P9-DEJ-402

FOURTH EDITION

ELEMENTARY DIFFERENTIAL EQUATIONS AND BOUNDARY VALUE PROBLEMS

FOURTH EDITION

ELEMENTARY DIFFERENTIAL EQUATIONS AND BOUNDARY VALUE PROBLEMS

WILLIAM E. BOYCE
RICHARD C. DiPRIMA
Rensselaer Polytechnic Institute

JOHN WILEY & SONS
NEW YORK CHICHESTER BRISBANE
TORONTO SINGAPORE

Library of Congress Cataloging in Publication Data:

Boyce, William E.
Elementary differential equations and boundary value problems.

Includes index.
1. Differential equations. 2. Boundary value problems. I. DiPrima, Richard C. II. Title.

QA371.B773 1986 515.3'5 85-20244
ISBN 0-471-07895-6

Printed in the United States of America

10 9 8 7 6 5

In loving memory of our parents
Ethel and Clyde DiPrima
Marie and Edward Boyce

PREFACE

A course in elementary differential equations is an excellent vehicle for conveying to the student a feeling for the interrelationship of pure mathematics and the physical sciences or engineering. Before the engineer or scientist can proceed confidently to use differential equations, he or she must master the techniques of solution and have at least a rudimentary knowledge of the underlying theory. Conversely, the student of pure mathematics often benefits greatly from a knowledge of some of the ways in which the necessity for solving specific problems has stimulated work of a more abstract nature.

We wrote this book from the intermediate viewpoint of the applied mathematician, whose interest in differential equations may be both highly theoretical and intensely practical. We have sought to combine a sound and accurate (but not abstract) exposition of the elementary theory of differential equations with considerable material on methods of solution that have proved useful in a wide variety of applications. We have given principal attention to those methods that are capable of broad application and that can be extended to problems beyond the range of our book. We emphasize that these methods have a systematic and orderly structure and are not merely a miscellaneous collection of unrelated mathematical tricks. At some point, however, the student should realize that many problems cannot be satisfactorily solved by elementary analytical techniques. Therefore, we have also discussed questions related to the qualitative behavior of solutions and ways of numerically approximating solutions.

With the widespread availability of enormous computing power, including versatile symbolic computation packages, it is reasonable to inquire whether elementary analytical methods of solving differential equations remain a worthwhile object of study. We believe that they do, for at least two reasons. First, solving a difficult problem in differential equations often requires the use of a variety of tools, both analytical and numerical. The implementation of an efficient numerical procedure typically rests on a good deal of preliminary analysis—to determine the qualitative features of the solution as a guide to computation, to test limiting or special cases, and so forth. Second, gaining an understanding of a complex natural process is often accomplished by combining or building upon simpler and more basic models. The latter are often described by differential equations of an elementary type. Thus a thorough knowledge of elementary differential equations and the models they represent is the first and indispensable step toward the solution of more complex problems.

The main reason for most students to study differential equations is that they are useful in understanding natural phenomena. We provide numerous examples of applications, many drawn from classical areas of the physical sciences but some from newer areas such as population dynamics or ecological modeling. The discussions of applications are usually accompanied by a careful derivation of the relevant equations. Although we believe that some attention to applications is an important part of an elementary course on differential equations, the applications are generally discussed in separate sections so that each instructor can cover as much or as little of this material as desired.

When one is faced with a problem whose solution cannot be found simply, questions arise as to the existence and uniqueness of solutions. We answer these questions by stating the relevant theorems, although not necessarily in their most general form, and discussing their significance. In particular, the differences between linear and nonlinear equations, and between initial and boundary value problems, are examined.

Although most students at this level can begin to grasp the importance and applications of the various theorems, the proofs in many cases involve concepts unknown to them, such as uniform convergence. Where this is true, we have omitted the proofs without apology. For example, we discuss some of the problems involved in proving the fundamental existence and uniqueness theorem for a first-order initial value problem by the method of successive approximations; our approach, however, is rather intuitive, and the more difficult analytical details are avoided.

We have written this book primarily for students who have a knowledge of calculus gained from a normal two- or three-semester course; most of the book is well within the capabilities of such students. In Chapter 7, which deals with systems of linear differential equations, experience with matrix algebra will be helpful, although we include two sections on matrices that summarize the information necessary to read the rest of the chapter. Sections marked by an asterisk probably require greater mathematical sophistication (although, strictly speaking, no more knowledge) than the rest of the book.

There is a tendency for courses dealing mainly with methods of solving differential equations to be "cookbookish" in nature. We have made a special effort to combat this tendency by explaining not only which methods work for which problems, but also *why* they work. Whenever possible, we draw upon the student's prior knowledge to suggest a method of attack for a new type of problem.

In an elementary book, terseness is not a primary virtue, and we have aimed for clarity rather than conciseness. We hope that whatever may have been lost in elegance has been regained in readability.

This book has more than average flexibility for classroom use. Beginning with Chapter 4, the chapters are substantially independent of each other, although Chapter 11 logically follows Chapter 10, and Chapter 9 contains references to Chapter 7. Thus, after the necessary parts of the first three chapters are completed (roughly Section 1.1, Sections 2.1 through 2.4, and Sections 3.1 through 3.6.2), the selection of additional topics, and the order and depth in which they are covered, is left to the discretion of the instructor.

Almost every section is followed by a set of problems for the student. These range from the routine to the challenging; some of the latter extend the theory discussed in the text or provide an introduction to other areas of application. The more difficult problems are indicated by an asterisk. Answers to all problems are given at the end of the book. There is also a Solutions Manual, compiled by Charles W. Haines, which contains solutions for many problems worked out in detail.

Sections are numbered in decimal form. Theorems and figures are numbered consecutively within each chapter. Thus, Theorem 3.4 is the fourth theorem in Chapter 3, but it is not necessarily in Section 3.4. General references are given at the end of each chapter, and more specific ones appear occasionally in footnotes.

The scope of the book can be adequately judged from the table of contents, and readers familiar with the previous edition will find that this one follows the same general pattern. Nevertheless, in preparing this edition we have made many minor changes and several more significant ones, including the following:

1. There is more emphasis throughout the book on the geometrical interpretation of solutions. For example, computer-generated graphs have been included in several chapters to provide concrete illustrations of such things as the convergence of power or Fourier series, or the pattern of trajectories in the phase plane. Such figures are invaluable in providing insight into the meaning and interpretation of the complicated formulas that often occur in solving differential equations.
2. A new section (2.6) on the geometrical analysis and qualitative properties of solutions of first order equations has been added, with applications to problems in population dynamics. The important concepts of asymptotic stability and instability are introduced here in an intuitive way.
3. The geometrical interpretation of solutions of linear systems has been given more emphasis in Chapter 7, and the classification of critical points is introduced here in an informal manner. In addition, more attention is paid to solving nonhomogeneous systems.
4. Chapter 9, on the stability and qualitative theory of linear and nonlinear autonomous systems, has been largely rewritten, in part to accommodate the inclusion of related material in Chapters 2 and 7, as mentioned above. We have also used vector notation wherever it is natural and helpful in this

discussion, thereby making the entire treatment of systems more unified and compact.

5. The chapter on numerical methods has been updated and some additional problems included. As in previous editions, primary emphasis is placed on the principles that underlie effective numerical procedures and on the causes, effects, and control of errors.
6. The chapter on Laplace transforms has been expanded to include more examples and a good many more problems. Indeed, examples or problems have been added in many places throughout the book.
7. The sections on applications of first order equations (2.5 through 2.7) have been placed before the sections on exact equations, integrating factors, and so forth, thereby giving the former greater prominence and the latter less.
8. Sections 3.2 and 3.3, on fundamental sets of solutions, linear independence, and Wronskians, have been rewritten so as to clarify the relationships among these important concepts.
9. Two topics that appeared in the preceding edition have, regretfully, been omitted from this one: the solution of linear systems by transformation to a single equation of higher order, and the discussion of limit cycles and orbital stability.

Dick DiPrima participated fully in the planning of this edition and contributed substantially to the revision of the early chapters, until the course of his final illness left him without the strength to continue. His death took from us a dedicated and inspiring teacher, an ingenious researcher, a thorough administrator, and a loyal friend.

WILLIAM E. BOYCE
Troy, New York
September, 1985

ACKNOWLEDGMENTS

In preparing this revision we have received assistance of various kinds from many individuals, and we now gratefully thank each one.

The computer-generated figures were produced using an innovative and powerful software package developed by Garrett M. Odell. Richard Welty also wrote several programs and provided much valuable advice.

The revision of Chapter 8 on numerical methods drew heavily on the comments offered by Joseph E. Flaherty and Bart Ng.

While drafting the Solutions Manual, Charles W. Haines carefully read the entire manuscript and suggested many corrections and improvements. Others who reviewed all or parts of the book include John E. Derwent and John H. George. We also appreciate the many comments from past users of the book who have taken the time to communicate with us. Each comment has been carefully considered, and many have been incorporated in the present volume.

In the actual production of the book, Helen D. Hayes, as in the past, skillfully typed the revised manuscript. John Reynders and John McAuley provided valuable assistance in reading and correcting the galley proofs. John Wiley's editorial and production staff smoothly and promptly brought the book from manuscript to printed volume.

Finally, I thank my wife Elsa for her continued support and encouragement while this book was in preparation, as well as for her perceptive technical and linguistic advice.

W. E. B.

CONTENTS

1.1 Classification of Differential Equations

Many important and significant problems in engineering, the physical sciences, and the social sciences, when formulated in mathematical terms, require the determination of a function satisfying an equation containing one or more derivatives of the unknown function. Such equations are called *differential equations*. Perhaps the most familiar example is Newton's law

$$m\frac{d^2u(t)}{dt^2} = F\left[t, u(t), \frac{du(t)}{dt}\right] \tag{1}$$

for the position $u(t)$ of a particle acted on by a force F, which may be a function of time t, the position $u(t)$, and the velocity $du(t)/dt$. To determine the motion of a particle acted on by a given force F it is necessary to find a function u satisfying Eq. (1).

The main purpose of this book is to discuss some of the properties of solutions of differential equations, and to describe some of the methods that have proved effective in finding solutions, or in some cases approximating them. To provide a framework for our presentation we first mention several useful ways of classifying differential equations.

ORDINARY AND PARTIAL DIFFERENTIAL EQUATIONS. One of the more obvious classifications is based on whether the unknown function depends on a single independent variable or on several independent variables. In the first case only ordinary derivatives appear in the differential equation, and it is said to be an *ordinary differential equation*. In the second case the derivatives are partial derivatives, and the equation is called a *partial differential equation*.

Two examples of ordinary differential equations, in addition to Eq. (1), are

$$L\frac{d^2Q(t)}{dt^2} + R\frac{dQ(t)}{dt} + \frac{1}{C}Q(t) = E(t), \tag{2}$$

for the charge $Q(t)$ on a condenser in a circuit with capacitance C, resistance R, inductance L, and impressed voltage $E(t)$; and the equation governing the decay with time of an amount $R(t)$ of a radioactive substance, such as radium,

$$\frac{dR(t)}{dt} = -kR(t), \tag{3}$$

where k is a known constant. Typical examples of partial differential equations are the potential equation

$$\frac{\partial^2 u(x,y)}{\partial x^2} + \frac{\partial^2 u(x,y)}{\partial y^2} = 0, \tag{4}$$

the diffusion or heat conduction equation

$$\alpha^2\frac{\partial^2 u(x,t)}{\partial x^2} = \frac{\partial u(x,t)}{\partial t}, \tag{5}$$

and the wave equation

$$a^2\frac{\partial^2 u(x,t)}{\partial x^2} = \frac{\partial^2 u(x,t)}{\partial t^2}. \tag{6}$$

Here α^2 and a^2 are certain constants. The potential equation, the diffusion equation, and the wave equation arise in a variety of problems in the fields of electricity and magnetism, elasticity, and fluid mechanics. Each is typical of distinct physical phenomena (note the names), and each is representative of a large class of partial differential equations. While we are primarily concerned with ordinary differential equations, we also consider partial differential equations, in particular the three important equations just mentioned, in Chapters 10 and 11.

SYSTEMS OF DIFFERENTIAL EQUATIONS. Another classification of differential equations depends on the number of unknown functions that are involved. If there is a single function to be determined, then one equation is sufficient. However, if there are two or more unknown functions, then a system of equations is required. For example, the Lotka-Volterra, or predator–prey, equations are important in ecological modeling. They have the form

$$\begin{aligned} dH/dt &= aH - \alpha HP, \\ dP/dt &= -cP + \gamma HP, \end{aligned} \tag{7}$$

where $H(t)$ and $P(t)$ are the respective populations of the prey and predator species. The constants a, α, c, and γ are based on empirical observations and depend on the particular species that are being studied. Systems of equations are discussed in Chapters 7 and 9; in particular, the Lotka-Volterra equations are examined in Section 9.4.

ORDER. The *order* of a differential equation is the order of the highest derivative that appears in the equation. Thus Eqs. (1) and (2) are second order ordinary differential equations, and Eq. (3) is a first order ordinary differential equation. Equations (4), (5), and (6) are second order partial differential equations. More generally, the equation

$$F\left[x, u(x), u'(x), \ldots, u^{(n)}(x)\right] = 0 \tag{8}$$

is an ordinary differential equation of the nth order. Equation (8) represents a relation between the independent variable x and the values of the function u and its first n derivatives $u', u'', \ldots, u^{(n)}$. It is convenient and customary in differential equations to write y for $u(x)$, with $y', y'', \ldots, y^{(n)}$ standing for $u'(x), u''(x), \ldots, u^{(n)}(x)$. Thus Eq. (8) is written as

$$F(x, y, y', \ldots, y^{(n)}) = 0. \tag{9}$$

For example,

$$y''' + 2e^x y'' + yy' = x^4 \tag{10}$$

is a third order differential equation for $y = u(x)$. Occasionally, other letters will be used instead of y; the meaning will be clear from the context.

We assume that it is always possible to solve a given ordinary differential equation for the highest derivative, obtaining

$$y^{(n)} = f(x, y, y', y'', \ldots, y^{(n-1)}). \tag{11}$$

We study only equations of the form (11). This is mainly to avoid the ambiguity that may arise because a single equation of the form (9) may correspond to several equations of the form (11). For example, the equation

$$y'^2 + xy' + 4y = 0$$

leads to the two equations

$$y' = \frac{-x + \sqrt{x^2 - 16y}}{2} \quad \text{or} \quad y' = \frac{-x - \sqrt{x^2 - 16y}}{2}.$$

SOLUTION. A *solution* of the ordinary differential equation (11) on the interval $\alpha < x < \beta$ is a function ϕ such that $\phi', \phi'', \ldots, \phi^{(n)}$ exist and satisfy

$$\phi^{(n)}(x) = f\left[x, \phi(x), \phi'(x), \ldots, \phi^{(n-1)}(x)\right] \tag{12}$$

for every x in $\alpha < x < \beta$. Unless stated otherwise, we assume that the function f of Eq. (11) is a real-valued function, and we are interested in obtaining real-valued solutions $y = \phi(x)$.

It is easily verified by direct substitution that the first order equation (3)

$$dR/dt = -kR$$

has the solution

$$R = \phi(t) = ce^{-kt}, \qquad -\infty < t < \infty, \tag{13}$$

where c is an arbitrary constant. Similarly, the functions $y_1(x) = \cos x$ and $y_2(x) = \sin x$ are solutions of

$$y'' + y = 0 \tag{14}$$

for all x. As a somewhat more complicated example, we verify that $\phi_1(x) = x^2 \ln x$ is a solution of

$$x^2y'' - 3xy' + 4y = 0, \qquad x > 0. \tag{15}$$

We have

$$\phi_1(x) = x^2 \ln x,$$

$$\phi_1'(x) = x^2(1/x) + 2x \ln x = x + 2x \ln x,$$

$$\phi_1''(x) = 1 + 2x(1/x) + 2 \ln x = 3 + 2 \ln x.$$

On substituting in the differential equation (15) we obtain

$$x^2(3 + 2 \ln x) - 3x(x + 2x \ln x) + 4(x^2 \ln x)$$

$$= 3x^2 - 3x^2 + (2 - 6 + 4)x^2 \ln x = 0,$$

which verifies that $\phi_1(x) = x^2 \ln x$ is a solution of Eq. (15). It can also be shown that $\phi_2(x) = x^2$ is a solution of Eq. (15); this is left as an exercise.

Although for the equations (3), (14), and (15) we are able to verify that certain simple functions are solutions, in general we do not have such solutions readily available. Thus a fundamental question is the following: Given an equation of the form (11), how can we tell whether it has a solution? This is the question of *existence* of a solution. The fact that we have written down an equation of the form (11) does not necessarily mean that there is a function $y = \phi(x)$ that satisfies it. Indeed, not all differential equations have solutions, nor is the question of existence a purely mathematical one. If a meaningful physical problem is correctly formulated mathematically as a differential equation, then the mathematical problem should have a solution. In this sense an engineer or scientist has some check on the validity of the mathematical formulation.

Second, assuming that a given equation has one solution, does it have other solutions? If so, what type of additional conditions must be specified to single out a particular solution? This is the question of *uniqueness*. Notice that there is an infinity of solutions of the first order equation (3) corresponding to the infinity of possible choices of the constant c in Eq. (13). If R is specified at some time t, this condition will determine a value for c; even so, however, we do not know yet that there may not be other solutions of Eq. (3) which also have the prescribed value of R at the prescribed time t. The questions of existence and uniqueness are difficult questions; they and related questions will be discussed as we proceed.

A third question, a more practical one, is: Given a differential equation of the form (11), how do we actually determine a solution? Note that if we find a solution of the given equation we have at the same time answered the question of the existence of a solution. On the other hand, without knowledge of existence theory we might, for example, use a computer to find a numerical approximation to a "solution" that does not exist. Even though we may know that a solution exists, it may be that the solution is not expressible in terms of the usual elementary functions—polynomial, trigonometric, exponential, logarithmic, and hyperbolic functions. Unfortunately, this is the situation for most differential equations. Thus, while we discuss elementary methods that can be used to obtain solutions of certain relatively simple problems, it is also important to consider methods of a more general nature that can be applied to more difficult problems.

LINEAR AND NONLINEAR EQUATIONS. A crucial classification of differential equations is according to whether they are linear or nonlinear. The ordinary differential equation

$$F(x, y, y', \dots, y^{(n)}) = 0$$

is said to be *linear* if F is a linear function of the variables $y, y', \dots, y^{(n)}$; a similar definition applies to partial differential equations. Thus the general linear ordinary differential equation of order n is

$$a_0(x)y^{(n)} + a_1(x)y^{(n-1)} + \cdots + a_n(x)y = g(x). \tag{16}$$

Equations (2), (3), (14), and (15) are linear equations. An equation that is not of the form (16) is a *nonlinear* equation. Equation (10) is nonlinear because of the term yy'. A simple physical problem that leads to a nonlinear differential equation is the oscillating pendulum. The angle θ that an oscillating pendulum of length l makes with the vertical direction (see Figure 1.1) satisfies the nonlinear equation

$$\frac{d^2\theta}{dt^2} + \frac{g}{l}\sin\theta = 0. \tag{17}$$

The mathematical theory and the techniques for solving linear equations are highly developed. In contrast, for nonlinear equations the situation is not as satisfactory. General techniques for solving nonlinear equations are largely lacking, and the theory associated with such equations is also more complicated than the theory for linear equations. In view of this it is fortunate that many significant problems lead to linear ordinary differential equations or, at least in the first approximation, to linear equations. For example, for the pendulum problem, if

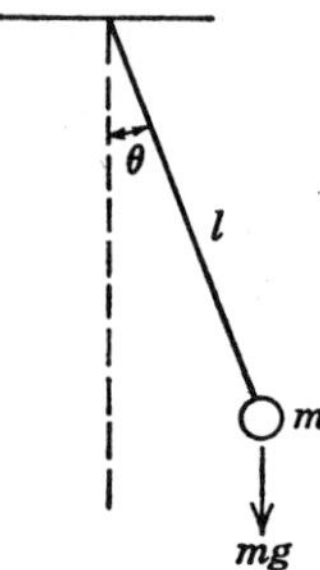

FIGURE 1.1 An oscillating pendulum.

the angle θ is small, then $\sin\theta \cong \theta$ and Eq. (17) can be replaced by the linear equation

$$\frac{d^2\theta}{dt^2} + \frac{g}{l}\theta = 0. \tag{18}$$

On the other hand, there are important physical phenomena, such as the ecological problems in Section 9.4, in which it is not possible to approximate the governing nonlinear differential equations by linear ones—the nonlinearity is crucial.

In an elementary text it is natural to emphasize the discussion of linear equations. The greater part of this book is therefore devoted to linear equations and various methods for solving them. However, Chapters 8 and 9, as well as a large part of Chapter 2, are concerned with nonlinear equations. Throughout the text we attempt to show why nonlinear equations are, in general, more difficult, and why many of the techniques that are useful in solving linear equations cannot be applied to nonlinear equations.

PROBLEMS

1. For each of the following ordinary differential equations determine its order and whether or not the equation is linear.

(a) $x^2\dfrac{d^2y}{dx^2} + x\dfrac{dy}{dx} + 2y = \sin x$ (b) $(1 + y^2)\dfrac{d^2y}{dx^2} + x\dfrac{dy}{dx} + y = e^x$

(c) $\dfrac{d^4y}{dx^4} + \dfrac{d^3y}{dx^3} + \dfrac{d^2y}{dx^2} + \dfrac{dy}{dx} + y = 1$ (d) $\dfrac{dy}{dx} + xy^2 = 0$

(e) $\dfrac{d^2y}{dx^2} + \sin(x + y) = \sin x$ (f) $\dfrac{d^3y}{dx^3} + x\dfrac{dy}{dx} + (\cos^2 x)y = x^3$

2. Verify in each of the following cases that the given function or functions is a solution of the differential equation.

(a) $y'' - y = 0$; $y_1(x) = e^x$, $y_2(x) = \cosh x$
(b) $y'' + 2y' - 3y = 0$; $y_1(x) = e^{-3x}$, $y_2(x) = e^x$

(c) $y'''' + 4y''' + 3y = x$; $\quad y_1(x) = x/3, \quad y_2(x) = e^{-x} + x/3$
(d) $2x^2y'' + 3xy' - y = 0, \quad x > 0$; $\quad y_1(x) = x^{1/2}, \quad y_2(x) = x^{-1}$
(e) $x^2y'' + 5xy' + 4y = 0, \quad x > 0$; $\quad y_1(x) = x^{-2}, \quad y_2(x) = x^{-2}\ln x$
(f) $y'' + y = \sec x, \quad 0 < x < \pi/2$; $\quad y = \phi(x) = (\cos x)\ln\cos x + x\sin x$
(g) $y' - 2xy = 1$; $\quad y = \phi(x) = e^{x^2}\int_0^x e^{-t^2}\,dt + e^{x^2}$

3. Determine for what values of r each of the following linear differential equations has solutions of the form $y = e^{rx}$.

(a) $y' + 2y = 0$ (b) $y'' - y = 0$
(c) $y'' + y' - 6y = 0$ (d) $y''' - 3y'' + 2y' = 0$

4. Determine for what values of r each of the following linear differential equations has solutions of the form $y = x^r$ for $x > 0$.

(a) $x^2y'' + 4xy' + 2y = 0$ (b) $x^2y'' - 4xy' + 4y = 0$

5. For each of the following partial differential equations determine the order and whether or not the equation is linear.

(a) $u_{xx} + u_{yy} + u_{zz} = 0$ (b) $\alpha^2 u_{xx} = u_t$
(c) $a^2 u_{xx} = u_{tt}$ (d) $u_{xx} + u_{yy} + uu_x + uu_y + u = 0$
(e) $u_{xxxx} + 2u_{xxyy} + u_{yyyy} = 0$ (f) $u_t + uu_x = 1 + u_{xx}$

6. Verify for each of the following partial differential equations that the given function or functions is a solution.

(a) $u_{xx} + u_{yy} = 0$; $\quad u_1(x, y) = \cos x\cosh y, \quad u_2(x, y) = \ln(x^2 + y^2)$
(b) $\alpha^2 u_{xx} = u_t$; $\quad u_1(x, t) = e^{-\alpha^2 t}\sin x, \quad u_2(x, t) = e^{-\alpha^2\lambda^2 t}\sin\lambda x$, λ a real constant
(c) $a^2 u_{xx} = u_{tt}$; $\quad u_1(x, t) = \sin\lambda x\sin\lambda at, \quad u_2(x, t) = \sin(x - at)$, λ a real constant

7. Verify for each of the following partial differential equations that the given function is a solution.

(a) $u_{xx} + u_{yy} + u_{zz} = 0$; $\quad u = \phi(x, y, z) = (x^2 + y^2 + z^2)^{-1/2}$, $(x, y, z) \neq (0, 0, 0)$
(b) $\alpha^2 u_{xx} = u_t$; $\quad u = \phi(x, t) = (\pi/t)^{1/2}e^{-x^2/4\alpha^2 t}, \quad t > 0$
(c) $a^2 u_{xx} = u_{tt}$; $\quad u = f(x - at) + g(x + at)$ where f and g are twice differentiable functions

1.2 Historical Remarks

Without knowing something about differential equations and methods of solving them, it is difficult to appreciate the history and development of this important branch of mathematics. Further, the development of differential equations is intimately interwoven with the general development of mathematics and cannot be divorced from it. In these brief comments we mention only the most prominent of the mathematicians of the seventeenth and eighteenth centuries who made important contributions to the theory, methods of solution, and applications of differential equations. Further information may be found in the books listed in the references at the end of the chapter.

The study of differential equations originated in the beginnings of the calculus with Isaac Newton (1642–1727) and Gottfried Wilhelm Leibniz (1646–1716) in the seventeenth century. Newton grew up in the English countryside, was educated at Trinity College, Cambridge, and became professor of mathematics there in 1669. His epochal discoveries of calculus and of the fundamental laws of mechanics date from 1665. They were circulated privately among his friends, but Newton was extremely sensitive to criticism, and did not begin to publish his results until 1687 with the appearance of his most famous book, *Philosophiae naturalis principia mathematica.* While Newton did relatively little work in differential equations per se, his development of the calculus and elucidation of the basic principles of mechanics provided a basis for their applications in the eighteenth century, most notably by Euler. Newton classified first order differential equations according to the forms $dy/dx = f(x)$, $dy/dx = f(y)$, and $dy/dx = f(x, y)$. For the latter equation he developed a method of solution using infinite series when $f(x, y)$ is a polynomial in x and y. Newton's active research career largely ended in the early 1690s, except for the solution of occasional challenge problems. He was appointed Warden of the British Mint in 1695 and resigned his professorship a few years later. He was knighted in 1705 and, upon his death, was buried in Westminster Abbey.

Leibniz was born at Leipzig and completed his doctorate in philosophy at the age of 20 at the University of Altdorf. Throughout his life he engaged in scholarly work in several different fields. He was mainly self-taught in mathematics, since his interest in this subject developed when he was in his twenties. Leibniz arrived at the fundamental results of calculus independently, although a little later than Newton, but was the first to publish them, in 1684. Leibniz was very conscious of the power of good mathematical notation, and our notation for the derivative (dy/dx) and the integral sign are due to him. He discovered the method of separation of variables (Section 2.4) in 1691, the reduction of homogeneous equations to separable ones (Section 2.10) in 1691, and the procedure for solving first order linear equations (Sections 2.1 and 2.2) in 1694. He spent his life as ambassador and adviser to several German royal families, which permitted him to travel widely and to carry on an extensive correspondence with other mathematicians, especially the Bernoulli brothers. In the course of this correspondence many problems in differential equations were solved during the latter part of the seventeenth century.

The brothers Jakob (1654–1705) and Johann (1667–1748) Bernoulli of Basel did much to develop methods of solving differential equations and to extend the range of their applications. Jakob became professor of mathematics at Basel in 1687, and Johann was appointed to the same position upon his brother's death in 1705. Both men were quarrelsome, jealous, and frequently embroiled in disputes, especially with each other. Nevertheless, both also made significant contributions to several areas of mathematics. With the aid of calculus they formulated as differential equations and solved a number of problems in mechanics. For example, Jakob Bernoulli solved the differential equation $y' = [a^3/(b^2y - a^3)]^{1/2}$ in 1690 and in the same paper first used the term "integral" in the modern sense. In 1694 Johann Bernoulli was able to solve the equation $dy/dx = y/ax$ even though it was not yet known that $d(\ln x) = dx/x$. One problem to which both brothers contributed, and which led to much friction between them, was the *brachistochrone* problem—the determination of the curve of fastest descent. In Problem 17 of Section 2.7 we will see that this problem leads to the nonlinear first order equation

$$y\left[1 + (y')^2\right] = c,$$

where c is a constant. The brachistochrone problem was also solved by Leibniz and Newton in addition to the Bernoulli brothers. It is said, perhaps apocryphally, that Newton learned of the problem late in the afternoon of a tiring day at the Mint, and solved it that evening after dinner. He published the solution anonymously, but on seeing it Johann Bernoulli exclaimed, "Ah, I know the lion by his paw."

Daniel Bernoulli (1700–1782), son of Johann, migrated to St. Petersburg as a young man to join the newly established St. Petersburg Academy, but returned to Basel in 1733 as professor of botany, and later, of physics. His interests were primarily in partial differential equations and their applications. For instance, it is his name that is associated with the famous Bernoulli equation in fluid mechanics. He was also the first to encounter the functions that a century later became known as Bessel functions.

The greatest mathematician of the eighteenth century, Leonhard Euler (1707–1783), grew up near Basel and was a student of Johann Bernoulli. He followed his friend Daniel Bernoulli to St. Petersburg in 1727. For the remainder of his life he was associated with the St. Petersburg Academy (1727–1741 and 1766–1783) and the Berlin Academy (1741–1766). Euler was the most prolific mathematician of all time; his collected works fill more than 70 large volumes. His interests ranged over all areas of mathematics and many fields of application. Even though he was blind during the last 17 years of his life, his work continued undiminished until the very day of his death. Of particular interest here is his formulation of problems in mechanics in mathematical language and his development of methods of solving these mathematical problems. Lagrange said of Euler's work in mechanics, "The first great work in which analysis is applied to the science of movement." Among other things, Euler identified the condition for exactness of first order differential equations (Section 2.8) in 1734–35, developed the theory of integrating factors (Section 2.9) in the same paper, and gave the general solution of homogeneous linear equations with constant coefficients (Sections 3.5, 3.5.1, and 5.3) in 1743. He extended the latter results to nonhomogeneous equations in 1750–51. Beginning about 1750, Euler made frequent use of power series (Chapter 4) in solving differential equations. He also proposed a numerical procedure (Section 8.2) in 1768–69, made important contributions in partial differential equations, and gave the first systematic treatment of the calculus of variations.

Joseph Louis Lagrange (1736–1813) became professor of mathematics in his native Turin at the age of 19. He succeeded Euler in the chair of mathematics at the Berlin Academy in 1766, and moved on to the Paris Academy in 1787. He is most famous for his monumental work *Mécanique analytique*, published in 1788, an elegant and comprehensive treatise on Newtonian mechanics. With respect to elementary differential equations, Lagrange showed in 1762–65 that the general solution of an nth order linear homogeneous differential equation is a linear combination of n independent solutions (Sections 3.2, 3.3, and 5.2). Later, in 1774–75, he gave a complete development of the method of variation of parameters (Sections 3.6.2 and 5.5). Lagrange is also known for fundamental work in partial differential equations and the calculus of variations.

The name of Pierre Simon de Laplace (1749–1827) is often linked with that of Lagrange, although the nature of his mathematical work was quite different. Laplace used mathematics as a tool in understanding nature, whereas Lagrange pursued mathematics for its own sake. Laplace lived in Normandy as a boy but came to Paris in 1768 and quickly made his mark in scientific circles, winning election to the Académie des Sciences in 1773. He was preeminent in the field of celestial mechanics; his greatest work, *Traité de mécanique céleste*, was published in five volumes between 1799 and 1825. The equation

$$u_{xx} + u_{yy} + u_{zz} = 0,$$

where subscripts denote partial derivatives, is known as Laplace's equation or as the potential equation (Section 10.8). It is fundamental in many branches of mathematical physics, and Laplace studied it extensively in connection with gravitational attraction. The Laplace transform (Chapter 6) is also named for him although its usefulness in solving differential equations was not recognized until much later.

By the latter part of the eighteenth century, most of the elementary methods of solving ordinary differential equations were known. Attention then turned to the development of a rigorous, systematic, and general theory. The goal was not so much to construct solutions of particular equations, but to investigate properties of solutions of classes of equations.

We conclude this short historical sketch with a remark that may bring some pleasure to the student who has noted with dismay how often phrases such as "It is clear..." or "It can be easily shown..." appear in mathematics texts. Nathaniel Bowditch (1773–1838), an American astronomer and mathematician, while translating Laplace's *Mécanique céleste* in the early 1800s, stated, "I never come across one of Laplace's 'Thus it plainly appears' without feeling sure that I have hours of hard work before me to fill up the chasm and find out and show how it plainly appears."

REFERENCES

For further reading in the history of mathematics one may consult books such as those listed below. Of the four books mentioned, the one by Kline is by far the most comprehensive.

Bell, E. T., *Men of Mathematics* (New York: Simon and Schuster, 1937).
Boyer, C. B., *A History of Mathematics* (New York: Wiley, 1968).
Eves, H., *An Introduction to the History of Mathematics* (3rd ed.) (New York: Holt, 1964),
Kline, M., *Mathematical Thought from Ancient to Modern Times* (New York: Oxford University Press, 1972).

A useful historical appendix also appears in

Ince, E. L., *Ordinary Differential Equations* (London: Longmans, Green, 1927; New York: Dover, 1956).

Several anthologies that include original source material as well as explanatory and historical commentary are available; for example,

Calinger, R., ed., *Classics of Mathematics* (Oak Park, Illinois: Moore Publishing Company, 1982).
Newman, J. R., ed., *The World of Mathematics* (4 vols.) (New York: Simon and Schuster, 1956).

Finally, an encyclopedic source of information about the lives and achievements of mathematicians of the past is

Gillespie, C. C., ed., *Dictionary of Scientific Biography* (15 vols.) (New York: Scribner's, 1971).

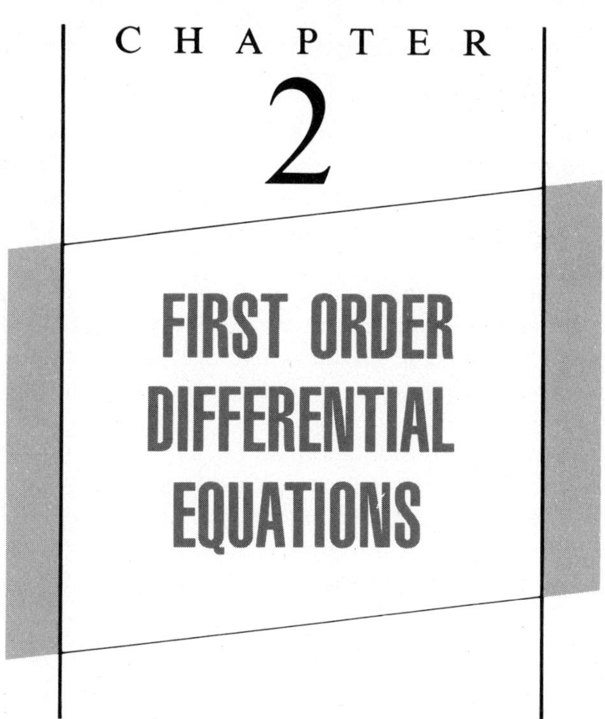

2.1 Linear Equations

This chapter deals with differential equations of first order, that is, equations of the form

$$y' = f(x, y), \tag{1}$$

where f is a given function of two variables. Any differentiable function $y = \phi(x)$ which identically satisfies Eq. (1) is called a solution, and our object is to determine whether such functions exist and, if so, how to find them.

The simplest type of first order differential equation occurs when f depends only on x. In this case

$$y' = f(x), \tag{2}$$

and we seek a function $y = \phi(x)$ whose derivative is the given function f. From elementary calculus we know that ϕ is an antiderivative of f, and we write

$$y = \phi(x) = \int^x f(t)\, dt + c, \tag{3}$$

where c is an arbitrary constant. For example, if

$$y' = \sin 2x,$$

then

$$y = \phi(x) = -\tfrac{1}{2}\cos 2x + c.$$

In Eq. (3) and elsewhere in this book we use the notation $\int^x f(t)\,dt$ to denote an antiderivative of the function f; that is, $F(x) = \int^x f(t)\,dt$ designates some particular representative of the class of functions whose derivatives are equal to f. All members of this class are included in the expression $F(x) + c$, where c is an arbitrary constant.

The very simple differential equation (2) illustrates two features that are typical of first order differential equations generally. First, a single integration process is required to eliminate the derivative of y and obtain y itself. Second, this integration process leads to an expression for the solution that involves an arbitrary constant.

We now return to a consideration of Eq. (1). Unfortunately, there is no satisfactory general method for writing down solutions of this equation. Instead, we will present several methods, each of which is applicable to a certain class of equations of the form (1). First, we suppose that $f(x, y)$ depends linearly on the dependent variable y. Then Eq. (1) becomes a linear first order equation, and can be written as

$$y' + p(x)y = g(x), \tag{4}$$

where p and g are given continuous functions on some interval $\alpha < x < \beta$. In this section we will be concerned with methods for solving Eq. (4). More theoretical questions involving the existence and uniqueness of solutions in general will be discussed in Section 2.2.

Let us begin with

$$y' + ay = 0, \tag{5}$$

where a is a real constant. This equation can be solved by inspection. We need a function y whose derivative y' is equal to $(-a)$ times y itself. Clearly, $y = e^{-ax}$ has this property, and therefore satisfies Eq. (5). Furthermore,

$$y = ce^{-ax}, \tag{6}$$

where c is an arbitrary constant, also does so. Since c is arbitrary, Eq. (6) represents infinitely many solutions of the differential equation (5). It is natural to ask whether Eq. (5) has any solutions other than those given by Eq (6). We show in the next section that there are no other solutions, but for the time being this question remains open.

Geometrically, Eq. (6) represents a one-parameter family of curves, called *integral curves* of Eq. (5). For $a = \frac{1}{2}$ several members of this family are sketched in Figure 2.1. Each integral curve is the geometric representation of the corresponding solution of the differential equation. Specifying a particular solution is equivalent to picking out a particular integral curve from the one-parameter family. It is usually convenient to do this by prescribing a point (x_0, y_0) through which the integral curve must pass; that is, we seek a solution $y = \phi(x)$ such that

$$\phi(x_0) = y_0.$$

Such a condition is called an *initial condition*. Since y stands for $\phi(x)$ we could

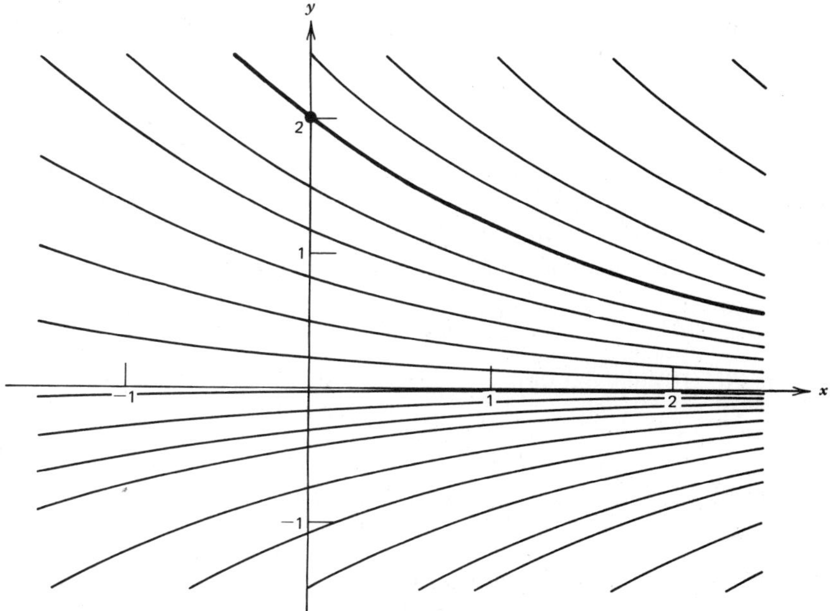

FIGURE 2.1 Integral curves of $y' + 0.5y = 0$.

also write

$$y = y_0 \quad \text{at} \quad x = x_0.$$

However, it is common practice to express the initial condition in the form

$$y(x_0) = y_0, \tag{7}$$

and this is the notation we will usually use in this book. A first order differential equation together with an initial condition form an *initial value problem*.[1]

For example, the differential equation (5),

$$y' + ay = 0,$$

and the initial condition

$$y(0) = 2, \tag{8}$$

form an initial value problem. As noted above, all solutions of the differential equation (5) are given by Eq. (6). The solution satisfying the initial condition (8) is found by substituting $x = 0$ and $y = 2$ in Eq. (6). Then $c = 2$ and the desired solution is

$$y = \phi(x) = 2e^{-ax}. \tag{9}$$

[1] This terminology is suggested by the fact that the independent variable often denotes time, the initial condition defines the situation at some fixed instant, and the solution of the initial value problem describes what happens later.

In particular, if $a = \frac{1}{2}$, then the solution of Eq. (5) passing through $(0, 2)$ is $y = 2e^{-x/2}$, and this solution is shown in Figure 2.1.

Now let us consider the equation

$$y' + ay = g(x). \tag{10}$$

If $a = 0$, then the left side of Eq. (10) is just the derivative of y, and the solution is given by Eq. (3) with f replaced by g. If $a \neq 0$, then the left side of Eq. (10) is a combination of terms involving y and y'. We ask whether these terms can be identified as the derivative of some function. In other words, can we write

$$\frac{dy}{dx} + ay = \frac{d}{dx}(\quad)? \tag{11}$$

If so, then Eq. (10) has the form

$$\frac{d}{dx}(\quad) = g(x),$$

and we can immediately solve it by integrating both sides.

A clue as to how to proceed can be found by looking again at the solution (6) of Eq. (5). Rewriting $y = ce^{-ax}$ in the form

$$ye^{ax} = c \tag{12}$$

and then differentiating both sides of Eq. (12), we obtain

$$\frac{d}{dx}(ye^{ax}) = 0,$$

or

$$\frac{dy}{dx}e^{ax} + ae^{ax}y = e^{ax}\left(\frac{dy}{dx} + ay\right) = 0.$$

Thus, if we multiply Eq. (10) by e^{ax}, then we can write the left side of the resulting equation as the derivative of the single function ye^{ax}:

$$\underbrace{e^{ax}y' + ae^{ax}y}_{} = e^{ax}g(x)$$

$$\frac{d}{dx}(e^{ax}y) = e^{ax}g(x). \tag{13}$$

Equation (13) can be integrated, and we obtain

$$e^{ax}y = \int^x e^{at}g(t)\,dt + c,$$

where c is an arbitrary constant. Hence, a one-parameter family of solutions of Eq. (10) is

$$y = \phi(x) = e^{-ax}\int^x e^{at}g(t)\,dt + ce^{-ax}. \tag{14}$$

Thus, for a given function g the problem of determining a solution of Eq. (10) is reduced to that of evaluating the antiderivative in Eq. (14). The difficulty involved in this depends on g; nevertheless, Eq. (14) gives an explicit formula for the solution $y = \phi(x)$. The constant c can be determined if an initial condition is prescribed.

EXAMPLE 1

Find the solution of the initial value problem

$$y' + 2y = e^{-x}, \tag{15}$$

$$y(0) = \tfrac{3}{4}. \tag{16}$$

To solve Eq. (15) we compare it with Eq. (10) and find that $a = 2$ in this problem. Thus, we multiply Eq. (15) by e^{2x}, thereby obtaining

$$e^{2x}y' + 2e^{2x}y = e^{2x}e^{-x}. \tag{17}$$

The left side of Eq. (17) is the derivative of $e^{2x}y$, so this equation can be written in the form

$$(e^{2x}y)' = e^{x}. \tag{18}$$

Integrating Eq. (18) yields

$$ye^{2x} = e^{x} + c;$$

hence

$$y = e^{-x} + ce^{-2x} \tag{19}$$

is a solution of Eq. (15) for any value of c. Some of the solutions (19) are shown in Figure 2.2. To satisfy the initial condition (16) we substitute $x = 0$ and $y = \frac{3}{4}$ in Eq. (19). This gives $c = -\frac{1}{4}$, and thus the solution of the initial value problem (15), (16) is

$$y = e^{-x} - \tfrac{1}{4}e^{-2x}. \tag{20}$$

This solution is also shown in Figure 2.2.

Now let us return to the general first order linear equation (4)

$$y' + p(x)y = g(x).$$

By analogy with the procedure used above, we would like to choose a function μ so that if Eq. (4) is multiplied by $\mu(x)$, then the left side of Eq. (4) can be written as the derivative[2] of the single function $\mu(x)y$. That is, we want to choose μ, if possible, so that

$$\begin{aligned}\mu(x)[y' + p(x)y] &= [\mu(x)y]' \\ &= \mu(x)y' + \mu'(x)y.\end{aligned}$$

[2]A function μ having this property is called an integrating factor. Integrating factors are discussed more fully in Section 2.9.

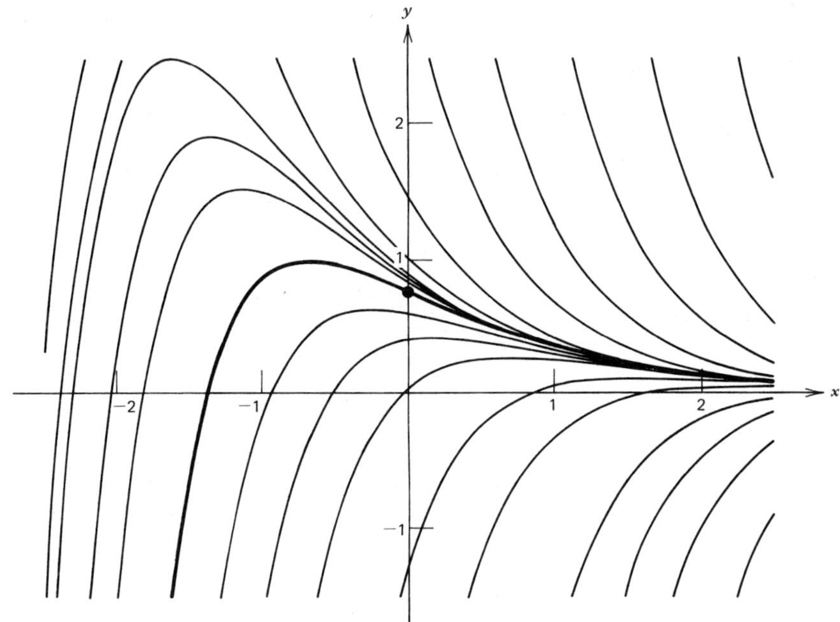

FIGURE 2.2 Integral curves of $y' + 2y = e^{-x}$.

Thus $\mu(x)$ must satisfy

$$\mu(x)p(x)y = \mu'(x)y.$$

Assuming for the moment that $\mu(x) > 0$, we obtain

$$\mu'(x)/\mu(x) = p(x).$$

Since $\mu'(x)/\mu(x)$ is the derivative of $\ln \mu(x)$, we have

$$\ln \mu(x) = \int^x p(t)\, dt,$$

and finally

$$\mu(x) = \exp\left[\int^x p(t)\, dt\right]. \tag{21}$$

Note that $\mu(x)$, as defined by Eq. (21), is indeed positive. Further, since $\int^x p(t)\, dt$ is determined only up to an arbitrary additive constant, $\mu(x)$ is determined only up to an arbitrary positive multiplicative constant.

Returning to Eq. (4) and multiplying by $\mu(x)$, we obtain

$$[\mu(x)y]' = \mu(x)g(x).$$

Therefore

$$\mu(x)y = \int^x \mu(s)g(s)\, ds + c,$$

or

$$y = \frac{1}{\mu(x)}\left[\int^x \mu(s)g(s)\,ds + c\right], \tag{22}$$

where $\mu(x)$ is given by Eq. (21). Equation (22) provides an explicit formula for the solution of the general first order linear equation (4), where p and g are given continuous functions. Two integrations are required, one to obtain $\mu(x)$ from Eq. (21) and the other to determine y from Eq. (22). The arbitrary constant c can be used to satisfy an initial condition.

Two additional things should be noted about the procedure we have developed for solving Eq. (4). First, before computing the integrating factor μ from Eq. (21), it is necessary to make sure that the differential equation is exactly in the form (4); specifically, the coefficient of y' must be one. Second, after finding $\mu(x)$ and multiplying Eq. (4) by it, one should make sure that the terms involving y are, in fact, the derivative of $\mu(x)y$, as they should be. This provides a check on the calculation of μ. Of course, once the solution y has been found, one can also check it by substituting it into the differential equation and initial condition.

EXAMPLE 2

Find the solution of the initial value problem

$$y' - 2xy = x, \qquad y(0) = 0. \tag{23}$$

To solve this equation we first determine the integrating factor:

$$\mu(x) = \exp\left(-\int^x 2t\,dt\right) = e^{-x^2}.$$

Hence

$$e^{-x^2}y' - 2xe^{-x^2}y = xe^{-x^2},$$

so that

$$\left(e^{-x^2}y\right)' = xe^{-x^2}.$$

Therefore,

$$ye^{-x^2} = \int^x te^{-t^2}\,dt + c = -\tfrac{1}{2}e^{-x^2} + c,$$

and finally

$$y = -\tfrac{1}{2} + ce^{x^2}.$$

To satisfy the initial condition $y(0) = 0$ we must choose $c = \frac{1}{2}$. Hence

$$y = -\tfrac{1}{2} + \tfrac{1}{2}e^{x^2} \tag{24}$$

is the solution of the given initial value problem. Some integral curves and the particular solution passing through the origin are shown in Figure 2.3.

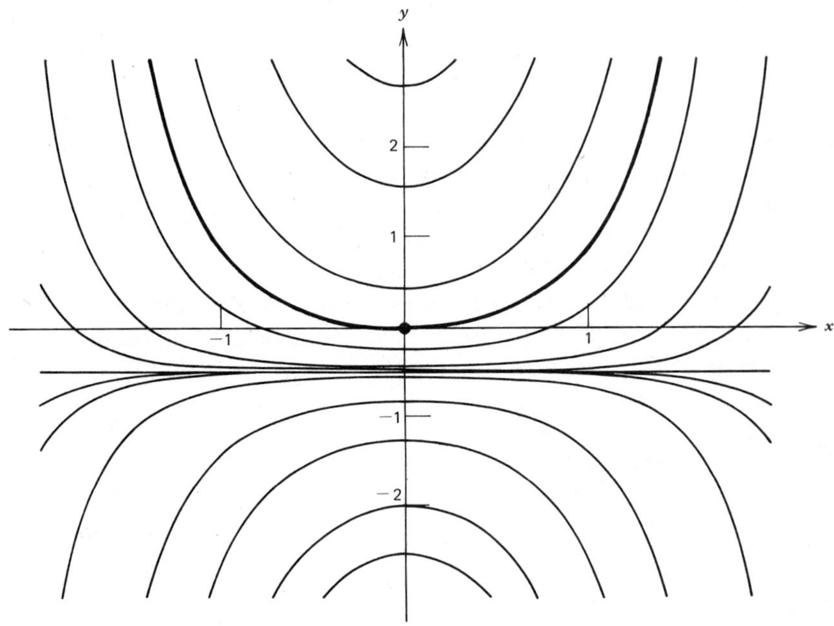

FIGURE 2.3 Integral curves of $y' - 2xy = x$.

PROBLEMS

In each of Problems 1 through 8 solve the given differential equation.

1. $y' + 3y = x + e^{-2x}$
2. $y' - 2y = x^2e^{2x}$
3. $y' + y = xe^{-x} + 1$
4. $y' + (1/x)y = 3\cos 2x, \quad x > 0$
5. $y' - y = 2e^x$
6. $xy' + 2y = \sin x, \quad x > 0$
7. $y' + 2xy = 2xe^{-x^2}$
8. $(1 + x^2)y' + 4xy = (1 + x^2)^{-2}$

In each of Problems 9 through 14 find the solution of the given initial value problem.

9. $y' - y = 2xe^{2x}, \quad y(0) = 1$
10. $y' + 2y = xe^{-2x}, \quad y(1) = 0$
11. $xy' + 2y = x^2 - x + 1, \quad y(1) = \frac{1}{2}, \quad x > 0$
12. $y' + \dfrac{2}{x}y = \dfrac{\cos x}{x^2}, \quad y(\pi) = 0, \quad x > 0$
13. $y' - 2y = e^{2x}, \quad y(0) = 2$
14. $xy' + 2y = \sin x, \quad y(\pi/2) = 1$

15. Find the solution of

$$\frac{dy}{dx} = \frac{1}{e^y - x}, \qquad y(1) = 0.$$

Hint: Consider x as the dependent variable instead of y.

16. (a) Show that $\phi(x) = e^{2x}$ is a solution of $y' - 2y = 0$ and that $y = c\phi(x)$ is also a solution of this equation for any value of the constant c.
(b) Show that $\phi(x) = 1/x$ is a solution of $y' + y^2 = 0$ for $x > 0$, but that $y = c\phi(x)$ is not a solution of this equation. Note that the equation of part (b) is nonlinear while that of part (a) is linear.

17. Show that if $y = \phi(x)$ is a solution of $y' + p(x)y = 0$ then $y = c\phi(x)$ is also a solution for any value of the constant c.

18. Let $y = y_1(x)$ be a solution of

$$y' + p(x)y = 0, \tag{i}$$

and let $y = y_2(x)$ be a solution of

$$y' + p(x)y = g(x). \tag{ii}$$

Show that $y = y_1(x) + y_2(x)$ is also a solution of Eq. (ii).

*19. Consider the following method of solving the general linear equation of first order:

$$y' + p(x)y = g(x). \tag{i}$$

(a) If $g(x)$ is identically zero, show that the solution is

$$y = A\exp\left[-\int^x p(t)\,dt\right], \tag{ii}$$

where A is a constant.
(b) If $g(x)$ is not identically zero, assume that the solution is of the form

$$y = A(x)\exp\left[-\int^x p(t)\,dt\right], \tag{iii}$$

where A is now a function of x. By substituting for y in the given differential equation show that $A(x)$ must satisfy the condition

$$A'(x) = g(x)\exp\left[\int^x p(t)\,dt\right]. \tag{iv}$$

(c) Find $A(x)$ from Eq. (iv). Then substitute for $A(x)$ in Eq. (iii) and determine y. Verify that the solution obtained in this manner agrees with that of Eq. (22) in the text. This technique is known as the method of *variation of parameters*; it is discussed in detail in Section 3.6.2 in connection with second order linear equations.

*20. Use the method of Problem 19 to solve each of the following differential equations.

(a) $y' - 2y = x^2e^{2x}$ (b) $y' + (1/x)y = 3\cos 2x, \quad x > 0$

2.2 *Further Discussion of Linear Equations*

In Section 2.1 we showed how to construct solutions of initial value problems for first order linear differential equations by changing the differential equation to a form that was directly integrable. Now we deal with some questions of a more general nature, namely:

1. Does such an initial value problem always have a solution?
2. May it have more than one solution?
3. Is the solution valid for all x, or only for some restricted interval about the initial point?

The following fundamental theorem responds to these questions.

Theorem 2.1. *If the functions p and g are continuous on an open interval $\alpha < x < \beta$ containing the point $x = x_0$, then there exists a unique function $y = \phi(x)$ that satisfies the differential equation*

$$y' + p(x)y = g(x) \tag{1}$$

for $\alpha < x < \beta$, and that also satisfies the initial condition

$$y(x_0) = y_0, \tag{2}$$

where y_0 is an arbitrary prescribed initial value.

The proof of this theorem is essentially contained in the discussion in the last section leading to the formula

$$y = \frac{1}{\mu(x)}\left[\int^x \mu(s)g(s)\,ds + c\right], \tag{3}$$

where

$$\mu(x) = \exp\int^x p(t)\,dt. \tag{4}$$

Assuming that Eq. (1) has a solution, the derivation in Section 2.1 shows that it must be of the form (3). Note that since p is continuous for $\alpha < x < \beta$, it follows that μ is defined in this interval, and is a nonzero differentiable function. This justifies the conversion of Eq. (1) into the form

$$[\mu(x)y]' = \mu(x)g(x). \tag{5}$$

The function μg has an antiderivative since μ and g are continuous, and Eq. (3) follows from Eq. (5). The initial assumption—that there is at least one solution of Eq. (1)—can now be verified by substituting the expression for y in Eq. (3) back into the differential equation. Finally, the initial condition (2) determines the constant c uniquely, thus completing the proof. Since Eq. (3) contains *all* solutions of Eq. (1), it is customary to call Eq. (3) the *general solution* of Eq. (1).

Equation (4) determines the integrating factor $\mu(x)$ only up to a multiplicative factor that depends on the lower limit of integration. If we choose this lower limit to be x_0, then

$$\mu(x) = \exp\int_{x_0}^x p(t)\,dt, \tag{6}$$

and it follows that $\mu(x_0) = 1$. Using the integrating factor given by Eq. (6) and choosing the lower limit of integration in Eq. (3) also to be x_0, we obtain the general solution of Eq. (1) in the form

$$y = \frac{1}{\mu(x)}\left[\int_{x_0}^{x} \mu(s)g(s)\,ds + c\right].$$

To satisfy the initial condition (2) we must choose $c = y_0$. Thus the solution of the initial value problem (1), (2) is

$$y = \frac{1}{\mu(x)}\left[\int_{x_0}^{x} \mu(s)g(s)\,ds + y_0\right], \tag{7}$$

where $\mu(x)$ is given by Eq. (6).

Several aspects of Theorem 2.1 should be noted. In the first place, it states that the given initial value problem *has* a solution and also that the problem has *only one* solution. In other words, the theorem asserts both the *existence* and *uniqueness* of the solution of the initial value problem (1), (2). Further, the solution $y = \phi(x)$ is a differentiable function and, in fact, since $y' = -p(x)y + g(x)$, the derivative $y' = \phi'(x)$ is continuous. Finally, the solution satisfies the differential equation (1) throughout the interval $\alpha < x < \beta$ in which the coefficients p and g are continuous. This means that the solution will break down, if at all, only at points where either p or g is discontinuous. Thus a certain amount of qualitative information about the solution is obtained merely by identifying points of discontinuity of p and g.

EXAMPLE 1

Solve the initial value problem

$$xy' + 2y = 4x^2, \tag{8}$$

$$y(1) = 2, \tag{9}$$

and determine the interval in which the solution is valid.

Proceeding as in Section 2.1, we rewrite Eq. (8) as

$$y' + \frac{2}{x}y = 4x \tag{10}$$

and seek a solution in an interval containing $x = 1$. Since the coefficients in Eq. (10) are continuous except at $x = 0$, it follows from Theorem 2.1 that the given initial value problem has a solution that is valid at least in the interval $0 < x < \infty$. To find this solution we first compute $\mu(x)$:

$$\mu(x) = \exp\left(\int^{x} \frac{2}{t}\,dt\right) = e^{2\ln x} = x^2. \tag{11}$$

Multiplying Eq. (10) by $\mu(x) = x^2$ gives

$$x^2y' + 2xy = 4x^3,$$

or

$$(x^2y)' = 4x^3.$$

Hence $x^2y = x^4 + c$, and

$$y = x^2 + \frac{c}{x^2}, \tag{12}$$

where c is arbitrary, is the general solution of Eq. (8). Integral curves of Eq. (8) for several values of c are sketched in Figure 2.4. To satisfy the initial condition (9) it is necessary that $c = 1$; thus

$$y = x^2 + \frac{1}{x^2}, \qquad x > 0 \tag{13}$$

is the solution of the initial value problem (8), (9). Observe that the function $y = x^2 + (1/x^2)$ for $x < 0$ is not part of the solution of this initial value problem.

Note that the solution (13) becomes unbounded as $x \to 0$. This is not surprising since $x = 0$ is a point of discontinuity of the coefficient of y in the differential equation (10). However, if the initial condition (9) is changed to

$$y(1) = 1, \tag{14}$$

then it follows from Eq. (12) that $c = 0$. Hence the solution of the initial value

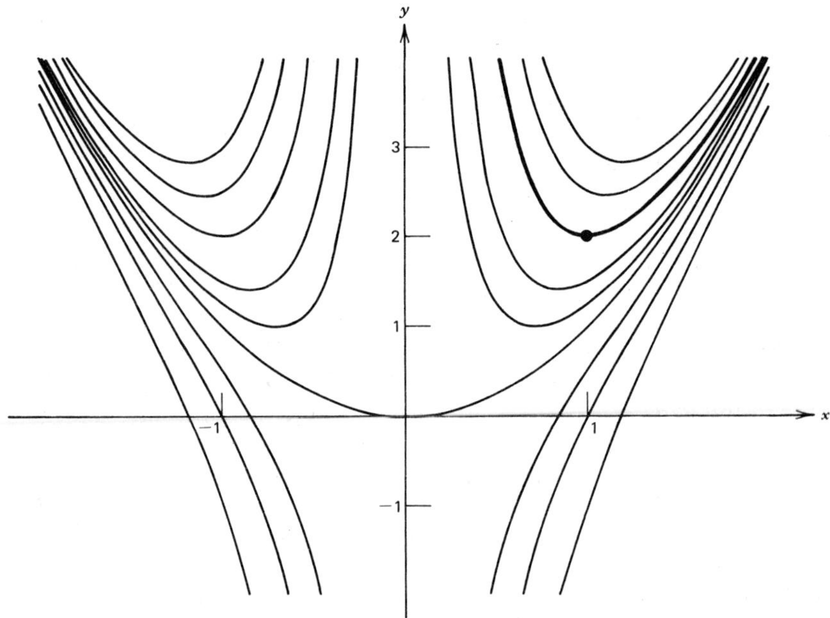

FIGURE 2.4 Integral curves of $xy' + 2y = 4x^2$.

problem (8), (14) is

$$y = x^2, \tag{15}$$

which is perfectly well behaved as $x \to 0$. This illustrates that Theorem 2.1 does *not* assert that the solution of an initial value problem must become singular whenever the functions p and g become discontinuous; rather, it asserts that the solution cannot become singular at other points.

The possible behavior of solutions of initial value problems for first order linear equations in the neighborhood of points where p and g are discontinuous is more varied than the previous discussion suggests. The possibilities are explored to some extent in Problems 12 through 15 and 20; a more detailed treatment appears in Chapter 4 in connection with second order linear equations.

EXAMPLE 2

Find the solution of the initial value problem

$$y' - 2xy = 1, \qquad y(0) = -\tfrac{1}{2}. \tag{16}$$

To solve the differential equation we observe that $\mu(x) = e^{-x^2}$; therefore

$$e^{-x^2}(y' - 2xy) = e^{-x^2}$$

and

$$ye^{-x^2} = \int^x e^{-t^2}\,dt + c. \tag{17}$$

To evaluate c it is convenient[3] to take the lower limit of integration as the initial point $x = 0$. Then, multiplying Eq. (17) by e^{x^2}, we obtain

$$y = e^{x^2}\int_0^x e^{-t^2}\,dt + ce^{x^2}.$$

To satisfy the initial condition $y(0) = -\frac{1}{2}$, we must choose $c = -\frac{1}{2}$; hence

$$y = e^{x^2}\int_0^x e^{-t^2}\,dt - \tfrac{1}{2}e^{x^2} \tag{18}$$

is the solution of the given initial value problem. Some of the integral curves and the particular solution passing through $(0, -\frac{1}{2})$ are shown in Figure 2.5.

Note that in the solution of Example 2 the integral of e^{-t^2} is not expressible as an elementary function. This illustrates the fact that it may be necessary to leave the solution of even a very simple problem in integral form. However, an

[3] The choice of the lower limit of integration is actually immaterial since the difference between $\int_a^x e^{-t^2}\,dt$ and $\int_b^x e^{-t^2}\,dt$ is merely a constant, which can be added to c.

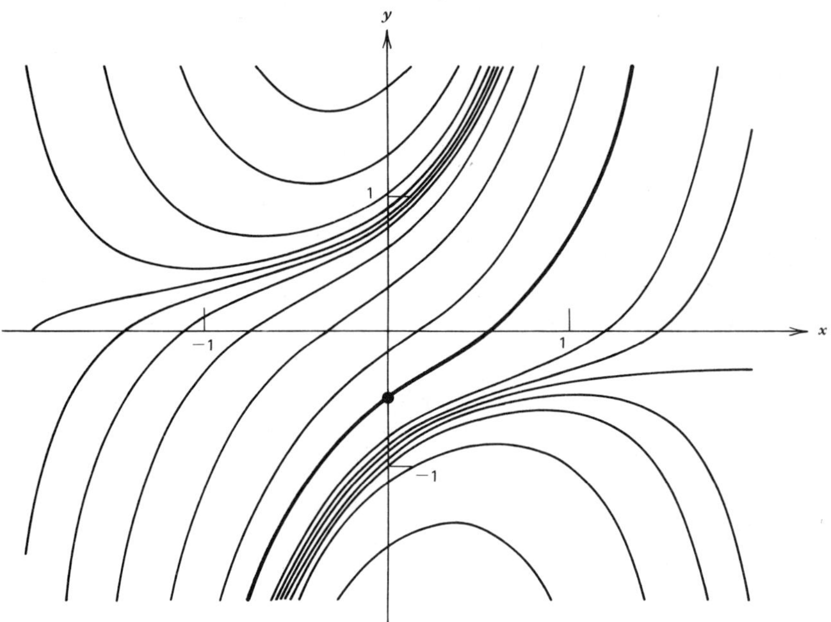

FIGURE 2.5 Integral curves of $y' - 2xy = 1$.

integral expression such as Eq. (18) can provide a starting point for the numerical calculation of the solution of an initial value problem. For a fixed value of x the integral in Eq. (18) is a *definite integral*, and its value can be determined numerically by Simpson's rule or some other numerical integration method. By repeating the numerical integration for different values of x, one can construct a table of values of the solution y. There are also other numerical procedures that use the differential equation directly to compute approximate values of the solution for a given set of values of x; these methods are discussed in Chapter 8.

In the case of Eq. (18) it also happens that the function

$$\operatorname{erf}(x) = \frac{2}{\sqrt{\pi}} \int_0^x e^{-t^2}\, dt, \tag{19}$$

known as the *error function*, has been extensively tabulated and can be regarded as a known function. Thus instead of Eq. (18) we can write

$$y = e^{x^2}\left[\frac{\sqrt{\pi}}{2}\operatorname{erf}(x) - \frac{1}{2}\right]. \tag{20}$$

To evaluate the right side of Eq. (20) for a given value of x we can consult a table of values of the error function, or resort to a numerical procedure as suggested above. In principle, working with the error function is no more difficult than working with the function e^{x^2}. The main difference is that many calculators have built-in routines for evaluating the exponential function but not the error function.

PROBLEMS

In each of Problems 1 through 4 find the general solution of the given differential equation.

1. $y' + (1/x)y = \sin x, \quad x > 0$
2. $x^2y' + 3xy = (\sin x)/x, \quad x < 0$
3. $y' + (\tan x)y = x \sin 2x, \quad -\pi/2 < x < \pi/2$
4. $xy' + 2y = e^x, \quad x > 0$

In each of Problems 5 through 11 find the solution of the given initial value problem. State the interval in which the solution is valid.

5. $xy' + 2y = x^2 - x + 1, \quad y(1) = \frac{1}{2}$
6. $xy' + y = e^x, \quad y(1) = 1$
7. $y' + (\cot x)y = 2 \csc x, \quad y(\pi/2) = 1$
8. $xy' + 2y = \sin x, \quad y(\pi) = 1/\pi$
9. $y' + (\cot x)y = 4 \sin x, \quad y(-\pi/2) = 0$
10. $x(2 + x)y' + 2(1 + x)y = 1 + 3x^2, \quad y(-1) = 1$
11. $y' + y = 1/(1 + x^2), \quad y(0) = 0$

Each of the equations in Problems 12 through 15 has at least one discontinuous coefficient at $x = 0$. Solve each equation for $x > 0$ and describe the behavior of the solution as $x \to 0$ for various values of the constant of integration. Sketch several members of the family of integral curves.

12. $y' + (2/x)y = 1/x^2$
13. $y' - (1/x)y = x$
14. $y' - (1/x)y = x^{1/2}$
15. $y' + (1/x)y = (\cos x)/x$
16. Use Simpson's rule (or any other numerical integration procedure that you know) to evaluate y from Eq. (18) for $x = 1$. Make sure that your answer is correct to at least three decimal places.
17. (a) Show that the solution of $y' - 2xy = 1$, $y(0) = y_0$ (see Example 2) can be written in the form

$$y = e^{x^2}\left[\frac{\sqrt{\pi}}{2}\operatorname{erf}(x) + y_0\right].$$

(b) From Figure 2.5 it appears that as $x \to \infty$ the solution grows without bound in the positive direction for some values of y_0 and in the negative direction for other values of y_0. It is also known that $\operatorname{erf}(x) \to 1$ as $x \to \infty$. Use this fact to find the critical value of y_0 which separates the solutions that grow positively from those that grow negatively.

18. (a) Show that the solution (3) of the general linear equation (1) can be written in the form

$$y = cy_1(x) + y_2(x), \tag{i}$$

where c is an arbitrary constant. Identify the functions y_1 and y_2.

(b) Show that y_1 is a solution of the differential equation

$$y' + p(x)y = 0, \tag{ii}$$

corresponding to $g(x) = 0$.
(c) Show that y_2 is a solution of the full linear equation (1). We see later (for example, in Section 3.6) that solutions of higher order linear equations have a pattern similar to Eq. (i).

19. Show that if a and λ are positive constants, and b is any real number, then every solution of the equation

$$y' + ay = be^{-\lambda x}$$

has the property that $y \to 0$ as $x \to \infty$.
Hint: Consider the cases $a = \lambda$ and $a \neq \lambda$ separately.

20. **Discontinuous Coefficients.** Linear differential equations sometimes occur in which one or both of the functions p and g have jump discontinuities. If x_0 is such a point of discontinuity, then it is necessary to solve the equation separately for $x < x_0$ and for $x > x_0$. Afterward the two solutions are matched so that y is continuous at x_0; this is accomplished by a proper choice of the arbitrary constants. The following two problems illustrate this situation. Note in each case that it is impossible also to make y' continuous at x_0.
(a) Solve the initial value problem

$$y' + 2y = g(x), \qquad y(0) = 0$$

where

$$g(x) = \begin{cases} 1, & 0 \leq x \leq 1, \\ 0, & x > 1. \end{cases}$$

(b) Solve the initial value problem

$$y' + p(x)y = 0, \qquad y(0) = 1$$

where

$$p(x) = \begin{cases} 2, & 0 \leq x \leq 1, \\ 1, & x > 1. \end{cases}$$

***Bernoulli Equations.** Sometimes it is possible to solve a nonlinear equation by making a change of the dependent variable that converts it into a linear equation. The most important class of such equations is of the form

$$y' + p(x)y = q(x)y^n.$$

Such equations are called Bernoulli equations after Jakob Bernoulli. Problems 21 through 25 deal with equations of this type.

21. (a) Solve Bernoulli's equation when $n = 0$; when $n = 1$.
(b) Show that if $n \neq 0, 1$, then the substitution $v = y^{1-n}$ reduces Bernoulli's equation to a linear equation. This method of solution was found by Leibniz in 1696.

In each of Problems 22 through 25 the given equation is a Bernoulli equation. In each case solve it by using the substitution mentioned in Problem 21(b).

22. $x^2y' + 2xy - y^3 = 0, \qquad x > 0$

23. $y' = ry - ky^2$, $r > 0$ and $k > 0$. This equation is important in population dynamics and is discussed in detail in Section 2.6.
24. $y' = \epsilon y - \sigma y^3$, $\epsilon > 0$ and $\sigma > 0$. This equation occurs in the study of the stability of fluid flow.
25. $dy/dt = (\Gamma \cos t + T)y - y^3$, where Γ and T are constants. This equation also occurs in the study of the stability of fluid flow.

2.3 Nonlinear Equations

We now turn to a study of differential equations of the form

$$y' = f(x, y), \tag{1}$$

subject to an initial condition

$$y(x_0) = y_0. \tag{2}$$

The differential equation (1) and the initial condition (2) together constitute an initial value problem. The basic questions to be considered are whether a solution of this initial value problem exists, whether such a solution is unique, over what interval a solution is defined, and how to construct a useful formula for the solution. All of these questions were answered with relative ease in Sections 2.1 and 2.2 for the case in which Eq. (1) is linear. However, if f is not a linear function of the dependent variable y, then our earlier treatment no longer applies. In this section we discuss in a general way some features of nonlinear initial value problems. In particular, we note several important differences between the nonlinear problem (1), (2) and the linear problem consisting of the differential equation

$$y' + p(x)y = g(x) \tag{3}$$

and the initial condition (2).

The reason first order linear differential equations are relatively simple is that there is a formula giving the solution of such an equation in all cases. In contrast, there is no corresponding general method for solving first order nonlinear equations. In fact, the analytic determination of the solution $y = \phi(x)$ of a nonlinear equation is often impossible.

The lack of a general formula for the solution of a nonlinear equation has at least two important consequences. In the first place, methods which yield approximate, perhaps numerical, solutions and qualitative information about solutions assume greater significance for nonlinear equations than for linear ones. Chapters 8 and 9 contain a discussion of some of these methods. Second, questions dealing with the existence and uniqueness of solutions must now be dealt with by indirect methods, since a direct construction of the solution cannot be carried out in general.

EXISTENCE AND UNIQUENESS. The following fundamental existence and uniqueness theorem is analogous to Theorem 2.1 for linear equations. However, its proof is a great deal more complicated and is postponed until Section 2.12.

> ***Theorem 2.2.*** *Let the functions f and $\partial f/\partial y$ be continuous in some rectangle $\alpha < x < \beta$, $\gamma < y < \delta$ containing the point (x_0, y_0). Then, in some interval $x_0 - h < x < x_0 + h$ contained in $\alpha < x < \beta$, there is a unique solution $y = \phi(x)$ of the initial value problem* (1), (2)
>
> $$y' = f(x, y), \qquad y(x_0) = y_0.$$

The conditions stated in Theorem 2.2 are sufficient to guarantee the existence of a unique solution of the initial value problem (1), (2) in some interval $x_0 - h < x < x_0 + h$. However, the determination of a value for h may be difficult. Moreover, even if f does not satisfy the hypotheses of the theorem, it is still possible that a unique solution may exist. Indeed, the conclusion of the theorem remains true if the hypothesis about the continuity of $\partial f/\partial y$ is replaced by certain weaker conditions. Further, the existence of a solution (but not its uniqueness) can be established on the basis of the continuity of f alone, without any additional hypotheses at all. The present form of the theorem, however, is satisfactory for most purposes.

It can be shown by means of examples that some conditions on f are essential in order to obtain the result stated in the theorem. For instance, the following example shows that the initial value problem (1), (2) may have more than one solution if the hypotheses of Theorem 2.2 are violated.

EXAMPLE 1

Consider the initial value problem

$$y' = y^{1/3}, \qquad y(0) = 0 \tag{4}$$

for $x \geq 0$.

This problem is easily solved by the method of Section 2.4. For the present we can verify that the function

$$y = \phi_1(x) = \left(\tfrac{2}{3}x\right)^{3/2}, \qquad x \geq 0$$

satisfies both of Eqs. (4). On the other hand, the function

$$y = \phi_2(x) = -\left(\tfrac{2}{3}x\right)^{3/2}, \qquad x \geq 0$$

is also a solution of the given initial value problem. Moreover, the function

$$y = \psi(x) = 0, \qquad x \geq 0$$

is yet another solution. Indeed, it is not hard to show that, for an arbitrary

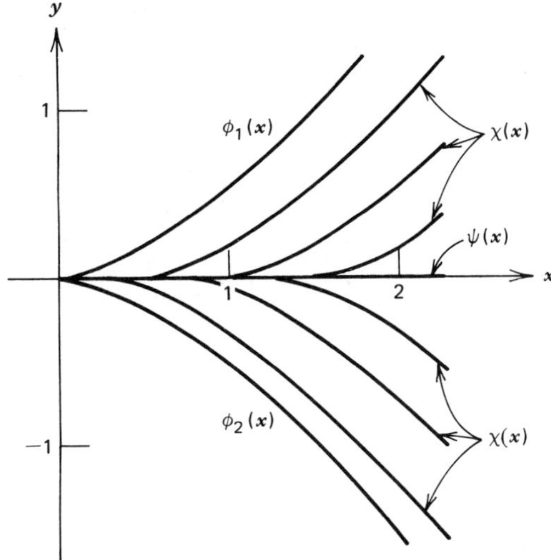

FIGURE 2.6 Several solutions of the initial value problem $y' = y^{1/3}$, $y(0) = 0$.

positive x_0, the functions

$$y = \chi(x) = \begin{cases} 0, & 0 \le x < x_0 \\ \pm\left[\frac{2}{3}(x - x_0)\right]^{3/2}, & x \ge x_0 \end{cases}$$

are continuous, differentiable (in particular at $x = x_0$), and are solutions of the initial value problem (4). Hence this problem has an infinite family of solutions; see Figure 2.6.

The nonuniqueness of the solutions of the problem (4) does not contradict the existence and uniqueness theorem since

$$\frac{\partial}{\partial y} f(x, y) = \frac{\partial}{\partial y}(y^{1/3}) = \tfrac{1}{3} y^{-2/3},$$

and this function is not continuous, or even defined, at any point where $y = 0$. Hence the theorem does not apply in any region containing any part of the x axis. If (x_0, y_0) is any point not on the x axis, however, then there is a unique solution of the differential equation $y' = y^{1/3}$ passing through (x_0, y_0).

INTERVAL OF DEFINITION. For the linear initial value problem (2), (3) the solution exists throughout any interval about $x = x_0$ in which the functions p and g are continuous. On the other hand, for the nonlinear initial value problem (1), (2), the interval in which a solution exists may be difficult to determine. The solution $y = \phi(x)$ exists as long as the point $[x, \phi(x)]$ remains within the region in which the hypotheses of the theorem are satisfied; however, since $\phi(x)$ is usually not known, it may be impossible to locate the point $[x, \phi(x)]$ with respect to this region. In any case, the interval in which a solution exists may have no simple relationship to the function f in Eq. (1). This is illustrated by the following example.

EXAMPLE 2

Consider the initial value problem

$$y' = y^2, \qquad y(0) = 1. \tag{5}$$

It can be readily verified by direct substitution that

$$y = \frac{1}{1 - x} \tag{6}$$

is the solution of this initial value problem. Clearly the solution becomes unbounded as $x \to 1$, and therefore it is valid only for $-\infty < x < 1$. There is no indication from the differential equation itself, however, that the point $x = 1$ is in any way remarkable. Moreover, if the initial condition is replaced by

$$y(0) = y_0, \tag{7}$$

it is again easy to verify that the solution of the differential equation (5) satisfying the initial condition (7) is

$$y = \frac{y_0}{1 - y_0 x}, \tag{8}$$

and that the solution now becomes unbounded as $x \to 1/y_0$. This illustrates another troublesome feature of initial value problems for nonlinear equations, namely, the singularities of the solution may depend on the initial condition.

GENERAL SOLUTION. Another way in which linear and nonlinear equations differ is in connection with the concept of a general solution. For a first order linear equation it is possible to obtain a solution containing one arbitrary constant, from which all possible solutions follow by specifying values for this constant. For nonlinear equations this may not be the case; even though a solution containing an arbitrary constant may be found, there may be other solutions that cannot be obtained by giving values to this constant. Specific examples are given in Problems 3 and 6. Thus we will use the term "general solution" only when discussing linear equations.

IMPLICIT SOLUTIONS. We recall again that for a first order linear equation there is an explicit formula [Eq. (22) of Section 2.1] for the solution $y = \phi(x)$. As long as the necessary antiderivatives can be determined, the value of the solution at any point can be found merely by substituting the appropriate value of x into the formula. For a nonlinear equation it is rarely possible to find such an explicit solution. Often the best that we can do is to eliminate the derivative that appears in Eq. (1), thereby obtaining in place of the differential equation a derivative-free equation of the form

$$\psi[x, \phi(x)] = 0 \tag{9}$$

which is satisfied by some, perhaps all, solutions $y = \phi(x)$ of Eq. (1). Even this cannot be accomplished in most cases for nonlinear equations. However, if a relation such as Eq. (9) can be found, it is customary to say that we have an *implicit formula* for the solutions of Eq. (1). An equation of the form (9) is also called an *integral* (or first integral) of Eq. (1).

For example, consider the simple nonlinear equation

$$y' = -x/y. \tag{10}$$

Using the methods of the next section, it is not hard to show that all solutions of Eq. (10) also satisfy the algebraic equation

$$x^2 + y^2 = c^2, \tag{11}$$

where c is an arbitrary positive constant. To verify that Eq. (11) is an implicit formula for solutions of Eq. (10), we can differentiate Eq. (11) with respect to x, thereby obtaining $2x + 2yy' = 0$, or $y' = -x/y$, which is Eq. (10). For $-c < x < c$, there are many functions $y = \phi(x)$ that satisfy Eq. (11). Some of these are:

$$y = \phi_1(x) = \sqrt{c^2 - x^2}, \qquad -c < x < c \tag{12}$$

$$y = \phi_2(x) = -\sqrt{c^2 - x^2}, \qquad -c < x < c \tag{13}$$

$$y = \phi_3(x) = \begin{cases} \sqrt{c^2 - x^2}, & -c < x \le 0 \\ -\sqrt{c^2 - x^2}, & 0 < x < c \end{cases} \tag{14}$$

$$y = \phi_4(x) = \begin{cases} -\sqrt{c^2 - x^2}, & -c < x \le -c/2 \\ \sqrt{c^2 - x^2}, & -c/2 < x < c/2 \\ -\sqrt{c^2 - x^2}, & c/2 \le x < c. \end{cases} \tag{15}$$

See Figure 2.7 for graphs of the functions given by Eqs. (12) to (15). However, only two of these functions satisfy the differential equation (10) over the *entire* interval $-c < x < c$, namely those given by Eqs. (12) and (13). If an initial condition is also given, then we want to select whichever of the functions ϕ_1 and ϕ_2 satisfies the specified condition, and also select the proper value of c. For example, if the initial condition is

$$y(0) = 3, \tag{16}$$

then we must discard ϕ_2, and keep ϕ_1 with c chosen to be 3. Hence

$$y = \sqrt{9 - x^2}, \qquad -3 < x < 3 \tag{17}$$

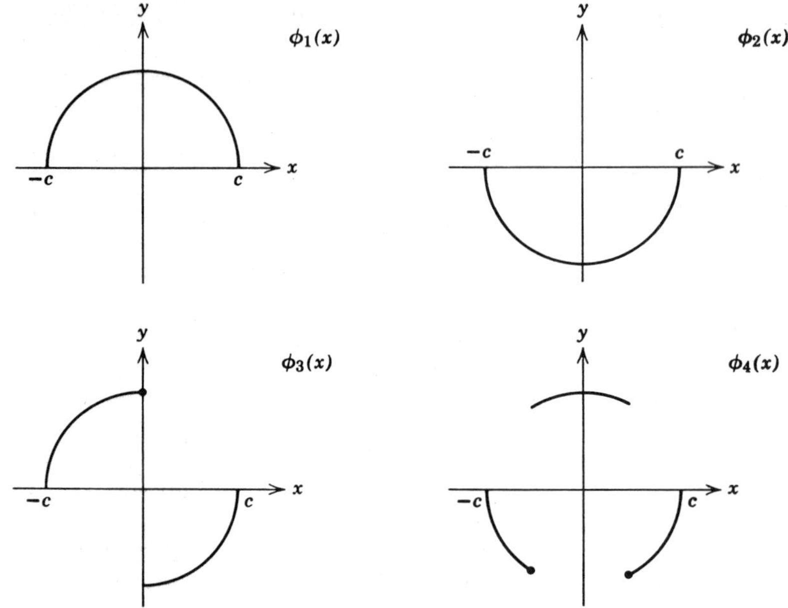

FIGURE 2.7 Several functions defined implicitly by $x^2 + y^2 = c^2$.

is a solution of the initial value problem (10) and (16). According to Theorem 2.2 there is no other solution of this problem on the interval $-3 < x < 3$.

In the foregoing example the explicit solutions given by Eqs. (12) and (13) were easily obtainable because the implicit relation (11) was quadratic in y. However, a little imagination suggests that an implicit relation (9), assuming it can be found, is often much more complicated than Eq. (11). If so, it is probably impossible to solve it (analytically) for y; it may be quite difficult even to determine the intervals in which solutions exist. Furthermore, it must be kept in mind that there may be solutions of the implicit relation (9) that do not satisfy the differential equation; also, in some cases the differential equation may have other solutions that do not satisfy the implicit relation.

GRAPHICAL OR NUMERICAL CONSTRUCTION OF INTEGRAL CURVES. Since there is no general way to obtain exact analytic solutions of nonlinear differential equations, methods that yield approximate solutions or other qualitative information about solutions may be of great importance. We reserve a discussion of numerical methods until Chapter 8; nevertheless, the student may wish to read at least the first few sections of that chapter in parallel with Chapter 2. Here we will say a little about the use of geometrical, or graphical, methods to investigate the behavior of solutions of the differential equation (1),

$$y' = f(x, y).$$

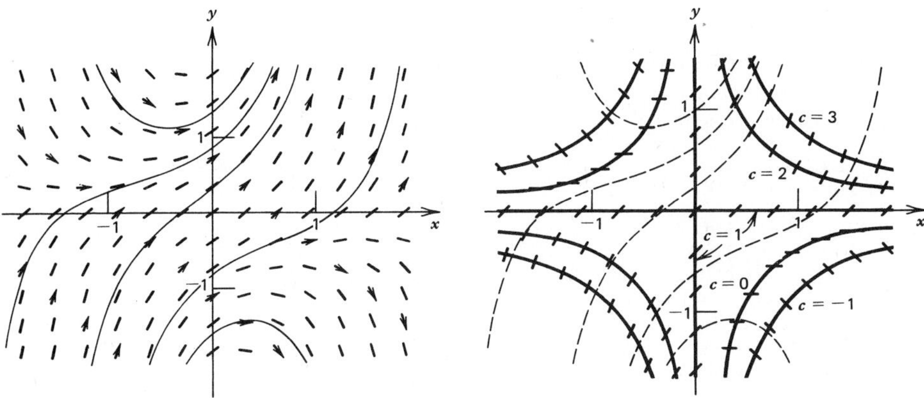

FIGURE 2.8 Direction field for $y' = 1 + 2xy$.

FIGURE 2.9 Isoclines and direction field for $y' = 1 + 2xy$.

Geometrically, the differential equation (1) says that at any point (x, y) the slope y' of the solution has the value $f(x, y)$. We can indicate this by drawing a line segment through (x, y) with slope $f(x, y)$. The collection of all such line segments is called the *direction field* for the differential equation (1). It is often helpful to visualize a direction field by drawing short line segments at some representative set of points. It is a comparatively simple task to use a computer to sketch the direction field for a given differential equation since it requires only the repeated evaluation of $f(x, y)$ for different values of x and y. For example, the direction field for

$$y' = 1 + 2xy \tag{18}$$

is shown in Figure 2.8. A solution $y = \phi(x)$ of the differential equation (1) has the property that at every point its graph is tangent to the element of the direction field at that point. For example, the integral curves of Eq. (18) were shown in Figure 2.5 of Section 2.2. They are superimposed lightly on the direction field in Figure 2.8. Thus, although a sketch of the direction field does not lead immediately to a formula for a solution, it does provide some qualitative information about solutions.

Note that for Eq. (1) the slope y' of the solution has the constant value c at all points on the curve $f(x, y) = c$; these curves are called *isoclines*. For simple equations one can often sketch the direction field by hand by drawing a few isoclines and the corresponding line segments tangent to the solution at several points on each one. For example, for the differential equation (18) the isoclines are the hyperbolas $xy = (c - 1)/2$. In Figure 2.9 we show the isoclines for $c = -1, 0, 1, 2,$ and 3 together with the corresponding tangent line segments. The reader may compare Figure 2.9 with Figure 2.8, in which the direction field for Eq. (18) was shown by means of line segments drawn on a uniformly spaced set of points.

PROBLEMS

1. For each of the following differential equations state the region in the xy plane where the existence of a unique solution through any specified point is guaranteed by the fundamental existence and uniqueness theorem.

   (a) $y' = \dfrac{x - y}{2x + 5y}$ (b) $y' = (1 - x^2 - y^2)^{1/2}$

   (c) $y' = 2xy/(1 + y^2)$ (d) $y' = 3(x + y)^{-2}$

   (e) $y' = \dfrac{\ln|xy|}{1 - x^2 + y^2}$ (f) $y' = (x^2 + y^2)^{3/2}$

2. Show that $y = \phi(x) = (1 - x^2)^{-1}$ is a solution of the initial value problem
$$y' = 2xy^2, \qquad y(0) = 1.$$
In what interval is this solution valid?

3. Show that $y = \phi(x) = [2(x + c)]^{-1/2}$, where c is an arbitrary constant, satisfies the differential equation
$$y' + y^3 = 0.$$
Find the solution that satisfies the initial condition $y(1) = 2$. Where is this solution valid? Note that $y = 0$ is also a solution of the differential equation, but that this solution cannot be obtained by assigning a value to c in $y = \phi(x)$.

4. Verify that
$$y = \phi(x) = \left[1 + \tfrac{2}{3}\ln(1 + x^3)\right]^{1/2}$$
is a solution of the initial value problem
$$y' = x^2/y(1 + x^3), \qquad y(0) = 1.$$
In what interval is this solution valid?

5. Verify that
$$y = (c^2 - 4x^2)^{1/2}, \qquad y = -(c^2 - 4x^2)^{1/2}$$
are solutions of the differential equation
$$y' = -4x/y.$$
For what parts of the xy plane are these solutions valid? Find the particular solution passing through the point $(0, 4)$; through the point $(1, -1)$.

6. (a) Verify that both $y_1(x) = 1 - x$ and $y_2(x) = -x^2/4$ are solutions of the initial value problem
$$y' = \frac{-x + (x^2 + 4y)^{1/2}}{2}, \qquad y(2) = -1.$$
Where are these solutions valid?

   (b) Explain why the existence of two solutions of the given problem does not contradict the uniqueness part of Theorem 2.2.

   (c) Show that $y = cx + c^2$, where c is an arbitrary constant, satisfies the differential equation in part (a) for $x \geq -2c$. If $c = -1$, the initial condition is also satisfied, and the solution $y = y_1(x)$ is obtained. Show that there is no choice of c that gives the second solution $y = y_2(x)$.

In each of Problems 7 through 14 determine the isoclines and use them to sketch the direction field. Then sketch a few integral curves.

7. $y' = x^2 + y^2$
8. $y' = x^2 - xy + y^2 - 1$
9. $y' = xy/(1 + x^2)$
10. $y' = (2x - 3y)/(x + y)$
11. $y' = -y(1 + y^2)$
12. $y' = y(1 - y^2)$
13. $y' = (1 - y)(2 - y)$
14. $y' + xy = 1$

2.4 Separable Equations

It is often convenient to write the equation

$$dy/dx = f(x, y) \tag{1}$$

in the form

$$M(x, y) + N(x, y)\frac{dy}{dx} = 0. \tag{2}$$

It is always possible to do this by setting $M(x, y) = -f(x, y)$ and $N(x, y) = 1$, but there may be other ways as well. In the event that M is a function of x only and N is a function of y only, then Eq. (2) takes the form

$$M(x) + N(y)\frac{dy}{dx} = 0. \tag{3}$$

For example, the equation

$$\frac{dy}{dx} = \frac{x^2}{1 + y^2}$$

can be written as

$$-x^2 + (1 + y^2)\frac{dy}{dx} = 0.$$

The differential equation (3) can also be written in the form

$$M(x)\,dx = -N(y)\,dy, \tag{4}$$

in which one side of the equation depends only on x, while the other side depends only on y. Such an equation is said to be *separable*. Now suppose that H_1 and H_2 are any functions such that

$$H_1'(x) = M(x), \qquad H_2'(y) = N(y); \tag{5}$$

then Eq. (3) becomes

$$H_1'(x) + H_2'(y)\frac{dy}{dx} = 0. \tag{6}$$

According to the chain rule

$$H_2'(y)\frac{dy}{dx} = \frac{d}{dx}H_2(y). \tag{7}$$

Consequently, if $y = \phi(x)$ is a solution of the differential equation (3), then Eq. (6) becomes

$$H_1'(x) + \frac{d}{dx}H_2[\phi(x)] = 0,$$

or

$$\frac{d}{dx}\{H_1(x) + H_2[\phi(x)]\} = 0. \tag{8}$$

Integrating Eq. (8) gives

$$H_1(x) + H_2[\phi(x)] = c, \tag{9}$$

or

$$H_1(x) + H_2(y) = c, \tag{10}$$

where c is an arbitrary constant. Thus the solution $y = \phi(x)$ of the differential equation (3) has been obtained in the implicit form (10). The functions H_1 and H_2 are any functions satisfying Eq. (5); that is, any antiderivatives of M and N, respectively. In practice the solution (10) is generally obtained from Eq. (4) by integrating the left side with respect to x and the right side with respect to y; the justification for this is the argument just given.

If, in addition to the differential equation, an initial condition

$$y(x_0) = y_0 \tag{11}$$

is prescribed, the solution of Eq. (3) satisfying this condition is obtained by setting $x = x_0$ and $y = y_0$ in Eq. (10). This gives

$$c = H_1(x_0) + H_2(y_0). \tag{12}$$

Substituting this value for c in Eq. (10), and noting that

$$H_1(x) - H_1(x_0) = \int_{x_0}^{x} M(t)\,dt, \qquad H_2(y) - H_2(y_0) = \int_{y_0}^{y} N(t)\,dt,$$

we obtain

$$\int_{x_0}^{x} M(t)\,dt + \int_{y_0}^{y} N(t)\,dt = 0. \tag{13}$$

Equation (13) is an implicit representation of the solution of the differential equation (3), which also satisfies the initial condition (11). The reader should bear in mind that the determination of the solution in explicit form, or even the determination of the precise interval in which the solution exists, generally requires that Eq. (13) be solved for y as a function of x. This may present formidable difficulties.

EXAMPLE 1

Solve the initial value problem

$$\frac{dy}{dx} = \frac{3x^2 + 4x + 2}{2(y - 1)}, \qquad y(0) = -1. \tag{14}$$

The differential equation can be written as

$$2(y-1)\,dy = (3x^2+4x+2)\,dx.$$

Integrating the left side with respect to y and the right side with respect to x gives

$$y^2 - 2y = x^3 + 2x^2 + 2x + c, \tag{15}$$

where c is an arbitrary constant. To determine the solution satisfying the prescribed initial condition we substitute $x = 0$ and $y = -1$ into Eq. (15), obtaining $c = 3$. Hence the solution of the initial value problem (14) is given implicitly by

$$y^2 - 2y = x^3 + 2x^2 + 2x + 3. \tag{16}$$

To obtain the solution explicitly, we must solve Eq. (16) for y in terms of x. This is a simple matter in this case since Eq. (16) is quadratic in y, and we obtain

$$y = 1 \pm \sqrt{x^3 + 2x^2 + 2x + 4}\,. \tag{17}$$

Thus Eq. (17) gives two solutions of the differential equation, only one of which, however, satisfies the given initial condition. This is the solution corresponding to the minus sign in Eq. (17), so that we finally obtain

$$y = \phi(x) = 1 - \sqrt{x^3 + 2x^2 + 2x + 4} \tag{18}$$

as the solution of the initial value problem (14). Note that if the plus sign is chosen by mistake in Eq. (17), then we obtain the solution of the same differential equation that satisfies the initial condition $y(0) = 3$. Finally, to determine the interval in which the solution (18) is valid, we must find the interval in which the quantity under the radical is positive. By plotting this expression as a function of x, the reader can show that the desired interval is $x > -2$. The solution of the initial value problem and some other integral curves of the differential equation are shown in Figure 2.10. Observe that the boundary of the interval of validity of the solution (18) is the point $(-2, 1)$ at which the tangent line is vertical.

EXAMPLE 2

Find the solution of the initial value problem

$$\frac{dy}{dx} = \frac{y\cos x}{1 + 2y^2}, \qquad y(0) = 1. \tag{19}$$

First we write the differential equation in the form

$$\frac{1 + 2y^2}{y}\,dy = \cos x\,dx. \tag{20}$$

Integrating the left side with respect to y and the right side with respect to x, we obtain

$$\ln|y| + y^2 = \sin x + c. \tag{21}$$

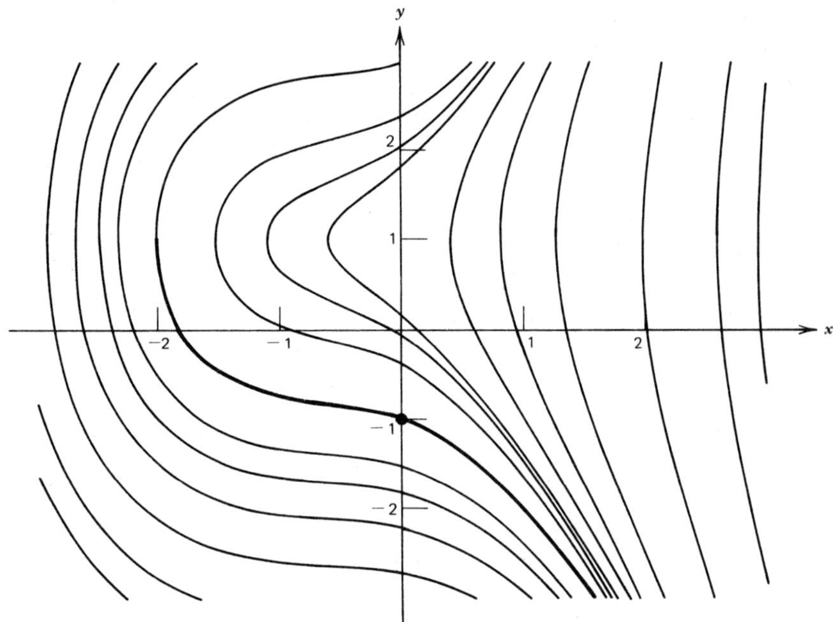

FIGURE 2.10 Integral curves of $y' = (3x^2 + 4x + 2)/2(y - 1)$.

To satisfy the initial condition we substitute $x = 0$ and $y = 1$ in Eq. (21); this gives $c = 1$. Hence the solution of the initial value problem (19) is given implicitly by

$$\ln|y| + y^2 = \sin x + 1. \tag{22}$$

Since Eq. (22) is not readily solved for y as a function of x, further analysis of this problem becomes more delicate. One fairly evident fact is that no solution crosses the x axis. To see this, observe that the left side of Eq. (22) becomes infinite if $y = 0$; however, the right side never becomes infinite, so no point on the x axis satisfies Eq. (22). Thus, for the solution of Eqs. (19) it follows that $y > 0$ always. Consequently, the absolute value bars in Eq. (22) can be dropped for this problem. It can also be shown that the interval of definition of the solution of the initial value problem (19) is the entire x axis, $-\infty < x < \infty$. Problem 21 indicates how to do this. Some solutions of the differential equation (19) are shown in Figure 2.11.

In Example 1 it was not difficult to solve explicitly for y as a function of x and to determine the exact interval in which the solution exists. However, this situation is exceptional, and often it will be necessary to leave the solution in implicit form, as in Example 2. Thus, in the problems below and in those following Sections 2.8 through 2.11, the terminology "Solve the following dif-

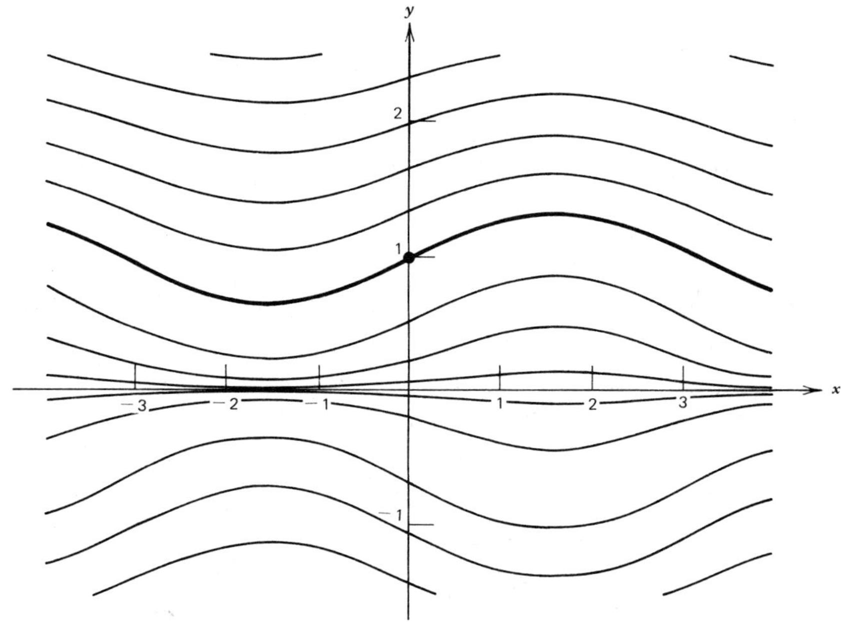

FIGURE 2.11 *Integral curves of* $y' = (y \cos x)/(1 + 2y^2)$.

ferential equation" means to find the solution explicitly if it is convenient to do so, but otherwise to find an implicit formula for the solution.

PROBLEMS

Solve each of the equations in Problems 1 through 8. State the regions of the xy plane in which the conditions of the fundamental existence and uniqueness theorem are satisfied.

1. $y' = x^2/y$
2. $y' = x^2/y(1 + x^3)$
3. $y' + y^2 \sin x = 0$
4. $y' = 1 + x + y^2 + xy^2$
5. $y' = (\cos^2 x)(\cos^2 2y)$
6. $xy' = (1 - y^2)^{1/2}$
7. $\dfrac{dy}{dx} = \dfrac{x - e^{-x}}{y + e^y}$
8. $\dfrac{dy}{dx} = \dfrac{x^2}{1 + y^2}$

For each of Problems 9 through 15 find the solution of the given initial value problem in explicit form, and determine (at least approximately) the interval in which it is defined.

9. $\sin 2x\, dx + \cos 3y\, dy = 0, \quad y(\pi/2) = \pi/3$
10. $x\, dx + ye^{-x}\, dy = 0, \quad y(0) = 1$
11. $dr/d\theta = r^2/\theta, \quad r(1) = 2$
12. $y' = 2x/(y + x^2 y), \quad y(0) = -2$
13. $y' = xy^3(1 + x^2)^{-1/2}, \quad y(0) = 1$
14. $y' = 2x/(1 + 2y), \quad y(2) = 0$
15. $y' = x(x^2 + 1)/4y^3, \quad y(0) = -1/\sqrt{2}$

16. Solve the initial value problem

$$y' = 3x^2/(3y^2 - 4), \qquad y(1) = 0$$

and determine the interval in which the solution is valid.
Hint: To find the interval of definition look for points where $dx/dy = 0$.

17. Solve the equation

$$y^2(1 - x^2)^{1/2}\,dy = \sin^{-1} x\,dx$$

in the interval $-1 < x < 1$.

18. Solve the equation

$$\frac{dy}{dx} = \frac{ax + b}{cx + d},$$

where a, b, c, and d are constants.

19. Solve the equation

$$\frac{dy}{dx} = \frac{ay + b}{cy + d},$$

where a, b, c, and d are constants.

*20. Show that the equation

$$\frac{dy}{dx} = \frac{y - 4x}{x - y}$$

is not separable, but that if the variable y is replaced by a new variable v defined by $v = y/x$, then the equation is separable in x and v. Find the solution of the given equation in this way. See Section 2.10 for a further discussion of this method.

*21. Consider again the initial value problem (19) of the text. Denote the right side of the differential equation by $f(x, y)$ and observe that

$$|f(x, y)| = \left|\frac{y}{1 + 2y^2}\right| |\cos x|.$$

(a) By finding the maximum and minimum values of $y/(1 + 2y^2)$, show that

$$\left|\frac{y}{1 + 2y^2}\right| \le \frac{1}{2\sqrt{2}}$$

for all y, and hence that $|f(x, y)| \le 1/2\sqrt{2}$ for all x and y.
(b) If $y = \phi(x)$ is the solution of the initial value problem (19), use the result of part (a) to show that $|\phi(x) - 1| \le |x|/2\sqrt{2}$ for all x. Hence conclude that the interval of definition of the solution ϕ is $-\infty < x < \infty$.

2.5 *Applications of First Order Linear Equations*

Differential equations are of interest to nonmathematicians primarily because of the possibility of using them to investigate a wide variety of problems in the physical, biological, and social sciences. Three identifiable steps in this process are present regardless of the specific field of application.

In the first place, it is necessary to translate the physical situation into mathematical terms. This is generally done by making assumptions about what is happening that appear to be consistent with the observed phenomena. For example, it has been observed that radioactive materials decay at a rate proportional to the amount of the material present, that heat passes from a warmer to a cooler body at a rate proportional to the temperature difference, that objects move about in accordance with Newton's laws of motion, and that isolated insect populations grow at a rate proportional to the current population. Each of these statements involves a rate of change (derivative) and consequently, when expressed mathematically, takes the form of a differential equation.

It is important to realize that the mathematical equations are almost always only an approximate description of the actual process because they are based on observations that are themselves approximations. For example, bodies moving at speeds comparable to the speed of light are not governed by Newton's laws, insect populations do not grow indefinitely as stated because of eventual limitations on their food supply, and heat transfer is affected by factors other than the temperature difference. Moreover, the process of formulating a physical problem mathematically often involves the conceptual replacement of a discrete process by a continuous one. For instance, the number of members in an insect population changes by discrete amounts; however, if the population is large, it seems reasonable to consider it as a continuous variable and even to speak of its derivative. Alternatively, one can adopt the point of view that the mathematical equations exactly describe the operation of a simplified model, which has been constructed (or conceived of) so as to embody the most important features of the actual process.

In any case, once the problem has been formulated mathematically, one is often faced with the problem of solving one or more differential equations or, failing that, of finding out as much as possible about the properties of the solution. It may happen that this mathematical problem is quite difficult and, if so, further approximations may be indicated at this stage to make the problem mathematically tractable. For example, a nonlinear equation may be approximated by a linear one, or a slowly varying function may be replaced by its average value. Naturally, any such approximations must also be examined from the physical point of view to make sure that the simplified mathematical problem still reflects the essential features of the physical process under investigation. At the same time, an intimate knowledge of the physics of the problem may suggest reasonable mathematical approximations that will make the mathematical problem more amenable to analysis. This interplay of understanding of physical phenomena and knowledge of mathematical techniques and their limitations is characteristic of applied mathematics at its best, and is indispensable in successfully constructing mathematical models of intricate physical processes.

Finally, having obtained the solution (or at least some information about it), one must interpret it in terms of the context in which the problem arose. In particular, one should always check as to whether the mathematical solution appears physically reasonable. This requires, at the very least, that the solution exist, that it be unique, and that it depend continuously on the data of the problem. This last consideration is important because the coefficients in the differential equation and initial conditions are often obtained as a result of measurements of some physical quantity and therefore are susceptible to small errors. If these small errors lead to large (or discontinuous) changes in the solution of the corresponding mathematical problem, which are not observed physically, then the relevance of the mathematical model to the physical problem must be reexamined. Of course, the fact that the mathematical solution appears to be reasonable does not guarantee that it is correct. However, if it is seriously inconsistent with careful observations of the physical system it purports to describe, this suggests either that errors have been made in solving the mathematical problem or that the mathematical model itself is too crude.

The examples in this section are typical of applications in which first order linear differential equations arise.

EXAMPLE 1

Radioactive Decay The radioactive isotope thorium-234 disintegrates at a rate proportional to the amount present. If 100 mg of this material is reduced to 82.04 mg in one week, find an expression for the amount present at any time. Also, find the time interval that must elapse for the mass to decay to one half of its original value.

Let $Q(t)$ be the amount of thorium-234 present at any time t, where Q is measured in milligrams and t in days. The physical observation that thorium-234 disintegrates at a rate proportional to the amount present means that the time rate of change dQ/dt is proportional to Q. Thus Q satisfies the differential equation

$$\frac{dQ}{dt} = -rQ, \tag{1}$$

where the constant $r > 0$ is known as the decay rate. We seek the solution of Eq. (1) that also satisfies the initial condition

$$Q(0) = 100 \tag{2}$$

as well as the condition

$$Q(7) = 82.04. \tag{3}$$

Equation (1) is a linear equation and also separable; its general solution is

$$Q(t) = ce^{-rt}, \tag{4}$$

where c is an arbitrary constant. The initial condition (2) requires that $c = 100$, and therefore

$$Q(t) = 100e^{-rt}. \tag{5}$$

To satisfy Eq. (3) we set $t = 7$ and $Q = 82.04$ in Eq. (5); this gives

$$82.04 = 100e^{-7r},$$

hence

$$r = -\frac{\ln 0.8204}{7} = 0.02828 \text{ days}^{-1}. \tag{6}$$

Thus the decay rate r has been determined. Using this value of r in Eq. (5), we obtain

$$Q(t) = 100e^{-0.02828t} \text{ mg}, \tag{7}$$

which gives the value of $Q(t)$ at any time.

The time period during which the mass is reduced to one-half of its original value is known as the *half-life* of the material. Let τ be the time at which $Q(t)$ is equal to 50 mg. Then, from Eq. (5),

$$50 = 100e^{-r\tau},$$

or

$$r\tau = \ln 2. \tag{8}$$

The relation (8) between the decay rate and the half-life is valid not only for thorium-234 but for any material that obeys the differential equation (1); using the value of r given by Eq. (6), we find that for thorium-234

$$\tau = \frac{\ln 2}{0.02828} \cong 24.5 \text{ days}. \tag{9}$$

EXAMPLE 2

Compound Interest Suppose that a sum of money S_0 is deposited in a bank or money market fund that pays interest at a rate r. The value $S(t)$ of the investment at the end of t years depends on the frequency with which the interest is compounded as well as on the interest rate.

If interest is compounded once a year, then

$$S(t) = S_0(1 + r)^t.$$

If interest is compounded twice a year, then at the end of six months the value of the investment is $S_0[1 + (r/2)]$, and at the end of one year it is $S_0[1 + (r/2)]^2$. Thus after t years we have

$$S(t) = S_0[1 + (r/2)]^{2t}.$$

In general, if interest is compounded m times per year, then

$$S(t) = S_0[1 + (r/m)]^{mt}. \tag{10}$$

Many financial institutions pay interest daily, corresponding to $m = 365$. This value of m is large enough so that it is of interest to consider what happens in the limiting case as $m \to \infty$, which means that interest is compounded *continuously*. By letting $m \to \infty$ in Eq. (10) and recalling from calculus that $[1 + (x/m)]^m \to e^x$ as $m \to \infty$, we obtain

$$S(t) = \lim_{m \to \infty} S_0\left(1 + \frac{r}{m}\right)^{mt} = S_0 e^{rt}. \tag{11}$$

Table 2.1 shows the effect of changing the frequency of compounding for an interest rate r of 8%. The second and third columns are calculated from Eq. (10) for quarterly and daily compounding, respectively, and the fourth column is calculated from Eq. (11), for continuous compounding. The results show that the frequency of compounding is not particularly important in most cases. For example, over a 10-year period the difference between quarterly and continuous compounding is \$17.50 per \$1000 invested, or less than \$2 per year. The difference would be somewhat greater for higher interest rates and less for lower rates.

From the first row in the table we see that, for the interest rate $r = 8\%$, the annual yield for quarterly compounding is 8.24% and for daily or continuous compounding it is 8.33%. Many banks advertise an annual yield even higher than that obtained by continuous compounding. This is accomplished by calculating a daily interest rate using a nominal year of 360 days, and then compounding this rate through the actual calendar year. Using this method for an interest rate r, and ignoring leap years, we find that

$$S(t) = S_0\left(1 + \frac{r}{360}\right)^{365t}. \tag{12}$$

Results from Eq. (12) are given in the last column of Table 2.1 for an interest rate of 8%. Observe that the effective annual yield is 8.45%.

Let us now set up an initial value problem that describes an investment program with continuous compounding. Let $S(t)$ be the balance in the account at

TABLE 2.1 GROWTH OF CAPITAL AT AN INTEREST RATE $r = 8\%$ FOR SEVERAL MODES OF COMPOUNDING

Years	$S(t)/S(t_0)$ from Eq. (10) $m = 4$	$S(t)/S(t_0)$ from Eq. (10) $m = 365$	$S(t)/S(t_0)$ from Eq. (11)	$S(t)/S(t_0)$ from Eq. (12)
1	1.0824	1.0833	1.0833	1.0845
2	1.1716	1.1735	1.1735	1.1761
5	1.4859	1.4918	1.4918	1.5001
10	2.2080	2.2253	2.2255	2.2502
20	4.8754	4.9522	4.9530	5.0634
30	10.7652	11.0202	11.0232	11.3937
40	23.7699	24.5238	24.5325	25.6382

time t. Then the rate of change dS/dt must be equal to the net rate at which funds are added to the account from all sources. The rate of accumulation due to interest is rS. Moreover, suppose that additional funds are continuously deposited at the constant rate k. Then

$$dS/dt = rS + k. \tag{13}$$

The initial condition is

$$S(0) = S_0. \tag{14}$$

We are assuming here that $k > 0$, but it is also possible for k to be negative. In the latter case, withdrawals are being made from the account, as in an annuity.

The general solution of Eq. (13) is

$$S(t) = ce^{rt} - (k/r), \tag{15}$$

where c is an arbitrary constant. To satisfy the initial condition (14) we must choose $c = S_0 + (k/r)$. Thus the solution of the initial value problem (13), (14) is

$$S(t) = S_0 e^{rt} + \frac{k}{r}(e^{rt} - 1). \tag{16}$$

The first term in expression (16) for $S(t)$ is the capital due to the initial investment S_0, while the second term gives the effect of the continuing investment program.

The attractiveness of stating the problem in a general way without specific values for S_0, r, and k lies in the generality of the resulting formula (16) for $S(t)$. With this formula we can readily compare the results of different investment programs and different interest rates.

For instance, suppose that one opens an Individual Retirement Account (IRA) at age 25 with an initial investment of \$2000, and then makes annual investments of \$2000 thereafter in a continuous manner. Assuming an interest rate of 8%, what will be the balance in the IRA at age 65? We have $S_0 = \$2000$, $r = 0.08$, and $k = \$2000$, and we wish to determine $S(40)$. From Eq. (18) we have

$$\begin{aligned} S(40) &= (2000)e^{3.2} + (25{,}000)(e^{3.2} - 1) \\ &= \$49{,}065 + \$588{,}313 = \$637{,}378. \end{aligned}$$

It is interesting to note that the total investment I is \$2000 + \$80,000, so the ratio $F(40)/I \simeq 7.8$.

Let us now examine the assumptions that have gone into the model. First, we have assumed that interest is compounded continuously. Over time periods of probable concern this does not lead to serious errors, as Table 2.1 shows, especially since many financial institutions compound interest daily. Second, we have assumed that additional capital is invested continuously. This is perhaps a more serious approximation, but the effect can be reduced if the investor makes deposits frequently, say monthly or even weekly. Finally, we have assumed that the interest rate r is constant for the entire period involved, whereas in fact interest rates have fluctuated considerably in recent years. Although we cannot reliably predict future interest rates, we can use expression (16) to determine the approximate effect of different interest rate projections. We can also take changes in r into account by using Eq. (16) over shorter time intervals. For example, the value $S(t_1)$ at the end of t_1 years with interest rate r_1 provides the initial condition for calculating $S(t_2)$, for the time period $t_2 - t_1$ with interest rate r_2, and so forth. In the same way one can also accommodate changes in the investment rate k.

EXAMPLE 3

Mixing At time $t = 0$ a tank contains Q_0 lb of salt dissolved in 100 gal of water; see Figure 2.12. Assume that water containing $\frac{1}{4}$ lb of salt per gallon is entering the tank at a rate of 3 gal/min, and that the well-stirred solution is leaving the tank at the same rate. Find an expression for the amount of salt $Q(t)$ in the tank at time t.

The rate of change of salt in the tank at time t, $Q'(t)$, must equal the rate at which salt enters the tank minus the rate at which it leaves. The rate at which salt enters is $\frac{1}{4}$ lb/gal times 3 gal/min. The rate at which salt leaves is $(Q(t)/100)$ lb/gal times 3 gal/min. Thus

$$Q'(t) = \tfrac{3}{4} - \tfrac{3}{100}Q(t) \tag{17}$$

is the differential equation governing this process. Equation (17) is linear and its general solution is

$$Q(t) = 25 + ce^{-0.03t}, \tag{18}$$

where c is arbitrary. To satisfy the initial condition

$$Q(0) = Q_0 \tag{19}$$

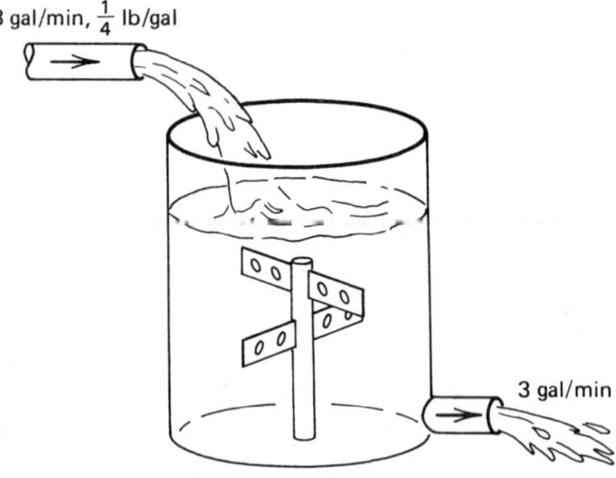

FIGURE 2.12 The water tank in Example 3.

we must take $c = Q_0 - 25$; hence

$$Q(t) = 25(1 - e^{-0.03t}) + Q_0 e^{-0.03t}. \tag{20}$$

The second term on the right side of Eq. (20) represents the portion of the original salt remaining in the tank at time t. This term becomes very small with the passage of time as the original solution is drained from the tank. The first term on the right side of Eq. (20) gives the amount of salt in the tank at time t due to the action of the flow processes. As t becomes large this term approaches the constant value 25 (pounds). It is also clear physically that this must be the limiting value of Q as the original solution in the tank is more and more completely replaced by that entering with a concentration of $\frac{1}{4}$lb/gal.

EXAMPLE 4

Determination of the Time of Death[4] In the investigation of a homicide or accidental death it is often important to estimate the time of death. Here we describe a mathematical way to approach this problem.

From experimental observations it is known that, to an accuracy satisfactory in many circumstances, the surface temperature of an object changes at a rate proportional to the difference between the temperature of the object and that of the surrounding environment (the ambient temperature). This is known as Newton's law of cooling. Thus, if $\theta(t)$ is the temperature of the object at time t, and T is the constant ambient temperature, then θ must satisfy the linear differential equation

$$d\theta/dt = -k(\theta - T), \tag{21}$$

where $k > 0$ is a constant of proportionality. The minus sign in Eq. (21) is due to the fact that if the object is warmer than its surroundings, ($\theta > T$), then it will become cooler with time. Thus $d\theta/dt < 0$ when $\theta - T > 0$.

Now suppose that at time $t = 0$ a corpse is discovered, and that its temperature is measured to be θ_0. We assume that at the time of death t_d the body temperature θ_d had the normal value of 98.6°F or 37°C. If we assume that Eq. (21) is valid in this situation, then our task is to determine t_d.

The solution of Eq. (21) subject to the initial condition $\theta(0) = \theta_0$ is

$$\theta(t) = T + (\theta_0 - T)e^{-kt}. \tag{22}$$

However, the cooling rate k that appears in this expression is as yet unknown. We can determine k by making a second measurement of the body's temperature at some later time t_1; suppose that $\theta = \theta_1$ when $t = t_1$. By substituting these values in Eq. (22) we find that

$$\theta_1 - T = (\theta_0 - T)e^{-kt_1},$$

[4]See J. F. Hurley, "An Application of Newton's Law of Cooling," *Mathematics Teacher 67* (1974), pp. 141–142, and David A. Smith, "The Homicide Problem Revisited," *The Two Year College Mathematics Journal 9* (1978), pp. 141–145.

hence

$$k = -\frac{1}{t_1} \ln \frac{\theta_1 - T}{\theta_0 - T}, \tag{23}$$

where θ_0, θ_1, T, and t_1 are known quantities.

Finally, to determine t_d we substitute $t = t_d$ and $\theta = \theta_d$ in Eq. (22) and then solve for t_d. We obtain

$$t_d = -\frac{1}{k} \ln \frac{\theta_d - T}{\theta_0 - T}, \tag{24}$$

where k is given by Eq. (23).

For example, suppose that the temperature of the corpse is 85°F when discovered and 74°F two hours later, and that the ambient temperature is 68°F. Then, from Eq. (23)

$$k = -\frac{1}{2} \ln \frac{74 - 68}{85 - 68} \simeq 0.5207 \text{ hr}^{-1},$$

and from Eq. (24)

$$t_d = -\frac{1}{0.5207} \ln \frac{98.6 - 68}{85 - 68} \simeq -1.129 \text{ hr}.$$

Thus we conclude that the body was discovered approximately 1 hr, 8 min after death.

PROBLEMS

1. The radioactive isotope plutonium-241 decays so as to satisfy the differential equation

$$dQ/dt = -0.0525Q,$$

where Q is measured in milligrams and t in years.
(a) Determine the half-life τ of plutonium-241.
(b) If 50 mg of plutonium are present today, how much will remain in 10 years?

2. Einsteinium-253 decays at a rate proportional to the amount present. Determine the half-life τ if this material loses one-third of its mass in 11.7 days.
3. Radium-226 has a half-life of 1620 years. Find the time period during which a body of this material is reduced to three-fourths of its original size.
4. Suppose that 100 mg of thorium-234 are initially present in a closed container, and that thorium-234 is added to the container at a constant rate of 1 mg/day.
(a) Find the amount $Q(t)$ of thorium-234 in the container at any time. Recall that the decay rate for thorium-234 was found in Example 1.
(b) Find the limiting amount Q_l of thorium-234 in the container as $t \to \infty$.
(c) How long a time period must elapse before the amount of thorium-234 in the container drops to within 0.5 mg of the limiting value Q_l?
(d) If thorium-234 is added to the container at a rate of k mg/day, find the value of k that is required to maintain a constant level of 100 mg of thorium-234.

5. **Radiocarbon Dating.** An important tool in archeological research is *radiocarbon dating*. This is a means of determining the age of certain wood and plant remains, hence of animal or human bones or artifacts found buried at the same levels. The procedure was developed by the American chemist Willard Libby (1908–1980) in the early 1950s and resulted in his winning the Nobel prize for chemistry in 1960. Radiocarbon dating is based on the fact that some wood or plant remains contain residual amounts of carbon-14, a radioactive isotope of carbon. This isotope is accumulated during the lifetime of the plant and begins to decay at its death. Since the half-life of carbon-14 is long (approximately 5568 years[5]), measurable amounts of carbon-14 remain after many thousands of years. Libby showed that if even a tiny fraction of the original amount of carbon-14 is still present, then by appropriate laboratory measurements the *proportion* of the original amount of carbon-14 that remains can be accurately determined. In other words, if $Q(t)$ is the amount of carbon-14 at time t and Q_0 is the original amount, then the ratio $Q(t)/Q_0$ can be determined, at least if this quantity is not too small. Present measurement techniques permit the use of this method for time periods up to about 100,000 years, after which the amount of carbon-14 remaining is only about 4×10^{-6} of the original amount.
   (a) Assuming that Q satisfies the differential equation $Q' = -rQ$, determine the decay constant r for carbon-14.
   (b) Find an expression for $Q(t)$ at any time t, if $Q(0) = Q_0$.
   (c) Suppose that certain remains are discovered in which the current residual amount of carbon-14 is 20% of the original amount. Determine the age of these remains.
6. Suppose that a sum S_0 is deposited in a bank that pays interest at an annual rate r compounded continuously.
   (a) Find the time T required for the original sum to double in value as a function of the interest rate r.
   (b) Determine T if $r = 7\%$.
   (c) Find the interest rate that must be paid if the initial investment is to double in eight years.
7. A young person with no initial capital invests k dollars per year at an annual interest rate r. Assume that investments are made continuously and that interest is compounded continuously.
   (a) Determine the sum $S(t)$ accumulated at any time t.
   (b) If $r = 7.5\%$ determine k so that one million dollars will be available for retirement in 40 years.
   (c) If $k = \$2000$ per year, determine the interest rate r that must be obtained to have one million dollars available in 40 years.
   Hint: Use Newton's method or some other appropriate numerical procedure in part (c).
8. The effect of a small change in the interest rate can be substantial for a long investment period. Confirm this by calculating the result for the IRA investment program in Example 2 if (a) $r = 7.5\%$; (b) $r = 9\%$.
9. Determine the sum accumulated in 20 years by each of the following investment programs. Suppose that 8 percent interest is compounded, and deposits made, continuously.

[5] The internationally agreed half-life of carbon-14 is 5568 ± 30 years, as reported in the *McGraw-Hill Encyclopedia of Science and Technology* (5th ed.) (New York: McGraw-Hill, 1982), Vol. 11, pp. 328–335.

(a) Nothing initially and \$1500 per year for 20 years.
(b) \$10,000 initially and \$1000 per year for 20 years.
(c) \$20,000 initially and \$500 per year for 20 years.
(d) \$30,000 initially and no additional deposits.
Note that in each case the total amount invested is \$30,000.

10. A retired person has a sum $S(t)$ invested so as to draw interest at an annual rate r compounded continuously. Withdrawals for living expenses are made at a rate of k dollars per year; assume that the withdrawals are made continuously.
(a) If the initial value of the investment is S_0, determine $S(t)$ at any time.
(b) Assuming that S_0 and r are fixed, determine the withdrawal rate k_0 at which $S(t)$ will remain constant.
(c) If k exceeds the value k_0 found in part (b), then $S(t)$ will decrease and ultimately become zero. Find the time T at which $S(t) = 0$.
(d) Determine T if $r = 8\%$ and $k = 2k_0$.
(e) Suppose that a person retiring with capital S_0 wishes to withdraw funds at an annual rate k for not more than T years. Determine the maximum possible rate of withdrawal.
(f) How large an initial investment is required to permit an annual withdrawal of \$12,000 for 20 years, assuming an interest rate of 8%?

11. In this problem we consider several 40-year investment programs using the notation of Example 2. In each case determine the accumulated capital $S(40)$ and the ratio $S(40)/I$, where I is the total amount invested.
(a) $S_0 = \$3000$, $k = \$1200/\text{year}$, $r = 8\%$ for the first 20 years and 10% for the next 20 years.
(b) $S_0 = \$3000$, $r = 8\%$, $k = \$1200/\text{year}$ for the first 20 years and \$1800/year for the next 20 years.
(c) $S_0 = \$3000$, $k = \$1200/\text{year}$, $r = 8\%$ for the first 20 years; $k = \$1800/\text{year}$, $r = 10\%$ for the next 20 years.

12. Assume the population of the earth changes at a rate proportional to the current population. (A more accurate hypothesis concerning population growth is considered in Section 2.6.) Further, it is estimated that at time $t = 0$ (A.D. 1650) the earth's population was 600 million (6.0×10^8); at time $t = 300$ (A.D. 1950) its population was 2.8 billion (2.8×10^9). Find an expression giving the population of the earth at any time. Assuming that the greatest population the earth can support is 25 billion (2.5×10^{10}), when will this limit be reached?

13. Suppose that the temperature of a cup of coffee obeys Newton's law of cooling. If the coffee has a temperature of 200°F when freshly poured, and one minute later has cooled to 190°F in a room at 70°F, determine when the coffee reaches a temperature of 150°F.

14. Find the interval between the time of death and the time of discovery of a corpse if the circumstances are as in Example 4, except that the ambient temperature is 32°F.

15. Suppose that a body with temperature 85°F is discovered at midnight, and that the ambient temperature is a constant 70°F. The body is removed quickly (assume instantly) to the morgue where the ambient temperature is maintained at 40°F. After one hour the body temperature is found to be 60°F. Estimate the time of death.

16. Assume that a spherical raindrop evaporates at a rate proportional to its surface area. If its radius originally is 3 mm, and one-half hour later has been reduced to 2 mm, find an expression for the radius of the raindrop at any time.

17. Consider a tank used in certain hydrodynamic experiments. After one experiment the tank contains 200 liters of a dye solution with a concentration of 1 g/liter. To prepare for the next experiment the tank is to be rinsed with fresh water flowing in at a rate of 2 liters/min, the well-stirred solution flowing out at the same rate. Find the time which will elapse before the concentration of dye in the tank reaches 1% of its original value.
18. A tank originally contains 100 gal of fresh water. Then water containing $\frac{1}{2}$ lb of salt per gallon is poured into the tank at a rate of 2 gal/min, and the mixture is allowed to leave at the same rate. After 10 min the process is stopped, and fresh water is poured into the tank at a rate of 2 gal/min, with the mixture again leaving at the same rate. Find the amount of salt in the tank at the end of 20 min.
19. A tank with a capacity of 500 gal originally contains 200 gal of water with 100 lb of salt in solution. Water containing 1 lb of salt per gallon is entering at a rate of 3 gal/min, and the mixture is allowed to flow out of the tank at a rate of 2 gal/min. Find the amount of salt in the tank at any time prior to the instant when the solution begins to overflow. Find the concentration (in pounds per gallon) of salt in the tank when it is on the point of overflowing. Compare this concentration with the theoretical limiting concentration if the tank had infinite capacity.
20. Suppose that a room containing 1200 ft^3 of air is originally free of carbon monoxide. Beginning at time $t = 0$ cigarette smoke, containing 4% carbon monoxide, is introduced into the room at a rate of 0.1 ft^3/min, and the well-circulated mixture is allowed to leave the room at the same rate.
(a) Find an expression for the concentration $x(t)$ of carbon monoxide in the room at any time $t > 0$.
(b) Extended exposure to a carbon monoxide concentration as low as 0.00012 is harmful to the human body. Find the time τ at which this concentration is reached.
21. Consider a lake of constant volume V containing at time t an amount $Q(t)$ of pollutant, evenly distributed throughout the lake with a concentration $c(t)$, where $c(t) = Q(t)/V$. Assume that water containing a concentration k of pollutant enters the lake at a rate r, and that water leaves the lake at the same rate. Suppose that pollutants are also added directly to the lake at a constant rate P. Note that the given assumptions neglect a number of factors that may, in some cases, be important; for example, the water added or lost by precipitation, absorption, and evaporation; the stratifying effect of temperature differences in a deep lake; the tendency of irregularities in the coastline to produce sheltered bays; and the fact that pollutants are not deposited evenly throughout the lake, but (usually) at isolated points around its periphery. The results below must be interpreted in the light of the neglect of such factors as these.

TABLE 2.2 VOLUME AND FLOW DATA FOR THE GREAT LAKES

Lake	V ($\text{km}^3 \times 10^{-3}$)	r (km^3/year)
Superior	12.2	65.2
Michigan	4.9	158
Erie	0.46	175
Ontario	1.6	209

(a) If at time $t = 0$ the concentration of pollutant is c_0, find an expression for the concentration $c(t)$ at any time. What is the limiting concentration as $t \to \infty$?
(b) If the addition of pollutants to the lake is terminated ($k = 0$ and $P = 0$ for $t > 0$), determine the time interval T that must elapse before the concentration of pollutants is reduced to 50% of its original value; to 10% of its original value.
(c) Table 2.2 contains data[6] for several of the Great Lakes. Using these data determine from part (b) the time T necessary to reduce the contamination of each of these lakes to 10% of the original value.

2.6 Population Dynamics and Some Related Problems

In a variety of applications, ranging from medicine to ecology to global economics, it is desirable to predict the future growth or decline of the population of a given species. In different situations we may be interested in a population of bacteria, insects, mammals, or even people. Similar equations also govern many other types of phenomena, and some are mentioned in the problems; for example, harvesting a renewable resource (Problems 18 and 19); epidemics (Problems 20 through 22); fluid bifurcation (Problem 23); and chemical reactions (Problem 24). Our main object in this section is to show how geometric methods can be used to obtain important qualitative information directly from the differential equation, without solving the equation. This will lead us at once to the very important concepts of stability and instability of solutions of differential equations. These ideas are introduced here, and examined in greater depth and in a more general setting in Chapter 9.

EXPONENTIAL GROWTH. Let $N(t)$ be the population of the given species at time t. The simplest hypothesis concerning the variation of $N(t)$ is that the rate of change of N is proportional[7] to the current value of N, that is,

$$dN/dt = rN, \tag{1}$$

where r is the constant of proportionality. The constant r is called the *rate of growth or decline*, depending on whether it is positive or negative. If $r < 0$, then the mathematical problem is the same as in radioactive decay, which was discussed in Section 2.5. Here we assume that $r > 0$, so that the population is growing.

Solving Eq. (1) subject to the initial condition

$$N(0) = N_0 \tag{2}$$

[6] This problem is based on R. H. Rainey, "Natural Displacement of Pollution from the Great Lakes," *Science 155* (1967) pp. 1242–1243; the information in the table was taken from that source.
[7] It was apparently the British economist Thomas Malthus (1766–1834) who first observed that many biological populations increase at a rate proportional to the population. His first paper on populations appeared in 1798.

we obtain

$$N(t) = N_0 e^{rt}. \tag{3}$$

Thus the mathematical model consisting of the initial value problem (1), (2) with $r > 0$ predicts that the population will grow exponentially for all time, as shown in Figure 2.13. Under ideal conditions Eq. (3) has been observed to be reasonably accurate for many populations, at least for limited periods of time. However, it is clear that such ideal conditions cannot continue indefinitely; eventually, limitations on space, food supply, or other resources will reduce the growth rate and bring an end to uninhibited exponential growth.

LOGISTIC GROWTH. To take account of the fact that the growth rate actually depends on the population, we replace the constant r in Eq. (1) by a function $f(N)$ and thereby obtain the modified equation

$$dN/dt = f(N)N. \tag{4}$$

We now want to choose $f(N)$ so that $f(N) \cong r > 0$ when N is small, $f(N)$ decreases as N grows larger, and $f(N) < 0$ when N is sufficiently large. The simplest function having these properties is $f(N) = r - aN$, where a is also a positive constant. Using this function in Eq. (4), we obtain

$$dN/dt = (r - aN)N. \tag{5}$$

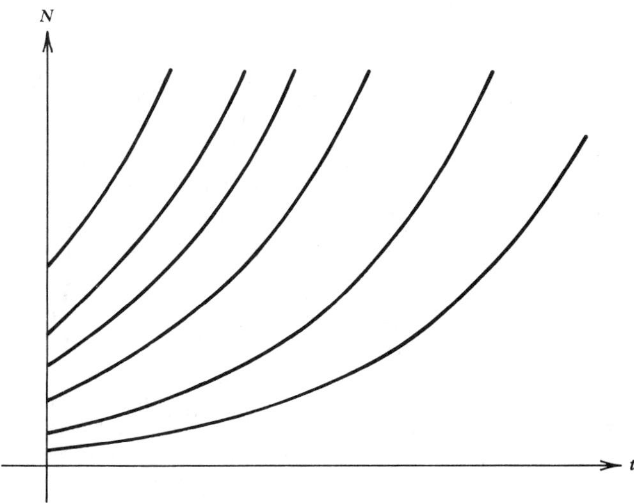

FIGURE 2.13 Exponential growth: N vs t for dN/dt = rN.

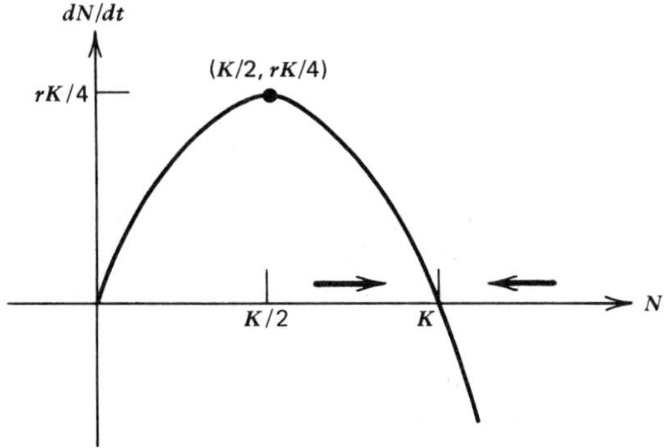

FIGURE 2.14 dN/dt vs N for $dN/dt = r(1 - N/K)N$.

Equation (5) is known as the *Verhulst*[8] equation or the *logistic* equation. It is often convenient to write the logistic equation in the equivalent form

$$dN/dt = r(1 - N/K)N, \tag{6}$$

where $K = r/a$. The constant r is called the *intrinsic growth rate*, that is, the growth rate in the absence of any limiting factors. The interpretation of K will be made clear shortly.

The solution of the logistic equation (6) can be easily found by the method of separation of variables,[9] and we derive the solution a little later. However, we now show that by using geometric reasoning the main features of the solution can be discovered directly from the differential equation itself, without solving it. This is important because the same methods can often be used on more complicated equations, whose solutions are more difficult to obtain.

In Figure 2.14 we show the graph of dN/dt versus N, where dN/dt is given by the right side of Eq. (6). The graph is a parabola with intercepts at $(0,0)$ and $(K,0)$, and with vertex at $(K/2, rK/4)$. For $0 < N < K$ we see that $dN/dt > 0$ and therefore N is an increasing function of t; this is indicated by the rightward pointing arrow near the N axis. Similarly, if $N > K$, then $dN/dt < 0$, hence $N(t)$ is decreasing, as indicated by the leftward pointing arrow. If $N = 0$ or $N = K$, then $dN/dt = 0$ and $N(t)$ does not change. The constant solutions $N = \phi_1(t) = 0$ and $N = \phi_2(t) = K$ are called *equilibrium solutions*. Corresponding to them, the

[8] P. F. Verhulst (1804–1849) was a Belgian mathematician who introduced Eq. (5) as a model for human population growth in 1838. He referred to it as logistic growth, hence Eq. (5) is often called the logistic equation. He was unable to test the accuracy of his model because of inadequate census data, and it did not receive much attention until many years later. Reasonable agreement with experimental data was demonstrated by R. Pearl (1930) for *Drosophila melanogaster* (fruit fly) populations, and by G. F. Gause (1935) for *Paramecium* and *Tribolium* (flour beetle) populations.

[9] The logistic equation is also a Bernoulli equation and can be solved by the substitution indicated in Problem 21 of Section 2.2.

points $N = 0$ and $N = K$ on the N axis are called *equilibrium points* or *critical points*.

Next, we wish to sketch the graphs of solutions $N(t)$ versus t for $t > 0$, $N > 0$, and for different initial values $N(0)$. For this purpose it is helpful to know the relationship between properties of the graph of dN/dt versus N and those of the graph of N versus t for any equation of the form

$$dN/dt = F(N).$$

This information is summarized in Table 2.3. The explanation of the entries in Table 2.3 is as follows. The graph of $N(t)$ vs t is increasing or decreasing, depending on whether dN/dt is positive or negative. To investigate concavity, note that if dN/dt is positive, then N and t increase or decrease together. Consequently, if dN/dt is positive and increasing as a function of N, then it is also increasing as a function of t, and the graph of N vs t is concave up. Similarly, if dN/dt is positive and decreasing, then the graph of N vs t is concave down.

The situation is reversed if dN/dt is negative, for then N decreases as t increases, and vice versa. Therefore, if dN/dt is negative and increasing as a function of N, then it is decreasing as a function of t and the graph of N vs t is concave down. Similarly, if dN/dt is negative and decreasing, then the graph of N vs t is concave up.

These results mean that the graphs of solutions of Eq. (6) must have the general shape shown in Figure 2.15, regardless of the values of r and K. The horizontal lines are the equilibrium solutions $\phi_1(t) = 0$ and $\phi_2(t) = K$. From Figure 2.14 we note that dN/dt is positive and increasing for $0 < N < K/2$, so the graph of $N(t)$ is increasing and concave up there. Similarly, dN/dt is positive and decreasing for $K/2 < N < K$, so the graph of $N(t)$ is increasing and concave down in this range. Thus, solutions that start below $K/2$ have the S-shaped or sigmoid character shown in Figure 2.15. On the other hand, for $N > K$, dN/dt is negative and decreasing, so the graph of $N(t)$ is decreasing and concave up for these values of N.

Finally, recall that Theorem 2.2, the fundamental existence and uniqueness theorem, guarantees that two different solutions never pass through the same

TABLE 2.3 RELATION BETWEEN THE GRAPHS OF dN/dt VS N AND OF N VS t, RESPECTIVELY

If dN/dt [or $F(N)$] is	then	$N(t)$ is
Positive and increasing		Increasing and concave up
Positive and decreasing		Increasing and concave down
Negative and increasing		Decreasing and concave down
Negative and decreasing		Decreasing and concave up

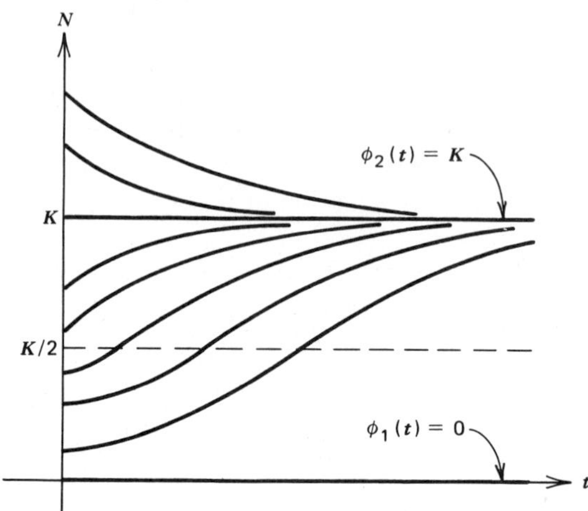

FIGURE 2.15 Logistic growth: N vs t for dN/dt = r(1 − N/K)N.

point. Hence, while solutions approach the equilibrium solution $N = K$ as $t \to \infty$, they do not attain this value at any finite time. Since K is the upper bound that is approached, but not exceeded, by growing populations starting below this value, it is natural to refer to K as the *saturation level*, or as the *environmental carrying capacity*, for the given species.

A comparison of Figures 2.13 and 2.15 reveals that solutions of the nonlinear equation (6) are strikingly different from those of the linear equation (1), at least for large values of t. Regardless of the value of K, that is, no matter how small the nonlinear term in Eq. (6), solutions of that equation approach a finite value as $t \to \infty$, whereas solutions of Eq. (1) grow (exponentially) without bound as $t \to \infty$. Thus, even a tiny nonlinear term in the differential equation has a decisive effect on the solution for large t.

In many situations it is sufficient to have the qualitative information about the solution $N(t)$ of Eq. (6) that is shown in Figure 2.15. We emphasize that this information was obtained entirely from the graph of dN/dt versus N, and without solving the differential equation (6). However, if we wish to have a more detailed description of logistic growth—for example, if we wish to know the value of the population at some particular time—then we must solve Eq. (6) subject to the initial condition (2). Provided that $N \neq 0$ and $N \neq K$, we can write Eq. (6) in the form

$$\frac{dN}{(1 - N/K)N} = r\,dt.$$

Using a partial fractions expansion on the left side we have

$$\left(\frac{1}{N} + \frac{1/K}{1 - N/K}\right) dN = r\,dt.$$

Then by integrating both sides we obtain

$$\ln|N| - \ln\left|1 - \frac{N}{K}\right| = rt + c, \tag{7}$$

where c is an arbitrary constant of integration to be determined from the initial condition $N(0) = N_0$. We have already noted that if $0 < N_0 < K$, then $N(t)$ remains in this interval for all time. Thus in this case we can remove the absolute value bars in Eq. (7), and by taking the exponential of both sides we find that

$$\frac{N}{1-(N/K)} = Ce^{rt}, \tag{8}$$

where $C = e^c$. To satisfy the initial condition $N(0) = N_0$ we must choose $C = N_0/[1 - (N_0/K)]$. Using this value for C in Eq. (8) and solving for N we obtain

$$N(t) = \frac{N_0 K}{N_0 + (K - N_0)e^{-rt}}. \tag{9}$$

We have derived the solution (9) under the assumption that $0 < N_0 < K$. If $N_0 > K$, then the details of dealing with Eq. (7) are only slightly different, and we leave it to the reader to show that Eq. (9) is also valid in this case. Finally, note that Eq. (9) also contains the equilibrium solutions $N = \phi_1(t) = 0$ and $N = \phi_2(t) = K$ corresponding to the initial conditions $N_0 = 0$ and $N_0 = K$, respectively.

All of the qualitative conclusions that we reached earlier by geometric reasoning can be confirmed by examining the solution (9). In particular, if $N_0 = 0$, then Eq. (9) requires that $N(t) \equiv 0$ for all t. If $N_0 > 0$, and if we let $t \to \infty$ in Eq. (9), then we obtain

$$\lim_{t\to\infty} N(t) = N_0 K/N_0 = K.$$

Thus for each $N_0 > 0$ the solution approaches the equilibrium solution $N = \phi_2(t) = K$ asymptotically (in fact, exponentially) as $t \to \infty$. Thus we say that the constant solution $\phi_2(t) = K$ is an *asymptotically stable solution* of Eq. (6), or that the point $N = K$ is an *asymptotically stable equilibrium or critical point*. This means that after a long time the population is close to the saturation level K regardless of the initial population size, as long as it is positive.

On the other hand, the situation for the equilibrium solution $N = \phi_1(t) = 0$ is quite different. Even solutions that start very near zero grow as t increases and, as we have seen, approach K as $t \to \infty$. We say that $\phi_1(t) = 0$ is an *unstable equilibrium solution* or that $N = 0$ is an *unstable equilibrium or critical point*. This means that the only way to guarantee that the solution remains near zero is to make sure that its initial value is *exactly* equal to zero.

These two cases may be visualized as follows. First, suppose that we are trying to perform a bacteria-free experiment in an environment where the growth of bacteria is governed by Eq. (6). Further, suppose that we have managed to reduce the bacteria population to an extremely low, but nonzero, level. Is it safe to carry out an experiment over an extended time period on the assumption that the bacteria population will remain low? No; as long as the initial population N_0 is nonzero the population eventually will approach the saturation level K in size. Of course, if the experiment is not too long and if N_0 is small enough, then the bacteria population may remain in an acceptable range for the duration of the experiment. Second, suppose that we are trying to perform an experiment at a bacteria level of K, but that from time to time contaminants are introduced that kill a few bacteria. Is it possible that this will cause the bacteria population to continue to decrease and eventually become extinct? No; if the population level varies slightly from K, whether above or below, it will tend to move back toward K as t increases.

EXAMPLE

The logistic model has been applied to the natural growth of the halibut population in certain areas of the Pacific Ocean.[10] Let $N(t)$, measured in kilograms, be the total mass, or biomass, of the halibut population at time t. The parameters in the logistic equation are estimated to have the values $r = 0.71/\text{year}$ and $K = 80.5 \times 10^6$ kg. If the initial biomass is $N_0 = 0.25K$, find the biomass two years later. Also find the time τ for which $N(\tau) = 0.75K$.

It is convenient to scale the solution (9) to the carrying capacity K; thus we write Eq. (9) in the form

$$\frac{N(t)}{K} = \frac{N_0/K}{(N_0/K) + [1 - (N_0/K)]e^{-rt}}. \tag{10}$$

Using the data given in the problem we find that

$$\frac{N(2)}{K} = \frac{0.25}{0.25 + 0.75e^{-1.42}} \cong 0.5797.$$

Consequently $N(2) \cong 46.7 \times 10^6$ kg.

To find τ we can first solve Eq. (10) for t. We obtain

$$e^{-rt} = \frac{(N_0/K)[1 - (N/K)]}{(N/K)[1 - (N_0/K)]};$$

hence

$$t = -\frac{1}{r}\ln\frac{(N_0/K)[1 - (N/K)]}{(N/K)[1 - (N_0/K)]}. \tag{11}$$

[10]A good source of information on the population dynamics and economics involved in making efficient use of a renewable resource, with particular emphasis on fisheries, is the book by Clark (see References at the end of the chapter). The parameter values used here are given on page 48, and were obtained as a result of a study by M. S. Mohring.

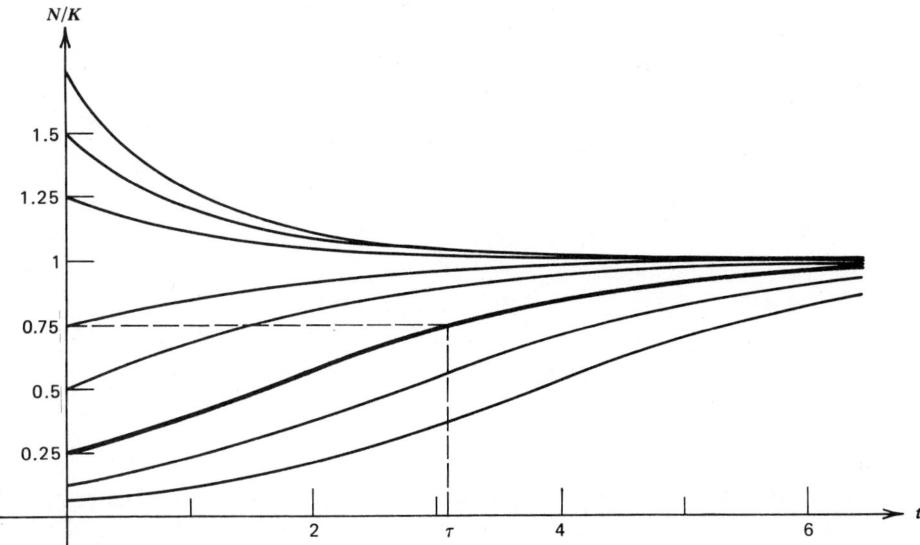

FIGURE 2.16 N/K *vs* t *for population model of halibut in the Pacific Ocean.*

Using the given values of r and N_0/K and setting $N/K = 0.75$, we find that

$$\tau = -\frac{1}{0.71}\ln\frac{(0.25)(0.25)}{(0.75)(0.75)} = \frac{1}{0.71}\ln 9 \cong 3.095 \text{ years.}$$

The graphs of $N(t)$ vs t for the given parameter values and for several initial conditions are shown in Figure 2.16.

A CRITICAL THRESHOLD. We now turn to a consideration of the equation

$$dN/dt = -r(1 - N/T)N, \tag{12}$$

where r and T are given positive constants. Observe that (except for replacing the parameter K by T) this equation differs from the logistic equation (6) only in the presence of the minus sign on the right side. However, as we will see, the solutions of Eq. (12) behave very differently from those of Eq. (6).

For Eq. (12) the graph of dN/dt versus N is the parabola shown in Figure 2.17. The intercepts on the N axis are the critical points $N = 0$ and $N = T$, corresponding to the equilibrium solutions $\phi_1(t) = 0$ and $\phi_2(t) = T$. If $0 < N < T$, then $dN/dt < 0$, and $N(t)$ decreases as t increases. On the other hand, if $N > T$, then $dN/dt > 0$, and $N(t)$ grows as t increases. Thus $\phi_1(t) = 0$ is an asymptotically stable equilibrium solution and $\phi_2(t) = T$ is an unstable one. Further, dN/dt is decreasing for $0 < N < T/2$ and increasing for $T/2 < N < T$, so the graph of $N(t)$ is concave up and concave down, respectively, in these

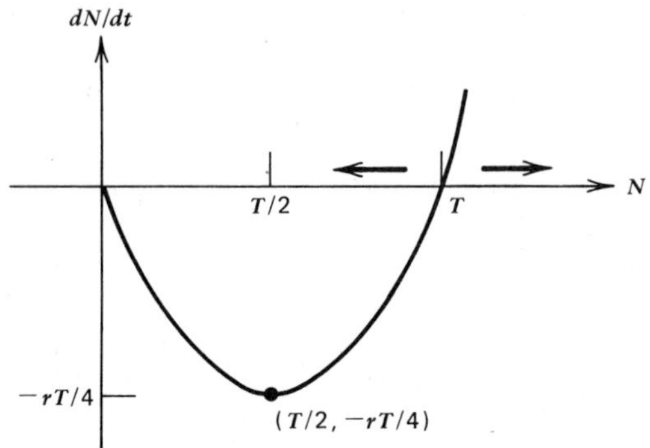

FIGURE 2.17 dN/dt *vs* N *for* $dN/dt = -r(1 - N/T)N$.

intervals (see Table 2.3). Also, dN/dt is increasing for $N > T$ so the graph of $N(t)$ is also concave up there. By making use of all of the information that we have obtained from Figure 2.17, we conclude that graphs of solutions of Eq. (12) for different values of N_0 must have the qualitative appearance shown in Figure 2.18. From this figure it is clear that as time increases $N(t)$ either approaches zero or grows without bound, depending on whether the initial value N_0 is less than or greater than T. Thus T is a *threshold level*, below which growth does not occur.

We can confirm the conclusions that we have reached through geometric reasoning by solving the differential equation (12). This can be done by separating the variables and integrating, just as we did for Eq. (6). However, if we note that Eq. (12) can be obtained from Eq. (6) by replacing K by T and r by $-r$, then we

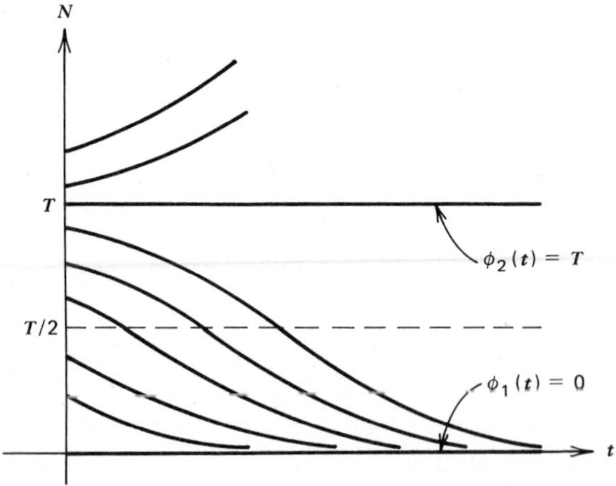

FIGURE 2.18 N *vs* t *for* $dN/dt = -r(1 - N/T)N$.

can make the same substitutions in the solution (9) and thereby obtain

$$N(t) = \frac{N_0 T}{N_0 + (T - N_0)e^{rt}}, \tag{13}$$

which is the solution of Eq. (12) subject to the initial condition $N(0) = N_0$.

If $N_0 < T$, then it is clear from Eq. (13) that $N(t) \to 0$ as $t \to \infty$. This agrees with our qualitative geometric analysis. If $N_0 > T$, then the denominator on the right side of Eq. (13) is zero for a certain finite value of t. We denote this value by t^*, and calculate it from

$$N_0 - (N_0 - T)e^{rt^*} = 0,$$

which gives

$$t^* = \frac{1}{r} \ln \frac{N_0}{N_0 - T}. \tag{14}$$

Thus if the initial population N_0 is above the threshold T, the threshold model predicts that the graph of $N(t)$ has a vertical asymptote at $t = t^*$; in other words, the population becomes unbounded in a finite time, which depends on the initial value N_0 and the threshold value T. The existence and location of this asymptote was not apparent from the geometric analysis, so in this case the explicit solution yields additional important qualitative, as well as quantitative, information.

The populations of some species exhibit the threshold phenomenon. If too few are present, the species cannot propagate itself successfully and the population becomes extinct. However, if a population larger than the threshold level can be brought together, then further growth occurs. Of course, the population cannot become unbounded, so eventually Eq. (12) must be modified to take this into account.

In fluid mechanics, equations of the form (6) or (12) often govern the evolution of a small disturbance N in a *laminar* (or smooth) fluid flow. For instance, if Eq. (12) holds and if $N < T$, then the disturbance is damped out and the laminar flow persists. However, if $N > T$, then the disturbance grows larger and the laminar flow breaks up into a turbulent one. In this case T is often referred to as the *critical amplitude*. Experimenters speak of keeping the disturbance level in a wind tunnel sufficiently low so that they can study laminar flow over an airfoil, for example.

The same type of situation can occur with automatic control devices. For example, suppose that N corresponds to the position of a flap on an airplane wing that is regulated by an automatic control. The desired position is $N = 0$. In the normal motion of the plane the changing aerodynamic forces on the flap will cause it to move from its set position, but then the automatic control will come into action to damp out the small deviation and return the flap to its desired position. However, if the airplane is caught in a high gust of wind, the flap may be deflected so much that the automatic control cannot bring it back to the set position (this would correspond to a deviation greater than T). Presumably the pilot would then take control and manually override the automatic system!

LOGISTIC GROWTH WITH A THRESHOLD. As we mentioned in the last subsection the threshold model (12) may need to be modified so that unbounded growth does not occur when N is above the threshold T. The simplest way to do this is to introduce another factor that will have the effect of making dN/dt negative when N is large. Thus we consider

$$dN/dt = -r(1 - N/T)(1 - N/K)N, \tag{15}$$

where $r > 0$ and $0 < T < K$.

The graph of dN/dt vs N is shown in Figure 2.19. In this problem there are three critical points: $N = 0$, $N = T$, and $N = K$, corresponding to the equilibrium solutions $\phi_1(t) = 0$, $\phi_2(t) = T$, and $\phi_3(t) = K$, respectively. From Figure 2.19 it is clear that $dN/dt > 0$ for $T < N < K$, and consequently $N(t)$ is increasing there. The reverse is true for $N < T$ and for $N > K$. Consequently, the equilibrium solutions $\phi_1(t)$ and $\phi_3(t)$ are stable, and the solution $\phi_2(t)$ is unstable. Graphs of $N(t)$ vs t have the qualitative appearance shown in Figure 2.20. If N starts below the threshold T, then N declines to ultimate extinction. On the other hand, if N starts above T, then $N(t)$ eventually approaches the carrying capacity K. The inflection points on the graphs of $N(t)$ vs t in Figure 2.20 correspond to the maximum and minimum points, N_1 and N_2, respectively, on the graph of dN/dt vs N in Figure 2.19. These values can be obtained by differentiating the right side of Eq. (15) with respect to N, setting the result equal to zero, and solving for N. We obtain

$$N_{1,2} = \left(K + T \pm \sqrt{K^2 - KT + T^2}\right)/3, \tag{16}$$

where the plus sign yields N_1 and the minus sign N_2.

A model of this general sort apparently governed the population of the passenger pigeon,[11] which was present in the United States in vast numbers until

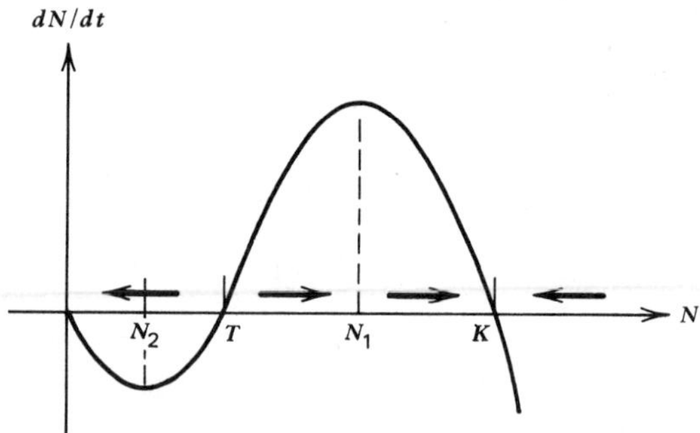

FIGURE 2.19 dN/dt *vs* N *for* $dN/dt = -r(1 - N/T)(1 - N/K)N$.

[11] See, for example, Oliver L. Austin, Jr., *Birds of the World* (New York: Golden Press, 1983), pp. 143–145.

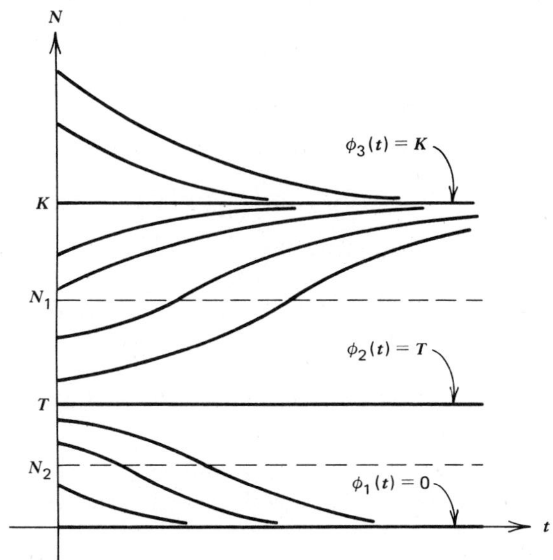

FIGURE 2.20 *N vs t for* $dN/dt = -r(1 - N/T)(1 - N/K)N$.

late in the nineteenth century. It was heavily hunted for food and for sport, and consequently its numbers were drastically reduced by the 1880s. Unfortunately, the passenger pigeon could apparently breed successfully only when present in a large concentration, corresponding to a relatively high threshold T. Although a reasonably large number of individual birds remained alive in the late 1880s, there were not enough in any one place to permit successful breeding, and the population rapidly declined to extinction. The last survivor died in 1914. The precipitous decline in the passenger pigeon population from huge numbers to extinction in scarcely more than three decades was one of the early factors contributing to concern for conservation in this country.

PROBLEMS

In each of Problems 1 through 6 sketch dN/dt vs N; determine the critical (equilibrium) points and classify each one as stable or unstable.

1. $dN/dt = aN + bN^2, \quad a > 0, \quad b > 0, \quad N_0 \geq 0$
2. $dN/dt = aN + bN^2, \quad a > 0, \quad b > 0, \quad -\infty < N_0 < \infty$
3. $dN/dt = N(N - 1)(N - 2), \quad N_0 \geq 0$
4. $dN/dt = e^N - 1, \quad -\infty < N_0 < \infty$
5. $dN/dt = e^{-N} - 1, \quad -\infty < N_0 < \infty$
6. $dN/dt = -2\,(\arctan N)/(1 + N^2), \quad -\infty < N_0 < \infty$

7. **Semistable equilibrium solutions.** Sometimes a constant equilibrium solution has the property that solutions lying on one side of the equilibrium solution tend to approach it whereas solutions lying on the other side recede from it (see Figure 2.21). In this case the equilibrium solution is said to be *semistable*.
(a) Consider the equation

$$dN/dt = k(1 - N)^2, \tag{i}$$

where k is a positive constant. Show that $N = 1$ is the only critical point, with the corresponding equilibrium solution $\phi(t) = 1$.
(b) Sketch dN/dt vs N. Show that N is increasing for $N < 1$ and also for $N > 1$. Thus solutions below the equilibrium solution approach it while those above it grow farther away. Thus $\phi(t) = 1$ is semistable.
(c) Solve Eq. (i) subject to the initial condition $N(0) = N_0$, and confirm the conclusions reached in part (b).

In each of Problems 8 through 13 sketch dN/dt vs N and determine the critical (equilibrium) points. Also classify each equilibrium point as stable, unstable, or semistable (see Problem 7).

8. $dN/dt = -k(N-1)^2, \quad k > 0, \quad -\infty < N_0 < \infty$
9. $dN/dt = N^2(N^2 - 1), \quad -\infty < N_0 < \infty$
10. $dN/dt = N(1 - N^2), \quad -\infty < N_0 < \infty$
11. $dN/dt = aN - b\sqrt{N}, \quad a > 0, \quad b > 0, \quad N_0 \geq 0$
12. $dN/dt = N^2(4 - N^2), \quad -\infty < N_0 < \infty$
13. $dN/dt = N^2(1 - N)^2, \quad -\infty < N_0 < \infty$

14. Consider the equation $dN/dt = F(N)$, and suppose that N_1 is a critical point, that is, $F(N_1) = 0$. Show that the constant equilibrium solution $\phi(t) = N_1$ is stable if $F'(N_1) < 0$ and unstable if $F'(N_1) > 0$.

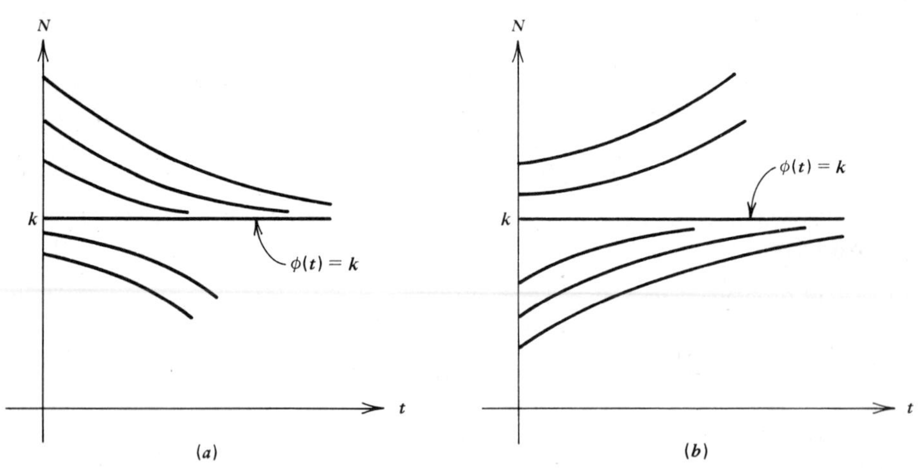

FIGURE 2.21 In both cases the equilibrium solution $\phi(t) = k$ is semistable. (a) $dN/dt < 0$; (b) $dN/dt \geq 0$.

15. Suppose that a certain population obeys the logistic equation $dN/dt = rN[1 - (N/K)]$.
(a) If $N_0 = K/3$, find the time τ at which the initial population has doubled. Find the value of τ corresponding to $r = 0.025$ per year.
(b) If $N_0/K = \alpha$, find the time T at which $N(T)/K = \beta$, where $0 < \alpha,\ \beta < 1$. Observe that $T \to \infty$ as $\alpha \to 0$ or as $\beta \to 1$. Find the value of T for $r = 0.025$ per year, $\alpha = 0.1$, and $\beta = 0.9$.
16. Another equation that has been used to model population growth is the Gompertz equation:

$$dN/dt = rN\ln(K/N),$$

where r and K are positive constants.
(a) Sketch the graph of dN/dt vs N, find the critical points, and determine whether each is stable or unstable.
(b) For $0 \le N \le K$ determine where the graph of N vs t is concave up and where it is concave down.
(c) For each N in $0 < N \le K$ show that dN/dt as given by the Gompertz equation is never less than dN/dt as given by the logistic equation.
17. (a) Solve the Gompertz equation

$$dN/dt = rN\ln(K/N),$$

subject to the initial condition $N(0) = N_0$.
Hint: You may wish to let $u = \ln(N/K)$.
(b) For the data given in the Example in the text [$r = 0.71$ per year, $K = 80.5 \times 10^6$ kg, $N_0/K = 0.25$], use the Gompertz model to find the predicted value of $N(2)$.
(c) For the same data as in part (b), use the Gompertz model to find the time τ at which $N(\tau) = 0.75K$.

Harvesting a Renewable Resource. Suppose that the population $N(t)$ of a certain species of fish (for example, tuna or halibut) in a given area of the ocean is described by the logistic equation

$$dN/dt = r(1 - N/K)N.$$

While it is desirable to utilize this source of food, it is intuitively clear that if too many fish are caught, then the fish population may be reduced below a useful level, and possibly even driven to extinction. Problems 18 and 19 explore some of the questions involved in formulating a rational strategy for managing the fishery.[12]

18. At a given level of effort, it is reasonable to assume that the rate at which fish are caught depends on the population N: the more fish there are, the easier it is to catch them. Thus we assume that the rate at which fish are caught, that is, the yield Y of the fishery, is given by $Y = EN$, where E is a positive constant, with dimensions 1/time, that measures the total effort made to harvest the given species of fish. To include this effect, the logistic equation is replaced by

$$dN/dt = r(1 - N/K)N - EN. \tag{i}$$

[12]An excellent treatment of this kind of problem, which goes far beyond what is outlined here, may be found in the book by Clark mentioned previously, especially in the first two chapters. Numerous additional references are given there.

This equation is known as the *Schaefer model* after the biologist, M. B. Schaefer, who applied it to fish populations.

(a) Show that if $E < r$, then there are two equilibrium points, $N_1 = 0$ and $N_2 = K[1 - E/r] > 0$.

(b) Show that $N = N_1$ is unstable and that $N = N_2$ is stable.

(c) Find the sustainable yield Y as a function of the effort E; the graph of this function is known as the yield–effort curve.

(d) Determine E so as to maximize Y, and thereby find the *maximum sustainable yield* Y_m.

19. In this problem we assume that fish are caught at a constant rate h independent of the size of the fish population. Then N satisfies

$$dN/dt = r(1 - N/K)N - h. \qquad \text{(i)}$$

The assumption of a constant catch rate h may be reasonable when N is large, but becomes less so when N is small.

(a) If $h < rK/4$, show that Eq. (i) has two equilibrium points N_1 and N_2 with $N_1 < N_2$; determine these points.

(b) Show that N_1 is unstable and N_2 is stable.

(c) From a plot of dN/dt vs N show that if the initial population $N_0 > N_1$, then $N(t) \to N_2$ as $t \to \infty$, but that if $N_0 < N_1$, then $N(t)$ decreases as t increases. Note that $N = 0$ is not an equilibrium point, so if $N_0 < N_1$, then extinction will be reached in a finite time.

(d) If $h > rK/4$, show that $N(t)$ decreases to zero as t increases regardless of the value of N_0.

(e) If $h = rK/4$, show that there is a single equilibrium point $N = K/2$, and that this point is semistable (see Problem 7). Thus the maximum sustainable yield is $h_m = rK/4$, corresponding to the equilibrium value $N = K/2$. Observe that h_m has the same value as Y_m in Problem 18(d). The fishery is considered to be overexploited if N is reduced to a level below $K/2$.

Epidemics. The use of mathematical methods to study the spread of contagious diseases goes back at least to some work by Daniel Bernoulli in 1760 on smallpox. In more recent years many mathematical models have been proposed and studied for many different diseases.[13] Problems 20 through 22 deal with a few of the simpler models and the conclusions that can be drawn from them. Similar models have also been used to describe the spread of rumors and of consumer products.

20. Suppose that a given population can be divided into two parts: those who have a given disease and can infect others, and those who do not have it but are susceptible. Let x be the proportion of susceptible individuals and y the proportion of infectious individuals; then $x + y = 1$. Assume that the disease spreads by contact between sick and well members of the population, and that the rate of spread dy/dt is proportional to the number of such contacts. Further, assume that members of both groups move about freely among each other, so that the number of contacts is proportional to the

[13]A standard source is the book by Bailey listed in the References. The models in Problems 20 through 22 are discussed by Bailey in Chapters 5, 10, and 20, respectively.

product of x and y. Since $x = 1 - y$, we obtain the initial value problem

$$dy/dt = \alpha y(1 - y), \qquad y(0) = y_0, \tag{i}$$

where α is a positive proportionality factor, and y_0 is the initial proportion of infectious individuals.

(a) Find the equilibrium points for the differential equation (i), and determine whether each is stable or unstable.

(b) Solve the initial value problem (i) and verify that the conclusions you reached in part (a) are correct. Show that $y(t) \to 1$ as $t \to \infty$, which means that ultimately the disease spreads through the entire population.

21. Some diseases (such as typhoid fever) are spread largely by *carriers*, individuals who can transmit the disease but who exhibit no overt symptoms. Let x and y, respectively, denote the proportion of susceptibles and carriers in the population. Suppose that carriers are identified and removed from the population at a rate β, so that

$$dy/dt = -\beta y. \tag{i}$$

Suppose also that the disease spreads at a rate proportional to the product of x and y; thus

$$dx/dt = -\alpha xy. \tag{ii}$$

(a) Determine y at any time t by solving Eq. (i) subject to the initial condition $y(0) = y_0$.

(b) Use the result of part (a) to find x at any time t by solving Eq. (ii) subject to the initial condition $x(0) = x_0$.

(c) Find the proportion of the population that escapes the epidemic by finding the limiting value of x as $t \to \infty$.

22. Daniel Bernoulli's work in 1760 had the goal of appraising the effectiveness of a controversial inoculation program against smallpox, which at that time was a major threat to public health. His model applies equally well to any other disease which, once contracted and survived, confers a lifetime immunity.

Consider the cohort of individuals born in a given year ($t = 0$), and let $n(t)$ be the number of these individuals surviving t years later. Let $x(t)$ be the number of members of this cohort who have not had smallpox by year t, and who are therefore still susceptible. Let β be the rate at which susceptibles contract smallpox, and let ν be the rate at which people who contract smallpox die from the disease. Finally, let $\mu(t)$ be the death rate from all causes other than smallpox. Then dx/dt, the rate at which the number of susceptibles declines, is given by

$$dx/dt = -[\beta + \mu(t)]x; \tag{i}$$

the first term on the right side of Eq. (i) is the rate at which susceptibles contract smallpox, while the second term is the rate at which they die from all other causes. Also

$$dn/dt = -\nu\beta x - \mu(t)n, \tag{ii}$$

where dn/dt is the death rate of the entire cohort, and the two terms on the right side are the death rates due to smallpox and to all other causes, respectively.

(a) Let $z = x/n$, and show that z satisfies the initial value problem

$$dz/dt = -\beta z(1 - \nu z), \qquad z(0) = 1. \tag{iii}$$

Observe that the initial value problem (iii) does not depend on $\mu(t)$.

(b) Find $z(t)$ by solving Eq. (iii).
(c) Bernoulli estimated that $\nu = \beta = \frac{1}{8}$. Using these values, determine the proportion of 20-year olds who have not had smallpox.
Note: Based on the model described above and on the best mortality data available at the time, Bernoulli calculated that if deaths due to smallpox could be eliminated ($\nu = 0$), then approximately 3 years could be added to the average life expectancy (in 1760) of 26 years 7 months. He therefore supported the inoculation program.

23. **Bifurcation theory.** In many physical problems some observable quantity, such as a velocity, waveform, or chemical reaction, depends on a parameter describing the physical state. As this parameter is increased a critical value is reached at which the velocity, or waveform, or reaction suddenly changes its character. For example, as the amount of one of the chemicals in a certain mixture is increased spiral wave patterns of varying color suddenly emerge in an originally quiescent fluid. In many such cases the mathematical analysis ultimately leads to an equation[14] of the form

$$dx/dt = (R - R_c)x - ax^3. \tag{i}$$

Here a and R_c are positive constants, and R is a parameter that may take on various values. For example, R may measure the amount of a certain chemical and x may measure a chemical reaction.
(a) If $R < R_c$, show that there is only one equilibrium solution $x = 0$, and that it is stable.
(b) If $R > R_c$, show that there are three equilibrium solutions, $x = 0$ and $x = \pm\sqrt{(R - R_c)/a}$, and that the first solution is unstable while the other two are stable.
(c) Draw a graph in the Rx plane showing all equilibrium solutions, and label each one as stable or unstable.
The point $R = R_c$ is called a *bifurcation point*. For $R < R_c$ one observes the stable equilibrium solution $x = 0$. However, this solution loses its stability as R passes through the value R_c, and for $R > R_c$ the stable (and hence the observable) solutions are $x = \sqrt{(R - R_c)/a}$ and $x = -\sqrt{(R - R_c)/a}$. Because of the way in which the solutions branch at R_c, this type of bifurcation is called a pitchfork bifurcation; your sketch should suggest that this name is appropriate.

24. **Chemical reactions.** A second order chemical reaction involves the interaction (collision) of one molecule of a substance P with one molecule of a substance Q to produce one molecule of a new substance X; this is denoted by $P + Q \to X$. Suppose that p and q, where $p \neq q$, are the initial concentrations of P and Q, respectively, and let $x(t)$ be the concentration of X at time t. Then $p - x(t)$ and $q - x(t)$ are the concentrations of P and Q at time t, and the rate at which the reaction occurs is given by the equation

$$dx/dt = \alpha(p - x)(q - x), \tag{i}$$

where α is a positive constant.

[14] In fluid mechanics Eq. (i) arises in the study of the transition from laminar to turbulent flow; there it is often called the Landau equation. L. D. Landau (1908–1968) was a Russian physicist who received the Nobel prize in 1962 for his contributions to the understanding of condensed states, particularly liquid helium. He was also the coauthor, with E. M. Lifschitz, of a well-known series of physics textbooks.

(a) If $x(0) = 0$, determine the limiting value of $x(t)$ as $t \to \infty$ without solving the differential equation. Then solve the initial value problem and find $x(t)$ for any t.
(b) If the substances P and Q are the same, then $p = q$ and Eq. (i) is replaced by

$$dx/dt = \alpha(p - x)^2. \tag{ii}$$

If $x(0) = 0$, determine the limiting value of $x(t)$ as $t \to \infty$ without solving the differential equation. Then solve the initial value problem and determine $x(t)$ for any t.

2.7 Elementary Mechanics

Some of the most important applications of first order differential equations are found in the realm of elementary mechanics. In this section we consider some problems involving the motion of a rigid body along a straight line. We assume that such bodies obey Newton's law of motion: *the product of the mass and the acceleration is equal to the external force*. In symbols,

$$F = ma, \tag{1}$$

where F is the external force, m is the mass of the body, and a is the acceleration in the direction of F.

Three systems of units are in common use. In the cgs system, the basic units are centimeters (length), grams (mass), and seconds (time). The unit of force, the dyne, is defined by Eq. (1) to be the force required to impart an acceleration of 1 cm/sec^2 to a mass of 1 g. Since one dyne is a very small force it is often preferable to use the mks system in which the basic units are meters (length), kilograms (mass), and seconds (time). The unit of force, the newton (N), is defined by Eq. (1) to be the force required to impart an acceleration of 1 m/sec^2 to a mass of 1 kg. Finally, in the English, or engineering, system the basic units are feet (length), pounds (force), and seconds (time). The unit of mass, the slug, is defined by Eq. (1) to be that mass to which a force of 1 lb gives an acceleration of 1 ft/sec^2. To compare the three systems note that 1 N is equal to 10^5 dynes and to 0.225 lb.

Now consider a body falling freely in a vacuum and close enough to the surface of the earth so that the only significant force acting on the body is its weight due to the earth's gravitational field. Equation (1) then takes the form

$$w = mg, \tag{2}$$

where w is the weight of the body and g denotes its acceleration due to gravity.

Even though the mass of the body remains constant, its weight and gravitational acceleration change with distance from the center of the earth's gravitational field. At sea level the value of g has been experimentally determined to be approximately 32 ft/sec^2 (English system), 980 cm/sec^2 (cgs system), or 9.8 m/sec^2 (mks system). It is customary to denote by g the gravitational acceleration at *sea level*; thus g is a constant. With this understanding Eq. (2) is exact only at sea level.

The general expression for the weight of a body of mass m is obtained from Newton's inverse-square law of gravitational attraction. If R is the radius of the earth, and x is the altitude above sea level, then

$$w(x) = \frac{K}{(R + x)^2}, \tag{3}$$

where K is a constant. At $x = 0$ (sea level), $w = mg$; hence $K = mgR^2$, and

$$w(x) = \frac{mgR^2}{(R + x)^2}. \tag{4}$$

Expanding $(R + x)^{-2}$ in a Taylor series about $x = 0$ leads to

$$w(x) = mg\left(1 - 2\frac{x}{R} + \cdots\right). \tag{5}$$

It follows that if terms of the order of magnitude of x/R can be neglected in comparison with unity, then Eq. (4) can be replaced by the simpler approximation, Eq. (2). Thus, for motions in the vicinity of the earth's surface, or if extreme accuracy is not required, it is usually sufficient to use Eq. (2). In other cases, such as problems involving space flight, it may be necessary to use Eq. (4), or at least a better approximation to it than Eq. (2).

In many cases the external force F will include effects other than the weight of the body. For example, frictional forces due to the resistance offered by the air or other surrounding medium may need to be considered. Similarly, the effects of mass changes may be important, as in the case of a space vehicle which burns its fuel during flight.

In the problems we consider here it is useful to describe the motion of the body with reference to some convenient coordinate system. We always choose the x axis to be the line along which the motion takes place; the positive direction on the x axis can be chosen arbitrarily. Once the positive x direction is specified, a positive value of x indicates a displacement in this direction, a positive value of $v = dx/dt$ indicates the body is moving in the direction of the positive x axis, a positive value for a force F indicates it acts in the positive x direction, and so forth. The following examples are concerned with some typical problems.

EXAMPLE 1

An object of mass m is dropped from rest in a medium that offers resistance proportional to $|v|$, the magnitude of the instantaneous velocity of the object.[15]

[15] In general, the resistance is a function of the speed, that is, the magnitude of the velocity. The assumption of a linear relation is reasonable only for comparatively low speeds. Where higher speeds are involved it may be necessary to assume that the resistance is proportional to some higher power of $|v|$, or even that it is given by some polynomial function of $|v|$.

Assuming the gravitational force to be constant, find the position and velocity of the object at any time t.

In this case it is convenient to draw the x axis positive downward with the origin at the initial position of the object (see Figure 2.22). The weight mg of the object then acts in the downward (positive) direction, but the resistance $k|v|$, where k is a positive constant, acts to impede the motion. When $v > 0$ the resistance is in the upward (negative) direction, and is thus given by $-kv$. When $v < 0$ the resistance acts downward and is still given by $-kv$. Thus in all cases Newton's law can be written as

$$m\frac{dv}{dt} = mg - kv$$

or

$$\frac{dv}{dt} + \frac{k}{m}v = g. \tag{6}$$

Equation (6) is a linear first order equation and has the integrating factor $e^{kt/m}$. The solution of Eq. (6) is

$$v = \frac{mg}{k} + c_1 e^{-kt/m}. \tag{7}$$

The initial condition $v(0) = 0$ requires that $c_1 = -mg/k$; thus

$$v = \frac{mg}{k}(1 - e^{-kt/m}). \tag{8}$$

To obtain the position x of the body we replace v by dx/dt in Eq. (8); integration and use of the second initial condition $x(0) = 0$ gives

$$x = \frac{mg}{k}t - \frac{m^2 g}{k^2}(1 - e^{-kt/m}). \tag{9}$$

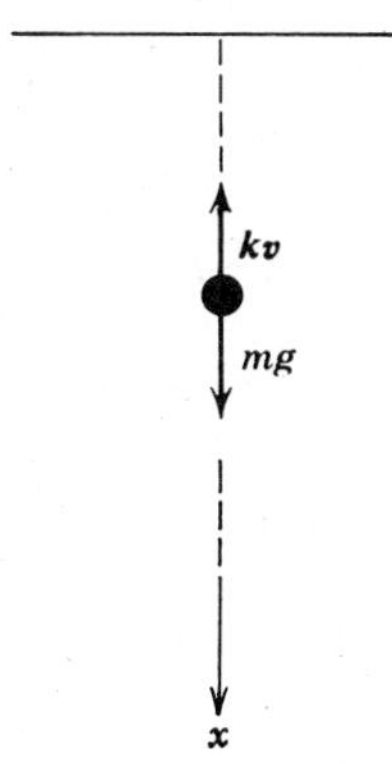

FIGURE 2.22 A mass falling in a resistive medium ($v > 0$).

The position and velocity of the body at any time t are given by Eqs. (9) and (8), respectively.

It is worth noting that as $t \to \infty$ in Eq. (7) the velocity approaches the limiting value

$$v_l = mg/k. \tag{10}$$

The limiting velocity can also be obtained directly from the differential equation (6) by setting $dv/dt = 0$ and then solving Eq. (6) for v. The limiting velocity v_l does not depend on the initial conditions, but only on the mass of the object and the coefficient of resistance of the medium. Given either v_l or k, Eq. (10) provides a formula for calculating the other. As k approaches zero (that is, as the resistance diminishes), the limiting velocity increases without bound.

EXAMPLE 2

A body of constant mass m is projected upward from the earth's surface with an initial velocity v_0. Assuming there is no air resistance, but taking into consideration the variation of the earth's gravitational field with altitude, find the smallest initial velocity for which the body will not return to the earth. This is the so-called "escape velocity."

Take the x axis positive upward, with the origin on the surface of the earth (see Figure 2.23). The only force acting on the body is its weight, given by Eq. (4), which acts downward. Thus the equation of motion is

$$m\frac{dv}{dt} = -\frac{mgR^2}{(R+x)^2}, \tag{11}$$

and the initial condition is

$$v(0) = v_0. \tag{12}$$

Since the force appearing on the right side of Eq. (11) is a function of x only, it is convenient to think of x, rather than t, as the independent variable. This requires that we express dv/dt in terms of dv/dx by the chain rule for differentiation; hence

$$\frac{dv}{dt} = \frac{dv}{dx}\frac{dx}{dt} = v\frac{dv}{dx}, \tag{13}$$

and Eq. (11) is replaced by

$$v\frac{dv}{dx} = -\frac{gR^2}{(R+x)^2}. \tag{14}$$

Separating the variables and integrating yields

$$\tfrac{1}{2}v^2 = \frac{gR^2}{R+x} + c. \tag{15}$$

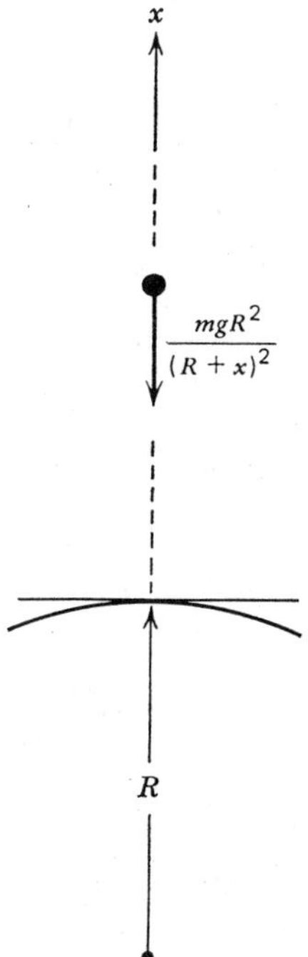

FIGURE 2.23 A body in the earth's gravitational field.

Since $x = 0$ when $t = 0$, the initial condition (12) at $t = 0$ can be replaced by the condition that $v = v_0$ when $x = 0$. Hence $c = \frac{1}{2}v_0^2 - gR$ and

$$v^2 = v_0^2 - 2gR + \frac{2gR^2}{R + x}. \tag{16}$$

The escape velocity is found by requiring that the velocity v given by Eq. (16) remain positive for all (positive) values of x. This condition will be met provided that $v_0^2 \geq 2gR$. Thus the escape velocity v_e is given by

$$v_e = (2gR)^{1/2}; \tag{17}$$

the magnitude of v_e is approximately 6.9 miles/sec or 11.1 km/sec.

In calculating the escape velocity $v_e = (2gR)^{1/2}$, air resistance has been neglected. The actual escape velocity, including the effect of air resistance, is somewhat higher. On the other hand, the effective escape velocity can be significantly reduced if the body is transported a considerable distance above sea level before being launched. Both gravitational and frictional forces are thereby reduced; air resistance, in particular, diminishes quite rapidly with increasing altitude.

PROBLEMS

1. A ball with mass 0.25 kg is thrown upward with initial velocity 20 m/sec from the roof of a building 30 m high. Neglect air resistance.
   (a) Find the maximum height above the ground that the ball reaches.
   (b) Assuming that the ball misses the building on the way down, find the time that it hits the ground.
2. Assume that the conditions are as in Problem 1 except that there is a force due to air resistance of $|v|/30$, where the velocity v is in meters per second.
   (a) Find the maximum height above the ground that the ball reaches.
   (b) Find the time that the ball hits the ground.
   Hint: Use Newton's method or some other appropriate numerical procedure in part (b).
3. A body of constant mass m is projected vertically upward with an initial velocity v_0. Assuming the gravitational attraction of the earth to be constant, and neglecting all other forces acting on the body, find
   (a) The maximum height attained by the body.
   (b) The time at which the maximum height is reached.
   (c) The time at which the body returns to its starting point.
4. A body is thrown vertically downward with an initial velocity v_0 in a medium offering resistance proportional to the magnitude of the velocity. Find the velocity v as a function of time t. Find the limiting velocity v_l approached after a long time.
5. An object of mass m is dropped from rest in a medium offering resistance proportional to the magnitude of the velocity. Find the time interval that elapses before the velocity of the object reaches 90% of its limiting value.
6. A boater and a motor boat together weigh 320 lb. If the thrust of the motor is equivalent to a constant force of 10 lb in the direction of motion, and if the resistance of the water to the motion is equal numerically to twice the speed in feet per second, and if the boat is initially at rest, find
   (a) The velocity of the boat at time t.
   (b) The limiting velocity.
7. A skydiver weighing 180 lb (including equipment) falls vertically downward from an altitude of 5000 ft, and opens the parachute after 10 sec of free fall. Assume that the force of air resistance is $0.75|v|$ when the parachute is closed and $12|v|$ when the parachute is open, where the velocity v is in feet per second.
   (a) Find the speed of the skydiver when the parachute opens.
   (b) Find the distance fallen before the parachute opens.
   (c) What is the limiting velocity v_l after the parachute opens?
   (d) Estimate how long the skydiver is in the air after the parachute opens.
8. A body with mass m is projected vertically downward with initial velocity v_0 in a medium offering resistance proportional to the square root of the magnitude of the velocity. Find the relation between the velocity v and the time t. Find the limiting velocity.
9. A body of mass m falls from rest in a medium offering resistance proportional to the square of the velocity. Find the relation between the velocity v and the time t. Find the limiting velocity.
10. A body of mass m falls in a medium offering resistance proportional to $|v|^r$, where r is a positive constant. Assuming the gravitational attraction to be constant, find the limiting velocity of the body.

11. A body of constant mass m is projected vertically upward with an initial velocity v_0 in a medium offering a resistance $k|v|$, where k is a constant. Neglect changes in the gravitational force.
(a) Find the maximum height x_m attained by the body and the time t_m at which this maximum height is reached.
(b) Show that if $kv_0/mg < 1$, then t_m and x_m can be expressed as

$$t_m = \frac{v_0}{g}\left[1 - \frac{1}{2}\frac{kv_0}{mg} + \frac{1}{3}\left(\frac{kv_0}{mg}\right)^2 - \cdots\right],$$

$$x_m = \frac{v_0^2}{2g}\left[1 - \frac{2}{3}\frac{kv_0}{mg} + \frac{1}{2}\left(\frac{kv_0}{mg}\right)^2 - \cdots\right].$$

12. A body of mass m is projected vertically upward with an initial velocity v_0 in a medium offering a resistance $k|v|$, where k is a constant. Assume that the gravitational attraction of the earth is constant.
(a) Find the velocity $v(t)$ of the body at any time.
(b) Use the result of part (a) to calculate the limit of $v(t)$ as $k \to 0$, that is, as the resistance approaches zero. Does this result agree with the velocity of a mass m projected upward with an initial velocity v_0 in a vacuum?
(c) Use the result of part (a) to calculate the limit of $v(t)$ as $m \to 0$, that is, as the mass approaches zero.
13. A body falling in a relatively dense fluid, oil for example, is acted on by three forces (see Figure 2.24): a resistive force R, a buoyant force B, and its weight w due to gravity. The buoyant force is equal to the weight of the fluid displaced by the object. For a slowly moving spherical body of radius a, the resistive force is given by Stokes'[16] law $R = 6\pi\mu a|v|$, where v is the velocity of the body, and μ is the coefficient of viscosity of the surrounding fluid.
(a) Find the limiting velocity of a solid sphere of radius a and density ρ falling freely in a medium of density ρ' and coefficient of viscosity μ.
(b) In 1910 the American physicist R. A. Millikan (1868–1953) determined the charge on an electron by studying the motion of tiny droplets of oil falling in an electric field. A field of strength E exerts a force Ee on a droplet with charge e. Assume that E has been adjusted so that the droplet is held stationary ($v = 0$), and that w and B are as given in part (a). Find a formula for e. Millikan was able to identify e as the charge on an electron and to determine that $e = 4.803 \times 10^{-10}$ esu.
14. A body of constant mass m is launched vertically upward from sea level with an initial velocity v_0 that does not exceed the escape velocity $v_e = (2gR)^{1/2}$. Neglecting air resistance, but considering the change of gravitational attraction with altitude, find the maximum altitude attained by the body (see Example 2).

[16]George Gabriel Stokes (1819–1903), professor at Cambridge, was one of the foremost applied mathematicians of the nineteenth century. The basic equations of fluid mechanics (the Navier–Stokes equations) are named partly in his honor, and one of the fundamental theorems of vector calculus bears his name. He was also one of the pioneers in the use of divergent (asymptotic) series, a subject of great interest and importance today.

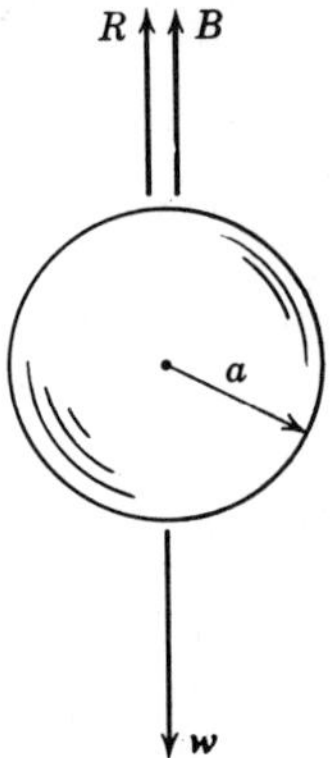

FIGURE 2.24 A body falling in a dense fluid.

15. Suppose that a rocket is launched straight up from the surface of the earth with initial velocity $v_0 = \sqrt{2gR}$, where R is the radius of the earth. Neglect air resistance.
(a) Find an expression for the velocity v in terms of the distance x from the surface of the earth.
(b) Find the time required for the rocket to go 240,000 miles (the approximate distance from the earth to the moon). Assume that $R = 4000$ miles.
16. Find the escape velocity for a body projected upward with an initial velocity v_0 from a point $x_0 = \xi R$ above the surface of the earth, where R is the radius of the earth and ξ is a constant. Neglect air resistance. Find the initial altitude from which the body must be launched in order to reduce the escape velocity to 85% of its value at the earth's surface.
17. **Brachistochrone Problem.** One of the famous problems in the history of mathematics is the brachistochrone problem: to find the curve along which a particle will slide without friction in the minimum time from one given point P to another Q, the second point being lower than the first but not directly beneath it (see Figure 2.25.). This problem was posed by Johann Bernoulli in 1696 as a challenge problem to the mathematicians of his day. Correct solutions were found by Johann Bernoulli and his brother Jakob Bernoulli, and by Isaac Newton, Gottfried Leibniz, and Marquis de

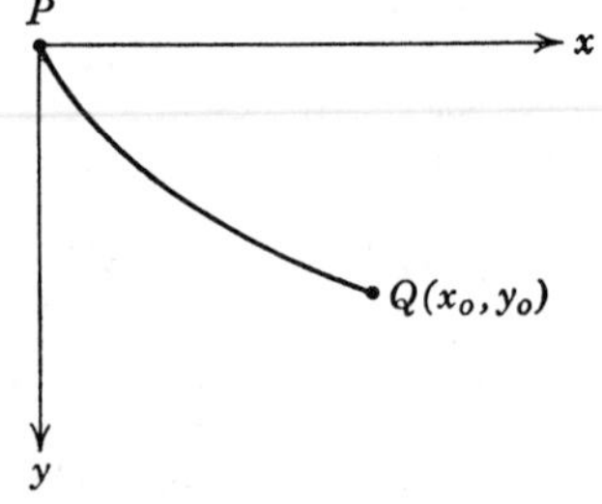

FIGURE 2.25 The brachistochrone.

L'Hôpital. The brachistochrone problem is important in the development of mathematics as one of the forerunners of the calculus of variations.

In solving this problem it is convenient to take the origin as the upper point P and to orient the axes as shown in Figure 2.25. The lower point Q has coordinates (x_0, y_0). It is then possible to show that the curve of minimum time is given by a function $y = \phi(x)$ that satisfies the differential equation

$$(1 + y'^2)y = k^2 \tag{i}$$

where k^2 is a certain positive constant to be determined later.
(a) Solve Eq. (i) for y'. Why is it necessary to choose the positive square root?
(b) Introduce the new variable t by the relation

$$y = k^2 \sin^2 t. \tag{ii}$$

Show that the equation found in part (a) then takes the form

$$2k^2 \sin^2 t\, dt = dx. \tag{iii}$$

(c) Letting $\theta = 2t$, show that the solution of Eq. (iii) for which $x = 0$ when $y = 0$ is given by

$$x = k^2(\theta - \sin\theta)/2, \qquad y = k^2(1 - \cos\theta)/2. \tag{iv}$$

Equations (iv) are parametric equations of the solution of Eq. (i) that passes through (0, 0). The graph of Eqs. (iv) is called a *cycloid*. If we make a proper choice of the constant k, then the cycloid also passes through the point (x_0, y_0), and is the solution of the brachistochrone problem. It is possible to eliminate θ and to obtain the solution in the form $y = \phi(x)$; however, it is easier to use the parametric equations.

2.8 Exact Equations

We have already mentioned that for first order nonlinear equations there are a number of integration methods that are applicable to various classes of problems. In Sections 2.8 through 2.10 and in some of the problems in Section 2.11 we describe the more important of these methods.

Let us first consider the equation

$$\psi(x, y) = c \tag{1}$$

where c is a constant. Assuming that Eq. (1) defines y implicitly as a differentiable function of x, we can differentiate Eq. (1) with respect to x and obtain[17]

$$\psi_x(x, y) + \psi_y(x, y)y' = 0. \tag{2}$$

Equation (2) is the differential equation whose solution is defined by Eq. (1).

Conversely, suppose that the differential equation

$$M(x, y) + N(x, y)y' = 0 \tag{3}$$

[17]Subscripts denote partial derivatives with respect to the variable indicated by the subscript.

is given. If there exists a function ψ such that

$$\psi_x(x, y) = M(x, y), \qquad \psi_y(x, y) = N(x, y), \tag{4}$$

and such that $\psi(x, y) = c$ defines $y = \phi(x)$ implicitly as a differentiable function of x, then

$$\begin{aligned} M(x, y) + N(x, y)y' &= \psi_x(x, y) + \psi_y(x, y)y' \\ &= \frac{d}{dx}\psi[x, \phi(x)]. \end{aligned} \tag{5}$$

As a result Eq. (3) becomes

$$\frac{d}{dx}\psi[x, \phi(x)] = 0. \tag{6}$$

In this case Eq. (3) is said to be an *exact differential equation*. The solution of Eq. (3), or the equivalent Eq. (6), is given implicitly by Eq. (1),

$$\psi(x, y) = c,$$

where c is an arbitrary constant.

In practice the differential equation (3) is often written in the symmetric differential form

$$M(x, y)\,dx + N(x, y)\,dy = 0. \tag{7}$$

EXAMPLE 1

Solve the differential equation

$$2xy^3 + 3x^2y^2\frac{dy}{dx} = 0. \tag{8}$$

By inspection it can be seen that the left side is the derivative of x^2y^3. Thus Eq. (8) can be rewritten as

$$\frac{d}{dx}(x^2y^3) = 0,$$

and the solution is given implicitly by $x^2y^3 = c$ and explicitly by

$$y = kx^{-2/3}, \qquad x < 0 \quad \text{or} \quad x > 0. \tag{9}$$

Note that the function $y = 0$, corresponding to $k = 0$, is a solution for all values of x. Also note that, for $x \neq 0$ and $y \neq 0$, we can divide Eq. (8) by xy^2 and obtain $2y + 3xy' = 0$, which can be solved either as a linear equation or as a separable one. The general solution is given by Eq. (9).

In this simple example it was easy to see that the differential equation was exact and, in fact, easy to find its solution, by recognizing that the left side was the derivative of x^2y^3. For more complicated equations it may not be possible to do this. A systematic procedure for determining whether a given differential equation is exact is provided by the following theorem.

Theorem 2.3. *Let the functions M, N, M_y, and N_x be continuous in the rectangular*[18] *region R: $\alpha < x < \beta$, $\gamma < y < \delta$. Then Eq.* (3),

$$M(x, y) + N(x, y)y' = 0,$$

is an exact differential equation in R if and only if

$$M_y(x, y) = N_x(x, y) \tag{10}$$

at each point of R. That is, there exists a function ψ satisfying Eqs. (4),

$$\psi_x(x, y) = M(x, y), \qquad \psi_y(x, y) = N(x, y),$$

if and only if M and N satisfy Eq. (10).

The proof of this theorem is in two parts. First we show that if there is a function ψ such that Eq. (4) is true, then it follows that Eq. (10) is satisfied. Computing M_y and N_x from Eq. (4) yields

$$M_y(x, y) = \psi_{xy}(x, y), \qquad N_x(x, y) = \psi_{yx}(x, y). \tag{11}$$

Since M_y and N_x are continuous, it follows that ψ_{xy} and ψ_{yx} are also continuous. This guarantees their equality,[19] and Eq. (10) follows.

We now show that if M and N satisfy Eq. (10), then Eq. (3) is exact. The proof involves the construction of a function ψ satisfying Eqs. (4),

$$\psi_x(x, y) = M(x, y), \qquad \psi_y(x, y) = N(x, y).$$

Integrating the first of Eqs. (4) with respect to x, holding y constant, gives

$$\psi(x, y) = \int^x M(t, y)\,dt + h(y). \tag{12}$$

The function h is an arbitrary function of y, playing the role of the arbitrary constant. Now we must show that it is always possible to choose $h(y)$ so that $\psi_y = N$. From Eq. (12)

$$\begin{aligned}\psi_y(x, y) &= \frac{\partial}{\partial y}\int^x M(t, y)\,dt + h'(y)\\ &= \int^x M_y(t, y)\,dt + h'(y).\end{aligned}$$

[18] It is not essential that the region R be rectangular, only that it be simply connected. In two dimensions this means that the region has no holes in its interior. Thus, for example, rectangular or circular regions are simply connected, but an annular region is not. More details can be found in most books on advanced calculus.

[19] The hypothesis of continuity is required, since otherwise ψ_{xy} and ψ_{yx} are not always equal. The exceptions are seldom encountered.

Setting $\psi_y = N$ and solving for $h'(y)$ gives

$$h'(y) = N(x, y) - \int^x M_y(t, y)\, dt. \tag{13}$$

To determine $h(y)$ from Eq. (13) it is essential that, despite its appearance, the right side of Eq. (13) be a function of y only. To establish this fact we can differentiate the quantity in question with respect to x, obtaining

$$N_x(x, y) - M_y(x, y),$$

which is zero on account of Eq. (10). Hence, despite its apparent form, the right side of Eq. (13) does not in fact depend on x, and a single integration then gives $h(y)$. Substituting for $h(y)$ in Eq. (12), we obtain as the solution of Eqs. (4)

$$\psi(x, y) = \int^x M(t, y)\, dt + \int^y \left[N(x, s) - \int^x M_s(t, s)\, dt \right] ds. \tag{14}$$

It should be noted that this proof contains a method for the computation of $\psi(x, y)$ and thus for solving the original differential equation (3). It is usually better to go through this process each time it is needed rather than to try to remember the result given in Eq. (14). Note also that the solution is obtained in implicit form; it may or may not be feasible to find the solution explicitly.

EXAMPLE 2

Solve

$$(y \cos x + 2xe^y) + (\sin x + x^2 e^y - 1)\, y' = 0. \tag{15}$$

It is clear that

$$M_y(x, y) = \cos x + 2xe^y = N_x(x, y),$$

and the given equation is exact. Thus there is a $\psi(x, y)$ such that

$$\psi_x(x, y) = y \cos x + 2xe^y,$$

$$\psi_y(x, y) = \sin x + x^2 e^y - 1.$$

Integrating the first of these equations, we obtain

$$\psi(x, y) = y \sin x + x^2 e^y + h(y). \tag{16}$$

Setting $\psi_y = N$ gives

$$\psi_y(x, y) = \sin x + x^2 e^y + h'(y) = \sin x + x^2 e^y - 1.$$

Thus

$$h'(y) = -1 \quad \text{and} \quad h(y) = -y.$$

The constant of integration can be omitted since any solution of the previous differential equation will suffice; we do not require the most general one. Substituting for $h(y)$ in Eq. (16) gives

$$\psi(x, y) = y \sin x + x^2 e^y - y.$$

Hence the solution of the original equation is given implicitly by

$$y \sin x + x^2 e^y - y = c. \tag{17}$$

EXAMPLE 3

Solve

$$(3x^2 + 2xy) + (x + y^2)y' = 0. \tag{18}$$

Here

$$M_y(x, y) = 2x, \qquad N_x(x, y) = 1;$$

since $M_y \neq N_x$, the given equation is not exact. To see that it cannot be solved by the procedure described above, let us seek a function ψ such that

$$\psi_x(x, y) = 3x^2 + 2xy, \qquad \psi_y(x, y) = x + y^2. \tag{19}$$

Integrating the first of Eqs. (19) gives

$$\psi(x, y) = x^3 + x^2 y + h(y), \tag{20}$$

where h is an arbitrary function of y only. To try to satisfy the second of Eqs. (19) we compute ψ_y from Eq. (20) and set it equal to N, obtaining

$$\psi_y(x, y) = x^2 + h'(y) = x + y^2$$

or

$$h'(y) = x + y^2 - x^2. \tag{21}$$

Since the right side of Eq. (21) depends on x as well as y, it is impossible to solve Eq. (21) for $h(y)$. Thus there is no $\psi(x, y)$ satisfying both of Eqs. (19).

PROBLEMS

Determine whether or not each of the equations in Problems 1 through 12 is exact. If it is exact, find the solution.

1. $(2x + 3) + (2y - 2)y' = 0$
2. $(2x + 4y) + (2x - 2y)y' = 0$
3. $(3x^2 - 2xy + 2)\,dx + (6y^2 - x^2 + 3)\,dy = 0$
4. $(2xy^2 + 2y) + (2x^2y + 2x)y' = 0$
5. $\dfrac{dy}{dx} = -\dfrac{ax + by}{bx + cy}$
6. $\dfrac{dy}{dx} = -\dfrac{ax - by}{bx - cy}$
7. $(e^x \sin y - 2y \sin x)\,dx + (e^x \cos y + 2\cos x)\,dy = 0$
8. $(e^x \sin y + 3y)\,dx - (3x - e^x \sin y)\,dy = 0$
9. $(ye^{xy} \cos 2x - 2e^{xy} \sin 2x + 2x)\,dx + (xe^{xy} \cos 2x - 3)\,dy = 0$
10. $(y/x + 6x)\,dx + (\ln x - 2)\,dy = 0, \qquad x > 0$
11. $(x \ln y + xy)\,dx + (y \ln x + xy)\,dy = 0; \qquad x > 0, \qquad y > 0$
12. $\dfrac{x\,dx}{(x^2 + y^2)^{3/2}} + \dfrac{y\,dy}{(x^2 + y^2)^{3/2}} = 0$

In each of Problems 13 and 14 solve the given initial value problem and determine at least approximately where the solution is valid.

13. $(2x - y)\,dx + (2y - x)\,dy = 0, \qquad y(1) = 3$
14. $(9x^2 + y - 1)\,dx - (4y - x)\,dy = 0, \qquad y(1) = 0$

15. Find the value of b for which each of the following equations is exact, and then solve it, using that value of b.
 (a) $(xy^2 + bx^2y)\,dx + (x + y)x^2\,dy = 0$
 (b) $(ye^{2xy} + x)\,dx + bxe^{2xy}\,dy = 0$
16. Consider the exact differential equation

$$M(x, y)\,dx + N(x, y)\,dy = 0.$$

 Find an implicit formula $\psi(x, y) = c$ for the solution analogous to Eq. (14) by first integrating the equation $\psi_y = N$, rather than $\psi_x = M$, as in the text.
17. Show that any equation which is separable, that is, of the form

$$M(x) + N(y)y' = 0,$$

 is also exact.

2.9 Integrating Factors

We now show how the methods of Section 2.8 can be extended to a somewhat larger class of problems. Consider an equation of the form

$$M(x, y)\,dx + N(x, y)\,dy = 0. \tag{1}$$

If this equation is not already exact, we will try to choose a function μ, which may depend on both x and y, so that the equation

$$\mu(M\,dx + N\,dy) = 0 \tag{2}$$

is exact. The function μ is called an *integrating factor*. Equation (2) can then be solved by the methods of Section 2.8, and its solutions will also satisfy the original equation (1). This approach is an extension of the method developed in Section 2.1 for linear equations.

To investigate the possibility of carrying out this procedure, we recall from Section 2.8 that Eq. (2) is exact if and only if

$$(\mu M)_y = (\mu N)_x. \tag{3}$$

Since M and N are given functions, the integrating factor μ must satisfy the first order partial differential equation

$$M\mu_y - N\mu_x + (M_y - N_x)\mu = 0. \tag{4}$$

If a function μ satisfying Eq. (4) can be found, then Eq. (2) will be exact. The solution of Eq. (2) can then be obtained by the method of Section 2.8 and is given implicitly in the form

$$\psi(x, y) = c.$$

This relation also defines the solution of Eq. (1), since the integrating factor μ can be canceled out of all terms in Eq. (2).

A partial differential equation of the form (4) may have more than one solution; if this is the case, any such solution may be used as an integrating factor of Eq. (1). This possible nonuniqueness of the integrating factor is illustrated in Example 2.

Unfortunately, Eq. (4), which determines the integrating factor, is ordinarily at least as difficult to solve as the original equation (1). Therefore, while in principle integrating factors are powerful tools for solving differential equations, in practice they can usually be found only in special cases. Some of these cases are indicated in the following examples and in Problems 5 and 6 at the end of the section.

EXAMPLE 1

Verify that $\mu(x, y) = (xy^2)^{-1}$ is an integrating factor for the differential equation

$$(y^2 + xy)\,dx - x^2\,dy = 0, \tag{5}$$

and then find its solution.

Since $M(x, y) = y^2 + xy$, and $N(x, y) = -x^2$, it follows that $M_y(x, y) = 2y + x$, and $N_x(x, y) = -2x$; thus the given equation is not exact. To show that $\mu(x, y) = (xy^2)^{-1}$ is an integrating factor for this equation, we multiply Eq. (5) by $\mu(x, y)$, thus obtaining

$$\left(\frac{1}{x} + \frac{1}{y}\right) dx - \frac{x}{y^2}\,dy = 0;$$

this latter equation is easily verified to be exact. By the method of Section 2.8 its solution is found to be given implicitly by

$$\ln|x| + (x/y) = c; \qquad x \neq 0, \quad y \neq 0. \tag{6}$$

Note that $y = 0$ is also a solution of Eq. (5), although it is not given by Eq. (6).

The two most important situations in which simple integrating factors can be found occur when μ is a function of only one of the variables x, y, instead of both. Let us determine necessary conditions on M and N so that $M\,dx + N\,dy = 0$ has an integrating factor μ, which depends on x only.

Assuming that μ is a function of x only, we have

$$(\mu M)_y = \mu M_y, \qquad (\mu N)_x = \mu N_x + N\frac{d\mu}{dx}.$$

Thus, if $(\mu M)_y$ is to equal $(\mu N)_x$, it is necessary that

$$\frac{d\mu}{dx} = \frac{M_y - N_x}{N}\mu. \tag{7}$$

If $(M_y - N_x)/N$ is a function of x only, then there is an integrating factor μ which also depends only on x; furthermore, $\mu(x)$ can be found by solving the first order *linear* equation (7).

A similar procedure can be used to determine a condition under which Eq. (1) has an integrating factor depending only on y (see Problem 5).

EXAMPLE 2

Find an integrating factor for the equation

$$(3xy + y^2) + (x^2 + xy)\frac{dy}{dx} = 0, \tag{8}$$

and then solve the equation.

On computing the quantity $(M_y - N_x)/N$ we find that

$$\frac{M_y(x, y) - N_x(x, y)}{N(x, y)} = \frac{3x + 2y - (2x + y)}{x^2 + xy} = \frac{1}{x}.$$

Thus, according to the preceding discussion, there is an integrating factor μ that is a function of x only and satisfies the differential equation

$$\frac{d\mu}{dx} = \frac{\mu}{x}.$$

Hence

$$\mu(x) = x. \tag{9}$$

Multiplying Eq. (8) by this integrating factor, we obtain

$$(3x^2y + xy^2) + (x^3 + x^2y)\frac{dy}{dx} = 0, \tag{10}$$

which is an exact equation. Its solution in easily found by the method of Section 2.8 to be given implicitly by

$$x^3y + \tfrac{1}{2}x^2y^2 = c. \tag{11}$$

Note that the quadratic equation (11) can be solved for y in terms of x. The reader may also verify, as in Example 1, that a second integrating factor of Eq. (8) is given by $\mu(x, y) = 1/xy(2x + y)$, and that the same solution is obtained, though with much greater difficulty, if this integrating factor is used (see Problem 14).

PROBLEMS

Show that the equations in Problems 1 through 4 are not exact, but become exact when multiplied by the given integrating factor. Then solve the equations.

1. $x^2y^3 + x(1 + y^2)y' = 0; \qquad \mu(x, y) = 1/xy^3$
2. $\left(\dfrac{\sin y}{y} - 2e^{-x}\sin x\right)dx + \left(\dfrac{\cos y + 2e^{-x}\cos x}{y}\right)dy = 0; \qquad \mu(x, y) = ye^x$
3. $y\,dx + (2x - ye^y)\,dy = 0; \qquad \mu(x, y) = y$
4. $(x + 2)\sin y\,dx + x\cos y\,dy = 0, \qquad \mu(x, y) = xe^x$
5. Show that if $(N_x - M_y)/M = Q$, where Q is a function of y only, then the differential equation

$$M + Ny' = 0$$

has an integrating factor of the form

$$\mu(y) = \exp\int^y Q(t)\,dt.$$

*6. Show that if $(N_x - M_y)/(xM - yN) = R$, where R depends on the quantity xy only, then the differential equation

$$M + Ny' = 0$$

has an integrating factor of the form $\mu(xy)$. Find a general formula for this integrating factor.

In each of Problems 7 through 13 find an integrating factor, and solve the given equation.

7. $(3x^2y + 2xy + y^3)\,dx + (x^2 + y^2)\,dy = 0$
8. $y' = e^{2x} + y - 1$
9. $dx + (x/y - \sin y)\,dy = 0$
10. $y\,dx + (2xy - e^{-2y})\,dy = 0$
11. $e^x\,dx + (e^x\cot y + 2y\csc y)\,dy = 0$
12. $[4(x^3/y^2) + (3/y)]\,dx + [3(x/y^2) + 4y]\,dy = 0$

*13. $\left(3x + \dfrac{6}{y}\right) + \left(\dfrac{x^2}{y} + 3\dfrac{y}{x}\right)\dfrac{dy}{dx} = 0$
Hint: See Problem 6.

*14. Solve the differential equation

$$(3xy + y^2) + (x^2 + xy)\,y' = 0$$

using the integrating factor $\mu(x, y) = [xy(2x + y)]^{-1}$. Verify that the solution is the same as that obtained in Example 2 with a different integrating factor.

2.10 Homogeneous Equations

The two main types of nonlinear first order equations that can be solved by direct processes of integration are those in which the variables separate and those which are exact. Section 2.9 dealt with the use of integrating factors to solve certain types of equations by reducing them to exact equations. This section introduces another technique, that of changes of variables, which can sometimes be used to simplify a differential equation and thus make its solution possible (or more convenient). Generally speaking, the appropriate substitution must be suggested by the repeated appearance of some combination of the variables, or some other peculiarity in the structure of the equation.

The most important class of equations for which a definite rule can be laid down is the class of homogeneous[20] differential equations. An equation of the form

$$dy/dx = f(x, y)$$

is said to be homogeneous whenever the function f does not depend on x and y separately, but only on their ratio y/x or x/y. Thus homogeneous equations are of the form

$$dy/dx = F(y/x). \tag{1}$$

Consider the following examples:

$$\text{(a)}\quad \frac{dy}{dx} = \frac{y^2 + 2xy}{x^2} = \left(\frac{y}{x}\right)^2 + 2\frac{y}{x};$$

$$\text{(b)}\quad \frac{dy}{dx} = \ln x - \ln y + \frac{x+y}{x-y} = \ln\frac{1}{y/x} + \frac{1+(y/x)}{1-(y/x)};$$

$$\text{(c)}\quad \frac{dy}{dx} = \frac{y^3 + 2xy}{x^2} = y\left(\frac{y}{x}\right)^2 + 2\frac{y}{x}.$$

Equations (a) and (b) are homogeneous, since the right side of each can be expressed as a function of y/x; since (c) cannot be so written, it is not homogeneous. Usually it is easy to tell by inspection whether or not a given equation is homogeneous; for complicated equations the criterion given in Problem 15 may be useful.

The form of a homogeneous equation suggests that it may be simplified by introducing a new variable, which we will denote by v, to represent the ratio of y to x. Thus

$$y = xv, \tag{2}$$

and Eq. (1) becomes

$$dy/dx = F(v). \tag{3}$$

Looking on v as the new dependent variable (replacing y), we must consider v as a function of x, and replace dy/dx in Eq. (3) by a suitable expression in terms of v. Differentiating Eq. (2) gives

$$\frac{dy}{dx} = x\frac{dv}{dx} + v,$$

[20] The word homogeneous is used in more than one way in the study of differential equations. The reader must be alert to distinguish these as they occur.

and hence Eq. (3) becomes

$$x\frac{dv}{dx} + v = F(v). \tag{4}$$

The most significant fact about Eq. (4) is that the variables x and v can *always* be separated, regardless of the form of the function F; in fact,

$$\frac{dx}{x} = \frac{dv}{F(v) - v}. \tag{5}$$

Solving Eq. (5) and then replacing v by y/x gives the solution of the original equation.

Thus any homogeneous equation can be transformed into one whose variables are separated by the substitution (2). As a practical matter, of course, it may or may not be possible to evaluate the integral required in solving Eq. (5) by elementary methods. Moreover, a homogeneous equation may also belong to one of the classes already discussed; it may be exact, or even linear, for example. In such cases there is a choice of methods for finding its solution.

EXAMPLE

Solve the differential equation

$$\frac{dy}{dx} = \frac{y^2 + 2xy}{x^2}. \tag{6}$$

Writing this equation as

$$\frac{dy}{dx} = \left(\frac{y}{x}\right)^2 + 2\frac{y}{x}$$

shows that it is homogeneous. The variables cannot be separated, nor is the equation exact, nor is an integrating factor obvious. Thus we are led to the substitution (2), which transforms the given equation into

$$x\frac{dv}{dx} + v = v^2 + 2v.$$

Hence

$$x\frac{dv}{dx} = v^2 + v$$

or, separating the variables,

$$\frac{dx}{x} = \frac{dv}{v(v+1)}.$$

Expanding the right side by partial fractions, we obtain

$$\frac{dx}{x} = \left(\frac{1}{v} - \frac{1}{v+1}\right) dv.$$

Integrating both sides then yields

$$\ln|x| + \ln|c| = \ln|v| - \ln|v + 1|,$$

where c is an arbitrary constant. Hence, combining the logarithms and taking the exponential of both sides, we obtain

$$cx = \frac{v}{v + 1}.$$

Finally, substituting for v in terms of y gives the solution of Eq. (6) in the form

$$cx = \frac{y/x}{(y/x) + 1} = \frac{y}{y + x}.$$

Solving for y we obtain

$$y = \frac{cx^2}{1 - cx}. \tag{7}$$

Sometimes it is helpful to transform both the independent and dependent variables. For example, in Problems 9, 10, and 11, a change of both variables is required to make the equation homogeneous, after which a further change of the dependent variable leads to a solution of the problem.

PROBLEMS

Show that the equations in Problems 1 through 8 are homogeneous, and find their solutions.

1. $\dfrac{dy}{dx} = \dfrac{x + y}{x}$
2. $2y\,dx - x\,dy = 0$
3. $\dfrac{dy}{dx} = \dfrac{x^2 + xy + y^2}{x^2}$
4. $\dfrac{dy}{dx} = \dfrac{x^2 + 3y^2}{2xy}$
5. $\dfrac{dy}{dx} = \dfrac{4y - 3x}{2x - y}$
6. $\dfrac{dy}{dx} = -\dfrac{4x + 3y}{2x + y}$
7. $\dfrac{dy}{dx} = \dfrac{x + 3y}{x - y}$
8. $(x^2 + 3xy + y^2)\,dx - x^2\,dy = 0$
9. (a) Find the solution of the equation

$$\frac{dy}{dx} = \frac{2y - x}{2x - y}.$$

(b) Find the solution of the equation

$$\frac{dy}{dx} = \frac{2y - x + 5}{2x - y - 4}.$$

Hint: To reduce the equation of part (b) to that of part (a), consider a preliminary substitution of the form $x = X - h$, $y = Y - k$. Choose the constants h and k so that the equation is homogeneous in the variables X and Y.

10. Solve $\dfrac{dy}{dx} = -\dfrac{4x + 3y + 15}{2x + y + 7}$. See Hint, Problem 9(b).

11. Solve $\dfrac{dy}{dx} = \dfrac{x + 3y - 5}{x - y - 1}$. See Hint, Problem 9(b).

12. Find the solution of the equation

$$(3xy + y^2)\, dx + (x^2 + xy)\, dy = 0$$

by the method of this section, and compare it with the solution obtained by other methods in Example 2 and Problem 14 of Section 2.9.

*13. Show that if

$$M(x, y)\, dx + N(x, y)\, dy = 0$$

is a homogeneous equation, then it has

$$\mu(x, y) = \frac{1}{xM(x, y) + yN(x, y)}$$

for an integrating factor.

*14. Use the result of Problem 13 to solve each of the following equations:
(a) The equation of Problem 2.
(b) The equation of Problem 4.

*15. Show that the equation $y' = f(x, y)$ is homogeneous if $f(x, y)$ is such that

$$f(x, tx) = f(1, t),$$

where t is a real parameter. Use this fact to determine whether each of the following equations is homogeneous.

(a) $y' = \dfrac{x^3 + xy + y^3}{x^2y + xy^2}$ (b) $y' = \ln x - \ln y + \dfrac{x + y}{x - y}$

(c) $y' = \dfrac{(x^2 + 3xy + 4y^2)^{1/2}}{x + 2y}$ (d) $y' = \dfrac{\sin(xy)}{x^2 + y^2}$

2.11 Miscellaneous Problems and Applications

This section consists of a list of problems. The first 32 problems can be solved by the methods of the previous sections. They are presented so that the reader may have some practice in identifying the method or methods applicable to a given equation. Next are a number of problems suggesting specialized techniques that are useful for certain types of equations. In particular, Problems 35 through 37 deal with Riccati equations. Problems 38 through 43 are concerned with some geometrical applications, while the remaining problems deal with some other applications of differential equations.

PROBLEMS

1. $\dfrac{dy}{dx} = \dfrac{x^3 - 2y}{x}$

2. $(x + y)\,dx - (x - y)\,dy = 0$

3. $\dfrac{dy}{dx} = \dfrac{2x + y}{3 + 3y^2 - x}, \quad y(0) = 0$

4. $(x + e^y)\,dy - dx = 0$

5. $\dfrac{dy}{dx} = -\dfrac{2xy + y^2 + 1}{x^2 + 2xy}$

6. $x\dfrac{dy}{dx} + xy = 1 - y, \quad y(1) = 0$

7. $\dfrac{dy}{dx} = \dfrac{x}{x^2y + y^3}$ *Hint:* Let $u = x^2$.

8. $x\dfrac{dy}{dx} + 2y = \dfrac{\sin x}{x}, \quad y(2) = 1$

9. $\dfrac{dy}{dx} = -\dfrac{2xy + 1}{x^2 + 2y}$

10. $(3y^2 + 2xy)\,dx - (2xy + x^2)\,dy = 0$

11. $(x^2 + y)\,dx + (x + e^y)\,dy = 0$

12. $\dfrac{dy}{dx} + y = \dfrac{1}{1 + e^x}$

13. $x\,dy - y\,dx = (xy)^{1/2}\,dx$

14. $(x + y)\,dx + (x + 2y)\,dy = 0, \quad y(2) = 3$

15. $(e^x + 1)\dfrac{dy}{dx} = y - ye^x$

16. $\dfrac{dy}{dx} = \dfrac{x^2 + y^2}{x^2}$

17. $\dfrac{dy}{dx} = e^{2x} + 3y$

18. $(2y + 3x)\,dx = -x\,dy$

19. $x\,dy - y\,dx = 2x^2y^2\,dy, \quad y(1) = -2$

20. $y' = e^{x+y}$

21. $xy' = y + xe^{y/x}$

22. $\dfrac{dy}{dx} = \dfrac{x^2 - 1}{y^2 + 1}, \quad y(-1) = 1$

23. $xy' + y - y^2e^{2x} = 0$

24. $2\sin y \cos x\,dx + \cos y \sin x\,dy = 0$

25. $\left(2\dfrac{x}{y} - \dfrac{y}{x^2 + y^2}\right)dx + \left(\dfrac{x}{x^2 + y^2} - \dfrac{x^2}{y^2}\right)dy = 0$

26. $(2y + 1)\,dx + \left(\dfrac{x^2 - y}{x}\right)dy = 0$

27. $(\cos 2y - \sin x)\,dx - 2\tan x \sin 2y\,dy = 0$

28. $\dfrac{dy}{dx} = \dfrac{3x^2 - 2y - y^3}{2x + 3xy^2}$

29. $\dfrac{dy}{dx} = \dfrac{2y + \sqrt{x^2 - y^2}}{2x}$

30. $\dfrac{dy}{dx} = \dfrac{y^3}{1 - 2xy^2}, \quad y(0) = 1$

31. $(x^2y + xy - y)\,dx + (x^2y - 2x^2)\,dy = 0$

32. $\dfrac{dy}{dx} = -\dfrac{3x^2y + y^2}{2x^3 + 3xy}, \quad y(1) = -2$

33. Show that an equation of the form

$$y = G(p),$$

where $p = dy/dx$, can be solved in the following manner.
(a) Differentiate with respect to x.
(b) Integrate the equation of part (a) to obtain x as a function of p. This equation and the original equation $y = G(p)$ form a parametric representation of the solution.

34. Solve the differential equation

$$y - \ln p = 0,$$

where $p = dy/dx$, by the method of Problem 33. Also solve this equation directly, and verify that the solutions are the same.

35. **Riccati equations.** The equation

$$\frac{dy}{dx} = q_1(x) + q_2(x)y + q_3(x)y^2$$

is known as a *Riccati*[21] *equation*. Suppose that some particular solution y_1 of this equation is known. A more general solution containing one arbitrary constant can be obtained through the substitution

$$y = y_1(x) + \frac{1}{v(x)}.$$

Show that $v(x)$ satisfies the first order *linear* equation

$$\frac{dv}{dx} = -(q_2 + 2q_3y_1)v - q_3.$$

Note that $v(x)$ will contain a single arbitrary constant.

36. Using the method of problem 35 and the given particular solution, solve each of the following Riccati equations.
(a) $y' = 1 + x^2 - 2xy + y^2; \qquad y_1(x) = x$

(b) $y' = -\dfrac{1}{x^2} - \dfrac{y}{x} + y^2; \qquad y_1(x) = 1/x$

(c) $\dfrac{dy}{dx} = \dfrac{2\cos^2 x - \sin^2 x + y^2}{2\cos x}; \qquad y_1(x) = \sin x$

37. The propagation of a single action in a large population (for example, drivers turning on headlights at sunset) often depends partly on external circumstances (gathering darkness) and partly on a tendency to imitate others who have already performed the action in question. In this case the proportion $y(t)$ of people who have performed the action can be described[22] by the equation

$$dy/dt = (1 - y)[x(t) + by], \qquad \text{(i)}$$

[21] Jacopo Francesco Riccati (1676–1754), a Venetian nobleman, declined university appointments in Italy, Austria, and Russia to pursue his mathematical studies privately at home. He studied the differential equation that now bears his name extensively; however, it was Euler (in 1760) who discovered the result stated in this problem.

[22] See Anatol Rapoport, "Contribution to the Mathematical Theory of Mass Behavior: I. The Propagation of Single Acts," *Bulletin of Mathematical Biophysics 14* (1952), pp. 159–169, and John Z. Hearon, "Note on the Theory of Mass Behavior," *Bulletin of Mathematical Biophysics*, *17* (1955), pp. 7–13.

where $x(t)$ measures the external stimulus and b is the imitation coefficient.
(a) Observe that Eq. (i) is a Riccati equation and that $y_1(t) = 1$ is one solution. Use the transformation suggested in Problem 35, and find the linear equation satisfied by $v(t)$.
(b) Find $v(t)$ in the case that $x(t) = at$, where a is a constant. Leave your answer in the form of an integral.

38. **Orthogonal Trajectories.** A fairly common geometrical problem is to find the family of curves that intersects a given family of curves orthogonally at each point. Such families of curves are said to be orthogonal trajectories of each other.
(a) Consider the family of parabolas

$$y = kx^2, \tag{i}$$

where k is a constant. Sketch the graph of Eq. (i) for several values of k. Find an expression for the slope of the parabola passing through a given point that involves the coordinates (x, y) of the point but not the parameter k.
Hint: Differentiate Eq. (i) and eliminate k.
(b) Making use of the fact that the slopes of orthogonal curves are negative reciprocals, write the differential equation for the orthogonal trajectories of Eq. (i).
(c) Solve the equation found in part (b) and determine the orthogonal trajectories. Sketch several numbers of this family of curves.

39. In each of the following cases use the method of Problem 38 to find the family of orthogonal trajectories of the given family of curves. Sketch both the given family and their orthogonal trajectories.
(a) The family of hyperbolas $xy = c$.
(b) The family of circles $(x - c)^2 + y^2 = c^2$.
(c) The family of ellipses $x^2 - xy + y^2 = c^2$.
(d) The family of parabolas $2cy + x^2 = c^2$, $c > 0$.

40. If two straight lines in the xy plane, having slopes m_1 and m_2, respectively, intersect at an angle θ, show that

$$(\tan\theta)(1 + m_1 m_2) = m_2 - m_1.$$

Using this fact, find the family of curves that intersects each of the following families at an angle of 45°. In each case sketch both families of curves.
(a) $x - 2y = c$ (b) $x^2 + y^2 = c^2$

41. Find all plane curves such that the tangent line at each point (x, y) passes through the fixed point (a, b).

42. The line normal to a given curve at each point (x, y) on the curve passes through the point $(2, 0)$. If the curve contains the point $(2, 3)$, find its equation.

43. Find all plane curves such that for each point on the curve the y axis bisects that part of the tangent line between the point of tangency and the x axis.

44. A certain man has a fortune that increases at a rate proportional to the square of his present wealth. If he had one million dollars a year ago, and has two million dollars today, how much will he be worth in 6 months? In 2 years?

45. A pond forms as water collects in a conical depression of radius a and depth h. Suppose that water flows in at a constant rate k, and is lost through evaporation at a rate proportional to the surface area.
(a) Show that the volume $V(t)$ of water in the pond at time t satisfies the differential

equation

$$dV/dt = k - \alpha\pi(3a/\pi h)^{2/3} V^{2/3},$$

where α is the coefficient of evaporation.
(b) Find the equilibrium depth of water in the pond. Is the equilibrium stable?
(c) Find a condition that must be satisfied if the pond is not to overflow.

46. Consider a cylindrical water tank of constant cross section A. Water is pumped into the tank at a constant rate k, and leaks out through a small hole of area a in the bottom of the tank. From Torricelli's theorem in hydrodynamics it follows that the rate at which water flows through the hole is $\alpha a\sqrt{2gh}$, where h is the current depth of water in the tank, g is the acceleration due to gravity, and α is a contraction coefficient which satisfies $0.5 \le \alpha \le 1.0$.
(a) Show that the depth of water in the tank at any time satisfies the equation

$$dh/dt = \left(k - \alpha a\sqrt{2gh}\right)/A.$$

(b) Determine the equilibrium depth h_e of water, and show that it is stable. Observe that h_e does not depend on A.

47. The growth of a cell depends on the flow of nutrients through its surface. Let $W(t)$ be the weight of the cell at time t, W_0 its weight at $t = 0$, and assume that dW/dt is proportional to the area of the cell surface.
(a) Give an argument to support the proposition that

$$\frac{dW}{dt} = \alpha W^{2/3},$$

where α is a constant of proportionality.
(b) Find the weight of the cell at any time t.

*2.12 The Existence and Uniqueness Theorem

In this section we discuss the proof of Theorem 2.2, the fundamental existence and uniqueness theorem for first order initial value problems. This theorem states that under certain conditions on $f(x, y)$, the initial value problem

$$y' = f(x, y), \tag{1a}$$

$$y(x_0) = y_0 \tag{1b}$$

has a unique solution in some interval containing the point x_0.

In some cases (for example, if the differential equation is linear) the existence of a solution of the initial value problem (1) can be established directly by actually solving the problem and exhibiting a formula for the solution. However, in general, this approach is not feasible because there is no method of solving Eq. (1a) that applies in all cases. Therefore, for the general case it is necessary to adopt an indirect approach that establishes the existence of a solution of Eqs. (1) but usually does not provide a practical means of finding it. The heart of this method is the construction of a sequence of functions that converges to a limit function satisfying the initial value problem, although the members of the sequence individually do not. As a rule, it is impossible to compute explicitly more than a few members of the sequence; therefore, the limit function can be found explicitly only in rare cases. Nevertheless, under the restrictions on $f(x, y)$ stated in Theorem 2.2, it is possible to show that the sequence in question converges and that the limit function has the desired properties. The argument is fairly intricate and depends, in part, on techniques and results that are usually encountered for the first time in a course on advanced calculus. Consequently, we do not go into all of the details of the proof here; we do however, indicate its main features and point out some of the difficulties involved.

First of all, we note that it is sufficient to consider the problem in which the initial point (x_0, y_0) is the origin; that is, the problem

$$y' = f(x, y), \tag{2a}$$

$$y(0) = 0. \tag{2b}$$

If some other initial point is given, then we can always make a preliminary change of variables, corresponding to a translation of the coordinate axes, that will take the given point (x_0, y_0) into the origin. Specifically, we introduce new dependent and independent variables w and s, respectively, defined by the equations

$$w = y - y_0, \qquad s = x - x_0. \tag{3}$$

Thinking of w as a function of s, we have by the chain rule

$$\frac{dw}{ds} = \frac{dw}{dx}\frac{dx}{ds} = \frac{d}{dx}(y - y_0)\frac{dx}{ds} = \frac{dy}{dx}.$$

Denoting $f(x, y) = f(s + x_0, w + y_0)$ by $F(s, w)$, the initial value problem (1) is converted into the form

$$w'(s) = F[s, w(s)], \tag{4a}$$

$$w(0) = 0. \tag{4b}$$

Except for the names of the variables, Eqs. (4) are the same as Eqs. (2). Hereafter, we consider the slightly simpler problem (2) rather than the original problem (1).

The existence and uniqueness theorem can now be stated in the following way.

Theorem 2.4. *If f and $\partial f/\partial y$ are continuous in a rectangle R: $|x| \le a$, $|y| \le b$, then there is some interval $|x| \le h \le a$ in which there exists a unique solution $y = \phi(x)$ of the initial value problem* (2)

$$y' = f(x, y), \qquad y(0) = 0.$$

To prove this theorem it is necessary to transform the initial value problem (2) into a more convenient form. If we suppose temporarily that there is a function $y = \phi(x)$ that satisfies the initial value problem, then $f[x, \phi(x)]$ is a continuous function of x only. Hence we can integrate Eq. (2a) from the initial point $x = 0$ to an arbitrary value of x, obtaining

$$\phi(x) = \int_0^x f[t, \phi(t)]\, dt, \tag{5}$$

where we have made use of the initial condition $\phi(0) = 0$.

Since Eq. (5) contains an integral of the unknown function ϕ, it is called an *integral equation*. This integral equation is not a formula for the solution of the initial value problem, but it does provide another relation satisfied by any solution of Eqs. (2). Conversely, suppose that there is a continuous function $y = \phi(x)$ that satisfies the integral equation (5); then this function also satisfies the initial value problem (2). To show this, we first substitute zero for x in Eq. (5), thus obtaining Eq. (2b). Further, since the integrand in Eq. (5) is continuous it follows from the fundamental theorem of calculus that $\phi'(x) = f[x, \phi(x)]$. Therefore the initial value problem and the integral equation are equivalent in the sense that any solution of one is also a solution of the other. It is more convenient to show that there is a unique solution of the integral equation in a certain interval $|x| \le h$. The same conclusion will then also hold for the initial value problem.

One method of showing that the integral equation (5) has a unique solution is known as the *method of successive approximations*, or Picard's[23] *iteration method*. In using this method we start by choosing an initial function ϕ_0, either arbitrarily or to approximate in some way the solution of the initial value problem. The simplest choice is

$$\phi_0(x) = 0; \tag{6}$$

then ϕ_0 at least satisfies the initial condition (2b) although presumably not the differential equation (2a). The next approximation ϕ_1 is obtained by substituting $\phi_0(t)$ for $\phi(t)$ in the right side of Eq. (5), and calling the result of this operation $\phi_1(x)$. Thus

$$\phi_1(x) = \int_0^x f[t, \phi_0(t)]\, dt. \tag{7}$$

Similarly, ϕ_2 is obtained from ϕ_1:

$$\phi_2(x) = \int_0^x f[t, \phi_1(t)]\, dt, \tag{8}$$

and in general,

$$\phi_{n+1}(x) = \int_0^x f[t, \phi_n(t)]\, dt. \tag{9}$$

[23] Charles Émile Picard (1856–1941), except for Henri Poincaré perhaps the most distinguished French mathematician of his generation, was appointed professor at the Sorbonne before the age of 30. He is known for important theorems in complex variables and algebraic geometry as well as in differential equations. A special case of the method of successive approximations was first published by Liouville in 1838. However, the method is usually credited to Picard, who established it in a general and widely applicable form in a series of papers beginning in 1890.

In this manner we generate the sequence of functions $\{\phi_n\} = \phi_0, \phi_1, \ldots, \phi_n, \ldots$. Each member of the sequence satisfies the initial condition (2b), but in general none satisfies the differential equation. However, if at some stage, say for $n = k$, we find that $\phi_{k+1}(x) = \phi_k(x)$, then it follows that ϕ_k is a solution of the integral equation (5). Hence ϕ_k is also a solution of the initial value problem (2), and the sequence is terminated at this point. In general, this will not occur, and it is necessary to consider the entire infinite sequence.

To establish Theorem 2.4 four principal questions must be answered:

1. Do all members of the sequence $\{\phi_n\}$ exist, or may the process break down at some stage?
2. Does the sequence converge?
3. What are the properties of the limit function? In particular, does it satisfy the integral equation (5), and hence the initial value problem (2)?
4. Is this the only solution, or may there be others?

We first show how these questions can be answered in a specific and relatively simple example, and then comment on some of the difficulties that may be encountered in the general case.

EXAMPLE

Consider the initial value problem

$$y' = 2x(1 + y), \qquad y(0) = 0. \tag{10}$$

To solve this problem by the method of successive approximations we note first that if $y = \phi(x)$, then the corresponding integral equation is

$$\phi(x) = \int_0^x 2t[1 + \phi(t)]\, dt. \tag{11}$$

If the initial approximation is $\phi_0(x) = 0$, it follows that

$$\phi_1(x) = \int_0^x 2t[1 + \phi_0(t)]\, dt = \int_0^x 2t\, dt = x^2. \tag{12}$$

Similarly

$$\phi_2(x) = \int_0^x 2t[1 + \phi_1(t)]\, dt = \int_0^x 2t[1 + t^2]\, dt = x^2 + \frac{x^4}{2}, \tag{13}$$

and

$$\phi_3(x) = \int_0^x 2t[1 + \phi_2(t)]\, dt = \int_0^x 2t\left[1 + t^2 + \frac{t^4}{2}\right] dt = x^2 + \frac{x^4}{2} + \frac{x^6}{2 \cdot 3}. \tag{14}$$

Equations (12), (13), and (14) suggest that

$$\phi_n(x) = x^2 + \frac{x^4}{2!} + \frac{x^6}{3!} + \cdots + \frac{x^{2n}}{n!} \tag{15}$$

for each $n \geq 1$, and this result can be established by mathematical induction. Clearly Eq. (15) is true for $n = 1$, and we must show that if it is true for $n = k$, then it also holds for $n = k + 1$. We have

$$\begin{aligned} \phi_{k+1}(x) &= \int_0^x 2t[1 + \phi_k(t)]\, dt \\ &= \int_0^x 2t\left(1 + t^2 + \frac{t^4}{2!} + \cdots + \frac{t^{2k}}{k!}\right) dt \\ &= x^2 + \frac{x^4}{2!} + \frac{x^6}{3!} + \cdots + \frac{x^{2k+2}}{(k+1)!}, \end{aligned} \tag{16}$$

and the inductive proof is complete.

It is clear from Eq. (15) that $\phi_n(x)$ is the nth partial sum of the infinite series

$$\sum_{k=1}^{\infty} x^{2k}/k!; \tag{17}$$

hence $\lim_{n\to\infty}\phi_n(x)$ exists if and only if the series (17) converges. Applying the ratio test, we see that for each x

$$\left|\frac{x^{2k+2}}{(k+1)!}\frac{k!}{x^{2k}}\right| = \frac{x^2}{k+1} \to 0 \quad \text{as} \quad k \to \infty; \tag{18}$$

thus the series (17) converges for all x, and its sum $\phi(x)$ is the limit[24] of the sequence $\{\phi_n(x)\}$. Further, since the series (17) is a Taylor series, it can be differentiated or integrated term by term as long as x remains within the interval of convergence, which in this case is the whole x axis. Therefore we can verify by direct computation that $\phi(x) = \sum_{k=1}^{\infty} x^{2k}/k!$ is a solution of the integral equation (11). Alternatively, by substituting $\phi(x)$ for y in Eq. (10) we can verify that this function also satisfies the initial value problem.

Finally, to deal with the question of uniqueness, let us suppose that the initial value problem has two solutions ϕ and ψ. Since ϕ and ψ both satisfy the integral equation (11), we have by subtraction that

$$\phi(x) - \psi(x) = \int_0^x 2t[\phi(t) - \psi(t)]\, dt.$$

Taking absolute values of both sides we have, if $x > 0$,

$$|\phi(x) - \psi(x)| \leq \int_0^x 2t|\phi(t) - \psi(t)|\, dt.$$

[24] In this case it is possible to identify ϕ in terms of elementary functions, namely, $\phi(x) = e^{x^2} - 1$. However, this is irrelevant to the discussion of existence and uniqueness.

If we restrict x to lie in the interval $0 \le x \le A/2$, where A is arbitrary, then $2t \le A$, and

$$|\phi(x) - \psi(x)| \le A \int_0^x |\phi(t) - \psi(t)|\, dt. \tag{19}$$

It is convenient at this point to introduce the function U defined by

$$U(x) = \int_0^x |\phi(t) - \psi(t)|\, dt. \tag{20}$$

Then it follows at once that

$$U(0) = 0, \tag{21}$$

$$U(x) \ge 0, \qquad \text{for} \quad x \ge 0. \tag{22}$$

Further, U is differentiable, and $U'(x) = |\phi(x) - \psi(x)|$. Hence, by Eq. (19),

$$U'(x) - AU(x) \le 0. \tag{23}$$

Multiplying Eq. (23) by the positive quantity e^{-Ax} gives

$$\left[e^{-Ax}U(x)\right]' \le 0. \tag{24}$$

Then, upon integrating Eq. (24) from zero to x and using Eq. (21), we obtain

$$e^{-Ax}U(x) \le 0 \qquad \text{for} \quad x \ge 0.$$

Hence $U(x) \le 0$ for $x \ge 0$, and in conjunction with Eq. (22), this requires that $U(x) = 0$ for each $x \ge 0$. Thus $U'(x) \equiv 0$, and therefore $\psi(x) \equiv \phi(x)$, which contradicts the original hypothesis. Consequently, there cannot be two different solutions of the initial value problem for $x \ge 0$. A slight modification of this argument leads to the same conclusion for $x \le 0$.

Returning now to the general problem of solving the integral equation (5), let us consider briefly each of the questions raised earlier.

1. Do all members of the sequence $\{\phi_n\}$ exist? In the example f and $\partial f/\partial y$ were continuous in the whole xy plane, and each member of the sequence could be explicitly calculated. In contrast, in the general case f and $\partial f/\partial y$ are assumed to be continuous only in the rectangle R: $|x| \le a$, $|y| \le b$ (see Figure 2.26). Furthermore, the members of the sequence cannot as a rule be explicitly determined. The danger is that at some stage, say for $n = k$, the graph of $y = \phi_k(x)$ may contain points that lie outside of the rectangle R. Hence at the next stage—in the computation of $\phi_{k+1}(x)$—it would be necessary to evaluate $f(x, y)$ at points where it is not known to be continuous or even to exist. Thus the calculation of $\phi_{k+1}(x)$ might be impossible.

   To avoid this danger it may be necessary to restrict x to a smaller interval than $|x| \le a$. To find such an interval we make use of the fact that a continuous function on a closed region is bounded. Hence f is bounded on R;

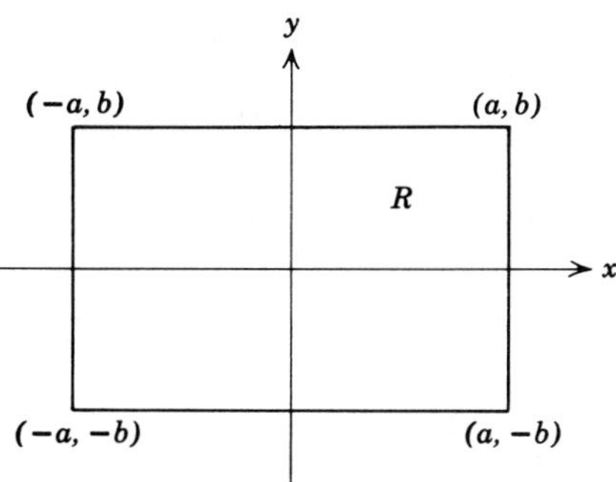

FIGURE 2.26 Region of definition for Theorem 2.4.

thus there exists a positive number M such that

$$|f(x, y)| \leq M, \qquad (x, y) \text{ in } R. \tag{25}$$

We have mentioned before that

$$\phi_n(0) = 0$$

for each n. Since $f[x, \phi_k(x)]$ is equal to $\phi'_{k+1}(x)$ the maximum absolute slope of the graph of the equation $y = \phi_{k+1}(x)$ is M. Since this graph contains the point $(0, 0)$, it must lie in the wedge-shaped shaded region in Figure 2.27. Hence the point $[x, \phi_{k+1}(x)]$ remains in R at least as long as R contains the wedge-shaped region, which is for $|x| \leq b/M$. We hereafter consider only the rectangle D: $|x| \leq h$, $|y| \leq b$, where h is equal either to a or to b/M, whichever is smaller. With this restriction all members of the sequence $\{\phi_n(x)\}$ exist. Note that if $b/M < a$, then a larger value of h can be obtained by

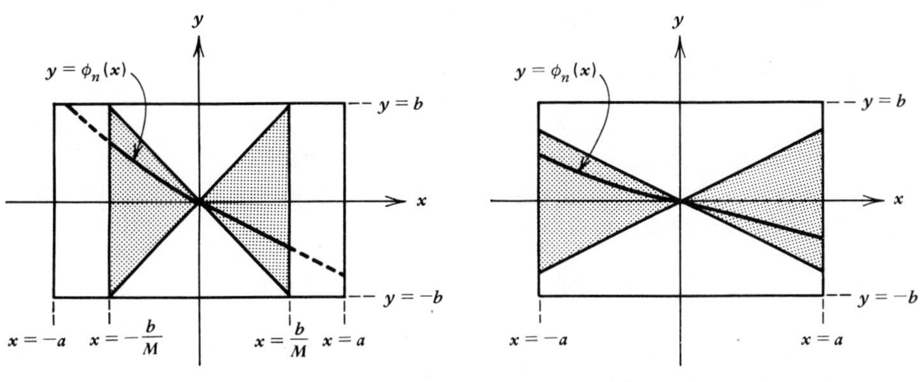

FIGURE 2.27 Region in which successive iterates lie.

finding a better bound for $|f(x, y)|$, provided that M is not already equal to the maximum value of $|f(x, y)|$.

2. Does the sequence $\{\phi_n(x)\}$ converge? As in the example, we can identify $\phi_n(x) = \phi_1(x) + [\phi_2(x) - \phi_1(x)] + \cdots + [\phi_n(x) - \phi_{n-1}(x)]$ as the nth partial sum of the series

$$\phi_1(x) + \sum_{k=1}^{\infty} [\phi_{k+1}(x) - \phi_k(x)]. \tag{26}$$

The convergence of the sequence $\{\phi_n(x)\}$ is established by showing that the series (26) converges. To do this it is necessary to estimate the magnitude $|\phi_{k+1}(x) - \phi_k(x)|$ of the general term. The argument by which this is done is indicated in Problems 3 through 6 and will be omitted here. Assuming that the sequence converges, we denote the limit function by ϕ, so that

$$\phi(x) = \lim_{n \to \infty} \phi_n(x). \tag{27}$$

3. What are the properties of the limit function ϕ? In the first place, we would like to know that ϕ is continuous. This is not, however, a necessary consequence of the convergence of the sequence $\{\phi_n(x)\}$, even though each member of the sequence is continuous itself. In Chapter 10 we encounter some problems in which a sequence of continuous functions converges to a limit function that is discontinuous. A simple example of this phenomenon is given in Problem 1. One way to show that ϕ is continuous is to show not only that the sequence $\{\phi_n\}$ converges but that it converges in a certain manner, known as uniform convergence. We do not take up this question here but note only that the argument referred to in paragraph 2 is sufficient to establish the uniform convergence of the sequence $\{\phi_n\}$ and hence the continuity of the limit function ϕ in the interval $|x| \leq h$.

Now let us return to Eq. (9),

$$\phi_{n+1}(x) = \int_0^x f[t, \phi_n(t)]\, dt.$$

Allowing n to approach ∞ on both sides, we obtain

$$\phi(x) = \lim_{n \to \infty} \int_0^x f[t, \phi_n(t)]\, dt. \tag{28}$$

We would like to interchange the operations of integration and taking the limit on the right side of Eq. (28), so as to obtain

$$\phi(x) = \int_0^x \lim_{n \to \infty} f[t, \phi_n(t)]\, dt. \tag{29}$$

In general, such an interchange is not permissible (see Problem 2, for example), but once again the fact that the sequence $\{\phi_n(x)\}$ not only converges but converges uniformly comes to the rescue and allows us to take the limiting operation inside the integral sign. Next we wish to take the limit inside the function f, which would give

$$\phi(x) = \int_0^x f\left[t, \lim_{n\to\infty} \phi_n(t)\right] dt \tag{30}$$

and hence

$$\phi(x) = \int_0^x f[t, \phi(t)]\, dt. \tag{31}$$

The statement that $\lim_{n\to\infty} f[t, \phi_n(t)] = f[t, \lim_{n\to\infty}\phi_n(t)]$ is equivalent to the statement that f is continuous in its second variable, which is known by hypothesis. Hence Eq. (31) is valid and the function ϕ satisfies the integral equation (5). Therefore ϕ is also a solution of the initial value problem (2).

4. Are there other solutions of the integral equation (5) besides $y = \phi(x)$? To show the uniqueness of the solution $y = \phi(x)$ we can proceed much as in the example. First assume the existence of another solution $y = \psi(x)$. It is then possible to show (see Problem 7) that the difference $\phi(x) - \psi(x)$ satisfies the inequality

$$|\phi(x) - \psi(x)| \le A \int_0^x |\phi(t) - \psi(t)|\, dt \tag{32}$$

for $0 \le x \le h$ and for a suitable positive number A. From this point the argument is identical to that given in the example, and we conclude that there is no solution of the initial value problem (2) other than that obtained by the method of successive approximations.

PROBLEMS

1. Let $\phi_n(x) = x^n$ for $0 \le x \le 1$, and show that

$$\lim_{n\to\infty} \phi_n(x) = \begin{cases} 0, & 0 \le x < 1, \\ 1, & x = 1 \end{cases}$$

This example shows that a sequence of continuous functions may converge to a limit function that is discontinuous.

2. Consider the sequence $\phi_n(x) = 2nxe^{-nx^2}$, $0 \le x \le 1$.
(a) Show that $\lim_{n\to\infty}\phi_n(x) = 0$ for $0 \le x \le 1$, hence that

$$\int_0^1 \lim_{n\to\infty} \phi_n(x)\, dx = 0.$$

(b) Show that $\int_0^1 2nxe^{-nx^2}\, dx = 1 - e^{-n}$, hence that

$$\lim_{n\to\infty} \int_0^1 \phi_n(x)\, dx = 1.$$

This example shows that it is not necessarily true that

$$\lim_{n\to\infty} \int_a^b \phi_n(x)\, dx = \int_a^b \lim_{n\to\infty} \phi_n(x)\, dx$$

even though $\lim_{n\to\infty}\phi_n(x)$ exists and is continuous.

In Problems 3 through 6 we indicate how to prove that the sequence $\{\phi_n(x)\}$, defined by Eqs. (6) through (9), converges.

3. If $\partial f/\partial y$ is continuous in the rectangle D, show that there is a positive constant K such that

$$|f(x, y_1) - f(x, y_2)| \le K|y_1 - y_2|$$

where (x, y_1) and (x, y_2) are any two points in D having the same x coordinate. *Hint:* Hold x fixed and use the mean value theorem on f as a function of y only. Choose K to be the maximum value of $|\partial f/\partial y|$ in D.

4. If $\phi_{n-1}(x)$ and $\phi_n(x)$ are members of the sequence $\{\phi_n(x)\}$, use the result of Problem 3 to show that

$$|f[x, \phi_n(x)] - f[x, \phi_{n-1}(x)]| \le K|\phi_n(x) - \phi_{n-1}(x)|.$$

5. (a) Show that if $|x| \le h$, then

$$|\phi_1(x)| \le M|x|$$

where M is chosen so that $|f(x, y)| \le M$ for (x, y) in D.
(b) Use the results of Problem 4 and part (a) of Problem 5 to show that

$$|\phi_2(x) - \phi_1(x)| \le \frac{MK|x|^2}{2}.$$

(c) Show, by mathematical induction, that

$$|\phi_n(x) - \phi_{n-1}(x)| \le \frac{MK^{n-1}|x|^n}{n!} \le \frac{MK^{n-1}h^n}{n!}.$$

6. Note that

$$\phi_n(x) = \phi_1(x) + [\phi_2(x) - \phi_1(x)] + \cdots + [\phi_n(x) - \phi_{n-1}(x)].$$

(a) Show that

$$|\phi_n(x)| \le |\phi_1(x)| + |\phi_2(x) - \phi_1(x)| + \cdots + |\phi_n(x) - \phi_{n-1}(x)|.$$

(b) Use the results of Problem 5 to show that

$$|\phi_n(x)| \le \frac{M}{K}\left[Kh + \frac{(Kh)^2}{2!} + \cdots + \frac{(Kh)^n}{n!}\right].$$

(c) Show that the sum in part (b) converges as $n \to \infty$, and hence show that the sum in part (a) also converges as $n \to \infty$. Conclude therefore that the sequence $\{\phi_n(x)\}$ converges, since it is the sequence of partial sums of a convergent infinite series.

7. In this problem we deal with the question of uniqueness of the solution of the integral equation (5),

$$\phi(x) = \int_0^x f[t, \phi(t)]\, dt.$$

(a) Suppose that ϕ and ψ are two solutions of Eq. (5). Show that

$$\phi(x) - \psi(x) = \int_0^x \{f[t, \phi(t)] - f[t, \psi(t)]\}\, dt.$$

(b) Show that

$$|\phi(x) - \psi(x)| \le \int_0^x |f[t,\phi(t)] - f[t,\psi(t)]|\, dt.$$

(c) Use the result of Problem 3 to show that

$$|\phi(x) - \psi(x)| \le K\int_0^x |\phi(t) - \psi(t)|\, dt,$$

where K is an upper bound for $|\partial f/\partial y|$ in D. This is the same as Eq. (32), and the rest of the proof may be constructed as indicated in the text.

REFERENCES

Two books mentioned in Section 2.6 are:

Bailey, N. T. J., *The Mathematical Theory of Infectious Diseases and Its Applications* (2nd ed.) (New York: Hafner Press, 1975).

Clark, Colin W., *Mathematical Bioeconomics* (New York: Wiley-Interscience, 1976).

An introduction to population dynamics in general is:

Frauenthal, J. C., *Introduction to Population Modeling* (Boston: Birkhauser, 1980).

A fuller discussion of the proof of the fundamental existence and uniqueness theorem can be found in many more advanced books on differential equations. Two that are reasonably accessible to elementary readers are:

Coddington, E. A., *An Introduction to Ordinary Differential Equations* (Englewood Cliffs, N.J.: Prentice-Hall, 1961).

Brauer, F., and Nohel, J., *Ordinary Differential Equations* (2nd ed.) (New York: Benjamin, 1973).

A useful catalog of differential equations and their solutions is contained in the following book:

Kamke, E., *Differentialgleichungen Lösungsmethoden und Lösungen* (New York: Chelsea, 1948).

Although the text is in German, very little knowledge of German is required to consult the list of solved problems.

A great deal of material relating to elementary applications of differential equations may be found in the library of UMAP Modules and the UMAP Journal produced by the Consortium for Mathematics and Its Application (COMAP), Lexington, Massachusetts.

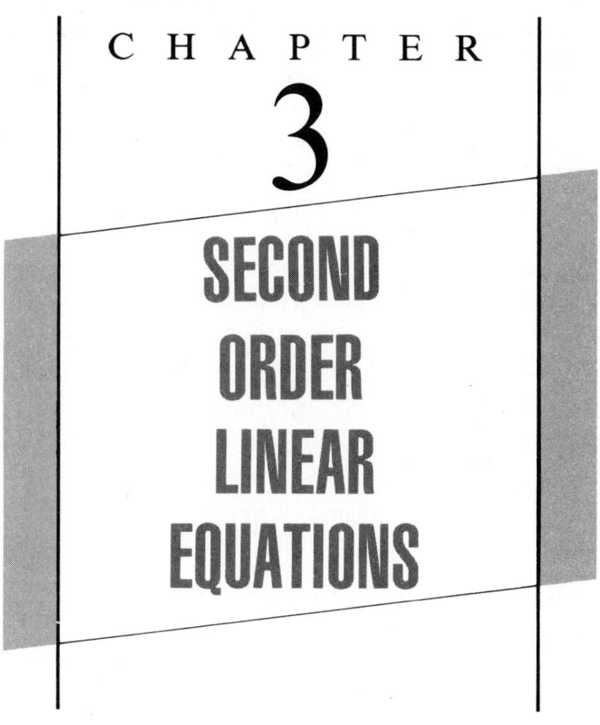

3.1 Introduction

We start our discussion by considering second order differential equations of the form

$$y'' = f(x, y, y'). \tag{1}$$

For the first order differential equation $y' = f(x, y)$ we found that there is a solution containing one arbitrary constant. Since a second order equation involves a second derivative and so, roughly speaking, two integrations are required to find a solution, it is natural to expect to find solutions of Eq. (1) containing two arbitrary constants. For example, the solution of

$$y'' = g(x) \tag{2}$$

is

$$y = \phi(x) = c_1 + c_2 x + \int^x \left[\int^t g(s)\, ds \right] dt, \tag{3}$$

where c_1 and c_2 are arbitrary constants. For a first order equation it was sufficient to specify the value of the solution at one point to determine a unique integral curve. Since we expect the solution of a second order equation to contain two arbitrary constants, we should also anticipate that to obtain a unique solution it is necessary to specify two conditions, for example, the value of the solution y_0 and its derivative y_0' at a point x_0. These conditions are referred to as *initial conditions*. Thus to determine uniquely an integral curve of a second order equation it is necessary to specify not only a point through which it passes, but also the slope of the curve at the point.

To insure the existence of a solution of Eq. (1) satisfying prescribed initial conditions it is necessary to postulate certain properties for the function f. The situation is governed by the following existence and uniqueness theorem.

Theorem 3.1. *If the functions f, f_y, and $f_{y'}$ are continuous in an open region R of the three-dimensional xyy' space, and if the point (x_0, y_0, y_0') is in R, then in some interval about x_0 there exists a unique solution $y = \phi(x)$ of the differential equation* (1)

$$y'' = f(x, y, y'),$$

that satisfies the prescribed initial conditions

$$y(x_0) = y_0, \qquad y'(x_0) = y_0'. \tag{4}$$

The functions f_y and $f_{y'}$ are the partial derivatives of f with respect to those variables. For example, if $y'' = f(x, y, y') = -y(y')^3 - x^2y^2$, then

$$f_y(x, y, y') = -(y')^3 - 2x^2y, \qquad f_{y'}(x, y, y') = -3y(y')^2. \tag{5}$$

The proof of this theorem is similar to that for the first order equation given in Section 2.12. In carrying out the proof it is usually convenient to replace the second order equation by an equivalent system of two first order equations (see Chapter 7). Proofs can be found in many more advanced books, for example, Coddington (Chapter 6) or Ince (Chapter 3). Even though the existence of a solution of Eq. (1) is guaranteed under the conditions of Theorem 3.1, it may not be possible to determine a convenient analytical expression for the solution unless f is a sufficiently simple function.

Just as in the case of first order equations, we distinguish between linear and nonlinear second order equations. The *general second order linear equation* has the form

$$P(x)\frac{d^2y}{dx^2} + Q(x)\frac{dy}{dx} + R(x)y = G(x), \tag{6}$$

where P, Q, R, and G are given functions.

A simple but important example of a second order linear differential equation is the equation governing the motion of a mass on a spring

$$m\frac{d^2u}{dt^2} + c\frac{du}{dt} + ku = F(t), \tag{7}$$

where m, c, and k are constants and F is a prescribed function. This equation is derived in Section 3.7. Other examples are Bessel's[1] equation of order ν

$$x^2y'' + xy' + (x^2 - \nu^2)y = 0, \tag{8}$$

and Legendre's[2] equation of order α

$$(1 - x^2)y'' - 2xy' + \alpha(\alpha + 1)y = 0, \tag{9}$$

where ν and α are constants, often integers. Bessel's equation arises in many different physical situations, particularly in problems involving circular geometry, such as the determination of the temperature distribution in a circular plate. Legendre's equation often occurs in physical situations involving spherical geometry.

If Eq. (1) is not of the form (6) it is said to be *nonlinear*. Although the theory of nonlinear second order differential equations is fairly difficult, there are two special cases where it is possible to simplify the general second order nonlinear equation (1). These occur when either the variable x or the variable y is missing in $f(x, y, y')$; that is, when Eq. (1) is of the form

$$y'' = f(x, y') \tag{10}$$

or

$$y'' = f(y, y'). \tag{11}$$

In these cases it is always possible to reduce Eq. (10) or Eq. (11) to a first order equation for $v = y'$. Provided that the first order equation is of a type discussed in Chapter 2, we can solve that equation for v, and one more integration yields the solution of the original differential equation. This is discussed in Problems 1, 2, and 3 at the end of this section.

The rest of this chapter and the next chapter are devoted to methods of solving second order linear differential equations. Although there is no specific formula for the solution of Eq. (6), as was the case for the first order linear equation, there is an extensive mathematical theory for second order linear equations. In the following discussion we assume, unless otherwise stated, that the functions P, Q, R, and G in Eq. (6) are continuous on some interval $\alpha < x < \beta$ (in some problems the interval may be unbounded, that is, α may be $-\infty$ and/or β may be $+\infty$), and that, further, the function P is nowhere zero in the interval.

[1]Friedrich Wilhelm Bessel (1784–1846) embarked on a career in business as a youth, but soon became interested in astronomy and mathematics. Largely self-taught, he was appointed director of the observatory at Königsberg in 1810, and held this position until his death. His study of planetary perturbations led him in 1824 to make the first systematic analysis of the solutions, known as Bessel functions, of Eq. (8). He is also famous for making the first accurate determination (1838) of the distance from the earth to a star.

[2]Adrien-Marie Legendre (1752–1833), a disciple of Euler and Lagrange, held various positions in the Académie des Sciences from 1783 onward. His primary work was in the fields of elliptic functions and number theory. The Legendre functions, solutions of Eq. (9), first appeared in 1784 in his study of the attraction of spheroids.

In this case we can divide Eq. (6) by $P(x)$ and obtain an equation of the form

$$\frac{d^2y}{dx^2} + p(x)\frac{dy}{dx} + q(x)y = g(x). \tag{12}$$

Notice that the above assumptions are fulfilled for Eq. (7) on $-\infty < x < \infty$; for Bessel's equation on any interval not including the origin; and for Legendre's equation on the intervals, $-1 < x < 1$, or $x > 1$, or $x < -1$.

If we write Eq. (12) in the form of Eq. (1), the function f is given by

$$f(x, y, y') = -p(x)y' - q(x)y + g(x). \tag{13}$$

Since $\partial f(x, y, y')/\partial y = -q(x)$ and $\partial f(x, y, y')/\partial y' = -p(x)$ are continuous, the existence and uniqueness of a solution of Eq. (12) satisfying the initial conditions $y(x_0) = y_0$, $y'(x_0) = y_0'$, $\alpha < x_0 < \beta$, in some interval about x_0 follows from Theorem 3.1. However, just as for first order equations, the existence and uniqueness theorem for second order equations can be stated in a stronger form for linear equations than for nonlinear ones.

Theorem 3.2. *If the functions p, q, and g are continuous on the open interval $\alpha < x < \beta$, then there exists one and only one function $y = \phi(x)$ satisfying the differential equation* (12),

$$y'' + p(x)y' + q(x)y = g(x),$$

on the entire interval $\alpha < x < \beta$, and the prescribed initial conditions (4)

$$y(x_0) = y_0, \qquad y'(x_0) = y_0',$$

at a particular point x_0 in the interval.

Note that any values whatever can be assigned to y_0 and y_0'. To illustrate one use of this theorem consider the following simple examples.

EXAMPLE 1

Find the solution of the differential equation

$$y'' + y = 0, \qquad -\infty < x < \infty,$$

satisfying the initial conditions $y(0) = 0$, $y'(0) = 1$.

It is easily verified that $\sin x$ and $\cos x$ are solutions of the differential equation. Further $y = \sin x$ satisfies $y(0) = 0$, $y'(0) = 1$; hence, according to Theorem 3.2, $y = \sin x$ is the unique solution of the problem.

EXAMPLE 2

What is the only solution of

$$y'' + p(x)y' + q(x)y = 0, \qquad \alpha < x < \beta,$$

satisfying the initial conditions $y(x_0) = 0$, $y'(x_0) = 0$, where p and q are continuous and x_0 is a point in the interval $\alpha < x < \beta$?

Since $y = 0$ satisfies both the differential equation and the initial conditions, it is the unique solution.

In Theorems 3.1 and 3.2 it is important to note that the initial conditions that determine a unique solution of Eq. (1) or Eq. (12) are conditions on the value of the solution and its first derivative at a fixed point in the interval. In contrast, the question of finding a solution of, say, Eq. (12) on the interval $x_0 < x < x_1$, and satisfying $y(x_0) = A$, $y(x_1) = B$ is not covered by this theorem. Indeed, the latter problem may not always have a solution. Such problems are known as *boundary value problems* and will be discussed in Chapters 10 and 11.

To solve the second order differential equation (12),

$$y'' + p(x)y' + q(x)y = g(x),$$

it is necessary to solve only the *homogeneous*,[3] or *reduced*, or *complementary*, equation

$$y'' + p(x)y' + q(x)y = 0 \tag{14}$$

obtained from Eq. (12) by setting $g(x) = 0$. Once the solution of the homogeneous equation (14) is known we can, by a general method, solve the nonhomogeneous equation (12). The next three sections contain some general results about the homogeneous equation (14). Then in Section 3.5 we show how Eq. (14) can be solved in the particular case that the functions p and q are constants. Even this case is of considerable practical importance; for example, Eq. (7) for the motion of a mass on a spring has constant coefficients. In Section 3.6 we turn our attention to the nonhomogeneous equation (12). The rest of the chapter is devoted to applications in the areas of mechanical vibrations and electrical networks.

The discussion of second order linear differential equations in Sections 3.2 through 3.6.2 is generalized in a straightforward manner in Chapter 5 to nth order linear differential equations. If desired, Chapter 5 can be read simultaneously with this chapter.

PROBLEMS

1. **Equations with y missing.** For a second order differential equation of the form $y'' = f(x, y')$, the substitution $v = y'$, $v' = y''$ leads to a first order equation of the form $v' = f(x, v)$. If this equation can be solved for v, then y can be obtained by integrating $dy/dx = v(x)$. Note that one arbitrary constant is obtained in solving the first order equation for v, and a second is obtained in the integration for y. Solve the

[3] Note that the use of the word homogeneous here is not related to its use in the discussion of first order homogeneous differential equations in Section 2.10.

following differential equations.

(a) $x^2y'' + 2xy' - 1 = 0, \quad x > 0$ (b) $xy'' + y' = 1, \quad x > 0$
(c) $y'' + x(y')^2 = 0$ (d) $2x^2y'' + (y')^3 = 2xy', \quad x > 0$
(e) $y'' + y' = e^{-x}$ (f) $x^2y'' = (y')^2, \quad x > 0$

2. **Equations with x missing.** Consider a second order differential equation of the form $y'' = f(y, y')$. The variable x does not appear explicitly, but only through the derivatives $y' = dy/dx$ and $y'' = d^2y/dx^2$. If we let v equal y', then we obtain $dv/dx = f(y,v)$. Since the right side depends on y and v, rather than on x and v, this equation is not of the form of the first order equations discussed in Chapter 2. However, if we think of y as the independent variable, then by the chain rule

$$\frac{dv}{dx} = \frac{dv}{dy}\frac{dy}{dx} = v\frac{dv}{dy}.$$

Hence the original differential equation can be written as

$$v\frac{dv}{dy} = f(y,v).$$

Provided that this first order equation can be solved, we obtain v as a function of y. A relation between y and x results from solving $dy/dx = v(y)$. Again, there will be two arbitrary constants in the final result. Solve the following differential equations.

(a) $yy'' + (y')^2 = 0$ (b) $y'' + y = 0$
(c) $y'' + y(y')^3 = 0$ (d) $2y^2y'' + 2y(y')^2 = 1$
(e) $yy'' - (y')^3 = 0$ (f) $y'' + (y')^2 = 2e^{-y}$
Hint for part (f): The transformed equation is a Bernoulli equation.

3. Using the techniques of Problems 1 and 2, solve the following differential equations. If initial conditions are prescribed, find the solution satisfying the stated conditions.

(a) $y'y'' = 2, \quad y(0) = 1, \quad y'(0) = 2$
(b) $y'' - 3y^2 = 0, \quad y(0) = 2, \quad y'(0) = 4$
(c) $(1 + x^2)y'' + 2xy' + 3x^{-2} = 0, \quad x > 0$
(d) $y'y'' - x = 0, \quad y(1) = 2, \quad y'(1) = 1$

4. Determine the intervals in which a unique solution of each of the following linear differential equations satisfying the initial conditions $y(x_0) = y_0$, $y'(x_0) = y_0'$, where x_0 is any point in the interval, is certain to exist.

(a) $xy'' + 3y = x$ (b) $y'' + 6y' + 7y = 2\sin x$
(c) $x(x - 1)y'' + 3xy' + 4y = 2$ (d) $y'' + (\cos x)y' + 3(\ln|x|)y = 0$
(e) $(1 + x^2)y'' + 4y' = e^x$ (f) $e^xy'' + x^2y' + y = \tan x$

5. Assuming that p and q are continuous on an open interval including the origin and that $y = \phi(x)$ is a solution of

$$y'' + p(x)y' + q(x)y = 0, \qquad y(0) = a_0, \quad y'(0) = a_1,$$

determine $\phi''(0)$. If p and q are polynomials it can be shown that the solution

$y = \phi(x)$ of the above differential equation can be differentiated infinitely many times. Assuming that p and q are polynomials, determine $\phi'''(0)$ in terms of a_0, a_1, $p(0)$, $q(0)$, $p'(0)$, and $q'(0)$. Can this process be continued indefinitely?

6. The solution of a second order equation of the form $y'' = f(x, y, y')$ generally involves two arbitrary constants. Conversely, a given family of functions involving two arbitrary constants can be shown to be the solution of some second order differential equation. By eliminating the constants c_1 and c_2 among y, y', and y'', find the differential equation satisfied by each of the following families of functions.

   (a) $y = c_1e^x + c_2e^{-x}$ (b) $y = c_1 \cos x + c_2 \sin x$
   (c) $y = c_1x + c_2 \sin x$ (d) $y = (c_1 + c_2x)e^x$
   (e) $y = c_1x + c_2x^2$ (f) $y = c_1 \cosh x + c_2 \sinh x$

3.2 Fundamental Solutions of the Homogeneous Equation

In developing the theory of linear differential equations, it is helpful to introduce a differential operator notation. Let p and q be continuous functions on an open interval $\alpha < x < \beta$. Then for any twice differentiable function ϕ on $\alpha < x < \beta$, we define the *differential operator* L by the equation

$$L[\phi] = \phi'' + p\phi' + q\phi. \tag{1}$$

Note that $L[\phi]$ is a function on $\alpha < x < \beta$. The value of the function $L[\phi]$ at the point x is

$$L[\phi](x) = \phi''(x) + p(x)\phi'(x) + q(x)\phi(x).$$

For example, if $p(x) = x^2$, $q(x) = 1 + x$, and $\phi(x) = \sin 3x$ then

$$\begin{aligned} L[\phi](x) &= (\sin 3x)'' + x^2(\sin 3x)' + (1 + x)\sin 3x \\ &= -9\sin 3x + 3x^2\cos 3x + (1 + x)\sin 3x. \end{aligned}$$

The operator L is often written as $L = D^2 + pD + q$, where D is the derivative operator.

In this section we study the second order linear homogeneous equation

$$L[\phi](x) = \phi''(x) + p(x)\phi'(x) + q(x)\phi(x) = 0, \tag{2}$$

where it is understood that the functions p and q are continuous on an open interval $\alpha < x < \beta$. Recall that it is customary to use the symbol y to denote $\phi(x)$. Thus we will often write in place of Eq. (2)

$$L[y] = y'' + p(x)y' + q(x)y = 0. \tag{3}$$

We emphasize that $L[\phi]$ as given in Eq. (1) is a function, while $L[y]$ as given in Eq. (3) is the value of the function $L[\phi]$ at the point x. Thus the symbol y is treated in a special way.

***Theorem 3.3.* (Superposition Principle).** *If $y = y_1(x)$ and $y = y_2(x)$ are solutions of the differential equation* (3),

$$L[y] = y'' + p(x)y' + q(x)y = 0,$$

then the linear combination $y = c_1 y_1(x) + c_2 y_2(x)$, where c_1 and c_2 are arbitrary constants, is also a solution of Eq. (3).

We must show that if $L[y_1] = y_1'' + py_1' + qy_1 = 0$,[4] and $L[y_2] = y_2'' + py_2' + qy_2 = 0$, then $L[c_1y_1 + c_2y_2] = 0$. But

$$\begin{aligned}
L[c_1y_1 + c_2y_2] &= (c_1y_1 + c_2y_2)'' + p(c_1y_1 + c_2y_2)' + q(c_1y_1 + c_2y_2)\\
&= c_1y_1'' + c_2y_2'' + p(c_1y_1' + c_2y_2') + q(c_1y_1 + c_2y_2)\\
&= c_1(y_1'' + py_1' + qy_1) + c_2(y_2'' + py_2' + qy_2)\\
&= c_1L[y_1] + c_2L[y_2]\\
&= 0,
\end{aligned}$$

which proves the theorem. If we set $c_2 = 0$ in the above theorem we obtain the result that if the function y_1 is a solution of Eq. (3), then any constant multiple of y_1 is also a solution of Eq. (3).

In the process of proving Theorem 3.3, we have shown that for any two functions u_1 and u_2 possessing continuous second derivatives, and for any two arbitrary constants c_1 and c_2,

$$L[c_1u_1 + c_2u_2] = c_1L[u_1] + c_2L[u_2].$$

An operator with this property is known as a *linear operator* and, in particular, the differential operator L is a second order linear differential operator.

The fact that a linear combination of solutions of a *linear*, *homogeneous* equation is also a solution of the equation, the superposition principle, is a result of fundamental importance. The theory of linear homogeneous equations, including higher order ordinary differential equations and partial differential equations, depends strongly on the superposition principle. This principle (and its limitations) are illustrated by the following simple examples.

EXAMPLE 1

Show that $y_1(x) = \cos x$ and $y_2(x) = \sin x$ are solutions of $y'' + y = 0$. Verify that the linear combination $\phi(x) = c_1 \cos x + c_2 \sin x$ is also a solution.

First, we have

$$\begin{aligned}
y_1''(x) + y_1(x) &= (\cos x)'' + \cos x\\
&= -\cos x + \cos x = 0,
\end{aligned}$$

[4] Notice that since $L[y_1]$ is a function, the 0 on the right side of the statement $L[y_1] = 0$ stands for the function that is identically zero on $\alpha < x < \beta$.

so $y_1(x)$ is a solution of the differential equation. In a similar manner we can show that $y_2(x)$ is a solution of the differential equation. Next, substituting $\phi(x)$ for y, we have

$$\begin{aligned}\phi''(x)+\phi(x) &= (c_1\cos x + c_2\sin x)'' + (c_1\cos x + c_2\sin x)\\ &= c_1\big[(\cos x)'' + \cos x\big] + c_2\big[(\sin x)'' + \sin x\big]\\ &= c_1[-\cos x + \cos x] + c_2[-\sin x + \sin x] = 0.\end{aligned}$$

EXAMPLE 2

Verify that $\phi(x) = x + 1$ is a solution of the differential equation $y'' + 3y' + y = x + 4$, but that $\psi(x) = 2\phi(x)$ is not a solution.

Since $\phi'(x) = 1$, $\phi''(x) = 0$, we have

$$\phi''(x) + 3\phi'(x) + \phi(x) = 0 + 3(1) + (x + 1) = x + 4.$$

However,

$$\psi''(x) + 3\psi'(x) + \psi(x) = 0 + 3(2) + 2(x + 1) \neq x + 4.$$

This is not a contradiction of Theorem 3.3, since the differential equation is not homogeneous.

EXAMPLE 3

Show that if the functions y_1 and y_2 are solutions of the equation $L[y] = y'' + y^2 = 0$, it does *not* necessarily follow that the linear combination $c_1y_1 + c_2y_2$ is a solution.

Substituting, we have

$$\begin{aligned}L[c_1y_1 + c_2y_2] &= (c_1y_1 + c_2y_2)'' + (c_1y_1 + c_2y_2)^2\\ &= c_1y_1'' + c_2y_2'' + c_1^2y_1^2 + 2c_1c_2y_1y_2 + c_2^2y_2^2,\end{aligned}$$

and this is not equal to $c_1L[y_1] + c_2L[y_2]$ for arbitrary values of c_1 and c_2. The superposition principle does not apply in this case because the differential equation is nonlinear.

We have seen that if the functions y_1 and y_2 are solutions of Eq. (3), then the linear combination $c_1y_1 + c_2y_2$ is a solution of Eq. (3) and, corresponding to the infinity of values we can assign to c_1 and c_2, we can construct an infinity of solutions of Eq. (3). We now ask: does this infinity of solutions include all possible solutions of Eq. (3)?

Two solutions y_1 and y_2 of Eq. (3) are said to form a *fundamental set* of solutions if every solution of Eq. (3) can be expressed as a linear combination of y_1 and y_2. To prove that two solutions, y_1 and y_2, form a fundamental set of solutions, we must show that for every solution $y = \phi(x)$ of Eq. (3) we can find constants c_1 and c_2 such that $\phi(x) = c_1 y_1(x) + c_2 y_2(x)$. We proceed as follows.

Let $y = \phi(x)$ be any solution of Eq. (3). Choose a point x_0 in the interval $\alpha < x < \beta$; then $\phi(x_0)$ and $\phi'(x_0)$, respectively, are the values of the chosen solution and its derivative at this point. We now seek constants c_1 and c_2 so that the function $c_1 y_1 + c_2 y_2$ and its derivative take on the same values $\phi(x_0)$ and $\phi'(x_0)$ at $x = x_0$. Thus c_1 and c_2 must satisfy the following system of linear algebraic equations:

$$\begin{aligned} c_1 y_1(x_0) + c_2 y_2(x_0) &= \phi(x_0), \\ c_1 y_1'(x_0) + c_2 y_2'(x_0) &= \phi'(x_0). \end{aligned} \tag{4}$$

The system of equations (4) has a unique solution for c_1 and c_2 provided that the determinant of coefficients is not zero; that is,

$$\begin{vmatrix} y_1(x_0) & y_2(x_0) \\ y_1'(x_0) & y_2'(x_0) \end{vmatrix} = y_1(x_0) y_2'(x_0) - y_1'(x_0) y_2(x_0) \neq 0. \tag{5}$$

Then with c_1 and c_2 determined by Eqs. (4), the functions $c_1 y_1 + c_2 y_2$ and ϕ satisfy the same initial conditions at $x = x_0$ as well as the same differential equation (3). Since Theorem 3.2 asserts that there is only one solution of this initial value problem, we conclude that $\phi(x) = c_1 y_1(x) + c_2 y_2(x)$ for each x in $\alpha < x < \beta$.

We will always be able to choose an x_0 such that Eq. (5) holds if we know that it holds for all x_0 in $\alpha < x < \beta$. In this case Eqs. (4) can always be solved regardless of the values x_0, $\phi(x_0)$, and $\phi'(x_0)$. Thus we have proved the following theorem.

Theorem 3.4. *If the functions p and q are continuous on the open interval $\alpha < x < \beta$ and if y_1 and y_2 are solutions of the differential equation* (3),

$$L[y] = y'' + p(x)y' + q(x)y = 0,$$

satisfying the condition

$$y_1(x) y_2'(x) - y_1'(x) y_2(x) \neq 0 \tag{6}$$

at every point in $\alpha < x < \beta$, then any solution of Eq. (3) *on the interval $\alpha < x < \beta$ can be expressed uniquely as a linear combination of y_1 and y_2.*

It is customary to call the linear combination $c_1 y_1 + c_2 y_2$, with c_1 and c_2 arbitrary, the *general solution* of Eq. (3).

EXAMPLE 4

Find a fundamental set of solutions of the equation

$$y'' + y = 0, \qquad -\infty < x < \infty. \tag{7}$$

We have shown in Example 1 that $y_1(x) = \cos x$ and $y_2(x) = \sin x$ are solutions of Eq. (7). Further,

$$y_1(x)y_2'(x) - y_1'(x)y_2(x) = \cos x \cos x - (-\sin x)\sin x = 1.$$

Hence if $y = \phi(x)$ is a solution of Eq. (7), then there exist constants c_1 and c_2 such that $\phi(x) = c_1 \cos x + c_2 \sin x$.

EXAMPLE 5

Show that $y_1(x) = x^{1/2}$ and $y_2(x) = x^{-1}$ form a fundamental set of solutions of

$$2x^2y'' + 3xy' - y = 0, \qquad x > 0. \tag{8}$$

First we verify by direct substitution that y_1 and y_2 are solutions of the differential equation. Since $y_1(x) = x^{1/2}$, $y_1'(x) = \frac{1}{2}x^{-1/2}$, and $y_1''(x) = -\frac{1}{4}x^{-3/2}$, we have

$$2x^2(-\tfrac{1}{4}x^{-3/2}) + 3x(\tfrac{1}{2}x^{-1/2}) - x^{1/2} = (-\tfrac{1}{2} + \tfrac{3}{2} - 1)x^{1/2} = 0.$$

Similarly, $y_2(x) = x^{-1}$, $y_2'(x) = -x^{-2}$, and $y_2''(x) = 2x^{-3}$, so

$$2x^2(2x^{-3}) + 3x(-x^{-2}) - x^{-1} = (4 - 3 - 1)x^{-1} = 0.$$

Next we calculate

$$\begin{aligned} y_1(x)y_2'(x) - y_1'(x)y_2(x) &= x^{1/2}(-x^{-2}) - \tfrac{1}{2}x^{-1/2}(x^{-1}) \\ &= -\tfrac{3}{2}x^{-3/2} \neq 0 \end{aligned} \tag{9}$$

for $x > 0$. Thus y_1 and y_2 are a fundamental set of solutions.

Given two functions y_1 and y_2 differentiable on some open interval, the function $y_1y_2' - y_1'y_2$ is referred to as the Wronskian[5] of y_1 and y_2, and is usually written as

$$W(y_1, y_2) = \begin{vmatrix} y_1 & y_2 \\ y_1' & y_2' \end{vmatrix} = y_1y_2' - y_1'y_2. \tag{10}$$

The value of the Wronskian of y_1 and y_2 at the point x is denoted by $W(y_1, y_2)(x)$ or simply $W(x)$ if it is clear what functions are being considered.

Now suppose we have found two solutions, y_1 and y_2, of Eq. (3). How do we decide whether the functions y_1 and y_2 form a fundamental set; that is, how do we decide whether $W(y_1, y_2)$ is ever zero in the interval $\alpha < x < \beta$? In Example 4, $W(\cos x, \sin x) = 1$, and hence it is clear that $\cos x$ and $\sin x$ form a fundamental set for the differential equation $y'' + y = 0$ on any interval. However, in general, $W(y_1, y_2)$ will not be constant, and it may be difficult to determine

[5]Wronskian determinants are named for Józef Maria Hoëné-Wroński (1776–1853), who was born in Poland but spent most of his life in France. A gifted but troubled man, Wronski's life was marked by repeated violent disputes with other individuals and institutions.

whether this function is zero at some point in an interval $\alpha < x < \beta$. Fortunately this question is answered by the following remarkable theorem.

Theorem 3.5. *If the functions p and q are continuous on the open interval $\alpha < x < \beta$, and if the functions y_1 and y_2 are solutions of the differential equation* (3),

$$L[y] = y'' + p(x)y' + q(x)y = 0,$$

on $\alpha < x < \beta$, then $W(y_1, y_2)$ either is identically zero, or else is never zero in $\alpha < x < \beta$.

The proof of this theorem can be based on an argument similar to that used in proving Theorem 3.4; however, we present an alternate proof in which we also derive a formula for $W(y_1, y_2)$. We make use of the fact that the functions y_1 and y_2 satisfy

$$\begin{aligned} y_1'' + py_1' + qy_1 &= 0, \\ y_2'' + py_2' + qy_2 &= 0. \end{aligned} \tag{11}$$

Multiplying the first equation by $-y_2$, the second by y_1, and adding gives

$$\left(y_1y_2'' - y_2y_1''\right) + p\left(y_1y_2' - y_1'y_2\right) = 0. \tag{12}$$

Letting $W(x) = W(y_1, y_2)(x)$, and noting that

$$W' = y_1y_2'' - y_1''y_2, \tag{13}$$

allows us to write Eq. (12) in the form

$$W' + pW = 0. \tag{14}$$

This is a separable equation (Section 2.4) and also a first order linear equation (Section 2.1), and it can be integrated immediately to give

$$W(x) = c\exp\left[-\int^x p(t)\,dt\right], \tag{15}$$

where c is a constant.[6] Since the exponential function is never zero, $W(x) = 0$ only if $c = 0$, and if $c = 0$ then $W(x)$ is identically zero, thus proving the theorem. In addition, Eq. (15) gives a formula for determining the Wronskian of a fundamental set of solutions of Eq. (3) up to a multiplicative constant without solving the equation. Also note that the Wronskians of any two fundamental sets of solutions of the same equation can differ only by a multiplicative constant.

[6] The result given in Eq. (15) was derived by the Norwegian mathematician N. H. Abel (1802–1829) in 1827, and is known as Abel's identity. He also showed that there is no general formula for solving a quintic equation in terms of explicit algebraic operations on the coefficients in the polynomial equation. His greatest contribution was in analysis in the area known as elliptic functions, although his memoir was not noticed until after his death. The distinguished French mathematician Legendre described it as "a monument more lasting than bronze."

EXAMPLE 6

For the differential equation of Example 5, verify that the Wronskian of $y_1(x) = x^{1/2}$ and $y_2(x) = x^{-1}$ is given by Eq. (15).

According to Eq. (9) we know that $W(y_1, y_2)(x) = -\frac{3}{2}x^{-3/2}$. To use Eq. (15) we must write the differential equation (8) in standard form:

$$y'' + \frac{3}{2x}y' - \frac{1}{2x^2}y = 0,$$

so $p(x) = 3/2x$. Hence

$$W(x) = c\exp\left[-\int^x \frac{3}{2t}\,dt\right] = c\exp\left(-\frac{3}{2}\ln x\right)$$
$$= cx^{-3/2}.$$

If we choose $c = -\frac{3}{2}$ we obtain the desired result.

By combining the results of Theorems 3.4 and 3.5, we can state the following theorem.

Theorem 3.6. *If the functions p and q are continuous on the open interval $\alpha < x < \beta$ and if the functions y_1 and y_2 are solutions of the differential equation* (3),

$$L[y] = y'' + p(x)y' + q(x)y = 0,$$

on the interval $\alpha < x < \beta$, and if there is at least one point in $\alpha < x < \beta$ where $W(y_1, y_2)$ is not zero, then every solution $y = \phi(x)$ of Eq. (3) *can be expressed in the form*

$$\phi(x) = c_1y_1(x) + c_2y_2(x).$$

Finally we must show that there actually does exist a fundamental set of solutions of Eq. (3); that is, we must prove the following theorem.

Theorem 3.7. *If the functions p and q are continuous on the open interval $\alpha < x < \beta$, then there always exists a fundamental set of solutions of the differential equation* (3),

$$L[y] = y'' + p(x)y' + q(x)y = 0,$$

on the interval $\alpha < x < \beta$.

Choose any point x_0 in $\alpha < x < \beta$. It follows from Theorem 3.2 that there exist unique solutions y_1 and y_2 of the initial value problems

$$y'' + p(x)y' + q(x)y = 0; \qquad y(x_0) = 1, \quad y'(x_0) = 0$$
$$y'' + p(x)y' + q(x)y = 0; \qquad y(x_0) = 0, \quad y'(x_0) = 1$$

on the interval $\alpha < x < \beta$. It is readily seen that $W(y_1, y_2)(x_0) = 1 \neq 0$. Hence it follows from Theorem 3.6 that y_1 and y_2 form a fundamental set of solutions of Eq. (3).

For example, we already know that $y_1(x) = \cos x$ and $y_2(x) = \sin x$ are a fundamental set of solutions of $y'' + y = 0$. Observe that these functions are solutions of the following initial value problems:

$$y_1(x) = \cos x: \qquad y'' + y = 0, \quad y(0) = 1, \quad y'(0) = 0$$

$$y_2(x) = \sin x: \qquad y'' + y = 0, \quad y(0) = 0, \quad y'(0) = 1.$$

We can summarize our discussion in this section as follows. To find the general solution of the differential equation

$$y'' + p(x)y' + q(x)y = 0, \qquad \alpha < x < \beta,$$

we must first find two functions y_1 and y_2 that satisfy the differential equation on $\alpha < x < \beta$, and then evaluate $W(y_1, y_2)$ at any convenient point x_0 in $\alpha < x < \beta$. If $W(y_1, y_2)(x_0) \neq 0$, then y_1 and y_2 form a fundamental set of solutions, and the general solution is

$$y = c_1 y_1(x) + c_2 y_2(x),$$

where c_1 and c_2 are arbitrary constants. If initial conditions $y(x_0) = y_0$ and $y'(x_0) = y_0'$ are prescribed at a point x_0 in $\alpha < x < \beta$, then the constants c_1 and c_2 can be chosen so that y satisfies the initial conditions.

PROBLEMS

1. Verify that e^x and e^{-2x} and the linear combination $c_1 e^x + c_2 e^{-2x}$, where c_1 and c_2 are arbitrary constants, are solutions of the differential equation

$$y'' + y' - 2y = 0.$$

2. In Problem 1 find the unique solution of the differential equation that satisfies the initial conditions $y(0) = 1$, $y'(0) = 0$. What is the unique solution of this problem if the initial conditions are $y(1) = 0$, $y'(1) = 0$?
3. Verify that e^x and e^{-x} are solutions of $y'' - y = 0$. Hence show that $\sinh x = (e^x - e^{-x})/2$ and $\cosh x = (e^x + e^{-x})/2$ are also solutions of this differential equation.
4. Assuming that $(cf)' = cf'$ where f is a real-valued function and c is a complex number, show that the linear combination $(1 + i)\sin x + (2 - i)\cos x$ is a solution of $y'' + y = 0$. Here i is the imaginary unit, $i^2 = -1$.
5. Verify that x^2 and x^{-1} and the linear combination $c_1 x^2 + c_2 x^{-1}$, where c_1 and c_2 are arbitrary constants, are solutions of the differential equation $x^2 y'' - 2y = 0$, $x > 0$.
6. Verify that 1 and $x^{1/2}$ are solutions of the differential equation $yy'' + (y')^2 = 0$, $x > 0$; but that the linear combination $c_1 + c_2 x^{1/2}$ is not, in general, a solution. Why?
7. Show that if $y = \phi(x)$ is a solution of the differential equation $y'' + p(x)y' + q(x)y = g(x)$, $g(x) \not\equiv 0$, then $y = c\phi(x)$, where c is any constant other than one, is not a solution. Why?

8. If $L[y] = ay'' + by' + cy$, where a, b, and c are constants, compute

(a) $L[x]$ (b) $L[\sin x]$
(c) $L[e^{rx}]$, r a constant (d) $L[x^r]$, r a constant

9. If $L[y] = ax^2y'' + bxy' + cy$, $x > 0$, where a, b, and c are constants, compute

(a) $L[x^2]$ (b) $L[e^{rx}]$, r a constant
(c) $L[x^r]$, r a constant (d) $L[\ln x]$

10. Compute the Wronskians of the following pairs of functions:

(a) e^{mx}, e^{nx}, where m and n are integers, and $m \neq n$
(b) $\sinh x$, $\cosh x$ (c) x, xe^x
(d) $e^x \sin x$, $e^x \cos x$ (e) $\cos^2 x$, $1 + \cos 2x$

11. In the following problems verify that the functions y_1 and y_2 are solutions of the given differential equation and determine in what intervals they form a fundamental set of solutions by computing $W(y_1, y_2)$.

(a) $y'' + \lambda^2 y = 0$; $y_1(x) = \sin \lambda x$, $y_2(x) = \cos \lambda x$, where λ is a real number
(b) $y'' - y' - 2y = 0$; $y_1(x) = e^{-x}$, $y_2(x) = e^{2x}$
(c) $y'' - 2y' + y = 0$; $y_1(x) = e^x$; $y_2(x) = xe^x$
(d) $x^2y'' - x(x+2)y' + (x+2)y = 0$; $y_1(x) = x$, $y_2(x) = xe^x$

Notice that if the equation of part (d) is put in the standard form $y'' + p(x)y' + q(x)y = 0$, the coefficients $p(x) = -(x+2)/x$ and $q(x) = (x+2)/x^2$ become unbounded as $x \to 0$, but the solutions x and xe^x are perfectly well behaved as $x \to 0$. Thus it does not necessarily follow that at a point where the coefficients are discontinuous the solution will be discontinuous—but this is often the case!

12. Verify that if ϕ, ϕ_1, and ϕ_2 are differentiable functions, then $W(\phi\phi_1, \phi\phi_2) = \phi^2 W(\phi_1, \phi_2)$.

13. In the following problems verify that the functions y_1 and y_2 form a fundamental set of solutions of the given differential equations and determine the solution satisfying the prescribed initial conditions.

(a) $y'' - y = 0$, $y(0) = 0$, $y'(0) = 1$; $y_1(x) = e^x$, $y_2(x) = e^{-x}$
(b) $y'' - y = 0$, $y(0) = 0$, $y'(0) = 1$; $y_1(x) = \sinh x$, $y_2(x) = \cosh x$
Compare with the result for part (a).
(c) $y'' + 5y' + 6y = 0$, $y(0) = 1$, $y'(0) = 1$; $y_1(x) = e^{-2x}$, $y_2(x) = e^{-3x}$
(d) $y'' + y' = 0$, $y(1) = 0$, $y'(1) = 1$; $y_1(x) = 2$, $y_2(x) = e^{-x}$

14. Given that y_1 and y_2 are linearly independent solutions of $y'' + 2x^{-1}y' + e^x y = 0$, and that $W(y_1, y_2)(1) = 2$, what is the value of $W(y_1, y_2)(5)$?

In Problems 15, 16, and 17 assume that p and q are continuous, and that the functions y_1 and y_2 are solutions of the differential equation $y'' + p(x)y' + q(x)y = 0$ on the interval $\alpha < x < \beta$.

15. Prove that if y_1 and y_2 vanish at the same point in $\alpha < x < \beta$, then they cannot form a fundamental set of solutions on that interval.

16. Prove that if y_1 and y_2 have maxima or minima at the same point in $\alpha < x < \beta$, then they cannot form a fundamental set of solutions on that interval.

*17. Prove that if y_1 and y_2 are a fundamental set of solutions, then they cannot have a common point of inflection in $\alpha < x < \beta$ unless p and q vanish simultaneously there.

Problems 18 through 21 deal with the concepts of an exact second order linear differential equation and the adjoint of a second order linear differential equation.

*18. The equation $P(x)y'' + Q(x)y' + R(x)y = 0$ is said to be *exact* if it can be written in the form $[P(x)y']' + [f(x)y]' = 0$, where $f(x)$ is to be determined in terms of $P(x)$, $Q(x)$, and $R(x)$. The latter equation can be integrated once immediately to give a first order linear equation for y which can be solved by the method of Section 2.1. Show by equating the coefficients of the above equations, and then eliminating $f(x)$, that a necessary condition for exactness is $P''(x) - Q'(x) + R(x) = 0$. It can also be shown that this is a sufficient condition for exactness. Determine whether each of the following equations is exact; if so, find its solution.

(a) $y'' + xy' + y = 0$
(b) $y'' + 3x^2y' + xy = 0$
(c) $xy'' - (\cos x)y' + (\sin x)y = 0, \quad x > 0$
(d) $x^2y'' + xy' - y = 0, \quad x > 0$

*19. If a second order linear homogeneous equation is not exact, it can be made exact by multiplying by an appropriate integrating factor $\mu(x)$. Thus we require that $\mu(x)$ be such that $\mu(x)P(x)y'' + \mu(x)Q(x)y' + \mu(x)R(x)y = 0$ can be written in the form $[\mu(x)P(x)y']' + [f(x)y]' = 0$. Show by equating coefficients in these two equations and then eliminating $f(x)$ that the function μ must satisfy

$$P\mu'' + (2P' - Q)\mu' + (P'' - Q' + R)\mu = 0.$$

This equation is known as the *adjoint* equation. It plays a very important role in the advanced theory of differential equations. In general the problem of solving the adjoint differential equation is as difficult as that of solving the original equation. Determine the adjoint equation for each of the following differential equations.
(a) The Bessel equation of order ν, $x^2y'' + xy' + (x^2 - \nu^2)y = 0$
(b) The Legendre equation of order α, $(1 - x^2)y'' - 2xy' + \alpha(\alpha + 1)y = 0$
(c) The Airy equation, $y'' - xy = 0$.

*20. Show for the second order linear equation $P(x)y'' + Q(x)y' + R(x)y = 0$ that the adjoint of the adjoint is the original equation.

*21. A second order linear equation $P(x)y'' + Q(x)y' + R(x)y = 0$ is said to be *self-adjoint* if its adjoint is identical with the original equation. Show that a necessary condition for this differential equation to be self-adjoint is $P'(x) = Q(x)$. Determine whether the equations of Problem 19 are self-adjoint.

3.3 Linear Independence

The concept of the general solution of a second order linear differential equation as a linear combination of two solutions whose Wronskian does not vanish is intimately related to the concept of linear independence of two functions. This is a very important concept and has significance far beyond the present context; we briefly discuss it in this section.

Two functions f and g are said to be *linearly dependent* on an interval $\alpha < x < \beta$ if there exist two constants k_1 and k_2, not both zero, such that

$$k_1 f(x) + k_2 g(x) = 0 \tag{1}$$

for all x in $\alpha < x < \beta$. Two functions f and g are said to be *linearly independent* on an interval $\alpha < x < \beta$ if they are not linearly dependent; that is, if Eq. (1) holds for all x in the interval only if $k_1 = k_2 = 0$. The same definitions apply on any interval, open or not.

EXAMPLE 1

The functions $\sin x$ and $\cos(x + \pi/2)$ are linearly dependent on any interval since

$$k_1 \sin x + k_2 \cos(x + \pi/2) = 0$$

for all x if we choose $k_1 = 1$, $k_2 = 1$.

EXAMPLE 2

Show that the functions e^x and e^{2x} are linearly independent on any interval. To establish this result we suppose that

$$k_1 e^x + k_2 e^{2x} = 0 \tag{2}$$

for all x in an interval $\alpha < x < \beta$; we must then show that $k_1 = k_2 = 0$. Choose two points x_0 and $x_1 \neq x_0$, and evaluate Eq. (2) at these points. This gives

$$\begin{aligned} k_1 e^{x_0} + k_2 e^{2x_0} &= 0, \\ k_1 e^{x_1} + k_2 e^{2x_1} &= 0. \end{aligned} \tag{3}$$

Since the determinant of coefficients is $\exp(x_0 + 2x_1) - \exp(2x_0 + x_1) \neq 0$, it follows that the only solution of Eqs. (3) is $k_1 = k_2 = 0$. Hence e^x and e^{2x} are linearly independent.

The following theorem relates linear dependence and independence to the Wronskian.

> ***Theorem 3.8.*** *If f and g are differentiable functions and if $W(f, g)(x_0) \neq 0$ for some x_0 in $\alpha < x < \beta$, then f and g are linearly independent on this interval. Alternatively, if f and g are linearly dependent, then $W(f, g)(x) = 0$ for all x in $\alpha < x < \beta$.*

We will prove the first statement of Theorem 3.8. Consider a linear combination $k_1 f(x) + k_2 g(x)$. Evaluating this expression and its derivative at x_0 we have

$$\begin{aligned} k_1 f(x_0) + k_2 g(x_0) &= 0, \\ k_1 f'(x_0) + k_2 g'(x_0) &= 0. \end{aligned} \tag{4}$$

The determinant of coefficients of Eqs. (4) is precisely $W(f, g)(x_0)$, which is not zero by hypothesis. Therefore the only solution of Eqs. (4) is $k_1 = k_2 = 0$, so f and g are linearly independent.

The second statement of Theorem 3.8 follows immediately from the first. Let f and g be linearly dependent, and suppose that the conclusion is false, namely, that $W(f, g)$ is not identically zero. Then there is a point x_0 such that $W(f, g)(x_0) \neq 0$; this implies that f and g are linearly independent, which is a contradiction.

We can apply this result to the two functions $f(x) = e^x$ and $g(x) = e^{2x}$, $-\infty < x < \infty$, discussed in Example 2. Let x_0 be some point in $(-\infty, \infty)$; then we have

$$W(f, g)(x_0) = \begin{vmatrix} e^{x_0} & e^{2x_0} \\ e^{x_0} & 2e^{2x_0} \end{vmatrix} = e^{3x_0} \neq 0. \tag{5}$$

Thus the functions e^x and e^{2x} are linearly independent on $-\infty < x < \infty$.

We emphasize that the converse of Theorem 3.8 is not true. That is, the functions f and g may be linearly independent on an interval containing x_0 even though $W(f, g)(x_0) = 0$. Indeed, it is even possible for f and g to be linearly independent when $W(f, g)(x) = 0$ for *all* x in $\alpha < x < \beta$. This is illustrated in Problem 13.

The situation is different if the functions f and g are restricted to be solutions y_1 and y_2, respectively, of a second order linear differential equation. In that case we have the following result.

Theorem 3.9. *Let y_1 and y_2 be solutions of*

$$y'' + p(x)y' + q(x)y = 0, \tag{6}$$

where p and q are continuous on $\alpha < x < \beta$. Then y_1 and y_2 are linearly dependent if and only if $W(y_1, y_2)(x)$ is identically zero on $\alpha < x < \beta$. Alternatively, y_1 and y_2 are linearly independent on $\alpha < x < \beta$ if and only if $W(y_1, y_2)(x)$ is not identically zero there.

Of course, we know by Theorem 3.5 in Section 3.2 that $W(y_1, y_2)(x)$ is either identically zero or nowhere zero. To prove Theorem 3.9 observe that if y_1 and y_2 are linearly dependent, then $W(y_1, y_2)(x)$ is identically zero by Theorem 3.8. On the other hand, if $W(y_1, y_2)(x)$ is identically zero, then $W(y_1, y_2)(x_0) = 0$ for any x_0 in $\alpha < x < \beta$. Consequently, the system of equations

$$\begin{aligned} c_1 y_1(x_0) + c_2 y_2(x_0) &= 0, \\ c_1 y_1'(x_0) + c_2 y_2'(x_0) &= 0, \end{aligned} \tag{7}$$

for c_1 and c_2 has a nontrivial solution. Using these values of c_1 and c_2, let $\phi(x) = c_1y_1(x) + c_2y_2(x)$. Then ϕ is a solution of Eq. (6) and, further, by Eqs. (7) it satisfies the initial conditions $\phi(x_0) = \phi'(x_0) = 0$. Therefore, by the existence and uniqueness theorem, $\phi(x) = 0$ for all x in $\alpha < x < \beta$ (see Example 2 of Section 3.1). Then $c_1y_1(x) + c_2y_2(x) = 0$ for all x in $\alpha < x < \beta$, which means that y_1 and y_2 are linearly dependent.

We can now summarize the facts about fundamental sets of solutions and linear independence in the following way. Let y_1 and y_2 be solutions of Eq. (6),

$$y'' + p(x)y' + q(x)y = 0,$$

where p and q are continuous on $\alpha < x < \beta$. Then the following four statements are equivalent, in the sense that each one implies the other three.

1. The functions y_1 and y_2 are a fundamental set of solutions on $\alpha < x < \beta$.
2. The functions y_1 and y_2 are linearly independent on $\alpha < x < \beta$.
3. $W(y_1, y_2)(x_0) \neq 0$ for some x_0 in $\alpha < x < \beta$.
4. $W(y_1, y_2)(x) \neq 0$ for all x in $\alpha < x < \beta$.

It is interesting to note the similarity between the theory of second order linear differential equations and two dimensional vector algebra. Two vectors **a** and **b** are said to be linearly dependent if there are two scalars k_1 and k_2, not both zero, such that $k_1\mathbf{a} + k_2\mathbf{b} = \mathbf{0}$; otherwise they are said to be linearly independent. Let **i** and **j** be unit vectors directed along the positive x and y axes, respectively. Since there are no scalars k_1 and k_2, not both zero, such that $k_1\mathbf{i} + k_2\mathbf{j} = \mathbf{0}$, the vectors **i** and **j** are linearly independent. Further we know that any vector with components a_1 and a_2 can be written as $a_1\mathbf{i} + a_2\mathbf{j}$; that is, as a linear combination of the two linearly independent vectors **i** and **j**. It is not difficult to show that any vector in two dimensions can be represented as a linear combination of any two linearly independent two-dimensional vectors (see Problem 11). Such a pair of linearly independent vectors is said to form a basis for the vector space of two-dimensional vectors.

The term vector space is also applied to other collections of mathematical objects that satisfy the same laws of addition and multiplication by scalars that geometric vectors do. For example, it can be shown that the set of functions that are twice differentiable on $\alpha < x < \beta$ forms a vector space. Similarly the set V of functions satisfying Eq. (6) also forms a vector space.

Since every member of V can be expressed as a linear combination of two linearly independent members y_1 and y_2, we say that such a pair forms a basis for V. This leads to the conclusion that V is two dimensional—Eq. (6) is second order—and analogous in many respects to the space of geometric vectors in a plane. Later we find that the set of solutions of an nth order linear homogeneous differential equation forms a vector space of dimension n, and that any set of n linearly independent solutions of the differential equation forms a basis for the space.

PROBLEMS

In each of Problems 1 through 4 prove that the functions y_1 and y_2 are linearly independent solutions of the given differential equation.

1. $y'' - y = 0; \quad y_1(x) = e^x, \quad y_2(x) = e^{-x}$
2. $y'' - y = 0; \quad y_1(x) = \cosh x, \quad y_2(x) = \sinh x$
3. $y'' - y' - 6y = 0; \quad y_1(x) = e^{-2x}, \quad y_2(x) = e^{3x}$
4. $x^2y'' + xy' - 4y = 0, \quad x > 0; \quad y_1(x) = x^2, \quad y_2(x) = x^{-2}$

In each of Problems 5 through 8 determine whether the given pair of functions is linearly independent.

5. $f(x) = e^{\lambda x}\cos\mu x, \quad g(x) = e^{\lambda x}\sin\mu x, \quad \mu \neq 0$
6. $f(x) = \cos 3x, \quad g(x) = 4\cos^3 x - 3\cos x$
7. $f(x) = x^2 + 5x, \quad g(x) = x^2 - 5x$
8. $f(x) = e^{3x}, \quad g(x) = e^{3(x-1)}$
9. Prove that if the functions y_1 and y_2 are linearly independent solutions of $y'' + p(x)y' + q(x)y = 0$, then c_1y_1 and c_2y_2 are linearly independent solutions provided neither c_1 nor c_2 equals 0.
10. Prove that if the functions y_1 and y_2 are linearly independent solutions of the differential equation $y'' + p(x)y' + q(x)y = 0$, then $y_3 = y_1 + y_2$, and $y_4 = y_1 - y_2$ also form a fundamental set of solutions. Conversely, if y_3 and y_4 are linearly independent solutions of the differential equation, show that y_1 and y_2 form a fundamental set of solutions.
11. (a) Prove that any two-dimensional vector can be written as a linear combination of the vectors $\mathbf{i} + \mathbf{j}$ and $\mathbf{i} - \mathbf{j}$.
    Hint: Any vector $\mathbf{a}$ can be written as $a_1\mathbf{i} + a_2\mathbf{j}$. Show that it is possible to determine k_1 and k_2 such that $\mathbf{a} = k_1(\mathbf{i} + \mathbf{j}) + k_2(\mathbf{i} - \mathbf{j})$ by equating the two expressions for $\mathbf{a}$ and solving for k_1 and k_2 in terms of a_1 and a_2.
    (b) Prove that if the vectors $\mathbf{x} = x_1\mathbf{i} + x_2\mathbf{j}$ and $\mathbf{y} = y_1\mathbf{i} + y_2\mathbf{j}$ are linearly independent, then any vector $\mathbf{z} = z_1\mathbf{i} + z_2\mathbf{j}$ can be expressed as a linear combination of $\mathbf{x}$ and $\mathbf{y}$. Note that if $\mathbf{x}$ and $\mathbf{y}$ are linearly independent then $x_1y_2 - y_1x_2 \neq 0$. Why?
12. Verify that x and x^2 are linearly independent on $-1 < x < 1$, but that $W(x, x^2)$ vanishes at $x = 0$. From this what can you conclude about the possibility of x and x^2 being solutions of Eq. (6) of the text? Show that x and x^2 are solutions of $x^2y'' - 2xy' + 2y = 0$. Does this contradict your conclusion? Does it contradict Theorem 3.5 of Section 3.2?
13. Show that the functions $f(x) = x|x|$ and $g(x) = x^2$ are linearly dependent on $0 < x < 1$ and on $-1 < x < 0$ but are linearly independent on $-1 < x < 1$. Note that while f and g are linearly independent, $W(f, g)$ is identically zero on $-1 < x < 1$; hence f and g cannot be solutions of Eq. (6) of the text.

*14. If the functions y_1 and y_2 are linearly independent solutions of the differential equation $y'' + p(x)y' + q(x)y = 0$, show that between consecutive zeros of y_1 there is one and only one zero of y_2.
Hint: Use Rolle's (1652–1719) theorem to prove the conclusion by contradiction. This result is often referred to as Sturm's (1803–1855) theorem. As a particular example, note that $\sin x$ and $\cos x$ are linearly independent solutions of $y'' + y = 0$, and that the zeros $(0, \pm\pi, \pm 2\pi, \ldots)$ of $\sin x$ and the zeros $(\pm\pi/2, \pm 3\pi/2, \ldots)$ of $\cos x$ are interlaced.

3.4 Reduction of Order

A very important and useful fact is the following: if one solution of a second order linear homogeneous differential equation is known, a second linearly independent solution (and hence a fundamental set of solutions) can be determined. The procedure, which is due to D'Alembert,[7] is usually referred to as the method of reduction of order.

Suppose we know one solution y_1, not identically zero, of

$$y'' + p(x)y' + q(x)y = 0. \tag{1}$$

Then cy_1, where c is any constant, is also a solution of Eq. (1). This suggests the following question: can we determine a function v, not a constant, such that $y = v(x)y_1(x)$ is a solution of Eq. (1)? The answer is yes, and further, we shall see that v can be determined in a straightforward manner. If we set

$$y = v(x)y_1(x), \tag{2}$$

then

$$y' = v(x)y_1'(x) + v'(x)y_1(x),$$
$$y'' = v(x)y_1''(x) + 2v'(x)y_1'(x) + v''(x)y_1(x).$$

Substituting for y, y', and y'' in Eq. (1) and collecting terms gives

$$v(y_1'' + py_1' + qy_1) + v'(2y_1' + py_1) + v''y_1 = 0. \tag{3}$$

Since y_1 is a solution of Eq. (1) the quantity in the first parentheses is zero. In any interval in which y_1 does not vanish, we can divide by y_1 obtaining

$$v'' + \left(p + 2\frac{y_1'}{y_1}\right)v' = 0. \tag{4}$$

Equation (4) for v' can be solved immediately either as a first order linear equation (Section 2.1) or as a separable equation (Section 2.4). The solution is

$$v'(x) = c\exp\left[-\int^x \left(p(s) + \frac{2y_1'(s)}{y_1(s)}\right) ds\right] = cu(x), \tag{5}$$

where c is an arbitrary constant, and

$$u(x) = \frac{1}{[y_1(x)]^2}\exp\left[-\int^x p(s)\,ds\right]. \tag{6}$$

[7]Jean d'Alembert (1717–1783), a French mathematician, was a contemporary of Euler and Daniel Bernoulli, and is known primarily for his work in mechanics and differential equations. D'Alembert's principle in mechanics and d'Alembert's paradox in hydrodynamics are named for him, and the wave equation (see Section 10.7) first appeared in his paper on vibrating strings in 1747. In his later years he devoted himself largely to philosophy and to his duties as science editor of Diderot's Encyclopédie.

Then

$$v(x) = c\int^x u(t)\,dt + k, \tag{7}$$

where k is also an arbitrary constant. However, we can omit the constant k since

$$\begin{aligned} y &= y_1(x)v(x) \\ &= cy_1(x)\int^x u(t)\,dt + ky_1(x), \end{aligned} \tag{8}$$

and hence it only adds a multiple of $y_1(x)$ to the second solution. Thus two solutions of Eq. (1) are

$$y = y_1(x) \quad \text{and} \quad y = y_1(x)\int^x u(t)\,dt. \tag{9}$$

Since the antiderivative of the function u cannot be a constant, the solutions are linearly independent. In using the method of reduction of order it is *not* important to try to memorize Eqs. (9) and (6); what is important to remember is that if one solution y_1 is known, a second solution of the form $y = v(x)y_1(x)$ can be found by the above procedure.

While the method of reduction of order does not tell us how to find the first solution of Eq. (1), it is encouraging to know that we have *reduced* the problem of solving Eq. (1) to that of finding just one solution.

It is also possible to derive the second linearly independent solution given in Eq. (9) by using Abel's formula for the Wronskian of two linearly independent solutions of Eq. (1). This is discussed in Problems 15 and 16.

The method of reduction of order is also useful for nonhomogeneous equations; this is discussed in Problem 17 of Section 3.6.2.

EXAMPLE 1

Given that $y_1(x) = x^{-1}$ is one solution of

$$2x^2y'' + 3xy' - y = 0, \qquad x > 0, \tag{10}$$

find a second linearly independent solution.

We set $y = x^{-1}v(x)$; then

$$y' = x^{-1}v' - x^{-2}v, \qquad y'' = x^{-1}v'' - 2x^{-2}v' + 2x^{-3}v.$$

Substituting for y, y', and y'' in Eq. (10) and collecting terms we obtain

$$\begin{aligned} &2x^2(x^{-1}v'' - 2x^{-2}v' + 2x^{-3}v) + 3x(x^{-1}v' - x^{-2}v) - x^{-1}v \\ &\quad = 2xv'' + (-4 + 3)v' + (4x^{-1} - 3x^{-1} - x^{-1})v \\ &\quad = 2xv'' - v' = 0. \end{aligned} \tag{11}$$

We note that the coefficient of v is zero, as it should be; this provides a good check on our algebra.

Using the method of separation of variables, we readily find from Eq. (11) that

$$v'(x) = cx^{1/2},$$

hence

$$v(x) = \tfrac{2}{3}cx^{3/2} + k.$$

Thus a second linearly independent solution of Eq. (10) is

$$y_2(x) = x^{-1}v(x) = \tfrac{2}{3}cx^{1/2} + kx^{-1}, \tag{12}$$

where c and k are arbitrary constants. As explained earlier, we can take $k = 0$, and for convenience we take $c = \frac{3}{2}$ so that $y_2(x) = x^{1/2}$.

EXAMPLE 2

Show that $y = x$ is a solution of the Legendre equation of order one

$$(1 - x^2)y'' - 2xy' + 2y = 0, \qquad -1 < x < 1, \tag{13}$$

and find a second linearly independent solution.

First if $y = x$, then $y' = 1$ and $y'' = 0$. Substituting for y, y', and y'' in Eq. (13) gives

$$(1 - x^2) \cdot 0 - 2x + 2x = 0,$$

so indeed $y = x$ is a solution. To find a second solution let $y = xv(x)$; then

$$y' = xv' + v, \qquad y'' = xv'' + 2v'.$$

Substituting for y, y', and y'' in Eq. (13) gives

$$(1 - x^2)(xv'' + 2v') - 2x(xv' + v) + 2xv = 0.$$

Collecting terms and dividing by $x(1 - x^2)$ we obtain

$$v'' + \left(\frac{2}{x} - \frac{2x}{1 - x^2}\right)v' = 0. \tag{14}$$

Note that the coefficient of v' is not defined at $x = 0$, the zero of $y_1(x)$. However, this will not cause any difficulty in the final solution. First we will obtain a solution of Eq. (14) in the intervals $-1 < x < 0$ and $0 < x < 1$.

Equation (14) is a first order linear equation for v', and the integrating factor is $x^2(1 - x^2)$; hence

$$[x^2(1 - x^2)v']' = 0,$$

$$x^2(1 - x^2)v' = c,$$

so

$$v(x) = c\int^x \frac{dt}{t^2(1-t^2)} = c\int^x \left(\frac{1}{t^2} + \frac{1}{1-t^2}\right) dt$$
$$= c\left(-\frac{1}{x} + \frac{1}{2}\ln\frac{1+x}{1-x}\right).$$

Consequently, a second solution (suppressing a constant multiplier without loss of generality) of Eq. (13) is

$$y_2(x) = xv(x) = 1 - \frac{x}{2}\ln\frac{1+x}{1-x}. \tag{15}$$

Although the function v is undefined at $x = 0$, clearly the limit of $y_2(x) = xv(x)$ as $x \to 0$ exists. The function y_2 defined by Eq. (15) on $-1 < x < 1$ actually satisfies the differential equation (13) on $-1 < x < 1$, not just on $-1 < x < 0$ and $0 < x < 1$. On the other hand, $y_2(x)$ becomes unbounded as $x \to \pm 1$; this is closely related to the fact that in Eq. (13) the coefficient of y'' is zero at $x = \pm 1$, while the coefficients of y' and y are nonzero. This will be discussed further in Chapter 4.

PROBLEMS

In Problems 1 through 11 find a second solution of the given differential equation by the method of reduction of order.

1. $y'' - 4y' - 12y = 0, \quad y_1(x) = e^{6x}$
2. $y'' + 2y' + y = 0, \quad y_1(x) = e^{-x}$
3. $x^2y'' + 2xy' = 0, \quad y_1(x) = 1$. For what range of x would you expect the solution to be valid?
4. $x^2y'' + 2xy' - 2y = 0, \quad y_1(x) = x$. For what range of x would you expect the solution to be valid?
5. $x^2y'' + 3xy' + y = 0, \quad x > 0; \quad y_1(x) = x^{-1}$
6. $x^2y'' - x(x+2)y' + (x+2)y = 0, \quad x > 0; \quad y_1(x) = x$
7. $(1 - x\cot x)y'' - xy' + y = 0, \quad y_1(x) = x$. Consider the interval $0 < x < \pi$.

   Hint: $$\int \frac{x\,dx}{1 - x\cot x} = \ln|x\cos x - \sin x|.$$

8. $xy'' - y' + 4x^3y = 0, \quad x > 0; \quad y_1(x) = \sin x^2$
9. $(x-1)y'' - xy' + y = 0, \quad x > 1; \quad y_1(x) = e^x$
10. $y'' - \left(\frac{1}{x} - \frac{3}{16x^2}\right)y = 0, \quad x > 0; \quad y_1(x) = x^{1/4}e^{2\sqrt{x}}$
11. $y'' - (1 - 2\,\text{sech}^2 x)y = 0, \quad y_1(x) = \text{sech}\, x$
    Hint: $\cosh 2x = 2\cosh^2 x - 1, \quad \sinh 2x = 2\sinh x \cosh x$.

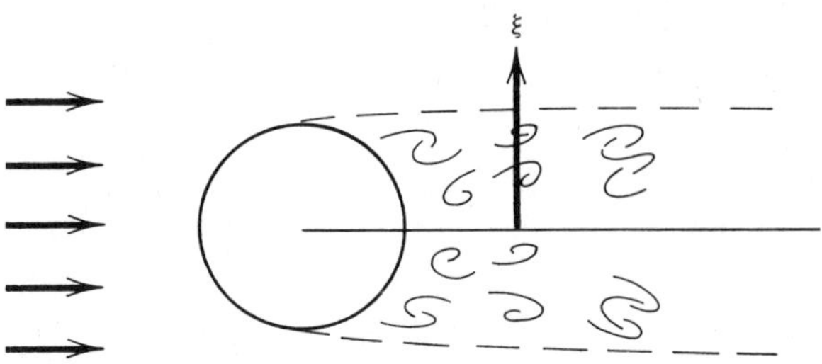

FIGURE 3.1 A cylinder in a uniform stream.

12. Verify that $y_1(x) = x^{-1/2}\sin x$ is one solution of Bessel's equation of order one-half,

$$x^2y'' + xy' + \left(x^2 - \tfrac{1}{4}\right)y = 0,$$

and determine a second solution. Consider the interval $0 < x < \infty$.

13. The differential equation

$$xy'' - (x + N)y' + Ny = 0,$$

where N is a nonnegative integer, has been discussed by several authors.[8] One reason it is of interest is because it has an exponential solution and a polynomial solution.
(a) Verify that one solution is $y_1(x) = e^x$.
(b) Show that a second solution has the form $y_2(x) = ce^x\int x^N e^{-x}\,dx$. Calculate $y_2(x)$ for $N = 1$ and $N = 2$ and convince yourself that, with $c = -1/N!$,

$$y_2(x) = 1 + \frac{x}{1!} + \frac{x^2}{2!} + \cdots + \frac{x^N}{N!}.$$

It is interesting to note that $y_2(x)$ is exactly the first $N + 1$ terms of the Taylor series for e^x.

14. Consider the two-dimensional flow of a uniform stream around a fixed circular cylinder whose axis is perpendicular to the stream. In the analysis of turbulent flow in the wake of the cylinder (see Figure 3.1), the differential equation

$$f'' + \delta(\xi f' + f) = 0$$

arises. Here ξ is a dimensionless space variable perpendicular both to the axis of the cylinder and to the stream, $f(\xi)$ measures the reduction of the stream velocity in the wake, a prime denotes differentiation with respect to ξ, and δ is a physical parameter that is inversely proportional to the turbulent diffusivity. Show that $f_1(\xi) = e^{-\delta\xi^2/2}$ is one solution and find an integral form for the general solution.

15. Suppose that y_1 is a nonvanishing solution of $y'' + p(x)y' + q(x)y = 0$ and that we wish to find a second linearly independent solution y_2. Show that $(y_2/y_1)' = W(y_1, y_2)/y_1^2$ and then use Abel's formula [Eq. (15) in Section 3.2] for $W(y_1, y_2)$ to obtain y_2.

16. Verify that $y_1(x) = x$ is a solution of $x^2y'' + 2xy' - 2y = 0$, $x > 0$, and then, using the result of Problem 15, determine a second linearly independent solution.

[8] T. A. Newton, "On Using a Differential Equation to Generate Polynomials," *American Mathematical Monthly*, *81* (1974), pp. 592–601. Also see the references given there.

3.5 Homogeneous Equations with Constant Coefficients

We now turn from the general theory of second order linear homogeneous equations to methods of actually solving such equations. In this section we consider the problem of finding the general solution of a second order linear homogeneous differential equation with constant coefficients. The corresponding problem with variable coefficients, which is much more difficult, is considered in Chapter 4.[9]

Consider the equation

$$\begin{aligned} L[y] &= ay'' + by' + cy \\ &= (aD^2 + bD + c)y = 0, \end{aligned} \tag{1}$$

where $a \neq 0$, b, and c are real numbers. Since a, b, and c are constants it follows immediately from the existence and uniqueness Theorem 3.2 that the solutions of Eq. (1) are valid on the interval $-\infty < x < \infty$.

A clue to a method of solving Eq. (1) can be found simply by reading the equation; that is, what function $y = \phi(x)$ satisfies the relationship that a times its second derivative plus b times its first derivative plus c times the function itself adds up to zero for all x? With only constant coefficients it is natural first to consider functions $y = \phi(x)$ such that y, y', and y'' differ only by constant multiplicative factors.

One function that has been studied extensively in the elementary calculus has just the property we wish—the exponential function e^{rx}. Hence we try to find solutions of Eq. (1) of the form e^{rx} for suitably chosen values of r. Substituting $y = e^{rx}$ in Eq. (1) leads to the equation

$$L[e^{rx}] = a(e^{rx})'' + b(e^{rx})' + ce^{rx} = 0 \tag{2}$$

or

$$e^{rx}(ar^2 + br + c) = 0. \tag{3}$$

Since e^{rx} is not zero, we must have

$$ar^2 + br + c = 0. \tag{4}$$

If r is a root of this quadratic equation, often called the *characteristic* or *auxiliary equation*, then e^{rx} is a solution of Eq. (1). Notice that the coefficients in the characteristic equation (4) are the same as those in the differential equation (1). The roots r_1 and r_2 of Eq. (4) are given by

$$r_1 = \frac{-b + (b^2 - 4ac)^{1/2}}{2a}, \qquad r_2 = \frac{-b - (b^2 - 4ac)^{1/2}}{2a}. \tag{5}$$

[9] In some cases it is possible to reduce a differential equation with variable coefficients to one with constant coefficients by a suitable change of variable (see Problems 18 through 21 of Section 3.5.1).

The nature of the solutions of Eq. (1) clearly depends on the values of r_1 and r_2, which in turn depend on the constant coefficients in the differential equation through the relations (5). Just as in elementary algebra we must examine separately the cases $b^2 - 4ac$ positive, zero, and negative.

The first two cases are considered in this section; the case $b^2 - 4ac < 0$ is considered in the next section.

REAL AND UNEQUAL ROOTS. For $b^2 - 4ac > 0$, Eqs. (5) give two real, unequal values for r_1 and r_2. Hence e^{r_1x} and e^{r_2x} are solutions of Eq. (1). To show that they are linearly independent solutions we compute $W(e^{r_1x}, e^{r_2x})$. We have

$$W(e^{r_1x}, e^{r_2x})(x) = \begin{vmatrix} e^{r_1x} & e^{r_2x} \\ r_1e^{r_1x} & r_2e^{r_2x} \end{vmatrix} = (r_2 - r_1)e^{(r_1+r_2)x},$$

which cannot be zero since $r_2 \neq r_1$. Hence the general solution of Eq. (1) is

$$y = c_1e^{r_1x} + c_2e^{r_2x}. \tag{6}$$

EXAMPLE 1

Find the solution of the differential equation $y'' + 5y' + 6y = 0$ satisfying the initial conditions $y(0) = 0$, $y'(0) = 1$.

Substituting $y = e^{rx}$ leads to

$$r^2 + 5r + 6 = (r + 3)(r + 2) = 0;$$

hence $r = -2, -3$. The general solution is

$$y = c_1e^{-2x} + c_2e^{-3x}.$$

To satisfy the initial conditions at $x = 0$ we must have

$$c_1 + c_2 = 0, \qquad -2c_1 - 3c_2 = 1.$$

Solving these equations gives

$$c_1 = 1, \qquad c_2 = -1.$$

Hence the solution of the differential equation satisfying the prescribed conditions is

$$y = e^{-2x} - e^{-3x}.$$

EXAMPLE 2

Find the solution of the differential equation $y'' + y' - 12y = 0$ satisfying the initial conditions $y(2) = 2$, $y'(2) = 0$.

Substituting $y = e^{rx}$ leads to

$$r^2 + r - 12 = (r - 3)(r + 4) = 0;$$

hence $r = 3, -4$. The general solution is

$$y = c_1 e^{3x} + c_2 e^{-4x}.$$

Since the initial conditions are given at $x = 2$, it is slightly more convenient to rewrite the general solution in the form

$$y = k_1 e^{3(x-2)} + k_2 e^{-4(x-2)}.$$

This corresponds to redefining c_1 and c_2 as $c_1 = k_1 e^{-6}$ and $c_2 = k_2 e^{8}$. To satisfy the initial conditions at $x = 2$ we must have

$$k_1 + k_2 = 2, \qquad 3k_1 - 4k_2 = 0.$$

Solving these equations, we obtain

$$k_1 = \tfrac{8}{7}, \qquad k_2 = \tfrac{6}{7}.$$

Hence the general solution of the differential equation satisfying the prescribed conditions is

$$y = \tfrac{8}{7} e^{3(x-2)} + \tfrac{6}{7} e^{-4(x-2)}.$$

REAL AND EQUAL ROOTS. When $b^2 - 4ac = 0$, it follows from Eqs. (5) that $r_1 = r_2 = -b/2a$, and we have only the one solution, $e^{-(b/2a)x}$. However, we can use the method of reduction of order (see Section 3.4) to reduce the order of the equation and find a second solution. Let

$$y = v(x) e^{-(b/2a)x};$$

then

$$y' = v'(x) e^{-(b/2a)x} - \frac{b}{2a} v(x) e^{-(b/2a)x},$$

$$y'' = \left[v''(x) - \frac{b}{a} v'(x) + \frac{b^2}{4a^2} v(x) \right] e^{-(b/2a)x}.$$

Substituting for y, y', and y'' in Eq. (1) and dividing out the common factor $e^{-(b/2a)x}$ gives the following equation for v:

$$a\left(v'' - \frac{b}{a} v' + \frac{b^2}{4a^2} v \right) + b\left(v' - \frac{b}{2a} v \right) + cv = 0.$$

After collecting terms we find that

$$av'' - \left(\frac{b^2}{4a} - c \right) v = 0.$$

Since $b^2 - 4ac = 0$ the last term drops out, and we obtain

$$v'' = 0.$$

Hence

$$v(x) = c_1 x + c_2,$$

where c_1 and c_2 are arbitrary constants. Consequently, a second solution of Eq. (1) is $(c_1 x + c_2)e^{-(b/2a)x}$. In particular, corresponding to $c_2 = 0$, $c_1 = 1$ we obtain $xe^{-(b/2a)x}$. We know from our earlier discussion of the method of reduction of order that $e^{-(b/2a)x}$ and $xe^{-(b/2a)x}$ are linearly independent solutions of Eq. (1). This can also be demonstrated by calculating their Wronskian (see Problem 16). Thus in the case $b^2 - 4ac = 0$ the general solution of Eq. (1) is

$$y = c_1 e^{r_1 x} + c_2 x e^{r_1 x}, \qquad r_1 = -b/2a. \tag{7}$$

EXAMPLE 3

Find the general solution of

$$y'' + 4y' + 4y = 0.$$

Substituting $y = e^{rx}$ leads to

$$r^2 + 4r + 4 = (r + 2)(r + 2) = 0;$$

hence $r = -2$ is a repeated root of the characteristic equation. One solution is e^{-2x}, and a second linearly independent solution, according to the above theory, is xe^{-2x}. Thus the general solution is

$$y = c_1 e^{-2x} + c_2 x e^{-2x}.$$

There is an interesting alternative way of determining the second solution when the roots of the characteristic equation are equal. Since r_1 is a repeated root of $ar^2 + br + c = 0$, it follows that $ar^2 + br + c = a(r - r_1)^2$. Then, for all r and x

$$L[e^{rx}] = a(e^{rx})'' + b(e^{rx})' + ce^{rx} = ae^{rx}(r - r_1)^2. \tag{8}$$

The right side of Eq. (8) is zero when $r = r_1$, which shows that $e^{r_1 x}$ is a solution of Eq. (1), as we already know. Taking the partial derivative of both sides of Eq. (8) with respect to r and interchanging differentiation with respect to r and x gives

$$L[xe^{rx}] = axe^{rx}(r - r_1)^2 + 2ae^{rx}(r - r_1). \tag{9}$$

When $r = r_1$ the right side of Eq. (9) is zero and hence $L[xe^{r_1 x}] = 0$; thus $xe^{r_1 x}$ is also a solution of Eq. (1).

PROBLEMS

In each of Problems 1 through 14 determine the general solution of the given differential equation. If initial conditions are given, find the solution satisfying the stated conditions.

1. $y'' + 2y' - 3y = 0$
2. $4y'' + 4y' + y = 0$
3. $6y'' - y' - y = 0$
4. $2y'' - 3y' + y = 0$
5. $y'' - y = 0$
6. $y'' - 2y' + y = 0$
7. $y'' + 5y' = 0$
8. $y'' - 9y' + 9y = 0$
9. $y'' - 2y' - 2y = 0$
10. $y'' + 2y' + y = 0$
11. $y'' + y' - 2y = 0, \quad y(0) = 1, \quad y'(0) = 1$
12. $y'' - 6y' + 9y = 0, \quad y(0) = 0, \quad y'(0) = 2$
13. $y'' + 8y' - 9y = 0, \quad y(1) = 1, \quad y'(1) = 0$
14. $y'' + 4y' + 4y = 0, \quad y(-1) = 2, \quad y'(-1) = 1$
15. Show that the general solution of $y'' - 4y = 0$ is
$$y = c_1 \sinh 2x + c_2 \cosh 2x.$$
16. Show that $W(e^{rx}, xe^{rx}) \neq 0$ for any value of r. Hence conclude that if $e^{r_1 x}$ and $xe^{r_1 x}$ are solutions of $ay'' + by' + cy = 0$, then they are a fundamental set of solutions.

*17. In this problem we present another argument that suggests the form of the second solution of $ay'' + by' + cy = 0$ when the roots r_1 and r_2 of the characteristic equation $ar^2 + br + c = 0$ are equal. First, suppose that $r_2 \neq r_1$ so that $e^{r_1 x}$ and $e^{r_2 x}$ are solutions of the differential equation. Verify that $\phi(x; r_1, r_2) = (e^{r_2 x} - e^{r_1 x})/(r_2 - r_1)$ is also a solution of the differential equation for $r_2 \neq r_1$. Now think of r_1 as fixed and evaluate the limit of $\phi(x; r_1, r_2)$ as $r_2 \to r_1$ by using L'Hôpital's rule.

3.5.1 Complex Roots

To complete our discussion of the equation

$$ay'' + by' + cy = 0, \tag{1}$$

we consider the case where the roots of the characteristic equation

$$ar^2 + br + c = 0 \tag{2}$$

are complex numbers of the form $\lambda + i\mu$ where λ and μ are real and $\mu \neq 0$. This immediately raises the question of what we mean by an expression of the form $e^{(\lambda + i\mu)x}$.

First recall that in elementary calculus it was shown that the Taylor series for e^x about $x = 0$ is

$$e^x = \sum_{n=0}^{\infty} \frac{x^n}{n!}, \qquad -\infty < x < \infty. \tag{3}$$

If we substitute ix for x in Eq. (3) and take the result as the definition of e^{ix}, we find that

$$
\begin{aligned}
e^{ix} &= \sum_{n=0}^{\infty} \frac{(ix)^n}{n!} \\
&= \sum_{n=0}^{\infty} \frac{(-1)^n x^{2n}}{(2n)!} + i \sum_{n=1}^{\infty} \frac{(-1)^{n-1} x^{2n-1}}{(2n-1)!},
\end{aligned} \tag{4}
$$

where we have made use of the fact that $i^2 = -1$, $i^3 = -i$, $i^4 = 1$, etc. The first series in Eq. (4) is precisely the Taylor series for $\cos x$, and the second is the Taylor series for $\sin x$. Hence we *define*

$$e^{ix} = \cos x + i \sin x. \tag{5}$$

This relationship is referred to as Euler's formula. Substituting $-x$ for x in Eq. (5) and making use of the relations $\cos(-x) = \cos x$, $\sin(-x) = -\sin x$ gives

$$e^{-ix} = \cos x - i \sin x. \tag{6}$$

The functions $\cos x$ and $\sin x$ can be expressed in terms of e^{ix} and e^{-ix} by adding and subtracting Eqs. (5) and (6), respectively. We obtain

$$\cos x = \frac{e^{ix} + e^{-ix}}{2}, \qquad \sin x = \frac{e^{ix} - e^{-ix}}{2i}.$$

Next we *define*

$$e^{(\lambda + i\mu)x} = e^{\lambda x} e^{i\mu x} = e^{\lambda x}(\cos \mu x + i \sin \mu x). \tag{7}$$

With this definition it can be shown that the usual laws of algebra hold. For example,

$$\frac{e^{(\lambda + i\mu)x}}{e^{(\alpha + i\beta)x}} = e^{(\lambda + i\mu)x - (\alpha + i\beta)x} = e^{(\lambda - \alpha)x + i(\mu - \beta)x},$$

and

$$\left[e^{(\lambda + i\mu)x}\right]^n = e^{n(\lambda + i\mu)x} = e^{n\lambda x + in\mu x},$$

where n is a positive or negative integer.

Finally, we must consider the question of differentiating $e^{(\lambda + i\mu)x}$ with respect to x. While we know that

$$\frac{d}{dx}(e^{rx}) = re^{rx} \tag{8}$$

for r a real number, we do not know that this is true if r is complex. However, it can be shown by direct computation using the definition given by Eq. (7) that Eq. (8) is true for r complex. Indeed, the usual laws of the elementary calculus are valid for complex exponentials with the understanding that e^{rx} is defined by Eq.

(7) for r complex. More generally for a complex-valued function $f(x) = u(x) + iv(x)$ where u and v are real-valued functions, and for a complex constant c, we have $f'(x) = u'(x) + iv'(x)$ and $[cf(x)]' = cf'(x)$.

Now we can return to the problem of solving Eq. (1) when the roots of the characteristic equation are complex. Since a, b, and c are real, the roots occur in conjugate pairs $r_1 = \lambda + i\mu$ and $r_2 = \lambda - i\mu$; hence the general solution of Eq. (1) is

$$y = c_1 e^{(\lambda + i\mu)x} + c_2 e^{(\lambda - i\mu)x}. \tag{9}$$

The general solution (9) has the disadvantage that the functions $e^{(\lambda+i\mu)x}$ and $e^{(\lambda-i\mu)x}$ are complex valued. Since the differential equation is real it would seem desirable, if possible, to express the general solution of Eq. (1) as a linear combination of real-valued solutions. This can be done as follows. Since $e^{(\lambda+i\mu)x}$ and $e^{(\lambda-i\mu)x}$ are solutions of Eq. (1), their sums and differences are also solutions; thus

$$\begin{aligned} e^{(\lambda+i\mu)x} + e^{(\lambda-i\mu)x} &= e^{\lambda x}(\cos\mu x + i\sin\mu x) + e^{\lambda x}(\cos\mu x - i\sin\mu x) \\ &= 2e^{\lambda x}\cos\mu x \end{aligned}$$

and

$$\begin{aligned} e^{(\lambda+i\mu)x} - e^{(\lambda-i\mu)x} &= e^{\lambda x}(\cos\mu x + i\sin\mu x) - e^{\lambda x}(\cos\mu x - i\sin\mu x) \\ &= 2ie^{\lambda x}\sin\mu x \end{aligned}$$

are solutions of Eq. (1). Hence, neglecting the constant multipliers 2 and $2i$, respectively, the functions

$$e^{\lambda x}\cos\mu x, \qquad e^{\lambda x}\sin\mu x \tag{10}$$

are real-valued solutions of Eq. (1). It can easily be shown that

$$W(e^{\lambda x}\cos\mu x, e^{\lambda x}\sin\mu x) = \mu e^{2\lambda x},$$

and since this is never zero these two functions form a fundamental set of real-valued solutions. Hence we can express the general solution of Eq. (1) in the form

$$y = c_1 e^{\lambda x}\cos\mu x + c_2 e^{\lambda x}\sin\mu x,$$

where c_1 and c_2 are arbitrary constants.

EXAMPLE 1

Find the general solution of

$$y'' + y' + y = 0. \tag{11}$$

The characteristic equation is

$$r^2 + r + 1 = 0,$$

and the roots are

$$r = \frac{-1 \pm (1-4)^{1/2}}{2} = -\frac{1}{2} \pm i\frac{\sqrt{3}}{2}.$$

Hence the general solution of Eq. (11) is

$$y = e^{-x/2}\left(c_1 \cos\frac{\sqrt{3}}{2}x + c_2 \sin\frac{\sqrt{3}}{2}x\right).$$

EXAMPLE 2

Find the solution of

$$y'' + 9y = 0 \tag{12}$$

satisfying the initial conditions $y(\pi/2) = A$ and $y'(\pi/2) = B$.

The characteristic equation is

$$r^2 + 9 = 0$$

with roots $r = 3i, -3i$. Hence the general solution of Eq. (12) is

$$y = c_1 \cos 3x + c_2 \sin 3x.$$

The initial conditions require that

$$c_1(0) + c_2(-1) = A, \qquad -3c_1(-1) + 3c_2(0) = B.$$

Hence $c_2 = -A$, $c_1 = B/3$ and the solution of Eq. (12) satisfying the prescribed initial conditions is

$$y = \tfrac{1}{3}B\cos 3x - A\sin 3x.$$

Let us consider again the real-valued solutions $e^{\lambda x}\cos\mu x$ and $e^{\lambda x}\sin\mu x$ of Eq. (1). Notice that these functions are the real and imaginary parts, respectively, of the complex-valued solution $e^{(\lambda+i\mu)x}$. This is just a special case of a general result that depends only on the fact that the coefficients in the differential equation are real-valued, whether or not they are constants. We state this result as a theorem.

Theorem 3.10. *Let the real-valued functions p and q be continuous on the open interval $\alpha < x < \beta$. Let $y = \phi(x) = u(x) + iv(x)$ be a complex-valued solution of the differential equation*

$$L[y] = y'' + p(x)y' + q(x)y = 0, \tag{13}$$

where u and v are real-valued functions. Then u and v are also solutions of the differential equation (13).

To prove this theorem, observe that

$$\begin{aligned} L[u+iv] &= (u+iv)'' + p(u+iv)' + q(u+iv) \\ &= (u'' + pu' + qu) + i(v'' + pv' + qv) \\ &= L[u] + iL[v] = 0. \end{aligned}$$

But a complex number is zero if and only if both its real and imaginary parts are zero; thus $L[u] = 0$ and $L[v] = 0$.

If the constants a, b, and c in Eq. (1) are complex numbers, it is still possible to find solutions of the form e^{rx} where r must satisfy

$$ar^2 + br + c = 0.$$

However, in general the roots of the characteristic equation will be complex numbers but not complex conjugates, and the corresponding solutions of Eq. (1) will be complex valued. This is discussed briefly in Problems 15, 16, and 17.

We can summarize the principal results of Sections 3.5 and 3.5.1 as follows. For the second order linear differential equation

$$ay'' + by' + cy = 0,$$

set $y = e^{rx}$ and determine the roots r_1 and r_2 of the characteristic equation

$$ar^2 + br + c = 0.$$

1. If r_1 and r_2 are real and $r_1 \neq r_2$, then the general solution is
$$y = c_1 e^{r_1 x} + c_2 e^{r_2 x}.$$
2. If r_1 and r_2 are real and $r_1 = r_2$, then the general solution is
$$y = c_1 e^{r_1 x} + c_2 x e^{r_1 x}.$$
3. If r_1 and r_2 are complex conjugate roots, $r_1, r_2 = \lambda \pm i\mu$ where $\mu \neq 0$, then the general solution is
$$y = c_1 e^{\lambda x}\cos \mu x + c_2 e^{\lambda x}\sin \mu x.$$

PROBLEMS

In each of Problems 1 through 10 determine the general solution of the given differential equation. If initial conditions are given, find the solution satisfying the stated conditions.

1. $y'' - 2y' + 2y = 0$
2. $y'' - 2y' + 6y = 0$
3. $y'' + 2y' - 8y = 0$
4. $y'' + 2y' + 2y = 0$
5. $9y'' - 6y' + y = 0$
6. $y'' + 6y' + 13y = 0$
7. $y'' + 4y = 0, \quad y(0) = 0, \quad y'(0) = 1$
8. $y'' + 4y' + 5y = 0, \quad y(0) = 1, \quad y'(0) = 0$
9. $y'' - 2y' + 5y = 0, \quad y(\pi/2) = 0, \quad y'(\pi/2) = 2$
10. $y'' + y = 0, \quad y(\pi/3) = 2, \quad y'(\pi/3) = -4$
11. Verify that $W(e^{\lambda x}\cos \mu x, e^{\lambda x}\sin \mu x) = \mu e^{2\lambda x}$.

Behavior of solutions as $x \to \infty$. Suppose that x represents time. Often it is desirable to study the behavior of solutions of a differential equation for large time ($x \to \infty$). Problems 12 through 14 are concerned with this question.

12. Show that if a, b, and c are positive constants, then all solutions of
$$ay'' + by' + cy = 0 \qquad \text{(i)}$$
approach zero as $x \to \infty$.
13. (a) If $a > 0$ and $c > 0$, but $b = 0$ in Eq. (i), show that the result of Problem 12 is no longer true, but that all solutions are bounded as $x \to \infty$.
(b) If $a > 0$ and $b > 0$ but $c = 0$ in Eq. (i), show that the result of Problem 12 is no longer true, but that all solutions approach a constant as $x \to \infty$. Determine this constant for the initial value problem $y(0) = y_0$, $y'(0) = y_0'$.
14. Show that $y = \sin x$ is a solution of
$$y'' + (k\sin^2 x)y' + (1 - k\cos x\sin x)y = 0$$
for any value of the constant k. Show that if $0 < k < 2$ then $(1 - k\cos x\sin x) > 0$ and $k\sin^2 x \geq 0$. Thus observe that even though the coefficients of this variable coefficient differential equation are nonnegative (and the coefficient of y' is zero only at the points $x = 0, \pi, 2\pi, \ldots$), it still has a solution that does not approach zero as $x \to \infty$. Thus we observe a not unusual situation in the theory of differential equations: apparently very similar equations can have quite different properties.

*15. Show that any complex number $a + ib$ can be written in the form $Ae^{i\theta}$, where A and θ are positive real numbers, and $0 \leq \theta < 2\pi$.
Hint: Expand $Ae^{i\theta}$ using Euler's formula and solve for A and θ in terms of a and b.

*16. Defining the two possible square roots of a complex number $Ae^{i\theta}$ as $\pm A^{1/2}e^{i\theta/2}$, compute the square roots of $1 + i, 1 - i, i$.

*17. Solve
(a) $y'' + iy' + 2y = 0$ (b) $y'' + 2y' + iy = 0$
Note that the roots of the characteristic equations are not complex conjugates. Does the real or imaginary part of either of the solutions satisfy the differential equation?

Change of Variables. Often a differential equation with variable coefficients,
$$y'' + p(x)y' + q(x)y = 0, \qquad \text{(i)}$$
can be put in a more suitable form for finding a solution by making a change of the independent and/or dependent variables. We explore these ideas in Problems 18 through 23. In particular, in Problem 18 we determine conditions under which Eq. (i) can be transformed into a differential equation with constant coefficients—a differential equation which we now know how to solve. This result is applied to several examples in Problems 19 through 21.

*18. In this problem we determine conditions on p and q such that Eq. (i) can be transformed into one with constant coefficients by a change of the independent variable. Let z be the new independent variable, $z = u(x)$, where for the moment the relationship between z and x is unspecified. Show that

(a) $$\frac{dy}{dx} = \frac{dz}{dx}\frac{dy}{dz}$$
$$\frac{d^2y}{dx^2} = \left(\frac{dz}{dx}\right)^2\frac{d^2y}{dz^2} + \frac{d^2z}{dx^2}\frac{dy}{dz}$$

(b) The differential equation becomes

$$\left(\frac{dz}{dx}\right)^2 \frac{d^2y}{dz^2} + \left(\frac{d^2z}{dx^2} + p(x)\frac{dz}{dx}\right)\frac{dy}{dz} + q(x)y = 0$$

(c) The equation of part (b) will have constant coefficients if we choose

$$z = u(x) = \int^x [q(t)]^{1/2}\,dt,$$

provided that

$$\frac{z'' + p(x)z'}{q(x)} = \frac{q'(x) + 2p(x)q(x)}{2[q(x)]^{3/2}}$$

is a constant. Notice that for the transformation $z = u(x)$ to be real valued it is necessary that $q(x) \geq 0$. This means that in the original differential equation $q(x)$ must be of one sign; if $q(x) < 0$ multiply the differential equation by -1.

Hence conclude that a necessary and sufficient condition for transforming $y'' + p(x)y' + q(x)y = 0$ into an equation with constant coefficients by a change of the independent variable is that the function $(q' + 2pq)/q^{3/2}$ be a constant.

*19. Using the result of Problem 18, determine whether each of the following equations can be transformed into an equation with constant coefficients by a change of the independent variable. If so, find the general solution of the given equation.

(a) $y'' + xy' + e^{-x^2}y = 0, \qquad -\infty < x < \infty$
(b) $y'' + 3xy' + x^2y = 0, \qquad -\infty < x < \infty$
(c) $xy'' + (x^2 - 1)y' + x^3y = 0, \qquad 0 < x < \infty$

*20. **Euler equations.** Using the result of Problem 18, show that it is always possible to transform an equation of the form

$$x^2y'' + \alpha xy' + \beta y = 0, \qquad x > 0,$$

where α and β are real constants, into an equation with constant coefficients by letting $z = \ln x$. An equation of this type is called an Euler equation; it is discussed in Section 4.4.

*21. Use the method of Problem 20 to solve each of the following Euler equations. Assume $x > 0$.

(a) $x^2y'' + xy' + y = 0$
(b) $x^2y'' + 4xy' + 2y = 0$
(c) $x^2y'' - 3xy' + 4y = 0$
(d) $x^2y'' - 4xy' - 6y = 0$

22. By suitable changes of the dependent variable we can establish a relation between the second order linear equation $y'' + p(x)y' + q(x)y = 0$ and a first order equation.
(a) Set $y'/y = -u$ and determine y in terms of u. Then show that u satisfies

$$du/dx = q(x) - p(x)u + u^2.$$

This equation is known as a Riccati equation; it is discussed in Problem 35 of Section 2.11. Note that the price for going from a second order to a first order equation is that of going from a linear to a nonlinear equation.
(b) Now consider the general Riccati equation

$$dw/dx = q_0(x) + q_1(x)w + q_2(x)w^2.$$

Show that the transformation $w = -y'/yq_2$ leads to the second order linear homogeneous equation

$$q_2(x)y'' - [q_2'(x) + q_1(x)q_2(x)]y' + q_2^2(x)q_0(x)y = 0.$$

*23. **Normal form.** In developing certain aspects of the theory of the second order linear equation

$$y'' + p(x)y' + q(x)y = 0 \tag{i}$$

it is often desirable to transform the dependent variable so that the first derivative term is missing. If we set $y = u(x)v(x)$ in Eq. (i), show that it is possible to choose $v(x)$ so that Eq. (i) takes the form $u'' + f(x)u = 0$. This is known as the normal form of a second order linear homogeneous equation. Determine $v(x)$ and $f(x)$ in terms of $p(x)$ and $q(x)$.

3.6 The Nonhomogeneous Problem

In the previous sections of this chapter we discussed the general theory of second order linear homogeneous differential equations, and in the case of constant coefficients showed how to construct solutions. We now turn our attention to solving the nonhomogeneous differential equation

$$L[y] = y'' + p(x)y' + q(x)y = g(x). \tag{1}$$

Throughout the discussion we assume that the functions p, q, and g are continuous on the interval of interest.

In modern engineering the problem of solving Eq. (1) is often referred to as an input–output problem. Supposing for the moment that the differential equation (1) represents a certain mechanical or electrical system, it is natural to refer to the solution $y = \phi(x)$ as the output of the system. The coefficients p and q are determined by the physical mechanism and the nonhomogeneous term g is the input function. The problem of solving Eq. (1) corresponds to determining the output of the system for different input functions.

Before considering particular cases of Eq. (1), we will prove several simple but very useful general results that will simplify our succeeding work.

Theorem 3.11. *The difference of any two solutions of the differential equation* (1),

$$L[y] = y'' + p(x)y' + q(x)y = g(x),$$

is a solution of the corresponding homogeneous differential equation

$$L[y] = y'' + p(x)y' + q(x)y = 0. \tag{2}$$

To prove this theorem suppose that the functions u_1 and u_2 are solutions of Eq. (1). Then

$$L[u_1] = g,$$

and

$$L[u_2] = g.$$

Subtracting the second equation from the first equation gives

$$L[u_1] - L[u_2] = 0,$$

or, since L is a linear operator,

$$L[u_1 - u_2] = 0,$$

which is the desired result. With the aid of this theorem we can prove the following important theorem.

Theorem 3.12. *Given one solution y_p of the nonhomogeneous linear differential equation* (1)

$$L[y] = y'' + p(x)y' + q(x)y = g(x),$$

then any solution $y = \phi(x)$ of this equation can be expressed as

$$\phi(x) = y_p(x) + c_1y_1(x) + c_2y_2(x), \tag{3}$$

where y_1 and y_2 are linearly independent solutions of the corresponding homogeneous equation.

To prove this theorem, observe that according to Theorem 3.11 $\phi - y_p$ is a solution of the homogeneous equation. Thus, according to Theorem 3.4 of Section 3.2, $\phi - y_p$ can be expressed as a linear combination of y_1 and y_2, which establishes the theorem.

It is customary to refer to the linear combination (3) as the *general solution* of Eq. (1).

Consequently, to find the general solution of Eq. (1) we must find the general solution of the homogeneous equation (2) and then find *any* solution of the nonhomogeneous equation. As we could have anticipated (see Section 3.1), the general solution of Eq. (1) involves two arbitrary constants; hence to specify a unique solution of Eq. (1) it is necessary to specify two additional conditions, namely, the initial conditions $y(x_0) = y_0$ and $y'(x_0) = y_0'$.

The general solution of the homogeneous equation (2) is often referred to as the *complementary solution* and denoted by y_c. Thus $y_c(x) = c_1y_1(x) + c_2y_2(x)$. A solution of the nonhomogeneous equation y_p is usually called a *particular solution.*[10] Notice that y_p is not uniquely determined since if Y is a solution of the nonhomogeneous equation then Y plus any multiple of y_1 or y_2 is still a solution of the nonhomogeneous equation. It follows from Theorem 3.12 that the

[10] This is rather an unfortunate usage, since the term particular solution may also refer to a solution satisfying prescribed initial conditions. The meaning is usually clear from the context.

general solution of the nonhomogeneous equation (1) is

$$y = y_c(x) + y_p(x). \tag{4}$$

EXAMPLE 1

Given the differential equation

$$y'' + 4y' + 4y = 8x^2 + 2, \tag{5}$$

verify that $y_p(x) = 2x^2 - 4x + \frac{7}{2}$ is a particular solution, and find the general solution.

To verify that y_p is a particular solution we substitute y_p in the differential equation:

$$\begin{aligned}
&y_p'' + 4y_p' + 4y_p \\
&\quad = \left(2x^2 - 4x + \tfrac{7}{2}\right)'' + 4\left(2x^2 - 4x + \tfrac{7}{2}\right)' + 4\left(2x^2 - 4x + \tfrac{7}{2}\right) \\
&\quad = 4 + 4(4x - 4) + 4\left(2x^2 - 4x + \tfrac{7}{2}\right) \\
&\quad = 8x^2 + (16 - 16)x + (4 - 16 + 14) \\
&\quad = 8x^2 + 2.
\end{aligned}$$

Hence y_p is a particular solution. Next we consider the corresponding homogeneous equation

$$y'' + 4y' + 4y = 0$$

and set $y = e^{rx}$. The characteristic equation is $r^2 + 4r + 4 = (r + 2)^2 = 0$ so $r = -2, -2$. Hence, the general solution of the homogeneous equation, or the complementary solution, is $y_c(x) = c_1 e^{-2x} + c_2 x e^{-2x}$.

The general solution of Eq. (5) is

$$y = c_1 e^{-2x} + c_2 x e^{-2x} + 2x^2 - 4x + \tfrac{7}{2}. \tag{6}$$

In many problems the nonhomogeneous term g may be very complicated; however, if g can be expressed as the sum of a finite number of functions we can make use of the linearity of the differential equation to replace the original problem by several simpler ones. For example, suppose that it is possible to write $g(x)$ as $g_1(x) + g_2(x) + \cdots + g_m(x)$; then Eq. (1) becomes

$$L[y] = y'' + p(x)y' + q(x)y = g_1(x) + g_2(x) + \cdots + g_m(x). \tag{7}$$

If we can find particular solutions y_{p_i} of the differential equations

$$L[y] = g_i(x), \qquad i = 1, 2, \ldots, m, \tag{8}$$

then it follows by direct substitution that

$$y_p(x) = y_{p_1}(x) + y_{p_2}(x) + \cdots + y_{p_m}(x) \tag{9}$$

is a particular solution of Eq. (7). Hence the general solution of Eq. (7) is

$$y = y_c(x) + y_{p_1}(x) + \cdots + y_{p_m}(x). \tag{10}$$

In general, it is easier to find solutions of Eqs. (8) and to add the results than to solve Eq. (7) as it stands. This method of constructing the solution of a complicated problem by adding solutions of simpler problems is known as the method of superposition.

EXAMPLE 2

Find the general solution of

$$y'' + 4y = 1 + x + \sin x. \tag{11}$$

First consider the corresponding homogeneous equation, $y'' + 4y = 0$. Substituting $y = e^{rx}$ yields the characteristic equation $r^2 + 4 = 0$. Thus the complementary solution is

$$y_c(x) = c_1 \cos 2x + c_2 \sin 2x.$$

To determine a particular solution of Eq. (11) we superimpose particular solutions of $y'' + 4y = 1$, $y'' + 4y = x$, and $y'' + 4y = \sin x$. It can be readily verified that $\frac{1}{4}$, $\frac{1}{4}x$, and $\frac{1}{3}\sin x$ are particular solutions of these equations, respectively. Thus the general solution of Eq. (11) is

$$y = c_1 \cos 2x + c_2 \sin 2x + \tfrac{1}{4} + \tfrac{1}{4}x + \tfrac{1}{3}\sin x.$$

In Sections 3.6.1 and 3.6.2 we show how to find particular solutions of nonhomogeneous equations.

3.6.1 The Method of Undetermined Coefficients

A number of methods can be used to obtain particular solutions of second order nonhomogeneous differential equations. When applicable, the method of undetermined coefficients is one of the simplest. Basically, the method consists of making an intelligent choice as to the form of the particular solution and then substituting this function, which in general involves one or more unknown coefficients, into the differential equation. For this method to be successful we must be able to determine the unknown coefficients so that the function actually does satisfy the differential equation.

Clearly such a method depends to a considerable extent on the ability to discover in advance the general form of a particular solution. For a completely arbitrary differential equation very little can be said. However, if the equation is of the form

$$ay'' + by' + cy = g(x), \tag{1}$$

where a, b, and c are real constants, and the nonhomogeneous term $g(x)$ is an exponential function ($e^{\alpha x}$), or a polynomial ($a_0 x^n + \cdots + a_n$), or sinusoidal in character ($\sin \beta x$ or $\cos \beta x$), then definite rules can be given for the determination of a particular solution by the method of undetermined coefficients. These rules also cover the more general case in which $g(x)$ is a product of terms of the above types, such as

$$g(x) = e^{\alpha x}\left(a_0 x^n + a_1 x^{n-1} + \cdots + a_n\right)\begin{cases} \cos \beta x \\ \sin \beta x. \end{cases} \tag{2}$$

Notice that any other product of exponentials, polynomials, and sines and cosines is equivalent to a sum of terms of the type (2). In particular, the product of two or more sinusoidal terms can always be reduced to a sum of individual sinusoidal terms by the use of trigonometric identities. We can treat the case in which $g(x)$ is a sum of terms of type (2) by using the method of superposition that was discussed in the last section. For more general nonhomogeneous terms than (2) or for equations with variable coefficients the method of undetermined coefficients is rarely useful. In this case we can use the method of variation of parameters discussed in the next section.

Before taking up the general procedure for the method of undetermined coefficients let us consider a few simple examples which will illustrate the method.

EXAMPLE 1

Find a particular solution of the differential equation

$$y'' - 3y' - 4y = 3e^{2x}. \tag{3}$$

We want to find a function y_p such that the sum of its second derivative minus three times its first derivative minus four times the function itself adds up to $3e^{2x}$. The obvious function to consider for $y_p(x)$ is e^{2x}. So we *assume* that $y_p(x)$ is of the form

$$y_p(x) = Ae^{2x},$$

where A is, for the moment, an unknown constant. Then

$$y_p'(x) = 2Ae^{2x} \quad \text{and} \quad y_p''(x) = 4Ae^{2x}.$$

Substituting for y, y', and y'' in Eq. (3) and collecting terms gives

$$(4A - 6A - 4A)e^{2x} = 3e^{2x}.$$

This equation will be satisfied if we choose $A = -\frac{1}{2}$. Hence a particular solution is

$$y_p(x) = -\tfrac{1}{2}e^{2x}.$$

EXAMPLE 2

Find a particular solution of the differential equation

$$y'' - 3y' - 4y = 2\sin x. \tag{4}$$

By analogy with Example 1, we assume $y_p(x) = A\sin x$ where A is a constant to be determined. On substituting in Eq. (4) we obtain

$$-A\sin x - 3A\cos x - 4A\sin x = 2\sin x.$$

However, because of the $\cos x$ term there is no choice of the constant A that satisfies this equation. This suggests that we must modify our initial choice for $y_p(x)$ to include a $\cos x$ term. Thus we assume that $y_p(x)$ is of the form

$$y_p(x) = A\sin x + B\cos x,$$

where A and B are unknown constants to be determined. Then

$$y_p'(x) = A\cos x - B\sin x \quad \text{and} \quad y_p''(x) = -A\sin x - B\cos x.$$

Substituting for y, y', and y'' in Eq. (4) and collecting terms gives

$$(-A + 3B - 4A)\sin x + (-B - 3A - 4B)\cos x = 2\sin x.$$

This equation will be satisfied identically if and only if

$$-5A + 3B = 2 \quad \text{and} \quad -3A - 5B = 0.$$

Hence $A = -\frac{5}{17}$, $B = \frac{3}{17}$, and a particular solution of Eq. (4) is

$$y_p(x) = \tfrac{1}{17}(3\cos x - 5\sin x).$$

EXAMPLE 3

Find a particular solution of the differential equation

$$y'' - 3y' - 4y = 4x^2. \tag{5}$$

It is natural to try $y_p(x) = Ax^2$ where A is a constant to be determined. Then $y_p'(x) = 2Ax$, $y_p''(x) = 2A$, and substituting in Eq. (5) gives

$$2A - 6Ax - 4Ax^2 = 4x^2.$$

If this equation is to be satisfied for all x the coefficients of like powers of x on each side of the equation must be identical. This leads to three equations, and there is no choice of A that will satisfy all three. Hence it is impossible to find a particular solution of Eq. (5) of the form Ax^2.

However, if we think of the nonhomogeneous term $4x^2$ in Eq. (5) as the polynomial $4x^2 + 0x + 0$ it now appears reasonable to assume that the form of $y_p(x)$ is $Ax^2 + Bx + C$, where A, B, and C are constants to be determined. Substituting for y, y', and y'' in Eq. (5) and equating like powers of x on both sides of the equation gives three simultaneous linear algebraic nonhomogeneous equations for the three unknowns A, B, and C. Their solution is $A = -1$, $B = \frac{3}{2}$, and $C = -\frac{13}{8}$; hence a particular solution of Eq. (5) is

$$y_p(x) = -x^2 + \tfrac{3}{2}x - \tfrac{13}{8}.$$

The reader may wonder what would happen if a higher degree polynomial, such as $Ax^4 + Bx^3 + Cx^2 + Dx + E$, were assumed for $y_p(x)$. The answer is that all coefficients beyond the quadratic term would turn out to be zero. Thus, with certain exceptions to be noted later, it is unnecessary to assume for $y_p(x)$ a polynomial of higher degree than the degree of the polynomial in the nonhomogeneous term.

EXAMPLE 4

Find a particular solution of the differential equation

$$y'' - 3y' - 4y = 3xe^{2x}. \tag{6}$$

Relying on the experience gained from Examples 1 and 3, we assume that $y_p(x) = e^{2x}(Ax + B)$. Note that we did not just take the term $e^{2x}Ax$, but that we also included the constant term B in the polynomial multiplying e^{2x}. Then

$$y_p'(x) = e^{2x}[A + 2(Ax + B)] = e^{2x}[2Ax + (A + 2B)]$$

and

$$y_p''(x) = e^{2x}[4Ax + 2(A + 2B) + 2A] = e^{2x}[4Ax + 4(A + B)].$$

Substituting for y_p, y_p', and y_p'' in Eq. (6) and collecting terms, we obtain

$$[4Ax + (4A + 4B) - 6Ax - 3(A + 2B) - 4(Ax + B)]e^{2x} = 3xe^{2x},$$
$$[-6Ax + (A - 6B)]e^{2x} = 3xe^{2x}.$$

Hence

$$-6A = 3 \quad \text{and} \quad A - 6B = 0,$$

so $A = -\frac{1}{2}$ and $B = -\frac{1}{12}$; and a particular solution of Eq. (6) is

$$y_p(x) = e^{2x}(-\tfrac{1}{2}x - \tfrac{1}{12}).$$

EXAMPLE 5

Find a particular solution of the differential equation

$$y'' - 3y' - 4y = e^{-x}. \tag{7}$$

It is natural to assume that $y_p(x) = Ae^{-x}$. Substituting in Eq. (7) gives

$$(A + 3A - 4A)e^{-x} = e^{-x},$$

or

$$0 \cdot Ae^{-x} = e^{-x},$$

and we cannot determine A. Hence the particular solution is not of the form Ae^{-x}. The difficulty stems from the fact that e^{-x} is a solution of the corresponding homogeneous equation. How can we modify our choice for $y_p(x)$ so that when we substitute it in the differential equation we will obtain an e^{-x}? The simplest function, other than e^{-x} itself, which on differentiation leads to an e^{-x} is xe^{-x}. Thus we might try $y_p(x) = e^{-x}(Ax + B)$. However, we do not need the Be^{-x} since this is a solution of the corresponding homogeneous equation and cannot help us in satisfying the nonhomogeneous differential equation. Hence we try $y_p(x) = Axe^{-x}$ so

$$y_p'(x) = e^{-x}(-Ax + A) \quad \text{and} \quad y_p''(x) = e^{-x}(Ax - 2A).$$

Substituting for y_p, y_p', and y_p'' in Eq. (7) and collecting terms, we obtain

$$[Ax - 2A - 3(-Ax + A) - 4Ax]e^{-x} = e^{-x}$$

$$[0 \cdot Ax - 5A]e^{-x} = e^{-x},$$

so $A = -\frac{1}{5}$. Hence a particular solution of Eq. (7) is $y_p(x) = -\frac{1}{5}xe^{-x}$.

From the preceding examples we can see the general idea of the method of undetermined coefficients. We assume a $y_p(x)$ such that $y_p'(x)$ and $y_p''(x)$ do not introduce any new types of terms and such that $g(x)$ is included in the set of terms generated by $ay_p'' + by_p' + cy_p$. Moreover, we try to keep the set of terms as small as possible. We encounter difficulties if our choice for $y_p(x)$ is a solution of the corresponding homogeneous equation. To produce a term that is not a solution of the homogeneous equation, we multiply our initial choice by x (as in Example 5) or, if necessary, by x^2. For a second order equation it is never necessary to multiply by a higher power than this. The procedure is summarized in Table 3.1. Notice that to determine whether our initial choice for y_p is a solution of the corresponding homogeneous equation, hence whether it must be multiplied by a power of x, we must know the general solution of the homogeneous equation. Thus the first step is always to solve the homogeneous problem.

In using the method of undetermined coefficients if one assumes too little, or the wrong terms, for y_p then a contradiction will soon be reached. Usually this

TABLE 3.1 THE PARTICULAR SOLUTION OF $ay'' + by' + cy = g(x)$

$g(x)$	$y_p(x)$
$P_n(x) = a_0x^n + a_1x^{n-1} + \cdots + a_n$	$x^s(A_0x^n + A_1x^{n-1} + \cdots + A_n)$
$P_n(x)e^{\alpha x}$	$x^s(A_0x^n + A_1x^{n-1} + \cdots + A_n)e^{\alpha x}$
$P_n(x)e^{\alpha x}\begin{cases}\sin\beta x\\ \cos\beta x\end{cases}$	$x^s\left[\left(A_0x^n + A_1x^{n-1} + \cdots + A_n\right)e^{\alpha x}\cos\beta x + \left(B_0x^n + B_1x^{n-1} + \cdots + B_n\right)e^{\alpha x}\sin\beta x\right]$

Notes. Here s is the smallest nonnegative integer ($s = 0$, 1, or 2) which will insure that no term in $y_p(x)$ is a solution of the corresponding homogeneous equation. Equivalently, for the three cases, s is the number of times 0 is a root of the characteristic equation, α is a root of the characteristic equation, and $\alpha + i\beta$ is a root of the characteristic equation, respectively.

contradiction points the way to correct the initial assumption. If one assumes too much, then the only harm is that it will be necessary to do some extra work and some coefficients will turn out to be zero, but one does obtain the correct answer. Thus in this sense the method of undetermined coefficients is self correcting.

These results, coupled with the fact that the principle of superposition can be used to find the corresponding particular solution when the nonhomogeneous term consists of the sum of several functions, allow us to solve fairly rapidly a wide class of linear constant coefficient nonhomogeneous differential equations.

EXAMPLE 6

Using the method of undetermined coefficients, determine the correct form for $y_p(x)$ for the differential equation

$$y'' + 4y = xe^x + x\sin 2x. \tag{8}$$

First we solve the corresponding homogeneous equation. We readily find that

$$y_c(x) = c_1 \cos 2x + c_2 \sin 2x.$$

To construct a particular solution y_p of Eq. (8) we use the superposition principle, and consider the problems

$$y'' + 4y = xe^x \tag{9}$$

and

$$y'' + 4y = x\sin 2x \tag{10}$$

separately. For the first problem we assume $y_p(x) = (A_0 x + A_1)e^x$. Since e^x is not a solution of the homogeneous equation it is not necessary to revise our initial guess. For the second problem we assume $y_p(x) = (B_0 x + B_1)\cos 2x + (C_0 x + C_1)\sin 2x$, but since $\cos 2x$ and $\sin 2x$ are solutions of the homogeneous equation it is necessary to multiply their polynomial coefficients by x. Hence the correct form for $y_p(x)$ for Eq. (8) is

$$y_p(x) = (A_0 x + A_1)e^x + \left(B_0 x^2 + B_1 x\right)\cos 2x + \left(C_0 x^2 + C_1 x\right)\sin 2x.$$

In determining the coefficients in this expression it is easier to compute the particular solution corresponding to xe^x, Eq. (9), and $x\sin 2x$, Eq. (10), separately than to try to do all the algebra at one time.

We now establish the results given in Table 3.1 by considering the general case in which $g(x)$ has one of the following forms:

$$g(x) = \begin{cases} P_n(x) = a_0 x^n + a_1 x^{n-1} + \cdots + a_n \\ e^{\alpha x} P_n(x) \\ e^{\alpha x} P_n(x) \sin \beta x \\ e^{\alpha x} P_n(x) \cos \beta x. \end{cases}$$

$g(x) = P_n(x)$. In this case Eq. (1) becomes

$$ay'' + by' + cy = a_0x^n + a_1x^{n-1} + \cdots + a_n. \tag{11}$$

To obtain a particular solution we assume

$$y_p(x) = A_0x^n + A_1x^{n-1} + \cdots + A_{n-2}x^2 + A_{n-1}x + A_n. \tag{12}$$

Substituting in Eq. (11) we obtain

$$a\left[n(n-1)A_0x^{n-2} + \cdots + 2A_{n-2}\right] + b\left(nA_0x^{n-1} + \cdots + A_{n-1}\right)$$
$$+c\left(A_0x^n + A_1x^{n-1} + \cdots + A_n\right) = a_0x^n + \cdots + a_n. \tag{13}$$

Equating the coefficients of like powers of x gives

$$cA_0 = a_0,$$
$$cA_1 + nbA_0 = a_1,$$
$$\vdots$$
$$cA_n + bA_{n-1} + 2aA_{n-2} = a_n.$$

Provided $c \neq 0$ the solution of the first equation is $A_0 = a_0/c$, and the remaining equations determine $A_1, A_2, \ldots, A_n$ successively. If $c = 0$, but $b \neq 0$, the polynomial on the left side of Eq. (13) is of degree $n - 1$, and we cannot satisfy Eq. (13). To insure that $ay_p''(x) + by_p'(x)$ will be a polynomial of degree n we must choose $y_p(x)$ to be a polynomial of degree $n + 1$. Hence we assume

$$y_p(x) = x\left(A_0x^n + \cdots + A_n\right).$$

There is no constant term in this expression for $y_p(x)$, but there is no need to include such a term since when $c = 0$ a constant is a solution of the homogeneous differential equation. Since $b \neq 0$ we have $A_0 = a_0/b(n+1)$, and the coefficients $A_1, \ldots, A_n$ can be determined similarly. If both c and b are zero, we assume

$$y_p(x) = x^2\left(A_0x^n + \cdots + A_n\right).$$

The term $ay_p''(x)$ gives rise to a term of degree n, and we can proceed as before.[11] Again the constant and linear terms in $y_p(x)$ are omitted since in this case they are both solutions of the homogeneous equation.

$g(x) = e^{\alpha x}P_n(x)$. The problem of determining a particular solution of

$$ay'' + by' + cy = e^{\alpha x}P_n(x) \tag{14}$$

[11] The student should note that in the case $c = 0,\ b \neq 0$, Eq. (11) can be integrated once, giving a first order linear equation whose nonhomogeneous term is a polynomial of degree $n + 1$. A particular solution of this equation could then be found by the method of undetermined coefficients with $y_p(x)$ a polynomial of degree $n + 1$. If both c and b vanish, then Eq. (11) can be integrated immediately, the solution being a polynomial of degree $n + 2$.

can be reduced to the one just solved.[12] Let

$$y_p(x) = e^{\alpha x}u(x);$$

then

$$y_p'(x) = e^{\alpha x}\left[u'(x) + \alpha u(x)\right]$$

and

$$y_p''(x) = e^{\alpha x}\left[u''(x) + 2\alpha u'(x) + \alpha^2 u(x)\right].$$

Substituting for y, y', and y'' in Eq. (14), canceling the factor $e^{\alpha x}$, and collecting terms give

$$au''(x) + (2a\alpha + b)u'(x) + (a\alpha^2 + b\alpha + c)u(x) = P_n(x). \tag{15}$$

The determination of a particular solution of Eq. (15) is precisely the problem that we just solved. If $a\alpha^2 + b\alpha + c$ is not zero we assume $u(x) = (A_0x^n + \cdots + A_n)$; hence a particular solution of Eq. (14) is of the form

$$y_p(x) = e^{\alpha x}\left(A_0x^n + A_1x^{n-1} + \cdots + A_n\right). \tag{16}$$

On the other hand, if $a\alpha^2 + b\alpha + c$ is zero but $(2a\alpha + b)$ is not, we must take $u(x)$ to be of the form $x(A_0x^n + \cdots + A_n)$. The corresponding form for $y_p(x)$ is x times the expression in the right member of Eq. (16). It should be noted that the vanishing of $a\alpha^2 + b\alpha + c$ implies that $e^{\alpha x}$ is a solution of the homogeneous equation. If both $a\alpha^2 + b\alpha + c$ and $2a\alpha + b$ are zero (and this implies that both $e^{\alpha x}$ and $xe^{\alpha x}$ are solutions of the homogeneous equation) then the correct form for $u(x)$ is $x^2(A_0x^n + \cdots + A_n)$; hence $y_p(x)$ is x^2 times the expression on the right side of Eq. (16).

$g(x) = e^{\alpha x}P_n(x)\cos\beta x$ or $e^{\alpha x}P_n(x)\sin\beta x$. The two cases are similar; consider the latter case. We can reduce this problem to the previous one by recalling that $\sin\beta x = (e^{i\beta x} - e^{-i\beta x})/2i$. Hence $g(x)$ is of the form

$$g(x) = P_n(x)\frac{e^{(\alpha+i\beta)x} - e^{(\alpha-i\beta)x}}{2i}$$

and we should choose

$$y_p(x) = e^{(\alpha+i\beta)x}\left(A_0x^n + \cdots + A_n\right) + e^{(\alpha-i\beta)x}\left(B_0x^n + \cdots + B_n\right),$$

or equivalently

$$y_p(x) = e^{\alpha x}\left(A_0x^n + \cdots + A_n\right)\cos\beta x + e^{\alpha x}\left(B_0x^n + \cdots + B_n\right)\sin\beta x.$$

Usually the latter form is preferred. If $\alpha \pm i\beta$ satisfy the characteristic equation corresponding to the homogeneous equation we must, of course, multiply each of the polynomials by x to increase their degree by one.

[12] One of a mathematician's favorite devices is to reduce a new problem to one that has already been solved.

If the nonhomogeneous term involves expressions such as $e^{\alpha x}\cos\beta x$ and $e^{\alpha x}\sin\beta x$ it is convenient to treat these together, since each one individually gives rise to the same form for a particular solution. For example, if $g(x) = x\sin x + 2\cos x$, the form for $y_p(x)$ would be

$$(A_0 x + A_1)\sin x + (B_0 x + B_1)\cos x$$

provided that $\sin x$ and $\cos x$ were not solutions of the homogeneous equation.

PROBLEMS

Using the method of undetermined coefficients to find a particular solution of the nonhomogeneous equation, find the general solution of the following differential equations. Where specified find the solution satisfying the given initial conditions.

1. $y'' + y' - 2y = 2x, \quad y(0) = 0, \quad y'(0) = 1$
2. $2y'' - 4y' - 6y = 3e^{2x}$
3. $y'' + 4y = x^2 + 3e^x, \quad y(0) = 0, \quad y'(0) = 2$
4. $y'' + 2y' = 3 + 4\sin 2x$
5. $y'' + 9y = x^2 e^{3x} + 6$
6. $y'' - 2y' + y = xe^x + 4; \quad y(0) = 1, \quad y'(0) = 1$
7. $2y'' + 3y' + y = x^2 + 3\sin x$
8. $y'' + y = 3\sin 2x + x\cos 2x$
9. $y'' + 2y' + y = e^x\cos x$
10. $u'' + \omega_0^2 u = \cos\omega t, \quad \omega^2 \neq \omega_0^2$
11. $u'' + \omega_0^2 u = \cos\omega_0 t$
12. $u'' + \mu u' + \omega_0^2 u = \cos\omega t, \quad \mu^2 - 4\omega_0^2 < 0, \quad \mu > 0$
13. $y'' + y' + y = \sin^2 x$
14. $y'' + y' + 4y = 2\sinh x \qquad$ *Hint:* $\sinh x = (e^x - e^{-x})/2$
15. $y'' - y' - 2y = \cosh 2x \qquad$ *Hint:* $\cosh x = (e^x + e^{-x})/2$

In Problems 16 through 22 determine a suitable form for $y_p(x)$ if the method of undetermined coefficients is to be used. Do not evaluate the constants.

16. $y'' + 3y' = 2x^4 + x^2e^{-3x} + \sin 3x$
17. $y'' + y = x(1 + \sin x)$
18. $y'' - 5y' + 6y = e^x\cos 2x + e^{2x}(3x + 4)\sin x$
19. $y'' + 2y' + 2y = 3e^{-x} + 2e^{-x}\cos x + 4e^{-x}x^2\sin x$
20. $y'' - 4y' + 4y = 2x^2 + 4xe^{2x} + x\sin 2x$
21. $y'' + 4y = x^2\sin 2x + (6x + 7)\cos 2x$
22. $y'' + 3y' + 2y = e^x(x^2 + 1)\sin 2x + 3e^{-x}\cos x + 4e^x$
23. Determine the general solution of

$$y'' + \lambda^2 y = \sum_{m=1}^{N} a_m \sin m\pi x,$$

where $\lambda > 0$ and $\lambda \neq m\pi$, $m = 1, 2, \ldots, N$.

24. In many physical problems the input function (that is, the nonhomogeneous term) may be specified by different formulas in different time periods. As a simple example of such a problem determine the solution $y = \phi(t)$ of

$$y'' + y = \begin{cases} t, & 0 \le t \le \pi, \\ \pi e^{\pi - t}, & t > \pi, \end{cases}$$

satisfying the initial conditions $y(0) = 0$ and $y'(0) = 1$, and the requirement that y and y' be continuous for all t. Plot the nonhomogeneous term and the solution $y = \phi(t)$ as a function of time.
Hint: First solve the initial value problem for $t \leq \pi$, then solve for $t > \pi$ determining the arbitrary constants so that y and y' are continuous at $t = \pi$.

Behavior of solutions as $x \to \infty$. In Problems 25 and 26 we continue the discussion started with Problems 12, 13, and 14 of Section 3.5.1. We study the differential equation

$$ay'' + by' + cy = g(x), \tag{i}$$

where a, b, and c are positive.

25. If $Y_1(x)$ and $Y_2(x)$ are solutions of Eq. (i), show that $Y_1(x) - Y_2(x) \to 0$ as $x \to \infty$. Is this result true if $b = 0$?
26. If $g(x) = d$, a constant, show that every solution of Eq. (i) approaches d/c as $x \to \infty$. A stronger result, namely that if $g(x) \to d$ as $x \to \infty$ then $y \to d/c$ as $x \to \infty$, can be established but the proof is beyond the level of this text.

*27. In this problem we sketch an alternate procedure[13] for solving the differential equation

$$y'' + by' + cy = (D^2 + bD + c)y = g(x). \tag{i}$$

Let r_1 and r_2 be the characteristic roots of the corresponding homogeneous differential equation. Note that r_1 and r_2 may be real and unequal, real and equal, or complex conjugate numbers.
(a) Verify that Eq. (i) can be written in the factored form

$$(D - r_1)(D - r_2)y = g(x),$$

where $r_1 + r_2 = -b$ and $r_1 r_2 = c$.
(b) Let $(D - r_2)y = u$ and show that solving Eq. (i) is equivalent to solving sequentially the following two first order linear equations:

$$(D - r_1)u = g(x), \tag{ii}$$

$$(D - r_2)y = u(x). \tag{iii}$$

(c) Show that y is given by

$$y = c_2 e^{r_2 x} + e^{r_2 x} \int^x e^{-r_2 t} u(t)\, dt$$

where

$$u(x) = c_1 e^{r_1 x} + e^{r_1 x} \int^x e^{-r_1 t} g(t)\, dt.$$

*28. Use the method of Problem 27 to solve each of the following equations.

(a) $y'' - 3y' - 4y = 3e^{2x}$ (see Example 1)
(b) $2y'' - 4y' - 6y = 3e^{2x}$ (see Problem 2)
(c) $y'' + 2y' = 3 + 4\sin 2x$ (see Problem 4)
(d) $y'' + 2y' + y = e^x \cos x$ (see Problem 9)

[13] R. S. Luthar, "Another Approach to a Standard Differential Equation," *Two Year College Mathematics Journal 10* (1979) pp. 200–201. Also see D. C. Sandell and F. M. Stein, "Factorization of Operators of Second Order Linear Homogeneous Ordinary Differential Equations," *Two Year College Mathematics Journal 8* (1977), pp. 132–141, for a more general discussion of factoring operators.

3.6.2 The Method of Variation of Parameters

In Section 3.6.1 we discussed a simple method for determining particular solutions of nonhomogeneous differential equations with constant coefficients provided the nonhomogeneous term is of a suitable form. In this section we consider a *general method* for determining a particular solution of the equation

$$y'' + p(x)y' + q(x)y = g(x) \tag{1}$$

where the functions p, q, and g are continuous on the interval of interest. To use this method, known as the method of *variation of parameters*, it is necessary to know a fundamental set of solutions of the corresponding homogeneous equation

$$y'' + p(x)y' + q(x)y = 0. \tag{2}$$

Suppose y_1 and y_2 are linearly independent solutions of the homogeneous equation (2). Then the general solution of Eq. (2) is

$$y_c(x) = c_1y_1(x) + c_2y_2(x).$$

The method of variation of parameters involves the replacement of the constants c_1 and c_2 by functions u_1 and u_2. We then seek to determine the two functions u_1 and u_2 so that

$$y_p(x) = u_1(x)y_1(x) + u_2(x)y_2(x) \tag{3}$$

satisfies the nonhomogeneous differential equation (1). The importance of this method is due to the fact that it is possible to determine the functions u_1 and u_2 in a simple manner. To determine the two functions u_1 and u_2 we need two conditions. One condition on u_1 and u_2 stems from the fact that y_p must satisfy Eq. (1). A second condition can be imposed arbitrarily, and is selected so as to facilitate the calculations. Differentiating Eq. (3) gives

$$y_p' = (u_1'y_1 + u_2'y_2) + (u_1y_1' + u_2y_2'). \tag{4}$$

As just indicated we can simplify this by requiring that u_1 and u_2 satisfy

$$u_1'y_1 + u_2'y_2 = 0. \tag{5}$$

With this condition on u_1' and u_2', y_p'' is given by

$$y_p'' = u_1'y_1' + u_2'y_2' + u_1y_1'' + u_2y_2''. \tag{6}$$

Notice that as a result of imposing condition (5), only *first derivatives* of u_1 and u_2 appear in the expression for y_p''. Substituting for y_p' and y_p'' in Eq. (1) and rearranging terms gives

$$u_1(y_1'' + py_1' + qy_1) + u_2(y_2'' + py_2' + qy_2) + u_1'y_1' + u_2'y_2' = g.$$

The terms in parentheses are zero since y_1 and y_2 are solutions of the homogeneous equation (2); hence the condition that y_p satisfy Eq. (1) leads to the requirement

$$u_1'y_1' + u_2'y_2' = g. \tag{7}$$

Rewriting Eqs. (5) and (7), we have the following system of two equations

$$u_1' y_1 + u_2' y_2 = 0$$

$$u_1' y_1' + u_2' y_2' = g$$

for the derivatives u_1' and u_2' of the two unknown functions. Solving this system of equations gives

$$u_1' = \frac{-y_2 g}{W(y_1, y_2)}, \qquad u_2' = \frac{y_1 g}{W(y_1, y_2)}, \tag{8}$$

where $W(y_1, y_2) = y_1 y_2' - y_1' y_2$. Division by $W(y_1, y_2)$ is permissible since y_1 and y_2 are linearly independent solutions of the homogeneous equation; hence $W(y_1, y_2)$ cannot vanish in the interval. Integrating Eqs. (8) and substituting in Eq. (3) gives a particular solution of the nonhomogeneous equation (1). We state this result in a theorem.

Theorem 3.13. *If the functions p, q, and g are continuous on $\alpha < x < \beta$, and if the functions y_1 and y_2 are linearly independent solutions of the homogeneous equation associated with the differential equation* (1),

$$y'' + p(x)y' + q(x)y = g(x),$$

then a particular solution of Eq. (1) *is given by*

$$y_p(x) = -y_1(x)\int^x \frac{y_2(t)g(t)}{W(y_1, y_2)(t)}\,dt + y_2(x)\int^x \frac{y_1(t)g(t)}{W(y_1, y_2)(t)}\,dt. \tag{9}$$

Equation (9) can also be written as

$$y_p(x) = \int^x \frac{y_1(t)\,y_2(x) - y_1(x)\,y_2(t)}{y_1(t)\,y_2'(t) - y_1'(t)\,y_2(t)}\,g(t)\,dt. \tag{10}$$

Note that in Eq. (10) t is a dummy variable of integration and x is the independent variable.

In using the method of variation of parameters for a particular problem it is usually safer to substitute $y_p(x) = u_1(x)y_1(x) + u_2(x)y_2(x)$ and to proceed as above rather than to try to remember either formula (9) or (10). If one of these formulas is used care should be taken that the differential equation is written in the standard form (1); otherwise the function g will not be correctly identified. It should also be mentioned that while Eqs. (9) and (10) provide formulas for computing $y_p(x)$, it may not always be easy or even possible to evaluate these integrals in closed form. Nevertheless, even in these cases the formulas for $y_p(x)$ provide a starting point for the numerical evaluation of $y_p(x)$, and that may be the best that can be done. From a numerical point of view it is usually advantageous to have an integral form of the solution, since numerical evaluation of an integral is usually, but not always, considerably easier than direct numerical integration of a differential equation.

Again we emphasize that in computing a particular solution by the method of variation of parameters it is not necessary that the coefficients in the differential equation be constants; all that is required is that we know two linearly independent solutions of the homogeneous equation.

EXAMPLE 1

Determine the general solution of the differential equation

$$y'' + y = \sec x, \qquad 0 < x < \pi/2. \tag{11}$$

Two linearly independent solutions of the homogeneous equation are $y_1(x) = \cos x$, $y_2(x) = \sin x$. Even though Eq. (11) has constant coefficients and we know $y_c(x)$, we cannot use the method of undetermined coefficients because the right side of Eq. (11) is not of the form $e^{\alpha x}P_n(x)\cos\beta x$ or $e^{\alpha x}P_n(x)\sin\beta x$. Instead we use the method of variation of parameters. We write

$$y_p(x) = u_1(x)\cos x + u_2(x)\sin x.$$

Then

$$y_p'(x) = [-u_1(x)\sin x + u_2(x)\cos x] + [u_1'(x)\cos x + u_2'(x)\sin x].$$

Setting the second term in brackets equal to zero, differentiating again, and substituting in Eq. (11) gives

$$u_1'(x)\cos x + u_2'(x)\sin x = 0,$$
$$-u_1'(x)\sin x + u_2'(x)\cos x = \sec x.$$

Solving for u_1' and u_2', we obtain

$$u_1'(x) = -\tan x, \qquad u_2'(x) = 1;$$

hence

$$u_1(x) = \ln\cos x, \qquad u_2(x) = x.$$

Consequently, a particular solution of Eq. (11) is

$$y_p(x) = x\sin x + (\cos x)\ln\cos x,$$

and the general solution is

$$y = c_1\cos x + c_2\sin x + x\sin x + (\cos x)\ln\cos x.$$

EXAMPLE 2

Determine the general solution of

$$y'' - 2y' + y = e^x/(1 + x^2), \qquad -\infty < x < \infty. \tag{12}$$

Two linearly independent solutions of the homogeneous equation are $y_1(x) = e^x$ and $y_2(x) = xe^x$. We look for a particular solution of the form

$$y_p(x) = u_1(x)e^x + u_2(x)xe^x. \tag{13}$$

In this example we illustrate the use of Eqs. (8) rather than proceeding from basic principles. We have

$$W(y_1, y_2)(x) = \begin{vmatrix} e^x & xe^x \\ e^x & e^x(x+1) \end{vmatrix} = e^{2x}(x + 1 - x) = e^{2x}.$$

With $g(x) = e^x/(1 + x^2)$, Eqs. (8) give

$$u_1'(x) = \frac{-xe^x}{e^{2x}} \frac{e^x}{1+x^2} = \frac{-x}{1+x^2} \quad \text{so} \quad u_1(x) = -\tfrac{1}{2}\ln(1+x^2),$$

$$u_2'(x) = \frac{e^x}{e^{2x}} \frac{e^x}{1+x^2} = \frac{1}{1+x^2} \quad \text{so} \quad u_2(x) = \arctan x.$$

Substituting for u_1 and u_2 in Eq. (13) we find that the general solution of Eq. (12) is

$$y = c_1e^x + c_2xe^x - \tfrac{1}{2}e^x \ln(1 + x^2) + xe^x \arctan x.$$

We can now summarize the results of the preceding sections of this chapter, which are concerned with methods of solving the differential equation (1)

$$y'' + p(x)y' + q(x)y = g(x),$$

where the functions p, q, and g are continuous on $\alpha < x < \beta$.

1. If p and q are constants it is always possible to find two linearly independent solutions of the homogeneous equation. A particular solution can then be obtained by the method of variation of parameters; hence the complete solution is known. If g is of the appropriate form it will probably be easier to use the method of undetermined coefficients rather than the method of variation of parameters to determine the particular solution.
2. If p and q are not constants, there is no general way to solve Eq. (1) in terms of a finite number of elementary functions.[14] However, if one solution of the homogeneous equation can be found, a second solution can be obtained, in principle, by reducing the order of the equation. Then a particular solution can be obtained by the method of variation of parameters; hence the complete solution is known. Actually both calculations can be done together. If we know one solution of the homogeneous equation, we can solve the nonhomogeneous equation by the method of reduction of order and obtain both a particular

[14] Methods involving infinite series, which are useful for a large class of problems involving variable coefficients, are discussed in Chapter 4. Numerical methods are discussed in Chapter 8.

solution and a second linearly independent solution of the homogeneous equation. This is illustrated in Problems 17 and 18. Thus, in general, the problem of solving Eq. (1) rests on the possibility of finding *one* solution of the corresponding homogeneous equation.

PROBLEMS

In each of Problems 1 through 7 determine a particular solution using the method of variation of parameters.

1. $y'' - 5y' + 6y = 2e^x$
2. $y'' - y' - 2y = 2e^{-x}$
3. $y'' + 2y' + y = 3e^{-x}$
4. $y'' + y = \tan x, \quad 0 < x < \pi/2$
5. $y'' + 9y = 9\sec^2 3x, \quad 0 < x < \pi/6$
6. $y'' + 4y' + 4y = x^{-2}e^{-2x}, \quad x > 0$
7. $y'' + 4y = 3\csc 2x, \quad 0 < x < \pi/2$
8. Verify that x and xe^x are solutions of the homogeneous equation corresponding to
$$x^2y'' - x(x+2)y' + (x+2)y = 2x^3, \qquad x > 0,$$
and find the general solution.
9. Verify that $(1 + x)$ and e^x are solutions of the homogeneous equation corresponding to
$$xy'' - (1+x)y' + y = x^2e^{2x}, \qquad x > 0,$$
and find the general solution.
10. Two linearly independent solutions of Bessel's equation of order one-half,
$$x^2y'' + xy' + \left(x^2 - \tfrac{1}{4}\right)y = 0, \qquad x > 0,$$
are $x^{-1/2}\sin x$ and $x^{-1/2}\cos x$. Find the general solution of
$$x^2y'' + xy' + \left(x^2 - \tfrac{1}{4}\right)y = 3x^{3/2}\sin x, \qquad x > 0.$$
11. Verify that e^x and x are solutions of the homogeneous equation corresponding to
$$(1-x)y'' + xy' - y = 2(x-1)^2e^{-x}, \qquad 0 < x < 1,$$
and find the general solution.
12. Find a formula for a particular solution of the differential equation
$$y'' - 5y' + 6y = g(x).$$
13. Find a formula for a particular solution of the differential equation
$$x^2y'' + xy' + \left(x^2 - \tfrac{1}{4}\right)y = g(x), \qquad x > 0.$$
(see Problem 10)
14. Show that the general solution of
$$y'' = g(x), \qquad y(0) = y_0, \quad y'(0) = y_0'$$
is
$$y = y_0 + xy_0' + \int_0^x \left(\int_0^s g(t)\,dt\right) ds.$$

*15. Consider the initial value problem

$$y'' + p(x)y' + q(x)y = g(x), \qquad y(x_0) = y_0, \quad y'(x_0) = y_0'. \tag{i}$$

(a) Using formulas (9) and (10) of the text show that

$$y_p(x) = \int_{x_0}^{x} \frac{y_1(t)y_2(x) - y_1(x)y_2(t)}{y_1(t)y_2'(t) - y_1'(t)y_2(t)} g(t)\,dt$$

is a particular solution satisfying the conditions $y_p(x_0) = 0$, $y_p'(x_0) = 0$.

(b) Let y_1 and y_2 be linearly independent solutions of the homogeneous equation that satisfy the conditions

$$y_1(x_0) = 1, \qquad y_1'(x_0) = 0 \quad \text{and} \quad y_2(x_0) = 0, \qquad y_2'(x_0) = 1.$$

Show that the general solution of Eq. (i) is

$$y = y_0 y_1(x) + y_0' y_2(x) + \int_{x_0}^{x} \frac{y_1(t)y_2(x) - y_1(x)y_2(t)}{y_1(t)y_2'(t) - y_1'(t)y_2(t)} g(t)\,dt.$$

Note that this equation gives a formula for computing the solution of the original initial value problem for any given nonhomogeneous term $g(t)$.

*16. Consider the initial value problem

$$y'' + y = g(x), \qquad y(0) = 0, \qquad y'(0) = 0.$$

Use the result of Problem 15 to show that

$$y = \phi(x) = \int_0^x g(s)\sin(x - s)\,ds.$$

If we think of x as representing time, then this formula shows the relation between the input $g(x)$ and the output $\phi(x)$. Further we see that the output at time x depends only on the behavior of the input from the initial time 0 to the time of interest x. This integral is referred to as the *convolution* of $\sin x$ and $g(x)$.

Show that if the initial conditions are $y(0) = y_0$ and $y'(0) = y_0'$, then the solution of the initial value problem is

$$y = \phi(x) = y_0 \cos x + y_0' \sin x + \int_0^x g(s)\sin(x - s)\,ds.$$

17. Provided that one solution of the homogeneous equation is known, the nonhomogeneous equation

$$y'' + p(x)y' + q(x)y = g(x) \tag{i}$$

can also be solved by the method of reduction of order (see Section 3.4). Suppose that y_1 is a known solution of the homogeneous equation.

(a) Show that $y = y_1(x)v(x)$ is a solution of Eq. (i) provided that v satisfies

$$y_1 v'' + (2y_1' + py_1)v' = g. \tag{ii}$$

(b) Equation (ii) is a first order linear equation for v'. Show that the solution is

$$y_1^2(x)h(x)v'(x) = \int^x y_1(t)h(t)g(t)\,dt + c_2,$$

where $h(x) = \exp[\int^x p(t)\,dt]$ and c_2 is a constant.

(c) Using part (b) show that the general solution of Eq. (i) is

$$y = c_1 y_1(x) + c_2 y_1(x) \int^x \frac{ds}{y_1^2(s) h(s)}$$

$$+ y_1(x) \int^x \frac{1}{y_1^2(s) h(s)} \left[\int^s y_1(t) h(t) g(t)\, dt \right] ds.$$

18. Using the procedure developed in Problem 17 solve the following differential equations.
(a) $x^2 y'' - 2xy' + 2y = 4x^2, \quad x > 0; \qquad y_1(x) = x$
(b) $x^2 y'' + 7xy' + 5y = x, \quad x > 0; \qquad y_1(x) = x^{-1}$

3.7 Mechanical Vibrations

In Sections 3.1 through 3.6.2 we have developed methods for solving initial value problems of the form

$$ay'' + by' + cy = g(t), \qquad y(0) = y_0, \quad y'(0) = y_0', \tag{1}$$

where a, b, and c are constants, t is the time, and $g(t)$ is a given function. Two important areas of application of this theory are in the fields of mechanical and electrical oscillations. For example, the motion of a mass on a vibrating spring, the torsional oscillation of a shaft with a flywheel, the flow of electric current in a simple series circuit, and other problems all lead to initial value problems of the form (1). This illustrates a fundamental relationship between mathematics and physics: *many physical problems, when formulated mathematically, are identical.* Thus, once we know how to solve the initial value problem (1) it is only necessary to make appropriate interpretations of the constants a, b, and c, and the functions y and g to obtain solutions of different physical problems. In this section and in Sections 3.7.1 and 3.7.2 we discuss the vibration of a mass on a vertical spring, and in Section 3.8 we discuss the flow of current in a simple (LRC) electric circuit.

The derivation of the initial value problem for the motion of a mass at the end of a vertical spring is worth considering in detail since the principles are common to many problems. First, consider the case of the static elongation of a spring of natural length l due to the addition of a mass m. Let Δl denote the elongation (see Figure 3.2). The forces acting on the mass are the force of gravity acting downward, $mg = w$, where g is the acceleration due to gravity and w is the weight of the mass, and the spring force acting upward. Since the mass is in equilibrium in this position these forces are numerically equal. If the displacement Δl is small compared to the length l, the spring force, according to Hooke's law,[15] is proportional to Δl and hence has magnitude $k\,\Delta l$. The constant k depends on

[15] Robert Hooke (1635–1703) first published his law in 1676 as an anagram: *ceiiinosssttuv*; and in 1678 gave the solution *ut tensio sic vis*, which means, roughly, "as the force so is the displacement."

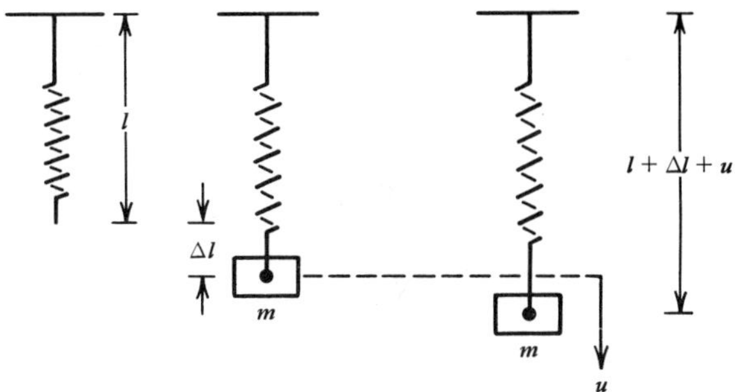

FIGURE 3.2 A spring–mass system.

the spring and is called the spring constant. For a known weight w it can be computed by measuring Δl and using

$$k\,\Delta l = mg. \tag{2}$$

Notice that k has the dimensions of $mg/\Delta l$—that is, force/length.

In the corresponding dynamic problem we are interested in studying the motion of the mass when it is acted on by an external force or is initially displaced. Let u, measured positive downward, denote the displacement of the mass from the equilibrium position. The displacement u depends on t and is related to the forces acting on the mass through Newton's law, which states that md^2u/dt^2 must have the same magnitude and be in the same direction as the net resultant force acting on the mass m. In deriving the equation governing the motion of the mass, care must be exercised to determine the correct signs of the various forces that act on the mass. In the derivation below we adopt the convention that a force in the downward direction is positive, while a force in the upward direction is negative. Four forces act on the mass:

1. Its weight $w = mg$, which always acts downward.
2. The force due to the spring F_s, which is proportional to the elongation $\Delta l + u$ and always acts to restore the spring to its natural position. If $\Delta l + u > 0$, the spring is extended and the force, which is directed upward, is given by $F_s(t) = -k(\Delta l + u)$. Note that the magnitude of the force on the mass due to the spring is $k(\Delta l + u)$; the minus sign tells us that this force is directed upward in this particular case. If $u + \Delta l < 0$ (that is, $-u > \Delta l$), the spring is compressed a distance[16] $(-u) - (\Delta l)$ and the force, which is directed downward, is $k[(-u) - (\Delta l)]$. Hence for all u

$$F_s(t) = -k(\Delta l + u). \tag{3}$$

[16]Remember that if the mass is above the equilibrium position its *distance* above the equilibrium position is $-u$, not u.

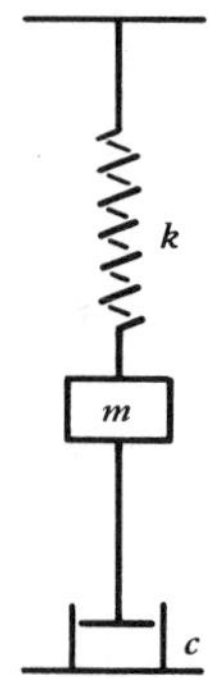

FIGURE 3.3 A spring–mass–dashpot system.

We emphasize again that Eq. (3) gives not only the magnitude but also the direction of the force exerted on the mass by the spring at any position u.

3. The damping or resisting force $F_d(t)$. This force may be due to the viscous properties of the fluid in which the mass is moving (air resistance, for instance); or the moving mass may be attached to an oil or dashpot mechanism (see Figure 3.3). In either case the damping force will act in a direction opposite to the direction of motion of the mass. We know from experimental evidence that as long as the speed of the mass is not too large the resistance may be taken to be proportional to the speed $|du/dt|$. If the mass is moving downward, u is increasing and hence du/dt is positive, so $F_d(t)$, which is directed upward, is given by $-c(du/dt)$. The positive constant c is referred to as the damping constant; it has the units of force/velocity. If the mass is moving upward, u is decreasing, du/dt is negative, and $F_d(t)$, which is now directed downward, is again given by $-c(du/dt)$. Hence for all u
$$F_d(t) = -c\,du/dt. \tag{4}$$
4. An applied force $F(t)$, which is directed downward or upward as $F(t)$ is positive or negative. This could be a force due to the motion of the mount to which the spring is attached, or it could be a force applied directly to the mass.

According to Newton's law
$$mg + F_s(t) + F_d(t) + F(t) = m\frac{d^2u}{dt^2}.$$
Substituting for $F_s(t)$ and $F_d(t)$ gives
$$mg - k(\Delta l + u) - c\frac{du}{dt} + F(t) = m\frac{d^2u}{dt^2}. \tag{5}$$
Since $k\,\Delta l = mg$, Eq. (5) reduces to
$$m\ddot{u} + c\dot{u} + ku = F(t). \tag{6}$$

In writing this equation we have followed a common convention of indicating time derivatives by dots. Note that Eq. (6) has the same form as Eq. (1).

It is important to remember that in Eq. (6), the mass m, the damping constant c, and the spring constant k are all *positive* constants. It is also important to recognize that Eq. (6) represents only an approximate equation for the determination of u. In reality the relationship between $F_s(t)$ and $\Delta l + u$ is not linear; however, for most of the usual materials, such as steel, and for $(\Delta l + u)/l \ll 1$, the relationship is, to a high degree of accuracy, approximately linear. Similarly, the expression for the damping force is an approximate one. Also we have neglected the mass of the spring compared to the mass of the attached body in deriving Eq. (6). Fortunately, for a wide variety of applications, Eq. (6) is a satisfactory mathematical model of the physical system.

The complete formulation of the vibration problem requires the specification of two initial conditions—the initial displacement u_0, and the initial velocity $\dot{u}_0$. It may not be obvious from physical reasoning that we can specify only two initial conditions, or that these are the two that should be specified. However, it does follow from the existence and uniqueness Theorem 3.2 that these conditions give a mathematical problem that has a unique solution. The solution of Eq. (6) satisfying these initial conditions gives the position of the mass as a function of time.

EXAMPLE

Derive the initial value problem for the following spring–dashpot–mass system. It is known that a 5-lb weight stretches the spring 1 in. The dashpot mechanism exerts a force of 0.02 lb for a velocity of 2 in./sec. A mass weighing 2 lb is attached to the spring and released from a position 2 in. below equilibrium.

In setting up the differential equation we must be consistent in expressing lengths, time, and forces in the same units. We remind the reader that we discussed the English and metric systems of units in Section 2.7. We have

$$m = \frac{w}{g} = \frac{2 \text{ lb}}{32 \text{ ft/sec}^2} = \frac{1}{16}\,\frac{\text{lb-sec}^2}{\text{ft}},$$

$$c = \frac{0.02 \text{ lb}}{2 \text{ in./sec}} = \frac{0.02 \text{ lb}}{(1/6) \text{ ft/sec}} = 0.12\frac{\text{lb-sec}}{\text{ft}},$$

$$k = \frac{5 \text{ lb}}{1 \text{ in.}} = \frac{5 \text{ lb}}{(1/12) \text{ ft}} = 60\frac{\text{lb}}{\text{ft}};$$

hence

$$\tfrac{1}{16}\ddot{u} + 0.12\dot{u} + 60u = 0,$$

or

$$\ddot{u} + 1.92\dot{u} + 960u = 0,$$

where u is measured in feet and t in seconds. The initial conditions are $u(0) = \frac{1}{6}$, $\dot{u}(0) = 0$.

3.7.1 Free Vibrations

UNDAMPED FREE VIBRATIONS. If there is no external force and no damping, then Eq. (6) from the previous section reduces to

$$m\ddot{u} + ku = 0. \tag{1}$$

The solution of this equation is

$$u = A\cos\omega_0 t + B\sin\omega_0 t, \tag{2}$$

where

$$\omega_0^2 = k/m, \tag{3}$$

and ω_0, called the circular frequency, clearly has the units of 1/time. For a particular problem, the constants A and B are determined by the prescribed initial conditions.

In discussing the solution of Eq. (1) it is convenient to write Eq. (2) in the form

$$u = R\cos(\omega_0 t - \delta). \tag{4}$$

Regardless of the values of A and B this can always be done by solving $A = R\cos\delta$ and $B = R\sin\delta$ for R and δ. It follows that $R = (A^2 + B^2)^{1/2}$ and $\tan\delta = B/A$. Note that in calculating δ care must be taken to choose the correct quadrant; this can be done by checking the signs of $\cos\delta$ and $\sin\delta$ from the equations $A = R\cos\delta$ and $B = R\sin\delta$.

Because of the periodic character of the cosine function, Eq. (4) represents a periodic motion, or simple harmonic motion, of period

$$T = \frac{2\pi}{\omega_0} = 2\pi\left(\frac{m}{k}\right)^{1/2}. \tag{5}$$

To distinguish between problems of free vibrations without damping and problems of damped vibrations or forced vibrations, the circular frequency ω_0 and period $2\pi/\omega_0$ of the undamped, unforced motion are usually referred to as the natural circular frequency and natural period of the vibration.

Since $|\cos(\omega_0 t - \delta)| \leq 1$, u always lies between the lines $u = \pm R$. The maximum displacement occurs at the times $\omega_0 t - \delta = 0,\ \pi, 2\pi, \ldots$ or $t = (\delta + n\pi)/\omega_0$, $n = 0, 1, 2, 3, \ldots$. The constants R and δ are referred to as the amplitude and phase angle, respectively, of the motion. Note that the amplitude R is the maximum displacement of the mass from the equilibrium position. A sketch of the motion represented by Eq. (4) is shown in Figure 3.4. Notice that this periodic motion does not die out with increasing time. This is a consequence of neglecting the damping force, which would allow a means of dissipating the energy supplied to the system by the original displacement and velocity.

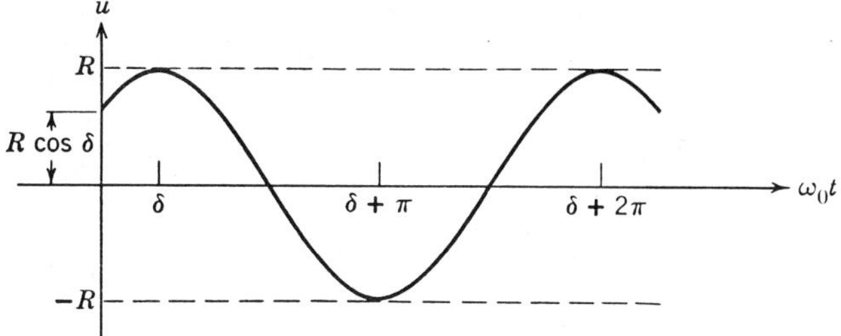

FIGURE 3.4 Simple harmonic motion.

EXAMPLE 1

Find the natural period of a spring–mass system if the mass weighs 10 lb and stretches a steel spring 2 in. If the spring is stretched an additional 2 in. and then released, determine the subsequent motion.

The spring constant is $k = 10$ lb/2 in. $= 60$ lb/ft. The mass $m = w/g = \frac{10}{32}$ lb/(ft/sec^2). Hence

$$T = 2\pi\left(\frac{m}{k}\right)^{1/2} = 2\pi\left[\frac{10}{(32)60}\right]^{1/2} = \frac{\pi\sqrt{3}}{12} \text{ sec.}$$

The equation of motion is

$$\tfrac{10}{32}\ddot{u} + 60u = 0;$$

hence

$$u = A\cos(8\sqrt{3}\,t) + B\sin(8\sqrt{3}\,t).$$

The solution satisfying the initial conditions $u(0) = \frac{1}{6}$ ft and $\dot{u}(0) = 0$ ft/sec is

$$u = \tfrac{1}{6}\cos(8\sqrt{3}\,t),$$

where t is measured in seconds and u is measured in feet.

DAMPED FREE VIBRATIONS. If we include the effect of damping, the differential equation governing the motion of the mass is

$$m\ddot{u} + c\dot{u} + ku = 0. \tag{6}$$

The roots of the corresponding characteristic equation are

$$r_1, r_2 = \frac{-c \pm \sqrt{c^2 - 4km}}{2m} = -\frac{c}{2m} \pm \frac{c}{2m}\left(1 - \frac{4km}{c^2}\right)^{1/2}. \tag{7}$$

Since c, k, and m are positive, $c^2 - 4km$ is always less than c^2; hence if $c^2 - 4km \geq 0$ the values of r_1 and r_2 given by Eq. (7) are *negative*. If $c^2 - 4km < 0$ the values of r_1 and r_2 given by Eq. (7) are complex, but with *negative* real part. In tabular form the solution of Eq. (6) is

$$c^2 - 4km > 0, \qquad u = Ae^{r_1 t} + Be^{r_2 t}, \qquad r_1 \text{ and } r_2 < 0; \tag{8}$$

$$c^2 - 4km = 0, \qquad u = (A + Bt)e^{-(c/2m)t}; \tag{9}$$

$$c^2 - 4km < 0, \qquad u = e^{-(c/2m)t}(A\cos\mu t + B\sin\mu t),$$

$$\mu = (4km - c^2)^{1/2}/2m > 0. \tag{10}$$

In all three cases, regardless of the initial conditions, that is, regardless of the values of A and B, $u \to 0$ as $t \to \infty$; hence the motion dies out with increasing time. Thus the solutions of Eqs. (1) and (6) confirm what we would intuitively expect—without damping the motion always continues unabated and with damping the motion must decay to zero with increasing time.

The first two cases, Eqs. (8) and (9), which are referred to as *overdamped* and *critically damped*, respectively, represent motions in which the originally displaced mass "creeps" back to its equilibrium position. Depending on the initial conditions, it may be possible to overshoot the equilibrium position once, but this is not a "vibratory motion." Two typical examples of critically damped motions are sketched in Figure 3.5. Note that in one case the mass passes through the equilibrium position once, but in the other case it never passes through the equilibrium position. The theory is discussed further in Problems 16 and 17.

The third case, which is referred to as an *underdamped motion*, often occurs in mechanical systems and represents a "damped vibration." To see this, we let $A = R\cos\delta$ and $B = R\sin\delta$ in Eq. (10), and we obtain

$$u = Re^{-(c/2m)t}\cos(\mu t - \delta). \tag{11}$$

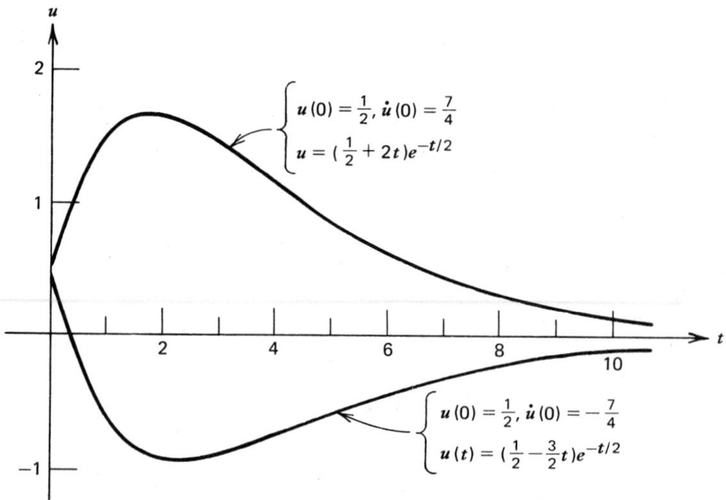

FIGURE 3.5 Critically damped motions: $\ddot{u} + \dot{u} + \frac{1}{4}u = 0$, $u = (A + Bt)e^{-t/2}$.

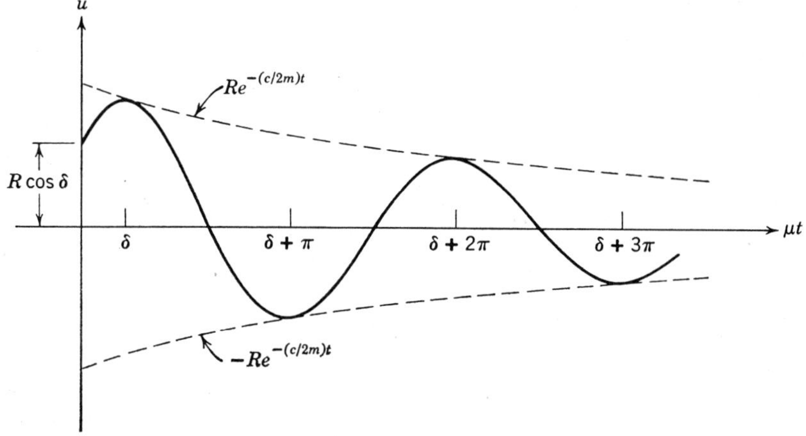

FIGURE 3.6 Damped vibration.

The displacement u must lie between the curves $u = \pm Re^{-(c/2m)t}$; hence it resembles a cosine curve with decreasing amplitude. A typical example is sketched in Figure 3.6.

While the motion is not truly periodic, we can define a *quasicircular frequency* $\omega_d = \mu$. The *quasiperiod* $T_d = 2\pi/\mu$ is the time between successive maxima or minima of the position of the mass. It is interesting to note the relation between T_d and T. We have, from Eq. (10),

$$T_d = \frac{2\pi}{\mu} = \frac{4\pi m}{(4km - c^2)^{1/2}} = 2\pi\left(\frac{k}{m}\right)^{-1/2}\left(1 - \frac{c^2}{4km}\right)^{-1/2}$$

$$= T\left(1 - \frac{c^2}{4km}\right)^{-1/2}. \tag{12}$$

Hence if $c^2/4km$ is small,

$$T_d \cong T\left(1 + \frac{c^2}{8km}\right).$$

Consequently, when the damping is very small (that is, when the dimensionless quantity $c^2/4km \ll 1$), we can neglect damping in computing the quasiperiod of vibration. In many physical problems this is precisely the situation. On the other hand, if we want to study the detailed motion of the mass for all time, we can *never* neglect the damping force, no matter how small. The importance of the damping force can be seen even more emphatically in the next section where we discuss forced vibrations.

EXAMPLE 2

The motion of a certain spring–mass system is governed by the differential equation

$$\ddot{u} + \tfrac{1}{8}\dot{u} + u = 0, \tag{13}$$

where u is measured in feet and t in seconds. If $u(0) = 2$ and $\dot{u}(0) = 0$ determine the subsequent motion, the quasiperiod, and the first time at which the mass passes through the equilibrium position.

The solution of Eq. (13) is

$$u = e^{-t/16}\left(A\cos\frac{\sqrt{255}}{16}t + B\sin\frac{\sqrt{255}}{16}t\right).$$

To satisfy the initial conditions we must choose $A = 2$ and $B = 2/\sqrt{255}$; hence the solution of the initial value problem is

$$\begin{aligned} u &= e^{-t/16}\left(2\cos\frac{\sqrt{255}}{16}t + \frac{2}{\sqrt{255}}\sin\frac{\sqrt{255}}{16}t\right) \\ &= \frac{32}{\sqrt{255}}e^{-t/16}\cos\left(\frac{\sqrt{255}}{16}t - \delta\right) \end{aligned}$$

where $\tan\delta = 1/\sqrt{255}$ so $\delta \simeq 0.06254$. The displacement of the mass as a function of time is sketched in Figure 3.7. For the purpose of comparison we also show the motion if the damping term is neglected.

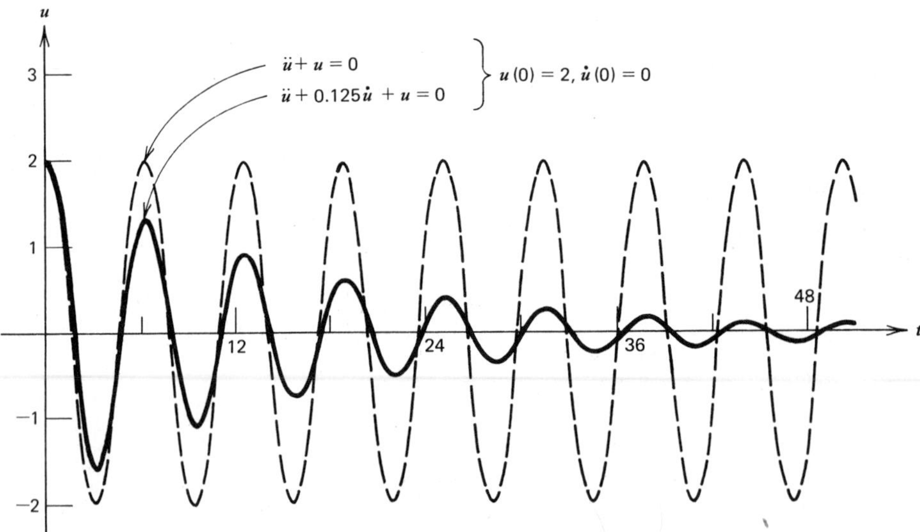

FIGURE 3.7 Vibration with light damping (solid curve) and without damping (dashed curve).

The quasiperiod is $T_d = 2\pi/(\sqrt{255}/16) = 32\pi/\sqrt{255} \cong 6.295$ sec. Since $c^2/4km = (\frac{1}{8})^2/4 \cdot 1 \cdot 1 = 1/256 \ll 1$ the quasiperiod is nearly equal to the natural period as is evident from Figure 3.7. The first time the mass passes through the equilibrium position is given by $\sqrt{255}\,t/16 - \delta = \pi/2$ or $t = 16(\pi/2 + \delta)/\sqrt{255} \simeq 1.637$ sec.

PROBLEMS

1. A mass weighing 2 lb stretches a 2-ft spring 6 in. If the mass is pulled down an additional 3 in. and released, determine the subsequent motion neglecting air resistance. What are the amplitude, circular frequency, and period of the motion?
2. A mass of 100 g is attached to a steel spring of natural length 50 cm. The spring is stretched 5 cm by the addition of this mass. If the mass is started in motion with a velocity of 10 cm/sec in the downward direction, determine the subsequent motion. Neglect air resistance. At what time does the mass pass through its equilibrium position?
3. A mass weighing 3 lb stretches a spiral spring 3 in. If the mass is pushed upward, contracting the spring a distance 1 in., and then released with a velocity in the downward direction of 2 ft/sec, determine the subsequent motion. Neglect air resistance. What are the amplitude, circular frequency, and period of the motion?
4. Show that the period of motion of an undamped vibration of a mass hanging from a vertical spring is $2\pi\sqrt{\Delta l/g}$ where Δl is the elongation due to the weight w.
5. In Problem 3 suppose air resistance exerts a damping force which is given by $0.3\dot{u}$ lb. Compare the quasiperiod of the damped motion with the natural period of the undamped motion. What is the percentage change based on the natural period?
6. In Problem 5, how large would the damping coefficient have to be to increase the quasiperiod by 50% relative to the natural period?
7. Show that $A\cos\omega_0 t + B\sin\omega_0 t$ can be written in the form $r\sin(\omega_0 t - \theta)$. Determine r and θ in terms of A and B. What is the relationship between r, R, θ, and δ if $R\cos(\omega_0 t - \delta) = r\sin(\omega_0 t - \theta)$?
8. Determine the solution of the differential equation $m\ddot{u} + ku = 0$ satisfying the following initial conditions. Express your answer in the form $R\cos(\omega_0 t - \delta)$.

   (a) $u(0) = u_0,\quad \dot{u}(0) = 0$; (b) $u(0) = 0,\quad \dot{u}(0) = \dot{u}_0$; (c) $u(0) = u_0,\quad \dot{u}(0) = \dot{u}_0$.

9. Determine the natural period of oscillation of the pendulum of length l shown in Figure 3.8. Neglect air resistance and the weight of the string compared to the weight of the mass. Assume that θ is small enough that $\sin\theta \simeq \theta$.
   Hint: Write Newton's law for the tangential component of the motion.
10. A cubic block of side l and mass density ρ per unit volume is floating in a fluid of mass density $\hat{\rho}$ per unit volume. If the block is slightly depressed and then released it oscillates in the vertical direction. Assuming that the viscous damping of the fluid and air can be neglected, derive the differential equation of motion and determine the period of the motion.
    Hint: In deriving the governing differential equation use Archimedes' principle: an object that is completely or partially submerged in a fluid is acted on by an upward (buoyant) force equal to the weight of the displaced fluid.

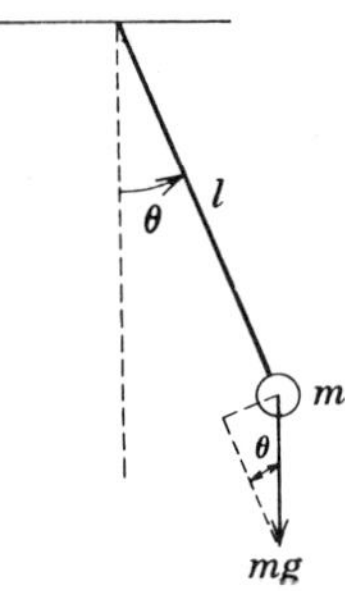

FIGURE 3.8 An oscillating pendulum.

11. A mass of 20 g stretches a spiral spring a distance of 5 cm. If the spring is attached to an oil dashpot damper with a damping constant of 400 dyne-sec/cm, determine the subsequent motion if the mass is pulled down an additional 2 cm and released.
12. A mass weighing 16 lb stretches a spring 3 in. The mass is attached to a dashpot mechanism with a damping constant of 2 lb-sec/ft. Determine the subsequent motion if the mass is released from its equilibrium position with a velocity of 3 in./sec in the downward direction. At what time does the mass pass through its equilibrium position?
13. A mass weighing 8 lb is attached to a steel spring of length 15 in. The mass extends the spring a distance of $\frac{1}{8}$ ft. If the mass is also attached to an oil dashpot damper with a damping constant c, determine the values of c for which the motion will be overdamped, critically damped, and underdamped. Be sure to give the units for c.
14. A spring is stretched 10 cm by a force of 3 newtons. A mass of 2 kg is attached to the spring which is also attached to a dashpot damping mechanism with a damping constant of 0.6 newton-sec/m. Determine the subsequent motion if the mass is pulled down a distance of 5 cm and released with a downward velocity of 10 cm/sec. Round all calculations to three nonzero digits.
15. Determine a solution of the differential equation $m\ddot{u} + c\dot{u} + ku = 0$ in the case $c^2 - 4km < 0$ satisfying the following initial conditions. Express your answer in the form $Re^{-(c/2m)t}\cos(\mu t - \delta)$.

    (a) $u(0) = u_0$, $\dot{u}(0) = 0$; (b) $u(0) = 0$, $\dot{u}(0) = \dot{u}_0$; (c) $u(0) = u_0$, $\dot{u}(0) = \dot{u}_0$

16. Show that if the damping constant c in the equation $m\ddot{u} + c\dot{u} + ku = 0$ is such that the motion is overdamped or critically damped, then the mass can pass through its zero displacement position at most once, regardless of the initial conditions. *Hint:* Determine all possible values of t such that $u = 0$.
17. For the case of critical damping, find the general solution of Eq. (6) satisfying the initial conditions $u(0) = u_0$, $\dot{u}(0) = \dot{u}_0$. If $\dot{u}_0 = 0$, show that $u \to 0$ as $t \to \infty$, but $u \neq 0$ for any finite value of t. Assuming u_0 is positive, determine a condition on $\dot{u}_0$ that will insure that the mass passes through the zero displacement position after it is released.
18. Show that in the limit $c \to 0$ an underdamped vibratory motion approaches a simple harmonic motion.

19. Show that the quasicircular frequency of a damped vibration is always less than the corresponding circular frequency of the undamped vibration.

*20. For the damped oscillation given by Eq. (11) the time between successive maxima is $T_d = 2\pi/\mu$. The ratio of the displacement at two successive maxima (time t and time $t + T_d$) is given by $e^{(c/2m)T_d}$. Thus successive displacements at time intervals T_d form a geometric progression with ratio $e^{(c/2m)T_d}$. The natural logarithm of this ratio, denoted by $\Delta = (c/2m)T_d = (\pi c/m\mu)$, is referred to as the *logarithmic decrement.* Since m, μ, and Δ are quantities which can easily be measured for a mechanical system, this equation provides a convenient and *practical* method of computing the damping constant of the system. In particular, for the motion of a vibrating mass in a viscous fluid the damping constant depends on the viscosity of the fluid; for simple geometric shapes the form of this dependence is known, and the above relation allows the determination of the viscosity experimentally. This is one of the most accurate ways of determining the viscosity of a gas at high pressure.
(a) In Problem 12, what is the logarithmic decrement?
(b) In Problem 13 if the measured logarithmic decrement is 3, and $T_d = 0.3$ sec, determine the damping constant c.

3.7.2 Forced Vibrations

Consider now the case in which a periodic external force, say $F_0 \cos \omega t$, is applied to a spring–mass system. In this case the equation of motion is

$$m\ddot{u} + c\dot{u} + ku = F_0 \cos \omega t. \tag{1}$$

First suppose there is no damping; then Eq. (1) reduces to

$$m\ddot{u} + ku = F_0 \cos \omega t. \tag{2}$$

Provided $\omega_0 = \sqrt{k/m} \neq \omega$, the general solution of Eq. (2) is

$$u = c_1 \cos \omega_0 t + c_2 \sin \omega_0 t + \frac{F_0}{m(\omega_0^2 - \omega^2)} \cos \omega t. \tag{3}$$

The constants c_1 and c_2 are determined by the initial conditions. The resultant motion is, in general, the sum of two periodic motions of different frequencies (ω_0 and ω) and amplitudes. We consider two different situations.

BEATS. Suppose the mass is initially at rest, that is, $u(0) = 0$, $\dot{u}(0) = 0$. Then it turns out that the constants c_1 and c_2 in Eq. (3) are given by

$$c_1 = \frac{-F_0}{m(\omega_0^2 - \omega^2)}, \qquad c_2 = 0, \tag{4}$$

and the solution of Eq. (2) is

$$u = \frac{F_0}{m(\omega_0^2 - \omega^2)}(\cos \omega t - \cos \omega_0 t). \tag{5}$$

This is the sum of two periodic functions of different periods but the same amplitude. Making use of the trigonometric identities $\cos(A \pm B) = \cos A \cos B \mp \sin A \sin B$ with $A = (\omega_0 + \omega)t/2$ and $B = (\omega_0 - \omega)t/2$, we can write Eq. (5) in the form

$$u = \left[\frac{2F_0}{m(\omega_0^2 - \omega^2)} \sin \frac{(\omega_0 - \omega)t}{2}\right] \sin \frac{(\omega_0 + \omega)t}{2}. \tag{6}$$

If $|\omega_0 - \omega|$ is small, then $(\omega_0 + \omega) \gg |\omega_0 - \omega|$, and consequently $\sin(\omega_0 + \omega)t/2$ is a rapidly oscillating function compared to $\sin(\omega_0 - \omega)t/2$. Then the motion is a rapid oscillation with circular frequency $(\omega_0 + \omega)/2$, but with a slowly varying sinusoidal amplitude

$$\frac{2F_0}{m(\omega_0^2 - \omega^2)} \sin \frac{(\omega_0 - \omega)t}{2}.$$

This type of motion, possessing a periodic variation of amplitude, exhibits what is called a *beat*. Such a phenomenon occurs in acoustics when two tuning forks of nearly equal frequency are sounded simultaneously. In this case the periodic variation of amplitude is quite apparent to the unaided ear. In electronics the variation of the amplitude with time is called *amplitude modulation*. The function $u = \phi(t)$ given in Eq. (6) is sketched in Figure 3.9.

RESONANCE. As a second example, consider the case $\omega = \omega_0$; that is, the period of the forcing function is the same as the natural period of the system. Then the nonhomogeneous term $F_0 \cos \omega t$ is a solution of the homogeneous equation. In this case, the solution of Eq. (2) is

$$u = c_1 \cos \omega_0 t + c_2 \sin \omega_0 t + \frac{F_0}{2m\omega_0} t \sin \omega_0 t. \tag{7}$$

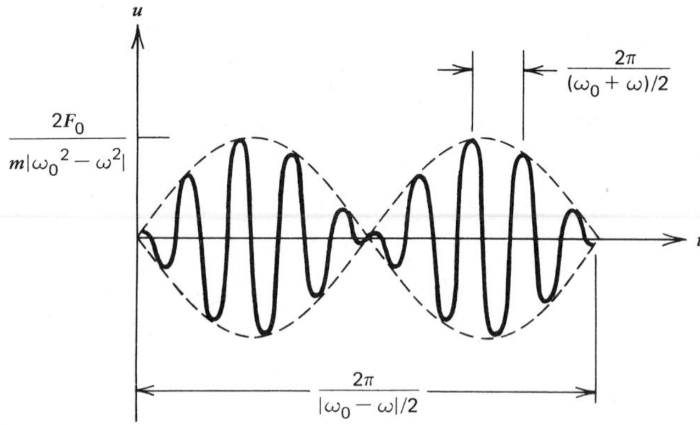

FIGURE 3.9 The phenomenon of beats.

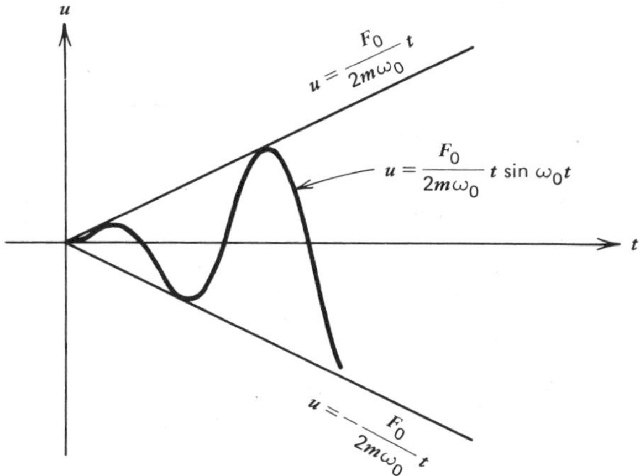

FIGURE 3.10 The phenomenon of resonance.

Because of the presence of the term $t \sin \omega_0 t$ in Eq. (7) it is clear, regardless of the values of c_1 and c_2, that the motion will become unbounded as $t \to \infty$ (see Figure 3.10). This is the phenomenon known as *resonance*. In actual practice, however, the spring would probably break. Of course, as soon as u becomes large, our theory is no longer valid since we have assumed that u is small when we used a linear relation to determine the spring constant. If damping is included in the system the motion will remain bounded; however, there may still be a large response to the input function $F_0 \cos \omega t$ if the damping is small and ω is close to ω_0.

The phenomenon of resonance can create serious difficulties in designing structures, where resonance can produce instabilities leading to destruction of the structure. Most of us are familiar with soldiers breaking step when crossing a bridge in order to eliminate the *periodic* force of their marching which could resonate a natural frequency of the bridge. A classic example of bridge failure is that of the Tacoma Narrows bridge on November 7, 1940. In this case the *periodic* external force was due to the periodic shedding of Karman vortices by the bridge in a transverse wind. A more recent example is the design of the high-pressure fuel turbopump for the space shuttle main engine. The turbopump was unstable and could not be operated over 20,000 rpm as compared to the design speed of 39,000 rpm. This difficulty led to a shutdown of the space shuttle program for 6 months at an estimated cost of $500,000 per day.[17]

FORCED VIBRATIONS WITH DAMPING. The motion of the spring–mass system with damping and the forcing function $F_0 \cos \omega t$ can be determined in a

[17] F. Ehrich and D. Childs, "Self-Excited Vibration in High-Performance Turbomachinery," *Mechanical Engineering* (May 1984), p. 66.

straightforward manner. Although the computations are rather lengthy they are not difficult. The solution of Eq. (1) is

$$u = c_1 e^{r_1 t} + c_2 e^{r_2 t} + \frac{F_0}{\sqrt{m^2(\omega_0^2 - \omega^2)^2 + c^2\omega^2}} \cos(\omega t - \delta), \tag{8}$$

where δ is given by $\cos\delta = m(\omega_0^2 - \omega^2)/\Delta$ and $\sin\delta = c\omega/\Delta$ with $\Delta = \sqrt{m^2(\omega_0^2 - \omega^2)^2 + c^2\omega^2}$. Here r_1 and r_2 are the roots of the characteristic equation associated with Eq. (1). As was shown in the last section both $e^{r_1 t}$ and $e^{r_2 t}$ approach zero as $t \to \infty$. Hence as $t \to \infty$,

$$u \to u_p(t) = \frac{F_0}{\sqrt{m^2(\omega_0^2 - \omega^2)^2 + c^2\omega^2}} \cos(\omega t - \delta). \tag{9}$$

For this reason $u_c(t) = c_1 e^{r_1 t} + c_2 e^{r_2 t}$ is often called the *transient solution* and $u_p(t)$ is called the *steady state solution*. To speak roughly, the transient solution allows us to satisfy the imposed initial conditions; with increasing time the energy put into the system by the initial displacement and velocity is dissipated through the damping force, and the motion then represents the response of the system to the external force $F(t)$. Without damping (which, of course, is impossible in any physical system) the effect of the initial conditions would persist for all time.

In Figure 3.11 we show a specific example of a forced vibration with damping. We also show the graph of the forcing function $3\cos 2t$. (The corresponding damped vibration without the forcing function is shown in Figure 3.7 of Section 3.7.1.) Notice the initial transient motion that decays as t increases. The steady state response of the system is

$$u_p(t) = \frac{3}{\sqrt{1(1 - 2^2)^2 + (\frac{1}{8})^2 2^2}} \cos(2t - \delta) \simeq 0.977 \cos(2t - \delta),$$

where the phase angle $\delta \simeq 3.058$ rad ($\simeq \pi$ rad). Since the period of the forcing function is π it follows that the steady state response of the system is almost exactly out of phase with the forcing function. This is evident in Figure 3.11 where we see that the response for large time is a maximum at the time the forcing is nearly a minimum and vice versa. In this regard we mention that it is desirable to express the steady state response in terms of the same trigonometric function (with a phase angle) as that of the forcing function so that the phase relationship between the forcing function and the response is made more obvious.

Notice that $m^2(\omega_0^2 - \omega^2)^2 + c^2\omega^2$ is never zero, even for $\omega = \omega_0$; hence with damping the motion is always bounded. The amplitude of the steady state solution is a maximum when $m^2(\omega_0^2 - \omega^2)^2 + c^2\omega^2$ is a minimum. Using the methods of the calculus we find, for fixed values of k, c, and m with $c^2/2km < 1$

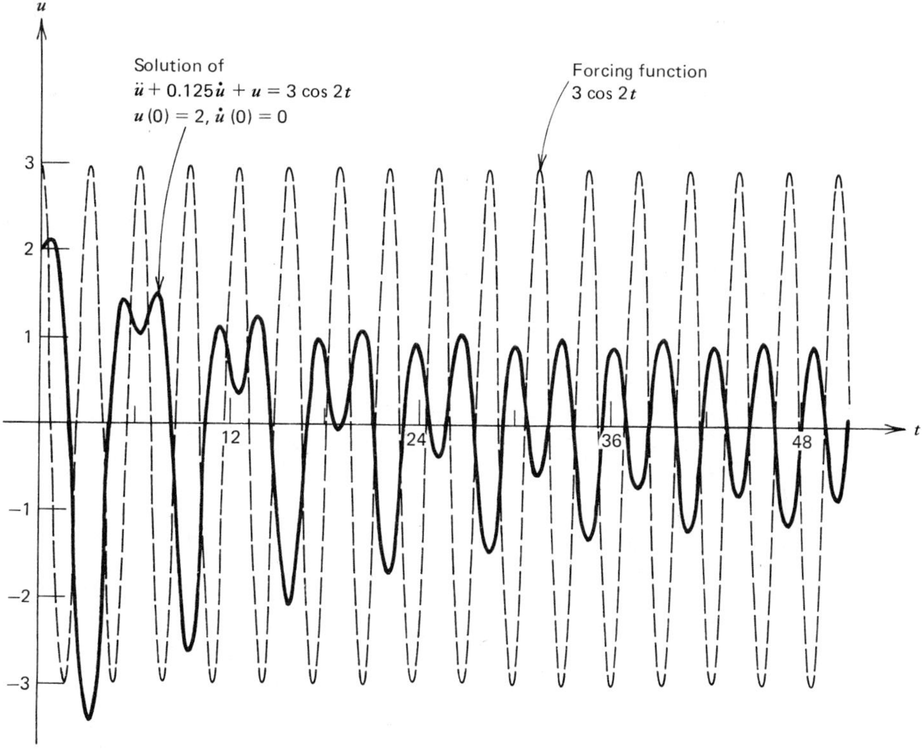

FIGURE 3.11. A forced vibration with damping.

(the damping is small), that the minimum corresponds to choosing

$$\omega^2 = \omega_{\max}^2 = \omega_0^2 - \frac{1}{2}\left(\frac{c}{m}\right)^2 = \frac{k}{m} - \frac{c^2}{2m^2}$$

$$= \frac{k}{m}\left(1 - \frac{c^2}{2km}\right). \tag{10}$$

Note that $\omega_{\max} < \omega_0$. For $c^2/2km \geq 1$ the minimum occurs for $\omega = 0$ and $m^2(\omega_0^2 - \omega^2)^2 + c^2\omega^2$ is a monotone increasing function of ω. Representative graphs of $M(\omega) = [m^2(\omega_0^2 - \omega^2)^2 + c^2\omega^2]^{-1/2}$ are shown in Figure 3.12.

In designing a spring–mass system[18] to detect periodic forces in a narrow frequency range around a given value of ω, it is clear that we would want to choose k, c, and m in such a way that Eq. (10) is satisfied or nearly satisfied. In

[18]Such instruments are usually referred to as seismic instruments since they are similar in principle to the seismograph used to detect motions of the earth's surface. The periodic motion of the mount to which the spring is attached is equivalent to a problem in which a periodic force acts on the mass.

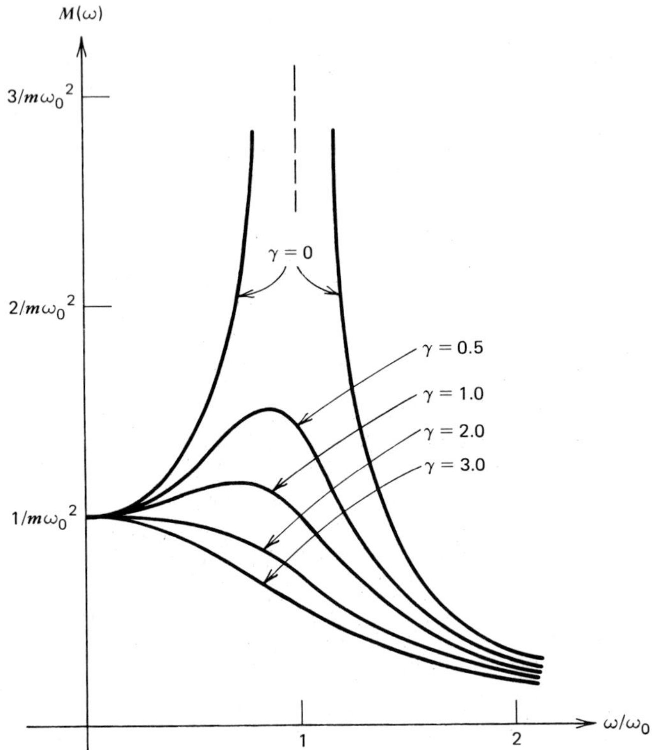

FIGURE 3.12 Forced vibration with damping: amplitude of steady state response versus frequency of driving force. $\gamma = c^2 / m^2\omega_0^2$.

this way we obtain the maximum response of the system to such forces thus making their detection easier. A similar situation occurs in problems involving the detection of electrical signals in electrical networks.

PROBLEMS

1. Express $\cos 9t - \cos 7t$ in the form $A \sin \alpha t \sin \beta t$.
2. Express $\sin 7t - \sin 6t$ in the form $A \sin \alpha t \cos \beta t$.
3. A spiral spring is stretched $\frac{1}{8}$ ft by a mass that weighs 4 lb. The spring is acted on by an external force $2 \cos 3t$ lb. If the mass is displaced a distance 2 in. from its position of equilibrium and then released, determine the subsequent motion. Neglect damping. If the external force is $4 \sin \omega t$ lb, for what value of ω will resonance occur?
4. If an undamped spring–mass system with a mass that weighs 6 lb and a spring constant 1 lb/in. is suddenly set in motion at $t = 0$ by an external force of $4 \cos 7t$ lb, determine the subsequent motion and draw a graph of the displacement versus t.

5. A mass that weighs 8 lb stretches a spring 6 in. The spring system is acted on by an external force $8\sin 8t$ lb. If the mass is pulled down a distance $\frac{1}{4}$ ft and released from rest determine the subsequent motion. Determine the first four times at which the velocity of the mass is zero.
6. Determine the solution of the differential equation $m\ddot{u} + ku = F_0\cos\omega t$, $\omega \neq \sqrt{k/m}$, satisfying the following initial conditions:

   (a) $u(0) = u_0$, $\dot{u}(0) = 0$; (b) $u(0) = 0$, $\dot{u}(0) = \dot{u}_0$; (c) $u(0) = u_0$, $\dot{u}(0) = \dot{u}_0$.

*7. Determine the motion $u = \phi(t)$ of a mass m hanging on a spring with spring constant k if the mass is acted on by a force

$$F(t) = F_0\begin{cases} t, & 0 \leq t \leq \pi \\ 2\pi - t, & \pi < t \leq 2\pi \\ 0, & 2\pi < t. \end{cases}$$

   For convenience assume $\omega_0^2 = k/m = 1$. Neglect air resistance and assume that the mass is initially at rest.
   Hint: Treat each time interval separately, and match the solutions in the different intervals by requiring that ϕ and $\dot{\phi}$ be continuous functions of t.
8. A spiral spring is stretched 6 in. by a mass that weighs 8 lb. The mass is attached to a dashpot mechanism which has a damping constant of 0.25 lb-sec/ft. If the mass is subjected to an external force of the form $4\cos 2t$ lb, determine the steady state response of the system.
9. In Problem 8 what is the optimum choice of the mass m in order to maximize the steady state response of the system to the external force?
10. A spring–mass system has spring constant 3 N/m. A mass of 2 kg is attached to the spring and the motion takes place in a viscous fluid which offers a resistance numerically equal to the magnitude of the instantaneous velocity. If the system is driven by an external force of $3\cos 3t - 2\sin 3t$ N, determine the steady state motion. Express your answer in the form $R\cos(\omega t - \delta)$.
11. Show that $M(\omega) = [m^2(\omega_0^2 - \omega^2)^2 + c^2\omega^2]^{-1/2}$ has a maximum at $\omega = 0$ if $c^2 > 2km$ and at $\omega^2 = \omega_0^2 - \frac{1}{2}(c/m)^2$ if $c^2 < 2km$.
12. Determine the solution of the differential equation $m\ddot{u} + c\dot{u} + ku = F_0\sin\omega t$ satisfying the following initial conditions. Assume $c^2 - 4km < 0$.

    (a) $u(0) = u_0$, $\dot{u}(0) = 0$; (b) $u(0) = 0$, $\dot{u}(0) = \dot{u}_0$; (c) $u(0) = u_0$, $\dot{u}(0) = \dot{u}_0$.

3.8 Electrical Networks

As a second example of the application of the theory of linear second order differential equations with constant coefficients, we consider the flow of electric current in a simple series circuit. The circuit is depicted in Figure 3.13. The current $I = \phi(t)$ (measured in amperes) is a function of the time t. The resistance R (ohms), the capacitance C (farads), and the inductance L (henrys) are all positive, and in general may depend on the time t and the current I. For a wide variety of applications this dependence can be neglected, and we assume that R, C, and L are known constants. The impressed voltage E (volts) is a given function of time—often of the form $E_0\cos\omega t$. Another physical quantity that

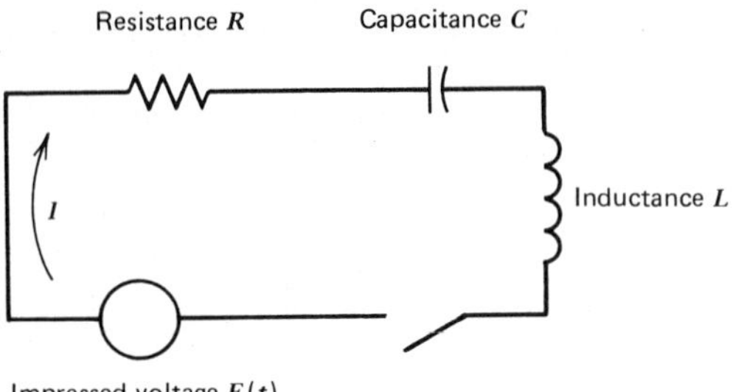

FIGURE 3.13 A simple electric circuit.

appears in our discussion is the total charge Q (coulombs) on the capacitor at the time t. The charge Q is related to the current I by $I = dQ/dt$.

The flow of current in the circuit under consideration is governed by Kirchhoff's[19] second law: *In a closed circuit the impressed voltage is equal to the sum of the voltage drops in the rest of the circuit.*

According to the elementary laws of electricity we know that

$$\text{The voltage drop across the resistance} = IR,$$
$$\text{The voltage drop across the capacitor} = Q/C$$
$$\text{The voltage drop across the inductance} = L\,dI/dt.$$

Hence

$$L\frac{dI}{dt} + RI + \frac{1}{C}Q = E(t). \tag{1}$$

The units have been chosen so that 1 volt = 1 ohm · 1 ampere = 1 coulomb/1 farad = 1 henry · 1 ampere/1 second.

Since $I = dQ/dt$ we can substitute for I in Eq. (1), obtaining the following second order linear nonhomogeneous equation for Q,

$$L\ddot{Q} + R\dot{Q} + (1/C)Q = E(t), \tag{2}$$

where dots denote differentiation with respect to t. The initial conditions that must be specified, say at time $t = 0$, are

$$Q(0) = Q_0, \qquad \dot{Q}(0) = I(0) = I_0. \tag{3}$$

Thus we must know the initial charge on the capacitor and the initial current in the circuit.

Alternatively, we can obtain a second order differential equation for the current I by differentiating Eq. (1) with respect to t, and then substituting for

[19]Gustav Kirchhoff (1824–1887), a German physicist, is famous for his contributions in electricity (the laws that bear his name) and for his work in spectroscopy.

TABLE 3.2 THE ANALOGY BETWEEN MECHANICAL AND ELECTRICAL VIBRATIONS

Mechanical System		Electric Circuit	
$m\ddot{u} + c\dot{u} + ku = F(t)$		$L\ddot{Q} + R\dot{Q} + \frac{1}{C}Q = E(t)$	
Displacement	u	Charge	Q
Velocity	$\dot{u}$	Current	$I = \dot{Q}$
Mass	m	Inductance	L
Damping	c	Resistance	R
Spring constant	k	Elastance	$1/C$
Impressed force	$F(t)$	Impressed voltage	$E(t)$

dQ/dt. This gives

$$L\ddot{I} + R\dot{I} + (1/C)I = \dot{E}(t). \tag{4}$$

The initial conditions that must be specified are

$$I(0) = I_0, \qquad \dot{I}(0) = \dot{I}_0. \tag{5}$$

Notice from Eq. (1) that

$$\dot{I}_0 = \frac{1}{L}\left[E(0) - RI_0 - \frac{1}{C}Q_0\right]. \tag{6}$$

Hence $\dot{I}_0$ is known if we specify $I(0)$ and $Q(0)$. The initial charge and current are physically measurable quantities and a knowledge of these quantities allows us to determine the initial conditions (3) corresponding to Eq. (2) or the initial conditions (5) corresponding to Eq. (4).

It is interesting to note the analogy between the mechanical vibration problem discussed earlier and the present electric circuit problem. This is clearly shown in Table 3.2. This analogy between the two systems allows us to solve problems in mechanical vibrations by building the corresponding electrical circuit and then measuring the charge Q. This is one of the basic ideas in the field of analog computers.[20]

Let us now consider Eq. (2) when the impressed voltage is a periodic function, $E_0 \cos \omega t$. Just as for the spring–mass system discussed in Sections 3.7.1 and 3.7.2 we can develop the concepts of a natural frequency, critical damping, resonance, beats, transient solution, and steady state solution for the series circuit. This is done in the following example and in the problems.

EXAMPLE

Determine the behavior of the solution of Eq. (2) as $t \to \infty$ if $E(t) = E_0 \cos \omega t$ and $R \neq 0$.

[20] For a particularly simple introduction to the theory of analog computers, see T. D. Truitt and A. E. Rogers.

First, since L, R, and C are positive it follows that all solutions of the homogeneous equation corresponding to Eq. (2) approach zero as $t \to \infty$ (see Section 3.7.1). To find a particular solution of

$$L\ddot{Q} + R\dot{Q} + (1/C)Q = E_0 \cos \omega t \tag{7}$$

we use the method of undetermined coefficients. Substituting

$$Q_p(t) = A \cos \omega t + B \sin \omega t \tag{8}$$

we find that A and B must satisfy

$$\left(\frac{1}{C} - L\omega^2\right)A + \omega RB = E_0,$$

$$-\omega RA + \left(\frac{1}{C} - L\omega^2\right)B = 0.$$

Solving for A and B and substituting for them in Eq. (8) gives

$$Q_p(t) = \frac{E_0(1/C - L\omega^2)\cos \omega t + \omega R E_0 \sin \omega t}{(1/C - L\omega^2)^2 + \omega^2 R^2}. \tag{9}$$

Equation (9) can also be written as

$$Q_p(t) = \frac{E_0 \sin(\omega t + \delta)}{\sqrt{(1/C - L\omega^2)^2 + \omega^2 R^2}}, \tag{10}$$

where δ is given by $\sin \delta = (1/C - L\omega^2)/\Delta$, $\cos \delta = \omega R/\Delta$ and $\Delta = \sqrt{(1/C - L\omega^2)^2 + \omega^2 R^2}$. Since $Q = Q_c(t) + Q_p(t)$ and $Q_c(t) \to 0$ as $t \to \infty$, it follows that $Q \to Q_p(t)$ as $t \to \infty$. For this reason $Q_p(t)$ is often referred to as the steady state solution of Eq. (7).

Generally, we wish to know the steady state current in the circuit, and this can be found by differentiating $Q_p(t)$. We obtain

$$I_p(t) = \frac{E_0 \cos(\omega t + \delta)}{\sqrt{R^2 + [\omega L - 1/(\omega C)]^2}}. \tag{11}$$

Notice that the steady state current has the same frequency as the impressed voltage. The term $\omega L - 1/(\omega C)$ that appears in Eq. (11) is known as the *reactance* of the circuit, and the expression $\sqrt{R^2 + [\omega L - 1/(\omega C)]^2}$ is called the *impedance* of the circuit.

PROBLEMS

1. If there is no resistance present in a simple series circuit, show that in the absence of an impressed voltage the charge Q on the capacitor is periodic in time with circular frequency $\omega_0 = \sqrt{1/LC}$. The quantity $\sqrt{1/LC}$ is referred to as the natural frequency of the circuit.
2. Show that if there is no resistance in the circuit and the impressed voltage is of the form $E_0 \cos \omega t$, then the charge on the capacitor will become unbounded as $t \to \infty$ if $\omega = \sqrt{1/LC}$. This is the phenomenon of resonance. Show that the charge will always be bounded, no matter what the choice of ω, provided that there is some resistance, no matter how small, in the circuit.
3. Show that the solution of $L\ddot{I} + R\dot{I} + (1/C)I = 0$ is of the form

$$I = \begin{cases} Ae^{r_1 t} + Be^{r_2 t}, & \text{for } R^2 - 4L/C > 0; \\ (A + Bt)e^{\lambda t}, & \text{for } R^2 - 4L/C = 0; \\ e^{\lambda t}(A \cos \mu t + B \sin \mu t) & \text{for } R^2 - 4L/C < 0, \end{cases}$$

   The circuit is said to be overdamped, critically damped, or underdamped corresponding to these three cases, respectively.
4. Suppose that a series circuit consisting of an inductor, a resistor, and a capacitor is open, and there is an initial charge $Q_0 = 10^{-6}$ on the capacitor. Determine the variation of the charge and the current after the switch is closed for the following cases:

   (a) $L = 0.2$, $C = 10^{-5}$, $R = 3 \times 10^2$
   (b) $L = 1$, $C = 4 \times 10^{-6}$, $R = 10^3$
   (c) $L = 2$, $C = 10^{-5}$, $R = 4 \times 10^2$

5. If $L = 0.2$ henry and $C = 0.8 \times 10^{-6}$ farads, determine the resistance R so that the circuit is critically damped.
6. A series circuit has a capacitor of 0.25×10^{-6} farad, a resistor of 5×10^3 ohms, and an inductor of 1 henry. The initial charge on the capacitor is zero. If a 12-volt battery is connected to the circuit and the circuit is closed at $t = 0$, determine the charge on the capacitor at $t = 0.001$ sec, at $t = 0.01$ sec, and the steady state charge.
7. Determine the steady state current in a series circuit if the impressed voltage is $E(t) = 110 \cos 120\pi t$ volts, and $L = 10$ henrys, $R = 3 \times 10^3$ ohms, and $C = 0.25 \times 10^{-5}$ farad.
8. For a series circuit with given values of L, R, and C and an impressed voltage $E_0 \cos \omega t$, for what value of ω will the steady state current be a maximum?

*9. In electrical engineering practice it is often convenient to think of $E_0 \cos \omega t$ as the real part of $E_0 e^{j\omega t}$ ($j = \sqrt{-1}$, which is standard notation in electrical engineering). Then corresponding to Eq. (7) of the text we would consider

$$L\ddot{Q} + R\dot{Q} + \frac{1}{C}Q = E_0 e^{j\omega t}, \tag{i}$$

where it is understood that when we finish our computations we must take the real part of the solution.[21] Assume that $R \neq 0$.
(a) Show that the steady state solution of Eq. (i) is

$$Q = \frac{E_0}{j\omega R - \omega^2 L + 1/C} e^{j\omega t}.$$

Hint: Substitute $Q = Ae^{j\omega t}$.
(b) Hence show that the steady state current is

$$I = \frac{E_0}{R + j(\omega L - 1/\omega C)} e^{j\omega t}.$$

(c) Using the fact that $\alpha + i\beta = \sqrt{\alpha^2 + \beta^2}\, e^{j\gamma}$, where $\tan\gamma = \beta/\alpha$, show that

$$I = \frac{E_0}{\sqrt{R^2 + (\omega L - 1/\omega C)^2}} e^{j(\omega t - \delta)},$$

where $\tan\delta = (\omega L - 1/\omega C)/R$. Finally, show that the real part of this expression is identical with that derived in the example of this section.
(d) What is the steady state current if the impressed voltage is $E(t) = E_0 \sin\omega t$?
(e) The complex parameter $Z = R + j(\omega L - 1/\omega C)$ is known as the *complex impedance*. The reciprocal of Z is called the *admittance*, and the real and imaginary parts of $1/Z$ are called the *conductance* and *susceptance*. Determine the admittance, conductance, and susceptance.

REFERENCES

Coddington, E. A., *An Introduction to Ordinary Differential Equations* (Englewood Cliffs, N.J.; Prentice-Hall, 1961).

Ince, E. L., *Ordinary Differential Equations* (London: Longmans, Green, 1927; New York: Dover, 1956).

[21] The use of complex notation for electrical circuit problems was pioneered by the mathematician and inventor Charles P. Steinmetz (1865–1923). The use of his symbolic notation in studying alternating current phenomena was largely responsible for the rapid progress made in the commercial introduction of alternating current apparatus. He also discovered the law of hysteresis and made important contributions to the study of lightning phenomena. As a young man he was a socialist agitator who left Germany in 1888 just ahead of the police, and without getting his doctoral degree. He came to the United States in 1889 and was employed by what is now General Electric Company.

Two books on analog computers are:

Truitt, T. D., and Rogers, A. E., *Basics of Analog Computers* (New York: Rider, 1960).

Jenners, Roger R., *Analog Computation and Simulation: Laboratory Approach* (Boston: Allyn and Bacon, 1965).

There are many books on mechanical vibrations and electric circuits. A classic book on mechanical vibrations is:

Den Hartog, J. P., *Mechanical Vibrations* (New York: McGraw-Hill, 1947).

More recent intermediate level books are:

Thomson, W. T., *Theory of Vibrations with Applications* (Englewood Cliffs, N.J.: Prentice-Hall, 1972).

Vierck, R. K., *Vibration Analysis* (Scranton, Pa.: International, 1967).

An elementary book with material on electrical circuits is:

Smith, J. R. J., *Circuits, Devices, and Systems* (2nd ed.) (New York: Wiley, 1976).

4.1 *Review of Power Series*

In this chapter we discuss the use of power series to construct fundamental sets of solutions of second order linear differential equations whose coefficients are functions of the independent variable. We begin by summarizing very briefly the pertinent results about infinite series, and in particular power series, that we need. Readers who are familiar with power series may go on to Section 4.2. Those who need more details than are presented here should consult a book on calculus.

1. A power series $\sum_{n=0}^{\infty} a_n(x - x_0)^n$ is said to converge at a point x if

$$\lim_{m \to \infty} \sum_{n=0}^{m} a_n(x - x_0)^n$$

exists. It is clear that the series converges for $x = x_0$; it may converge for all x, or it may converge for some values of x and not for others.

2. The series $\sum_{n=0}^{\infty} a_n(x - x_0)^n$ is said to converge absolutely at a point x if the series

$$\sum_{n=0}^{\infty} |a_n(x - x_0)^n|$$

converges. It can be shown that if the series converges absolutely, then the series also converges; however, the converse is not necessarily true.

3. One of the most useful tests for the absolute convergence of a power series is the ratio test. If $a_n \neq 0$, and if for a fixed value of x

$$\lim_{n\to\infty}\left|\frac{a_{n+1}(x-x_0)^{n+1}}{a_n(x-x_0)^n}\right| = |x-x_0|\lim_{n\to\infty}\left|\frac{a_{n+1}}{a_n}\right| = l,$$

then the power series converges absolutely at that value of x if $l < 1$, and diverges if $l > 1$. If $l = 1$ the test is inconclusive.

EXAMPLE 1

For which values of x does the power series

$$\sum_{n=1}^{\infty}(-1)^{n+1}n(x-2)^n$$

converge?

To test for convergence we use the ratio test. We have

$$\lim_{n\to\infty}\left|\frac{(-1)^{n+2}(n+1)(x-2)^{n+1}}{(-1)^{n+1}n(x-2)^n}\right| = |x-2|\lim_{n\to\infty}\frac{n+1}{n} = |x-2|.$$

According to Statement 3 the series converges absolutely for $|x-2| < 1$, or $1 < x < 3$, and diverges for $|x-2| > 1$. The values of x corresponding to $|x-2| = 1$ are $x = 1$ and $x = 3$. The series diverges for each of these values of x since the nth term of the series does not approach zero as $n \to \infty$.

4. If the power series $\sum_{n=0}^{\infty} a_n(x-x_0)^n$ converges at $x = x_1$, it converges absolutely for $|x-x_0| < |x_1-x_0|$; and if it diverges at $x = x_1$, it diverges for $|x-x_0| > |x_1-x_0|$.

5. There is a number $\rho \geq 0$, called the *radius of convergence*, such that $\sum_{n=0}^{\infty} a_n(x-x_0)^n$ converges absolutely for $|x-x_0| < \rho$ and diverges for $|x-x_0| > \rho$. For a series that converges nowhere except at x_0, we define ρ to

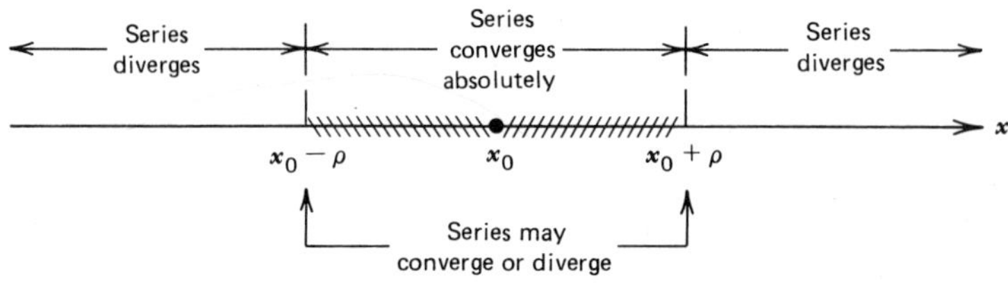

FIGURE 4.1 The interval of convergence of a power series.

be zero; for a series that converges for all x, we say that ρ is infinite. If $\rho > 0$, then the interval $|x - x_0| < \rho$ is called the *interval of convergence*; it is indicated by the hatched lines in Figure 4.1. The series may either converge or diverge when $|x - x_0| = \rho$.

EXAMPLE 2

Determine the radius of convergence of the power series

$$\sum_{n=1}^{\infty} \frac{(x+1)^n}{n2^n}.$$

We apply the ratio test:

$$\lim_{n\to\infty} \left| \frac{(x+1)^{n+1}}{(n+1)2^{n+1}} \frac{n2^n}{(x+1)^n} \right| = |x+1| \lim_{n\to\infty} \frac{n}{2(n+1)} = \frac{|x+1|}{2}.$$

Thus the series converges absolutely for $|x + 1| < 2$, or $-3 < x < 1$, and diverges for $|x + 1| > 2$. The radius of convergence of the power series is $\rho = 2$. Finally, we check the end points of the interval of convergence. At $x = -3$ we have

$$\sum_{n=1}^{\infty} \frac{(-3+1)^n}{n2^n} = \sum_{n=1}^{\infty} \frac{(-1)^n}{n}$$

which converges, but does not converge absolutely. The series is said to converge conditionally at $x = -3$. At $x = 1$ the series becomes

$$\sum_{n=1}^{\infty} \frac{1}{n},$$

which diverges. To summarize, the series converges for $-3 \le x < 1$, converges absolutely for $-3 < x < 1$, and has radius of convergence 2.

If $\sum_{n=0}^{\infty} a_n(x - x_0)^n$ and $\sum_{n=0}^{\infty} b_n(x - x_0)^n$ converge to $f(x)$ and $g(x)$, respectively, for $|x - x_0| < \rho$, $\rho > 0$, then the following are true for $|x - x_0| < \rho$.

6. The series can be added or subtracted termwise, and

$$f(x) \pm g(x) = \sum_{n=0}^{\infty} (a_n \pm b_n)(x - x_0)^n.$$

7. The series can be formally multiplied, and

$$f(x)g(x) = \left[\sum_{n=0}^{\infty} a_n(x - x_0)^n\right]\left[\sum_{n=0}^{\infty} b_n(x - x_0)^n\right] = \sum_{n=0}^{\infty} c_n(x - x_0)^n$$

where $c_n = a_0 b_n + a_1 b_{n-1} + \cdots + a_n b_0$. Further, if $g(x_0) \neq 0$, the series can be formally divided and

$$\frac{f(x)}{g(x)} = \sum_{n=0}^{\infty} d_n(x - x_0)^n,$$

although the calculation of the d_n is somewhat complicated. Also, in the case of division the radius of convergence of the resulting power series may be less than ρ.

8. The function f is continuous and has derivatives of all orders for $|x - x_0| < \rho$. Further, $f', f'', \ldots$ can be computed by differentiating the series termwise; that is,

$$f'(x) = a_1 + 2a_2(x - x_0) + \cdots + na_n(x - x_0)^{n-1} + \cdots$$
$$= \sum_{n=1}^{\infty} na_n(x - x_0)^{n-1},$$
$$f''(x) = 2a_2 + 6a_3(x - x_0) + \cdots + n(n-1)a_n(x - x_0)^{n-2} + \cdots$$
$$= \sum_{n=2}^{\infty} n(n-1)a_n(x - x_0)^{n-2},$$

and so forth, and each of the series converges absolutely for $|x - x_0| < \rho$.

9. The value of a_n is given by

$$a_n = \frac{f^{(n)}(x_0)}{n!}.$$

The series is called the Taylor[1] series for the function f about $x = x_0$.

10. If $\sum_{n=0}^{\infty} a_n(x - x_0)^n = \sum_{n=0}^{\infty} b_n(x - x_0)^n$ for each x, then $a_n = b_n$, $n = 0, 1, 2, \ldots$. In particular, if $\sum_{n=0}^{\infty} a_n(x - x_0)^n = 0$ for each x, then $a_0 = a_1 = \cdots = a_n = \cdots = 0$.

A function f that has a Taylor series expansion about $x = x_0$,

$$f(x) = \sum_{n=0}^{\infty} \frac{f^{(n)}(x_0)}{n!}(x - x_0)^n,$$

with a radius of convergence $\rho > 0$ is said to be *analytic* at $x = x_0$. According to the above statements, if f and g are analytic at x_0 then $f \pm g$, $f \cdot g$, and f/g (provided $g(x_0) \neq 0$) are analytic at $x = x_0$. The result that will be used most often in the following sections is that a polynomial is analytic at every point; thus sums, differences, products, and quotients (except at zeros of the

[1] Brook Taylor (1685–1731) was the leading English mathematician in the generation following Newton. In 1715 he published a general statement of the expansion theorem that is named for him, a result that is fundamental in all branches of analysis. He was also one of the founders of the calculus of finite differences, and was the first to recognize the existence of singular solutions of differential equations.

denominator) of polynomials are analytic at every point.

SHIFT OF INDEX OF SUMMATION. The index of summation in an infinite series is a dummy parameter just as the integration variable in a definite integral is a dummy variable. For example,

$$\sum_{n=0}^{\infty} \frac{2^n x^n}{n!} = \sum_{j=0}^{\infty} \frac{2^j x^j}{j!}.$$

Just as we make changes of the variable of integration in a definite integral, we find it convenient to make changes of summation indices in calculating series solutions of differential equations. We illustrate by several examples how to shift the summation index.

1.
$$\begin{aligned}\sum_{n=2}^{\infty} a_n x^n &= a_2x^2 + a_3x^3 + a_4x^4 + \cdots + a_nx^n + \cdots \\ &= a_{0+2}x^{0+2} + a_{1+2}x^{1+2} + \cdots + a_{m+2}x^{m+2} + \cdots \\ &= \sum_{m=0}^{\infty} a_{m+2}x^{m+2} = \sum_{n=0}^{\infty} a_{n+2}x^{n+2}.\end{aligned}$$

We have shifted the counting index by $+2(n \to n+2)$ and have started counting 2 lower.

2.
$$\sum_{n=2}^{\infty} (n+2)(n+1)a_n(x-x_0)^{n-2}.$$

It may be desirable to write this series as a series with generic term $(x - x_0)^n$ rather than $(x - x_0)^{n-2}$. To accomplish this we shift the index by $+2(n \to n + 2)$ and start counting 2 lower as in the previous example. We obtain

$$\sum_{n=0}^{\infty} (n+4)(n+3)a_{n+2}(x-x_0)^n.$$

It can be readily verified that the terms in the two series are exactly the same.

3.
$$x^2 \sum_{n=0}^{\infty} (n+r)a_n x^{n+r-1}.$$

First we take the x^2 inside the summation obtaining

$$\sum_{n=0}^{\infty} (n+r)a_n x^{n+r+1}.$$

To write this series as a new series with generic term x^{n+r} we shift the index down by 1 ($n \to n - 1$) and start counting 1 higher. Thus

$$\sum_{n=0}^{\infty} (n + r)a_n x^{n+r+1} = \sum_{n=1}^{\infty} (n + r - 1)a_{n-1} x^{n+r}.$$

Again, it can be readily verified that the terms in the two series are exactly the same.

4. As a final example we consider the implication of the equation

$$\sum_{n=1}^{\infty} n a_n x^{n-1} = \sum_{n=0}^{\infty} a_n x^n.$$

We wish to use Statement 10 to determine the a_n. To equate coefficients of like powers of x, it is helpful to have the generic term in each series involve x^n. To accomplish this we shift the index of summation on the left side by the rule $n \to n + 1$ and start counting one lower. We obtain

$$\sum_{n=0}^{\infty} (n + 1)a_{n+1} x^n = \sum_{n=0}^{\infty} a_n x^n.$$

According to Statement 10 we conclude that

$$(n + 1)a_{n+1} = a_n, \qquad n = 0, 1, 2, 3, \ldots,$$

or

$$a_{n+1} = a_n/(n + 1).$$

Hence

$$a_1 = a_0, \qquad a_2 = a_1/2 = a_0/2, \qquad a_3 = a_2/3 = a_0/2 \cdot 3 = a_0/3!,$$

and in general

$$a_n = a_0/n!, \qquad n = 1, 2, 3, \ldots .$$

Thus the original relation determines all of the coefficients in terms of a_0. Moreover,

$$\sum_{n=0}^{\infty} a_n x^n = \sum_{n=0}^{\infty} a_0 \frac{x^n}{n!} = a_0 \sum_{n=0}^{\infty} \frac{x^n}{n!},$$

and this is precisely $a_0 e^x$. In writing the above series we have used the common convention that $0! = 1$.

PROBLEMS

1. Determine the radius of the convergence of each of the following power series.

(a) $\sum_{n=0}^{\infty} (x - 3)^n$ (b) $\sum_{n=0}^{\infty} \frac{n}{2^n} x^n$ (c) $\sum_{n=0}^{\infty} \frac{x^{2n}}{n!}$

(d) $\sum_{n=0}^{\infty} 2^n x^n$ (e) $\sum_{n=1}^{\infty} \frac{(2x + 1)^n}{n^2}$ (f) $\sum_{n=1}^{\infty} \frac{(x - x_0)^n}{n}$

(g) $\sum_{n=1}^{\infty} \frac{(-1)^n n^2 (x + 2)^n}{3^n}$ (h) $\sum_{n=0}^{\infty} \frac{(-1)^n (n!)^2 x^{2n+1}}{(2n)!}$ (i) $\sum_{n=1}^{\infty} \frac{n! x^n}{n^n}$

2. Determine the Taylor series about the point x_0 for each of the following functions. Also determine the radius of convergence of the series.

(a) $\sin x, \quad x_0 = 0$ (b) $e^x, \quad x_0 = 0$
(c) $x, \quad x_0 = 1$ (d) $x^2, \quad x_0 = -1$
(e) $\ln x, \quad x_0 = 1$ (f) $\dfrac{1}{1+x}, \quad x_0 = 0$
(g) $\dfrac{1}{1-x}, \quad x_0 = 0$ (h) $\dfrac{1}{1-x}, \quad x_0 = 2$

3. Given that $y = \sum_{n=0}^{\infty} nx^n$, compute y' and y'' and write out the first four terms of each series as well as the coefficient of x^n in the general term.
4. Given that $y = \sum_{n=0}^{\infty} a_n x^n$, compute y' and y'' and write out the first four terms of each series as well as the coefficient of x^n in the general term. Show that if $y'' = y$, then the coefficients a_0 and a_1 are arbitrary, and determine a_2 and a_3 in terms of a_0 and a_1. Show that $a_{n+2} = a_n/(n+2)(n+1)$, $n = 0, 1, 2, 3, \ldots$.
5. Verify the following:

(a) $\sum_{n=0}^{\infty} a_n (x-1)^{n+1} = \sum_{n=1}^{\infty} a_{n-1}(x-1)^n$

(b) $\sum_{n=2}^{\infty} n(n-1)a_n x^{n-2} = \sum_{n=0}^{\infty} (n+2)(n+1)a_{n+2}x^n$

(c) $\sum_{n=0}^{\infty} a_n x^{n+2} = \sum_{n=2}^{\infty} a_{n-2}x^n$

(d) $\sum_{n=k}^{\infty} a_{n+m}x^{n+p} = \sum_{n=0}^{\infty} a_{n+m+k}x^{n+p+k}, \qquad x > 0,$

where k and m are given integers and p is a constant.

(e) $x^2 \sum_{n=0}^{\infty} (n+r)(n+r-1)a_n x^{n+r-2} + \alpha x \sum_{n=0}^{\infty} (n+r)a_n x^{n+r-1} + \beta x^2 \sum_{n=0}^{\infty} a_n x^{n+r}$

$$= [r(r-1) + \alpha r]a_0 x^r + [(r+1)r + \alpha(r+1)]a_1 x^{r+1}$$

$$+ \sum_{n=2}^{\infty} [(n+r)(n+r-1)a_n + \alpha(n+r)a_n + \beta a_{n-2}]x^{n+r}, \qquad x > 0,$$

where r is a constant.

6. Determine the a_n so that the equation

$$\sum_{n=1}^{\infty} na_n x^{n-1} + 2\sum_{n=0}^{\infty} a_n x^n = 0$$

is satisfied. Try to identify the function represented by the series $\sum_{n=0}^{\infty} a_n x^n$.

4.2 Series Solutions near an Ordinary Point, Part I

In Chapter 3 we described methods of solving second order linear differential equations with constant coefficients. We now consider methods of solving second order linear equations when the coefficients are functions of the independent variable. It is sufficient to consider the homogeneous equation

$$P(x)\frac{d^2y}{dx^2} + Q(x)\frac{dy}{dx} + R(x)y = 0, \tag{1}$$

since the procedure for the corresponding nonhomogeneous equation is similar.

A wide class of problems in mathematical physics leads to equations of the form (1) having polynomial coefficients; for example, the Bessel equation

$$x^2y'' + xy' + (x^2 - \nu^2)y = 0,$$

where ν is a constant, and the Legendre equation

$$(1 - x^2)y'' - 2xy' + \alpha(\alpha + 1)y = 0,$$

where α is a constant. For this reason, as well as to simplify the algebraic computations, we primarily consider the case in which the functions P, Q, and R are polynomials. However, as we will see, the method of solution is applicable for a class of functions more general than polynomials.

For the present, then, suppose that P, Q, and R are polynomials, and that they have no common factors. Suppose also that we wish to solve Eq. (1) in the neighborhood of a point x_0. The solution of Eq. (1) in an interval containing x_0 is closely associated with the behavior of P in that interval.

A point x_0 such that $P(x_0) \neq 0$ is called an *ordinary point*. Since P is continuous, it follows that there is an interval about x_0 in which $P(x)$ is never zero. In that interval we can divide Eq. (1) by $P(x)$ to obtain

$$y'' + p(x)y' + q(x)y = 0, \tag{2}$$

where $p(x) = Q(x)/P(x)$ and $q(x) = R(x)/P(x)$ are continuous functions. Hence, according to the existence and uniqueness Theorem 3.2, there exists in that interval a unique solution of Eq. (1) that also satisfies the initial conditions $y(x_0) = y_0$, $y'(x_0) = y_0'$ for arbitrary values of y_0 and y_0'. In this and the following section we discuss the solution of Eq. (1) in the neighborhood of an ordinary point.

On the other hand, if $P(x_0) = 0$, then x_0 is called a *singular point* of Eq. (1). In this case at least one of $Q(x_0)$ and $R(x_0)$ is not zero. Consequently, at least one of the coefficients p and q in Eq. (2) becomes unbounded as $x \to x_0$, and therefore Theorem 3.2 does not apply in this case. Sections 4.3 through 4.7 deal with finding solutions of Eq. (1) in the neighborhood of a singular point.

We now take up the problem of solving Eq. (1) in the neighborhood of an ordinary point. We look for solutions of the form

$$y = a_0 + a_1(x - x_0) + \cdots + a_n(x - x_0)^n + \cdots = \sum_{n=0}^{\infty} a_n(x - x_0)^n. \quad (3)$$

We must consider two questions. The first is whether we can formally determine[2] the a_n so that y as given by Eq. (3) satisfies Eq. (1). The second is whether the series thus determined actually converges, and if so, for what values of $x - x_0$. If we can show that the series does converge for $|x - x_0| < \rho$, $\rho > 0$, then all the formal procedures such as termwise differentiation can be justified, and we will have constructed a solution of Eq. (1) that is valid for $|x - x_0| < \rho$.

We postpone until Section 4.2.1 all consideration of more theoretical matters such as the radius of convergence of the series (3). For the present we simply assume that such a solution exists, and show how to determine the a_n. The most practical way to do this is to substitute the series (3) and its derivatives for y, y', and y'' in Eq. (1); the a_n are then determined so that the differential equation is formally satisfied. The following examples illustrate the procedure. The differential equations are also of considerable importance in their own right.

EXAMPLE 1

Find a series solution of the equation

$$y'' + y = 0, \qquad -\infty < x < \infty. \quad (4)$$

As we well know, two linearly independent solutions of this equation are $\sin x$ and $\cos x$. To obtain practice in finding a series solution we illustrate how the $\sin x$ and $\cos x$ solutions can be obtained using power series. For Eq. (4), $P(x) = 1$, $Q(x) = 0$, and $R(x) = 1$; hence every point is an ordinary point, in particular, $x = 0$.

We look for a solution of the form

$$y = a_0 + a_1x + a_2x^2 + \cdots + a_nx^n + a_{n+1}x^{n+1} + a_{n+2}x^{n+2} + \cdots$$

$$= \sum_{n=0}^{\infty} a_nx^n. \quad (5)$$

[2] By "formally determine" we mean "carry out all the algebraic computations necessary for the determination of a solution without justifying at each step that it is permissible to perform the computation."

Differentiating term by term yields

$$y' = a_1 + 2a_2 x + \cdots + na_n x^{n-1} + (n+1)a_{n+1}x^n + (n+2)a_{n+2}x^{n+1} + \cdots = \sum_{n=1}^{\infty} na_n x^{n-1}, \tag{6}$$

$$y'' = 2a_2 + \cdots + n(n-1)a_n x^{n-2} + (n+1)na_{n+1}x^{n-1} + (n+2)(n+1)a_{n+2}x^n + \cdots = \sum_{n=2}^{\infty} n(n-1)a_n x^{n-2}. \tag{7}$$

Substituting the series (5) and (7) for y and y'' in Eq. (4) gives

$$\sum_{n=2}^{\infty} n(n-1)a_n x^{n-2} + \sum_{n=0}^{\infty} a_n x^n = 0.$$

Next, in the first sum, we shift the index of summation by replacing n by $n+2$ and starting the sum at 0 rather than 2. We obtain

$$\sum_{n=0}^{\infty} (n+2)(n+1)a_{n+2}x^n + \sum_{n=0}^{\infty} a_n x^n = 0$$

or

$$\sum_{n=0}^{\infty} [(n+2)(n+1)a_{n+2} + a_n]x^n = 0.$$

For this equation to be satisfied for all x it is necessary that the coefficient of each power of x be zero; hence, we conclude that

$$(n+2)(n+1)a_{n+2} + a_n = 0, \qquad n = 0, 1, 2, 3, \ldots. \tag{8}$$

Equation (8) is referred to as a *recurrence relation*. The successive coefficients can be evaluated one by one by writing the recurrence relation first for $n = 0$, then for $n = 1$, and so forth. In this example Eq. (8) relates each coefficient to the second one before it. Thus the even-numbered coefficients ($a_0, a_2, a_4, \ldots$) and the odd-numbered ones ($a_1, a_3, a_5, \ldots$) are determined separately. For the even-numbered coefficients we have

$$a_2 = -\frac{a_0}{2\cdot 1} = -\frac{a_0}{2!}, \qquad a_4 = -\frac{a_2}{4\cdot 3} = +\frac{a_0}{4!},$$

$$a_6 = -\frac{a_4}{6\cdot 5} = -\frac{a_0}{6!}, \quad \ldots.$$

These results suggest[3] that in general, if $n = 2k$, then

$$a_n = a_{2k} = \frac{(-1)^k}{(2k)!}a_0, \qquad k = 1, 2, 3, \ldots. \tag{9}$$

[3] The result given in Eq. (9) and other similar formulas in this chapter can be proved by mathematical induction. We assume that these results are plausible and omit the inductive argument.

Similarly, for the odd-numbered coefficients

$$a_3 = -\frac{a_1}{2 \cdot 3} = -\frac{a_1}{3!}, \qquad a_5 = -\frac{a_3}{5 \cdot 4} = +\frac{a_1}{5!},$$

$$a_7 = -\frac{a_5}{7 \cdot 6} = -\frac{a_1}{7!}, \quad \dots,$$

and in general, if $n = 2k + 1$, then

$$a_n = a_{2k+1} = \frac{(-1)^k}{(2k+1)!} a_1, \qquad k = 1, 2, 3, \dots \tag{10}$$

Thus the series (5) takes the form

$$\begin{aligned} y &= a_0 + a_1 x - \frac{a_0}{2!}x^2 - \frac{a_1}{3!}x^3 + \frac{a_0}{4!}x^4 + \frac{a_1}{5!}x^5 \\ &\quad + \cdots + \frac{(-1)^n a_0}{(2n)!}x^{2n} + \frac{(-1)^n a_1}{(2n+1)!}x^{2n+1} + \cdots \\ &= a_0\left[1 - \frac{x^2}{2!} + \frac{x^4}{4!} + \cdots + \frac{(-1)^n}{(2n)!}x^{2n} + \cdots\right] \\ &\quad + a_1\left[x - \frac{x^3}{3!} + \frac{x^5}{5!} + \cdots + \frac{(-1)^n}{(2n+1)!}x^{2n+1} + \cdots\right] \\ &= a_0 \sum_{n=0}^{\infty} \frac{(-1)^n}{(2n)!}x^{2n} + a_1 \sum_{n=0}^{\infty} \frac{(-1)^n}{(2n+1)!}x^{2n+1} \end{aligned} \tag{11}$$

The first series is exactly the Taylor series for $\cos x$; the second series is the Taylor series for $\sin x$. Thus, as we expected at the outset, we obtain the solution $y = a_0 \cos x + a_1 \sin x$. Notice that no conditions are imposed on a_0 and a_1; hence they are arbitrary. This is true in general for a series solution at an ordinary point; upon reflection this is not surprising. From Eqs. (5) and (6) we see that y and y' evaluated at $x = 0$ are a_0 and a_1, respectively. Since the initial conditions $y(0)$ and $y'(0)$ can be chosen arbitrarily, it follows that a_0 and a_1 should be arbitrary until specific initial conditions are stated.

Figures 4.2 and 4.3 show how the partial sums of the series in Eq. (11) approximate $\cos x$ and $\sin x$. As the number of terms increases, the interval over which the approximation is satisfactory becomes longer, and for each x in this interval the accuracy of the approximation improves.

In Example 1 we knew from the start that $\sin x$ and $\cos x$ form a fundamental set of solutions of Eq. (4). However, if we had not known this and had simply solved Eq. (4) using series methods, we would still have obtained the solution (11). In recognition of the fact that the differential equation (4) often occurs in applications we might decide to give the two solutions of Eq. (11) special names;

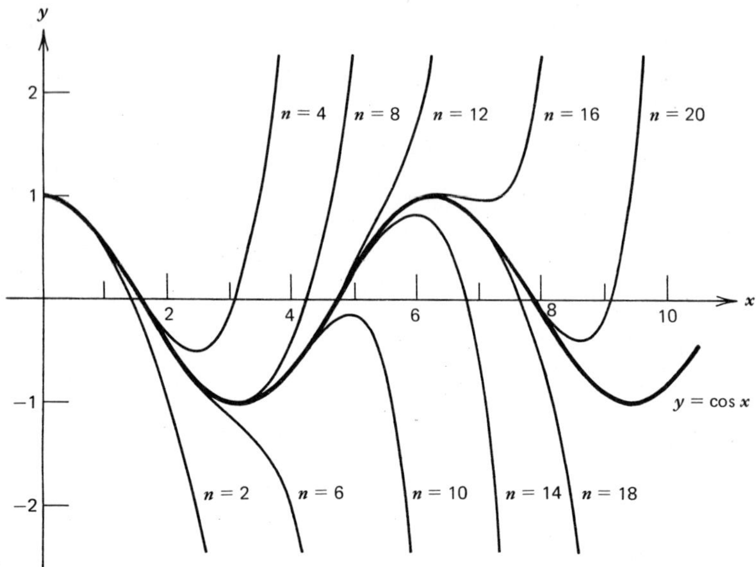

FIGURE 4.2 Polynomial approximations to $\cos x$*. The value of* n *is the degree of the approximating polynomial.*

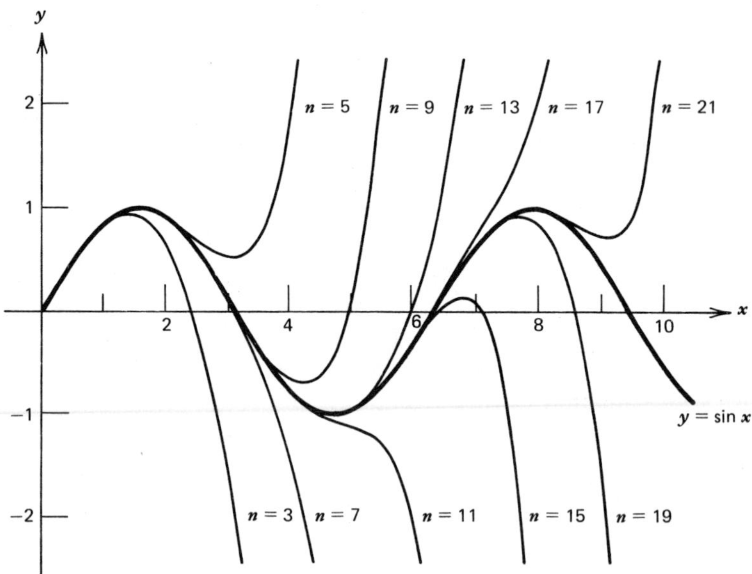

FIGURE 4.3 Polynomial approximations to $\sin x$*. The value of* n *is the degree of the approximating polynomial.*

perhaps,

$$\cos x \quad \text{for} \sum_{n=0}^{\infty} \frac{(-1)^n}{(2n)!} x^{2n}, \qquad \sin x \quad \text{for} \sum_{n=0}^{\infty} \frac{(-1)^n}{(2n+1)!} x^{2n+1}.$$

Then we might ask what properties these functions have. For instance, it is reasonably obvious from the series that $\sin 0 = 0$, $\cos 0 = 1$, $\cos(-x) = \cos x$, and $\sin(-x) = -\sin x$. Another formula that is easily derived is

$$\frac{d}{dx} \sin x = \frac{d}{dx} \sum_{n=0}^{\infty} \frac{(-1)^n}{(2n+1)!} x^{2n+1} = \sum_{n=0}^{\infty} \frac{(-1)^n (2n+1)}{(2n+1)!} x^{2n}$$

$$= \sum_{n=0}^{\infty} \frac{(-1)^n}{(2n)!} x^{2n} = \cos x.$$

Similarly, $d(\cos x)/dx = -\sin x$. Moreover, calculating with the infinite series, we can derive[4] all of the usual analytical and algebraic properties of the sine and cosine functions without any recourse to the area in which most students first learn about them—plane trigonometry.

It is interesting that these very important functions in mathematical physics can be defined as solutions of a simple, but frequently occurring, second order linear differential equation. To be precise, the function $\sin x$ can be defined as the solution of the initial value problem $y'' + y = 0$, $y(0) = 0$, $y'(0) = 1$; similarly, the function $\cos x$ can be defined as the solution of the initial value problem $y'' + y = 0$, $y(0) = 1$, $y'(0) = 0$. Other initial value problems can also be used to define the same functions; see Problem 12.

EXAMPLE 2

Find a series solution in powers of x of Airy's[5] equation

$$y'' = xy, \qquad -\infty < x < \infty. \tag{12}$$

For this equation $P(x) = 1$, $Q(x) = 0$, and $R(x) = -x$; hence every point is an ordinary point, in particular, $x = 0$. We assume that

$$y = \sum_{n=0}^{\infty} a_n x^n. \tag{13}$$

The series for y'' is given by Eq. (7); as explained in the previous example, we can

[4] Such an analysis is given in Section 24 of K. Knopp, *Theory and Applications of Infinite Series* (New York: Hafner, 1951).

[5] Sir George Airy (1801–1892), an English astronomer and mathematician, was director of the Greenwich Observatory from 1835 to 1881. One reason Airy's equation is of interest is that for x negative the solutions are oscillatory, similar to trigonometric functions, and for x positive they are monotonic, similar to hyperbolic functions. Can you explain why it is reasonable to expect such behavior?

rewrite it as

$$y'' = \sum_{n=0}^{\infty} (n+2)(n+1)a_{n+2}x^n. \tag{14}$$

Substituting the series (13) and (14) for y and y'' in Eq. (12), we obtain

$$\sum_{n=0}^{\infty} (n+2)(n+1)a_{n+2}x^n = x\sum_{n=0}^{\infty} a_n x^n = \sum_{n=0}^{\infty} a_n x^{n+1}. \tag{15}$$

Next, we shift the index of summation in the series on the right side of the above equation by replacing n by $n-1$ and starting the summation at 1 rather than zero. Thus, we have

$$2\cdot 1a_2 + \sum_{n=1}^{\infty} (n+2)(n+1)a_{n+2}x^n = \sum_{n=1}^{\infty} a_{n-1}x^n.$$

Again, for this equation to be satisfied for all x it is necessary that the coefficients of like powers of x be equal; hence $a_2 = 0$, and we obtain the recurrence relation

$$(n+2)(n+1)a_{n+2} = a_{n-1} \qquad \text{for} \quad n = 1, 2, 3, \ldots . \tag{16}$$

Since a_{n+2} is given in terms of a_{n-1}, it is clear that the a's are determined in steps of three. Thus a_0 determines a_3, which in turn determines $a_6, \ldots$; a_1 determines a_4, which in turn determines $a_7, \ldots$; and a_2 determines a_5, which in turn determines $a_8, \ldots$. Since $a_2 = 0$, we immediately conclude that $a_5 = a_8 = a_{11} = \cdots = 0$.

For the sequence $a_0, a_3, a_6, a_9, \ldots$ we set $n = 1, 4, 7, 10, \ldots$ in the recurrence relation:

$$a_3 = \frac{a_0}{3\cdot 2}, \qquad a_6 = \frac{a_3}{6\cdot 5} = \frac{a_0}{(6\cdot 5)(3\cdot 2)},$$

$$a_9 = \frac{a_6}{(9\cdot 8)} = \frac{a_0}{(9\cdot 8)(6\cdot 5)(3\cdot 2)}, \ldots .$$

For this sequence of coefficients it is convenient to write a formula for a_{3n}, $n = 1, 2, 3, \ldots$. The above results suggest the general formula

$$a_{3n} = \frac{1}{[(3n)(3n-1)][(3n-3)(3n-4)]\cdots[6\cdot 5][3\cdot 2]}a_0, \qquad n = 1, 2, \ldots .$$

This expression can also be obtained by rewriting the recurrence relation (16) as follows. First replace n by $n-2$ to obtain $n(n-1)a_n = a_{n-3}$; next, to generate the coefficients corresponding to a_0, replace n by $3n$, which gives $(3n)(3n-1)a_{3n} = a_{3n-3}$. Repeated application of this formula gives the above formula for a_{3n}.

For the sequence $a_1, a_4, a_7, a_{10}, \ldots$, we set $n = 2, 5, 8, 11, \ldots$ in the recurrence relation:

$$a_4 = \frac{a_1}{4\cdot 3}, \qquad a_7 = \frac{a_4}{7\cdot 6} = \frac{a_1}{(7\cdot 6)(4\cdot 3)},$$

$$a_{10} = \frac{a_7}{10\cdot 9} = \frac{a_1}{(10\cdot 9)(7\cdot 6)(4\cdot 3)}, \ldots .$$

In the same way as before we find that

$$a_{3n+1} = \frac{1}{[(3n+1)(3n)][(3n-2)(3n-3)]\cdots[7\cdot 6][4\cdot 3]}a_1,$$
$$n = 1, 2, 3, \ldots$$

The solution of Airy's equation is

$$y = a_0\left[1 + \frac{x^3}{3\cdot 2} + \frac{x^6}{6\cdot 5\cdot 3\cdot 2} + \cdots + \frac{x^{3n}}{(3n)(3n-1)\cdots 3\cdot 2} + \cdots\right]$$
$$+ a_1\left[x + \frac{x^4}{4\cdot 3} + \frac{x^7}{7\cdot 6\cdot 4\cdot 3} + \cdots\right.$$
$$\left. + \frac{x^{3n+1}}{(3n+1)(3n)(3n-2)(3n-3)\cdots 4\cdot 3} + \cdots\right]$$
$$= a_0\left[1 + \sum_{n=1}^{\infty}\frac{x^{3n}}{(3n)(3n-1)(3n-3)(3n-4)\cdots 3\cdot 2}\right]$$
$$+ a_1\left[x + \sum_{n=1}^{\infty}\frac{x^{3n+1}}{(3n+1)(3n)(3n-2)(3n-3)\cdots 4\cdot 3}\right]. \tag{17}$$

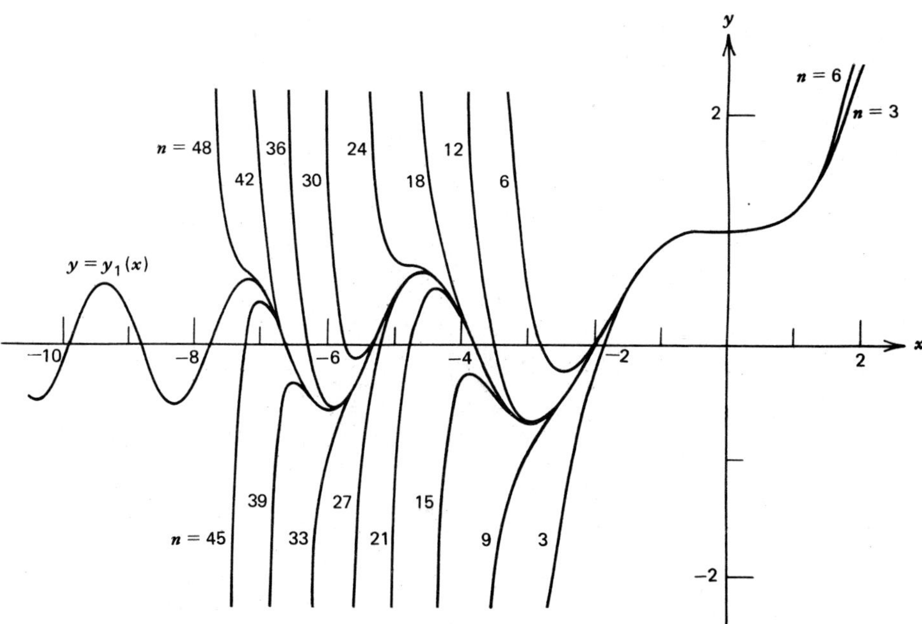

FIGURE 4.4 Polynomial approximations to the solution $y_1(x)$ of Airy's equation. The value of n is the degree of the approximating polynomial.

Because of the rapid growth of the denominators of the terms in the series (17), we might expect these series to have a large radius of convergence. Indeed, as we show in the next section, both of these series converge for all x. Also see Problem 11.

Assume for the moment that the series do converge for all x, and let y_1 and y_2 denote the functions defined by the first and second sets of brackets, respectively, in Eq. (17). Then, by first choosing $a_0 = 1$, $a_1 = 0$ and then $a_0 = 0$, $a_1 = 1$, it is clear that y_1 and y_2 are individually solutions of Eq. (12). Notice that y_1 satisfies the initial conditions $y_1(0) = 1$, $y_1'(0) = 0$ and y_2 satisfies the initial conditions $y_2(0) = 0$, $y_2'(0) = 1$. Thus $W(y_1, y_2)(0) = 1 \neq 0$, and consequently y_1 and y_2 are linearly independent. Hence the general solution of Airy's equation is

$$y = a_0 y_1(x) + a_1 y_2(x), \qquad -\infty < x < \infty.$$

In Figures 4.4 and 4.5, respectively, we show the graphs of the solutions y_1 and y_2 of Airy's equation, as well as graphs of several partial sums of the two series in Eq. (17). Observe that both y_1 and y_2 are monotone for $x > 0$ and oscillatory for $x < 0$. One can also see from the figures that the oscillations are not uniform, but decay in amplitude and increase in frequency as the distance from the origin increases.

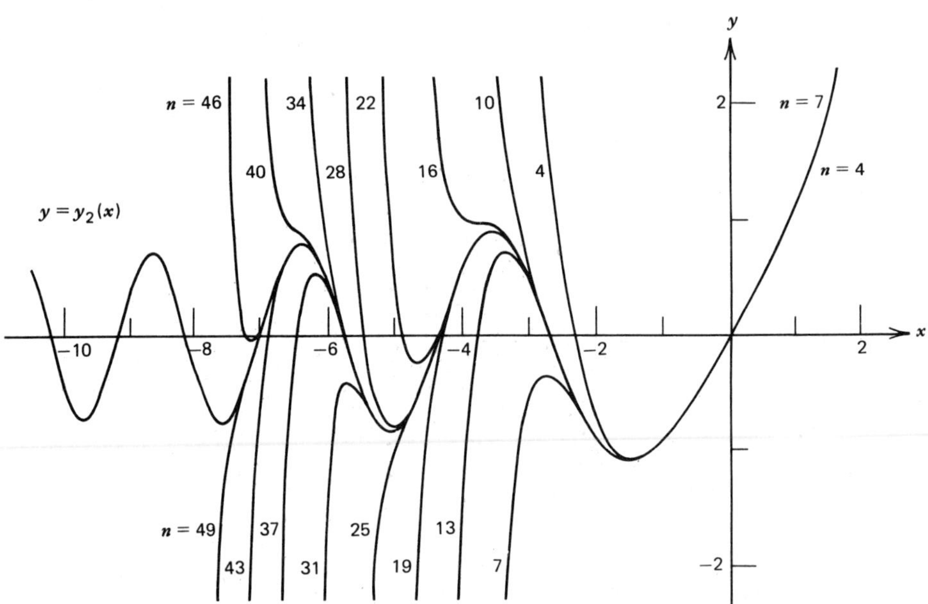

FIGURE 4.5 Polynomial approximations to the solution $y_2(x)$ of Airy's equation. The value of n is the degree of the approximating polynomial.

EXAMPLE 3

Find a solution of Airy's equation in powers of $x - 1$.

The point $x = 1$ is an ordinary point of Eq. (12), and thus we look for a solution of the form

$$y = \sum_{n=0}^{\infty} a_n(x - 1)^n.$$

Then

$$y' = \sum_{n=1}^{\infty} na_n(x - 1)^{n-1} = \sum_{n=0}^{\infty} (n + 1)a_{n+1}(x - 1)^n,$$

and

$$y'' = \sum_{n=2}^{\infty} n(n - 1)a_n(x - 1)^{n-2} = \sum_{n=0}^{\infty} (n + 2)(n + 1)a_{n+2}(x - 1)^n.$$

Substituting for y and y'' in Eq. (12) gives

$$\sum_{n=0}^{\infty} (n + 2)(n + 1)a_{n+2}(x - 1)^n = x \sum_{n=0}^{\infty} a_n(x - 1)^n. \tag{18}$$

Now to equate the coefficients of like powers of $(x - 1)$ we must express x, the coefficient of y in Eq. (12), in powers of $x - 1$; that is, we write $x = 1 + (x - 1)$. Note that this is precisely the Taylor series for x about $x = 1$. Then Eq. (18) takes the form

$$\sum_{n=0}^{\infty} (n + 2)(n + 1)a_{n+2}(x - 1)^n = [1 + (x - 1)] \sum_{n=0}^{\infty} a_n(x - 1)^n$$

$$= \sum_{n=0}^{\infty} a_n(x - 1)^n + \sum_{n=0}^{\infty} a_n(x - 1)^{n+1}.$$

Shifting the index of summation in the second series on the right gives

$$\sum_{n=0}^{\infty} (n + 2)(n + 1)a_{n+2}(x - 1)^n = \sum_{n=0}^{\infty} a_n(x - 1)^n + \sum_{n=1}^{\infty} a_{n-1}(x - 1)^n.$$

Equating coefficients of like powers of $x - 1$ we obtain

$$\begin{aligned} 2a_2 &= a_0, \\ (3 \cdot 2)a_3 &= a_1 + a_0, \\ (4 \cdot 3)a_4 &= a_2 + a_1, \\ (5 \cdot 4)a_5 &= a_3 + a_2, \\ &\vdots \end{aligned}$$

The general recurrence relation is

$$(n + 2)(n + 1)a_{n+2} = a_n + a_{n-1} \qquad \text{for} \quad n \geq 1. \tag{19}$$

Solving for the first few a_n in terms of a_0 and a_1 gives

$$a_2 = \frac{a_0}{2}, \qquad a_3 = \frac{a_1}{6} + \frac{a_0}{6},$$

$$a_4 = \frac{a_2}{12} + \frac{a_1}{12} = \frac{a_0}{24} + \frac{a_1}{12},$$

$$a_5 = \frac{a_3}{20} + \frac{a_2}{20} = \frac{a_0}{30} + \frac{a_1}{120},$$

Hence

$$y = a_0\left[1 + \frac{(x-1)^2}{2} + \frac{(x-1)^3}{6} + \frac{(x-1)^4}{24} + \frac{(x-1)^5}{30} + \cdots\right]$$
$$+ a_1\left[(x-1) + \frac{(x-1)^3}{6} + \frac{(x-1)^4}{12} + \frac{(x-1)^5}{120} + \cdots\right]. \tag{20}$$

In general, when the recurrence relation has three terms, such as the one given in Eq. (19), the determination of the coefficients in the series solution will be fairly complicated. In this example the formula for a_n in terms of a_0 and a_1 is not readily apparent. However, even without knowing the formula for a_n, we shall see in Section 4.2.1 that it is possible to establish that the series in Eq. (20) converge for all x, and further define functions y_3 and y_4, which are linearly independent solutions of the Airy equation (12). Thus

$$y = a_0 y_3(x) + a_1 y_4(x)$$

is the general solution of Airy's equation for $-\infty < x < \infty$.

It is worth emphasizing, as we saw in Example 3, that if we look for a solution of Eq. (1) of the form $y = \sum_{n=0}^{\infty} a_n(x - x_0)^n$, then the coefficients $P(x)$, $Q(x)$, and $R(x)$ in Eq. (1) must also be expressed in powers of $(x - x_0)$. Alternatively we can make the change of variable $x - x_0 = t$, obtaining a new differential equation for y as a function of t and then look for solutions of this new equation of the form $\sum_{n=0}^{\infty} a_n t^n$. When we have finished the calculations we replace t by $x - x_0$ (see Problem 10).

Another point of interest is the following. The functions y_1 and y_2 defined by the series in Eq. (17) are linearly independent solutions of Eq. (12) for all x, and this is similarly true for the functions y_3 and y_4 defined by the series in Eq. (20). According to the general theory of second order linear equations each of the first two functions can be expressed as a linear combination of the latter two functions and vice versa—a result that is certainly not obvious from an examination of the series alone.

EXAMPLE 4

The Hermite[6] equation is

$$y'' - 2xy' + \lambda y = 0, \qquad -\infty < x < \infty, \tag{21}$$

where λ is a constant. This equation is important in many branches of mathematical physics; for example, in quantum mechanics the Hermite equation arises in the investigation of the Schrödinger (1887–1961) equation for a harmonic oscillator.

To find a solution of Eq. (21) in the neighborhood of the ordinary point $x = 0$, we substitute the series (5), (6), and (14) for y, y', and y'' in Eq. (21), obtaining

$$\sum_{n=0}^{\infty} (n+2)(n+1)a_{n+2}x^n - \sum_{n=1}^{\infty} 2na_n x^n + \sum_{n=0}^{\infty} \lambda a_n x^n = 0$$

or

$$(2a_2 + \lambda a_0) + \sum_{n=1}^{\infty} [(n+2)(n+1)a_{n+2} - 2na_n + \lambda a_n]x^n = 0. \tag{22}$$

Hence $a_2 = -\lambda a_0/2$, and the general recurrence relation is

$$a_{n+2} = \frac{2n - \lambda}{(n+2)(n+1)} a_n, \qquad n \geq 1. \tag{23}$$

Note that the recurrence relation (23) actually includes the result $a_2 = -\lambda a_0/2$ corresponding to $n = 0$. It is clear from Eq. (23) that a_0 determines a_2, which in turn determines a_4, and so on. Similarly, the coefficients of the odd powers of x are determined in terms of a_1. The formal series solution for the Hermite equation is

$$\begin{aligned} y = a_0 &\left[1 - \frac{\lambda}{2!}x^2 - \frac{(4-\lambda)\lambda}{4!}x^4 - \frac{(8-\lambda)(4-\lambda)\lambda}{6!}x^6 - \cdots\right] \\ &+ a_1\left[x + \frac{(2-\lambda)}{3!}x^3 + \frac{(6-\lambda)(2-\lambda)}{5!}x^5 \right. \\ &\qquad \left. + \frac{(10-\lambda)(6-\lambda)(2-\lambda)}{7!}x^7 + \cdots\right] \\ = {} & a_0 y_1(x) + a_1 y_2(x). \end{aligned} \tag{24}$$

Again, the results of Section 4.2.1 establish the convergence of these series for all x. If λ is a nonnegative even integer, one or the other of the above series terminates. In particular for $\lambda = 0, 2, 4, 6$, respectively, one solution of the Hermite equation is 1, x, $1 - 2x^2$, and $x - 2x^3/3$. The polynomial solution

[6] Charles Hermite (1822–1901) was an influential French analyst and algebraist. He introduced the Hermite functions in 1864, and showed in 1873 that e is a transcendental number (that is, e is not a root of any polynomial equation with rational coefficients). His name is also associated with Hermitian matrices (see Section 7.3), some of whose properties he discovered.

corresponding to $\lambda = 2n$, after being multiplied by a proper constant,[7] is known as the Hermite polynomial $H_n(x)$.

PROBLEMS

In each of Problems 1 through 8 find the recurrence relation and two linearly independent solutions in powers of $x - x_0$ for the given differential equation.

1. $y'' - y = 0, \quad x_0 = 0$
2. $y'' - xy' - y = 0, \quad x_0 = 0$
3. $y'' - xy' - y = 0, \quad x_0 = 1$
4. $y'' + k^2x^2y = 0; \quad x_0 = 0$, k a constant
5. $(1 - x)y'' + y = 0, \quad x_0 = 0$
6. $(2 + x^2)y'' - xy' + 4y = 0, \quad x_0 = 0$
7. $y'' + xy' + 2y = 0, \quad x_0 = 0$
8. $xy'' + y' + xy = 0, \quad x_0 = 1$

9. Find the solution of Problem 2 that satisfies the initial conditions $y(0) = 0$, $y'(0) = 1$; the initial conditions $y(0) = 2$, $y'(0) = 1$.
10. By making the change of variable $x - 1 = t$, find two linearly independent series solutions of

$$y'' + (x - 1)^2 y' + (x^2 - 1) y = 0$$

in powers of $x - 1$. Show that you can obtain the same result by expanding $x^2 - 1$ in a Taylor series about $x = 1$, and then substituting

$$y = \sum_{n=0}^{\infty} a_n (x - 1)^n.$$

11. Show directly, using the ratio test, that the series solutions of Airy's equation about $x = 0$ converge for all x. See Eq. (17) of the text.
12. Consider the initial value problem $(y')^2 = 1 - y^2$, $y(0) = 0$ with the additional condition $y'(0) > 0$. This additional condition is necessary to determine a unique solution since only the squares of y and y' appear in the differential equation; hence if $\phi(x)$ is a solution of the differential equation, then so is $-\phi(x)$.
(a) Show that $y = \sin x$ is the solution of this initial value problem.
(b) Look for a series solution of the initial value problem of the form $y = a_0 + a_1 x + a_2 x^2 + \cdots$, and show that $a_0 = 0$, $a_1 = 1$, $a_2 = 0$, and $a_3 = -1/3!$. Note that the first two nonzero terms in the series solution are the first two terms in the Taylor series expansion of $\sin x$ about $x = 0$.

[7] This process is usually referred to as normalization. The constant is often chosen so that the solution has a specified value at a particular point, or so that a certain integral of the solution over a prescribed range has a definite value. For example, the Hermite polynomial of order n is uniquely specified by the statement that it is a polynomial solution of degree n of the Hermite equation with $\lambda = 2n$, and with the coefficient of x^n equal to 2^n.

4.2.1 Series Solutions near an Ordinary Point, Part II

In the previous section we considered the problem of finding solutions of

$$P(x)y'' + Q(x)y' + R(x)y = 0, \tag{1}$$

where P, Q, and R are polynomials, in the neighborhood of an ordinary point x_0. Assuming that Eq. (1) does have a solution $y = \phi(x)$, and that ϕ has a Taylor series

$$y = \phi(x) = \sum_{n=0}^{\infty} a_n(x - x_0)^n \tag{2}$$

which converges for $|x - x_0| < \rho$, $\rho > 0$, we found that the a_n can be determined by directly substituting the series (2) for y in Eq. (1).

Let us now consider how we might justify the statement that if x_0 is an ordinary point of Eq. (1), then there exist solutions of the form (2). We also consider the question of the radius of convergence of such a series. In doing this we are led to a generalization of the definition of an ordinary point.

Suppose, then, that there is a solution of Eq. (1) of the form (2). By differentiating Eq. (2) m times and setting x equal to x_0 it follows that

$$m!a_m = \phi^{(m)}(x_0).$$

Hence, to compute the a_n in the series (2), we must show that we can determine $\phi^{(n)}(x_0)$ for $n = 0, 1, 2, \ldots$ from the differential equation (1).

Suppose that $y = \phi(x)$ is a solution of Eq. (1) satisfying the initial conditions $y(x_0) = y_0$, $y'(x_0) = y_0'$. Then $a_0 = y_0$ and $a_1 = y_0'$. If we are solely interested in finding a solution of Eq. (1) without specifying any initial conditions, then a_0 and a_1 will be arbitrary. To determine $\phi^{(n)}(x_0)$ and the corresponding a_n for $n = 2, 3, \ldots$, we turn to Eq. (1). Since ϕ is a solution of Eq. (1) we have

$$P(x)\phi''(x) + Q(x)\phi'(x) + R(x)\phi(x) = 0.$$

For the interval about x_0 for which P is nonvanishing we can write this equation in the form

$$\phi''(x) = -p(x)\phi'(x) - q(x)\phi(x) \tag{3}$$

where $p(x) = Q(x)/P(x)$ and $q(x) = R(x)/P(x)$. Setting x equal to x_0 in Eq. (3) gives

$$\phi''(x_0) = -p(x_0)\phi'(x_0) - q(x_0)\phi(x_0).$$

Hence a_2 is given by

$$2!a_2 = \phi''(x_0) = -p(x_0)a_1 - q(x_0)a_0. \tag{4}$$

To determine a_3 we differentiate Eq. (3) and then set x equal to x_0, obtaining

$$3!a_3 = \phi'''(x_0) = -[p\phi'' + (p' + q)\phi' + q'\phi]\big|_{x=x_0}$$

$$= -2!p(x_0)a_2 - [p'(x_0) + q(x_0)]a_1 - q'(x_0)a_0. \qquad (5)$$

Substituting for a_2 from Eq. (4) gives a_3 in terms of a_1 and a_0. Since P, Q, and R are polynomials and $P(x_0) \neq 0$, all of the derivatives of p and q exist at x_0. Hence, we can continue to differentiate Eq. (3) indefinitely, determining after each differentiation the successive coefficients $a_4, \ldots,$ etc., by setting x equal to x_0.

Notice that the important property that we used in determining the a_n was that we could compute infinitely many derivatives of the functions p and q. It might seem reasonable to relax our assumption that the functions p and q are ratios of polynomials, and simply require that they be infinitely differentiable in the neighborhood of x_0. Unfortunately, this condition is too weak to insure that we can prove the convergence of the resulting series expansion for $y = \phi(x)$. However, there is a more general condition we can require of p and q. It allows us to carry out the above computations for the a_n and also to prove the convergence of the series solution. This condition is that the functions p and q be *analytic* at x_0; that is, that they have Taylor series expansions that converge in some interval about the point x_0:

$$p(x) = p_0 + p_1(x - x_0) + \cdots + p_n(x - x_0)^n + \cdots = \sum_{n=0}^{\infty} p_n(x - x_0)^n, \qquad (6)$$

$$q(x) = q_0 + q_1(x - x_0) + \cdots + q_n(x - x_0)^n + \cdots = \sum_{n=0}^{\infty} q_n(x - x_0)^n. \qquad (7)$$

As mentioned in Section 4.1 if each of the functions p and q is the quotient of polynomials, then they have series expansions[8] of the form (6) and (7). With this idea in mind we can generalize the definition of an ordinary point and a singular point of Eq. (1) as follows: If the functions $p = Q/P$ and $q = R/P$ are analytic at x_0, then the point x_0 is said to be an *ordinary point* of the differential equation (1); otherwise it is a *singular point*.

Now let us turn to the question of the interval of convergence of the series solution. One possibility (but not an appealing one) is actually to compute the series solution for each problem and then to apply one of the tests for convergence of an infinite series to determine the radius of convergence of the series

[8] The careful reader has probably reached the conclusion (which is correct) that if p is of the form (6) then p has infinitely many derivatives at x_0, but that the converse does not follow. An example of a function with infinitely many derivatives at $x = 0$, but which does not have a Taylor series expansion, is given in Problem 8.

solution. Fortunately, the question can be answered at once for a wide class of problems by the following theorem.

> ***Theorem 4.1.*** *If x_0 is an ordinary point of the differential equation* (1)
>
> $$P(x)y'' + Q(x)y' + R(x)y = 0;$$
>
> *that is, if $p = Q/P$ and $q = R/P$ are analytic at x_0, then the general solution of Eq.* (1) *is*
>
> $$y = \sum_{n=0}^{\infty} a_n(x - x_0)^n = a_0 y_1(x) + a_1 y_2(x), \tag{8}$$
>
> *where a_0 and a_1 are arbitrary, and y_1 and y_2 are linearly independent series solutions which are analytic at x_0. Further the radius of convergence for each of the series solutions y_1 and y_2 is at least as large as the minimum of the radii of convergence of the series for p and q. The coefficients in the series solutions are determined by substituting the series* (2) *for y in Eq.* (1).

Notice from the form of the series solution that $y_1(x) = 1 + b_2(x - x_0)^2 + \cdots$, and $y_2(x) = (x - x_0) + c_2(x - x_0)^2 + \cdots$. Hence y_1 is the solution satisfying the initial conditions $y_1(x_0) = 1$, $y_1'(x_0) = 0$, and y_2 is the solution satisfying the initial conditions $y_2(x_0) = 0$, $y_2'(x_0) = 1$. Also note that while the calculation of the coefficients by successively differentiating the differential equation is excellent in theory, it is not in general a practical computational procedure. Rather, one should substitute the series (2) for y in the differential equation (1) and determine the coefficients so that the differential equation is satisfied as in the examples in the previous section.

We will not prove this theorem, which in a slightly more general form is due to Fuchs.[9] What is important for our purposes is that there is a series solution of the form (2), and that the radius of convergence of the series solution cannot be less than the smaller of the radii of convergence of the series for p and q; hence we need only determine these.

This can be done in one of two ways. Again, one possibility is simply to compute the power series for p and q, and then to determine the radii of convergence by using one of the convergence tests for infinite series. However, there is an easier way when P, Q, and R are polynomials. It is shown in the theory of functions of a complex variable that the ratio of two polynomials, say Q/P, has a convergent power series expansion about a point $x = x_0$ if $P(x_0) \neq 0$.

[9] Immanuel Lazarus Fuchs (1833–1902) was a student and later a professor at the University of Berlin. He proved the result of Theorem 4.1 in 1866. His most important research was on singular points of linear differential equations. He recognized the significance of regular singular points (Section 4.3), and equations whose only singularities, including the point at infinity, are regular singular points are known as Fuchsian equations.

Further, assuming that any factors common to Q and P have been canceled, the radius of convergence of the power series for Q/P about the point x_0 is precisely the distance from x_0 to the nearest zero of P. In determining this distance we must remember that $P(x) = 0$ may have complex roots, and these must also be considered.

EXAMPLE 1

What is the radius of convergence of the Taylor series for $(1 + x^2)^{-1}$ about $x = 0$?

For $x^2 < 1$ we know that a convergent power series expansion for $(1 + x^2)^{-1}$ is

$$\frac{1}{1 + x^2} = 1 - x^2 + x^4 - x^6 + \cdots + (-1)^n x^{2n} + \cdots.$$

Clearly this series diverges for $x^2 \geq 1$ since then the nth term does not approach zero as $n \to \infty$. Hence the radius of convergence is $\rho = 1$. (This can be verified by the ratio test.)

To determine the radius of convergence using the theory that was just discussed, we note that the polynomial $1 + x^2$ has zeros at $x = \pm i$; since the distance from 0 to i in the complex plane is 1, the radius of convergence of the power series about $x = 0$ is 1.

EXAMPLE 2

What is the radius of convergence of the Taylor series for $(x^2 - 2x + 2)^{-1}$ about $x = 0$? About $x = 1$?

First notice that

$$x^2 - 2x + 2 = 0$$

has solutions $x = 1 \pm i$. The distance from $x = 0$ to either $x = 1 + i$ or $x = 1 - i$ in the complex plane is $\sqrt{2}$; hence the radius of convergence of the Taylor series expansion $\sum_{n=0}^{\infty} a_n x^n$ about $x = 0$ is $\sqrt{2}$.

The distance from $x = 1$ to either $x = 1 + i$ or $x = 1 - i$ is 1; hence the radius of convergence of the Taylor series expansion $\sum_{n=0}^{\infty} b_n(x - 1)^n$ about $x = 1$ is 1.

According to Theorem 4.1 the series solutions of the Airy equation in Examples 2 and 3, and of the Hermite equation in Example 4 of the previous section, converge for all values of x, $x - 1$, and x, respectively, since in each problem $P(x) = 1$ and hence is never zero.

It should be emphasized that a series solution may converge for a wider range of x than indicated by Theorem 4.1, so the theorem actually gives only a lower bound on the radius of convergence of the series solution. This is illustrated by the Legendre polynomial solution of the Legendre equation given in the next example.

EXAMPLE 3

Determine a lower bound for the radius of convergence of series solutions about $x = 0$ for the Legendre equation

$$(1 - x^2)y'' - 2xy' + \alpha(\alpha + 1)y = 0,$$

where α is a constant.

Note that $P(x) = 1 - x^2$, $Q(x) = -2x$, and $R(x) = \alpha(\alpha + 1)$ are polynomials, and that the zeros of P, $x = \pm 1$, are a distance 1 from $x = 0$. Hence a series solution of the form $\sum_{n=0}^{\infty} a_n x^n$ will converge for $|x| < 1$ at least, and possibly for larger values of x. Indeed, it can be shown that if α is a positive integer, one of the series solutions terminates after a finite number of terms and hence converges not just for $|x| < 1$ but for all x. For example, if $\alpha = 1$ the polynomial solution is $y = x$. See Problems 10 through 16 at the end of this section for a more complete discussion of the Legendre equation.

EXAMPLE 4

Determine a lower bound for the radius of convergence of series solutions of the differential equation

$$(1 + x^2)y'' + 2xy' + 4x^2y = 0 \tag{9}$$

about the point $x = 0$; about the point $x = -\frac{1}{2}$.

Again, P, Q, and R are polynomials, and P has zeros at $x = \pm i$. The distance in the complex plane from 0 to $\pm i$ is 1, and from $-\frac{1}{2}$ to $\pm i$ is $\sqrt{1 + \frac{1}{4}} = \sqrt{5}/2$. Hence in the first case the series $\sum_{n=0}^{\infty} a_n x^n$ converges at least for $|x| < 1$, and in the second case the series $\sum_{n=0}^{\infty} b_n(x + \frac{1}{2})^n$ converges at least for $|x + \frac{1}{2}| < \sqrt{5}/2$.

An interesting observation that we can make about Eq. (9) follows from the existence and uniqueness Theorem 3.2 and Theorem 4.1 that was just stated. Suppose that initial conditions $y(0) = y_0$ and $y'(0) = y_0'$ are given. Since $1 + x^2 \neq 0$ for all x, we know from Theorem 3.2 that there exists a unique solution of the initial value problem on $-\infty < x < \infty$. On the other hand, Theorem 4.1 only guarantees a series solution of the form $\sum_{n=0}^{\infty} a_n x^n$ ($a_0 = y_0$, $a_1 = y_0'$) for

$-1 < x < 1$. The unique solution on the interval $-\infty < x < \infty$ may not have a power series about $x = 0$ that converges for all x.

EXAMPLE 5

Can we determine a series solution about $x = 0$ for the differential equation

$$y'' + (\sin x)y' + (1 + x^2)y = 0,$$

and if so, what is the radius of convergence?

For this differential equation, $p(x) = \sin x$ and $q(x) = 1 + x^2$. Recall from Section 4.1 that $\sin x$ has a Taylor series expansion about $x = 0$, which converges for all x. Further, q also has a Taylor series expansion about $x = 0$, namely $q(x) = 1 + x^2$, which converges for all x. Thus there is a series solution of the form $y = \sum_{n=0}^{\infty} a_n x^n$ with a_0 and a_1 arbitrary, and the series converges for all x.

PROBLEMS

1. Determine $\phi''(x_0)$, $\phi'''(x_0)$, and $\phi^{\text{iv}}(x_0)$ for the given point x_0 if $y = \phi(x)$ is a solution of the given initial value problem.

   (a) $y'' + xy' + y = 0; \quad y(0) = 1, \quad y'(0) = 0$
   (b) $y'' + (\sin x)y' + (\cos x)y = 0; \quad y(0) = 0, \quad y'(0) = 1$
   (c) $x^2y'' + (1 + x)y' + 3(\ln x)y = 0; \quad y(1) = 2, \quad y'(1) = 0$
   (d) $y'' + x^2y' + (\sin x)y = 0; \quad y(0) = a_0, \quad y'(0) = a_1$
   In part (d) express $\phi''(0)$, $\phi'''(0)$, and $\phi^{\text{iv}}(0)$ in terms of a_0 and a_1.

2. Determine a lower bound for the radius of convergence of series solutions about each given point x_0 for each of the following differential equations.

   (a) $y'' + 4y' + 6xy = 0; \quad x_0 = 0, \quad x_0 = 4$
   (b) $(x^2 - 2x - 3)y'' + xy' + 4y = 0; \quad x_0 = 4, \quad x_0 = -4, \quad x_0 = 0$
   (c) $(1 + x^3)y'' + 4xy' + y = 0; \quad x_0 = 0, \quad x_0 = 2$
   (d) $xy'' + y = 0; \quad x_0 = 1$

3. Determine a lower bound for the radius of convergence of series solutions for each of the differential equations in Problems 1 through 8 of Section 4.2.
4. The Tchebycheff (1821–1894) differential equation is

$$(1 - x^2)y'' - xy' + \alpha^2 y = 0,$$

   where α is a constant.
   (a) Determine two linearly independent solutions in powers of x for $|x| < 1$.
   (b) Show that if α is a nonnegative integer n, then there is a polynomial solution of degree n. These polynomials, when properly normalized, are called the Tchebycheff polynomials. They are very useful in problems requiring a polynomial approximation to a function defined on $-1 \le x \le 1$.
   (c) Find a polynomial solution for each of the cases $\alpha = n = 0$, 1, 2, and 3.

5. Find the first three terms in each of two linearly independent power series solutions in powers of x of

$$y'' + (\sin x)y = 0.$$

Hint: Expand $\sin x$ in a Taylor series about $x = 0$, and retain a sufficient number of terms to compute the necessary coefficients in $y = \sum_{n=0}^{\infty} a_n x^n$.

6. Find the first three terms in each of two linearly independent power series solutions in powers of x of

$$e^x y'' + xy = 0.$$

What is the radius of convergence of each series solution?
Hint: Expand e^x or xe^{-x} in a power series about $x = 0$.

7. Suppose that you are told that x and x^2 are solutions of a differential equation $P(x)y'' + Q(x)y' + R(x)y = 0$. What can you say about the point $x = 0$: Is it an ordinary point or a singular point?
Hint: Use Theorem 3.2, and note the value of x and x^2 at $x = 0$.

*8. To show that it is possible to have a function that is infinitely differentiable at a point but does not have a Taylor series expansion about the point, consider the function

$$f(x) = \begin{cases} e^{-1/x^2}, & x \neq 0, \\ 0, & x = 0. \end{cases}$$

Show, using the definition of a derivative as the limit of a difference quotient, that $f'(0) = f''(0) = 0$. It can be shown that f is infinitely differentiable at the origin, and that $f^{(n)}(0) = 0$ for all n. Hence the series

$$\sum_{n=0}^{\infty} \frac{f^{(n)}(0)}{n!} x^n$$

is zero for all x, and converges to $f(x)$ only at $x = 0$.

9. The series methods discussed in this section are directly applicable to the first order linear differential equation $P(x)y' + Q(x)y = 0$ at a point x_0, if the function $p = Q/P$ has a Taylor series expansion about that point. Such a point is called an ordinary point, and further, the radius of convergence of the series $y = \sum_{n=0}^{\infty} a_n(x - x_0)^n$ is at least as large as the radius of convergence of the series for Q/P. Solve the following differential equations by series in powers of x and verify that a_0 is arbitrary in each case. Problems (e) and (f) involve nonhomogeneous differential equations to which the series methods can be easily extended.

   (a) $y' - y = 0$ (b) $y' - xy = 0$
   (c) $y' = e^{x^2}y$, three terms (d) $(1 - x)y' = y$
   *(e) $y' - y = x^2$ *(f) $y' + xy = 1 + x$

   Where possible, compare the series solution with the solution obtained by using the methods of Chapter 2.

Legendre Equation. Problems 10 through 16 deal with the Legendre equation

$$(1 - x^2)y'' - 2xy' + \alpha(\alpha + 1)y = 0.$$

As indicated in Example 3, the point $x = 0$ is an ordinary point of this equation, and the distance from the origin to the nearest zero of $P(x) = 1 - x^2$ is 1. Hence the radius of convergence of series solutions about $x = 0$ is at least 1. Also notice that it is only necessary to consider $\alpha > -1$ because if $\alpha \leq -1$, then the substitution $\alpha = -(1 + \gamma)$ where $\gamma \geq 0$ leads to the Legendre equation $(1 - x^2)y'' - 2xy' + \gamma(\gamma + 1)y = 0$.

10. Show that two linearly independent solutions of the Legendre equation for $|x| < 1$ are

$$y_1(x) = 1 + \sum_{m=1}^{\infty} (-1)^m \times \frac{\alpha(\alpha - 2)(\alpha - 4) \cdots (\alpha - 2m + 2)(\alpha + 1)(\alpha + 3) \cdots (\alpha + 2m - 1)}{(2m)!} x^{2m},$$

$$y_2(x) = x + \sum_{m=1}^{\infty} (-1)^m \times \frac{(\alpha - 1)(\alpha - 3) \cdots (\alpha - 2m + 1)(\alpha + 2)(\alpha + 4) \cdots (\alpha + 2m)}{(2m + 1)!} x^{2m+1}.$$

11. Show that if α is zero or a positive even integer $2n$ the series solution for y_1 reduces to a polynomial of degree $2n$ containing only even powers of x. Show that corresponding to $\alpha = 0$, 2, and 4 these polynomials are

$$1, \qquad 1 - 3x^2, \qquad 1 - 10x^2 + \tfrac{35}{3}x^4.$$

Show that if α is a positive odd integer $2n + 1$ then the series solution for y_2 reduces to a polynomial of degree $2n + 1$ containing only odd powers of x and that corresponding to $\alpha = 1, 3, 5$ these polynomials are

$$x, \qquad x - \tfrac{5}{3}x^3, \qquad x - \tfrac{14}{3}x^3 + \tfrac{21}{5}x^5.$$

12. The Legendre polynomial P_n is defined as the polynomial solution of the Legendre equation with $\alpha = n$ satisfying the condition $P_n(1) = 1$. Using the results of Problem 11, show that

$$P_0(x) = 1, \qquad P_1(x) = x, \qquad P_2(x) = \tfrac{1}{2}(3x^2 - 1), \qquad P_3(x) = \tfrac{1}{2}(5x^3 - 3x).$$

It can be shown that the general formula is

$$P_n(x) = \frac{1}{2^n} \sum_{k=0}^{[n/2]} \frac{(-1)^k (2n - 2k)!}{k!(n - k)!(n - 2k)!} x^{n-2k},$$

where $[n/2]$ denotes the greatest integer less than or equal to $n/2$. By observing the form of $P_n(x)$ for n even and n odd show that $P_n(-1) = (-1)^n$.

13. The Legendre polynomials play an important role in mathematical physics. For example, in solving Laplace's equation (the potential equation) in spherical coordinates we encounter the equation

$$\frac{d^2F(\varphi)}{d\varphi^2} + \cot\varphi \frac{dF(\varphi)}{d\varphi} + n(n + 1)F(\varphi) = 0, \qquad 0 < \varphi < \pi,$$

where n is a positive integer. Show that the change of variable $x = \cos\varphi$ leads to the Legendre equation with $\alpha = n$ for $y = f(x) = F(\cos^{-1} x)$.

14. Show that for $n = 0, 1, 2, 3$ the corresponding Legendre polynomial is given by

$$P_n(x) = \frac{1}{2^n n!} \frac{d^n}{dx^n}(x^2 - 1)^n.$$

This formula, known as Rodrigues' (1794–1851) formula, is true for all positive integers n.

15. Show that the Legendre equation can also be written as

$$[(1 - x^2)y']' = -\alpha(\alpha + 1)y.$$

Then it follows that $[(1 - x^2)P_n'(x)]' = -n(n+1)P_n(x)$ and $[(1 - x^2)P_m'(x)]' = -m(m+1)P_m(x)$. Show, by multiplying the first equation by $P_m(x)$ and the second equation by $P_n(x)$, and then integrating by parts, that

$$\int_{-1}^{1} P_n(x)P_m(x)\,dx = 0 \qquad \text{if} \quad n \neq m.$$

This property of the Legendre polynomials is known as the orthogonality property. If $m = n$, it can be shown that the value of the above integral is $2/(2n+1)$.

16. Given a polynomial f of degree n, it is possible to express f as a linear combination of $P_0, P_1, P_2, \ldots, P_n$:

$$f(x) = \sum_{k=0}^{n} a_k P_k(x).$$

Using the result of Problem 15, show that

$$a_k = \frac{2k+1}{2} \int_{-1}^{1} f(x)P_k(x)\,dx.$$

4.3 Regular Singular Points

In this section we will consider the equation

$$P(x)y'' + Q(x)y' + R(x)y = 0 \tag{1}$$

in the neighborhood of a singular point x_0. Recall that if the functions P, Q, and R are polynomials having no common factors, the singular points of Eq. (1) are the points for which $P(x) = 0$.

EXAMPLE 1

Determine the singular points and ordinary points of the Bessel equation of order ν

$$x^2y'' + xy' + (x^2 - \nu^2)y = 0. \tag{2}$$

The point $x = 0$ is clearly a singular point since $P(x) = x^2$ is zero there. All other points are ordinary points of Eq. (2).

EXAMPLE 2

Determine the singular points and ordinary points of the Legendre equation

$$(1 - x^2)y'' - 2xy' + \alpha(\alpha + 1)y = 0, \tag{3}$$

where α is a constant.

The singular points are the zeros of $P(x) = 1 - x^2$, namely the points $x = \pm 1$. All other points are ordinary points.

Unfortunately, if we attempt to use the methods of the preceding two sections to solve Eq. (1) in the neighborhood of a singular point x_0, we find that these methods fail. This is because the solution of Eq. (1) is often not analytic at x_0, and consequently cannot be represented by a Taylor series in powers of $x - x_0$. Instead, we must use a more general series expansion.

Since the singular points of a differential equation are usually few in number, we might ask whether we can simply ignore them, especially since we already know how to construct solutions about ordinary points. However, this is not feasible because the singular points determine the principal features of the solution to a much larger extent than one might at first suspect. In the neighborhood of a singular point the solution often becomes large in magnitude, experiences rapid changes in magnitude, or is peculiar in some manner. Thus the behavior of a physical system modeled by a differential equation frequently is most interesting in the neighborhood of a singular point. Often geometrical singularities in a physical problem, such as corners or sharp edges, lead to singular points in the corresponding differential equation. Thus, while at first we might want to avoid the few points where a differential equation is singular, it is precisely at these points that it is necessary to study the solution most carefully.

As an alternative to analytical methods, one can consider the use of numerical methods, which are discussed in Chapter 8. However, these methods are ill-suited for the study of solutions near a singular point. Thus, even if one adopts a numerical approach, it is necessary to combine it with the analytical methods of this chapter in order to examine the behavior of solutions near singular points.

Without any additional information about the behavior of Q/P and R/P in the neighborhood of the singular point, it is impossible to describe the behavior of the solutions of Eq. (1) near $x = x_0$. It may be that there are two linearly independent solutions of Eq. (1) that remain bounded as $x \to x_0$, or there may be only one with the other becoming unbounded as $x \to x_0$, or they may both become unbounded as $x \to x_0$. To illustrate this consider the following examples.

EXAMPLE 3

The differential equation

$$x^2y'' - 2y = 0 \tag{4}$$

has a singular point at $x = 0$. It can be easily verified by direct substitution that for $x > 0$ or $x < 0$, $y = x^2$ and $y = 1/x$ are linearly independent solutions of Eq. (4). Thus in any interval not containing the origin the general solution of Eq. (4) is $y = c_1x^2 + c_2x^{-1}$. The only solution of Eq. (4) that is bounded as $x \to 0$ is $y = c_1x^2$. Indeed, this solution is analytic at the origin even though if Eq. (4) is put in the standard form, $y'' - (2/x^2)y = 0$, the function $q(x) = -2/x^2$ is not analytic at $x = 0$, and Theorem 4.1 is not applicable. On the other hand, notice that the solution $y = x^{-1}$ does not have a Taylor series expansion about the origin (is not analytic at $x = 0$); therefore, the method of Section 4.2 would fail in this case.

EXAMPLE 4

The differential equation

$$x^2y'' - 2xy' + 2y = 0 \tag{5}$$

also has a singular point at $x = 0$. It can be verified that $y_1(x) = x$ and $y_2(x) = x^2$ are linearly independent solutions of Eq. (5), and both are analytic at $x = 0$. Nevertheless, it is still not proper to pose an initial value problem with initial conditions at $x = 0$. It is impossible to prescribe arbitrary initial conditions at $x = 0$, since any linear combination of x and x^2 is zero at $x = 0$.

It is also possible to construct a differential equation with a singular point x_0, such that every solution of the differential equation becomes unbounded as $x \to x_0$. Even if Eq. (1) does not have a solution that remains bounded as $x \to x_0$, it is often important to determine how the solutions of Eq. (1) behave as $x \to x_0$; for example, does $y \to \infty$ in the same manner as $(x - x_0)^{-1}$ or $|x - x_0|^{-1/2}$, or in some other manner?

To develop a reasonably simple mathematical theory for solving Eq. (1) in the neighborhood of a singular point x_0, it is necessary to restrict ourselves to cases in which the singularities in the functions Q/P and R/P at $x = x_0$ are not too severe, that is, to what we might call "weak singularities." A priori, it is not clear exactly what is an acceptable singularity. However, we show later (see Section 4.5.1, Problems 12 and 13) that if P, Q, and R are polynomials, the conditions that distinguish "weak singularities" are

$$\lim_{x \to x_0} (x - x_0)\frac{Q(x)}{P(x)} \quad \text{is finite} \tag{6}$$

and

$$\lim_{x \to x_0} (x - x_0)^2\frac{R(x)}{P(x)} \quad \text{is finite.} \tag{7}$$

This means that the singularity in Q/P can be no worse than $(x - x_0)^{-1}$ and the singularity in R/P can be no worse than $(x - x_0)^{-2}$. Such a point is called a *regular singular point* of Eq. (1). For more general functions than polynomials, x_0 is a *regular singular point* of Eq. (1) if it is a singular point, and if both[10]

$$(x - x_0)\frac{Q(x)}{P(x)} \quad \text{and} \quad (x - x_0)^2\frac{R(x)}{P(x)} \tag{8}$$

have convergent Taylor series about x_0; that is, if the functions in Eq. (8) are analytic at $x = x_0$. Equations (6) and (7) imply that this will be the case when P, Q, and R are polynomials. Any singular point of Eq. (1) that is not a regular singular point is called an *irregular singular point* of Eq. (1).

In the following sections we discuss how to solve Eq. (1) in the neighborhood of a regular singular point. A discussion of the solutions of differential equations in the neighborhood of irregular singular points is beyond the scope of an elementary text.

EXAMPLE 5

In Example 2 we observed that the singular points of the Legendre equation

$$(1 - x^2)y'' - 2xy' + \alpha(\alpha + 1)y = 0$$

are $x = \pm 1$. Determine whether these singular points are regular or irregular singular points.

We consider the point $x = 1$ first and also observe that on dividing by $1 - x^2$ the coefficients of y' and y are $-2x/(1 - x^2)$ and $\alpha(\alpha + 1)/(1 - x^2)$, respectively. Thus we calculate

$$\lim_{x \to 1}(x - 1)\frac{-2x}{1 - x^2} = \lim_{x \to 1}\frac{(x - 1)(-2x)}{(1 - x)(1 + x)} = \lim_{x \to 1}\frac{2x}{1 + x} = 1$$

and

$$\begin{aligned}\lim_{x \to 1}(x - 1)^2\frac{\alpha(\alpha + 1)}{1 - x^2} &= \lim_{x \to 1}\frac{(x - 1)^2\alpha(\alpha + 1)}{(1 - x)(1 + x)}\\ &= \lim_{x \to 1}\frac{(x - 1)(-\alpha)(\alpha + 1)}{1 + x} = 0.\end{aligned}$$

Since these limits are finite, the point $x = 1$ is a regular singular point. It can be shown in a similar manner that $x = -1$ is also a regular singular point.

EXAMPLE 6

Determine the singular points of the differential equation

$$2(x - 2)^2xy'' + 3xy' + (x - 2)y = 0,$$

and classify them as regular or irregular.

[10] The functions given in Eq. (8) may not be defined at x_0, in which case their values at x_0 are to be taken equal to their limits as $x \to x_0$.

Dividing the differential equation by $2(x-2)^2x$, we have

$$y'' + \frac{3}{2(x-2)^2}y' + \frac{1}{2(x-2)x}y = 0,$$

so $p(x) = Q(x)/P(x) = 3/2(x-2)^2$ and $q(x) = R(x)/P(x) = 1/2x(x-2)$. The singular points are $x = 0$ and $x = 2$. Consider $x = 0$. We have

$$\lim_{x\to 0} xp(x) = \lim_{x\to 0} x\frac{3}{2(x-2)^2} = 0$$

$$\lim_{x\to 0} x^2q(x) = \lim_{x\to 0} x^2\frac{1}{2x(x-2)} = 0.$$

Since these limits are finite, $x = 0$ is a regular singular point. For $x = 2$ we have

$$\lim_{x\to 2} (x-2)p(x) = \lim_{x\to 2} (x-2)\frac{3}{2(x-2)^2}$$

does not exist; hence $x = 2$ is an irregular singular point.

PROBLEMS

In Problems 1 through 8 determine whether each of the points -1, 0, and 1 is an ordinary point, a regular singular point, or an irregular singular point for the given differential equation.

1. $xy'' + (1-x)y' + xy = 0$
2. $x^2(1-x^2)y'' + 2xy' + 4y = 0$
3. $2x^4(1-x^2)y'' + 2xy' + 3x^2y = 0$
4. $x^2(1-x^2)y'' + (2/x)y' + 4y = 0$
5. $(1-x^2)^2y'' + x(1-x)y' + (1+x)y = 0$
6. $y'' + \left(\frac{x}{1-x}\right)^2 y' + 3(1+x)^2y = 0$
7. $(x+3)y'' - 2xy' + (1-x^2)y = 0$
8. $x(1-x^2)^3y'' + (1-x^2)^2y' + 2(1+x)y = 0$
9. Determine whether the singular point $x = 0$ of the Bessel equation
$$x^2y'' + xy' + (x^2 - \nu^2)y = 0$$
is a regular or irregular singular point.
10. For the following differential equation classify the point $x = 0$ as an ordinary point, a regular singular point, or an irregular singular point:
$$xy'' + e^xy' + (3\cos x)y = 0.$$
11. For the following differential equations classify the point $x = 0$ as an ordinary point, a regular singular point, or an irregular singular point:

    (a) $x^2y'' + 2(e^x - 1)y' + (e^{-x}\cos x)y = 0$
    (b) $x^2y'' - 3(\sin x)y' + (1+x^2)y = 0$

    Hint: Consider the power series for e^x and $\sin x$ in parts (a) and (b), respectively.

12. Assuming that the function f defined by $f(x) = x^{-1}\sin x$ for $x \neq 0$ and $f(0) = 1$ is analytic at $x = 0$, show that $x = 0$ is a regular singular point of the differential equation

$$(\sin x)\,y'' + xy' + 4y = 0,$$

and an irregular singular point of the differential equation

$$(x\sin x)\,y'' + 3y' + xy = 0.$$

*13. **Singularities at infinity.** The definitions of an ordinary point and a regular singular point given in the preceding sections apply only if the point x_0 is finite. In more advanced work in differential equations it is often necessary to discuss the point at infinity. This is done by making the change of variable $\xi = 1/x$ and studying the resulting equation at $\xi = 0$. Show that for the differential equation $P(x)y'' + Q(x)y' + R(x)y = 0$ the point at infinity is an ordinary point if

$$\frac{1}{P(1/\xi)}\left[\frac{2P(1/\xi)}{\xi} - \frac{Q(1/\xi)}{\xi^2}\right] \quad \text{and} \quad \frac{R(1/\xi)}{\xi^4 P(1/\xi)}$$

have Taylor series expansions about $\xi = 0$. Show also that the point at infinity is a regular singular point if one or the other of the above functions does not have a Taylor series expansion, but both

$$\frac{\xi}{P(1/\xi)}\left[\frac{2P(1/\xi)}{\xi} - \frac{Q(1/\xi)}{\xi^2}\right] \quad \text{and} \quad \frac{R(1/\xi)}{\xi^2 P(1/\xi)}$$

do have such expansions.

*14. For the following differential equations determine whether the point at infinity is an ordinary point, a regular singular point, or an irregular singular point.

(a) $(1 - x^2)y'' - 2xy' + \alpha(\alpha + 1)y = 0$ (Legendre equation)
(b) $x^2y'' + xy' + (x^2 - \nu^2)y = 0$ (Bessel equation)
(c) $y'' - 2xy' + \lambda y = 0$ (Hermite equation)
(d) $y'' - xy = 0$ (Airy equation)

4.4 *Euler Equations*

The simplest example of a differential equation having a regular singular point is the Euler equation or equidimensional equation

$$L[y] = x^2y'' + \alpha xy' + \beta y = 0, \tag{1}$$

where α and β are constants. Unless otherwise stated we shall consider α and β to be real. It is easy to see that $x = 0$ is a regular singular point of Eq. (1). Because the solution of the Euler equation is typical of the solutions of all differential equations with a regular singular point, it is worthwhile considering this equation in detail before discussing the more general problem.

In any interval not including the origin, Eq. (1) has a general solution of the form $y = c_1y_1(x) + c_2y_2(x)$ where y_1 and y_2 are linearly independent. For convenience we first consider the interval $x > 0$, extending our results later to the interval $x < 0$. First, we note that $(x^r)' = rx^{r-1}$ and $(x^r)'' = r(r-1)x^{r-2}$. Hence, if we assume that Eq. (1) has a solution of the form[11]

$$y = x^r, \tag{2}$$

then we obtain

$$\begin{aligned} L[x^r] &= x^2(x^r)'' + \alpha x(x^r)' + \beta x^r \\ &= x^r F(r), \end{aligned} \tag{3}$$

where

$$F(r) = r(r-1) + \alpha r + \beta. \tag{4}$$

If r is a root of the quadratic equation $F(r) = 0$, then $L[x^r]$ is zero, and $y = x^r$ is a solution of Eq. (1). The roots of $F(r) = 0$ are

$$\begin{aligned} r_1 &= \frac{-(\alpha-1) + \sqrt{(\alpha-1)^2 - 4\beta}}{2}, \\ r_2 &= \frac{-(\alpha-1) - \sqrt{(\alpha-1)^2 - 4\beta}}{2}, \end{aligned} \tag{5}$$

and $F(r) = (r - r_1)(r - r_2)$. Just as in the case of the second order linear differential equation with constant coefficients (see Section 3.5), we must examine separately the cases $(\alpha - 1)^2 - 4\beta$ positive, zero, and negative. Indeed, the entire theory presented in this section is similar to the theory for second order linear equations with constant coefficients, e^{rx} being replaced by x^r (see Problems 16 and 17). Also, we note that in solving a specific Euler equation it is much easier to substitute $y = x^r$, to obtain the quadratic equation (4), and calculate the roots r_1 and r_2, rather than to memorize the general formula (5).

***CASE 1.* $(\alpha - 1)^2 - 4\beta > 0$.** In this case Eqs. (5) give two real unequal roots r_1 and r_2. Since $W(x^{r_1}, x^{r_2}) = (r_2 - r_1)x^{r_1+r_2-1}$ is nonvanishing for $r_1 \neq r_2$ and $x > 0$, it follows that the general solution of Eq. (1) is

$$y = c_1 x^{r_1} + c_2 x^{r_2}, \qquad x > 0. \tag{6}$$

Note that if r_1 is not a rational number, then x^{r_1} is defined by $x^{r_1} = e^{r_1 \ln x}$.

EXAMPLE 1

Solve

$$2x^2y'' + 3xy' - y = 0, \qquad x > 0. \tag{7}$$

Substituting $y = x^r$ gives

$$x^r[2r(r-1) + 3r - 1] = x^r(2r^2 + r - 1) = x^r(2r-1)(r+1) = 0.$$

[11] This assumption is also suggested by the fact that the first order Euler equation $xy' + ky = 0$ has the solution $y = x^{-k}$.

Hence

$$r_1 = \tfrac{1}{2}, \qquad r_2 = -1,$$

and

$$y = c_1 x^{1/2} + c_2 x^{-1}, \qquad x > 0. \tag{8}$$

CASE 2. $(\alpha - 1)^2 - 4\beta = 0$. In this case, according to Eqs. (5), $r_1 = r_2 = -(\alpha - 1)/2$, and we have only one solution, $y_1(x) = x^{r_1}$, of the differential equation. A second solution can be obtained by the method of reduction of order, but for the purpose of our future discussion we consider an alternative method. Since $r_1 = r_2$, it follows that $F(r) = (r - r_1)^2$. Thus in this case not only does $F(r_1) = 0$, but also $F'(r_1) = 0$. This suggests differentiating Eq. (3) with respect to r and then setting r equal to r_1. Differentiating Eq. (3) with respect to r gives

$$\frac{\partial}{\partial r} L[x^r] = \frac{\partial}{\partial r}[x^r F(r)].$$

Substituting for $F(r)$, interchanging differentiation with respect to x and r, and noting that $\partial(x^r)/\partial r = x^r \ln x$, we obtain

$$L[x^r \ln x] = (r - r_1)^2 x^r \ln x + 2(r - r_1)x^r. \tag{9}$$

The right side of Eq. (9) is zero for $r = r_1$; consequently,

$$y_2(x) = x^{r_1} \ln x, \qquad x > 0, \tag{10}$$

is a second solution of Eq. (1). It is easy to show that $W(x^{r_1}, x^{r_1} \ln x) = x^{2r_1 - 1}$. Hence x^{r_1} and $x^{r_1} \ln x$ are linearly independent for $x > 0$, and the general solution of Eq. (1) is

$$y = (c_1 + c_2 \ln x)x^{r_1}, \qquad x > 0. \tag{11}$$

EXAMPLE 2

Solve

$$x^2 y'' + 5xy' + 4y = 0, \qquad x > 0. \tag{12}$$

Substituting $y = x^r$ gives

$$x^r[r(r - 1) + 5r + 4] = x^r(r^2 + 4r + 4) = 0.$$

Hence $r_1 = r_2 = -2$, and

$$y = x^{-2}(c_1 + c_2 \ln x), \qquad x > 0. \tag{13}$$

CASE 3. $(\alpha - 1)^2 - 4\beta < 0$. In this case the roots r_1 and r_2 are complex conjugates, say $r_1 = \lambda + i\mu$ and $r_2 = \lambda - i\mu$, with $\mu \neq 0$. We must now explain what is meant by x^r when r is complex. Remembering that when $x > 0$ and r is real we can write

$$x^r = e^{r \ln x}, \tag{14}$$

we can use this equation to *define* x^r when r is complex. Then

$$\begin{aligned} x^{(\lambda + i\mu)} = e^{(\lambda + i\mu)\ln x} &= e^{\lambda \ln x} e^{i\mu \ln x} \\ &= x^{\lambda} e^{i\mu \ln x} \\ &= x^{\lambda}[\cos(\mu \ln x) + i \sin(\mu \ln x)], \qquad x > 0. \end{aligned} \tag{15}$$

With this definition of x^r for complex values of r, it can be verified that the usual laws of algebra and the differential calculus hold, and hence x^{r_1} and x^{r_2} are indeed solutions of Eq. (1). The general solution of Eq. (1) is

$$y = c_1 x^{\lambda + i\mu} + c_2 x^{\lambda - i\mu}. \tag{16}$$

The disadvantage of this expression is that the functions $x^{\lambda + i\mu}$ and $x^{\lambda - i\mu}$ are complexed valued. Recall that we had a similar situation for the second order differential equation with constant coefficients when the roots of the characteristic equation were complex. In the same way as we did then we observe that the real and imaginary parts of $x^{\lambda + i\mu}$,

$$x^{\lambda}\cos(\mu \ln x) \quad \text{and} \quad x^{\lambda}\sin(\mu \ln x), \tag{17}$$

are also solutions of Eq. (1). A straightforward calculation shows that

$$W[x^{\lambda}\cos(\mu \ln x), x^{\lambda}\sin(\mu \ln x)] = \mu x^{2\lambda - 1}.$$

Hence these solutions are also linearly independent for $x > 0$, and the general solution of Eq. (1) is

$$y = c_1 x^{\lambda}\cos(\mu \ln x) + c_2 x^{\lambda}\sin(\mu \ln x), \qquad x > 0. \tag{18}$$

EXAMPLE 3

Solve

$$x^2 y'' + x y' + y = 0. \tag{19}$$

Substituting $y = x^r$ gives

$$x^r[r(r-1) + r + 1] = x^r(r^2 + 1) = 0$$

Hence $r = \pm i$, and the general solution is

$$y = c_1 \cos(\ln x) + c_2 \sin(\ln x), \qquad x > 0. \tag{20}$$

Now let us consider the interval $x < 0$. The difficulty here is in explaining what is meant by x^r when x is negative, and r is not an integer; similarly $\ln x$ is not defined for $x < 0$. The solutions of the Euler equation that we have given for $x > 0$ can be shown to be valid for $x < 0$ but they are in general complex valued. Thus in Example 1 the solution $x^{1/2}$ is imaginary for $x < 0$.

It is always possible to obtain real-valued solutions of the Euler equation (1) in the interval $x < 0$ by making the following change of variable. Let $x = -\xi$ where $\xi > 0$, and let $y = u(\xi)$. Then we have

$$\frac{dy}{dx} = \frac{du}{d\xi}\frac{d\xi}{dx} = -\frac{du}{d\xi}, \qquad \frac{d^2y}{dx^2} = \frac{d}{d\xi}\left[-\frac{du}{d\xi}\right]\frac{d\xi}{dx} = \frac{d^2u}{d\xi^2}. \tag{21}$$

Thus Eq. (1) takes the form

$$\xi^2\frac{d^2u}{d\xi^2} + \alpha\xi\frac{du}{d\xi} + \beta u = 0, \qquad \xi > 0. \tag{22}$$

But this is exactly the problem that we have just solved; from Eqs. (6), (11), and (18) we have

$$u(\xi) = \begin{cases} c_1\xi^{r_1} + c_2\xi^{r_2} \\ (c_1 + c_2 \ln \xi)\xi^{r_1} \\ c_1\xi^{\lambda}\cos(\mu \ln \xi) + c_2\xi^{\lambda}\sin(\mu \ln \xi) \end{cases} \tag{23}$$

depending on whether $(\alpha - 1)^2 - 4\beta$ is positive, zero, or negative. To obtain u in terms of x we replace ξ by $-x$ in Eqs. (23).

We can combine our results for $x > 0$ and $x < 0$ as follows. We have $|x| = x$ for $x > 0$ and $|x| = -x$ for $x < 0$. Thus we need only replace x by $|x|$ in Eqs. (6), (11), and (18) to obtain real-valued solutions valid in any interval not containing the origin (also see Problems 18 and 19). These results are summarized in the following theorem.

Theorem 4.2. *To solve the Euler equation* (1)

$$x^2y'' + \alpha xy' + \beta y = 0$$

in any interval not containing the origin, substitute $y = x^r$, *and compute the roots* r_1 *and* r_2 *of the equation*

$$F(r) = r^2 + (\alpha - 1)r + \beta = 0.$$

If the roots are real and unequal, then

$$y = c_1|x|^{r_1} + c_2|x|^{r_2}. \tag{24}$$

If the roots are real and equal, then

$$y = (c_1 + c_2 \ln|x|)|x|^{r_1}. \tag{25}$$

If the roots are complex, then

$$y = |x|^{\lambda}[c_1\cos(\mu \ln|x|) + c_2\sin(\mu \ln|x|)], \tag{26}$$

where $r_1, r_2 = \lambda \pm i\mu$.

For an Euler equation of the form

$$(x - x_0)^2 y'' + \alpha(x - x_0) y' + \beta y = 0, \tag{27}$$

the situation is exactly the same. One looks for solutions of the form $y = (x - x_0)^r$. The general solution is given by either Eq. (24), (25), or (26) with x replaced by $x - x_0$. Alternatively, we can reduce Eq. (27) to the form of Eq. (1) by making the change of independent variable $t = x - x_0$.

The situation for a general second order differential equation with a regular singular point is similar to that for an Euler equation. We consider that problem in the next section.

PROBLEMS

In each of Problems 1 through 12 determine the general solution of the given second order differential equation that is valid in any interval not including the singular point.

1. $x^2y'' + 4xy' + 2y = 0$
2. $(x + 1)^2y'' + 3(x + 1)y' + \frac{3}{4}y = 0$
3. $x^2y'' - 3xy' + 4y = 0$
4. $x^2y'' + 3xy' + 5y = 0$
5. $x^2y'' - xy' + y = 0$
6. $(x - 1)^2y'' + 8(x - 1)y' + 12y = 0$
7. $x^2y'' + 6xy' - y = 0$
8. $2x^2y'' - 4xy' + 6y = 0$
9. $x^2y'' - 5xy' + 9y = 0$
10. $(x - 2)^2y'' + 5(x - 2)y' + 8y = 0$
11. $x^2y'' + 2xy' + 4y = 0$
12. $x^2y'' - 4xy' + 4y = 0$

13. Using the method of reduction of order, show that if r_1 is a repeated root of $r(r - 1) + \alpha r + \beta = 0$, then x^{r_1} and $x^{r_1} \ln x$ are solutions of

$$x^2y'' + \alpha xy' + \beta y = 0, \qquad x > 0.$$

14. Find all values of α for which all solutions of $x^2y'' + \alpha xy' + \frac{5}{2}y = 0$ approach zero as $x \to 0$.
15. Find all values of β for which all solutions of $x^2y'' + \beta y = 0$ approach zero as $x \to 0$.
16. The second order Euler equation $x^2y'' + \alpha xy' + \beta y = 0$ can be reduced to a second order linear equation with constant coefficients by an appropriate change of the independent variable. Remembering that x^r plays the same role for the Euler equation as $e^{rx} = (e^x)^r$ plays for the equation with constant coefficients, it would appear reasonable to make the change of variable $x = e^z$, or $z = \ln x$. Consider only the interval $x > 0$.
(a) Show that

$$\frac{dy}{dx} = \frac{1}{x}\frac{dy}{dz} \quad \text{and} \quad \frac{d^2y}{dx^2} = \frac{1}{x^2}\frac{d^2y}{dz^2} - \frac{1}{x^2}\frac{dy}{dz}.$$

(b) Show that the Euler equation becomes

$$\frac{d^2y}{dz^2} + (\alpha - 1)\frac{dy}{dz} + \beta y = 0.$$

Letting r_1 and r_2 denote the roots of $r^2 + (\alpha - 1)r + \beta = 0$, show that

(c) if r_1 and r_2 are real and unequal, then

$$y = c_1 e^{r_1 z} + c_2 e^{r_2 z} = c_1 x^{r_1} + c_2 x^{r_2},$$

(d) if r_1 and r_2 are equal, then

$$y = (c_1 + c_2 z)e^{r_1 z} = (c_1 + c_2 \ln x)x^{r_1},$$

(e) if r_1 and r_2 are complex conjugates, $r_1 = \lambda + i\mu$, then

$$y = e^{\lambda z}(c_1 \cos \mu z + c_2 \sin \mu z) = x^{\lambda}[c_1 \cos(\mu \ln x) + c_2 \sin(\mu \ln x)].$$

17. Using the method of Problem 16, solve the following differential equations for $x > 0$.

(a) $x^2 y'' - 2y = 0$
(b) $x^2 y'' - 3xy' + 4y = \ln x$
(c) $x^2 y'' + 7xy' + 5y = x$
(d) $x^2 y'' - 2xy' + 2y = 3x^2 + 2\ln x$
(e) $x^2 y'' + xy' + 4y = \sin(\ln x)$
(f) $3x^2 y'' + 12xy' + 9y = 0$

18. Show that if $L[y] = x^2 y'' + \alpha xy' + \beta y$, then

$$L[(-x)^r] = (-x)^r F(r)$$

for $x < 0$, where $F(r) = r(r-1) + \alpha r + \beta$. Hence conclude that if $r_1 \neq r_2$ are roots of $F(r) = 0$, then linearly independent solutions of $L[y] = 0$ for $x < 0$ are $(-x)^{r_1}$ and $(-x)^{r_2}$.

19. Suppose that x^{r_1} and x^{r_2} are solutions of an Euler equation, where $r_1 \neq r_2$, and r_1 is an integer. According to Eq. (24) the general solution in any interval not containing the origin is $y = c_1|x|^{r_1} + c_2|x|^{r_2}$. Show that the general solution can also be written as $y = k_1 x^{r_1} + k_2|x|^{r_2}$.
Hint: Show by a proper choice of constants that the expressions are identical for $x > 0$; and by a different choice of constants that they are identical for $x < 0$.

20. If the constants α and β in the Euler equation $x^2 y'' + \alpha xy' + \beta y = 0$ are complex numbers, it is still possible to obtain solutions of the form x^r. However, in general, the solutions are no longer real valued. Determine the general solution of each of the following equations for the interval $x > 0$.

(a) $x^2 y'' + 2ixy' - iy = 0$
(b) $x^2 y'' + (1-i)xy' + 2y = 0$
(c) $x^2 y'' + xy' - 2iy = 0$
Hint: See Problems 15, 16, and 17 of Section 3.5.1

4.5 Series Solutions near a Regular Singular Point, Part I

We now consider the question of solving the general second order linear equation

$$P(x)y'' + Q(x)y' + R(x)y = 0 \tag{1}$$

in the neighborhood of a regular singular point $x = x_0$. For convenience we assume that $x_0 = 0$. If $x_0 \neq 0$ the equation can be transformed into one for which the regular singular point is at the origin by letting $x - x_0$ equal z.

The fact that $x = 0$ is a regular singular point of Eq. (1) means that $xQ(x)/P(x) = xp(x)$ and $x^2R(x)/P(x) = x^2q(x)$ have finite limits as $x \to 0$, and are analytic at $x = 0$. Thus they have power series expansions of the form

$$xp(x) = \sum_{n=0}^{\infty} p_n x^n, \qquad x^2q(x) = \sum_{n=0}^{\infty} q_n x^n, \tag{2}$$

which are convergent for some interval $|x| < \rho$, $\rho > 0$, about the origin. To make the quantities $xp(x)$ and $x^2q(x)$ appear in Eq. (1), it is convenient to divide Eq. (1) by $P(x)$ and then to multiply by x^2, obtaining

$$x^2y'' + x[xp(x)]y' + [x^2q(x)]y = 0, \tag{3a}$$

or

$$x^2y'' + x(p_0 + p_1x + \cdots + p_nx^n + \cdots)y' + (q_0 + q_1x + \cdots + q_nx^n + \cdots)y = 0. \tag{3b}$$

If all of the p_n and q_n are zero except

$$p_0 = \lim_{x\to 0} \frac{xQ(x)}{P(x)}, \qquad q_0 = \lim_{x\to 0} \frac{x^2R(x)}{P(x)}, \tag{4}$$

then Eq. (3) reduces to the Euler equation

$$x^2y'' + p_0xy' + q_0y = 0, \tag{5}$$

which was discussed in the previous section. In general, of course, some of the p_n and q_n, $n \geq 1$, are not zero. However, the essential character of solutions of Eq. (3) is identical to that of solutions of the Euler equation. The presence of the terms $p_1x + \cdots + p_nx^n + \cdots$ and $q_1x + \cdots + q_nx^n + \cdots$ merely complicates the calculations.

We primarily restrict our discussion to the interval $x > 0$. The interval $x < 0$ can be treated, just as for the Euler equation, by making the change of variable $x = -\xi$ and then solving the resulting equation for $\xi > 0$.

Since the coefficients in Eq. (3) are "Euler coefficients" times power series, it is natural to seek solutions of the form of "Euler solutions" times power series:

$$y = x^r(a_0 + a_1x + \cdots + a_nx^n + \cdots) = x^r \sum_{n=0}^{\infty} a_nx^n = \sum_{n=0}^{\infty} a_nx^{r+n}. \tag{6}$$

As part of our problem we have to determine

1. The values of r for which Eq. (1) has a solution of the form (6).
2. The recurrence relation for the a_n.
3. The radius of convergence of the series $\sum_{n=0}^{\infty} a_nx^n$.

We will find that the possible values of r are precisely those associated with the Euler equation (5); hence they are roots of the quadratic equation

$$F(r) = r(r-1) + p_0r + q_0 = 0, \tag{7}$$

where p_0 and q_0 are given by Eqs. (4). Equation (7) is called the *indicial equation* for the differential equation (1) and the two corresponding values of r are called the *exponents of the singularity* at the regular singular point $x = 0$. Note that the exponent r determines the qualitative behavior of the solution (6) in the neighborhood of the singular point, and that the possible exponents are easily found from Eqs. (4) and (7).

We will also find that Eq. (1) always has a solution of the form (6) with a_0 arbitrary, for at least one of the possible exponents. Often there is a second solution of the form (6) for the second exponent. If there is not a second solution of the form (6), the second solution will involve a logarithmic term just as did the second solution of the Euler equation when the roots were equal. In theory, of course, once we obtain the first solution, which is of the form (6), a second solution can be obtained by the method of reduction of order. Unfortunately, this is not always a convenient method for obtaining a second solution; we give an alternative method in Sections 4.6 and 4.7.

The general theory is due to Frobenius[12] and is fairly complicated. Rather than trying to present this theory we simply assume in this and the next three sections that there does exist a solution of the stated form. In particular we assume that any power series in an expression for a solution has a nonzero radius of convergence, and concentrate on showing how to determine the coefficients in such a series.

Let us first illustrate the method of Frobenius for a specific differential equation with two solutions of the form (6). Consider the equation

$$2x^2y'' - xy' + (1 + x)y = 0. \tag{8}$$

By comparison with Eq. (3), it is clear that $x = 0$ is a regular singular point of Eq. (8). Also by comparison with Eqs. (3a) and (3b) we see that $xp(x) = -\frac{1}{2}$ and $x^2q(x) = (1 + x)/2$. Thus $p_0 = -\frac{1}{2}$, $q_0 = \frac{1}{2}$, $q_1 = \frac{1}{2}$, and all other p's and q's are zero. The corresponding Euler equation is

$$2x^2y'' - xy' + y = 0. \tag{9}$$

We formally try to find a solution of Eq. (8) of the form (6) for $x > 0$. Then y' and y'' are given by

$$y' = \sum_{n=0}^{\infty} a_n(r + n)x^{r+n-1}, \tag{10}$$

and

$$y'' = \sum_{n=0}^{\infty} a_n(r + n)(r + n - 1)x^{r+n-2}. \tag{11}$$

[12] Georg Ferdinand Frobenius (1849–1917) was (like Fuchs) a student and eventually a professor at the University of Berlin. He showed how to construct series solutions about regular singular points in 1874. His most distinguished work, however, was in algebra where he was one of the foremost early developers of group theory.

Hence

$$2x^2y'' - xy' + (1 + x)y = \sum_{n=0}^{\infty} 2a_n(r+n)(r+n-1)x^{r+n}$$
$$- \sum_{n=0}^{\infty} a_n(r+n)x^{r+n} + \sum_{n=0}^{\infty} a_n x^{r+n} + \sum_{n=0}^{\infty} a_n x^{r+n+1}.$$

Noting that the last sum can be written in the form $\sum_{n=1}^{\infty} a_{n-1}x^{r+n}$ and combining terms gives

$$2x^2y'' - xy' + (1 + x)y = a_0[2r(r-1) - r + 1]x^r$$
$$+ \sum_{n=1}^{\infty} \{[2(r+n)(r+n-1) - (r+n) + 1]a_n + a_{n-1}\}x^{r+n} = 0. \tag{12}$$

If Eq. (12) is to be identically satisfied, the coefficient of each power of x in Eq. (12) must be zero. Corresponding to x^r we obtain, since $a_0 \neq 0$,[13]

$$2r(r-1) - r + 1 = 2r^2 - 3r + 1 = 0. \tag{13}$$

From the coefficient of x^{r+n} we have

$$[2(r+n)(r+n-1) - (r+n) + 1]a_n + a_{n-1} = 0,$$

or

$$a_n = -\frac{1}{2(r+n)^2 - 3(r+n) + 1}a_{n-1}, \qquad n \geq 1. \tag{14}$$

Equation (13) is the indicial equation for Eq. (8). Note that it is exactly the equation we would obtain for the Euler equation (9) associated with Eq. (8). Factoring Eq. (13) gives $(2r-1)(r-1) = 0$; hence the roots of the indicial equation are

$$r_1 = 1, \qquad r_2 = \tfrac{1}{2}. \tag{15}$$

In the following, whenever the roots of the indicial equation are real and unequal, the larger one is denoted by r_1.

Equation (14) gives a recurrence relation for the coefficients a_n. Corresponding to the factored form of the indicial equation we can write the recurrence relation as

$$a_n = \frac{-1}{[2(r+n)-1][(r+n)-1]}a_{n-1}, \qquad n \geq 1. \tag{16}$$

[13] Nothing is gained by taking $a_0 = 0$, since this would simply mean that the series would start with the term a_1x^{r+1} or, if $a_1 = 0$, the term a_2x^{r+2}, and so on. Suppose the first nonzero term is a_mx^{r+m}; this is equivalent to using the series (6) with r replaced by $r + m$ and a_0 replaced by a_m. Indeed, r is chosen to be the lowest power of x appearing in the series and we take $a_0 \neq 0$ as its coefficient.

For each root r_1 and r_2 of the indicial equation we use the recurrence relation to determine a set of coefficients $a_1, a_2, \ldots$. For $r = r_1 = 1$, Eq. (16) becomes

$$a_n = -\frac{a_{n-1}}{(2n+1)n}, \qquad n \geq 1.$$

Thus

$$a_1 = -\frac{a_0}{3 \cdot 1},$$

$$a_2 = -\frac{a_1}{5 \cdot 2} = \frac{a_0}{(3 \cdot 5)(1 \cdot 2)},$$

and

$$a_3 = -\frac{a_2}{7 \cdot 3} = -\frac{a_0}{(3 \cdot 5 \cdot 7)(1 \cdot 2 \cdot 3)}.$$

In general we have

$$a_n = \frac{(-1)^n}{[3 \cdot 5 \cdot 7 \cdots (2n+1)]n!} a_0, \qquad n \geq 1. \tag{17}$$

Hence, omitting the constant multiplier a_0, one solution of Eq. (8) is

$$y_1(x) = x\left[1 + \sum_{n=1}^{\infty} \frac{(-1)^n x^n}{[3 \cdot 5 \cdot 7 \cdots (2n+1)]n!}\right], \qquad x > 0. \tag{18}$$

To determine the radius of convergence of the series in Eq. (18) we use the ratio test:

$$\lim_{n \to \infty} \left| \frac{a_{n+1}x^{n+1}}{a_n x^n} \right| = \lim_{n \to \infty} \frac{|x|}{(2n+3)(n+1)} = 0$$

for all x. Thus the series converges for all x.

Corresponding to the second root $r = r_2 = \frac{1}{2}$, we proceed similarly. From Eq. (16) we have

$$a_n = -\frac{a_{n-1}}{2n(n-\frac{1}{2})} = -\frac{a_{n-1}}{n(2n-1)}, \qquad n \geq 1.$$

Hence

$$a_1 = -\frac{a_0}{1 \cdot 1},$$

$$a_2 = -\frac{a_1}{2 \cdot 3} = \frac{a_0}{(1 \cdot 2)(1 \cdot 3)},$$

$$a_3 = -\frac{a_2}{3 \cdot 5} = -\frac{a_0}{(1 \cdot 2 \cdot 3)(1 \cdot 3 \cdot 5)},$$

and in general

$$a_n = \frac{(-1)^n}{n![1 \cdot 3 \cdot 5 \cdots (2n-1)]}, \qquad n \geq 1. \tag{19}$$

Again omitting the constant multiplier a_0, we obtain the second solution

$$y_2(x) = x^{1/2}\left[1 + \sum_{n=1}^{\infty} \frac{(-1)^n x^n}{n![1 \cdot 3 \cdot 5 \cdots (2n-1)]}\right], \qquad x > 0. \qquad (20)$$

As before, we can show that the series in Eq. (20) converges for all x. Since the leading terms in the series solutions y_1 and y_2 are x and $x^{1/2}$, respectively, it is clear that the solutions are linearly independent. Hence the general solution of Eq. (8) is

$$y = c_1 y_1(x) + c_2 y_2(x), \qquad x > 0.$$

The preceding example is illuminating in several respects. Let us reconsider our calculations from a more general viewpoint. Let

$$\begin{aligned} F(r) &= 2r(r-1) - r + 1 \\ &= (2r-1)(r-1); \end{aligned} \qquad (21)$$

then the indicial equation is $F(r) = 0$, and the recurrence relation (14) becomes

$$F(r+n)a_n + a_{n-1} = 0, \qquad n \geq 1. \qquad (22)$$

Equation (22) is typical of the general case except that the term a_{n-1} is, in general, replaced by a linear combination of $a_0, a_1, a_2, \ldots, a_{n-1}$. However, we can use Eq. (22) as a model for the discussion that follows with the understanding that the discussion is not limited to the specific example just completed.

There are two possible sources of trouble. First, if the two roots r_1 and r_2 of the indicial equation are equal, then we will obtain only one series solution of the form (6). Second, there is the possibility that for a given root r_1 or r_2 we may not be able to solve for one of the a_n from Eq. (22) because $F(r+n)$ may be zero. When can this happen? The only zeros of $F(r)$ are r_1 and r_2, and if $r_1 > r_2$, then clearly $r_1 + n > r_1 > r_2$ for all n and $F(r_1 + n) \neq 0$ for all n. Thus there will never be any difficulty in computing the series solution corresponding to the larger of the roots of the indicial equation. On the other hand consider $F(r_2 + n)$, and suppose that $r_1 - r_2 = N$, a positive integer. Then for $n = N$ we have

$$F(r_2 + N)a_N + a_{N-1} = 0;$$

but $F(r_2 + N) = F(r_1) = 0$ and we cannot solve for a_N unless also $a_{N-1} = 0$, and hence cannot obtain a second series solution. In more general problems, as we see in the next section, the recurrence relation will be more complicated, and even though $F(r_2 + N)$ may vanish, it may be possible to determine a second solution of the form (6) corresponding to $r = r_2$. In any event the cases $r_1 = r_2$ and $r_1 - r_2 = N$, a positive integer, require special attention.

Finally, we note that if the roots of the indicial equation are complex, then this type of difficulty cannot occur; on the other hand, the solutions corresponding to r_1 and r_2 will be complex valued functions of x. However, just as for the Euler equation, it is possible to obtain real valued solutions by taking the real and imaginary parts of the complex valued solutions.

Before turning to a general discussion and the special cases $r_1 = r_2$ and $r_1 - r_2 = N$, a positive integer, a practical point should be mentioned. If P, Q, and R are polynomials, it is much better to work directly with Eq. (1) than with Eq. (3). This avoids the necessity of expressing $xQ(x)/P(x)$ and $x^2R(x)/P(x)$ as power series. For example, it is more convenient to consider the equation

$$x(1 + x)y'' + 2y' + xy = 0$$

than to write it in the form

$$x^2y'' + x\frac{2}{1 + x}y' + \frac{x^2}{1 + x}y = 0,$$

and then to expand $2/(1 + x)$ and $x^2/(1 + x)$, which gives

$$x^2y'' + x\left[2(1 - x + x^2 - \cdots)\right]y' + (x^2 - x^3 + x^4 - \cdots)y = 0.$$

PROBLEMS

In each of Problems 1 through 6 show that the given differential equation has a regular singular point at $x = 0$. Determine the indicial equation, the recurrence relation, and the roots of the indicial equation. Find the series solution ($x > 0$) corresponding to the larger root. If the roots are unequal and do not differ by an integer, find the series solution corresponding to the smaller root.

1. $2xy'' + y' + xy = 0$
2. $x^2y'' + xy' + (x^2 - \frac{1}{9})y = 0$
3. $xy'' + y = 0$
4. $xy'' + y' - y = 0$
5. $3x^2y'' + 2xy' + x^2y = 0$
6. $x^2y'' + xy' + (x - 2)y = 0$
7. The Legendre equation of order α is
$$(1 - x^2)y'' - 2xy' + \alpha(\alpha + 1)y = 0.$$
The solution of this equation near the ordinary point $x = 0$ was discussed in Problems 10 and 11 of Section 4.2.1. In Example 5 of Section 4.3 it was shown that $x = \pm 1$ are regular singular points. Determine the indicial equation and its roots for the point $x = 1$. Find a series solution in powers of $x - 1$ for $x - 1 > 0$.
Hint: Write $1 + x = 2 + (x - 1)$, and $x = 1 + (x - 1)$. Alternatively, make the change of variable $x - 1 = z$ and determine a series solution in powers of z.
8. The Tchebycheff equation is
$$(1 - x^2)y'' - xy' + \alpha^2y = 0,$$
where α is a constant; see Problem 4 of Section 4.2.1.
(a) Show that $x = 1$ and $x = -1$ are regular singular points, and find the exponents at each of these singularities.
(b) Find two linearly independent solutions about $x = 1$.
9. The Laguerre (1834–1886) differential equation is
$$xy'' + (1 - x)y' + \lambda y = 0.$$
Show that $x = 0$ is a regular singular point. Determine the indicial equation, its roots, the recurrence relation, and one solution ($x > 0$). Show that if $\lambda = m$, a positive integer, this solution reduces to a polynomial. When properly normalized this polynomial is known as the Laguerre polynomial, $L_m(x)$.

10. The Bessel equation of order zero is

$$x^2y'' + xy' + x^2y = 0.$$

Show that $x = 0$ is a regular singular point; that the roots of the indicial equation are $r_1 = r_2 = 0$; and that one solution for $x > 0$ is

$$J_0(x) = 1 + \sum_{n=1}^{\infty} \frac{(-1)^n x^{2n}}{2^{2n}(n!)^2}.$$

Notice that the series converges for all x, not just $x > 0$. In particular, $J_0(x)$ is bounded as $x \to 0$. The function J_0 is known as the Bessel function of the first kind of order zero.

11. Referring to Problem 10 show, using the method of reduction of order, that the second solution of the Bessel equation of order zero contains a logarithmic term.
Hint: If $y_2(x) = J_0(x)v(x)$, then

$$y_2(x) = J_0(x)\int^x \frac{ds}{s[J_0(s)]^2}.$$

Find the first term in the series expansion for $1/s[J_0(s)]^2$.

12. The Bessel equation of order one is

$$x^2y'' + xy' + (x^2 - 1)y = 0.$$

(a) Show that $x = 0$ is a regular singular point; that the roots of the indicial equation are $r_1 = 1$ and $r_2 = -1$; and that one solution for $x > 0$ is

$$J_1(x) = \frac{x}{2} \sum_{n=0}^{\infty} \frac{(-1)^n x^{2n}}{(n+1)!n!2^{2n}}.$$

Notice as in Problem 10 that the series actually converges for all x and that $J_1(x)$ is bounded as $x \to 0$. The function J_1 is known as the Bessel function of the first kind of order one.

(b) Show that it is impossible to determine a second solution of the form

$$x^{-1} \sum_{n=0}^{\infty} b_n x^n, \qquad x > 0.$$

4.5.1 Series Solutions near a Regular Singular Point, Part II

Now let us consider the general problem of determining a solution of the equation

$$x^2y'' + x[xp(x)]y' + [x^2q(x)]y = 0 \tag{1}$$

where

$$xp(x) = \sum_{n=0}^{\infty} p_n x^n \quad \text{and} \quad x^2q(x) = \sum_{n=0}^{\infty} q_n x^n,$$

and both series have nonzero radii of convergence (see Eq. (3) of Section 4.5). The point $x = 0$ is a regular singular point. For convenience, we first suppose that

$x > 0$, and look for a solution of the form

$$y = \sum_{n=0}^{\infty} a_n x^{r+n}. \tag{2}$$

Substituting in Eq. (1) gives

$$\begin{aligned} &a_0 r(r-1)x^r + a_1(r+1)rx^{r+1} + \cdots + a_n(r+n)(r+n-1)x^{r+n} + \cdots \\ &\quad + (p_0 + p_1 x + \cdots + p_n x^n + \cdots) \\ &\qquad \times \left[a_0 r x^r + a_1(r+1)x^{r+1} + \cdots + a_n(r+n)x^{r+n} + \cdots\right] \\ &\quad + (q_0 + q_1 x + \cdots + q_n x^n + \cdots) \\ &\qquad \times \left(a_0 x^r + a_1 x^{r+1} + \cdots + a_n x^{r+n} + \cdots\right) = 0. \end{aligned} \tag{3}$$

Multiplying the infinite series together and then collecting terms, we obtain

$$\begin{aligned} &a_0 F(r)x^r + [a_1 F(r+1) + a_0(p_1 r + q_1)]x^{r+1} \\ &\quad + \{a_2 F(r+2) + a_0(p_2 r + q_2) + a_1[p_1(r+1) + q_1]\}x^{r+2} \\ &\quad + \cdots + \{a_n F(r+n) + a_0(p_n r + q_n) + a_1[p_{n-1}(r+1) + q_{n-1}] \\ &\quad + \cdots + a_{n-1}[p_1(r+n-1) + q_1]\}x^{r+n} + \cdots = 0, \end{aligned}$$

or in a more compact form

$$a_0 F(r)x^r + \sum_{n=1}^{\infty} \left\{ F(r+n)a_n + \sum_{k=0}^{n-1} a_k[(r+k)p_{n-k} + q_{n-k}] \right\} x^{r+n} = 0, \tag{4}$$

where

$$F(r) = r(r-1) + p_0 r + q_0. \tag{5}$$

If Eq. (4) is to be satisfied identically, the coefficient of each power of x must be zero.

Since $a_0 \neq 0$, the term involving x^r yields the indicial equation $F(r) = 0$. We emphasize again that the indicial equation $F(r) = 0$ is exactly the equation we would obtain in looking for solutions x^r of the Euler equation $x^2y'' + p_0xy' + q_0y = 0$ corresponding to Eq. (1). Let us denote the roots of the indicial equation by r_1 and r_2 with $r_1 \geq r_2$ if the roots are real. If the roots are complex the ordering is immaterial. Only for these values of r can we expect to find solutions of Eq. (1) of the form (2).

Setting the coefficient of x^{r+n} in Eq. (4) equal to zero gives the recurrence relation

$$F(r+n)a_n + \sum_{k=0}^{n-1} a_k[(r+k)p_{n-k} + q_{n-k}] = 0, \qquad n \geq 1. \tag{6}$$

Equation (6) shows that, in general, a_n depends on the value of r and all of the preceding coefficients $a_0, a_1, \ldots, a_{n-1}$. It also shows that we can successively compute $a_1, a_2, \ldots, a_n, \ldots,$ in terms of the coefficients in the series for $xp(x)$ and $x^2q(x)$ provided that $F(r+1), F(r+2), \ldots, F(r+n), \ldots$ are not zero. The only values of r for which $F(r) = 0$ are $r = r_1$ and $r = r_2$; since $r_1 + n$ is not equal to r_1 or r_2 for $n \geq 1$, it follows that $F(r_1 + n) \neq 0$ for $n \geq 1$. Hence we can always determine one solution

$$y_1(x) = x^{r_1}\left[1 + \sum_{n=1}^{\infty} a_n(r_1)x^n\right]. \tag{7}$$

Here we have introduced the notation $a_n(r_1)$ to indicate that a_n has been determined from Eq. (6) with $r = r_1$. Also for definiteness we have taken a_0 to be one.

If r_2 is not equal to r_1, and $r_1 - r_2$ is not a positive integer, then $r_2 + n$ is not equal to r_1 for any value of $n \geq 1$; hence $F(r_2 + n) \neq 0$, and we can obtain a second solution

$$y_2(x) = x^{r_2}\left[1 + \sum_{n=1}^{\infty} a_n(r_2)x^n\right]. \tag{8}$$

Before continuing, we make several remarks about the series solutions (7) and (8). Within their radii of convergence, the power series $\sum_{n=0}^{\infty} a_n(r_1)x^n$ and $\sum_{n=0}^{\infty} a_n(r_2)x^n$ define functions that are analytic at $x = 0$. Thus the singular behavior, if there is any, of the solutions y_1 and y_2 is determined by the terms x^{r_1} and x^{r_2} which multiply these two analytic functions, respectively. Next, to obtain real-valued solutions for $x < 0$, we can make the substitution $x = -\xi$ with $\xi > 0$. As we might expect from our discussion of the Euler equation it turns out that we need only replace x^{r_1} in Eq. (7) and x^{r_2} in Eq. (8) by $|x|^{r_1}$ and $|x|^{r_2}$, respectively. Finally, we note that if r_1 and r_2 are complex numbers, then they are necessarily complex conjugates and $r_2 \neq r_1 + N$. Thus we can compute two series solutions of the form (2); however, they are complex-valued functions of x. Real-valued solutions can be obtained by taking the real and imaginary parts of the complex-valued solutions.

If $r_2 = r_1$, we obtain only one solution of the form (2). The second solution contains a logarithmic term. A procedure for determining a second solution is discussed in Section 4.6 and an example is given in Section 4.7.

If $r_1 - r_2 = N$, where N is a positive integer, then $F(r_2 + N) = F(r_1) = 0$, and we cannot determine $a_N(r_2)$ from Eq. (6) unless it happens that the sum in Eq. (6) also is zero for $n = N$. If this is so, it can be shown that a_N is arbitrary and we can obtain a second series solution; if not, there is only one series solution of the form (2). If the latter is the case, then the second solution contains a logarithmic term. The procedure is discussed briefly in Section 4.6, and examples of both cases are given in Section 4.7.

The next question to consider is the convergence of the infinite series in the formal solution. As we might expect, this is related to the radii of convergence of the power series for $xp(x)$ and $x^2q(x)$. We summarize the results of our discussion and the additional conclusions about the convergence of the series solution in the following theorem.

Theorem 4.3. *Consider the differential equation* (1)

$$x^2y'' + x[xp(x)]y' + [x^2q(x)]y = 0,$$

where $x = 0$ is a regular singular point. Then $xp(x)$ and $x^2q(x)$ are analytic at $x = 0$ with convergent power series expansions

$$xp(x) = \sum_{n=0}^{\infty} p_n x^n, \qquad x^2q(x) = \sum_{n=0}^{\infty} q_n x^n$$

for $|x| < \rho$, where $\rho > 0$ is the minimum of the radii of convergence of the power series for $xp(x)$ and $x^2q(x)$. Let r_1 and r_2 be the roots of the indicial equation

$$F(r) = r(r-1) + p_0 r + q_0 = 0,$$

with $r_1 \geq r_2$ if r_1 and r_2 are real. Then in either of the intervals $-\rho < x < 0$ or $0 < x < \rho$, there exists a solution of the form

$$y_1(x) = |x|^{r_1}\left[1 + \sum_{n=1}^{\infty} a_n(r_1)x^n\right] \tag{9}$$

where the $a_n(r_1)$ are given by the recurrence relation (6) *with $a_0 = 1$ and $r = r_1$. If $r_1 - r_2$ is not zero or a positive integer, then in either of the intervals $-\rho < x < 0$ or $0 < x < \rho$, there exists a second linearly independent solution of the form*

$$y_2(x) = |x|^{r_2}\left[1 + \sum_{n=1}^{\infty} a_n(r_2)x^n\right]. \tag{10}$$

The $a_n(r_2)$ are also determined by the recurrence relation (6) *with $a_0 = 1$ and $r = r_2$. The power series in Eqs.* (9) *and* (10) *converge at least for $|x| < \rho$.*

It is possible that the series in Eqs. (9) and (10) may have radii of convergence greater than ρ; if so, the solutions (9) and (10) will be valid in correspondingly larger intervals. While it is beyond the scope of this text to prove this theorem, we can make several observations. To determine the roots r_1 and r_2 of the indicial equation, which play such an important role in the present theory, we need only determine p_0 and q_0 and then solve the equation $r(r-1) + p_0 r + q_0 = 0$. These coefficients are given by

$$p_0 = \lim_{x\to 0} xp(x), \qquad q_0 = \lim_{x\to 0} x^2q(x). \tag{11}$$

In practice, rather than remember the recurrence formula (6) for the a_n we usually substitute the series (2) for y in Eq. (1), determine the values of r, and then explicitly calculate the a_n.

Finally we note that if $x = 0$ is a regular singular point of the equation

$$P(x)y'' + Q(x)y' + R(x)y = 0, \tag{12}$$

where the functions P, Q, and R are polynomials, then $xp(x) = xQ(x)/P(x)$

and $x^2q(x) = x^2R(x)/P(x)$. Thus

$$p_0 = \lim_{x\to 0} x\frac{Q(x)}{P(x)}, \qquad q_0 = \lim_{x\to 0} x^2\frac{R(x)}{P(x)}. \tag{13}$$

Further, the radii of convergence for the series in Eqs. (9) and (10) are at least equal to the distance from the origin to the nearest zero of P.

EXAMPLE

Discuss the nature of the solutions of the equation

$$2x(1+x)y'' + (3+x)y' - xy = 0$$

in the neighborhood of the singular points.

This equation is of the form (12) with $P(x) = 2x(1+x)$, $Q(x) = 3+x$, and $R(x) = -x$. Clearly the points $x = 0$ and $x = -1$ are singular points. The point $x = 0$ is a regular singular point, since

$$\lim_{x\to 0} x\frac{Q(x)}{P(x)} = \lim_{x\to 0} x\frac{3+x}{2x(1+x)} = \frac{3}{2},$$

$$\lim_{x\to 0} x^2\frac{R(x)}{P(x)} = \lim_{x\to 0} x^2\frac{-x}{2x(1+x)} = 0.$$

Further, from Eq. (13), $p_0 = \frac{3}{2}$ and $q_0 = 0$. Thus the indicial equation is $r(r-1) + \frac{3}{2}r = 0$, and the roots are $r_1 = 0$, $r_2 = -\frac{1}{2}$. Since these roots are not equal and do not differ by an integer there are two linearly independent solutions of the form

$$y_1(x) = 1 + \sum_{n=1}^{\infty} a_n(0)x^n \quad \text{and} \quad y_2(x) = |x|^{-1/2}\left[1 + \sum_{n=1}^{\infty} a_n(-\tfrac{1}{2})x^n\right]$$

for $0 < |x| < \rho$. A lower bound for the radius of convergence of each series is 1, the distance from $x = 0$ to $x = -1$, the other zero of $P(x)$. Note that the solution y_1 is bounded as $x \to 0$, indeed is analytic there. On the other hand, the second solution y_2 is unbounded as $x \to 0$.

The point $x = -1$ is also a regular singular point, since

$$\lim_{x\to -1} (x+1)\frac{Q(x)}{P(x)} = \lim_{x\to -1} \frac{(x+1)(3+x)}{2x(1+x)} = -1,$$

$$\lim_{x\to -1} (x+1)^2\frac{R(x)}{P(x)} = \lim_{x\to -1} \frac{(x+1)^2(-x)}{2x(1+x)} = 0.$$

In this case $p_0 = -1$, $q_0 = 0$ so the indicial equation is $r(r-1) - r = 0$. The roots of the indicial equation are $r_1 = 2$ and $r_2 = 0$. Corresponding to the larger root there is a solution of the form

$$y_1(x) = (x+1)^2\left[1 + \sum_{n=1}^{\infty} a_n(2)(x+1)^n\right].$$

The series converges at least for $|x+1| < 1$. Since the two roots differ by a positive integer there may or may not be a second solution of the form

$$y_2(x) = \left[1 + \sum_{n=1}^{\infty} a_n(0)(x+1)^n\right].$$

We cannot say more without further analysis.

PROBLEMS

In each of Problems 1 through 8 find all the regular singular points of the given differential equation. Determine the indicial equation and the exponents of the singularity at each regular singular point.

1. $xy'' + 2xy' + 6e^x y = 0$
2. $x^2y'' - x(2+x)y' + (2+x^2)y = 0$
3. $x(x-1)y'' + 6x^2y' + 3y = 0$
4. $y'' + 4xy' + 6y = 0$
5. $x^2y'' + 3(\sin x)y' - 2y = 0$
6. $2x(x+2)y'' + y' - xy = 0$
7. $x^2y'' + \frac{1}{2}(x + \sin x)y' + y = 0$
8. $(x+1)^2y'' + 3(x^2-1)y' + 3y = 0$
9. Show that

$$x^2y'' + (\sin x)y' - (\cos x)y = 0$$

has a regular singular point at $x = 0$, and that the roots of the indicial equation are ± 1. Determine the first three nonzero terms in the series corresponding to the larger root.

10. Show that

$$(\ln x)y'' + \tfrac{1}{2}y' + y = 0$$

has a regular singular point at $x = 1$. Determine the roots of the indicial equation at $x = 1$. Determine the first three nonzero terms in the series $\sum_{n=0}^{\infty} a_n(x-1)^{r+n}$ corresponding to the larger root. Take $x - 1 > 0$. What would you expect the radius of convergence of the series to be?

11. In several problems in mathematical physics (for example, the Schrödinger equation for a hydrogen atom) it is necessary to study the differential equation

$$x(1-x)y'' + [\gamma - (1+\alpha+\beta)x]y' - \alpha\beta y = 0, \tag{i}$$

where α, β, and γ are constants. This equation is known as the *hypergeometric* equation.

(a) Show that $x = 0$ is a regular singular point, and that the roots of the indicial equation are 0 and $1 - \gamma$.

(b) Show that $x = 1$ is a regular singular point, and that the roots of the indicial equation are 0 and $\gamma - \alpha - \beta$.

(c) Assuming that $1 - \gamma$ is not a positive integer, show that in the neighborhood of $x = 0$ one solution of (i) is

$$y_1(x) = 1 + \frac{\alpha\beta}{\gamma \cdot 1!}x + \frac{\alpha(\alpha+1)\beta(\beta+1)}{\gamma(\gamma+1)2!}x^2 + \cdots.$$

What would you expect the radius of convergence of this series to be?

(d) Assuming that $1 - \gamma$ is not an integer or zero, show that a second solution for $0 < x < 1$ is

$$y_2(x) = x^{(1-\gamma)}\left[1 + \frac{(\alpha-\gamma+1)(\beta-\gamma+1)}{(2-\gamma)1!}x + \frac{(\alpha-\gamma+1)(\alpha-\gamma+2)(\beta-\gamma+1)(\beta-\gamma+2)}{(2-\gamma)(3-\gamma)2!}x^2 + \cdots\right].$$

*(e) Show that the point at infinity is a regular singular point, and that the roots of the indicial equation are α and β. See Problem 13 of Section 4.3.

12. Consider the differential equation

$$x^3y'' + \alpha xy' + \beta y = 0$$

where $\alpha \neq 0$ and β are real constants.
(a) Show that $x = 0$ is an irregular singular point.
(b) Show that if we attempt to determine a solution of the form $x^r \sum_{n=0}^{\infty} a_n x^n$, the indicial equation for r will be linear, and as a consequence there will be only one formal solution of this form.

13. Consider the differential equation

$$y'' + \frac{\alpha}{x^s}y' + \frac{\beta}{x^t}y = 0, \qquad \text{(i)}$$

where $\alpha \neq 0$ and $\beta \neq 0$ are real numbers, and s and t are positive integers that for the moment are arbitrary.
(a) Show that if $s > 1$ or $t > 2$, then the point $x = 0$ is an irregular singular point.
(b) Suppose we try to find a solution of Eq. (i) of the form

$$y = \sum_{n=0}^{\infty} a_n x^{r+n}, \qquad x > 0. \qquad \text{(ii)}$$

Show that if $s = 2$ and $t = 2$, then there is a only one possible value of r for which there is a formal solution of Eq. (i) of the form (ii).
(c) Show that if $s = 1$ and $t = 3$ there are no solutions of Eq. (i) of the form (ii).
(d) Show that the maximum values of s and t for which the indicial equation is quadratic in r [and hence we can hope to find two solutions of the form (ii)] are $s = 1$ and $t = 2$. These are precisely the conditions that distinguish a "weak singularity," or a regular singular point from an irregular singular point, as we defined them in Section 4.3.
As a note of caution we should point out that while it is sometimes possible to obtain a formal series solution of the form (ii) at an irregular singular point, the series may not converge.

*4.6 Series Solutions near a Regular Singular Point; $r_1 = r_2$ and $r_1 - r_2 = N$

We now consider the cases in which the roots r_1 and r_2 of the indicial equation are equal or differ by a positive integer, $r_1 - r_2 = N$. If $r_1 = r_2$, we would expect by analogy with the Euler equation that the second solution will contain a logarithmic term. This may also be the case when the roots differ by an integer. The general situation, including the case in which r_1 and r_2 do not differ by an integer, is summarized in the following theorem.

Theorem 4.4. *Consider the differential equation*

$$L[y] = x^2y'' + x[xp(x)]y' + [x^2q(x)]y = 0, \tag{1}$$

where $x = 0$ is a regular singular point. Then the functions $xp(x)$ and $x^2q(x)$ are analytic at $x = 0$ with power series representations

$$xp(x) = \sum_{n=0}^{\infty} p_n x^n, \qquad x^2q(x) = \sum_{n=0}^{\infty} q_n x^n, \tag{2}$$

which converge for $|x| < \rho$, where $\rho > 0$ is the minimum of the radii of convergence of the power series for $xp(x)$ and $x^2q(x)$. Let r_1 and r_2, where $r_1 \geq r_2$ if they are real, be the roots of the indicial equation

$$F(r) = r(r-1) + p_0 r + q_0 = 0. \tag{3}$$

Then in either of the intervals $-\rho < x < 0$ or $0 < x < \rho$, Eq. (1) *has two linearly independent solutions y_1 and y_2 of the following form.*

1. *If $r_1 - r_2$ is not an integer or zero, then*

$$y_1(x) = |x|^{r_1}\left[1 + \sum_{n=1}^{\infty} a_n(r_1)x^n\right], \tag{4a}$$

$$y_2(x) = |x|^{r_2}\left[1 + \sum_{n=1}^{\infty} a_n(r_2)x^n\right]. \tag{4b}$$

2. *If $r_1 = r_2$, then*

$$y_1(x) = |x|^{r_1}\left[1 + \sum_{n=1}^{\infty} a_n(r_1)x^n\right], \tag{5a}$$

$$y_2(x) = y_1(x)\ln|x| + |x|^{r_1}\sum_{n=1}^{\infty} b_n(r_1)x^n. \tag{5b}$$

3. *If $r_1 - r_2 = N$, a positive integer, then*

$$y_1(x) = |x|^{r_1}\left[1 + \sum_{n=1}^{\infty} a_n(r_1)x^n\right], \tag{6a}$$

$$y_2(x) = ay_1(x)\ln|x| + |x|^{r_2}\left[1 + \sum_{n=1}^{\infty} c_n(r_2)x^n\right]. \tag{6b}$$

The coefficients $a_n(r_1), a_n(r_2), b_n(r_1), c_n(r_2)$, and the constant a can be determined by substituting the form of the series solutions for y in Eq. (1)*. The constant a may turn out to be zero. Each of the series in Eqs.* (4), (5), *and* (6) *converges at least for $|x| < \rho$ and defines a function that is analytic at $x = 0$.*

The case in which $r_1 - r_2$ is not an integer was discussed in Section 4.5.1. We turn now to the cases $r_1 = r_2$ and $r_1 - r_2 = N$, and assume that $x > 0$. Real-valued solutions for $x < 0$ can be obtained, as before, by letting $x = -\xi$.

$r_1 = r_2$. The method of finding the second solution is essentially the same as that which we used in finding the second solution of the Euler equation (see Section 4.4) when the roots of the indicial equation were equal. We seek a solution of Eq. (1) of the form

$$y = \phi(r, x) = x^r \sum_{n=0}^{\infty} a_n x^n = x^r\left(a_0 + \sum_{n=1}^{\infty} a_n x^n\right), \tag{7}$$

where we have written $y = \phi(r, x)$ to emphasize the fact that ϕ depends on the choice of r. If we substitute the series (2) for $xp(x)$ and $x^2q(x)$, compute and substitute for y' and y'' from Eq. (7), and finally collect terms, we obtain (see Eqs. (3) and (4) of Section 4.5.1)

$$L[\phi](r, x) = a_0 F(r) x^r + \sum_{n=1}^{\infty}\left\{a_n F(r+n) + \sum_{k=0}^{n-1} a_k[(r+k)p_{n-k} + q_{n-k}]\right\}x^{r+n}, \tag{8}$$

where

$$F(r) = r(r-1) + p_0 r + q_0. \tag{9}$$

If the roots of the indicial equation are both equal to r_1, we can obtain one solution by setting r equal to r_1 and choosing the a_n so that each term of the series in Eq. (8) is zero.

To determine a second solution we think of r as a continuous variable, and determine a_n as a function of r by requiring that the coefficient of x^{r+n}, $n \geq 1$, in Eq. (8) be zero. Then assuming $F(r+n) \neq 0$ we have

$$a_n(r) = -\frac{\sum_{k=0}^{n-1} a_k[(r+k)p_{n-k} + q_{n-k}]}{F(r+n)}, \qquad n \geq 1. \tag{10}$$

With this choice for $a_n(r)$ for $n \geq 1$, Eq. (8) reduces to

$$L[\phi](r, x) = a_0 x^r F(r). \tag{11}$$

Since r_1 is a repeated root of $F(r)$, it follows from Eq. (9) that $F(r) = (r - r_1)^2$. Setting $r = r_1$ in Eq. (11) we find that $L[\phi](r_1, x) = 0$; hence, as we already know,

$$y_1(x) = \phi(r_1, x) = x^{r_1}\left[a_0 + \sum_{n=1}^{\infty} a_n(r_1) x^n\right], \qquad x > 0, \tag{12}$$

is one solution of Eq. (1). But more important, it also follows from Eq. (11), just as for the Euler equation, that

$$L\left[\frac{\partial\phi}{\partial r}\right](r_1, x) = a_0\,\frac{\partial}{\partial r}\left[x^r(r-r_1)^2\right]\Big|_{r=r_1}$$
$$= a_0\left[(r-r_1)^2 x^r \ln x + 2(r-r_1)x^r\right]\Big|_{r=r_1}$$
$$= 0. \tag{13}$$

Hence, a second solution of Eq. (1) is

$$y_2(x) = \frac{\partial\phi(r,x)}{\partial r}\Big|_{r=r_1} = \frac{\partial}{\partial r}\left\{x^r\left[a_0 + \sum_{n=1}^{\infty} a_n(r)x^n\right]\right\}\Big|_{r=r_1}$$
$$= (x^{r_1}\ln x)\left[a_0 + \sum_{n=1}^{\infty} a_n(r_1)x^n\right] + x^{r_1}\sum_{n=1}^{\infty} a_n'(r_1)x^n$$
$$= y_1(x)\ln x + x^{r_1}\sum_{n=1}^{\infty} a_n'(r_1)x^n, \qquad x > 0, \tag{14}$$

where $a_n'(r_1)$ denotes da_n/dr evaluated at $r = r_1$ and $a_n(r)$ is given by Eq. (10).

Note that in addition to the assumptions we have already made, we have also assumed that it is permissible to differentiate the series $\sum_{n=1}^{\infty} a_n(r)x^n$ term by term with respect to r. Further, it is certainly not a trivial matter to determine $a_n(r_1)$ and $a_n'(r_1)$; however, the determination of two solutions of a second order differential equation with a regular singular point is not a trivial problem.

In summary, there are three ways that we can proceed to find a second solution when $r_1 = r_2$. First, we can compute $b_n(r_1)$ directly by substituting the expression (5b) for y in Eq. (1). Second, we can compute $b_n(r_1) = a_n'(r_1)$ by first determining $a_n(r)$ and then calculating $a_n'(r_1)$. Notice that in calculating the first solution it is necessary to know $a_n(r_1)$ so it may be just as easy to calculate the general expression for $a_n(r)$ from which we can calculate both $a_n(r_1)$ and $a_n'(r_1)$. The first procedure may be more convenient if only a few terms in the series for $y_2(x)$ are needed or if the formula for $a_n(r)$ is very complicated or difficult to obtain. The third alternative is to use the method of reduction of order, but this will not avoid the use of series.

$r_1 - r_2 = N$. For this case the detailed arguments to derive the form (6b) of the second solution are considerably more complicated and will not be given. However, we note that if we assume $y_2(x) = \sum_{n=0}^{\infty} a_n x^{r_2+n}$, then there may be difficulty in computing $a_N(r_2)$ from Eq. (10) since $F(r+N) = (r+N-r_1)(r+N-r_2) = (r-r_2)(r+N-r_2)$ vanishes for $r = r_2$. This difficulty is circumvented by choosing a_0, which can be chosen arbitrarily, to be $r - r_2$. Then each of the a's will be proportional to $r - r_2$, so there will be a factor of $r - r_2$ in the numerator of Eq. (10) to cancel the corresponding factor in the denominator that occurs when $n = N$. Following an analysis similar to that for the case $r_1 = r_2$, it turns out that the solution is of the form (6b) with

$$c_n(r_2) = \frac{d}{dr}[(r - r_2)a_n(r)]\bigg|_{r=r_2}, \qquad n = 1, 2, \ldots \tag{15}$$

where $a_n(r)$ is given by Eq. (10) with $a_0 = 1$, and with

$$a = \lim_{r \to r_2} (r - r_2)a_N(r). \tag{16}$$

If $a_N(r_2)$ is finite, then $a = 0$ and there is no logarithmic term in y_2. The procedure is discussed in Coddington (Chapter 4).

In practice the best way to determine whether a is zero in the second solution is simply to try to compute the a_n corresponding to the root r_2 and to see whether it is possible to determine $a_N(r_2)$. If so, there is no further problem. If not, we must use the form (6b) with $a \neq 0$.

Again, in summary, there are three ways to find a second solution when $r_1 - r_2 = N$. First, we can calculate a and $c_n(r_2)$ directly by substituting the expression (6b) for y in Eq. (1). Second, we can calculate $c_n(r_2)$ and a of Eq. (6b) using the formulas (15) and (16). If this is the planned procedure, then in calculating the solution corresponding to $r = r_1$ be sure to obtain the general formula for $a_n(r)$ rather than just $a_n(r_1)$. The third alternative is to use the method of reduction of order.

In the next section we give three examples that illustrate the cases $r_1 = r_2$, $r_1 - r_2 = N$ with $a = 0$, and $r_1 - r_2 = N$ with $a \neq 0$.

*4.7 Bessel's Equation

In this section we consider three special cases of Bessel's equation,

$$x^2y'' + xy' + (x^2 - \nu^2)y = 0, \tag{1}$$

where ν is a constant, which illustrate the theory discussed in Section 4.6. Clearly $x = 0$ is a regular singular point. For simplicity we consider only the case $x > 0$.

BESSEL EQUATION OF ORDER ZERO. This example illustrates the situation in which the roots of the indicial equation are equal. Setting $\nu = 0$ in Eq. (1) gives

$$L[y] = x^2y'' + xy' + x^2y = 0. \tag{2}$$

Substituting

$$y = \phi(r, x) = a_0x^r + \sum_{n=1}^{\infty} a_nx^{r+n}, \tag{3}$$

we obtain

$$L[\phi](r,x) = \sum_{n=0}^{\infty} a_n[(n+r)(n+r-1)+(n+r)]x^{r+n} + \sum_{n=0}^{\infty} a_n x^{r+n+2}$$
$$= a_0[r(r-1)+r]x^r + a_1[(r+1)r+(r+1)]x^{r+1}$$
$$+ \sum_{n=2}^{\infty} \{a_n[(n+r)(n+r-1)+(n+r)] + a_{n-2}\}x^{r+n} = 0. \quad (4)$$

The roots of the indicial equation $F(r) = r(r-1) + r = 0$ are $r_1 = 0$ and $r_2 = 0$; hence we have the case of equal roots. The recurrence relation is

$$a_n(r) = \frac{-a_{n-2}(r)}{(n+r)(n+r-1)+(n+r)} = -\frac{a_{n-2}(r)}{(n+r)^2}, \qquad n \geq 2. \quad (5)$$

To determine $y_1(x)$ we set r equal to 0. Then from Eq. (4) it follows that for the coefficient of x^{r+1} to be zero we must choose $a_1 = 0$. Hence from Eq. (5), $a_3 = a_5 = a_7 = \cdots = a_{2n+1} = \cdots = 0$. Further

$$a_n(0) = -a_{n-2}(0)/n^2, \qquad n = 2, 4, 6, 8, \ldots,$$

or letting $n = 2m$,

$$a_{2m}(0) = -a_{2m-2}(0)/(2m)^2, \qquad m = 1, 2, 3, \ldots.$$

Thus

$$a_2(0) = -\frac{a_0}{2^2}$$

$$a_4(0) = -\frac{a_2(0)}{(2\cdot 2)^2} = +\frac{a_0}{(2\cdot 2)^2 2^2} = \frac{a_0}{2^4 2^2}$$

$$a_6(0) = -\frac{a_4(0)}{(2\cdot 3)^2} = -\frac{a_0}{(2\cdot 3)^2 2^4 2^2} = -\frac{a_0}{2^6(3\cdot 2)^2}$$

$$\vdots$$

$$a_{2m}(0) = \frac{(-1)^m a_0}{2^{2m}(m!)^2}, \qquad m = 1, 2, 3, \ldots. \quad (6)$$

Hence

$$y_1(x) = a_0\left[1 + \sum_{m=1}^{\infty} \frac{(-1)^m x^{2m}}{2^{2m}(m!)^2}\right], \qquad x > 0. \quad (7)$$

The function in brackets is known as the *Bessel function of the first kind of order zero*, and is denoted by $J_0(x)$. It follows from Theorem 4.4 that the series converges for all x, and that J_0 is analytic at $x = 0$. Some of the important properties of J_0 are discussed in the problems. Figure 4.6 shows the graphs of $y = J_0(x)$ and of some of the partial sums of the series (7).

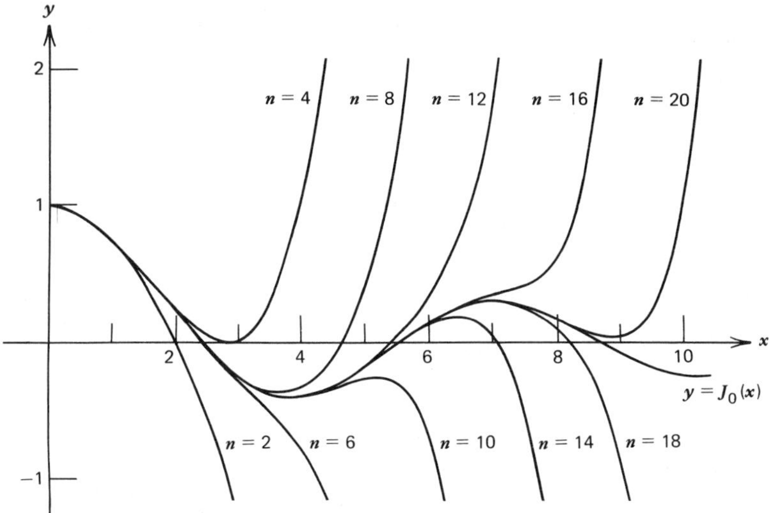

FIGURE 4.6 Polynomial approximations to $J_0(x)$. The value of n is the degree of the approximating polynomial.

In this example we determine $y_2(x)$ by computing $a_n'(0)$. The alternative procedure in which we simply substitute the form (5b) of Section 4.6 in Eq. (2), and then determine the b_n is discussed in Problem 7. First we note from the coefficient of x^{r+1} in Eq. (4) that $(r+1)^2a_1(r) = 0$. It follows that not only does $a_1(0) = 0$, but also $a_1'(0) = 0$. It is easy to deduce from the recurrence relation (5) that $a_3'(0) = a_5'(0) = \cdots = a_{2n+1}'(0) = \cdots = 0$; hence we need only compute $a_{2m}'(0)$, $m = 1, 2, 3, \ldots$. From Eq. (5)

$$a_{2m}(r) = -a_{2m-2}(r)/(2m+r)^2, \qquad m = 1, 2, 3, \ldots.$$

Hence

$$a_2(r) = -\frac{a_0}{(2+r)^2}$$

$$a_4(r) = -\frac{a_2(r)}{(4+r)^2} = +\frac{a_0}{(4+r)^2(2+r)^2}$$

$$a_6(r) = -\frac{a_4}{(6+r)^2} = -\frac{a_0}{(6+r)^2(4+r)^2(2+r)^2}$$

$$\vdots$$

$$a_{2m}(r) = \frac{(-1)^m a_0}{(2m+r)^2(2m-2+r)^2\cdots(4+r)^2(2+r)^2}, \quad m = 1, 2, 3, \ldots. \tag{8}$$

The computation of $a'_{2m}(r)$ can be carried out most conveniently by noting that if

$$f(x) = (x-\alpha_1)^{\beta_1}(x-\alpha_2)^{\beta_2}(x-\alpha_3)^{\beta_3}\cdots(x-\alpha_n)^{\beta_n},$$

then

$$f'(x) = \beta_1(x-\alpha_1)^{\beta_1-1}\left[(x-\alpha_2)^{\beta_2}\cdots(x-\alpha_n)^{\beta_n}\right]$$
$$+\beta_2(x-\alpha_2)^{\beta_2-1}\left[(x-\alpha_1)^{\beta_1}(x-\alpha_3)^{\beta_3}\cdots(x-\alpha_n)^{\beta_n}\right]+\cdots.$$

Hence for x not equal to $\alpha_1, \alpha_2, \ldots, \alpha_n$

$$\frac{f'(x)}{f(x)} = \frac{\beta_1}{x-\alpha_1} + \frac{\beta_2}{x-\alpha_2} + \cdots + \frac{\beta_n}{x-\alpha_n}.$$

Thus, from Eq. (8)

$$\frac{a'_{2m}(r)}{a_{2m}(r)} = -2\left(\frac{1}{2m+r} + \frac{1}{2m-2+r} + \cdots + \frac{1}{2+r}\right),$$

and setting r equal to 0 we obtain

$$a'_{2m}(0) = -2\left[\frac{1}{2m} + \frac{1}{2(m-1)} + \frac{1}{2(m-2)} + \cdots + \frac{1}{2}\right]a_{2m}(0).$$

Substituting for $a_{2m}(0)$ from Eq. (6), and letting

$$H_m = \frac{1}{m} + \frac{1}{m-1} + \cdots + \frac{1}{2} + 1, \tag{9}$$

we obtain finally

$$a'_{2m}(0) = -H_m\frac{(-1)^m a_0}{2^{2m}(m!)^2}, \qquad m = 1, 2, 3, \ldots.$$

The second solution of the Bessel equation of order zero is obtained by setting $a_0 = 1$, and substituting for $y_1(x)$ and $a'_{2m}(0) = b_{2m}(0)$ in Eq. (5b) of Section 4.6. We obtain

$$y_2(x) = J_0(x)\ln x + \sum_{m=1}^{\infty}\frac{(-1)^{m+1}H_m}{2^{2m}(m!)^2}x^{2m}, \qquad x > 0. \tag{10}$$

In place of y_2, the second solution is usually taken to be a certain linear combination of J_0 and y_2. It is known as the Bessel function of the second kind of order zero, and is denoted by Y_0. Following Copson (Chapter 12), we define[14]

$$Y_0(x) = \frac{2}{\pi}\left[y_2(x) + (\gamma - \ln 2)J_0(x)\right]. \tag{11}$$

[14] Other authors use other definitions for Y_0. The present choice for Y_0 is also known as the Weber (1842–1913) function.

Here γ is a constant, known as the Euler-Máscheroni (1750–1800) constant; it is defined by the equation

$$\gamma = \lim_{n\to\infty} (H_n - \ln n) \cong 0.5772. \tag{12}$$

Substituting for $y_2(x)$ in Eq. (11), we obtain

$$Y_0(x) = \frac{2}{\pi}\left[\left(\gamma + \ln\frac{x}{2}\right)J_0(x) + \sum_{m=1}^{\infty}\frac{(-1)^{m+1}H_m}{2^{2m}(m!)^2}x^{2m}\right], \qquad x > 0. \tag{13}$$

The general solution of the Bessel equation of order zero for $x > 0$ is

$$y = c_1 J_0(x) + c_2 Y_0(x).$$

Note that since $J_0(x) \to 1$ as $x \to 0$, $Y_0(x)$ has a logarithmic singularity at $x = 0$; that is, $Y_0(x)$ behaves as $(2/\pi)\ln x$ when $x \to 0$ through positive values. Thus if we are interested in solutions of Bessel's equation of order zero that are finite at the origin, which is often the case, we must discard Y_0. The graphs of the functions J_0 and Y_0 are shown in Figure 4.7.

It is interesting to note from Figure 4.7 that for x large both $J_0(x)$ and $Y_0(x)$ are oscillatory. Such a behavior might be anticipated from the original equation; indeed it is true for the solutions of the Bessel equation of order ν. If we divide Eq. (1) by x^2, we obtain

$$y'' + \frac{1}{x}y' + \left(1 - \frac{\nu^2}{x^2}\right)y = 0.$$

For x very large it is reasonable to suspect that the terms $(1/x)y'$ and $(\nu^2/x^2)y$ are small and hence can be neglected. If this is true, then the Bessel equation of

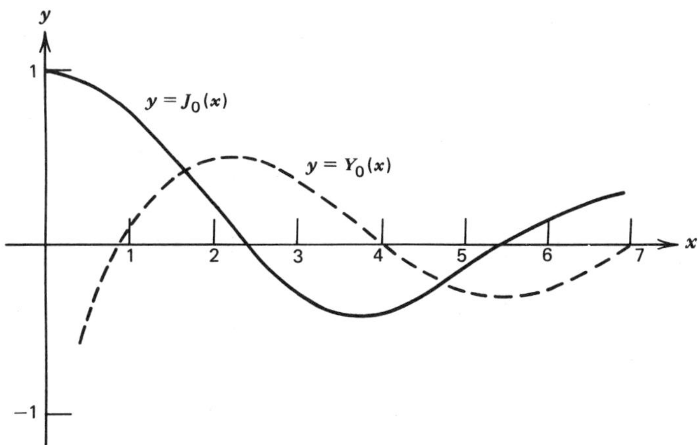

FIGURE 4.7 The Bessel functions of order zero.

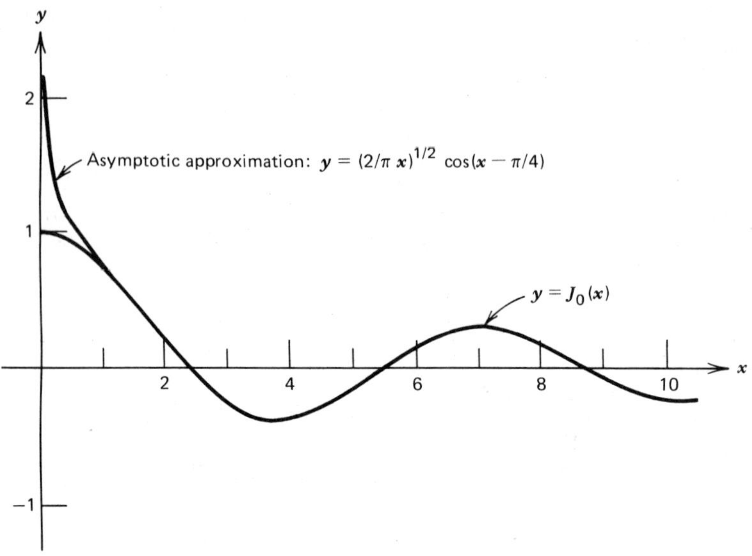

FIGURE 4.8 Asymptotic approximation to $J_0(x)$.

order ν can be approximated by

$$y'' + y = 0.$$

The solutions of this equation are $\sin x$ and $\cos x$; thus we might anticipate that the solutions of Bessel's equation for x large are similar to linear combinations of $\sin x$ and $\cos x$. This is correct insofar as the Bessel functions are oscillatory; however, it is only partly correct. For x large the functions J_0 and Y_0 also decay as x increases; thus the equation $y'' + y = 0$ does not provide an adequate approximation to the Bessel equation for large x, and a more delicate analysis is required. In fact, it is possible to show that

$$J_0(x) \cong (2/\pi x)^{1/2} \cos(x - \pi/4) \qquad \text{as} \quad x \to \infty. \tag{14}$$

As can be seen from Figure 4.8, this asymptotic approximation is actually reasonably accurate for all $x \geq 1$. Thus to approximate $J_0(x)$ over the entire range from zero to infinity, one can use two or three terms of the series (7) for $x \leq 1$ and the asymptotic approximation (14) for $x \geq 1$.

BESSEL EQUATION OF ORDER ONE-HALF. This example illustrates the situation in which the roots of the indicial equation differ by a positive integer, but there is no logarithmic term in the second solution. Setting $\nu = \frac{1}{2}$ in Eq. (1) gives

$$L[y] = x^2y'' + xy' + \left(x^2 - \tfrac{1}{4}\right)y = 0. \tag{15}$$

If we substitute the series (3) for $y = \phi(r, x)$, we obtain

$$L[\phi](r,x) = \sum_{n=0}^{\infty} \left[(r+n)(r+n-1) + (r+n) - \tfrac{1}{4}\right] a_n x^{r+n} + \sum_{n=0}^{\infty} a_n x^{r+n+2}$$
$$= \left(r^2 - \tfrac{1}{4}\right) a_0 x^r + \left[(r+1)^2 - \tfrac{1}{4}\right] a_1 x^{r+1}$$
$$+ \sum_{n=2}^{\infty} \left\{\left[(r+n)^2 - \tfrac{1}{4}\right] a_n + a_{n-2}\right\} x^{r+n} = 0. \tag{16}$$

The roots of the indicial equation are $r_1 = \frac{1}{2}$, $r_2 = -\frac{1}{2}$; hence the roots differ by an integer. The recurrence relation is

$$\left[(r+n)^2 - \tfrac{1}{4}\right] a_n = -a_{n-2}, \qquad n \geq 2. \tag{17}$$

Corresponding to the larger root $r_1 = \frac{1}{2}$ we find from the coefficient of x^{r+1} in Eq. (16) that $a_1 = 0$. Hence, from Eq. (17), $a_3 = a_5 = \cdots = a_{2n+1} = \cdots = 0$. Further, for $r = \frac{1}{2}$

$$a_n = -\frac{a_{n-2}}{n(n+1)}, \qquad n = 2, 4, 6, \ldots,$$

or letting $n = 2m$,

$$a_{2m} = -\frac{a_{2m-2}}{(2m+1)2m}, \qquad m = 1, 2, 3, \ldots.$$

Thus

$$a_2 = -\frac{a_0}{3 \cdot 2} = -\frac{a_0}{3!}$$
$$a_4 = -\frac{a_2}{5 \cdot 4} = +\frac{a_0}{5 \cdot 4 \cdot 3!} = \frac{a_0}{5!}$$
$$\vdots$$
$$a_{2m} = \frac{(-1)^m a_0}{(2m+1)!}, \qquad m = 1, 2, 3, \ldots. \tag{18}$$

Hence, taking $a_0 = 1$, we obtain

$$y_1(x) = x^{1/2}\left[1 + \sum_{m=1}^{\infty} \frac{(-1)^m x^{2m}}{(2m+1)!}\right]$$
$$= x^{-1/2} \sum_{m=0}^{\infty} \frac{(-1)^m x^{2m+1}}{(2m+1)!}, \qquad x > 0. \tag{19}$$

The power series in Eq. (19) is precisely the Taylor series for $\sin x$; hence one solution of the Bessel equation of order one-half is $x^{-1/2} \sin x$. The Bessel function of the first kind of order one-half, $J_{1/2}$, is defined as $(2/\pi)^{1/2} y_1$. Thus

$$J_{1/2}(x) = \left(\frac{2}{\pi x}\right)^{1/2} \sin x, \qquad x > 0. \tag{20}$$

Corresponding to the root $r_2 = -\frac{1}{2}$ it is possible that we may have difficulty in computing a_1 since $N = r_1 - r_2 = 1$. However, it is clear from Eq. (16) that for $r = -\frac{1}{2}$ the coefficients of a_0 and a_1 are zero, and hence a_0 and a_1 can be chosen arbitrarily. Then corresponding to a_0 we obtain $a_2, a_4, \ldots$ from the recurrence relation (17); and corresponding to a_1 we obtain $a_3, a_5, a_7, \ldots$. Hence the second solution does not involve a logarithmic term. It is left as an exercise for the student to show from Eq. (17) that for $r = -\frac{1}{2}$,

$$a_{2n} = \frac{(-1)^n a_0}{(2n)!}, \qquad n = 1, 2, \ldots,$$

and

$$a_{2n+1} = \frac{(-1)^n a_1}{(2n+1)!}, \qquad n = 1, 2, \ldots.$$

Hence

$$y_2(x) = x^{-1/2}\left[a_0 \sum_{n=0}^{\infty} \frac{(-1)^n x^{2n}}{(2n)!} + a_1 \sum_{n=0}^{\infty} \frac{(-1)^n x^{2n+1}}{(2n+1)!}\right]$$

$$= a_0 \frac{\cos x}{x^{1/2}} + a_1 \frac{\sin x}{x^{1/2}}, \qquad x > 0. \tag{21}$$

The constant a_1 simply introduces a multiple of $y_1(x)$. The second linearly independent solution of the Bessel equation of order one-half is usually taken to be the solution generated by a_0 with $a_0 = (2/\pi)^{1/2}$. It is denoted by $J_{-1/2}$. Then

$$J_{-1/2}(x) = \left(\frac{2}{\pi x}\right)^{1/2} \cos x, \qquad x > 0. \tag{22}$$

The general solution of Eq. (15) is $y = c_1 J_{1/2}(x) + c_2 J_{-1/2}(x)$.

BESSEL EQUATION OF ORDER ONE. This example illustrates the situation in which the roots of the indicial equation differ by a positive integer and the second solution involves a logarithmic term. Setting $\nu = 1$ in Eq. (1) gives

$$L[y] = x^2 y'' + xy' + (x^2 - 1)y = 0. \tag{23}$$

If we substitute for $y = \phi(r, x)$ the series (3), and collect terms as in the previous examples, we obtain

$$L[\phi](r, x) = a_0(r^2 - 1)x^r + a_1\left[(r+1)^2 - 1\right]x^{r+1}$$

$$+ \sum_{n=2}^{\infty} \left\{\left[(r+n)^2 - 1\right]a_n + a_{n-2}\right\}x^{r+n} = 0. \tag{24}$$

The roots of the indicial equation are $r_1 = 1$ and $r_2 = -1$. The recurrence relation is

$$\left[(r+n)^2 - 1\right]a_n(r) = -a_{n-2}(r), \qquad n \geq 2. \tag{25}$$

Corresponding to the larger root $r = 1$ the recurrence relation is

$$a_n = -\frac{a_{n-2}}{(n+2)n}, \qquad n = 2, 3, 4, \ldots .$$

We also find from the coefficient of x^{r+1} in Eq. (24) that $a_1 = 0$; hence from the recurrence relation $a_3 = a_5 = \cdots = 0$. For even values of n, let $n = 2m$; then

$$a_{2m} = -\frac{a_{2m-2}}{(2m+2)(2m)} = -\frac{a_{2m-2}}{2^2(m+1)m}, \qquad m = 1, 2, 3, \ldots .$$

Hence

$$a_2 = -\frac{a_0}{2^2(2)(1)}$$

$$a_4 = -\frac{a_2}{2^2(3)(2)} = +\frac{a_0}{2^4(3 \cdot 2)(2 \cdot 1)} = \frac{a_0}{2^4 3!2!}$$

$$a_6 = -\frac{a_4}{2^2(4)(3)} = -\frac{a_0}{2^6 4!3!}$$

$$\vdots$$

$$a_{2m} = \frac{(-1)^m a_0}{2^{2m}(m+1)!m!}, \qquad m = 1, 2, 3, \ldots . \tag{26}$$

With $a_0 = 1$ we have

$$y_1(x) = x \sum_{m=0}^{\infty} \frac{(-1)^m x^{2m}}{2^{2m}(m+1)!m!}. \tag{27}$$

The Bessel function of the first kind of order one is usually taken to be $\frac{1}{2}y_1$ and is denoted by J_1:

$$J_1(x) = \tfrac{1}{2}y_1(x) = \frac{x}{2} \sum_{m=0}^{\infty} \frac{(-1)^m x^{2m}}{2^{2m}(m+1)!m!}. \tag{28}$$

The series converges absolutely for all x; hence the function J_1 is defined for all x.

In determining a second solution of Bessel's equation of order one, we illustrate the method of direct substitution. The calculation is difficult, and inexperienced students are not expected to perform similar calculations in their entirety. However, it is feasible for a student to obtain a and the first few c_n of Eq. (29) given below. According to Theorem 4.4 we assume that

$$y_2(x) = aJ_1(x)\ln x + x^{-1}\left[1 + \sum_{n=1}^{\infty} c_n x^n\right], \qquad x > 0. \tag{29}$$

Computing $y_2'(x)$, $y_2''(x)$, substituting in Eq. (23), and making use of the fact that

J_1 is a solution of Eq. (23) gives

$$2axJ_1'(x) + \sum_{n=0}^{\infty} [(n-1)(n-2)c_n + (n-1)c_n - c_n]x^{n-1} + \sum_{n=0}^{\infty} c_n x^{n+1} = 0 \tag{30}$$

where $c_0 = 1$. Substituting for $J_1(x)$ from Eq. (28), shifting the indices of summation in the two series, and carrying out several steps of algebra gives

$$-c_1 + [0 \cdot c_2 + c_0]x + \sum_{n=2}^{\infty} [(n^2-1)c_{n+1} + c_{n-1}]x^n$$
$$= -a\left[x + \sum_{m=1}^{\infty} \frac{(-1)^m(2m+1)x^{2m+1}}{2^{2m}(m+1)!m!}\right]. \tag{31}$$

From Eq. (31) we observe first that $c_1 = 0$, and $a = -c_0 = -1$. Next since there are only odd powers of x on the right, the coefficient of each even power of x on the left must be zero. Thus, since $c_1 = 0$, we have $c_3 = c_5 = \cdots = 0$. Corresponding to the odd powers of x we obtain the recurrence relation (let $n = 2m+1$ in the series on the left side of Eq. (31))

$$\left[(2m+1)^2 - 1\right]c_{2m+2} + c_{2m} = \frac{(-1)(-1)^m(2m+1)}{2^{2m}(m+1)!m!}, \qquad m = 1,2,3,\ldots. \tag{32}$$

When we set $m = 1$ in Eq. (32) we obtain

$$(3^2-1)c_4 + c_2 = (-1)3/(2^2 \cdot 2!).$$

Notice that c_2 can be selected *arbitrarily*, and then this equation determines c_4. Also notice that in the equation for the coefficient of x, c_2 appeared multiplied by 0, and that equation was used to determine a. That c_2 is arbitrary is not surprising, since c_2 is the coefficient of x in the expression $x^{-1}[1 + \sum_{n=1}^{\infty} c_n x^n]$. Consequently, c_2 simply generates a multiple of J_1, and y_2 is only determined up to an additive multiple of J_1. In accord with the usual practice we choose $c_2 = 1/2^2$. Then we obtain

$$c_4 = \frac{-1}{2^4 \cdot 2}\left[\frac{3}{2} + 1\right] = \frac{-1}{2^4 2!}\left[\left(1 + \frac{1}{2}\right) + 1\right]$$
$$= \frac{(-1)}{2^4 \cdot 2!}(H_2 + H_1).$$

Although it is not an easy task to show (and we do not expect the student to perform this calculation), the solution of the recurrence relation (32) is

$$c_{2m} = \frac{(-1)(-1)^m(H_m + H_{m-1})}{2^{2m}m!(m-1)!}, \qquad m = 1,2,\ldots$$

with the understanding that $H_0 = 0$. Thus

$$y_2(x) = -J_1(x)\ln x + \frac{1}{x}\left[1 - \sum_{m=1}^{\infty} \frac{(-1)^m(H_m + H_{m-1})}{2^{2m}m!(m-1)!}x^{2m}\right], \qquad x > 0. \tag{33}$$

The calculation of $y_2(x)$ using the alternative procedure [see Eqs. (15) and (16) of Section 4.6] in which we determine the $c_n(r_2)$ is slightly easier. In particular the latter procedure yields the general formula for c_{2m} without the necessity of solving a recurrence relation of the form (32) (see Problem 8). In this regard the reader may also wish to compare the calculations of the second solution of Bessel's equation of order zero in the text and in Problem 7.

The second solution of Eq. (23), the Bessel function of the second kind of order one, Y_1, is usually taken to be a certain linear combination of J_1 and y_2. Following Copson (Chapter 12), Y_1 is defined as

$$Y_1(x) = \frac{2}{\pi}[-y_2(x) + (\gamma - \ln 2)J_1(x)], \tag{34}$$

where γ is defined in Eq. (12). The general solution of Eq. (23) for $x > 0$ is

$$y = c_1J_1(x) + c_2Y_1(x).$$

Notice that while J_1 is analytic at $x = 0$, the second solution Y_1 becomes unbounded in the same manner as $1/x$ as $x \to 0$.

PROBLEMS

1. Show that each of the following differential equations has a regular singular point at $x = 0$, and determine two linearly independent solutions for $x > 0$.

   (a) $x^2y'' + 2xy' + xy = 0$
   (b) $x^2y'' + 3xy' + (1 + x)y = 0$
   (c) $x^2y'' + xy' + 2xy = 0$
   (d) $x^2y'' + 4xy' + (2 + x)y = 0$

2. Find two linearly independent solutions of the Bessel equation of order $\frac{3}{2}$,

$$x^2y'' + xy' + (x^2 - \tfrac{9}{4})y = 0, \qquad x > 0.$$

3. Show that the Bessel equation of order one-half,

$$x^2y'' + xy' + (x^2 - \tfrac{1}{4})y = 0, \qquad x > 0,$$

   can be reduced to the equation

$$v'' + v = 0$$

   by the change of dependent variable $y = x^{-1/2}v(x)$. From this conclude that $y_1(x) = x^{-1/2}\cos x$ and $y_2(x) = x^{-1/2}\sin x$ are solutions of the Bessel equation of order one-half.

4. Show directly that the series for $J_0(x)$, Eq. (7), converges absolutely for all x.
5. Show directly that the series for $J_1(x)$, Eq. (28), converges absolutely for all x and that $J_0'(x) = -J_1(x)$.

6. Consider the Bessel equation of order ν

$$x^2y'' + xy' + (x^2 - \nu^2)y = 0, \qquad x > 0.$$

Take ν real and greater than zero.
(a) Show that $x = 0$ is a regular singular point, and that the roots of the indicial equation are ν and $-\nu$.
(b) Corresponding to the larger root ν, show that one solution is

$$y_1(x) = x^{\nu}\left[1 + \sum_{m=1}^{\infty} \frac{(-1)^m}{m!(m+\nu)(m+\nu-1)\cdots(2+\nu)(1+\nu)}\left(\frac{x}{2}\right)^{2m}\right].$$

(c) If 2ν is not an integer show that a second solution is

$$y_2(x) = x^{-\nu}\left[1 + \sum_{m=1}^{\infty} \frac{(-1)^m}{m!(m-\nu)(m-\nu-1)\cdots(2-\nu)(1-\nu)}\left(\frac{x}{2}\right)^{2m}\right].$$

Note that $y_1(x)$ is analytic at $x = 0$, and that $y_2(x)$ is unbounded as $x \to 0$.
(d) Verify by direct methods that the power series in the expressions for $y_1(x)$ and $y_2(x)$ converge absolutely for all x. Also verify that y_2 is a solution provided only that ν is not an integer.

7. In this section we showed that one solution of Bessel's equation of order zero,

$$L[y] = x^2y'' + xy' + x^2y = 0$$

is J_0, where $J_0(x)$ is given by Eq. (7) with $a_0 = 1$. According to Theorem 4.4 a second solution has the form $(x > 0)$

$$y_2(x) = J_0(x)\ln x + \sum_{n=1}^{\infty} b_n x^n.$$

(a) Show that

$$L[y_2](x) = \sum_{n=2}^{\infty} n(n-1)b_n x^n + \sum_{n=1}^{\infty} nb_n x^n + \sum_{n=1}^{\infty} b_n x^{n+2} + 2xJ_0'(x). \qquad \text{(i)}$$

(b) Substituting the series representation for $J_0(x)$ in Eq. (i), show that

$$b_1 x + 2^2 b_2 x^2 + \sum_{n=3}^{\infty}(n^2 b_n + b_{n-2})x^n = -2\sum_{n=1}^{\infty} \frac{(-1)^n 2nx^{2n}}{2^{2n}(n!)^2}. \qquad \text{(ii)}$$

(c) Note that only even powers of x appear on the right side of Eq. (ii). Show that $b_1 = b_3 = b_5 = \cdots = 0$, $b_2 = 1/2^2(1!)^2$, and that

$$(2n)^2 b_{2n} + b_{2n-2} = -2(-1)^n(2n)/2^{2n}(n!)^2, \qquad n = 2, 3, 4, \ldots.$$

Deduce that

$$b_4 = \frac{-1}{2^2 4^2}\left(1 + \frac{1}{2}\right) \quad \text{and} \quad b_6 = \frac{1}{2^2 4^2 6^2}\left(1 + \frac{1}{2} + \frac{1}{3}\right).$$

The general solution of the recurrence relation is $b_{2n} = (-1)^{n+1}H_n/2^{2n}(n!)^2$. Substituting for b_n in the expression for $y_2(x)$ we obtain the solution given in Eq. (11).

8. Find a second solution of Bessel's equation of order one by computing the $c_n(r_2)$ and a of Eq. (6b) of Section 4.6 according to the formulas (15) and (16) of that section.

Some guidelines along the way of this calculation are the following. First, use Eq. (24) to show that $a_1(-1)$ and $a_1'(-1) = 0$. Then show that $c_1(-1) = 0$ and, from the recurrence relation, that $c_n(-1) = 0$ for $n = 3, 5, \ldots$. Finally, use Eq. (25) to show that

$$a_{2m}(r) = \frac{(-1)^m a_0}{(2m+r+1)(2m+r-1)^2(2m+r-3)^2 \cdots (r+3)^2(r+1)}$$

$m = 1, 2, 3, \ldots$, and calculate $c_{2m}(-1) = (-1)^{m+1}(H_m + H_{m-1})/2^{2m}m!(m-1)!$.

9. By a suitable change of variables it is often possible to transform a differential equation with variable coefficients into a Bessel equation of a certain order. For example, show that a solution of

$$x^2y'' + \left(\alpha^2\beta^2x^{2\beta} + \tfrac{1}{4} - \nu^2\beta^2\right)y = 0, \qquad x > 0,$$

is given by $y = x^{1/2}f(\alpha x^\beta)$ where $f(\xi)$ is a solution of the Bessel equation of order ν.

10. Using the result of Problem 9 show that the general solution of the Airy equation

$$y'' - xy = 0, \qquad x > 0,$$

is $y = x^{1/2}[c_1 f_1(\tfrac{2}{3}ix^{3/2}) + c_2 f_2(\tfrac{2}{3}ix^{3/2})]$ where $f_1(\xi)$ and $f_2(\xi)$ are linearly independent solutions of the Bessel equation of order one-third.

11. It can be shown that J_0 has infinitely many zeros for $x > 0$. In particular the first three zeros are approximately 2.405, 5.520, and 8.653 (see Figure 4.6). Letting λ_j, $j = 1, 2, \ldots$, denote the zeros of J_0, it follows that

$$J_0(\lambda_j x) = \begin{cases} 1, & x = 0, \\ 0, & x = 1. \end{cases}$$

Verify that $y = J_0(\lambda_j x)$ satisfies the differential equation

$$y'' + \frac{1}{x}y' + \lambda_j^2 y = 0, \qquad x > 0.$$

Hence show that

$$\int_0^1 xJ_0(\lambda_i x)J_0(\lambda_j x)\,dx = 0 \qquad \text{if} \quad \lambda_i \neq \lambda_j.$$

This important property of $J_0(\lambda_i x)$, known as the orthogonality property, is useful in solving boundary value problems (see Sections 11.5 and 11.6.).

Hint: Write the differential equation for $J_0(\lambda_i x)$. Multiply it by $xJ_0(\lambda_j x)$ and subtract it from $xJ_0(\lambda_i x)$ times the differential equation for $J_0(\lambda_j x)$. Then integrate from 0 to 1.

REFERENCES

Coddington, E. A., *An Introduction to Ordinary Differential Equations*, (Englewood Cliffs, N.J.: Prentice-Hall, 1961).

Copson, E. T., *An Introduction to the Theory of Functions of a Complex Variable* (Oxford: Oxford University, 1935).

Proofs of Theorems 4.1, 4.3, and 4.4 can be found in intermediate or advanced books; for example, see Chapters 3 and 4 of Coddington, or Chapters 3 and 4 of

Rainville, E. D., *Intermediate Differential Equations* (2nd ed.) (New York: Macmillan, 1964).

Also see these texts for a discussion of the point at infinity, which was mentioned in Problem 13 of Section 4.3. The behavior of solutions near an irregular singular point is an even more advanced topic; a brief discussion can be found in Chapter 5 of

Coddington, E. A., and Levinson, N., *Theory of Ordinary Differential Equations* (New York: McGraw-Hill, New York, 1955).

More complete discussions of the Bessel equation, the Legendre equation, and many of the other named equations can be found in advanced books on differential equations, methods of applied mathematics, and special functions. A text dealing with special functions such as the Legendre polynomials, the Bessel functions, etc., is

Hochstadt, H., *Special Functions of Mathematical Physics* (New York: Holt, 1961).

An excellent compilation of formulas, graphs, and tables of Bessel functions, Legendre functions, and other special functions of mathematical physics may be found in

Abramowitz, M., and Stegun, I. A. (Eds.), *Handbook of Mathematical Functions* (New York: Dover, 1965); originally published by the National Bureau of Standards, Washington, D.C., 1964.

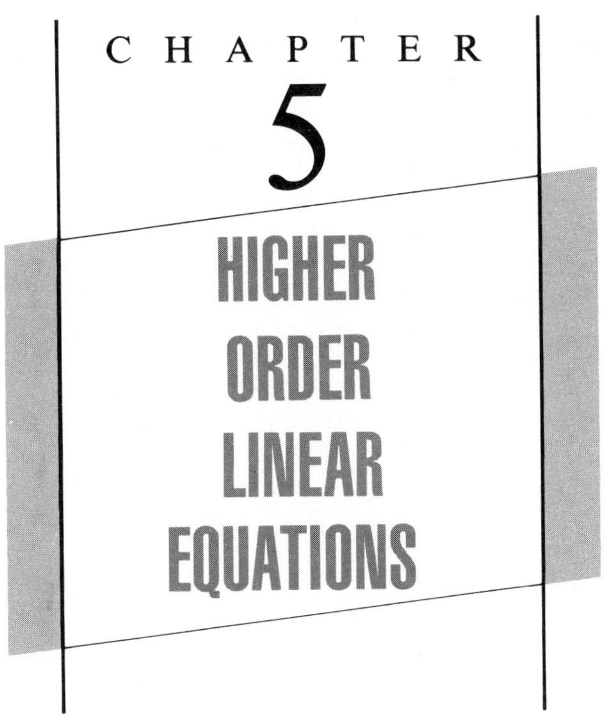

5.1 Introduction

In this chapter we extend the theory of second order linear equations developed in Chapter 3 to higher order linear equations. An nth order linear differential equation is an equation of the form

$$P_0(x)\frac{d^n y}{dx^n} + P_1(x)\frac{d^{n-1}y}{dx^{n-1}} + \cdots + P_{n-1}(x)\frac{dy}{dx} + P_n(x)y = G(x). \quad (1)$$

We assume, unless otherwise stated, that the functions $P_0, \ldots, P_n$, and G are continuous real-valued functions on some interval $\alpha < x < \beta$, and that P_0 is nowhere zero in the interval. Then dividing Eq. (1) by $P_0(x)$ we obtain

$$L[y] = \frac{d^n y}{dx^n} + p_1(x)\frac{d^{n-1}y}{dx^{n-1}} + \cdots + p_{n-1}(x)\frac{dy}{dx} + p_n(x)y = g(x), \quad (2)$$

where, using the same notation as in Chapter 3, we have introduced the linear differential operator L. The mathematical theory associated with Eq. (2) is completely analogous to that for the second order linear equation; for this reason we simply state the results for the nth order linear equation. The proofs of most of the results are also similar to those for the second order linear equation and are usually left as exercises.

First we note that since Eq. (2) involves the nth derivative of y with respect to x, it will, so to speak, require n integrations to solve Eq. (2). Each of these integrations introduces an arbitrary constant. Hence we can expect that to obtain a unique solution it is necessary to specify n initial conditions,

$$y(x_0) = y_0, y'(x_0) = y_0', \ldots, y^{(n-1)}(x_0) = y_0^{(n-1)}, \tag{3}$$

where x_0 may be any point in the interval $\alpha < x < \beta$ and $y_0, y_0', \ldots, y_0^{(n-1)}$ is any set of prescribed real constants. That there does exist such a solution and that it is unique is assured by the following existence and uniqueness theorem.

> ***Theorem 5.1.*** *If the functions $p_1, p_2, \ldots, p_n$, and g are continuous on the open interval $\alpha < x < \beta$, then there exists one and only one function $y = \phi(x)$ satisfying Eq.* (2) *on the interval $\alpha < x < \beta$ and the prescribed initial conditions* (3).

The proof of this theorem will not be given here; however, we note that if the functions $p_1, p_2, \ldots, p_n$ are constants we can construct the solution of Eq. (2) satisfying the initial conditions (3) (see Sections 5.3–5.5). Even though we know a solution in this case, we do not know that it is unique without the use of Theorem 5.1. A proof of the complete theorem can be found in Ince (Section 3.32) or Coddington (Chapter 6).

As in the corresponding second order problem, we first discuss the problem of solving the homogeneous or complementary equation

$$y^{(n)} + p_1(x)y^{(n-1)} + \cdots + p_{n-1}(x)y' + p_n(x)y = 0. \tag{4}$$

The general theory of the homogeneous equation (4) is discussed in Section 5.2, and a method of solving Eq. (4) when the p_j are constants is given in Section 5.3. Next we consider the problem of finding a particular solution y_p of the nonhomogeneous equation (2). The method of undetermined coefficients and the method of variation of parameters for determining a particular solution of Eq. (2) are discussed in Sections 5.4 and 5.5, respectively.

PROBLEMS

1. Assuming that the functions $p_1, \ldots, p_n$ are continuous on an interval including the origin, and that $y = \phi(x)$ is a solution of the initial value problem

$$y^{(n)} + p_1(x)y^{(n-1)} + \cdots + p_n(x)y = 0, \qquad y(0) = y_0, \ldots, y^{(n-1)}(0) = y_0^{(n-1)},$$

   determine $\phi^{(n)}(0)$. Show that if $p_1, \ldots, p_n$ are differentiable at $x = 0$, then $\phi^{(n+1)}(0)$ can be determined in terms of the initial data.
2. We can expect by analogy with the first order and second order differential equations that under suitable conditions on f the nth order differential equation $y^{(n)} = f(x, y, y', y'', \ldots, y^{(n-1)})$ has a solution involving n arbitrary constants. Conversely, a family of functions involving n arbitrary constants can be shown to be the solution of

an nth order differential equation. By eliminating the constants $c_1, \ldots, c_n$, determine the differential equation satisfied by each of the following functions.

(a) $y = c_1 + c_2 x + c_3 x^2 + \sin x$
(b) $y = c_1 + c_2 \cos x + c_3 \sin x$
(c) $y = c_1 e^x + c_2 e^{-x} + c_3 e^{2x}$
(d) $y = c_1 x + c_2 x^2 + c_3 x^3$
(e) $y = x + c_1 + c_2 \cos x + c_3 \sin x$
(f) $y = c_1 + c_2 x + c_3 \sinh x + c_4 \cosh x$

3. For each of the following linear equations determine intervals in which a solution is sure to exist.

(a) $y^{\text{iv}} + 4y''' + 3y = x$
(b) $xy''' + (\sin x)y'' + 3y = \cos x$
(c) $x(x-1)y^{\text{iv}} + e^x y'' + 4x^2 y = 0$
(d) $y''' + xy'' + x^2 y' + x^3 y = \ln x$

5.2 General Theory of nth Order Linear Equations

If the functions $y_1, y_2, \ldots, y_n$ are solutions of the nth order linear homogeneous differential equation

$$L[y] = y^{(n)} + p_1(x) y^{(n-1)} + \cdots + p_n(x) y = 0, \tag{1}$$

then it follows by direct computation that the linear combination

$$y = c_1 y_1(x) + c_2 y_2(x) + \cdots + c_n y_n(x), \tag{2}$$

where $c_1, \ldots, c_n$ are arbitrary constants, is also a solution of Eq. (1). It is natural to ask whether every solution of Eq. (1) can be expressed as a linear combination of $y_1, y_2, \ldots, y_n$. This will be true if, regardless of the initial conditions

$$y(x_0) = y_0,\ y'(x_0) = y_0', \ldots, y^{(n-1)}(x_0) = y_0^{(n-1)} \tag{3}$$

that are specified, it is possible to choose the constants $c_1, \ldots, c_n$ so that the linear combination (2) satisfies the initial conditions. Specifically, for any choice of the point x_0 in $\alpha < x < \beta$, and for any choice of $y_0, y_0', y_0'', \ldots, y_0^{(n-1)}$, we must be able to determine $c_1, \ldots, c_n$ so that the equations

$$\begin{aligned} c_1 y_1(x_0) + \cdots + c_n y_n(x_0) &= y_0 \\ c_1 y_1'(x_0) + \cdots + c_n y_n'(x_0) &= y_0' \\ &\ \vdots \\ c_1 y_1^{(n-1)}(x_0) + \cdots + c_n y_n^{(n-1)}(x_0) &= y_0^{(n-1)} \end{aligned} \tag{4}$$

are satisfied. Equations (4) can always be solved for the constants $c_1, \ldots, c_n$, provided that the determinant of the coefficients is not zero. On the other hand, if the determinant of the coefficients is zero, it is always possible to choose values of $y_0, y_0', \ldots, y_0^{(n-1)}$ such that Eqs. (4) do not have a solution. Hence a necessary and sufficient condition for the existence of a solution of Eqs. (4) for arbitrary values of $y_0, y_0', \ldots, y_0^{(n-1)}$ is that the Wronskian

$$W(y_1, y_2, \ldots, y_n) = \begin{vmatrix} y_1 & y_2 & \cdots & y_n \\ y_1' & y_2' & \cdots & y_n' \\ \vdots & \vdots & & \vdots \\ y_1^{(n-1)} & y_2^{(n-1)} & \cdots & y_n^{(n-1)} \end{vmatrix} \tag{5}$$

does not vanish at $x = x_0$. Since x_0 can be any point in the interval $\alpha < x < \beta$ it is necessary and sufficient that $W(y_1, y_2, \ldots, y_n)$ be nonzero at every point in the interval. Just as for the second order linear equation, it can be shown that if $y_1, y_2, \ldots, y_n$ are solutions of Eq. (1), then $W(y_1, y_2, \ldots, y_n)$ is either identically zero on the interval $\alpha < x < \beta$ or else never vanishes (see Problem 2). Hence we have the following theorem.

Theorem 5.2. *If the functions $p_1, p_2, \ldots, p_n$ are continuous on the open interval $\alpha < x < \beta$, if the functions $y_1, y_2, \ldots, y_n$ are solutions of Eq.* (1), *and if $W(y_1, \ldots, y_n)(x) \neq 0$ at least at one point in $\alpha < x < \beta$, then any solution of Eq.* (1) *can be expressed as a linear combination of the solutions $y_1, y_2, \ldots, y_n$.*

Such a set of solutions $y_1, y_2, \ldots, y_n$ of Eq. (1) is referred to as a *fundamental set of solutions* of Eq. (1). That a fundamental set of solutions exists can be shown in precisely the same way as for the second order linear equation (see Theorem 3.7). Since all solutions of Eq. (1) are of the form (2), it is customary to use the term *general solution* to refer to an arbitrary linear combination of any fundamental set of solutions of Eq. (1).

The discussion of linear dependence and independence given in Section 3.3 can also be generalized. The functions $f_1, \ldots, f_n$ are said to be *linearly dependent* on $\alpha < x < \beta$ if there exists a set of constants $k_1, \ldots, k_n$, not all zero, such that

$$k_1 f_1(x) + k_2 f_2(x) + \cdots + k_n f_n(x) = 0 \tag{6}$$

for all x in $\alpha < x < \beta$. The functions $f_1, \ldots, f_n$ are said to be *linearly independent* if they are not linearly dependent. If $y_1, \ldots, y_n$ are solutions of Eq. (1), it can be shown that a necessary and sufficient condition for them to be linearly independent is that $W(y_1, \ldots, y_n)(x_0) \neq 0$ for some x_0 in $\alpha < x < \beta$ (see Problem 3). Hence a fundamental set of solutions of Eq. (1) is linearly independent, and a linearly independent set of n solutions of Eq. (1) forms a fundamental set of solutions of Eq. (1).

NONHOMOGENEOUS PROBLEM. Now consider the nonhomogeneous equation

$$L[y] = y^{(n)} + p_1(x) y^{(n-1)} + \cdots + p_n(x) y = g(x). \tag{7}$$

If y_{p_1} and y_{p_2} are any two particular solutions of Eq. (7), it follows immediately from the linearity of the operator L that $L[y_{p_1} - y_{p_2}] = g - g = 0$. Hence the difference of any two solutions of the nonhomogeneous equation (7) is a solution of the homogeneous equation (1). Since any solution of the homogeneous equation can be expressed as a linear combination of a fundamental set of solutions $y_1, y_2, \ldots y_n$, it follows that any solution of Eq. (7) can be written as

$$\begin{aligned} y &= y_c(x) + y_p(x) \\ &= c_1 y_1(x) + c_2 y_2(x) + \cdots + c_n y_n(x) + y_p(x), \end{aligned} \tag{8}$$

where y_p is some particular solution of the nonhomogeneous equation (7). The linear combination (8) is usually referred to as the *general solution* of the nonhomogeneous equation (7).

The primary problem is to determine a fundamental set of solutions $y_1, y_2, \ldots, y_n$. If the coefficients in the differential equation are constants this is a fairly simple problem; it is discussed in the next section. If the coefficients are not constants, it is usually necessary to use numerical methods (Chapter 8) or series methods similar to those used for the second order linear differential equation with variable coefficients.[1]

In conclusion, it can be shown that the method of reduction of order also applies to nth order linear differential equations. Thus if y_1 is one solution of Eq. (1), then the substitution $y = y_1(x)v(x)$ leads to a linear differential equation of order $n - 1$ for v' (see Problems 7 and 8). Corresponding to y_1 and the $n - 1$ linearly independent solutions $v_1, \ldots, v_{n-1}$ of the reduced equation we obtain the fundamental set of solutions $y_1, y_1v_1, \ldots, y_1v_{n-1}$ of Eq. (1). If one solution of the equation for v' is known, then the method of reduction of order can again be used to obtain a linear differential equation of order $n - 2$ and so on until a first order equation is obtained. However, in practice the method of reduction of order is seldom useful for equations of higher than second order. If $n \geq 3$ the reduced equation is itself at least of second order, and only rarely will it be simpler to solve than the original equation. On the other other hand, as we have seen in Section 3.4, if $n = 2$ the reduced equation is of first order, and hence considerable simplification may be achieved.

PROBLEMS

1. Verify that the differential operator L defined by $L[y] = y^{(n)} + p_1(x)y^{(n-1)} + \cdots + p_n(x)y$ is a linear differential operator. That is, show that

$$L[c_1y_1 + c_2y_2] = c_1L[y_1] + c_2L[y_2],$$

where y_1 and y_2 are n times differentiable functions and c_1 and c_2 are arbitrary constants. Hence, show that if $y_1, y_2, \ldots, y_n$ are solutions of $L[y] = 0$, then the linear combination $c_1y_1 + \cdots + c_ny_n$ is also a solution of $L[y] = 0$.

[1]A discussion of series methods for higher order equations can be found in Ince (Chapter 16) or Coddington and Levinson (Chapter 4).

2. In Section 3.2 it was shown that the Wronskian $W(y_1, y_2)$ of two solutions of $y'' + p_1(x)y' + p_2(x)y = 0$ can be written as $W(y_1, y_2)(x) = c\exp\left[-\int^x p_1(t)\,dt\right]$, where c is a constant. More generally, it can be shown that if $y_1, y_2, \ldots, y_n$ are solutions of $y^{(n)} + p_1(x)y^{(n-1)} + \cdots + p_n(x)y = 0$ for $\alpha < x < \beta$, then

$$W(y_1, y_2, \ldots, y_n)(x) = c\exp\left[-\int^x p_1(t)\,dt\right].$$

This is known as Abel's identity. To show this result for $n = 3$ we can proceed as follows.
(a) Show that

$$W' = \begin{vmatrix} y_1 & y_2 & y_3 \\ y_1' & y_2' & y_3' \\ y_1''' & y_2''' & y_3''' \end{vmatrix},$$

where $W(y_1, y_2, y_3) = W$.
Hint: The derivative of a 3 by 3 determinant is the sum of three 3 by 3 determinants with the first, second, and third rows differentiated, respectively.
(b) Substitute for y_1''', y_2''', and y_3''' from the differential equation; multiply the first row by p_3, the second by p_2, and add these to the last row to obtain

$$W' = -p_1W.$$

The desired result follows from this equation. The proof for the general case is similar. Since the exponential function is never zero, this result shows that $W(y_1, y_2, \ldots, y_n)$ either is identically zero or else is nowhere zero on $\alpha < x < \beta$.

3. The purpose of this problem is to show that if $W(y_1, y_2, \ldots, y_n)(x_0) \neq 0$ for some x_0 in $\alpha < x < \beta$, then $y_1, y_2, \ldots, y_n$ are linearly independent; and if they are linearly independent and solutions of

$$L[y] = y^{(n)} + p_1(x)y^{(n-1)} + \cdots + p_n(x)y = 0, \qquad \alpha < x < \beta, \tag{i}$$

then $W(y_1, y_2, \ldots, y_n)$ is nowhere zero on $\alpha < x < \beta$.
(a) Suppose $W(y_1, y_2, \ldots, y_n)(x_0) \neq 0$. To show that $y_1, y_2, \ldots, y_n$ are linearly independent we must show that it is impossible to find constants $c_1, c_2, \ldots, c_n$ (not all zero) such that

$$c_1y_1(x) + c_2y_2(x) + \cdots + c_ny_n(x) = 0 \tag{ii}$$

for all x in $\alpha < x < \beta$. Assume that Eq. (ii) is true for all x in $\alpha < x < \beta$. By writing the equations for the first, second, ..., and $(n - 1)$st derivatives of Eq. (ii) at x_0, show that the c's must all be zero. Hence $y_1, y_2, \ldots, y_n$ are linearly independent.
(b) Suppose that $y_1, y_2, \ldots, y_n$ are linearly independent solutions of Eq. (i). To show that $W(y_1, y_2, \ldots, y_n)$ is nowhere zero on $\alpha < x < \beta$, assume that $W(y_1, y_2, \ldots, y_n)(x_0) = 0$ and show that this leads to a contradiction.
Hint: If $W(y_1, y_2, \ldots, y_n)(x_0) = 0$, there exists a nonzero solution of Eq. (i) satisfying the initial conditions $y = y' = y'' = \cdots = y^{(n-1)} = 0$ at x_0; then use the existence and uniqueness theorem.

4. Verify that the given functions are solutions of the differential equation, and compute the Wronskian of the solutions.

   (a) $y''' + y' = 0; \quad 1, \cos x, \quad \sin x$
   (b) $y^{iv} + y'' = 0; \quad 1, \ x, \quad \cos x, \quad \sin x$
   (c) $y''' + 2y'' - y' - 2y = 0; \quad e^x, \quad e^{-x}, \quad e^{-2x}$
   (d) $y^{iv} + 2y''' + y'' = 0; \quad 1, \quad x, \quad e^{-x}, \quad xe^{-x}$
   (e) $xy''' - y'' = 0; \quad 1, \quad x, \quad x^3$

5. Show that $W(5, \sin^2 x, \cos 2x) = 0$ for all x. Can you establish this result directly without evaluating the Wronskian?
6. Let the linear differential operator L be defined by

$$L[y] = a_0 y^{(n)} + a_1 y^{(n-1)} + a_2 y^{(n-2)} + \cdots + a_n y,$$

   where $a_0, a_1, \ldots, a_n$ are real constants.
   (a) Find $L[x^n]$.
   (b) Find $L[e^{rx}]$.
   (c) Determine four solutions of the equation $y^{iv} - 5y'' + 4y = 0$. Do you think the four solutions form a fundamental set of solutions? Why?
7. Show that if y_1 is a solution of

$$y''' + p_1(x)y'' + p_2(x)y' + p_3(x)y = 0,$$

   then the substitution $y = y_1(x)v(x)$ leads to the following second order linear equation for v':

$$y_1 v''' + (3y_1' + p_1 y_1)v'' + (3y_1'' + 2p_1 y_1' + p_2 y_1)v' = 0.$$

8. Using the method of reduction of order, solve the following differential equations.

   (a) $x^3 y''' - 3x^2 y'' + 6xy' - 6y = 0, \quad x > 0; \qquad y_1(x) = x$
   (b) $x^2(x+3)y''' - 3x(x+2)y'' + 6(1+x)y' - 6y = 0, \quad x > 0; \qquad y_1(x) = x^2,$ $y_2(x) = x^3$

5.3 Homogeneous Equations with Constant Coefficients

Consider the nth order linear homogeneous differential equation

$$L[y] = a_0 y^{(n)} + a_1 y^{(n-1)} + \cdots + a_{n-1} y' + a_n y = 0, \tag{1}$$

where $a_0, a_1, \ldots, a_n$ are real constants. It is natural to anticipate from our knowledge of second order linear equations with constant coefficients that $y = e^{rx}$ is a solution of Eq. (1) for suitable values of r. Indeed,

$$\begin{aligned} L[e^{rx}] &= e^{rx}(a_0 r^n + a_1 r^{n-1} + \cdots + a_{n-1} r + a_n) \\ &= e^{rx} Z(r) \end{aligned} \tag{2}$$

for all r. For those values of r for which $Z(r) = 0$, it follows that $L[e^{rx}] = 0$ and

$y = e^{rx}$ is a solution of Eq. (1). The polynomial $Z(r)$ is referred to as the *characteristic polynomial* or *auxiliary polynomial* and the equation $Z(r) = 0$ as the *characteristic equation* or *auxiliary equation* of the differential equation (1). A polynomial of degree n has n zeros, say $r_1, r_2, \ldots, r_n$; hence we can write the characteristic polynomial in the form

$$Z(r) = a_0(r - r_1)(r - r_2) \cdots (r - r_n). \tag{3}$$

REAL AND UNEQUAL ROOTS. If the roots of the characteristic equation are real and no two are equal, then we have n distinct solutions $e^{r_1 x}, e^{r_2 x}, \ldots, e^{r_n x}$ of Eq. (1). To show in this case that the general solution of Eq. (1) is of the form

$$y = c_1 e^{r_1 x} + c_2 e^{r_2 x} + \cdots + c_n e^{r_n x}, \tag{4}$$

we must show that the functions $e^{r_1 x}, \ldots, e^{r_n x}$ are linearly independent on the interval $-\infty < x < \infty$. One way to do this is to show that their Wronskian is nonzero. However, it is not always easy to evaluate an $n \times n$ determinant, so we will use a different argument. Let us assume that $e^{r_1 x}, \ldots, e^{r_n x}$ are linearly dependent and show that this leads to a contradiction. Then there exist constants $c_1, c_2, \ldots, c_n$, not all zero, such that $c_1 e^{r_1 x} + c_2 e^{r_2 x} + \cdots + c_n e^{r_n x} = 0$ for each x in $-\infty < x < \infty$. Multiplying by $e^{-r_1 x}$ gives $c_1 + c_2 e^{(r_2 - r_1)x} + \cdots + c_n e^{(r_n - r_1)x} = 0$ and, differentiating, we obtain

$$(r_2 - r_1)c_2 e^{(r_2 - r_1)x} + (r_3 - r_1)c_3 e^{(r_3 - r_1)x} + \cdots + (r_n - r_1)c_n e^{(r_n - r_1)x} = 0$$

for $-\infty < x < \infty$. Multiplying this last result by $e^{-(r_2 - r_1)x}$ and then differentiating gives

$$(r_3 - r_2)(r_3 - r_1)c_3 e^{(r_3 - r_2)x} + \cdots + (r_n - r_2)(r_n - r_1)c_n e^{(r_n - r_2)x} = 0$$

for $-\infty < x < \infty$. Continuing in this manner, we finally obtain

$$(r_n - r_{n-1})(r_n - r_{n-2}) \cdots (r_n - r_2)(r_n - r_1)c_n e^{(r_n - r_{n-1})x} = 0 \tag{5}$$

for $-\infty < x < \infty$. Since the exponential function does not vanish and the r_i are unequal, we have $c_n = 0$. Thus $c_1 e^{r_1 x} + c_2 e^{r_2 x} + \cdots + c_{n-1} e^{r_{n-1} x} = 0$ for each x in $-\infty < x < \infty$. Repeating the argument we have just given, we obtain $c_{n-1} = 0$. Similarly $c_{n-2}, \ldots, c_1 = 0$, and this is a contradiction of the assumption that $e^{r_1 x}, \ldots, e^{r_n x}$ are linearly dependent.

COMPLEX ROOTS. If the characteristic equation has complex roots, they must occur in conjugate pairs, $\lambda \pm i\mu$, since the coefficients $a_0, \ldots, a_n$ are real numbers. Provided that none of the roots is repeated, the general solution of Eq. (1) is still of the form[2] (4). However, following the same procedure as for the second order equation (Section 3.5.1) in place of the complex-valued solutions $e^{(\lambda + i\mu)x}$

[2] The linear independence of the solutions $e^{r_1 x}, \ldots, e^{r_n x}$, in the case that some of the r's are complex numbers, follows by a generalization of the argument just given for the case of the r's real.

and $e^{(\lambda - i\mu)x}$ we normally use the real-valued solutions

$$e^{\lambda x} \cos \mu x, \qquad e^{\lambda x} \sin \mu x \tag{6}$$

obtained as the real and imaginary parts of $e^{(\lambda + i\mu)x}$. Thus, even though some of the roots of the characteristic equation are complex, it is still possible to express the general solution of Eq. (1) as a linear combination of real-valued solutions.

EXAMPLE 1

Find the general solution of

$$y^{\text{iv}} - y = 0. \tag{7}$$

Substituting e^{rx} for y, we find that the characteristic equation is

$$r^4 - 1 = (r^2 - 1)(r^2 + 1) = 0.$$

The roots are $r = 1, -1, i, -i$; hence the general solution of Eq. (7) is

$$y = c_1 e^x + c_2 e^{-x} + c_3 \cos x + c_4 \sin x.$$

REPEATED ROOTS. If the roots of the characteristic equation are not distinct, that is, if some of the roots are repeated, then the solution (4) is clearly not the general solution of Eq. (1). Recalling that if r_1 is a repeated root for the second order linear equation $a_0 y'' + a_1 y' + a_2 y = 0$, then the two linearly independent solutions are $e^{r_1 x}$ and $xe^{r_1 x}$, it seems reasonable to expect that if a root of $Z(r) = 0$, say $r = r_1$, is repeated s times ($s \leq n$) then

$$e^{r_1 x}, xe^{r_1 x}, x^2 e^{r_1 x}, \ldots, x^{s-1} e^{r_1 x} \tag{8}$$

are solutions of Eq. (1). To prove this, we observe that if r_1 is an s-fold zero of $Z(r)$, then Eq. (2) can be written as

$$\begin{aligned} L[e^{rx}] &= e^{rx} a_0 (r - r_1)^s (r - r_{s+1}) \cdots (r - r_n) \\ &= e^{rx} (r - r_1)^s H(r) \end{aligned} \tag{9}$$

for all values of r, where $H(r_1) \neq 0$. Next, we use the facts that $\partial e^{rx}/\partial r = xe^{rx}$ and that partial differentiation of e^{rx} with respect to x and r can be interchanged. Then on differentiating Eq. (9) with respect to r we obtain

$$L[xe^{rx}] = e^{rx}\left[x(r - r_1)^s H(r) + s(r - r_1)^{s-1} H(r) + (r - r_1)^s H'(r)\right]. \tag{10}$$

Since $s \geq 2$ the right side of Eq. (10) is zero for $r = r_1$ and hence $xe^{r_1 x}$ is also a solution of Eq. (1). If $s \geq 3$, differentiating Eq. (10) again with respect to r and setting r equal to r_1 shows that $x^2 e^{r_1 x}$ is also a solution of Eq. (1). This process can be continued through $s - 1$ differentiations, which gives the desired result. Notice that the sth derivative of the right side of Eq. (9) is not zero for $r = r_1$

since the sth derivative of $(r - r_1)^s$ is a constant and $H(r_1) \neq 0$. It is reasonable to expect that $e^{r_1 x}, xe^{r_1 x}, \ldots, x^{s-1}e^{r_1 x}$ are linearly independent, and we accept this fact without proof.

Finally, if a complex root $\lambda + i\mu$ is repeated s times the complex conjugate $\lambda - i\mu$ is also repeated s times. Corresponding to these $2s$ complex-valued solutions we can find $2s$ real-valued solutions by noting that the real and imaginary parts of $e^{(\lambda + i\mu)x}, xe^{(\lambda + i\mu)x}, \ldots, x^{s-1}e^{(\lambda + i\mu)x}$ are also linearly independent solutions:

$$e^{\lambda x}\cos\mu x, \qquad e^{\lambda x}\sin\mu x, \qquad xe^{\lambda x}\cos\mu x, \qquad xe^{\lambda x}\sin\mu x,$$
$$\ldots, x^{s-1}e^{\lambda x}\cos\mu x, \qquad x^{s-1}e^{\lambda x}\sin\mu x.$$

Hence the general solution of Eq. (1) can always be expressed as a linear combination of n real-valued solutions. Consider the following example.

EXAMPLE 2

Find the general solution of

$$y^{\text{iv}} + 2y'' + y = 0. \tag{11}$$

The characteristic equation is

$$r^4 + 2r^2 + 1 = (r^2 + 1)(r^2 + 1) = 0.$$

The roots are $r = i, i, -i, -i$, and the general solution of Eq. (11) is

$$y = c_1 \cos x + c_2 \sin x + c_3 x \cos x + c_4 x \sin x.$$

In determining the roots of the characteristic equation it is often necessary to compute the cube roots, or fourth roots, or even higher roots of a (possibly complex) number. This can usually be done most conveniently by using Euler's formula $e^{ix} = \cos x + i \sin x$, and the algebraic laws given in Section 3.5.1. This is illustrated in the following example, and is also discussed in Problems 1 and 2.

EXAMPLE 3

Find the general solution of

$$y^{\text{iv}} + y = 0. \tag{12}$$

The characteristic equation is

$$r^4 + 1 = 0.$$

In this case the polynomial is not readily factored. We must compute the fourth roots of -1. Now -1, thought of as a complex number, is $-1 + 0i$. It has magnitude 1 and polar angle π. Thus

$$-1 = \cos\pi + i\sin\pi = e^{i\pi}.$$

Moreover, the angle is determined only up to a multiple of 2π. Thus

$$-1 = \cos(\pi + 2m\pi) + i\sin(\pi + 2m\pi) = e^{i(\pi+2m\pi)},$$

where m is zero or any positive or negative integer. Thus

$$(-1)^{1/4} = e^{i(\pi/4+m\pi/2)} = \cos\left(\frac{\pi}{4} + \frac{m\pi}{2}\right) + i\sin\left(\frac{\pi}{4} + \frac{m\pi}{2}\right).$$

The four fourth roots of -1 are obtained by setting $m = 0$, 1, 2, and 3; they are

$$\frac{1+i}{\sqrt{2}}, \qquad \frac{-1+i}{\sqrt{2}}, \qquad \frac{-1-i}{\sqrt{2}}, \qquad \frac{1-i}{\sqrt{2}}.$$

It is easy to verify that for any other value of m we obtain one of these four roots. For example, corresponding to $m = 4$ we obtain $(1 + i)/\sqrt{2}$. The general solution of Eq. (12) is

$$y = e^{x/\sqrt{2}}\left(c_1\cos\frac{x}{\sqrt{2}} + c_2\sin\frac{x}{\sqrt{2}}\right) + e^{-x/\sqrt{2}}\left(c_3\cos\frac{x}{\sqrt{2}} + c_4\sin\frac{x}{\sqrt{2}}\right).$$

In conclusion, we mention that the problem of determining the roots of $Z(r) = 0$ for $n > 2$ may be difficult if they cannot be found by inspection or by a simple process of trial and error. One result that can assist in the search for rational roots is the following. Suppose that the characteristic equation $Z(r) = a_0r^n + a_1r^{n-1} + \cdots + a_{n-1}r + a_n = 0$ has integer coefficients. Then, if $r = p/q$ is a rational root, where p/q is in lowest terms, q must be a factor of a_0 and p must be a factor of a_n. For example, consider the cubic polynomial equation

$$r^3 - 2r^2 + r - 2 = 0.$$

The factors of $a_0 = 1$ are ± 1 and the factors of $a_3 = -2$ are ± 1 and ± 2. Thus the only possible rational roots are ± 1 and ± 2. It is not difficult to confirm by direct calculation that $r = 2$ is the only rational root. Once we find the root $r = 2$, we can divide out the factor $(r - 2)$, obtaining the product $(r - 2)(r^2 + 1)$. So the roots of the polynomial equation are 2, i, and $-i$. Note that if $a_0 = 1$, then one need only consider the factors of a_n in the search for rational roots.

Although there are formulas similar to the quadratic formula for the roots of cubic and quartic polynomial equations, there is no formula[3] for $n > 4$. For irrational roots, even for third and fourth degree polynomial equations, it is usually more efficient to use numerical methods, such as Newton's method, rather than the exact formula for determining the roots. Of course, computer routines are available for carrying out such calculations.

[3] The formula for solving the cubic equation is usually attributed to Cardano (1501–1576), and that for the quartic equation to his pupil Ferrari (1522–1565). That it is impossible to express the roots of a general algebraic equation of degree higher than four by a formula involving only rational operations (addition, multiplication, etc.) and root extractions was established by Abel and Galois (1811–1832). A discussion of methods of solving algebraic equations can be found in Uspensky.

If the constants $a_0, a_1, \ldots, a_n$ in Eq. (1) are complex numbers, the solution of Eq. (1) is still of the form (4). In this case, however, the roots of the characteristic equation are, in general, complex numbers; and it is no longer true that the complex conjugate of a root is also a root. The corresponding solutions are complex valued.

PROBLEMS

1. Express each of the following complex numbers in the form

$$R(\cos\theta + i\sin\theta) = Re^{i\theta}.$$

Note that $e^{i(\theta+2m\pi)} = e^{i\theta}$, $m = 0, \pm 1, \pm 2, \ldots$.

(a) $1 + i$ (b) $-1 + i\sqrt{3}$ (c) -3
(d) $-i$ (e) $\sqrt{3} - i$ (f) $-1 - i$

2. Observing that $e^{i(\theta+2m\pi)} = e^{i\theta}$ for m an integer, and that

$$[e^{i(\theta+2m\pi)}]^{1/n} = e^{i[(\theta+2m\pi)/n]} = \cos\left(\frac{\theta}{n} + \frac{2m\pi}{n}\right) + i\sin\left(\frac{\theta}{n} + \frac{2m\pi}{n}\right),$$

determine the indicated roots of each of the following complex numbers.

(a) $1^{1/3}$ (b) $(1 - i)^{1/2}$
(c) $1^{1/4}$ (d) $\left[2\left(\cos\frac{\pi}{3} + i\sin\frac{\pi}{3}\right)\right]^{1/2}$

In each of Problems 3 through 18 determine the general solution of the given differential equation. Where specified, find the solution satisfying the given initial conditions.

3. $y''' - y'' - y' + y = 0$
4. $y''' - 3y'' + 3y' - y = 0$
5. $2y''' - 4y'' - 2y' + 4y = 0$
6. $y^{\text{iv}} - 4y''' + 4y'' = 0$
7. $y^{\text{vi}} + y = 0$
8. $y^{\text{iv}} - 5y'' + 4y = 0$
9. $y^{\text{vi}} - 3y^{\text{iv}} + 3y'' - y = 0$
10. $y^{\text{vi}} - y'' = 0$
11. $y^{\text{v}} - 3y^{\text{iv}} + 3y''' - 3y'' + 2y' = 0$
12. $y^{\text{iv}} - 8y' = 0$
13. $y^{\text{viii}} + 8y^{\text{iv}} + 16y = 0$
14. $y^{\text{iv}} + 2y'' + y = 0$
15. $y''' + y' = 0$; $y(0) = 0$, $y'(0) = 1$, $y''(0) = 2$
16. $y^{\text{iv}} - y = 0$; $y(0) = 1$, $y'(0) = 0$, $y''(0) = -1$, $y'''(0) = 0$
17. $y^{\text{iv}} - 4y''' + 4y'' = 0$; $y(1) = -1$, $y'(1) = 2$, $y''(1) = 0$, $y'''(1) = 0$
18. $y''' - y'' + y' - y = 0$; $y(\pi/2) = 2$, $y'(\pi/2) = 1$, $y''(\pi/2) = 0$
19. Show that the general solution of the differential equation

$$y^{\text{iv}} - y = 0$$

can be written as

$$y = c_1\cos x + c_2\sin x + c_3\cosh x + c_4\sinh x.$$

Determine the solution satisfying the initial conditions $y(0) = 0$, $y'(0) = 0$, $y''(0) = 1$, $y'''(0) = 1$. Why is it convenient to use $\cosh x$ and $\sinh x$ rather than e^x and e^{-x}?

Problems 20 through 23 deal with the nth order Euler equation.

*20. The nth order Euler or equidimensional equation is

$$L[y] = x^n y^{(n)} + a_1 x^{n-1} y^{(n-1)} + \cdots + a_{n-1} x y' + a_n y = 0 \qquad \text{(i)}$$

where $a_1, a_2, \ldots, a_n$ are real constants. Consider only the interval $x > 0$.
(a) Show that $L[x^r] = x^r F(r)$ where

$$F(r) = r(r-1)\cdots(r-n+1) + a_1[r(r-1)\cdots(r-n+2)] + \cdots + a_{n-1} r + a_n$$

is a polynomial of degree n in r. The functions $x^{r_1}, x^{r_2}, \ldots, x^{r_n}$ corresponding to the roots $r_1, r_2, \ldots, r_n$ of $F(r) = 0$ are solutions of Eq. (i).
(b) Show that if r_1 is an s-fold root of $F(r) = 0$ then $x^{r_1}, x^{r_1} \ln x, x^{r_1}(\ln x)^2, \ldots, x^{r_1}(\ln x)^{s-1}$ are solutions of Eq. (i). When the roots of $F(r) = 0$ are complex, real-valued solutions can be obtained in the manner described for the second order Euler equation. See Section 4.4.

*21. Using the results of Problem 20 determine the general solution of each of the following differential equations. Consider only the interval $x > 0$.

(a) $x^3 y''' + x^2 y'' - 2xy' + 2y = 0$
(b) $x^3 y''' + xy' - y = 0$
(c) $x^3 y''' + 2x^2 y'' + xy' - y = 0$
(d) $x^4 y^{\text{iv}} + 5x^3 y''' + 7x^2 y'' + 8xy' = 0$
(e) $x^3 y''' + x^2 y'' = 0$
(f) $x^3 y''' + 4x^2 y'' - 2xy' - 4y = 0$

*22. In determining the drag on a very small sphere of radius a placed in a uniform viscous flow it is necessary to solve the differential equation

$$\rho^3 f^{\text{iv}}(\rho) + 8\rho^2 f'''(\rho) + 8\rho f''(\rho) - 8f'(\rho) = 0, \qquad \rho > a,$$

where ρ is the distance from the center of the sphere. Note that this is an Euler equation for f' and show, using the results of Problem 20, that the general solution is

$$f(\rho) = \frac{A}{\rho^3} + \frac{B}{\rho} + C + D\rho^2$$

where A, B, C, and D are constants. Show that the solution satisfying the boundary conditions $f(a) = 0$, $f'(a) = 0$, and $f(\rho) \to U$ (the velocity at ∞) as $\rho \to \infty$ is $f(\rho) = Ua^3/2\rho^3 - 3Ua/2\rho + U$. The formula for the drag (Stokes' formula) on the sphere turns out to be $6\pi\mu a U$ where μ is the viscosity of the fluid. This result was used by R. A. Millikan in his famous experiment to measure the charge on an electron. Also see Problem 13 of Section 2.7.

*23. Show, for $x > 0$, that the change of variable $x = e^z$ reduces the third order Euler equation $x^3 y''' + a_1 x^2 y'' + a_2 xy' + a_3 y = 0$ to a third order linear equation with constant coefficients. This transformation also reduces the nth order Euler equation to an nth order linear equation with constant coefficients. Solve Problem 21b by this method.

*24. It is often of importance in engineering applications to know whether all solutions of a linear homogeneous equation approach zero as x approaches infinity. If so, the equation is said to be *asymptotically stable*. If x represents time and y the response of a physical system, the condition that the differential equation is asymptotically stable means that regardless of the initial conditions the response of the system to the initial conditions eventually decays to zero as x becomes large. For the second order equation $ay'' + by' + cy = 0$, it was shown in Section 3.7.1 (also see Section 3.5.1, Problem 12) that a sufficient condition for asymptotic stability is that a, b, and c be positive. More generally the equation

$$a_0 y^{(n)} + a_1 y^{(n-1)} + \cdots + a_n y = 0, \tag{i}$$

where $a_0, a_1, \ldots, a_n$ are real, is asymptotically stable if all the roots of the corresponding characteristic equation have negative real parts. A necessary and sufficient condition for this to be true has been given by Hurwitz (1859–1919). For $n = 4$ the Hurwitz stability criterion can be stated as follows: Eq. (i) is asymptotically stable if and only if for $a_0 > 0$,

$$a_1, \quad \begin{vmatrix} a_1 & a_0 \\ a_3 & a_2 \end{vmatrix}, \quad \begin{vmatrix} a_1 & a_0 & 0 \\ a_3 & a_2 & a_1 \\ 0 & a_4 & a_3 \end{vmatrix}, \quad \begin{vmatrix} a_1 & a_0 & 0 & 0 \\ a_3 & a_2 & a_1 & a_0 \\ 0 & a_4 & a_3 & a_2 \\ 0 & 0 & 0 & a_4 \end{vmatrix}$$

are all positive. For $n = 3$ the condition just applies to the first three expressions with $a_4 = 0$, and for $n = 2$ to the first two expressions with $a_3 = 0$. A complete discussion of the Hurwitz stability criterion can be found in Guillemin (Chapter 6, Article 26). Determine whether each of the following equations is asymptotically stable, and verify your result if possible by computing the general solution.

(a) $y''' + 3y'' + 3y' + y = 0$ (b) $y''' - y = 0$
(c) $y''' + y'' - y' + y = 0$ (d) $y^{\text{iv}} + 2y'' + y = 0$
(e) $y''' + 0.1y'' + 1.2y' - 0.4y = 0$ (f) $y''' + 3.2y'' + 2.41y' + 0.21y = 0$

5.4 The Method of Undetermined Coefficients

A particular solution of the nonhomogeneous nth order linear equation with constant coefficients

$$L[y] = a_0 y^{(n)} + a_1 y^{(n-1)} + \cdots + a_{n-1} y' + a_n y = g(x) \tag{1}$$

can be obtained by the method of undetermined coefficients provided that $g(x)$ is of an appropriate form. While the method of undetermined coefficients is not as general as the method of variation of parameters described in the next section, it is usually much easier to use when applicable.

Just as for the second order linear equation, it is clear that when the constant coefficient linear differential operator L is applied to a polynomial $A_0 x^m + A_1 x^{m-1} + \cdots + A_m$, an exponential function $e^{\alpha x}$, or a sine function $\sin \beta x$, or a cosine function $\cos \beta x$, the result is a polynomial, an exponential function, or a linear combination of sine and cosine functions, respectively. Hence if $g(x)$ is a sum of polynomials, exponentials, sines and cosines, or even products of such functions, we can expect that it is possible to find $y_p(x)$ by choosing a suitable combination of polynomials, exponentials, etc., with a number of undetermined constants. The constants are then determined so that Eq. (1) is satisfied.

First consider the case that $g(x)$ is a polynomial of degree m

$$g(x) = b_0x^m + b_1x^{m-1} + \cdots + b_m, \tag{2}$$

where $b_0, b_1, \ldots, b_m$ are given constants. It is natural to look for a particular solution of the form

$$y_p(x) = A_0x^m + A_1x^{m-1} + \cdots + A_m. \tag{3}$$

Substituting for y in Eq. (1), and equating the coefficients of like powers of x, we find from the terms in x^m that $a_nA_0 = b_0$. Provided that $a_n \neq 0$ we have $A_0 = b_0/a_n$. The constants $A_1, \ldots, A_m$ are determined from the coefficients of the terms in $x^{m-1}, x^{m-2}, \ldots, x^0$.

If $a_n = 0$, that is, if a constant is a solution of the homogeneous equation, we cannot solve for A_0; in this case it is necessary to assume for $y_p(x)$ a polynomial of degree $m + 1$ to obtain a term in $L[y_p](x)$ to balance against b_0x^m. However, it is not necessary to carry the constant in the assumed form for $y_p(x)$. More generally, it is easy to verify that if zero is an s-fold root of the characteristic polynomial, in which case $1, x, x^2, \ldots, x^{s-1}$ are solutions of the homogeneous equation, then a suitable form for $y_p(x)$ is

$$y_p(x) = x^s(A_0x^m + A_1x^{m-1} + \cdots + A_m). \tag{4}$$

As a second problem suppose that $g(x)$ is of the form

$$g(x) = e^{\alpha x}(b_0x^m + b_1x^{m-1} + \cdots + b_m). \tag{5}$$

Then we would expect $y_p(x)$ to be of the form

$$y_p(x) = e^{\alpha x}(A_0x^m + A_1x^{m-1} + \cdots + A_m), \tag{6}$$

provided that $e^{\alpha x}$ is not a solution of the homogeneous equation. If α is an s-fold root of the characteristic equation, a suitable form for $y_p(x)$ is

$$y_p(x) = x^se^{\alpha x}(A_0x^m + A_1x^{m-1} + \cdots + A_m). \tag{7}$$

These results can be proved, as for the second order linear nonhomogeneous equation, by reducing this problem to the previous one by the substitution $y = e^{\alpha x}u(x)$. The function u will satisfy an nth order linear nonhomogeneous equation with constant coefficients—the nonhomogeneous term being precisely the polynomial (2) (see Problem 18).

Similarly, if $g(x)$ is of the form

$$g(x) = e^{\alpha x}(b_0x^m + b_1x^{m-1} + \cdots + b_m)\begin{cases}\sin\beta x,\\ \cos\beta x,\end{cases} \tag{8}$$

then a suitable form for $y_p(x)$, provided that $\alpha + i\beta$ is not a root of the characteristic equation, is

$$\begin{aligned} y_p(x) &= e^{\alpha x}(A_0x^m + A_1x^{m-1} + \cdots + A_m)\cos\beta x \\ &\quad + e^{\alpha x}(B_0x^m + B_1x^{m-1} + \cdots + B_m)\sin\beta x. \end{aligned} \tag{9}$$

TABLE 5.1 THE PARTICULAR SOLUTION OF $a_0y^{(n)} + a_1y^{(n-1)} + \cdots + a_{n-1}y' + a_ny = g(x)$

$g(x)$	$y_p(x)$
$P_m(x) = b_0x^m + b_1x^{m-1} + \cdots + b_m$	$x^s(A_0x^m + \cdots + A_m)$
$P_m(x)e^{\alpha x}$	$x^s(A_0x^m + \cdots + A_m)e^{\alpha x}$
$P_m(x)e^{\alpha x}\begin{cases}\sin\beta x\\ \cos\beta x\end{cases}$	$x^s[(A_0x^m + \cdots + A_m)e^{\alpha x}\cos\beta x + (B_0x^m + \cdots + B_m)e^{\alpha x}\sin\beta x]$

Notes. Here s is the smallest nonnegative integer for which every term in $y_p(x)$ differs from every term in the complementary function $y_c(x)$. Equivalently, for the three cases, s is the number of times 0 is a root of the characteristic equation, α is a root of the characteristic equation, and $\alpha + i\beta$ is a root of the characteristic equation, respectively.

If $\alpha + i\beta$ is an s-fold root of the characteristic equation, then it is necessary to multiply the right side of Eq. (9) by x^s.

These results are summarized in Table 5.1.

If $g(x)$ is a sum of terms of the form (2), (5), and (8), it is usually easier in practice to compute separately the particular solution corresponding to each term in $g(x)$. From the principle of superposition (since the differential equation is linear), the particular solution of the complete problem is the sum of the particular solutions of the individual problems. This is illustrated in the following example.

EXAMPLE

Find a particular solution of

$$y''' - 4y' = x + 3\cos x + e^{-2x}. \tag{10}$$

First we solve the homogeneous equation. The characteristic equation is $r^3 - 4r = 0$, and the roots are $0, \pm 2$; hence

$$y_c(x) = c_1 + c_2e^{2x} + c_3e^{-2x}.$$

Using the superposition principle, we can write a particular solution of Eq. (10) as the sum of particular solutions corresponding to the differential equations

$$y''' - 4y' = x, \qquad y''' - 4y' = 3\cos x, \qquad y''' - 4y' = e^{-2x}.$$

Our initial choice for a particular solution, y_{p_1}, of the first equation is $A_0x + A_1$; but since a constant is a solution of the homogeneous equation, we multiply by x. Thus

$$y_{p_1}(x) = x(A_0x + A_1).$$

For the second equation we choose

$$y_{p_2}(x) = B\cos x + C\sin x,$$

and there is no need to modify this initial choice since $\cos x$ and $\sin x$ are not solutions of the homogeneous equation. Finally, for the third equation, since e^{-2x} is a solution of the homogeneous equation, we assume that

$$y_{p_3}(x) = Exe^{-2x}.$$

The constants are determined by substituting into the individual differential equations; they are $A_0 = -\frac{1}{8}$, $A_1 = 0$, $B = 0$, $C = -\frac{3}{5}$, and $E = \frac{1}{8}$. Hence a particular solution of Eq. (10) is

$$y_p(x) = -\tfrac{1}{8}x^2 - \tfrac{3}{5}\sin x + \tfrac{1}{8}xe^{-2x}.$$

The method of undetermined coefficients can be used whenever it is possible to guess the correct form for $y_p(x)$. However, this is usually impossible for other than constant coefficient differential equations and for other than nonhomogeneous terms of the type described above. For more complicated problems we can use the method of variation of parameters, which is discussed in the next section.

PROBLEMS

In each of Problems 1 through 11 determine the general solution of the given differential equation. Where specified find the solution satisfying the given initial conditions.

1. $y''' - y'' - y' + y = 2e^{-x} + 3$
2. $y^{\text{iv}} - y = 3x + \cos x$
3. $y''' + y'' + y' + y = e^{-x} + 4x$
4. $y''' - y' = 2\sin x$
5. $y^{\text{iv}} - 4y'' = x^2 + e^x$
6. $y^{\text{iv}} + 2y'' + y = 3 + \cos 2x$
7. $y^{\text{vi}} + y''' = x$
8. $y^{\text{iv}} + y''' = \sin 2x$
9. $y''' + 4y' = x, \quad y(0) = y'(0) = 0, \quad y''(0) = 1$
10. $y^{\text{iv}} + 2y'' + y = 3x + 4, \quad y(0) = y'(0) = 0, \quad y''(0) = y'''(0) = 1$
11. $y''' - 3y'' + 2y' = x + e^x, \quad y(0) = 1, \quad y'(0) = -\frac{1}{4}, \quad y''(0) = -\frac{3}{2}$

In each of Problems 12 through 17 determine a suitable form for $y_p(x)$ if the method of undetermined coefficients is to be used. Do not evaluate the constants.

12. $y''' - 2y'' + y' = x^3 + 2e^x$
13. $y''' - y' = xe^{-x} + 2\cos x$
14. $y^{\text{iv}} - 2y'' + y = e^x + \sin x$
15. $y^{\text{iv}} + 4y'' = \sin 2x + xe^x + 4$
16. $y^{\text{iv}} - y''' - y'' + y' = x^2 + 4 + x\sin x$
17. $y^{\text{iv}} + 2y''' + 2y'' = 3e^x + 2xe^{-x} + e^{-x}\sin x$

18. Consider the nonhomogeneous nth order linear differential equation

$$a_0y^{(n)} + a_1y^{(n-1)} + \cdots + a_ny = g(x)$$

where $a_0, \ldots, a_n$ are constants. Verify that if $g(x)$ is of the form

$$e^{\alpha x}(b_0x^m + \cdots + b_m),$$

then the substitution $y = e^{\alpha x}u(x)$ reduces the above equation to the form

$$t_0 u^{(n)} + t_1 u^{(n-1)} + \cdots + t_n u = b_0 x^m + \cdots + b_m,$$

where $t_0, \ldots, t_n$ are constants. Determine t_0 and t_n in terms of the a's and α. Thus the problem of determining a particular solution of the original equation is reduced to the simpler problem of determining a particular solution of an equation with constant coefficients and a polynomial for the nonhomogeneous term.

Method of Annihilators. In Problems 19 through 21 we consider another way of arriving at the proper form of $y_p(x)$ for use in the method of undetermined coefficients. The procedure is based on the observation that exponential, polynomial, or sinusoidal terms (or sums and products of such terms) can be viewed as solutions of certain linear homogeneous differential equations with constant coefficients. It is convenient to use the symbol D for d/dx. Then, for example, e^{-x} is a solution of $(D + 1)y = 0$; the differential operator $D + 1$ is said to *annihilate*, or to be an *annihilator* of, e^{-x}. Similarly, $D^2 + 4$ is an annihilator of $\sin 2x$ or $\cos 2x$, $(D - 3)^2 = D^2 - 6D + 9$ is an annihilator of e^{3x} or xe^{3x}, and so forth.

19. Show that linear differential operators with constant coefficients obey the commutative law, that is

$$(D - a)(D - b)f = (D - b)(D - a)f$$

for any twice differentiable function f, and any constants a and b. The result extends at once to any finite number of factors.

20. Consider the problem of finding the form of the particular solution $y_p(x)$ of

$$(D - 2)^3(D + 1)y_p = 3e^{2x} - xe^{-x}, \tag{i}$$

where the left side of the equation is written in a form corresponding to the factorization of the characteristic polynomial.

(a) Show that $D - 2$ and $(D + 1)^2$, respectively, are annihilators of the terms on the right side of Eq. (i), and that the combined operator $(D - 2)(D + 1)^2$ annihilates both terms on the right side of Eq. (i) simultaneously.

(b) Apply the operator $(D - 2)(D + 1)^2$ to Eq. (i), and use the result of Problem 19 to obtain

$$(D - 2)^4(D + 1)^3 y_p = 0. \tag{ii}$$

Thus y_p is a solution of the homogeneous equation (ii). By solving Eq. (ii), show that

$$y_p(x) = c_1 e^{2x} + c_2 xe^{2x} + c_3 x^2 e^{2x} + c_4 x^3 e^{2x}$$
$$+ c_5 e^{-x} + c_6 xe^{-x} + c_7 x^2 e^{-x}, \tag{iii}$$

where $c_1, \ldots, c_7$ are constants, as yet undetermined.

(c) Observe that e^{2x}, xe^{2x}, x^2e^{2x}, and e^{-x} are solutions of the homogeneous equation corresponding to Eq. (i); hence these terms are not useful in solving the nonhomogeneous equation. Therefore choose c_1, c_2, c_3, and c_5 to be zero in Eq. (iii), so that

$$y_p(x) = c_4 x^3 e^{2x} + c_6 xe^{-x} + c_7 x^2 e^{-x}. \tag{iv}$$

This is the form of the particular solution y_p of Eq. (i). The values of the coefficients c_4, c_6, and c_7 can be found by substituting from Eq. (iv) in the differential equation (i).

Summary. Suppose that

$$L(D)y = g(x), \tag{v}$$

where $L(D)$ is a linear differential operator with constant coefficients, and $g(x)$ is a sum or product of exponential, polynomial, or sinusoidal terms. To find the form of the particular solution of Eq. (v) one can proceed as follows.
(a) Find a differential operator $H(D)$ with constant coefficients that annihilates $g(x)$; that is, an operator such that $H(D)g(x) = 0$.
(b) Apply $H(D)$ to Eq. (v), obtaining

$$H(D)L(D)y = 0, \tag{vi}$$

which is a homogeneous equation of higher order.
(c) Solve Eq. (vi).
(d) Eliminate from the solution found in step (c) the terms that also appear in the solution of $L(D)y = 0$. The remaining terms constitute the correct form of the particular solution of Eq. (v).

21. Use the method of annihilators to find the form of the particular solution $y_p(x)$ for each of the equations in Problems 12 through 17. Do not evaluate the coefficients.

5.5 The Method of Variation of Parameters

The method of variation of parameters for determining a particular solution of the nonhomogeneous nth order linear differential equation

$$L[y] = y^{(n)} + p_1(x)y^{(n-1)} + \cdots + p_{n-1}(x)y' + p_n(x)y = g(x) \tag{1}$$

is a direct extension of the theory for the second order differential equation (see Section 3.6.2). As before, to use the method of variation of parameters it is first necessary to solve the corresponding homogeneous differential equation. In general, this may be difficult unless the coefficients are constants. However, the method of variation of parameters is still more general than the method of undetermined coefficients in the following sense. The method of undetermined coefficients is usually applicable only for constant coefficient equations and a limited class of functions g; for constant coefficient equations the homogeneous equation can be solved, and hence a particular solution for *any* continuous function g can be determined by the method of variation of parameters. Suppose then that we know a fundamental set of solutions $y_1, y_2, \ldots, y_n$ of the homogeneous equation. Then

$$y_c(x) = c_1y_1(x) + c_2y_2(x) + \cdots + c_ny_n(x). \tag{2}$$

The method of variation of parameters for determining a particular solution of Eq. (1) rests on the possibility of determining n functions $u_1, u_2, \ldots, u_n$ such that $y_p(x)$ is of the form

$$y_p(x) = u_1(x)y_1(x) + u_2(x)y_2(x) + \cdots + u_n(x)y_n(x). \tag{3}$$

Since we have n functions to determine, we will have to specify n conditions. One of these is clearly that y_p satisfy Eq. (1). The other $n-1$ conditions are chosen so as to facilitate the calculations. Since we can hardly expect a simplification in determining y_p if we must solve high order differential equations for the u_i, $i = 1, 2, \ldots, n$, it is natural to impose conditions to suppress the terms that lead to higher derivatives of the u_i. From Eq. (3) we obtain

$$y_p' = \left(u_1y_1' + u_2y_2' + \cdots + u_ny_n'\right) + \left(u_1'y_1 + u_2'y_2 + \cdots + u_n'y_n\right). \tag{4}$$

Thus the first condition that we impose on the u_i is that

$$u_1'y_1 + u_2'y_2 + \cdots + u_n'y_n = 0. \tag{5}$$

Continuing this process in a similar manner through $n-1$ derivatives of y_p gives

$$y_p^{(m)} = u_1y_1^{(m)} + u_2y_2^{(m)} + \cdots + u_ny_n^{(m)}, \qquad m = 0, 1, 2, \ldots, n-1, \tag{6}$$

and the following $n-1$ conditions on the functions $u_1, \ldots, u_n$:

$$u_1'y_1^{(m-1)} + u_2'y_2^{(m-1)} + \cdots + u_n'y_n^{(m-1)} = 0, \qquad m = 1, 2, \ldots, n-1. \tag{7}$$

The nth derivative of y_p is

$$y_p^{(n)} = \left(u_1y_1^{(n)} + \cdots + u_ny_n^{(n)}\right) + \left(u_1'y_1^{(n-1)} + \cdots + u_n'y_n^{(n-1)}\right). \tag{8}$$

Finally, we impose the condition that y_p be a solution of Eq. (1). On substituting for the derivatives of y_p from Eqs. (6) and (8), collecting terms, and making use of the fact that $L[y_i] = 0$, $i = 1, 2, \ldots, n$, we obtain

$$u_1'y_1^{(n-1)} + u_2'y_2^{(n-1)} + \cdots + u_n'y_n^{(n-1)} = g. \tag{9}$$

Equation (9), coupled with the $n-1$ equations (7), give n simultaneous linear nonhomogeneous equations for $u_1', u_2', \ldots, u_n'$:

$$\begin{aligned} y_1u_1' + y_2u_2' + \cdots + y_nu_n' &= 0, \\ y_1'u_1' + y_2'u_2' + \cdots + y_n'u_n' &= 0, \\ y_1''u_1' + y_2''u_2' + \cdots + y_n''u_n' &= 0, \\ &\vdots \\ y_1^{(n-1)}u_1' + \cdots + y_n^{(n-1)}u_n' &= g. \end{aligned} \tag{10}$$

A sufficient condition for the existence of a solution of the system of equations (10) is that the determinant of the coefficients is nonzero for each value of x. However, the determinant of the coefficients is precisely $W(y_1, y_2, \ldots, y_n)$, and it is nowhere zero since $y_1, \ldots, y_n$ are linearly independent solutions of the homogeneous equation. Hence it is possible to determine $u_1', \ldots, u_n'$. Using

Cramer's rule, we find that the solution of the system of equations (10) is

$$u'_m(x) = \frac{g(x)W_m(x)}{W(x)}, \qquad m = 1, 2, \ldots, n. \tag{11}$$

Here $W(x) = W(y_1, y_2, \ldots, y_n)(x)$ and W_m is the determinant obtained from $W(y_1, y_2, \ldots, y_n)$ by replacing the mth column by the column $(0, 0, \ldots, 0, 1)$. With this notation a particular solution of Eq. (1) is given by

$$y_p(x) = \sum_{m=1}^{n} y_m(x) \int^x \frac{g(t)W_m(t)}{W(t)}\, dt. \tag{12}$$

While the procedure is straightforward, the computational problem of evaluating $y_p(x)$ from Eq. (12) for $n > 2$ is not a trivial one. The calculation may be simplified to some extent by using Abel's identity:

$$W(x) = W(y_1, y_2, \ldots, y_n)(x) = c \exp\left[-\int^x p_1(t)\, dt\right].$$

(See Problem 2 of Section 5.2.) The constant c can be determined by evaluating $W(y_1, y_2, \ldots, y_n)$ at a conveniently chosen point.

EXAMPLE

Given that $y_1(x) = e^x$, $y_2(x) = xe^x$, and $y_3(x) = e^{-x}$ are solutions of the homogeneous equation corresponding to

$$y''' - y'' - y' + y = g(x),$$

determine a particular solution in terms of an integral.

We use formula (12). First, we have

$$W(e^x, xe^x, e^{-x})(x) = \begin{vmatrix} e^x & xe^x & e^{-x} \\ e^x & (x+1)e^x & -e^{-x} \\ e^x & (x+2)e^x & e^{-x} \end{vmatrix}.$$

Factoring an e^x from each of the first two columns, an e^{-x} from the third column, and then adding the second row to the first and third rows, we obtain $W(e^x, xe^x, e^{-x}) = 4e^x$. Next

$$W_1(x) = \begin{vmatrix} 0 & xe^x & e^{-x} \\ 0 & (x+1)e^x & -e^{-x} \\ 1 & (x+2)e^x & e^{-x} \end{vmatrix} = x(-1) - (x+1) = -1 - 2x.$$

In a similar manner we find that $W_2(x) = 2$ and $W_3(x) = e^{2x}$. Substituting these results in formula (12), we obtain

$$\begin{aligned} y_p(x) &= e^x \int^x \frac{g(t)(-1-2t)}{4e^t}\, dt + xe^x \int^x \frac{g(t)(2)}{4e^t}\, dt + e^{-x} \int^x \frac{g(t)e^{2t}}{4e^t}\, dt \\ &= \frac{1}{4} \int^x \left\{e^{x-t}[-1 + 2(x-t)] + e^{-(x-t)}\right\} g(t)\, dt. \end{aligned}$$

PROBLEMS

In each of Problems 1 through 3 use the method of variation of parameters to determine a particular solution of the given differential equation.

1. $y''' + y' = \tan x, \quad 0 < x < \pi/2$
2. $y''' - y' = x$
3. $y''' - 2y'' - y' + 2y = e^{4x}$
4. Given that x, x^2, and $1/x$ are solutions of the homogeneous equation corresponding to

$$x^3y''' + x^2y'' - 2xy' + 2y = 2x^4, \qquad x > 0,$$

determine a particular solution.

5. Find a formula involving integrals for a particular solution of the differential equation

$$y''' - y'' + y' - y = g(x).$$

6. Find a formula involving integrals for a particular solution of the differential equation

$$y^{\text{iv}} - y = g(x).$$

Hint: The functions $\sin x, \cos x, \sinh x, \cosh x$ form a fundamental set of solutions of the homogeneous equation.

7. Find a formula involving integrals for a particular solution of the differential equation

$$y''' - 3y'' + 3y' - y = g(x).$$

If $g(x) = x^{-2}e^x$, determine $y_p(x)$.

8. Find a formula involving integrals for a particular solution of the differential equation

$$x^3y''' - 3x^2y'' + 6xy' - 6y = g(x), \qquad x > 0.$$

Hint: Verify that x, x^2, and x^3 are solutions of the homogeneous equation.

REFERENCES

Coddington, E. A., *An Introduction to Ordinary Differential Equations* (Englewood Cliffs, N.J.: Prentice-Hall, 1961).

Coddington, E. A., and Levinson, N., *Theory of Ordinary Differential Equations* (New York: McGraw-Hill, 1955).

Guillemin, E. A., *The Mathematics of Circuit Analysis* (New York: Wiley, 1949; Cambridge, Mass.: MIT Press, 1969).

Ince, E. L., *Ordinary Differential Equations* (London: Longmans, Green, 1927; New York: Dover, 1956).

Uspensky, J. V., *Theory of Equations* (New York: McGraw-Hill, 1948).

6.1 Definition of the Laplace Transform

Among the tools that are very useful in solving linear differential equations are *integral transforms*. An integral transform is a relation of the form

$$F(s) = \int_{\alpha}^{\beta} K(s,t) f(t)\, dt, \tag{1}$$

wherein a given function f is transformed into another function F by means of an integral. The function F is said to be the transform of f, and the function K is called the *kernel* of the transformation. The general idea is to use the relation (1) to transform a problem for f into a simpler problem for F, to solve this simpler problem, and then to recover the desired function f from its transform F. By making a suitable choice of the kernel K and the integration limits α and β, it is often possible to simplify substantially a problem involving a linear differential equation. Several integral transformations are widely used, each being appropriate for certain types of problems.

In this chapter we discuss the properties and some of the applications of the Laplace[1] transform. This transform is defined in the following way. Let $f(t)$ be

[1] The Laplace transform is named for the eminent French mathematician P. S. Laplace, who studied the relation (2) in 1782. However, the techniques described in this chapter were not developed until a century or more later. They are due mainly to Oliver Heaviside (1850–1925), an innovative but unconventional English electrical engineer, who made significant contributions to the development and application of electromagnetic theory.

given for $t \geq 0$, and suppose that f satisfies certain conditions to be stated a little later. Then the Laplace transform of f, which we will denote by $\mathscr{L}\{f(t)\}$ or by $F(s)$, is defined by the equation

$$\mathscr{L}\{f(t)\} = F(s) = \int_0^\infty e^{-st} f(t)\, dt. \tag{2}$$

This transform makes use of the kernel $K(s,t) = e^{-st}$ and is associated particularly with linear differential equations with constant coefficients. It is especially useful in solving problems with nonhomogeneous terms of a discontinuous or impulsive nature. Such problems are relatively awkward to handle by the methods previously discussed, which involve piecing together solutions valid in different intervals. The Laplace transform is particularly valuable in circuit analysis, where discontinuous or impulsive forcing terms are common, but it is also important in other applications.

Since the Laplace transform is defined by an integral over the range from zero to infinity, it is useful to mention first some basic facts about such integrals. In the first place, an integral over an unbounded interval is called an *improper integral*, and is defined as a limit of integrals over finite intervals; thus

$$\int_a^\infty f(t)\, dt = \lim_{A \to \infty} \int_a^A f(t)\, dt, \tag{3}$$

where A is a positive real number. If the integral from a to A exists for each $A > a$, and if the limit as $A \to \infty$ exists, then the improper integral is said to *converge* to that limiting value. Otherwise the integral is said to *diverge*, or to fail to exist. The following examples illustrate both possibilities.

EXAMPLE 1

Let $f(t) = e^{ct}$, $t \geq 0$, where c is a real nonzero constant. Then

$$\int_0^\infty e^{ct}\, dt = \lim_{A \to \infty} \int_0^A e^{ct}\, dt = \lim_{A \to \infty} \left. \frac{e^{ct}}{c} \right|_0^A$$

$$= \lim_{A \to \infty} \frac{1}{c}(e^{cA} - 1).$$

It follows that the improper integral converges if $c < 0$, and diverges if $c > 0$. If $c = 0$ the integrand is unity, and the integral again diverges.

EXAMPLE 2

Let $f(t) = 1/t, t \geq 1$. Then

$$\int_1^\infty \frac{dt}{t} = \lim_{A \to \infty} \int_1^A \frac{dt}{t} = \lim_{A \to \infty} \ln A.$$

Since $\lim_{A \to \infty} \ln A = \infty$, the improper integral diverges.

EXAMPLE 3

Let $f(t) = t^{-p}, t \geq 1$, where p is a real constant and $p \neq 1$; the case $p = 1$ was considered in Example 2. Then

$$\int_1^\infty t^{-p}\,dt = \lim_{A\to\infty} \int_1^A t^{-p}\,dt = \lim_{A\to\infty} \frac{1}{1-p}(A^{1-p} - 1).$$

As $A \to \infty$, $A^{1-p} \to 0$ if $p > 1$, but $A^{1-p} \to \infty$ if $p < 1$. Hence $\int_1^\infty t^{-p}\,dt$ converges for $p > 1$, but (incorporating the result of Example 2) diverges for $p \leq 1$. These results are analogous to those for the infinite series $\sum_{n=1}^{\infty} n^{-p}$.

Before discussing the possible existence of $\int_a^\infty f(t)\,dt$, it is helpful to define certain terms. A function f is said to be *piecewise continuous* on an interval $\alpha \leq t \leq \beta$ if the interval can be partitioned by a finite number of points $\alpha = t_0 < t_1 < \cdots < t_n = \beta$ so that

1. f is continuous on each open subinterval $t_{i-1} < t < t_i$;
2. f approaches a finite limit as the end points of each subinterval are approached from within the subinterval.

In other words, f is piecewise continuous on $\alpha \leq t \leq \beta$ if it is continuous there except for a finite number of jump discontinuities. If f is piecewise continuous on $\alpha \leq t \leq \beta$ for every $\beta > \alpha$, then f is said to be piecewise continuous on $t \geq \alpha$. An example of a piecewise continuous function is shown in Fig. 6.1.

If f is piecewise continuous on the interval $a \leq t \leq A$, then it can be shown that $\int_a^A f(t)\,dt$ exists. Hence, if f is piecewise continuous for $t \geq a$, then $\int_a^A f(t)\,dt$ exists for each $A > a$. However, piecewise continuity is not enough to insure convergence of the improper integral $\int_a^\infty f(t)\,dt$, as the above examples show.

If f cannot be easily integrated in terms of elementary functions, the definition of convergence of $\int_a^\infty f(t)\,dt$ may be difficult to apply. Frequently, the most convenient way to test the convergence or divergence of an improper integral is by the following comparison theorem, which is analogous to a similar theorem for infinite series.

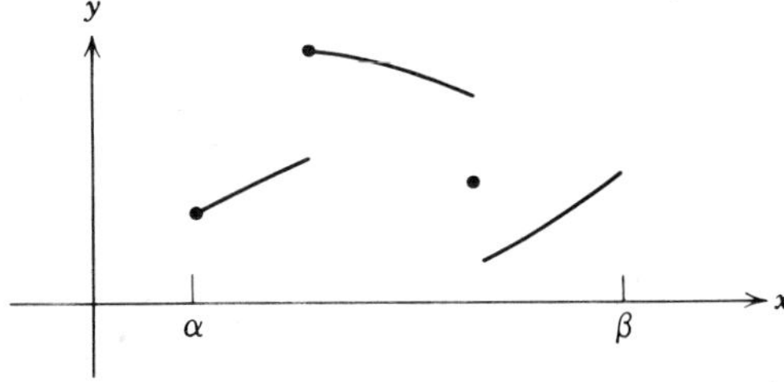

FIGURE 6.1 A piecewise continuous function.

Theorem 6.1. *If f is piecewise continuous for $t \geq a$, if $|f(t)| \leq g(t)$ when $t \geq M$ for some positive constant M, and if $\int_M^\infty g(t)\,dt$ converges, then $\int_a^\infty f(t)\,dt$ also converges. On the other hand, if $f(t) \geq g(t) \geq 0$ for $t \geq M$, and if $\int_M^\infty g(t)\,dt$ diverges, then $\int_a^\infty f(t)\,dt$ also diverges.*

The proof of this result from the calculus will not be given here. It is made plausible, however, by comparing the areas represented by $\int_M^\infty g(t)\,dt$ and $\int_M^\infty |f(t)|\,dt$. The functions most useful for comparison purposes are e^{ct} and t^{-p}, which were considered in Examples 1, 2, and 3.

We now return to a consideration of the Laplace transform $\mathscr{L}\{f(t)\}$ or $F(s)$, which is defined by Eq. (2) whenever this improper integral converges. In general, the parameter s may be complex, but for our discussion we need consider only real values of s. According to the foregoing discussion of integrals the function f must satisfy certain conditions in order for its Laplace transform F to exist. The simplest and most useful of these conditions are stated in the following theorem.

Theorem 6.2. *Suppose that*

1. *f is piecewise continuous on the interval $0 \leq t \leq A$ for any positive A;*
2. *$|f(t)| \leq Ke^{at}$ when $t \geq M$. In this inequality K, a, and M are real constants, K and M necessarily positive.*

Then the Laplace transform $\mathscr{L}\{f(t)\} = F(s)$, defined by Eq. (2), exists for $s > a$.

To establish this theorem it is necessary to show only that the integral in Eq. (2) converges for $s > a$. Splitting the improper integral into two parts, we have

$$\int_0^\infty e^{-st}f(t)\,dt = \int_0^M e^{-st}f(t)\,dt + \int_M^\infty e^{-st}f(t)\,dt. \tag{4}$$

The first integral on the right side of Eq. (4) exists by hypothesis (1) of the theorem; hence the existence of $F(s)$ depends on the convergence of the second integral. By hypothesis (2) we have, for $t \geq M$,

$$|e^{-st}f(t)| \leq Ke^{-st}e^{at} = Ke^{(a-s)t},$$

and thus, by Theorem 6.1, $F(s)$ exists provided $\int_M^\infty e^{(a-s)t}\,dt$ converges. Referring to Example 1 with c replaced by $a - s$, we see that this latter integral converges when $a - s < 0$, which establishes Theorem 6.2.

Unless the contrary is specifically stated, in this chapter we deal only with functions satisfying the conditions of Theorem 6.2. Such functions are described as piecewise continuous, and of *exponential order* as $t \to \infty$. The Laplace transforms of some important elementary functions are given in the examples below.

EXAMPLE 4

Let $f(t) = 1, t \geq 0$. Then

$$\mathscr{L}\{1\} = \int_0^\infty e^{-st}\,dt = \frac{1}{s}, \qquad s > 0.$$

EXAMPLE 5

Let $f(t) = e^{at}, t \geq 0$. Then

$$\mathscr{L}\{e^{at}\} = \int_0^\infty e^{-st}e^{at}\,dt = \int_0^\infty e^{-(s-a)t}\,dt$$
$$= \frac{1}{s-a}, \qquad s > a.$$

EXAMPLE 6

Let $f(t) = \sin at, t \geq 0$. Then

$$\mathscr{L}\{\sin at\} = F(s) = \int_0^\infty e^{-st}\sin at\,dt, \qquad s > 0.$$

Since

$$F(s) = \lim_{A\to\infty}\int_0^A e^{-st}\sin at\,dt,$$

upon integrating by parts we obtain

$$F(s) = \lim_{A\to\infty}\left[-\left.\frac{e^{-st}\cos at}{a}\right|_0^A - \frac{s}{a}\int_0^A e^{-st}\cos at\,dt\right]$$
$$= \frac{1}{a} - \frac{s}{a}\int_0^\infty e^{-st}\cos at\,dt.$$

A second integration by parts then yields

$$F(s) = \frac{1}{a} - \frac{s^2}{a^2}\int_0^\infty e^{-st}\sin at\,dt$$
$$= \frac{1}{a} - \frac{s^2}{a^2}F(s).$$

Hence, solving for $F(s)$, we have

$$F(s) = \frac{a}{s^2 + a^2}, \qquad s > 0.$$

Now let us suppose that f_1 and f_2 are two functions whose Laplace transforms exist for $s > a_1$ and $s > a_2$, respectively. Then, for s greater than the maximum of a_1 and a_2,

$$\begin{aligned}\mathscr{L}\{c_1 f_1(t) + c_2 f_2(t)\} &= \int_0^\infty e^{-st}[c_1 f_1(t) + c_2 f_2(t)]\,dt \\ &= c_1 \int_0^\infty e^{-st} f_1(t)\,dt + c_2 \int_0^\infty e^{-st} f_2(t)\,dt;\end{aligned}$$

hence

$$\mathscr{L}\{c_1 f_1(t) + c_2 f_2(t)\} = c_1 \mathscr{L}\{f_1(t)\} + c_2 \mathscr{L}\{f_2(t)\}. \tag{5}$$

Equation (5) is a statement of the fact that the Laplace transform is a *linear operator*. This property is of paramount importance, and we make frequent use of it later.

PROBLEMS

1. Sketch the graph of each of the following functions. In each case determine whether f is continuous, piecewise continuous, or neither on the interval $0 \le t \le 3$.

(a) $f(t) = \begin{cases} t^2, & 0 \le t \le 1 \\ 2 + t, & 1 < t \le 2 \\ 6 - t, & 2 < t \le 3 \end{cases}$ (b) $f(t) = \begin{cases} t^2, & 0 \le t \le 1 \\ (t-1)^{-1}, & 1 < t \le 2 \\ 1, & 2 < t \le 3 \end{cases}$

(c) $f(t) = \begin{cases} t^2, & 0 \le t \le 1 \\ 1, & 1 < t \le 2 \\ 3 - t, & 2 < t \le 3 \end{cases}$ (d) $f(t) = \begin{cases} t, & 0 \le t \le 1 \\ 3 - t, & 1 < t \le 2 \\ 1, & 2 < t \le 3 \end{cases}$

2. Find the Laplace transform of each of the following functions.

(a) t (b) t^2 (c) t^n, where n is a positive integer

3. Find the Laplace transform of $f(t) = \cos at$, where a is a real constant.
4. Recalling that $\cosh bt = (e^{bt} + e^{-bt})/2$ and $\sinh bt = (e^{bt} - e^{-bt})/2$, find the Laplace transform of each of the following functions; a and b are real constants.

(a) $\cosh bt$ (b) $\sinh bt$
(c) $e^{at} \cosh bt$ (d) $e^{at} \sinh bt$

5. Recall that $\cos bt = (e^{ibt} + e^{-ibt})/2$ and $\sin bt = (e^{ibt} - e^{-ibt})/2i$. Assuming that the necessary elementary integration formulas extend to this case, find the Laplace transform of each of the following functions; a and b are real constants.

(a) $\sin bt$ (b) $\cos bt$
(c) $e^{at} \sin bt$ (d) $e^{at} \cos bt$

6. Using integration by parts, find the Laplace transform of each of the following functions; n is a positive integer and a is a real constant.

(a) te^{at} (b) $t \sin at$ (c) $t \cosh at$
(d) $t^n e^{at}$ (e) $t^2 \sin at$ (f) $t^2 \sinh at$

7. Determine whether each of the following integrals converges or diverges.

(a) $\int_0^\infty (t^2 + 1)^{-1}\,dt$ (b) $\int_0^\infty te^{-t}\,dt$ (c) $\int_1^\infty t^{-2}e^t\,dt$

8. Suppose that f and f' are continuous for $t \geq 0$, and of exponential order as $t \to \infty$. Show by integration by parts that if $F(s) = \mathscr{L}\{f(t)\}$, then $\lim_{s\to\infty} F(s) = 0$. The result is actually true under less restrictive conditions, such as those of Theorem 6.2.
9. **The Gamma Function.** The Gamma function is denoted by $\Gamma(p)$ and is defined by the integral

$$\Gamma(p+1) = \int_0^\infty e^{-x}x^p\,dx. \tag{i}$$

This integral converges at infinity for all p. For $p < 0$ it is also improper because the integrand becomes unbounded as $x \to 0$. However, the integral can be shown to converge at $x = 0$ for $p > -1$.
(a) Show that for $p > 0$

$$\Gamma(p+1) = p\Gamma(p).$$

(b) Show that $\Gamma(1) = 1$.
(c) If p is a positive integer n, show that

$$\Gamma(n+1) = n!.$$

Since $\Gamma(p)$ is also defined when p is not an integer, this function provides an extension of the factorial function to nonintegral values of the independent variable. Note that it is also consistent to define $0! = 1$.
(d) Show that for $p > 0$

$$p(p+1)(p+2)\cdots(p+n-1) = \Gamma(p+n)/\Gamma(p).$$

Thus $\Gamma(p)$ can be determined for all positive values of p if $\Gamma(p)$ is known in a single interval of unit length, say $0 < p \leq 1$. It is possible to show that $\Gamma(\frac{1}{2}) = \sqrt{\pi}$. Find $\Gamma(\frac{3}{2})$ and $\Gamma(\frac{11}{2})$.
10. Consider the Laplace transform of t^p, where $p > -1$.
(a) Referring to Problem 9, show that

$$\mathscr{L}\{t^p\} = \int_0^\infty e^{-st}t^p\,dt = \frac{1}{s^{p+1}}\int_0^\infty e^{-x}x^p\,dx$$
$$= \Gamma(p+1)/s^{p+1}, \qquad s > 0.$$

(b) If p is a positive integer n, show that

$$\mathscr{L}\{t^n\} = n!/s^{n+1}, \qquad s > 0.$$

(c) Show that

$$\mathscr{L}\{t^{-1/2}\} = \frac{2}{\sqrt{s}}\int_0^\infty e^{-x^2}\,dx, \qquad s > 0.$$

It is possible to show that

$$\int_0^\infty e^{-x^2}\,dx = \frac{\sqrt{\pi}}{2};$$

hence

$$\mathcal{L}\{t^{-1/2}\} = \sqrt{\pi/s}, \qquad s > 0.$$

(d) Show that

$$\mathcal{L}\{t^{1/2}\} = \sqrt{\pi}/2s^{3/2}, \qquad s > 0.$$

6.2 Solution of Initial Value Problems

In this section we show how the Laplace transform can be used to solve initial value problems for linear differential equations with constant coefficients. The usefulness of the Laplace transform in this connection rests primarily on the fact that the transform of f' is related in a simple way to the transform of f. The relationship is expressed in the following theorem.

Theorem 6.3. *Suppose that f is continuous and that f' is piecewise continuous on any interval $0 \le t \le A$. Suppose further that there exist constants K, a, and M such that $|f(t)| \le Ke^{at}$ for $t \ge M$. Then $\mathcal{L}\{f'(t)\}$ exists for $s > a$, and moreover*

$$\mathcal{L}\{f'(t)\} = s\mathcal{L}\{f(t)\} - f(0). \tag{1}$$

To prove this theorem we consider the integral

$$\int_0^A e^{-st}f'(t)\,dt.$$

Letting $t_1, t_2, \ldots, t_n$ be the points in the interval $0 \le t \le A$ where f' is discontinuous, we obtain

$$\int_0^A e^{-st}f'(t)\,dt = \int_0^{t_1} e^{-st}f'(t)\,dt + \int_{t_1}^{t_2} e^{-st}f'(t)\,dt + \cdots + \int_{t_n}^{A} e^{-st}f'(t)\,dt.$$

Integrating each term on the right by parts yields

$$\int_0^A e^{-st}f'(t)\,dt = e^{-st}f(t)\Big|_0^{t_1} + e^{-st}f(t)\Big|_{t_1}^{t_2} + \cdots + e^{-st}f(t)\Big|_{t_n}^{A}$$
$$+ s\left[\int_0^{t_1} e^{-st}f(t)\,dt + \int_{t_1}^{t_2} e^{-st}f(t)\,dt + \cdots + \int_{t_n}^{A} e^{-st}f(t)\,dt\right].$$

Since f is continuous, the contributions of the integrated terms at $t_1, t_2, \ldots, t_n$ cancel. Combining the integrals gives

$$\int_0^A e^{-st}f'(t)\,dt = e^{-sA}f(A) - f(0) + s\int_0^A e^{-st}f(t)\,dt.$$

As $A \to \infty, e^{-sA}f(A) \to 0$ whenever $s > a$. Hence, for $s > a$,

$$\mathcal{L}\{f'(t)\} = s\mathcal{L}\{f(t)\} - f(0),$$

which establishes the theorem.

If f' and f'' satisfy the same conditions that are imposed on f and f', respectively, in Theorem 6.3, then it follows that the Laplace transform of f'' also exists for $s > a$ and is given by

$$\mathscr{L}\{f''(t)\} = s^2\mathscr{L}\{f(t)\} - sf(0) - f'(0). \tag{2}$$

Indeed, provided the function f and its derivatives satisfy suitable conditions, an expression for the transform of the nth derivative $f^{(n)}$ can be derived by successive applications of this theorem. The result is given in the following corollary.

Corollary. *Suppose that the functions $f, f', \ldots, f^{(n-1)}$ are continuous, and that $f^{(n)}$ is piecewise continuous on any interval $0 \le t \le A$. Suppose further that there exist constants K, a, and M such that $|f(t)| \le Ke^{at}, |f'(t)| \le Ke^{at}, \ldots, |f^{(n-1)}(t)| \le Ke^{at}$ for $t \ge M$. Then $\mathscr{L}\{f^{(n)}(t)\}$ exists for $s > a$ and is given by*

$$\mathscr{L}\{f^{(n)}(t)\} = s^n\mathscr{L}\{f(t)\} - s^{n-1}f(0) - \cdots - sf^{(n-2)}(0) - f^{(n-1)}(0). \tag{3}$$

We now show how the Laplace transform can be used to solve initial value problems. First we consider the differential equation

$$y'' - y' - 2y = 0 \tag{4}$$

and the initial conditions

$$y(0) = 1, \qquad y'(0) = 0. \tag{5}$$

This simple problem is easily solved by the methods of Section 3.5. The characteristic equation is

$$r^2 - r - 2 = (r - 2)(r + 1) = 0, \tag{6}$$

and consequently the general solution of Eq. (4) is

$$y = c_1e^{-t} + c_2e^{2t}. \tag{7}$$

To satisfy the initial conditions (5) we must have $c_1 + c_2 = 1$ and $-c_1 + 2c_2 = 0$; hence $c_1 = \frac{2}{3}$ and $c_2 = \frac{1}{3}$, so that the solution of the initial value problem (4) and (5) is

$$y = \phi(t) = \tfrac{2}{3}e^{-t} + \tfrac{1}{3}e^{2t}. \tag{8}$$

Now let us solve the given problem by means of the Laplace transform. To do this we must assume that the problem has a solution $y = \phi(t)$, which with its first two derivatives satisfies the conditions of the corollary. Then, taking the Laplace transform of the differential equation (4), we obtain

$$\mathscr{L}\{y''\} - \mathscr{L}\{y'\} - 2\mathscr{L}\{y\} = 0, \tag{9}$$

where we have used the linearity of the transform to write the transform of a sum as the sum of the separate transforms. Upon using the corollary to express

$\mathscr{L}\{y''\}$ and $\mathscr{L}\{y'\}$ in terms of $\mathscr{L}\{y\}$, we find that Eq. (9) then takes the form

$$s^2\mathscr{L}\{y\} - sy(0) - y'(0) - [s\mathscr{L}\{y\} - y(0)] - 2\mathscr{L}\{y\} = 0,$$

or

$$(s^2 - s - 2)Y(s) + (1 - s)y(0) - y'(0) = 0, \tag{10}$$

where $Y(s) = \mathscr{L}\{y\}$. Substituting for $y(0)$ and $y'(0)$ in Eq. (10) from the initial conditions (5), and then solving for $Y(s)$, we obtain

$$Y(s) = \frac{s-1}{s^2 - s - 2} = \frac{s-1}{(s-2)(s+1)}. \tag{11}$$

We have thus obtained an expression for the Laplace transform $Y(s)$ of the solution $y = \phi(t)$ of the given initial value problem. To determine the function ϕ we must find the function whose Laplace transform is $Y(s)$, as given by Eq. (11).

This can be done most easily by expanding the right side of Eq. (11) in partial fractions. Thus we write

$$Y(s) = \frac{s-1}{(s-2)(s+1)} = \frac{a}{s-2} + \frac{b}{s+1} = \frac{a(s+1) + b(s-2)}{(s-2)(s+1)}. \tag{12}$$

Then, by equating numerators of the second and fourth members of Eq. (12) we obtain

$$s - 1 = a(s+1) + b(s-2),$$

an equation that must hold for all s. In particular, if we set $s = 2$, then it follows that $a = \frac{1}{3}$. Similarly, if we set $s = -1$, then we find that $b = \frac{2}{3}$. By substituting these values for a and b, respectively, we have

$$Y(s) = \frac{1/3}{s-2} + \frac{2/3}{s+1}. \tag{13}$$

Finally, using the result of Example 5 of Section 6.1, it follows that $\frac{1}{3}e^{2t}$ has the transform $\frac{1}{3}(s-2)^{-1}$; similarly $\frac{2}{3}e^{-t}$ has the transform $\frac{2}{3}(s+1)^{-1}$. Hence, by the linearity of the Laplace transform,

$$y = \phi(t) = \tfrac{1}{3}e^{2t} + \tfrac{2}{3}e^{-t}$$

has the transform (13). This is the same solution, of course, as that obtained earlier in a different way.

The same procedure can be applied to the general second order linear equation with constant coefficients,

$$ay'' + by' + cy = f(t). \tag{14}$$

Assuming that the solution $y = \phi(t)$ satisfies the conditions of the corollary for $n = 2$, we can take the transform of Eq. (14) and thereby obtain

$$a[s^2Y(s) - sy(0) - y'(0)] + b[sY(s) - y(0)] + cY(s) = F(s), \tag{15}$$

where $F(s)$ is the transform of $f(t)$. By solving Eq. (15) for $Y(s)$ we find that

$$Y(s) = \frac{(as+b)y(0) + ay'(0)}{as^2 + bs + c} + \frac{F(s)}{as^2 + bs + c}. \tag{16}$$

The problem is then solved, provided we can find the function $y = \phi(t)$ whose transform is $Y(s)$.

Even at this early stage of our discussion we can point out some of the essential features of the transform method. In the first place, the transform $Y(s)$ of the unknown function $y = \phi(t)$ is found by solving an *algebraic equation* rather than a *differential equation*, Eq. (10) rather than Eq. (4), or in general Eq. (15) rather than Eq. (14). This is the key to the usefulness of Laplace transforms for solving ordinary differential equations—the problem is reduced from a differential equation to an algebraic one. Next, the solution satisfying given initial conditions is automatically found, so that the task of determining appropriate values for the arbitrary constants in the general solution does not arise. Further, as indicated in Eq. (15), nonhomogeneous equations are handled in exactly the same way as homogeneous ones; it is not necessary to solve the corresponding homogeneous equation first. Finally, the method can be applied in the same way to higher order equations, as long as we assume that the solution satisfies the conditions of the corollary for the appropriate value of n.

Observe that the polynomial $as^2 + bs + c$ in the denominator on the right side of Eq. (16) is precisely the characteristic polynomial associated with Eq. (14). Since the use of a partial fraction expansion of $Y(s)$ to determine $\phi(t)$ requires that this polynomial be factored, the use of Laplace transforms does not avoid the necessity of finding the roots of the characteristic equation. For equations of higher than second order this may be a difficult algebraic problem, particularly if the roots are irrational or complex.

The main difficulty that occurs in solving initial value problems by the transform technique lies in the problem of determining the function $y = \phi(t)$ corresponding to the transform $Y(s)$. This problem is known as the inversion problem for the Laplace transform; $\phi(t)$ is called the inverse transform corresponding to $Y(s)$, and the process of finding $\phi(t)$ from $Y(s)$ is known as "inverting the transform." We also use the notation $\mathscr{L}^{-1}\{Y(s)\}$ to denote the inverse transform of $Y(s)$. There is a general formula for the inverse Laplace transform, but its use requires a knowledge of the theory of functions of a complex variable, and we do not consider it in this book. However, it is still possible to develop many important properties of the Laplace transform, and to solve many interesting problems, without the use of complex variables.

In solving the initial value problem (4), (5) we did not consider the question of whether there may be functions other than that given by Eq. (8) that also have the transform (13). In fact, it can be shown that if f is a continuous function with the Laplace transform F, then there is no other continuous function having the same transform. In other words, there is essentially a one-to-one correspondence between functions and their Laplace transforms. This fact suggests the compila-

TABLE 6.1 ELEMENTARY LAPLACE TRANSFORMS

$f(t) = \mathscr{L}^{-1}\{F(s)\}$	$F(s) = \mathscr{L}\{f(t)\}$	Notes
1	$\dfrac{1}{s}, \quad s > 0$	Sec. 6.1; Ex. 4
e^{at}	$\dfrac{1}{s-a}, \quad s > a$	Sec. 6.1; Ex. 5
t^n, $n =$ positive integer	$\dfrac{n!}{s^{n+1}}, \quad s > 0$	Sec. 6.1; Prob. 10
$t^p, p > -1$	$\dfrac{\Gamma(p+1)}{s^{p+1}}, \quad s > 0$	Sec. 6.1; Prob. 10
$\sin at$	$\dfrac{a}{s^2+a^2}, \quad s > 0$	Sec. 6.1; Ex. 6
$\cos at$	$\dfrac{s}{s^2+a^2}, \quad s > 0$	Sec. 6.1; Prob. 3
$\sinh at$	$\dfrac{a}{s^2-a^2}, \quad s > \lvert a \rvert$	Sec. 6.1; Prob. 4
$\cosh at$	$\dfrac{s}{s^2-a^2}, \quad s > \lvert a \rvert$	Sec. 6.1; Prob. 4
$e^{at}\sin bt$	$\dfrac{b}{(s-a)^2+b^2}, \quad s > a$	Sec. 6.1; Prob. 5
$e^{at}\cos bt$	$\dfrac{s-a}{(s-a)^2+b^2}, \quad s > a$	Sec. 6.1; Prob. 5
$t^n e^{at}$, $n =$ positive integer	$\dfrac{n!}{(s-a)^{n+1}}, \quad s > a$	Sec. 6.1; Prob. 6
$u_c(t)$	$\dfrac{e^{-cs}}{s}, \quad s > 0$	Sec. 6.3
$u_c(t)f(t-c)$	$e^{-cs}F(s)$	Sec. 6.3
$e^{ct}f(t)$	$F(s-c)$	Sec. 6.3
$f(ct)$	$\dfrac{1}{c}F\left(\dfrac{s}{c}\right), \quad c > 0$	Sec. 6.3; Prob. 4
$\int_0^t f(t-\tau)g(\tau)\,d\tau$	$F(s)G(s)$	Sec. 6.5
$\delta(t-c)$	e^{-cs}	Sec. 6.4
$f^{(n)}(t)$	$s^nF(s) - s^{n-1}f(0) - \cdots - f^{(n-1)}(0)$	Sec. 6.2
$(-t)^n f(t)$	$F^{(n)}(s)$	Sec. 6.2; Prob. 28

tion of a table (somewhat analogous to a table of integrals) giving the transforms of functions frequently encountered, and vice versa (see Table 6.1).[2] The entries in the second column are the transforms of those in the first column. Perhaps more important, the functions in the first column are the inverse transforms of those in the second column. Thus, for example, if the transform of the solution of a differential equation is known, the solution itself can often be found merely by looking it up in the table. Some of the entries in Table 6.1 have been used as examples, or appear as problems in Section 6.1, while others will be developed later in the chapter. The third column of the table indicates where the derivation of the given transforms may be found.

Frequently, a Laplace transform $F(s)$ is expressible as a sum of several terms,

$$F(s) = F_1(s) + F_2(s) + \cdots + F_n(s). \tag{17}$$

Suppose that $f_1(t) = \mathscr{L}^{-1}\{F_1(s)\}, \ldots, f_n(t) = \mathscr{L}^{-1}\{F_n(s)\}$. Then the function

$$f(t) = f_1(t) + \cdots + f_n(t)$$

has the Laplace transform $F(s)$. By the uniqueness property stated above there is no other continuous function f having the same transform. Thus

$$\mathscr{L}^{-1}\{F(s)\} = \mathscr{L}^{-1}\{F_1(s)\} + \cdots + \mathscr{L}^{-1}\{F_n(s)\}; \tag{18}$$

that is, the inverse Laplace transform is also a linear operator.

In many problems it is convenient to make use of this property by decomposing a given transform into a sum of functions whose inverse transforms are already known or can be found in the table. Partial fraction expansions are particularly useful in this connection, and a general result covering many cases is given in Problem 33. Other useful properties of Laplace transforms are derived later in this chapter.

As further illustrations of the technique of solving initial value problems by means of the Laplace transform and partial fraction expansions, consider the following examples.

EXAMPLE 1

Find the solution of the differential equation

$$y'' + y = \sin 2t, \tag{19}$$

satisfying the initial conditions

$$y(0) = 0, \qquad y'(0) = 1. \tag{20}$$

We assume that this initial value problem has a solution $y = \phi(t)$, which with its first two derivatives satisfies the conditions of the corollary to Theorem

[2] Larger tables are also available. See the references at the end of the chapter.

6.3. Then, taking the Laplace transform of the differential equation, we have

$$s^2Y(s) - sy(0) - y'(0) + Y(s) = 2/(s^2 + 4),$$

where the transform of $\sin 2t$ has been obtained from Table 6.1. Substituting for $y(0)$ and $y'(0)$ from the initial conditions and solving for $Y(s)$, we obtain

$$Y(s) = \frac{s^2 + 6}{(s^2 + 1)(s^2 + 4)}. \tag{21}$$

Using partial fractions we can write $Y(s)$ in the form

$$Y(s) = \frac{as + b}{s^2 + 1} + \frac{cs + d}{s^2 + 4} = \frac{(as + b)(s^2 + 4) + (cs + d)(s^2 + 1)}{(s^2 + 1)(s^2 + 4)}. \tag{22}$$

By expanding the numerator on the right side of Eq. (22) and equating it to the numerator in Eq. (21) we find that

$$s^2 + 6 = (a + c)s^3 + (b + d)s^2 + (4a + c)s + (4b + d)$$

for all s. Then, comparing coefficients of like powers of s, we have

$$a + c = 0, \qquad b + d = 1,$$
$$4a + c = 0, \qquad 4b + d = 6.$$

Consequently, $a = 0$, $c = 0$, $b = \frac{5}{3}$, and $d = -\frac{2}{3}$, from which it follows that

$$Y(s) = \frac{5/3}{s^2 + 1} - \frac{2/3}{s^2 + 4}. \tag{23}$$

From Example 6 of Section 6.1, or from Table 6.1, the solution of the given initial value problem is

$$y = \phi(t) = \tfrac{5}{3}\sin t - \tfrac{1}{3}\sin 2t. \tag{24}$$

EXAMPLE 2

Find the solution of the initial value problem

$$y^{\text{iv}} - y = 0, \tag{25}$$

$$y(0) = 0, \qquad y'(0) = 1, \qquad y''(0) = 0, \qquad y'''(0) = 0. \tag{26}$$

In this problem we need to assume that the solution $y = \phi(t)$ satisfies the conditions of the corollary to Theorem 6.3 for $n = 4$. The Laplace transform of the differential equation (25) is

$$s^4Y(s) - s^3y(0) - s^2y'(0) - sy''(0) - y'''(0) - Y(s) = 0.$$

Then, using the initial conditions (26) and solving for $Y(s)$, we have

$$Y(s) = \frac{s^2}{s^4 - 1}. \tag{27}$$

The partial fraction expansion of $Y(s)$ is

$$Y(s) = \frac{as + b}{s^2 - 1} + \frac{cs + d}{s^2 + 1}$$

and it follows that

$$(as + b)(s^2 + 1) + (cs + d)(s^2 - 1) = s^2 \tag{28}$$

for all s. By setting $s = 1$ and $s = -1$, respectively, in Eq. (28) we obtain the pair of equations

$$2(a + b) = 1, \qquad 2(-a + b) = 1,$$

and therefore $a = 0$ and $b = \frac{1}{2}$. If we set $s = 0$ in Eq. (28) then $b - d = 0$, so $d = \frac{1}{2}$. Finally, equating the coefficients of the quadratic terms on each side of Eq. (28), we find that $a + c = 0$, so $c = 0$. Thus

$$Y(s) = \frac{1/2}{s^2 - 1} + \frac{1/2}{s^2 + 1}, \tag{29}$$

and from Table 6.1 the solution of the initial value problem (25), (26) is

$$y = \phi(t) = (\sinh t + \sin t)/2. \tag{30}$$

The most important elementary applications of the Laplace transform are in the study of mechanical vibrations and in the analysis of electric circuits; the governing equations were derived in Sections 3.7 and 3.8, respectively. A vibrating spring–mass system has the equation of motion

$$m\frac{d^2u}{dt^2} + c\frac{du}{dt} + ku = F(t), \tag{31}$$

where m is the mass, c the damping coefficient, k the spring constant, and $F(t)$ the applied external force. The equation describing an electric circuit containing an inductance L, a resistance R, and a capacitance C (an LRC circuit) is

$$L\frac{d^2Q}{dt^2} + R\frac{dQ}{dt} + \frac{1}{C}Q = E(t), \tag{32}$$

where $Q(t)$ is the charge on the condenser and $E(t)$ is the applied voltage. In terms of the current $I(t) = dQ(t)/dt$ we can write Eq. (32) in the form

$$L\frac{d^2I}{dt^2} + R\frac{dI}{dt} + \frac{1}{C}I = \frac{dE}{dt}(t). \tag{33}$$

Suitable initial conditions on u, Q, or I must also be prescribed.

We have noted previously in Section 3.8 that Eq. (31) for the spring–mass system and Eqs. (32) or (33) for the electric circuit are identical mathematically, differing only in the interpretation of the constants and variables appearing in them. Other physical problems also lead to the same differential equation. Thus, once the mathematical problem is solved, its solution can be interpreted in terms of whichever corresponding physical problem is of immediate interest.

In the problem lists following this and other sections in this chapter are numerous initial value problems for second order linear differential equations with constant coefficients. Many can be interpreted as models of particular physical systems, but usually we do not point this out explicitly.

PROBLEMS

In each of Problems 1 through 10 find the inverse Laplace transform of the given function.

1. $\dfrac{3}{s^2+4}$
2. $\dfrac{4}{(s-1)^3}$
3. $\dfrac{2}{s^2+3s-4}$
4. $\dfrac{3s}{s^2-s-6}$
5. $\dfrac{2s+2}{s^2+2s+5}$
6. $\dfrac{2s-3}{s^2-4}$
7. $\dfrac{2s+1}{s^2-2s+2}$
8. $\dfrac{8s^2-4s+12}{s(s^2+4)}$
9. $\dfrac{1-2s}{s^2+4s+5}$
10. $\dfrac{2s-3}{s^2+2s+10}$

In each of Problems 11 through 23 use the Laplace transform to solve the given initial value problem.

11. $y''-y'-6y=0;\quad y(0)=1,\quad y'(0)=-1$
12. $y''+3y'+2y=0;\quad y(0)=1,\quad y'(0)=0$
13. $y''-2y'+2y=0;\quad y(0)=0,\quad y'(0)=1$
14. $y''-4y'+4y=0;\quad y(0)=1,\quad y'(0)=1$
15. $y''-2y'-2y=0;\quad y(0)=2,\quad y'(0)=0$
16. $y''+2y'+5y=0;\quad y(0)=2,\quad y'(0)=-1$
17. $y^{\text{iv}}-4y'''+6y''-4y'+y=0;\quad y(0)=0,\ y'(0)=1,\ y''(0)=0,\ y'''(0)=1$
18. $y^{\text{iv}}-y=0;\quad y(0)=1,\quad y'(0)=0,\quad y''(0)=1,\quad y'''(0)=0$
19. $y^{\text{iv}}-4y=0;\quad y(0)=1,\quad y'(0)=0,\quad y''(0)=-2,\quad y'''(0)=0$
20. $y''+\omega^2y=\cos 2t,\quad \omega^2\neq 4;\quad y(0)=1,\quad y'(0)=0$
21. $y''-2y'+2y=\cos t;\quad y(0)=1,\quad y'(0)=0$
22. $y''-2y'+2y=e^{-t};\quad y(0)=0,\quad y'(0)=1$
23. $y''+2y'+y=4e^{-t};\quad y(0)=2,\quad y'(0)=-1$

In each of Problems 24 through 26 find the Laplace transform $Y(s)=\mathscr{L}\{y\}$ of the solution of the given initial value problem. A method for determining the inverse transforms is developed in Section 6.3.

24. $y''+4y=\begin{cases}1, & 0\le t<\pi,\\ 0, & \pi\le t<\infty;\end{cases}\quad y(0)=1,\quad y'(0)=0$

25. $y'' + y = \begin{cases} t, & 0 \le t < 1, \\ 0, & 1 \le t < \infty; \end{cases} \qquad y(0) = 0, \quad y'(0) = 0$

26. $y'' + 4y = \begin{cases} t, & 0 \le t < 1, \\ 1, & 1 \le t < \infty; \end{cases} \qquad y(0) = 0, \quad y'(0) = 0$

*27. The Laplace transforms of certain functions can be conveniently found from their Taylor series expansions.
(a) Using the Taylor series for $\sin t$,

$$\sin t = \sum_{n=0}^{\infty} \frac{(-1)^n t^{2n+1}}{(2n+1)!},$$

and assuming that the Laplace transform of this series can be computed term by term, verify that

$$\mathscr{L}\{\sin t\} = \frac{1}{s^2+1}, \qquad s > 1.$$

(b) Let

$$f(t) = \begin{cases} (\sin t)/t, & t \neq 0, \\ 1, & t = 0. \end{cases}$$

Find the Taylor series for f about $t = 0$. Assuming that the Laplace transform of this function can be computed term by term, verify that

$$\mathscr{L}\{f(t)\} = \arctan 1/s, \qquad s > 1.$$

(c) The Bessel function of the first kind of order zero J_0 has the Taylor series (see Section 4.7)

$$J_0(t) = \sum_{n=0}^{\infty} \frac{(-1)^n t^{2n}}{2^{2n}(n!)^2}.$$

Assuming that the following Laplace transforms can be computed term by term, verify that

$$\mathscr{L}\{J_0(t)\} = (s^2+1)^{-1/2}, \qquad s > 1,$$

and that

$$\mathscr{L}\{J_0(\sqrt{t})\} = s^{-1}e^{-1/4s}, \qquad s > 0.$$

Problems 28 through 31 are concerned with differentiation of the Laplace transform.

28. Let

$$F(s) = \int_0^{\infty} e^{-st} f(t)\, dt.$$

It is possible to show that as long as f satisfies the conditions of Theorem 6.2, it is legitimate to differentiate under the integral sign with respect to the parameter s when $s > a$.
(a) Show that $F'(s) = \mathscr{L}\{-tf(t)\}$.
(b) Show that $F^{(n)}(s) = \mathscr{L}\{(-t)^n f(t)\}$; hence differentiating the Laplace transform corresponds to multiplying the original function by $-t$.

29. Use the result of Problem 28 to find the Laplace transform of each of the following functions; a and b are real numbers and n is a positive integer.

(a) te^{at} (b) $t^2 \sin bt$
(c) t^n (d) $t^n e^{at}$
(e) $te^{at} \sin bt$ (f) $te^{at} \cos bt$

*30. Consider Bessel's equation of order zero

$$ty'' + y' + ty = 0.$$

Recall from Section 4.3 that $t = 0$ is a regular singular point for this equation, and therefore solutions may become unbounded as $t \to 0$. However, let us try to determine whether there are any solutions that remain finite at $t = 0$ and have finite derivatives there. Assuming that there is such a solution $y = \phi(t)$, let $Y(s) = \mathscr{L}\{\phi(t)\}$.
(a) Show that $Y(s)$ satisfies

$$(1 + s^2)Y'(s) + sY(s) = 0.$$

(b) Show that $Y(s) = c(1 + s^2)^{-1/2}$, where c is an arbitrary constant.
(c) Expanding $(1 + s^2)^{-1/2}$ in a binominal series valid for $s > 1$, and assuming that it is permissible to take the inverse transform term by term, show that

$$y = c \sum_{n=0}^{\infty} \frac{(-1)^n t^{2n}}{2^{2n}(n!)^2} = cJ_0(t),$$

where J_0 is the Bessel function of the first kind of order zero. Note that $J_0(0) = 1$, and that J_0 has finite derivatives of all orders at $t = 0$. It was shown in Section 4.7 that the second solution of this equation becomes unbounded as $t \to 0$.

31. For each of the following initial value problems use the results of Problem 28 to find the differential equation satisfied by $Y(s) = \mathscr{L}\{\phi(t)\}$, where $y = \phi(t)$ is the solution of the given initial value problem.

(a) $y'' - ty = 0$; $y(0) = 1$, $y'(0) = 0$ (Airy's equation)
(b) $(1 - t^2)y'' - 2ty' + \alpha(\alpha + 1)y = 0$; $y(0) = 0$, $y'(0) = 1$ (Legendre's equation)

Note that the differential equation for $Y(s)$ is of first order in part (a), but of second order in part (b). This is due to the fact that t appears at most to the first power in the equation of part (a), whereas it appears to the second power in that of part (b). This illustrates that the Laplace transform is not often useful in solving differential equations with variable coefficients, unless all of the coefficients are at most linear functions of the independent variable.

32. Suppose that

$$g(t) = \int_0^t f(\tau)\, d\tau.$$

If $G(s)$ and $F(s)$ are the Laplace transforms of $g(t)$ and $f(t)$, respectively, show that

$$G(s) = F(s)/s.$$

33. In this problem we show how a general partial fraction expansion can be used to calculate many inverse Laplace transforms. Suppose that

$$F(s) = P(s)/Q(s),$$

where $Q(s)$ is a polynomial of degree n with distinct zeros $r_1, \ldots, r_n$ and $P(s)$ is a polynomial of degree less than n. In this case it is possible to show that $P(s)/Q(s)$ has a partial fraction expansion of the form

$$\frac{P(s)}{Q(s)} = \frac{A_1}{s - r_1} + \cdots + \frac{A_n}{s - r_n}, \tag{i}$$

where the coefficients $A_1, \ldots, A_n$ must be determined.

(a) Show that

$$A_k = P(r_k)/Q'(r_k), \qquad k = 1, \ldots, n. \tag{ii}$$

Hint: One way to do this is to multiply Eq. (i) by $s - r_k$ and then to take the limit as $s \to r_k$.

(b) Show that

$$\mathscr{L}^{-1}\{F(s)\} = \sum_{k=1}^{n} \frac{P(r_k)}{Q'(r_k)} e^{r_k t}. \tag{iii}$$

6.3 Step Functions

In Section 6.2 the general procedure involved in solving initial value problems by means of the Laplace transform was outlined. Some of the most interesting elementary applications of the transform method occur in the solution of linear differential equations with discontinuous or impulsive forcing functions. Equations of this type frequently arise in the analysis of the flow of current in electric circuits or the vibrations of mechanical systems. In this section and the following ones we develop some additional properties of the Laplace transform that are useful in the solution of such problems. Unless a specific statement is made to the contrary, all functions appearing below will be assumed to be piecewise continuous and of exponential order, so that their Laplace transforms exist, at least for s sufficiently large.

To deal effectively with functions having jump discontinuities, it is very helpful to introduce a function known as the *unit step function*. This function will be denoted by u_c, and is defined by

$$u_c(t) = \begin{cases} 0, & t < c, \\ 1, & t \geq c, \end{cases} \qquad c \geq 0. \tag{1}$$

The graph of $y = u_c(t)$ is shown in Figure 6.2. The step can also be negative. For instance, Figure 6.3 shows the graph of $y = 1 - u_c(t)$; this function can be thought of as representing a rectangular pulse.

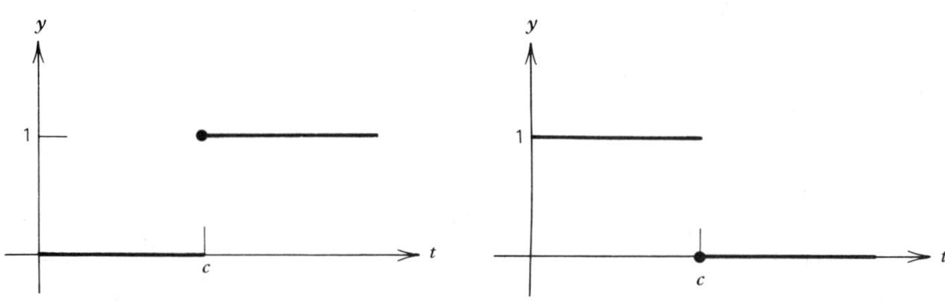

FIGURE 6.2 $y = u_c(t)$.

FIGURE 6.3 $y = 1 - u_c(t)$.

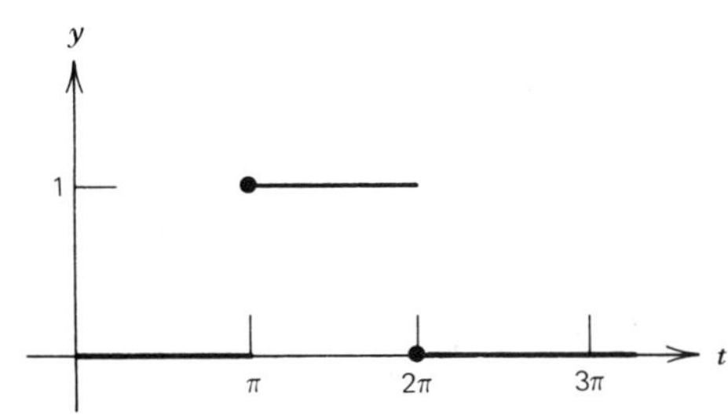

FIGURE 6.4 $y = u_\pi(t) - u_{2\pi}(t)$.

EXAMPLE 1

Sketch the graph of $y = h(t)$, where

$$h(t) = u_\pi(t) - u_{2\pi}(t), \qquad t \geq 0.$$

From the definition of $u_c(t)$ in Eq. (1) we have

$$h(t) = \begin{cases} 0 - 0 = 0, & 0 \leq t < \pi, \\ 1 - 0 = 1, & \pi \leq t < 2\pi, \\ 1 - 1 = 0, & 2\pi \leq t < \infty. \end{cases}$$

Thus the equation $y = h(t)$ has the graph shown in Figure 6.4.

For a given function f it is often necessary to consider the related function g defined by

$$y = g(t) = \begin{cases} 0, & t < c, \\ f(t - c), & t \geq c, \end{cases}$$

which represents a translation of f a distance c in the positive t direction; see Figure 6.5. In terms of the unit step function we can write $g(t)$ in the convenient form

$$g(t) = u_c(t) f(t - c).$$

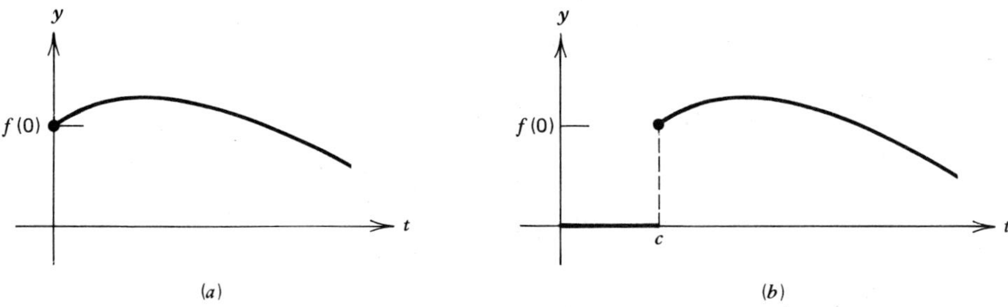

FIGURE 6.5 A translation of a given function. (*a*) $y = f(t)$; (*b*) $y = u_c(t)f(t-c)$.

The Laplace transform of u_c is easily determined:

$$\begin{aligned}\mathcal{L}\{u_c(t)\} &= \int_0^\infty e^{-st}u_c(t)\,dt\\ &= \int_c^\infty e^{-st}\,dt\\ &= \frac{e^{-cs}}{s}, \qquad s > 0.\end{aligned} \tag{2}$$

The unit step function is particularly important in transform theory because of the following relation between the transform of $f(t)$ and that of its translation $u_c(t)f(t-c)$.

Theorem 6.4. *If* $F(s) = \mathcal{L}\{f(t)\}$ *exists for* $s > a \geq 0$, *and if* c *is a positive constant, then*

$$\mathcal{L}\{u_c(t)f(t-c)\} = e^{-cs}\mathcal{L}\{f(t)\} = e^{-cs}F(s), \qquad s > a. \tag{3}$$

Conversely, if $f(t) = \mathcal{L}^{-1}\{F(s)\}$, *then*

$$u_c(t)f(t-c) = \mathcal{L}^{-1}\{e^{-cs}F(s)\}. \tag{4}$$

Theorem 6.4 simply states that the translation of $f(t)$ a distance c in the positive t direction corresponds to the multiplication of $F(s)$ by e^{-cs}. To prove Theorem 6.4 it is sufficient to compute the transform of $u_c(t)f(t-c)$:

$$\begin{aligned}\mathcal{L}\{u_c(t)f(t-c)\} &= \int_0^\infty e^{-st}u_c(t)f(t-c)\,dt\\ &= \int_c^\infty e^{-st}f(t-c)\,dt.\end{aligned}$$

Introducing a new integration variable $\xi = t - c$, we have

$$\begin{aligned}\mathcal{L}\{u_c(t)f(t-c)\} &= \int_0^\infty e^{-(\xi+c)s}f(\xi)\,d\xi = e^{-cs}\int_0^\infty e^{-s\xi}f(\xi)\,d\xi\\ &= e^{-cs}F(s).\end{aligned}$$

Thus Eq. (3) is established; Eq. (4) follows by taking the inverse transform of both sides of Eq. (3).

A simple example of this theorem occurs if we take $f(t) = 1$. Recalling that $\mathscr{L}\{1\} = 1/s$, we immediately have from Eq. (3) that $\mathscr{L}\{u_c(t)\} = e^{-cs}/s$. This result agrees with that of Eq. (2). Examples 2 and 3 illustrate further how Theorem 6.4 can be used in the calculation of transforms and inverse transforms.

EXAMPLE 2

If the function f is defined by

$$f(t) = \begin{cases} \sin t, & 0 \le t < \pi/4, \\ \sin t + \cos(t - \pi/4), & t \ge \pi/4, \end{cases}$$

find $\mathscr{L}\{f(t)\}$.

Note that $f(t) = \sin t + g(t)$, where

$$g(t) = \begin{cases} 0, & t < \pi/4, \\ \cos(t - \pi/4), & t \ge \pi/4. \end{cases}$$

Thus

$$g(t) = u_{\pi/4}(t)\cos(t - \pi/4),$$

and

$$\begin{aligned} \mathscr{L}\{f(t)\} &= \mathscr{L}\{\sin t\} + \mathscr{L}\{u_{\pi/4}(t)\cos(t - \pi/4)\} \\ &= \mathscr{L}\{\sin t\} + e^{-\pi s/4}\mathscr{L}\{\cos t\}. \end{aligned}$$

Introducing the transforms of $\sin t$ and $\cos t$, we obtain

$$\begin{aligned} \mathscr{L}\{f(t)\} &= \frac{1}{s^2 + 1} + e^{-\pi s/4}\frac{s}{s^2 + 1} \\ &= \frac{1 + se^{-\pi s/4}}{s^2 + 1}. \end{aligned}$$

The reader should compare this method with the calculation of $\mathscr{L}\{f(t)\}$ directly from the definition.

EXAMPLE 3

Find the inverse transform of

$$F(s) = \frac{1 - e^{-2s}}{s^2}.$$

From the linearity of the inverse transform we have

$$f(t) = \mathscr{L}^{-1}\{F(s)\} = \mathscr{L}^{-1}\left\{\frac{1}{s^2}\right\} - \mathscr{L}^{-1}\left\{\frac{e^{-2s}}{s^2}\right\}$$
$$= t - u_2(t)(t-2).$$

The function f may also be written as

$$f(t) = \begin{cases} t, & 0 \le t < 2, \\ 2, & t \ge 2. \end{cases}$$

The following theorem contains another very useful property of Laplace transforms that is somewhat analogous to that given in Theorem 6.4.

Theorem 6.5. *If $F(s) = \mathscr{L}\{f(t)\}$ exists for $s > a \ge 0$, and if c is a constant, then*

$$\mathscr{L}\{e^{ct}f(t)\} = F(s-c), \qquad s > a + c. \tag{5}$$

Conversely, if $f(t) = \mathscr{L}^{-1}\{F(s)\}$, then

$$e^{ct}f(t) = \mathscr{L}^{-1}\{F(s-c)\}. \tag{6}$$

According to Theorem 6.5, multiplication of $f(t)$ by e^{ct} results in a translation of the transform $F(s)$ a distance c in the positive s direction, and conversely. The proof of this theorem requires merely the evaluation of $\mathscr{L}\{e^{ct}f(t)\}$. Thus

$$\mathscr{L}\{e^{ct}f(t)\} = \int_0^\infty e^{-st}e^{ct}f(t)\,dt = \int_0^\infty e^{-(s-c)t}f(t)\,dt$$
$$= F(s-c),$$

which is Eq. (5). The restriction $s > a + c$ follows from the observation that, according to hypothesis (ii) of Theorem 6.2, $|f(t)| \le Ke^{at}$; hence $|e^{ct}f(t)| \le Ke^{(a+c)t}$. Equation (6) follows by taking the inverse transform of Eq. (5), and the proof is complete.

The principal application of Theorem 6.5 is in the evaluation of certain inverse transforms, as illustrated by Example 4.

EXAMPLE 4

Find the inverse transform of

$$G(s) = \frac{1}{s^2 - 4s + 5}.$$

By completing the square in the denominator we can write

$$G(s) = \frac{1}{(s-2)^2 + 1} = F(s-2),$$

where $F(s) = (s^2+1)^{-1}$. Since $\mathcal{L}^{-1}\{F(s)\} = \sin t$, it follows from Theorem 6.5 that

$$g(t) = \mathcal{L}^{-1}\{G(s)\} = e^{2t}\sin t.$$

The results of this section are often useful in solving differential equations, particularly those having discontinuous forcing functions. The next section is devoted to examples illustrating this fact.

PROBLEMS

1. Sketch the graph of each of the following functions on the interval $t \geq 0$.

   (a) $u_1(t) + 2u_3(t) - 6u_4(t)$
   (b) $(t-3)u_2(t) - (t-2)u_3(t)$
   (c) $f(t-\pi)u_\pi(t)$, where $f(t) = t^2$
   (d) $f(t-3)u_3(t)$, where $f(t) = \sin t$
   (e) $f(t-1)u_2(t)$, where $f(t) = 2t$
   (f) $f(t) = (t-1)u_1(t) - 2(t-2)u_2(t) + (t-3)u_3(t)$

2. Find the Laplace transform of each of the following functions.

   (a) $f(t) = \begin{cases} 0, & t < 2 \\ (t-2)^2, & t \geq 2 \end{cases}$
   (b) $f(t) = \begin{cases} 0, & t < 1 \\ t^2 - 2t + 2, & t \geq 1 \end{cases}$

   (c) $f(t) = \begin{cases} 0, & t < \pi \\ t - \pi, & \pi \leq t < 2\pi \\ 0, & t \geq 2\pi \end{cases}$
   (d) $f(t) = u_1(t) + 2u_3(t) - 6u_4(t)$

   (e) $f(t) = (t-3)u_2(t) - (t-2)u_3(t)$
   (f) $f(t) = t - u_1(t)(t-1), \quad t \geq 0$

3. Find the inverse Laplace transform of each of the following functions.

   (a) $F(s) = \dfrac{3!}{(s-2)^4}$
   (b) $F(s) = \dfrac{e^{-2s}}{s^2 + s - 2}$

   (c) $F(s) = \dfrac{2(s-1)e^{-2s}}{s^2 - 2s + 2}$
   (d) $F(s) = \dfrac{2e^{-2s}}{s^2 - 4}$

   (e) $F(s) = \dfrac{(s-2)e^{-s}}{s^2 - 4s + 3}$
   (f) $F(s) = \dfrac{e^{-s} + e^{-2s} - e^{-3s} - e^{-4s}}{s}$

4. Suppose that $F(s) = \mathcal{L}\{f(t)\}$ exists for $s > a \geq 0$.
   (a) Show that if c is a positive constant, then

$$\mathcal{L}\{f(ct)\} = \frac{1}{c}F\left(\frac{s}{c}\right), \qquad s > ca.$$

(b) Show that if k is a positive constant, then

$$\mathscr{L}^{-1}\{F(ks)\} = \frac{1}{k} f\left(\frac{t}{k}\right).$$

(c) Show that if a and b are constants with $a > 0$, then

$$\mathscr{L}^{-1}\{F(as+b)\} = \frac{1}{a} e^{-bt/a} f\left(\frac{t}{a}\right).$$

5. Using the results of Problem 4, find the inverse Laplace transform of each of the following functions.

(a) $F(s) = \dfrac{2^{n+1} n!}{s^{n+1}}$ (b) $F(s) = \dfrac{2s+1}{4s^2+4s+5}$

(c) $F(s) = \dfrac{1}{9s^2 - 12s + 3}$ (d) $F(s) = \dfrac{e^2 e^{-4s}}{2s-1}$

6. Find the Laplace transform of each of the following functions. In part (d) assume that term by term integration of the infinite series is permissible.

(a) $f(t) = \begin{cases} 1, & 0 \le t < 1 \\ 0, & t \ge 1 \end{cases}$ (b) $f(t) = \begin{cases} 1, & 0 \le t < 1 \\ 0, & 1 \le t < 2 \\ 1, & 2 \le t < 3 \\ 0, & t \ge 3 \end{cases}$

(c) $f(t) = 1 - u_1(t) + \cdots + u_{2n}(t) - u_{2n+1}(t) = 1 + \sum_{k=1}^{2n+1} (-1)^k u_k(t)$

(d) $f(t) = 1 + \sum_{k=1}^{\infty} (-1)^k u_k(t)$. See Figure 6.6.

*7. Let f satisfy $f(t+T) = f(t)$ for all $t \ge 0$ and for some fixed positive number T; f is said to be periodic with period T on $0 \le t < \infty$. Show that

$$\mathscr{L}\{f(t)\} = \frac{\int_0^T e^{-st} f(t)\, dt}{1 - e^{-sT}}.$$

*8. Use the result of Problem 7 to find the Laplace transform of each of the following functions.

(a) $f(t) = \begin{cases} 1, & 0 \le t < 1, \\ 0, & 1 \le t < 2; \end{cases}$

$f(t+2) = f(t)$.

Compare with Problem 6(d).

(b) $f(t) = \begin{cases} 1, & 0 \le t < 1, \\ -1, & 1 \le t < 2; \end{cases}$

$f(t+2) = f(t)$.

See Figure 6.7.

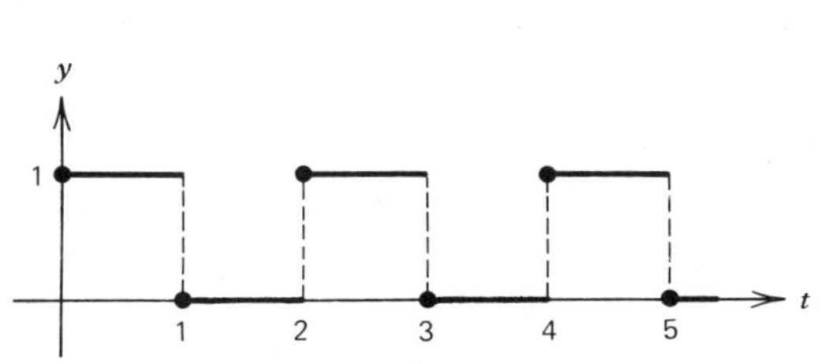

FIGURE 6.6 Square wave.

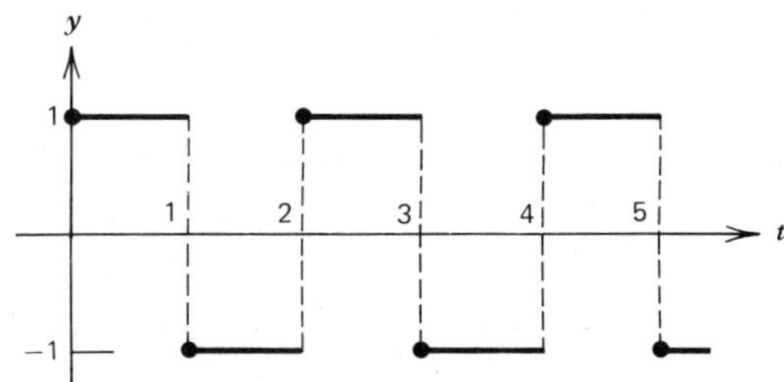

FIGURE 6.7 Square wave.

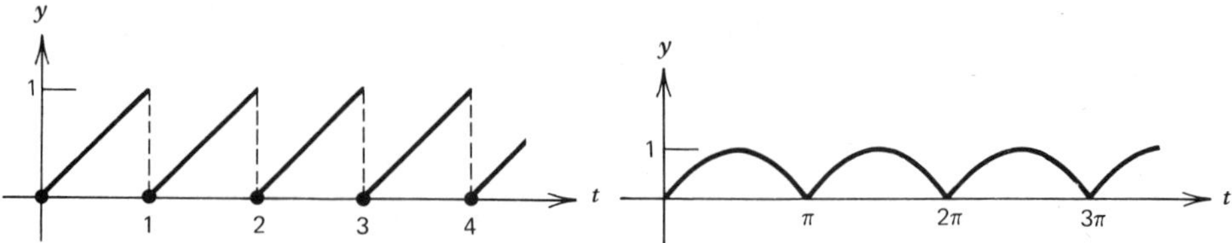

FIGURE 6.8 Sawtooth wave. FIGURE 6.9 Rectified sine wave.

(c) $f(t) = t, \quad 0 \le t < 1;$

$f(t+1) = f(t).$

See Figure 6.8.

(d) $f(t) = \sin t, \quad 0 \le t < \pi;$

$f(t+\pi) = f(t).$

See Figure 6.9.

*9. (a) If $f(t) = 1 - u_1(t)$, find $\mathscr{L}\{f(t)\}$; compare with Problem 6(a). Sketch the graph of $y - f(t)$.
(b) Let $g(t) = \int_0^t f(\xi)\,d\xi$, where the function f is defined in part (a). Sketch the graph of $y = g(t)$, and find $\mathscr{L}\{g(t)\}$.
(c) Let $h(t) = g(t) - u_1(t)g(t-1)$, where g is defined in part (b). Sketch the graph of $y = h(t)$, and find $\mathscr{L}\{h(t)\}$.

*10. Consider the function p defined by

$$p(t) = \begin{cases} t, & 0 \le t < 1, \\ 2 - t, & 1 \le t < 2; \end{cases} \qquad p(t+2) = p(t).$$

(a) Sketch the graph of $y = p(t)$.
(b) Find $\mathscr{L}\{p(t)\}$ by noting that p is the periodic extension of the function h in Problem 9(c) and then using the result of Problem 7.
(c) Find $\mathscr{L}\{p(t)\}$ by noting that

$$p(t) = \int_0^t f(t)\,dt,$$

where f is the function in Problem 8(b), and then using Theorem 6.3.

6.3.1 Differential Equations with Discontinuous Forcing Functions

In this section we turn our attention to some examples in which the nonhomogeneous term, or forcing function, is discontinuous.

EXAMPLE 1

Find the solution of the differential equation

$$y'' + y' + \tfrac{5}{4}y = g(t), \tag{1}$$

where

$$g(t) = 1 - u_\pi(t) = \begin{cases} 1, & 0 \le t < \pi, \\ 0, & t \ge \pi. \end{cases} \tag{2}$$

Assume that the initial conditions are

$$y(0) = 0, \qquad y'(0) = 0. \tag{3}$$

This problem governs the charge on the condenser in a simple electric circuit with an impressed voltage in the form of a rectangular pulse. Alternatively, y may represent the response of a damped oscillator subject to the applied force $g(t)$.

The Laplace transform of Eq. (1) is

$$\begin{aligned} s^2Y(s) - sy(0) - y'(0) + sY(s) - y(0) + \tfrac{5}{4}Y(s) &= \mathscr{L}\{1\} - \mathscr{L}\{u_\pi(t)\} \\ &= (1 - e^{-\pi s})/s. \end{aligned}$$

Introducing the initial values (3) and solving for $Y(s)$, we obtain

$$Y(s) = \frac{1 - e^{-\pi s}}{s\left(s^2 + s + \frac{5}{4}\right)}. \tag{4}$$

To find $y = \phi(t)$ it is convenient to write $Y(s)$ as

$$Y(s) = (1 - e^{-\pi s})H(s), \tag{5}$$

where

$$H(s) = 1/s\left(s^2 + s + \tfrac{5}{4}\right). \tag{6}$$

Then, if $h(t) = \mathscr{L}^{-1}\{H(s)\}$, we have

$$y = h(t) - u_\pi(t)h(t - \pi). \tag{7}$$

Observe that we have used Theorem 6.4 to write the inverse transform of $e^{-\pi s}H(s)$. Finally, to determine $h(t)$ we use the partial fraction expansion of $H(s)$:

$$H(s) = \frac{a}{s} + \frac{bs + c}{s^2 + s + \frac{5}{4}}. \tag{8}$$

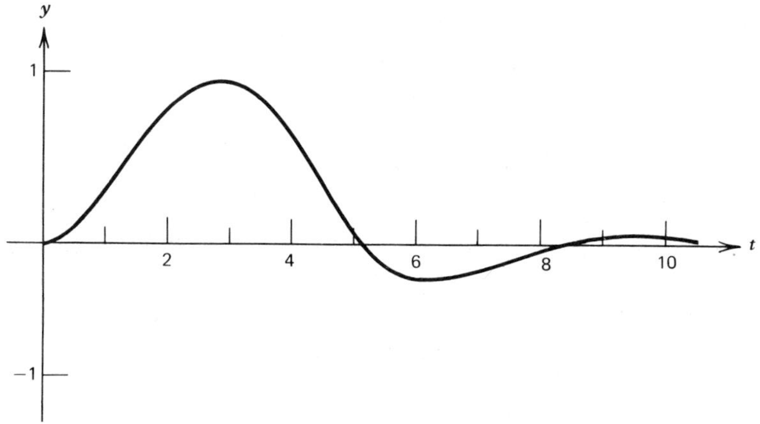

FIGURE 6.10 Solution of initial value problem (1), (2), (3).

Upon determining the coefficients we find that $a = \frac{4}{5}$, $b = -\frac{4}{5}$, and $c = -\frac{4}{5}$. Thus

$$H(s) = \frac{4/5}{s} - \frac{4}{5}\frac{s+1}{s^2+s+\frac{5}{4}}$$

$$= \frac{4/5}{s} - \frac{4}{5}\frac{s+\frac{1}{2}+\frac{1}{2}}{\left[s+\frac{1}{2}\right]^2+1}, \tag{9}$$

so that, by referring to Table 6.1, we obtain

$$h(t) = \tfrac{4}{5} - \tfrac{4}{5}\left(e^{-t/2}\cos t + \tfrac{1}{2}e^{-t/2}\sin t\right). \tag{10}$$

The graph of $y = \phi(t)$ from Eqs. (7) and (10) is shown in Figure 6.10.

The effect of the discontinuity in the forcing function can be seen if we examine the solution $\phi(t)$ of Example 1 more closely. By writing $\phi(t)$ without use of the unit step function, we have

$$\phi(t) = \begin{cases} \frac{4}{5} - \left(\frac{4}{5}e^{-t/2}\cos t + \frac{2}{5}e^{-t/2}\sin t\right), & t < \pi, \\ -(1+e^{\pi/2})\left(\frac{4}{5}e^{-t/2}\cos t + \frac{2}{5}e^{-t/2}\sin t\right), & t \ge \pi. \end{cases} \tag{11}$$

According to the existence and uniqueness Theorem 3.2 the solution ϕ and its first two derivatives are continuous except possibly at the point $t = \pi$. This can also be seen at once from Eq. (7). One can also show by direct computation from Eq. (11) that ϕ and ϕ' are continuous even at $t = \pi$. However, if we calculate ϕ'', we obtain

$$\phi''(t) = \begin{cases} e^{-t/2}\cos t - \frac{1}{2}e^{-t/2}\sin t, & t < \pi, \\ (1+e^{\pi/2})\left(e^{-t/2}\cos t - \frac{1}{2}e^{-t/2}\sin t\right), & t > \pi. \end{cases} \tag{12}$$

Consequently,

$$\lim_{t\to\pi-}\phi''(t) = -e^{-\pi/2}, \qquad \lim_{t\to\pi+}\phi''(t) = -(1+e^{\pi/2})e^{-\pi/2}.$$

Therefore at $t = \pi$ the function ϕ'' has a jump of

$$\lim_{t\to\pi+}\phi''(t) - \lim_{t\to\pi-}\phi''(t) = -(1+e^{\pi/2})e^{-\pi/2} + e^{-\pi/2} = -1.$$

Observe that this is exactly the same as the jump in the forcing function at the same point.

Consider now the general second order linear equation,

$$y'' + p(t)y' + q(t)y = g(t), \tag{13}$$

where p and q are continuous on some interval $\alpha < t < \beta$, but g is only piecewise continuous there. If $y = \psi(t)$ is a solution of Eq. (13) then ψ and ψ' will be continuous on $\alpha < t < \beta$, but ψ'' will have jump discontinuities at the same points as g. Similar remarks apply to higher order equations; the highest derivative of the solution appearing in the differential equation will have jump discontinuities at the same points as the forcing function, but the solution itself and its lower derivatives will be continuous even at those points.

An alternative way to solve the initial value problem (1), (2), and (3) is to solve the differential equation in the two separate intervals $0 < t < \pi$ and $t > \pi$. In the first of these intervals the arbitrary constants in the general solution are chosen as to satisfy the given initial conditions. In the second interval the constants are chosen so as to make the solution ϕ and its derivative ϕ' continuous at $t = \pi$. Note that there are not enough constants available to require that ϕ'' also be continuous there. The same general procedure can be used in other problems having piecewise continuous forcing functions. However, once the necessary theory is developed, it is usually easier to use the Laplace transform to solve such problems all at once.

EXAMPLE 2

Find the solution of the initial value problem

$$y'' + 4y = g(t), \tag{14}$$

$$y(0) = 0, \qquad y'(0) = 0, \tag{15}$$

where

$$g(t) = \begin{cases} t, & 0 \le t < \pi/2, \\ \pi/2, & t \ge \pi/2. \end{cases} \tag{16}$$

In this example the forcing function has the form shown in Figure 6.11 and is known as ramp loading. It is convenient to introduce the unit step function and

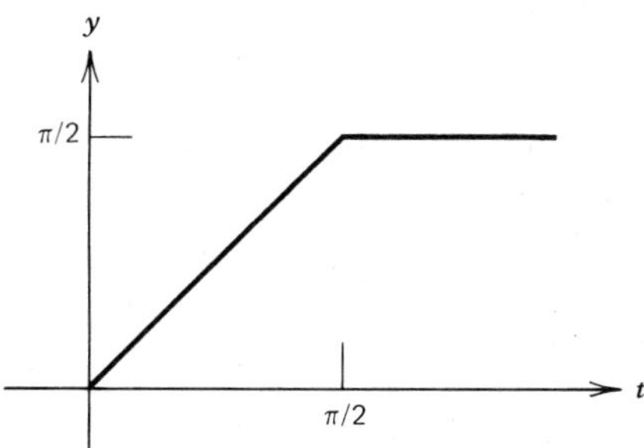

FIGURE 6.11 Ramp loading.

to write

$$g(t) = t - u_{\pi/2}(t)(t - \pi/2), \tag{17}$$

as the reader may verify. To solve the problem we take the Laplace transform of the differential equation and use the initial conditions, thereby obtaining

$$(s^2 + 4)Y(s) = (1 - e^{-\pi s/2})/s^2,$$

or

$$Y(s) = (1 - e^{-\pi s/2})H(s), \tag{18}$$

where

$$H(s) = 1/s^2(s^2 + 4). \tag{19}$$

Thus the solution of the initial value problem (14), (15) is

$$y = \phi(t) = h(t) - u_{\pi/2}(t)h(t - \pi/2), \tag{20}$$

where $h(t)$ is the inverse transform of $H(s)$. The partial fraction expansion of $H(s)$ is

$$H(s) = \frac{1/4}{s^2} - \frac{1/4}{s^2 + 4}, \tag{21}$$

and it then follows from Table 6.1 that

$$h(t) = \tfrac{1}{4}t - \tfrac{1}{8}\sin 2t. \tag{22}$$

The graph of $y = \phi(t)$ is shown in Figure 6.12.

Note that in this example the forcing function g is continuous but g' is discontinuous at $t = \pi/2$. It follows that the solution ϕ and its first two derivatives are continuous, but ϕ''' has a discontinuity at $t = \pi/2$ that matches the discontinuity in g' at that point.

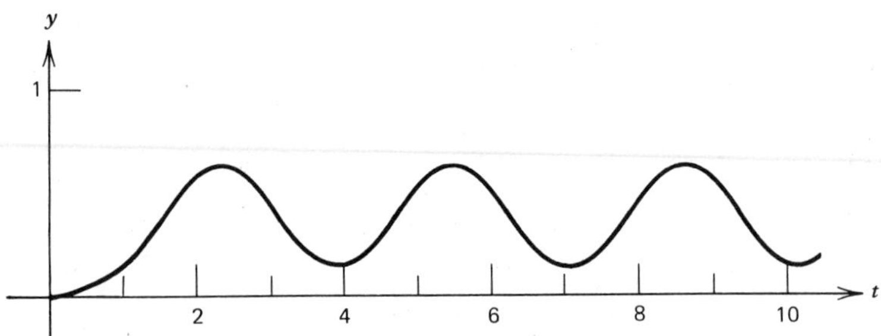

FIGURE 6.12 Solution of initial value problem (14), (15), (16).

PROBLEMS

In each of Problems 1 through 13 find the solution of the given initial value problem.

1. $y'' + y = f(t);\quad f(t) = \begin{cases} 1, & 0 \le t < \pi/2 \\ 0, & \pi/2 \le t < \infty \end{cases}$
   $y(0) = 0,\quad y'(0) = 1$

2. $y'' + 2y' + 2y = h(t);\quad h(t) = \begin{cases} 1, & \pi \le t < 2\pi \\ 0, & 0 \le t < \pi \text{ and } t \ge 2\pi \end{cases}$
   $y(0) = 0,\quad y'(0) = 1$

3. $y'' + 4y = \sin t - u_{2\pi}(t)\sin(t - 2\pi);\quad y(0) = 0,\quad y'(0) = 0$

4. $y'' + 4y = \sin t + u_{\pi}(t)\sin(t - \pi);\quad y(0) = 0,\quad y'(0) = 0$

5. $y'' + 2y' + y = f(t);\quad f(t) = \begin{cases} 1, & 0 \le t < 1 \\ 0, & t \ge 1 \end{cases}$
   $y(0) = 1,\quad y'(0) = 0$

6. $y'' + 3y' + 2y = u_2(t);\quad y(0) = 0,\quad y'(0) = 1$

7. $y'' + y = u_{\pi}(t);\quad y(0) = 1,\quad y'(0) = 0$

8. $y'' + y' + \frac{5}{4}y = t - u_{\pi/2}(t)(t - \pi/2);\quad y(0) = 0,\quad y'(0) = 0$

9. $y'' + y = g(t);\quad g(t) = \begin{cases} t, & 0 \le t < 1 \\ 1, & t \ge 1 \end{cases}$
   $y(0) = 0,\quad y'(0) = 1$

10. $y'' + y' + \frac{5}{4}y = g(t);\quad g(t) = \begin{cases} \sin t, & 0 \le t < \pi \\ 0, & t \ge \pi \end{cases}$
    $y(0) = 0,\quad y'(0) = 0$

11. $y'' + 4y = u_{\pi}(t) - u_{2\pi}(t);\quad y(0) = 1,\quad y'(0) = 0$

12. $y^{\text{iv}} - y = u_1(t) - u_2(t);\quad y(0) = 0,\quad y'(0) = 0,\quad y''(0) = 0,\quad y'''(0) = 0$

13. $y^{\text{iv}} + 5y'' + 4y = 1 - u_{\pi}(t);\quad y(0) = 0,\quad y'(0) = 0,\quad y''(0) = 0,\quad y'''(0) = 0$

*14. Let the function f be defined by

$$f(t) = \begin{cases} 1, & 0 \le t < \pi, \\ 0, & \pi \le t < 2\pi, \end{cases}$$

and for all other positive values of t so that $f(t + 2\pi) = f(t)$. That is, f is periodic with period 2π (see Problem 7 in Section 6.3). Find the solution of the initial value problem

$$y'' + y = f(t);\qquad y(0) = 1,\quad y'(0) = 0.$$

6.4 Impulse Functions

In some applications it is necessary to deal with phenomena of an impulsive nature; for example, voltages or forces of large magnitude that act over very short time intervals. Such problems often lead to differential equations of the form

$$ay'' + by' + cy = g(t), \tag{1}$$

where $g(t)$ is large during a short interval $t_0 - \tau < t < t_0 + \tau$ and is otherwise zero.

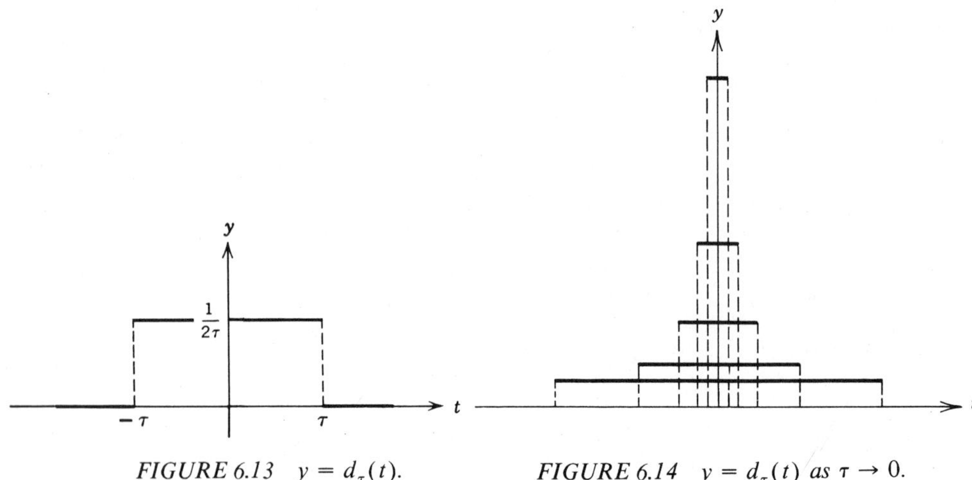

FIGURE 6.13 $y = d_\tau(t)$.

FIGURE 6.14 $y = d_\tau(t)$ as $\tau \to 0$.

The integral $I(\tau)$, defined by

$$I(\tau) = \int_{t_0-\tau}^{t_0+\tau} g(t)\, dt, \tag{2}$$

or, since $g(t) = 0$ outside of the interval $(t_0 - \tau, t_0 + \tau)$,

$$I(\tau) = \int_{-\infty}^{\infty} g(t)\, dt, \tag{3}$$

is a measure of the strength of the forcing function. In a mechanical system, where $g(t)$ is a force, then $I(\tau)$ is the total *impulse* of the force $g(t)$ over the time interval $(t_0 - \tau, t_0 + \tau)$. Similarly, if y is the current in an electric circuit and $g(t)$ is the time derivative of the voltage, then $I(\tau)$ represents the total voltage impressed on the circuit during the interval $(t_0 - \tau, t_0 + \tau)$.

In particular, let us suppose that t_0 is zero, and that $g(t)$ is given by

$$g(t) = d_\tau(t) = \begin{cases} 1/2\tau, & -\tau < t < \tau, \\ 0, & t \leq -\tau \text{ or } t \geq \tau, \end{cases} \tag{4}$$

where τ is a small positive constant (see Figure 6.13). According to Eq. (2) or (3) it follows immediately that in this case $I(\tau) = 1$ independent of the value of τ, as long as $\tau \neq 0$. Now let us idealize the forcing function d_τ by prescribing it to act over shorter and shorter time intervals; that is, we require that $\tau \to 0$, as indicated in Figure 6.14. As a result of this limiting operation we obtain

$$\lim_{\tau \to 0} d_\tau(t) = 0, \qquad t \neq 0. \tag{5}$$

Further, since $I(\tau) = 1$ for each $\tau \neq 0$, it follows that

$$\lim_{\tau \to 0} I(\tau) = 1. \tag{6}$$

Equations (5) and (6) can be used to define an idealized "unit impulse function" δ, which imparts an impulse of magnitude one at $t = 0$, but is zero for all values of t other than zero. That is, the "function" δ is defined to have the properties

$$\delta(t) = 0, \qquad t \neq 0; \tag{7}$$

$$\int_{-\infty}^{\infty} \delta(t)\, dt = 1. \tag{8}$$

Clearly, there is no ordinary function of the kind studied in elementary calculus that satisfies both of Eqs. (7) and (8). The "function" δ, defined by those equations, is an example of what are known as *generalized functions* and is usually called the Dirac[3] "delta function." It is frequently convenient to represent impulsive forcing functions by means of the delta function. Since $\delta(t)$ corresponds to a unit impulse at $t = 0$, a unit impulse at an arbitrary point $t = t_0$ is given by $\delta(t - t_0)$. From Eqs. (7) and (8) it follows that

$$\delta(t - t_0) = 0, \qquad t \neq t_0; \tag{9}$$

$$\int_{-\infty}^{\infty} \delta(t - t_0)\, dt = 1. \tag{10}$$

The delta function does not satisfy the conditions of Theorem 6.2, but its Laplace transform can nevertheless be formally defined. Since $\delta(t)$ is defined as the limit of $d_\tau(t)$ as $\tau \to 0$, it is natural to define the Laplace transform of δ as a similar limit of the transform of d_τ. In particular, we will assume that $t_0 > 0$, and define $\mathscr{L}\{\delta(t - t_0)\}$ by the equation

$$\mathscr{L}\{\delta(t - t_0)\} = \lim_{\tau \to 0} \mathscr{L}\{d_\tau(t - t_0)\}. \tag{11}$$

To evaluate the limit in Eq. (11) we first observe that if $\tau < t_0$, which must eventually be the case as $\tau \to 0$, then $t_0 - \tau > 0$. Since $d_\tau(t - t_0)$ is nonzero only in the interval from $t_0 - \tau$ to $t_0 + \tau$, we have

$$\begin{aligned} \mathscr{L}\{d_\tau(t - t_0)\} &= \int_0^{\infty} e^{-st} d_\tau(t - t_0)\, dt \\ &= \int_{t_0 - \tau}^{t_0 + \tau} e^{-st} d_\tau(t - t_0)\, dt. \end{aligned}$$

Substituting for $d_\tau(t - t_0)$ from Eq. (4), we obtain

$$\begin{aligned} \mathscr{L}\{d_\tau(t - t_0)\} &= \frac{1}{2\tau} \int_{t_0 - \tau}^{t_0 + \tau} e^{-st}\, dt = -\frac{1}{2\tau s} e^{-st} \Bigg|_{t = t_0 - \tau}^{t = t_0 + \tau} \\ &= \frac{1}{2\tau s} e^{-st_0}\left(e^{s\tau} - e^{-s\tau}\right) \end{aligned}$$

or

$$\mathscr{L}\{d_\tau(t - t_0)\} = \frac{\sinh s\tau}{s\tau} e^{-st_0}. \tag{12}$$

[3] Paul A. M. Dirac (1902–1984) was awarded the Nobel Prize (with Erwin Schrödinger) in 1933 for his work in quantum mechanics.

The quotient $(\sinh s\tau)/s\tau$ is indeterminate as $\tau \to 0$, but its limit can be evaluated by L'Hospital's rule. We obtain

$$\lim_{\tau \to 0} \frac{\sinh s\tau}{s\tau} = \lim_{\tau \to 0} \frac{s \cosh s\tau}{s} = 1.$$

Then from Eq. (11) it follows that

$$\mathscr{L}\{\delta(t - t_0)\} = e^{-st_0}. \tag{13}$$

Equation (13) defines $\mathscr{L}\{\delta(t - t_0)\}$ for any $t_0 > 0$. We extend this result, to allow t_0 to be zero, by letting $t_0 \to 0$ on the right side of Eq. (13); thus

$$\mathscr{L}\{\delta(t)\} = \lim_{t_0 \to 0} e^{-st_0} = 1. \tag{14}$$

In a similar way it is possible to define the integral of the product of the delta function and any continuous function f. We have

$$\int_{-\infty}^{\infty} \delta(t - t_0) f(t)\, dt = \lim_{\tau \to 0} \int_{-\infty}^{\infty} d_\tau(t - t_0) f(t)\, dt. \tag{15}$$

Using the definition (4) of $d_\tau(t)$ and the mean value theorem for integrals, we find that

$$\begin{aligned} \int_{-\infty}^{\infty} d_\tau(t - t_0) f(t)\, dt &= \frac{1}{2\tau} \int_{t_0 - \tau}^{t_0 + \tau} f(t)\, dt \\ &= \frac{1}{2\tau} \cdot 2\tau \cdot f(t^*) = f(t^*), \end{aligned}$$

where $t_0 - \tau < t^* < t_0 + \tau$. Hence, $t^* \to t_0$ as $\tau \to 0$, and it follows from Eq. (15) that

$$\int_{-\infty}^{\infty} \delta(t - t_0) f(t)\, dt = f(t_0). \tag{16}$$

It is often convenient to introduce the delta function when working with impulse problems, and to operate formally on it as though it were a function of the ordinary kind. This is illustrated in the example below. It is important to realize, however, that the ultimate justification of such procedures must rest on a careful analysis of the limiting operations involved. Such a rigorous mathematical theory has been developed, but it is too difficult to discuss here.

EXAMPLE

Find the solution of the initial value problem

$$y'' + 2y' + 2y = \delta(t - \pi), \tag{17}$$

$$y(0) = 0, \qquad y'(0) = 0. \tag{18}$$

This initial value problem might arise from the study of a certain electric circuit to which a unit voltage impulse is applied at time $t = \pi$.

To solve the given problem we take the Laplace transform of the differential equation, obtaining

$$(s^2 + 2s + 2)Y(s) = e^{-\pi s},$$

where the initial conditions (18) have been used. Thus

$$Y(s) = \frac{e^{-\pi s}}{s^2 + 2s + 2} = e^{-\pi s}\frac{1}{(s+1)^2 + 1}. \tag{19}$$

By Theorem 6.5 or from Table 6.1

$$\mathscr{L}^{-1}\left\{\frac{1}{(s+1)^2 + 1}\right\} = e^{-t}\sin t. \tag{20}$$

Hence, by Theorem 6.4 we have

$$y = \mathscr{L}^{-1}\{Y(s)\} = u_\pi(t)e^{-(t-\pi)}\sin(t - \pi), \tag{21}$$

which is the formal solution of the given problem. It is also possible to write y in the form

$$y = \begin{cases} 0, & t < \pi, \\ e^{-(t-\pi)}\sin(t - \pi), & t \geq \pi. \end{cases} \tag{22}$$

The graph of Eq. (22) is shown in Figure 6.15.

Since the initial conditions at $t = 0$ are homogeneous, and there is no external excitation until $t = \pi$, there is no response in the interval $0 < t < \pi$. The impulse at $t = \pi$ produces a response that persists indefinitely, although it decays exponentially in the absence of any further external excitation. The response is continuous at $t = \pi$ despite the singularity in the forcing function at that point. However, both the first and second derivatives of the solution have jump discontinuities at $t = \pi$. This is required by the differential equation (17) since a singularity on one side of the equation must be balanced by one on the other.

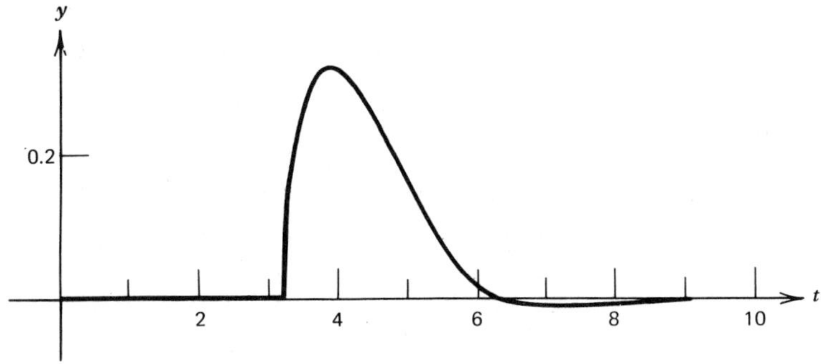

FIGURE 6.15 Solution of initial value problem (17), (18).

PROBLEMS

In each of Problems 1 through 12 find the solution of the given initial value problem by means of the Laplace transform.

1. $y'' + 2y' + 2y = \delta(t - \pi)$; $y(0) = 1$, $y'(0) = 0$
2. $y'' + 4y = \delta(t - \pi) - \delta(t - 2\pi)$; $y(0) = 0$, $y'(0) = 0$
3. $y'' + 2y' + y = \delta(t) + u_{2\pi}(t)$; $y(0) = 0$, $y'(0) = 1$
4. $y'' - y = 2\delta(t - 1)$; $y(0) = 1$, $y'(0) = 0$
5. $y'' + 2y' + 3y = \sin t + \delta(t - \pi)$; $y(0) = 0$, $y'(0) = 1$
6. $y'' + \omega^2 y = \delta(t - \pi/\omega)$; $y(0) = 1$, $y'(0) = 0$
7. $y'' + y = \delta(t - \pi)\cos t$; $y(0) = 0$, $y'(0) = 1$
8. $y'' + 4y = 2\delta(t - \pi/4)$; $y(0) = 0$, $y'(0) = 0$
9. $y'' + y = u_{\pi/2}(t) + \delta(t - \pi) - u_{3\pi/2}(t)$; $y(0) = 0$, $y'(0) = 0$
10. $y'' + 4y = 4\delta(t - \pi/6)\sin t$; $y(0) = 0$, $y'(0) = 0$
11. $y'' + 2y' + 2y = \cos t + \delta(t - \pi/2)$; $y(0) = 0$, $y'(0) = 0$
12. $y^{\text{iv}} - y = \delta(t - 1)$; $y(0) = 0$, $y'(0) = 0$, $y''(0) = 0$, $y'''(0) = 0$

*13. (a) Show by the method of variation of parameters that the solution of the initial value problem

$$y'' + 2y' + 2y = f(t); \qquad y(0) = 0, \quad y'(0) = 0$$

is

$$y = \int_0^t e^{-(t-\tau)} f(\tau)\sin(t - \tau)\, d\tau.$$

(b) Show that if $f(t) = \delta(t - \pi)$, then the solution of part (a) reduces to

$$y = u_\pi(t)e^{-(t-\pi)}\sin(t - \pi),$$

which agrees with Eq. (21) of the text.

6.5 *The Convolution Integral*

Sometimes it is possible to identify a Laplace transform $H(s)$ as the product of two other transforms $F(s)$ and $G(s)$, the latter transforms corresponding to known functions f and g, respectively. In this event, we might anticipate that $H(s)$ would be the transform of the product of f and g. However, this is not the case; in other words, the Laplace transform cannot be commuted with ordinary multiplication. On the other hand, if an appropriately defined "generalized product" is introduced, then the situation changes, as stated in the following theorem.

Theorem 6.6. *If* $F(s) = \mathscr{L}\{f(t)\}$ *and* $G(s) = \mathscr{L}\{g(t)\}$ *both exist for* $s > a \geq 0$, *then*

$$H(s) = F(s)G(s) = \mathscr{L}\{h(t)\}, \qquad s > a, \tag{1}$$

where

$$h(t) = \int_0^t f(t - \tau)g(\tau)\, d\tau = \int_0^t f(\tau)g(t - \tau)\, d\tau. \tag{2}$$

The function h is known as the convolution of f and g; the integrals in Eq. (2) are known as convolution integrals.

The equality of the two integrals in Eq. (2) follows by making the change of variable $t - \tau = \xi$ in the first integral. Before giving the proof of this theorem let us make some observations about the convolution integral. According to this theorem, the transform of the convolution of two functions, rather than the transform of their ordinary product, is given by the product of the separate transforms. It is conventional to emphasize that the convolution integral can be thought of as a "generalized product" by writing

$$h(t) = (f * g)(t). \tag{3}$$

In particular, the notation $(f * g)(t)$ serves to indicate the first integral appearing in Eq. (2).

The convolution $f * g$ has many of the properties of ordinary multiplication. For example, it is relatively simple to show that

$$f * g = g * f \qquad \text{(commutative law)} \tag{4}$$

$$f * (g_1 + g_2) = f * g_1 + f * g_2 \qquad \text{(distributive law)} \tag{5}$$

$$(f * g) * h = f * (g * h) \qquad \text{(associative law)} \tag{6}$$

$$f * 0 = 0 * f = 0. \tag{7}$$

The proofs of these properties are left to the reader. However, there are other properties of ordinary multiplication which the convolution integral does not have. For example, it is not true in general that $f * 1$ is equal to f. To see this, note that

$$(f * 1)(t) = \int_0^t f(t - \tau) \cdot 1\, d\tau = \int_0^t f(t - \tau)\, d\tau.$$

If, for example, $f(t) = \cos t$, then

$$\begin{aligned} (f * 1)(t) &= \int_0^t \cos(t - \tau)\, d\tau = -\sin(t - \tau)\Big|_{\tau=0}^{\tau=t} \\ &= -\sin 0 + \sin t \\ &= \sin t. \end{aligned}$$

Clearly, $(f * 1)(t) \neq f(t)$. Similarly, it may not be true that $f * f$ is nonnegative. See Problem 3 for an example.

Convolution integrals arise in various applications in which the behavior of the system at time t depends not only on its state at time t, but on its past history as well. Systems of this kind are sometimes called hereditary systems, and occur in such diverse fields as neutron transport, viscoelasticity, and population dynamics.

Turning now to the proof of Theorem 6.6, we note first that if

$$F(s) = \int_0^\infty e^{-s\xi} f(\xi)\, d\xi$$

and

$$G(s) = \int_0^\infty e^{-s\eta} g(\eta)\, d\eta,$$

then

$$F(s)G(s) = \int_0^\infty e^{-s\xi} f(\xi)\, d\xi \int_0^\infty e^{-s\eta} g(\eta)\, d\eta. \tag{8}$$

Since the integrand of the first integral does not depend on the integration variable of the second, we can write $F(s)G(s)$ as an iterated integral,

$$F(s)G(s) = \int_0^\infty g(\eta)\, d\eta \int_0^\infty e^{-s(\xi+\eta)} f(\xi)\, d\xi. \tag{9}$$

This expression can be put into a more convenient form by introducing new variables of integration. First let $\xi = t - \eta$, for fixed η. Then the integral with respect to ξ in Eq. (9) is transformed into one with respect to t; hence

$$F(s)G(s) = \int_0^\infty g(\eta)\, d\eta \int_\eta^\infty e^{-st} f(t-\eta)\, dt. \tag{10}$$

Next let $\eta = \tau$; then Eq. (10) becomes

$$F(s)G(s) = \int_0^\infty g(\tau)\, d\tau \int_\tau^\infty e^{-st} f(t-\tau)\, dt. \tag{11}$$

The integral on the right side of Eq. (11) is carried out over the shaded wedge-shaped region extending to infinity in the $t\tau$ plane shown in Figure 6.16. Assuming that the order of integration can be reversed, we finally obtain

$$F(s)G(s) = \int_0^\infty e^{-st}\, dt \int_0^t f(t-\tau) g(\tau)\, d\tau, \tag{12}$$

or

$$\begin{aligned} F(s)G(s) &= \int_0^\infty e^{-st} h(t)\, dt \\ &= \mathscr{L}\{h(t)\}, \end{aligned} \tag{13}$$

where $h(t)$ is defined by Eq. (2). This completes the proof of Theorem 6.6.

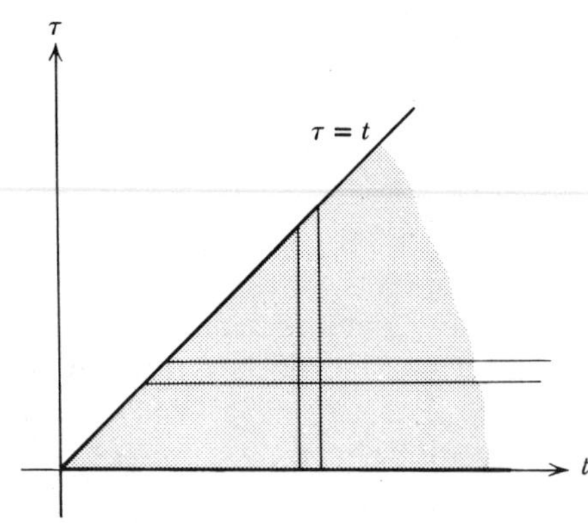

FIGURE 6.16 Region of integration in F(s)G(s).

EXAMPLE 1

Find the inverse transform of

$$H(s) = \frac{a}{s^2(s^2 + a^2)}. \tag{14}$$

It is convenient to think of $H(s)$ as the product of s^{-2} and $a/(s^2 + a^2)$, which, according to Table 6.1, are the transforms of t and $\sin at$, respectively. Hence, by Theorem 6.6, the inverse transform of $H(s)$ is

$$h(t) = \int_0^t (t - \tau) \sin a\tau \, d\tau = \frac{at - \sin at}{a^2}. \tag{15}$$

The reader may show that the same result is obtained if $h(t)$ is written in the alternate form

$$h(t) = \int_0^t \tau \sin a(t - \tau) \, d\tau,$$

which verifies Eq. (2) in this case. Of course, $h(t)$ can also be found by expanding $H(s)$ in partial fractions.

EXAMPLE 2

Find the solution of the initial value problem

$$y'' + 4y = g(t), \tag{16}$$

$$y(0) = 3, \qquad y'(0) = -1. \tag{17}$$

By taking the Laplace transform of the differential equation and using the initial conditions, we obtain

$$s^2Y(s) - 3s + 1 + 4Y(s) = G(s),$$

or

$$Y(s) = \frac{3s - 1}{s^2 + 4} + \frac{G(s)}{s^2 + 4}. \tag{18}$$

Observe that the first and second terms on the right side of Eq. (18) contain the dependence of $Y(s)$ on the initial conditions and forcing function, respectively. It is convenient to write $Y(s)$ in the form

$$Y(s) = 3\frac{s}{s^2 + 4} - \frac{1}{2}\frac{2}{s^2 + 4} + \frac{1}{2}\frac{2}{s^2 + 4}G(s). \tag{19}$$

Then, using Table 6.1 and Theorem 6.6, we obtain

$$y = 3\cos 2t - \tfrac{1}{2}\sin 2t + \tfrac{1}{2}\int_0^t \sin 2(t - \tau) g(\tau) \, d\tau. \tag{20}$$

If a specific forcing function g is given, then the integral in Eq. (20) can be evaluated (by numerical means, if necessary).

Example 2 illustrates the power of the convolution integral as a tool for writing the solution of an initial value problem in terms of an integral. In fact, it is possible to proceed in much the same way in more general problems. Consider the problem consisting of the differential equation

$$ay'' + by' + cy = g(t), \tag{21}$$

where a, b, and c are real constants and g is a given function, together with the initial conditions

$$y(0) = y_0, \qquad y'(0) = y_0'. \tag{22}$$

As we have noted previously, the initial value problem (21), (22) is often referred to as an input–output problem. The coefficients a, b, and c describe the properties of some physical system, and $g(t)$ is the input to the system. The values y_0 and y_0' describe the initial state, and the solution y is the output at timc t.

By taking the Laplace transform of Eq. (21), and using the initial conditions (22), we obtain

$$(as^2 + bs + c)Y(s) - (as + b)y_0 - ay_0' = G(s).$$

If we let

$$\Phi(s) = \frac{(as + b)y_0 + ay_0'}{as^2 + bs + c}, \qquad \Psi(s) = \frac{G(s)}{as^2 + bs + c}, \tag{23}$$

then we can write

$$Y(s) = \Phi(s) + \Psi(s). \tag{24}$$

Consequently,

$$y = \phi(t) + \psi(t), \tag{25}$$

where $\phi(t) = \mathcal{L}^{-1}\{\Phi(s)\}$ and $\psi(t) = \mathcal{L}^{-1}\{\Psi(s)\}$. Observe that $y = \phi(t)$ is the solution of the initial value problem

$$ay'' + by' + cy = 0, \qquad y(0) = y_0, \quad y'(0) = y_0' \tag{26}$$

obtained from Eqs. (21) and (22) by setting $g(t)$ equal to zero. Similarly $y = \psi(t)$ is the solution of

$$ay'' + by' + cy = g(t), \qquad y(0) = 0, \quad y'(0) = 0 \tag{27}$$

in which the initial values y_0 and y_0' are each replaced by zero.

Once specific values of a, b, and c are given, we can find $\phi(t) = \mathcal{L}^{-1}\{\Phi(s)\}$ by using Table 6.1, possibly in conjunction with a translation or a partial fractions expansion. To find $\psi(t) = \mathcal{L}^{-1}\{\Psi(s)\}$ it is convenient to write $\Psi(s)$ as

$$\Psi(s) = H(s)G(s), \tag{28}$$

where $H(s) = (as^2 + bs + c)^{-1}$. The function H is known as the *transfer func-*

tion,[4] and depends only on the properties of the system under consideration; that is, $H(s)$ is determined entirely by the coefficients a, b, and c. On the other hand, $G(s)$ depends only on the external excitation $g(t)$ that is applied to the system. By the convolution theorem we can write

$$\psi(t) = \mathscr{L}^{-1}\{H(s)G(s)\} = \int_0^t h(t-\tau)g(\tau)\,d\tau, \tag{29}$$

where $h(t) = \mathscr{L}^{-1}\{H(s)\}$, and $g(t)$ is the given forcing function.

To obtain a better understanding of the significance of $h(t)$, we consider the case in which $G(s) = 1$; consequently $g(t) = \delta(t)$ and $\Psi(s) = H(s)$. This means that $y = h(t)$ is the solution of the initial value problem

$$ay'' + by' + cy = \delta(t), \qquad y(0) = 0, \quad y'(0) = 0; \tag{30}$$

obtained from Eq. (27) by replacing $g(t)$ by $\delta(t)$. Thus $h(t)$ is the response of the system to a unit impulse applied at $t = 0$, and it is natural to call $h(t)$ the *impulse response* of the system. Equation (29) then says that $\psi(t)$ is the convolution of the impulse response and the forcing function.

Referring to Example 2, we note that in that case the transfer function is $H(s) = 1/(s^2 + 4)$, and the impulse response is $h(t) = (\sin 2t)/2$. Also, the first two terms on the right side of Eq. (20) constitute the function $\phi(t)$, the solution of the corresponding homogeneous equation that satisfies the given initial conditions.

PROBLEMS

1. Establish the commutative, distributive, and associative properties of the convolution integral.

   (a) $f * g = g * f$ (b) $f * (g_1 + g_2) = f * g_1 + f * g_2$
   (c) $f * (g * h) = (f * g) * h$

2. Find an example different from the one in the text showing that $(f * 1)(t)$ need not be equal to $f(t)$.
3. Show, by means of the example $f(t) = \sin t$, that $f * f$ is not necessarily nonnegative.
4. Find the Laplace transform of each of the following functions.

   (a) $f(t) = \int_0^t (t-\tau)^2 \cos 2\tau\, d\tau$ (b) $f(t) = \int_0^t e^{-(t-\tau)} \sin\tau\, d\tau$

   (c) $f(t) = \int_0^t (t-\tau)e^\tau\, d\tau$ (d) $f(t) = \int_0^t \sin(t-\tau)\cos\tau\, d\tau$

[4] This terminology arises from the fact that $H(s)$ is the ratio of the transforms of the output and the input of the problem (27).

5. Find the inverse Laplace transform of each of the following functions by using the convolution theorem.

(a) $F(s) = \dfrac{1}{s^4(s^2+1)}$ (b) $F(s) = \dfrac{s}{(s+1)(s^2+4)}$

(c) $F(s) = \dfrac{1}{(s+1)^2(s^2+4)}$ (d) $F(s) = \dfrac{G(s)}{s^2+1}$

In each of Problems 6 through 13 express the solution of the given initial value problem in terms of a convolution integral.

6. $y'' + \omega^2 y = g(t), \quad y(0) = 0, \quad y'(0) = 1$
7. $y'' + 2y' + 2y = \sin \alpha t, \quad y(0) = 0, \quad y'(0) = 0$
8. $4y'' + 4y' + 17y = g(t), \quad y(0) = 0, \quad y'(0) = 0$
9. $y'' + y' + \frac{5}{4}y = 1 - u_\pi(t), \quad y(0) = 1, \quad y'(0) = -1$
10. $y'' + 4y' + 4y = g(t), \quad y(0) = 2, \quad y'(0) = -3$
11. $y'' + 3y' + 2y = \cos \alpha t, \quad y(0) = 1, \quad y'(0) = 0$
12. $y^{\text{iv}} - y = g(t), \quad y(0) = 0, \quad y'(0) = 0, \quad y''(0) = 0, \quad y'''(0) = 0$
13. $y^{\text{iv}} + 5y'' + 4y = g(t), \quad y(0) = 1, \quad y'(0) = 0, \quad y''(0) = 0, \quad y'''(0) = 0$

14. Consider the equation

$$\phi(t) + \int_0^t k(t-\xi)\phi(\xi)\,d\xi = f(t),$$

in which f and k are known functions, and ϕ is to be determined. Since the unknown function ϕ appears under an integral sign, the given equation is called an *integral equation*; in particular, it belongs to a class of integral equations known as Volterra integral equations. Take the Laplace transform of the given integral equation, and obtain an expression for $\mathscr{L}\{\phi(t)\}$ in terms of the transforms $\mathscr{L}\{f(t)\}$ and $\mathscr{L}\{k(t)\}$ of the given functions f and k. The inverse transform of $\mathscr{L}\{\phi(t)\}$ is the solution of the original integral equation.

15. Consider the Volterra integral equation (see Problem 14)

$$\phi(t) + \int_0^t (t-\xi)\phi(\xi)\,d\xi = \sin 2t.$$

(a) Show that if u is a function such that $u''(t) = \phi(t)$, then

$$u''(t) + u(t) - tu'(0) - u(0) = \sin 2t.$$

(b) Show that the given integral equation is equivalent to the initial value problem

$$u''(t) + u(t) = \sin 2t; \qquad u(0) = 0, \quad u'(0) = 0.$$

(c) Solve the given integral equation by using the Laplace transform.

(d) Solve the initial value problem of part (b) and verify that the solution is the same as that obtained in (c).

*16. **The Tautochrone.** A problem of interest in the history of mathematics is that of finding the *tautochrone*—the curve down which a particle will slide freely under gravity alone, reaching the bottom in the same time regardless of its starting point on the curve. This problem arose in the construction of a clock pendulum whose period is independent of the amplitude of its motion. The tautochrone was found by Christian Huygens (1629–1695) in 1673 by geometrical methods, and later by Leibniz

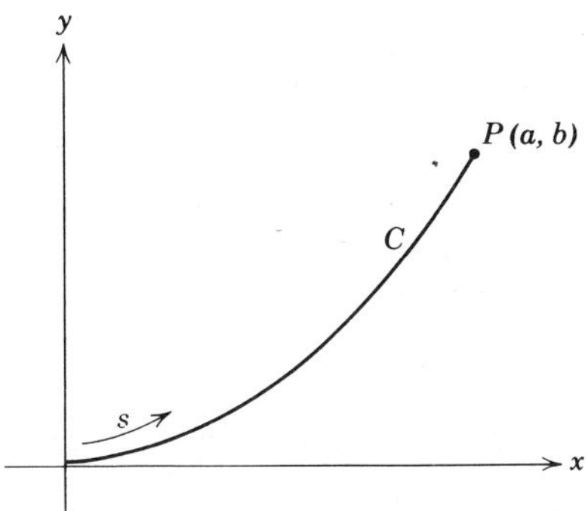

FIGURE 6.17 The tautochrone.

and Jakob Bernoulli using analytical arguments. Bernoulli's solution (in 1690) was one of the first occasions on which a differential equation was explicitly solved.

The geometrical configuration is shown in Figure 6.17. The starting point $P(a, b)$ is joined to the terminal point (0, 0) by the arc C. Arc length s is measured from the origin, and $f(y)$ denotes the rate of change of s with respect to y:

$$f(y) = \frac{ds}{dy} = \left[1 + \left(\frac{dx}{dy}\right)^2\right]^{1/2}. \tag{i}$$

Then it follows from the principle of conservation of energy that the time $T(b)$ required for a particle to slide from P to the origin is

$$T(b) = \frac{1}{\sqrt{2g}} \int_0^b \frac{f(y)}{\sqrt{b-y}}\, dy. \tag{ii}$$

(a) Assume that $T(b) = T_0$, a constant, for each b. By taking the Laplace transform of Eq. (ii) in this case, and using the convolution theorem, show that

$$F(s) = \sqrt{\frac{2g}{\pi}} \frac{T_0}{\sqrt{s}}; \tag{iii}$$

then show that

$$f(y) = \frac{\sqrt{2g}\, T_0}{\pi\sqrt{y}}. \tag{iv}$$

Hint: See Problem 10 of Section 6.1.

(b) Combining Eqs. (i) and (iv), show that

$$\frac{dx}{dy} = \sqrt{\frac{2\alpha - y}{y}}, \tag{v}$$

where $\alpha = gT_0^2/\pi^2$.

(c) Use the substitution $y = 2\alpha \sin^2(\theta/2)$ to solve Eq. (v), and show that

$$x = \alpha(\theta + \sin\theta), \qquad y = \alpha(1 - \cos\theta). \tag{vi}$$

Equations (vi) can be identified as parametric equations of a cycloid. Thus the tautochrone is an arc of a cycloid.

REFERENCES

The books listed below contain additional information on the Laplace transform and its applications:

Churchill, R. V., *Operational Mathematics* (3rd ed.) (New York: McGraw-Hill, 1971).

Doetsch, Gustave, *Guide to the Applications of Laplace Transforms* (London: Van Nostrand, 1967).

Kaplan, W., *Operational Methods for Linear Systems*, (Reading, Mass.: Addison-Wesley, 1962).

McLachlan, N. W., *Modern Operational Calculus* (New York: Dover, 1962).

Miles, J. W., *Integral Transforms in Applied Mathematics* (London: Cambridge University Press, 1971).

Rainville, E. D., *The Laplace Transform: an Introduction* (New York: Macmillan, 1963).

Each of the books mentioned above contains a table of transforms. Extensive tables are also available; see, for example:

Erdelyi, A. (Ed.), *Tables of Integral Transforms* (Vol. 1) (New York: McGraw-Hill, 1954).

Roberts, G. E., and Kaufman, H., *Table of Laplace Transforms* (Philadelphia: Saunders, 1966).

A further discussion of generalized functions can be found in:

Lighthill, M. J., *Fourier Analysis and Generalized Functions* (London: Cambridge University Press, 1958).

CHAPTER

7

SYSTEMS OF FIRST ORDER LINEAR EQUATIONS

7.1 Introduction

This chapter is devoted to a discussion of systems of simultaneous ordinary differential equations. Such systems arise naturally in problems involving several dependent variables, each of which is a function of a single independent variable. We will denote the independent variable by t, and let $x_1, x_2, x_3, \ldots$ represent dependent variables which are functions of t. Differentiation with respect to t will be denoted by a prime.

For example, the motion of a particle in space is governed by Newton's law in three dimensions,

$$\begin{aligned} m\frac{d^2x_1}{dt^2} &= F_1\left(t, x_1, x_2, x_3, \frac{dx_1}{dt}, \frac{dx_2}{dt}, \frac{dx_3}{dt}\right), \\ m\frac{d^2x_2}{dt^2} &= F_2\left(t, x_1, x_2, x_3, \frac{dx_1}{dt}, \frac{dx_2}{dt}, \frac{dx_3}{dt}\right), \\ m\frac{d^2x_3}{dt^2} &= F_3\left(t, x_1, x_2, x_3, \frac{dx_1}{dt}, \frac{dx_2}{dt}, \frac{dx_3}{dt}\right), \end{aligned} \tag{1}$$

where m is the mass of the particle; x_1, x_2, and x_3 are its spatial coordinates; and F_1, F_2, and F_3 are the forces acting on the particle in the x_1, x_2, and x_3

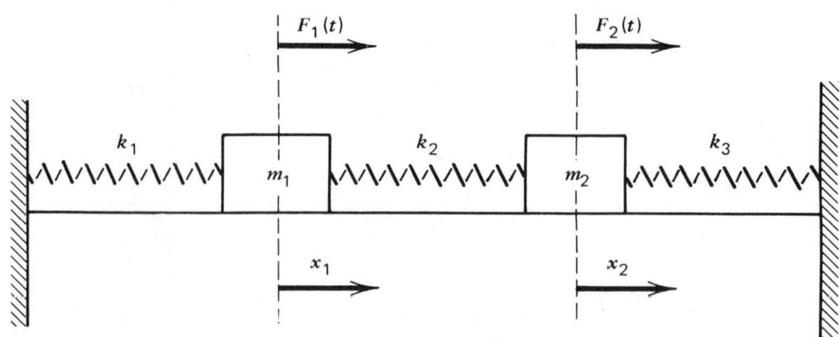

FIGURE 7.1 A two degrees of freedom spring–mass system.

directions, respectively. If the particle is an idealization of a space vehicle, for instance, then F_1, F_2, and F_3 include the gravitational forces exerted on the vehicle by neighboring celestial bodies, any forces produced by the vehicle's own propulsive system, and also drag forces if the vehicle is within the earth's atmosphere.

As another example, consider the spring–mass system shown in Figure 7.1. The two masses move on a frictionless surface under the influence of external forces $F_1(t)$ and $F_2(t)$, and they are also constrained by the three springs whose constants are k_1, k_2, and k_3, respectively. Using arguments similar to those in Section 3.7, we find the following equations for the coordinates x_1 and x_2 of the two masses:

$$\begin{aligned} m_1\frac{d^2x_1}{dt^2} &= k_2(x_2 - x_1) - k_1x_1 + F_1(t) \\ &= -(k_1 + k_2)x_1 + k_2x_2 + F_1(t), \\ m_2\frac{d^2x_2}{dt^2} &= -k_3x_2 - k_2(x_2 - x_1) + F_2(t) \\ &= k_2x_1 - (k_2 + k_3)x_2 + F_2(t). \end{aligned} \tag{2}$$

A derivation of Eqs. (2) is outlined in Problem 6.

Next, consider the parallel *LRC* circuit shown in Figure 7.2. Let V be the voltage drop across the capacitor and I the current through the inductance. Then, referring to Section 3.8 and to Problem 7 of this section, we can show that the voltage and current in this circuit are governed by the system of equations

$$\begin{aligned} \frac{dI}{dt} &= \frac{V}{L}, \\ \frac{dV}{dt} &= -\frac{I}{C} - \frac{V}{RC}, \end{aligned} \tag{3}$$

where L is the inductance, C the capacitance, and R the resistance.

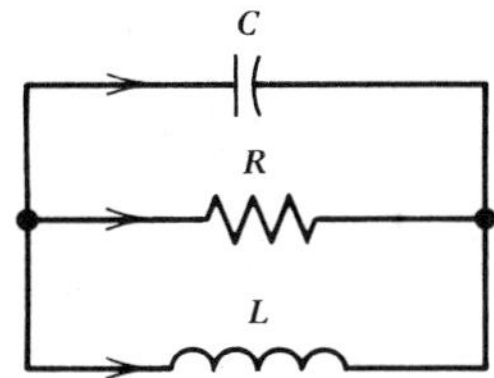

FIGURE 7.2 A parallel LRC *circuit.*

As a final example we mention the predator–prey problem, one of the fundamental problems of mathematical ecology. Let $H(t)$ and $P(t)$ denote the populations at time t of two species, one of which (P) preys upon the other (H). For example, $P(t)$ and $H(t)$ may be the number of foxes and rabbits, respectively, in a woods, or the number of bass and redear (which are eaten by bass) in a pond. Without the prey the predators will decrease, and without the predator the prey will increase. A mathematical model showing how an ecological balance can be maintained when both are present was proposed about 1925 by Lotka and Volterra. This model consists of the system of differential equations

$$\begin{aligned} dH/dt &= a_1 H - b_1 HP, \\ dP/dt &= -a_2 P + b_2 HP. \end{aligned} \tag{4}$$

In Eqs. (4) the coefficient a_1 is the birthrate of the population H; similarly, a_2 is the death rate of the population P. The coefficients b_1 and b_2 are the coefficients of interaction between predator and prey. Equations of the form (4) are sometimes called predator–prey equations. They are discussed in some detail in Section 9.4.

There is an important connection between systems of equations and single equations of arbitrary order. In fact, an nth order equation

$$y^{(n)} = F(t, y, y', \dots, y^{(n-1)}) \tag{5}$$

can always be reduced to a system of n first order equations having a rather special form. To show this we introduce the variables $x_1, x_2, \dots, x_n$ defined by

$$x_1 = y,\ x_2 = y',\ x_3 = y'', \dots, x_n = y^{(n-1)}. \tag{6}$$

It then follows immediately that

$$\begin{aligned} x_1' &= x_2, \\ x_2' &= x_3, \\ &\vdots \\ x_{n-1}' &= x_n, \end{aligned} \tag{7}$$

and from Eq. (5)

$$x_n' = F(t, x_1, x_2, \dots, x_n). \tag{8}$$

Equations (7) and (8) are a special case of the more general system

$$\begin{aligned} x_1' &= F_1(t, x_1, \ldots, x_n), \\ x_2' &= F_2(t, x_1, \ldots, x_n), \\ &\vdots \\ x_n' &= F_n(t, x_1, \ldots, x_n). \end{aligned} \tag{9}$$

In a similar way the systems (1) and (2) can also be reduced to systems of first order equations of the form (9) containing, respectively, six and four equations. The systems (3) and (4) are already in the form (9). In fact, systems of the form (9) include almost all cases of interest, and as a consequence much of the more advanced theory of differential equations is devoted to such systems.

The system of equations (9) is said to have a *solution* on the interval $\alpha < t < \beta$ if there exists a set of n functions

$$x_1 = \phi_1(t), \ldots, x_n = \phi_n(t), \tag{10}$$

which are differentiable at all points in $\alpha < t < \beta$ and which satisfy the system (9) at all points in this interval. In addition to the given system of differential equations, there may also be given initial conditions of the form

$$x_1(t_0) = x_1^0, \quad x_2(t_0) = x_2^0, \ldots, \quad x_n(t_0) = x_n^0, \tag{11}$$

where t_0 is a specified value of t in $\alpha < t < \beta$, and $x_1^0, \ldots, x_n^0$ are prescribed numbers. The differential equations (9) and initial conditions (11) together form an initial value problem.

A solution (10) can be viewed as a set of parametric equations in an n-dimensional space. For a given value of t Eqs. (10) give values for the coordinates $x_1, \ldots, x_n$ of a point in the space. As t changes, the coordinates in general also change. The collection of points corresponding to $\alpha < t < \beta$ form a curve in the space. It is often helpful to think of the curve as the trajectory or path of a particle moving in accordance with the system of differential equations (9). The initial conditions (10) determine the starting point of the moving particle.

To guarantee that the initial value problem (9) and (11) has a solution and that the solution is unique, it is necessary to impose certain conditions on the functions $F_1, F_2, \ldots, F_n$. The following theorem is analogous to the existence and uniqueness theorems for single equations of first and second order, Theorems 2.2 and 3.1, respectively.

Theorem 7.1. *Let each of the functions $F_1, \ldots, F_n$ and the partial derivatives $\partial F_1/\partial x_1, \ldots, \partial F_1/\partial x_n, \ldots, \partial F_n/\partial x_1, \ldots, \partial F_n/\partial x_n$ be continuous in a region R of $tx_1x_2 \cdots x_n$ space containing the point $(t_0, x_1^0, x_2^0, \ldots, x_n^0)$. Then there is an interval $|t - t_0| < h$ in which there exists a unique solution $x_1 = \phi_1(t), \ldots, x_n = \phi_n(t)$ of the system of differential equations* (9), *which also satisfies the initial conditions* (11).

The proof of this theorem can be constructed by generalizing the argument in Section 2.12, but we do not give it here. Note, however, that in the hypotheses of the theorem nothing is said about the partial derivatives of the functions $F_1, F_2, \ldots, F_n$ with respect to t. Also, in the conclusion, the length $2h$ of the interval in which the solution exists is not specified exactly, and in some cases may be very short. Finally, the theorem is not the most general one known, and the given conditions are sufficient, but not necessary, to ensure the conclusion.

If each of the functions $F_1, \ldots, F_n$ in Eqs. (9) is a linear function of the dependent variables $x_1, \ldots, x_n$, then the system of equations is said to be *linear*; otherwise it is *nonlinear*. Thus the most general system of n linear first order equations has the form

$$\begin{aligned} x_1' &= p_{11}(t)x_1 + \cdots + p_{1n}(t)x_n + g_1(t), \\ &\vdots \\ x_n' &= p_{n1}(t)x_1 + \cdots + p_{nn}(t)x_n + g_n(t). \end{aligned} \tag{12}$$

If each of the functions $g_1, \ldots, g_n$ is identically zero, then the system (12) is said to be *homogeneous*; otherwise it is *nonhomogeneous*. Observe that the systems (2) and (3) are both linear; the system (2) is nonhomogeneous unless $F_1(t) = F_2(t) = 0$, while the system (3) is homogeneous. The system (4) is nonlinear. The system (1) may be either linear or nonlinear, depending on the nature of the functions F_1, F_2, and F_3.

For the linear system (12) the existence and uniqueness theorem is simpler and also has a stronger conclusion. It is analogous to Theorems 2.1 and 3.2.

***Theorem** 7.2. If the functions $p_{11}, \ldots, p_{nn}, g_1, \ldots, g_n$ are continuous on an open interval $\alpha < t < \beta$, containing the point $t = t_0$, then there exists a unique solution $x_1 = \phi_1(t), \ldots, x_n = \phi_n(t)$ of the system of differential equations* (12), *which also satisfies the initial conditions* (11). *This solution is valid throughout the interval $\alpha < t < \beta$.*

Note that in contrast to the situation for a nonlinear system the existence and uniqueness of the solution of a linear system is guaranteed throughout the interval in which the hypotheses are satisfied. Furthermore, for a linear system the initial values $x_1^0, \ldots, x_n^0$ at $t = t_0$ are completely arbitrary, whereas in the nonlinear case the initial point must lie in the region R defined in Theorem 7.1.

The rest of this chapter is devoted to systems of linear first order equations (nonlinear systems are included in the discussion in Chapters 8 and 9). Our presentation makes use of matrix notation and assumes that the reader has some familiarity with the properties of matrices. The necessary facts about matrices are summarized in Sections 7.2 and 7.3; more details can be found in any elementary book on linear algebra.

PROBLEMS

1. Reduce each of the systems (1) and (2) to a system of first order equations of the form (9).
2. Consider the initial value problem $u'' + p(t)u' + q(t)u = g(t)$, $u(0) = u_0$, $u'(0) = u'_0$. Transform this problem into an initial value problem for two first order linear equations.
3. Show that if a_{11}, a_{12}, a_{21}, and a_{22} are constants with a_{12} and a_{21} not both zero, and if the functions g_1 and g_2 are differentiable, then the initial value problem
$$x'_1 = a_{11}x_1 + a_{12}x_2 + g_1(t), \qquad x_1(0) = x_1^0$$
$$x'_2 = a_{21}x_1 + a_{22}x_2 + g_2(t), \qquad x_2(0) = x_2^0$$
can be transformed into an initial value problem for a single second order equation. Can the same procedure be carried out if $a_{11}, \ldots, a_{22}$ are functions of t?
4. Consider the linear homogeneous system
$$x' = p_{11}(t)x + p_{12}(t)y,$$
$$y' = p_{21}(t)x + p_{22}(t)y.$$
Show that if $x = x_1(t)$, $y = y_1(t)$ and $x = x_2(t)$, $y = y_2(t)$ are two solutions of the given system, then $x = c_1x_1(t) + c_2x_2(t)$, $y = c_1y_1(t) + c_2y_2(t)$ is also a solution for any constants c_1 and c_2. This is the principle of superposition.
5. Let $x = x_1(t)$, $y = y_1(t)$ and $x = x_2(t)$, $y = y_2(t)$ be any two solutions of the linear nonhomogeneous system
$$x' = p_{11}(t)x + p_{12}(t)y + g_1(t),$$
$$y' = p_{21}(t)x + p_{22}(t)y + g_2(t).$$
Show that $x = x_1(t) - x_2(t)$, $y = y_1(t) - y_2(t)$ is a solution of the corresponding homogeneous system.
6. Equations (2) can be derived by drawing a free-body diagram showing the forces acting on each mass. Figure 7.3a shows the situation when the displacements x_1 and x_2 of the two masses are both positive (to the right) and $x_2 > x_1$. Then springs 1 and 2 are elongated and spring 3 is compressed, giving rise to forces as shown in Figure 7.3b. Use Newton's law ($F = ma$) to derive Eqs. (2).

Electric Circuits. The theory of electric circuits, such as that shown in Figure 7.2, consisting of inductances, resistors, and capacitors, is based on Kirchhoff's laws: (1) The net flow of current through each node (or junction) is zero, (2) the net voltage drop around each closed loop is zero. In addition to Kirchhoff's laws we also have the relation between the current I in amperes through each circuit element and the voltage drop V in volts across the element; namely,

$$V = RI; \qquad R = \text{resistance in ohms}$$
$$C\frac{dV}{dt} = I; \qquad C = \text{capacitance in farads}$$
$$L\frac{dI}{dt} = V; \qquad L = \text{inductance in henrys.}$$

Kirchhoff's laws and the current–voltage relation for each circuit element provide a system of algebraic and differential equations from which the voltage and current throughout the circuit can be determined. Problems 7 through 9 illustrate the procedure described above.

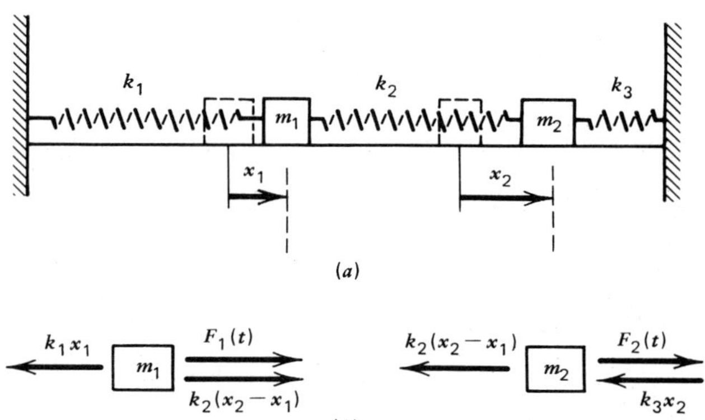

FIGURE 7.3 Free body diagram for spring–mass system.

7. Consider the circuit shown in Figure 7.2. Let I_1, I_2, and I_3 be the current through the capacitor, resistor, and inductance, respectively. Likewise, let V_1, V_2, and V_3 be the corresponding voltage drops. The arrows denote the arbitrarily chosen directions in which currents and voltage drops will be taken to be positive.
(a) Applying Kirchhoff's second law to the upper loop in the circuit, show that

$$V_1 - V_2 = 0. \tag{i}$$

In a similar way, show that

$$V_2 - V_3 = 0. \tag{ii}$$

(b) Applying Kirchhoff's first law to either node in the circuit, show that

$$I_1 + I_2 + I_3 = 0. \tag{iii}$$

(c) Use the current–voltage relation through each element in the circuit to obtain the equations

$$CV_1' = I_1, \qquad V_2 = RI_2, \qquad LI_3' = V_3. \tag{iv}$$

(d) Eliminate V_2, V_3, I_1, and I_2 among Eqs. (i) through (iv) to obtain

$$CV_1' = -I_3 - \frac{V_1}{R}, \qquad LI_3' = V_1. \tag{v}$$

Observe that if we omit the subscripts in Eqs. (v) then we have the system (3) of the text.

8. Consider the circuit shown in Figure 7.4. Use the method outlined in Problem 7 to show that the current I through the inductance and the voltage V across the capacitor satisfy the system of differential equations

$$\frac{dI}{dt} = -I - V,$$

$$\frac{dV}{dt} = 2I - V.$$

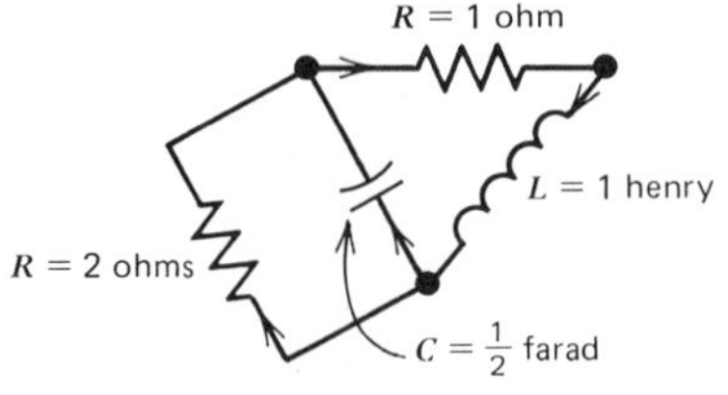

FIGURE 7.4 *The circuit in Problem 8.*

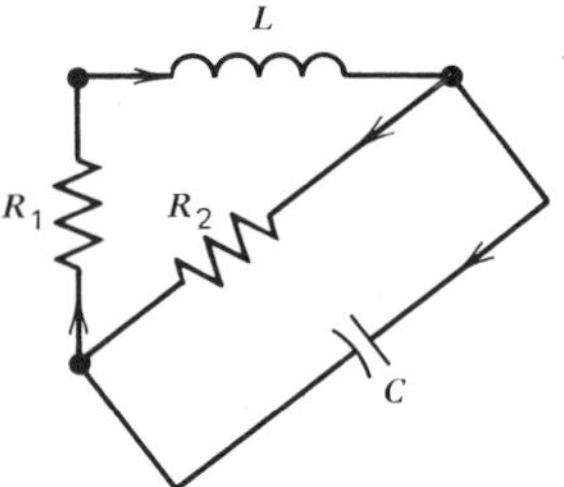

FIGURE 7.5 *The circuit in Problem 9.*

9. Consider the circuit shown in Figure 7.5. Use the method outlined in Problem 7 to show that the current I through the inductance and the voltage V across the capacitor satisfy the system of differential equations

$$L\frac{dI}{dt} = -R_1 I - V,$$

$$C\frac{dV}{dt} = I - \frac{V}{R_2}.$$

7.2 Review of Matrices

For both theoretical and computational reasons it is advisable to bring to bear some of the results of matrix theory on the initial value problem for a system of linear differential equations. For reference purposes this section and the next are devoted to a brief summary[1] of the facts that will be needed later. We assume, however, that the reader is familiar with determinants and how to evaluate them.

We designate matrices by bold-faced capitals $\mathbf{A}, \mathbf{B}, \mathbf{C}, \ldots,$ occasionally using bold-faced Greek capitals $\mathbf{\Phi}, \mathbf{\Psi}, \ldots$. A matrix $\mathbf{A}$ consists of a rectangular array of numbers, or elements, arranged in m rows and n columns, that is,

$$\mathbf{A} = \begin{pmatrix} a_{11} & a_{12} & \cdots & a_{1n} \\ a_{21} & a_{22} & \cdots & a_{2n} \\ \vdots & \vdots & & \vdots \\ a_{m1} & a_{m2} & \cdots & a_{mn} \end{pmatrix}. \tag{1}$$

We speak of $\mathbf{A}$ as an $m \times n$ matrix. Although later in the chapter we will often assume that the elements of certain matrices are real numbers, in this section we assume that the elements of matrices may be complex numbers. The element lying in the ith row and jth column is designated by a_{ij}, the first subscript identifying its row and the second its column. Sometimes the notation (a_{ij}) is used to denote the matrix whose generic element is a_{ij}.

[1]A representative list of books on matrix algebra is given at the end of the chapter.

Associated with each matrix $\mathbf{A}$ is the matrix $\mathbf{A}^T$, known as the *transpose* of $\mathbf{A}$, and obtained from $\mathbf{A}$ by interchanging the rows and columns of $\mathbf{A}$. Thus, if $\mathbf{A} = (a_{ij})$, then $\mathbf{A}^T = (a_{ji})$. Also, we will denote by $\bar{a}_{ij}$ the complex conjugate of a_{ij}, and by $\overline{\mathbf{A}}$ the matrix obtained from $\mathbf{A}$ by replacing each element a_{ij} by its conjugate $\bar{a}_{ij}$. The matrix $\overline{\mathbf{A}}$ is called the *conjugate* of $\mathbf{A}$. It will also be necessary to consider the transpose of the conjugate matrix $\overline{\mathbf{A}}^T$. This matrix is called the *adjoint* of $\mathbf{A}$ and will be denoted by $\mathbf{A}^*$.

For example, let

$$\mathbf{A} = \begin{pmatrix} 3 & 2 - i \\ 4 + 3i & -5 + 2i \end{pmatrix}.$$

Then

$$\mathbf{A}^T = \begin{pmatrix} 3 & 4 + 3i \\ 2 - i & -5 + 2i \end{pmatrix}, \qquad \overline{\mathbf{A}} = \begin{pmatrix} 3 & 2 + i \\ 4 - 3i & -5 - 2i \end{pmatrix},$$

$$\mathbf{A}^* = \begin{pmatrix} 3 & 4 - 3i \\ 2 + i & -5 - 2i \end{pmatrix}.$$

We are particularly interested in two somewhat special kinds of matrices: square matrices, which have the same number of rows and columns, that is, $m = n$; and vectors (or column vectors), which can be thought of as $n \times 1$ matrices, or matrices having only one column. Square matrices having n rows and n columns are said to be of order n. We denote (column) vectors by bold-faced lowercase letters $\mathbf{x}, \mathbf{y}, \boldsymbol{\xi}, \boldsymbol{\eta}, \ldots$. The transpose $\mathbf{x}^T$ of an $n \times 1$ column vector is a $1 \times n$ row vector; that is, the matrix consisting of one row whose elements are the same as the elements in the corresponding positions of $\mathbf{x}$.

ALGEBRAIC PROPERTIES. 1. **Equality.** Two $m \times n$ matrices $\mathbf{A}$ and $\mathbf{B}$ are said to be equal if corresponding elements are equal; that is, if $a_{ij} = b_{ij}$ for each i and j.

2. **Zero.** The symbol $\mathbf{0}$ will be used to denote the matrix (or vector) each of whose elements is zero.

3. **Addition.** The sum of two $m \times n$ matrices $\mathbf{A}$ and $\mathbf{B}$ is defined as the matrix obtained by adding corresponding elements:

$$\mathbf{A} + \mathbf{B} = (a_{ij}) + (b_{ij}) = (a_{ij} + b_{ij}). \tag{2}$$

With this definition it follows that matrix addition is commutative and associative, so that

$$\mathbf{A} + \mathbf{B} = \mathbf{B} + \mathbf{A}, \qquad \mathbf{A} + (\mathbf{B} + \mathbf{C}) = (\mathbf{A} + \mathbf{B}) + \mathbf{C}. \tag{3}$$

4. **Multiplication by a Number.** The product of a matrix $\mathbf{A}$ by a complex number α is defined as follows:

$$\alpha\mathbf{A} = \alpha(a_{ij}) = (\alpha a_{ij}). \tag{4}$$

The distributive laws

$$\alpha(\mathbf{A} + \mathbf{B}) = \alpha\mathbf{A} + \alpha\mathbf{B}, \qquad (\alpha + \beta)\mathbf{A} = \alpha\mathbf{A} + \beta\mathbf{A} \tag{5}$$

are satisfied for this type of multiplication. In particular, the negative of **A**, denoted by $-\mathbf{A}$, is defined by

$$-\mathbf{A} = (-1)\mathbf{A}. \tag{6}$$

5. **Subtraction.** The difference $\mathbf{A} - \mathbf{B}$ of two $m \times n$ matrices is defined by

$$\mathbf{A} - \mathbf{B} = \mathbf{A} + (-\mathbf{B}). \tag{7}$$

Thus

$$(a_{ij}) - (b_{ij}) = (a_{ij} - b_{ij}), \tag{8}$$

which is similar to Eq. (2).

6. **Multiplication.** The product **AB** of two matrices is defined whenever the number of columns of the first factor is the same as the number of rows in the second. If **A** and **B** are $m \times n$ and $n \times r$ matrices, respectively, then the product $\mathbf{C} = \mathbf{AB}$ is an $m \times r$ matrix. The element in the ith row and jth column of **C** is found by multiplying each element of the ith row of **A** by the corresponding element of the jth column of **B** and then adding the resulting products. Symbolically,

$$c_{ij} = \sum_{k=1}^{n} a_{ik} b_{kj}. \tag{9}$$

By direct calculation it can be shown that matrix multiplication satisfies the associative law

$$(\mathbf{AB})\mathbf{C} = \mathbf{A}(\mathbf{BC}) \tag{10}$$

and the distributive law

$$\mathbf{A}(\mathbf{B} + \mathbf{C}) = \mathbf{AB} + \mathbf{AC}. \tag{11}$$

However, in general, matrix multiplication is not commutative. For both products **AB** and **BA** to exist and to be of the same size, it is necessary that **A** and **B** be square matrices of the same order. Even in that case the two products need not be equal, so that, in general

$$\mathbf{AB} \neq \mathbf{BA}. \tag{12}$$

EXAMPLE 1

To illustrate the multiplication of matrices, and also the fact that matrix multiplication is not necessarily commutative, consider the matrices

$$\mathbf{A} = \begin{pmatrix} 1 & -2 & 1 \\ 0 & 2 & -1 \\ 2 & 1 & 1 \end{pmatrix}, \qquad \mathbf{B} = \begin{pmatrix} 2 & 1 & -1 \\ 1 & -1 & 0 \\ 2 & -1 & 1 \end{pmatrix}.$$

From the definition of multiplication given in Eq. (9) we have

$$\mathbf{AB} = \begin{pmatrix} 2-2+2 & 1+2-1 & -1+0+1 \\ 0+2-2 & 0-2+1 & 0+0-1 \\ 4+1+2 & 2-1-1 & -2+0+1 \end{pmatrix}$$

$$= \begin{pmatrix} 2 & 2 & 0 \\ 0 & -1 & -1 \\ 7 & 0 & -1 \end{pmatrix}.$$

Similarly, we find that

$$\mathbf{BA} = \begin{pmatrix} 0 & -3 & 0 \\ 1 & -4 & 2 \\ 4 & -5 & 4 \end{pmatrix}.$$

Clearly, $\mathbf{AB} \neq \mathbf{BA}$.

---

7. **Multiplication of Vectors.** Matrix multiplication as described above also applies as a special case if the matrices $\mathbf{A}$ and $\mathbf{B}$ are $1 \times n$ and $n \times 1$ row and column vectors, respectively. Denoting these vectors by $\mathbf{x}^T$ and $\mathbf{y}$ we have

$$\mathbf{x}^T\mathbf{y} = \sum_{i=1}^{n} x_i y_i. \tag{13}$$

The result of such an operation is a (complex) number, and it follows directly from Eq. (13) that

$$\mathbf{x}^T\mathbf{y} = \mathbf{y}^T\mathbf{x}, \qquad \mathbf{x}^T(\mathbf{y}+\mathbf{z}) = \mathbf{x}^T\mathbf{y} + \mathbf{x}^T\mathbf{z}, \qquad (\alpha\mathbf{x})^T\mathbf{y} = \alpha(\mathbf{x}^T\mathbf{y}) = \mathbf{x}^T(\alpha\mathbf{y}). \tag{14}$$

There is another very useful type of vector multiplication, which is also defined for any two vectors having the same number of components. This product, denoted by $(\mathbf{x}, \mathbf{y})$, is called the *scalar* or *inner product*, and is defined by

$$(\mathbf{x}, \mathbf{y}) = \sum_{i=1}^{n} x_i \bar{y}_i. \tag{15}$$

The scalar product is also a (complex) number, and by comparing Eqs. (13) and (15) we see that

$$(\mathbf{x}, \mathbf{y}) = \mathbf{x}^T\bar{\mathbf{y}}. \tag{16}$$

From Eq. (15) it follows that

$$\begin{aligned} (\mathbf{x}, \mathbf{y}) &= \overline{(\mathbf{y}, \mathbf{x})}, & (\mathbf{x}, \mathbf{y}+\mathbf{z}) &= (\mathbf{x}, \mathbf{y}) + (\mathbf{x}, \mathbf{z}), \\ (\alpha\mathbf{x}, \mathbf{y}) &= \alpha(\mathbf{x}, \mathbf{y}), & (\mathbf{x}, \alpha\mathbf{y}) &= \bar{\alpha}(\mathbf{x}, \mathbf{y}). \end{aligned} \tag{17}$$

Note that even if the vector **x** has elements with nonzero imaginary parts, the scalar product of **x** with itself yields a nonnegative real number,

$$(\mathbf{x}, \mathbf{x}) = \sum_{i=1}^{n} x_i \bar{x}_i = \sum_{i=1}^{n} |x_i|^2. \tag{18}$$

The nonnegative quantity $(\mathbf{x}, \mathbf{x})^{1/2}$, often denoted by $\|\mathbf{x}\|$, is called the *length*, or *magnitude*, of **x**. If $(\mathbf{x}, \mathbf{y}) = 0$, then the two vectors **x** and **y** are said to be *orthogonal*. For example, the unit vectors **i**, **j**, **k** of three dimensional vector geometry form an orthogonal set.

On the other hand, the matrix product

$$\mathbf{x}^T\mathbf{x} = \sum_{i=1}^{n} x_i^2 \tag{19}$$

may not be a real number. If all of the components of the second factor of Eqs. (13) and (15) are real, then the two products are identical, and both reduce to the dot product usually encountered in physical and geometrical contexts with $n = 3$.

For example, let

$$\mathbf{x} = \begin{pmatrix} i \\ -2 \\ 1+i \end{pmatrix}, \qquad \mathbf{y} = \begin{pmatrix} 2-i \\ i \\ 3 \end{pmatrix}.$$

Then

$$\begin{aligned}
\mathbf{x}^T\mathbf{y} &= (i)(2-i) + (-2)(i) + (1+i)(3) = 4 + 3i, \\
(\mathbf{x}, \mathbf{y}) &= (i)(2+i) + (-2)(-i) + (1+i)(3) = 2 + 7i, \\
\mathbf{x}^T\mathbf{x} &= (i)^2 + (-2)^2 + (1+i)^2 = 3 + 2i, \\
(\mathbf{x}, \mathbf{x}) &= (i)(-i) + (-2)(-2) + (1+i)(1-i) = 7.
\end{aligned}$$

8. **Identity.** The multiplicative identity, or simply the identity matrix **I**, is given by

$$\mathbf{I} = \begin{pmatrix} 1 & 0 & \cdots & 0 \\ 0 & 1 & \cdots & 0 \\ \vdots & \vdots & & \vdots \\ 0 & 0 & \cdots & 1 \end{pmatrix}. \tag{20}$$

From the definition of matrix multiplication we have

$$\mathbf{AI} = \mathbf{IA} = \mathbf{A} \tag{21}$$

for any (square) matrix **A**. Hence the commutative law does hold for square matrices if one of the matrices is the identity.

9. **Inverse.** To define an operation for square matrices analogous to division for numbers, we need to determine, for a given square matrix **A**, another matrix **B**

such that $\mathbf{AB} = \mathbf{I}$, where $\mathbf{I}$ is the identity. If $\mathbf{B}$ exists, it is called the multiplicative inverse, or simply the inverse, of $\mathbf{A}$, and we write $\mathbf{B} = \mathbf{A}^{-1}$. It is possible to show that if $\mathbf{A}^{-1}$ exists, then

$$\mathbf{AA}^{-1} = \mathbf{A}^{-1}\mathbf{A} = \mathbf{I}. \tag{22}$$

In other words, multiplication is commutative between any matrix and its inverse. If $\mathbf{A}$ has a multiplicative inverse $\mathbf{A}^{-1}$, then $\mathbf{A}$ is said to be *nonsingular*; otherwise $\mathbf{A}$ is called *singular*.

There are various ways to compute $\mathbf{A}^{-1}$ from $\mathbf{A}$, assuming that it exists. One way involves the use of determinants. Associated with each element a_{ij} of a given matrix is the minor M_{ij}, which is the determinant of the matrix obtained by deleting the ith row and jth column of the original matrix, that is, the row and column containing a_{ij}. Also associated with each element a_{ij} is the cofactor C_{ij} defined by the equation

$$C_{ij} = (-1)^{i+j}M_{ij}. \tag{23}$$

If $\mathbf{B} = \mathbf{A}^{-1}$, then it can be shown that the general element b_{ij} is given by

$$b_{ij} = \frac{C_{ji}}{\det \mathbf{A}}. \tag{24}$$

While Eq. (24) is not an efficient way[2] to calculate $\mathbf{A}^{-1}$, it does suggest a condition that $\mathbf{A}$ must satisfy for it to have an inverse. In fact, the condition is both necessary and sufficient: $\mathbf{A}$ is nonsingular if and only if $\det \mathbf{A} \neq 0$. If $\det \mathbf{A} = 0$, then $\mathbf{A}$ is singular.

Another and usually better way to compute $\mathbf{A}^{-1}$ is by means of elementary row operations. There are three such operations:

1. Interchange of two rows.
2. Multiplication of a row by a nonzero scalar.
3. Addition of any multiple of one row to another row.

Any nonsingular matrix $\mathbf{A}$ can be transformed into the identity $\mathbf{I}$ by a systematic sequence of these operations. It is possible to show that if the same sequence of operations is then performed upon $\mathbf{I}$, it is transformed into $\mathbf{A}^{-1}$. The transformation of a matrix by a sequence of elementary row operations is usually referred to as *row reduction*. The following example illustrates the process.

[2]For large n the number of multiplications required to evaluate $\mathbf{A}^{-1}$ by Eq. (24) is proportional to $n!$. If one uses more efficient methods, such as the row reduction procedure described below, the number of multiplications is proportional only to n^3. Even for small values of n (such as $n = 4$), determinants are not an economical tool in calculating inverses, and row reduction methods are preferred.

EXAMPLE 2

Find the inverse of $\mathbf{A} = \begin{pmatrix} 1 & -1 & -1 \\ 3 & -1 & 2 \\ 2 & 2 & 3 \end{pmatrix}$.

The matrix **A** can be transformed into **I** by the following sequence of operations. The result of each step appears in the right column.

(a) Obtain zeros in the off-diagonal positions in the first column by adding (-3) times the first row to the second row and adding (-2) times the first row to the third row.
$$\begin{pmatrix} 1 & -1 & -1 \\ 0 & 2 & 5 \\ 0 & 4 & 5 \end{pmatrix}$$

(b) Obtain a one in the diagonal position in the second column by multiplying the second row by $\frac{1}{2}$.
$$\begin{pmatrix} 1 & -1 & -1 \\ 0 & 1 & \frac{5}{2} \\ 0 & 4 & 5 \end{pmatrix}$$

(c) Obtain zeros in the off-diagonal positions in the second column by adding the second row to the first row and adding (-4) times the second row to the third row.
$$\begin{pmatrix} 1 & 0 & \frac{3}{2} \\ 0 & 1 & \frac{5}{2} \\ 0 & 0 & -5 \end{pmatrix}$$

(d) Obtain a one in the diagonal position in the third column by multiplying the third row by $(-\frac{1}{5})$.
$$\begin{pmatrix} 1 & 0 & \frac{3}{2} \\ 0 & 1 & \frac{5}{2} \\ 0 & 0 & 1 \end{pmatrix}$$

(e) Obtain zeros in the off-diagonal positions in the third column by adding $(-\frac{3}{2})$ times the third row to the first row and adding $(-\frac{5}{2})$ times the third row to the first row.
$$\begin{pmatrix} 1 & 0 & 0 \\ 0 & 1 & 0 \\ 0 & 0 & 1 \end{pmatrix}$$

If we perform the same sequence of operations in the same order on **I**, we obtain the following sequence of matrices:

$$\begin{pmatrix} 1 & 0 & 0 \\ 0 & 1 & 0 \\ 0 & 0 & 1 \end{pmatrix}, \quad \begin{pmatrix} 1 & 0 & 0 \\ -3 & 1 & 0 \\ -2 & 0 & 1 \end{pmatrix}, \quad \begin{pmatrix} 1 & 0 & 0 \\ -\frac{3}{2} & \frac{1}{2} & 0 \\ -2 & 0 & 1 \end{pmatrix}, \quad \begin{pmatrix} -\frac{1}{2} & \frac{1}{2} & 0 \\ -\frac{3}{2} & \frac{1}{2} & 0 \\ 4 & -2 & 1 \end{pmatrix},$$

$$\begin{pmatrix} -\frac{1}{2} & \frac{1}{2} & 0 \\ -\frac{3}{2} & \frac{1}{2} & 0 \\ -\frac{4}{5} & \frac{2}{5} & -\frac{1}{5} \end{pmatrix}, \quad \begin{pmatrix} \frac{7}{10} & -\frac{1}{10} & \frac{3}{10} \\ \frac{1}{2} & -\frac{1}{2} & \frac{1}{2} \\ -\frac{4}{5} & \frac{2}{5} & -\frac{1}{5} \end{pmatrix}.$$

The last of these matrices is $\mathbf{A}^{-1}$, a result that can be verified by direct multiplication with the original matrix **A**.

This example is made slightly simpler by the fact that the original matrix **A** had a one in the upper left corner ($a_{11} = 1$). If this is not the case, then the first step is to produce a one there by multiplying the first row by $1/a_{11}$, as long as $a_{11} \neq 0$. If $a_{11} = 0$, then the first row must be interchanged with some other row to bring a nonzero element into the upper left position before proceeding as above.

MATRIX FUNCTIONS. We sometimes need to consider vectors or matrices whose elements are functions of a real variable t. We write

$$\mathbf{x}(t) = \begin{pmatrix} x_1(t) \\ \vdots \\ x_n(t) \end{pmatrix}, \qquad \mathbf{A}(t) = \begin{pmatrix} a_{11}(t) & \cdots & a_{1n}(t) \\ \vdots & & \vdots \\ a_{n1}(t) & \cdots & a_{nn}(t) \end{pmatrix}, \tag{25}$$

respectively.

The matrix $\mathbf{A}(t)$ is said to be continuous at $t = t_0$ or on an interval $\alpha < t < \beta$ if each element of **A** is a continuous function at the given point or on the given interval. Similarly, $\mathbf{A}(t)$ is said to be differentiable if each of its elements in differentiable, and its derivative $d\mathbf{A}/dt$ is defined by

$$\frac{d\mathbf{A}}{dt} = \left(\frac{da_{ij}}{dt}\right); \tag{26}$$

that is, each element of $d\mathbf{A}/dt$ is the derivative of the corresponding element of **A**. In the same way the integral of a matrix function is defined as

$$\int_a^b \mathbf{A}(t)\,dt = \left(\int_a^b a_{ij}(t)\,dt\right). \tag{27}$$

For example, if

$$\mathbf{A}(t) = \begin{pmatrix} \sin t & t \\ 1 & \cos t \end{pmatrix},$$

then

$$\mathbf{A}'(t) = \begin{pmatrix} \cos t & 1 \\ 0 & -\sin t \end{pmatrix}, \qquad \int_0^\pi \mathbf{A}(t)\,dt = \begin{pmatrix} 2 & \pi^2/2 \\ \pi & 0 \end{pmatrix}.$$

Many of the rules of elementary calculus can be easily extended to matrix functions; in particular,

$$\frac{d}{dt}(\mathbf{CA}) = \mathbf{C}\frac{d\mathbf{A}}{dt}, \qquad \text{where } \mathbf{C} \text{ is a constant matrix}; \tag{28}$$

$$\frac{d}{dt}(\mathbf{A} + \mathbf{B}) = \frac{d\mathbf{A}}{dt} + \frac{d\mathbf{B}}{dt}; \tag{29}$$

$$\frac{d}{dt}(\mathbf{AB}) = \mathbf{A}\frac{d\mathbf{B}}{dt} + \frac{d\mathbf{A}}{dt}\mathbf{B}. \tag{30}$$

In Eqs. (28) and (30) care must be taken in each term to avoid carelessly interchanging the order of multiplication. The definitions expressed by Eqs. (26) and (27) also apply as special cases to vectors.

PROBLEMS

1. If $\mathbf{A} = \begin{pmatrix} 1 & -2 & 0 \\ 3 & 2 & -1 \\ -2 & 1 & 3 \end{pmatrix}$ and $\mathbf{B} = \begin{pmatrix} 4 & -2 & 3 \\ -1 & 5 & 0 \\ 6 & 1 & 2 \end{pmatrix}$,

   find

   (a) $2\mathbf{A} + \mathbf{B}$, (b) $\mathbf{A} - 4\mathbf{B}$, (c) $\mathbf{AB}$, (d) $\mathbf{BA}$.

2. If $\mathbf{A} = \begin{pmatrix} 1+i & -1+2i \\ 3+2i & 2-i \end{pmatrix}$ and $\mathbf{B} = \begin{pmatrix} i & 3 \\ 2 & -2i \end{pmatrix}$,

   find

   (a) $\mathbf{A} - 2\mathbf{B}$, (b) $3\mathbf{A} + \mathbf{B}$, (c) $\mathbf{AB}$, (d) $\mathbf{BA}$.

3. If $\mathbf{A} = \begin{pmatrix} -2 & 1 & 2 \\ 1 & 0 & -3 \\ 2 & -1 & 1 \end{pmatrix}$ and $\mathbf{B} = \begin{pmatrix} 1 & 2 & 3 \\ 3 & -1 & -1 \\ -2 & 1 & 0 \end{pmatrix}$,

   find

   (a) $\mathbf{A}^T$, (b) $\mathbf{B}^T$, (c) $\mathbf{A}^T + \mathbf{B}^T$, (d) $(\mathbf{A} + \mathbf{B})^T$.

4. If $\mathbf{A} = \begin{pmatrix} 3-2i & 1+i \\ 2-i & -2+3i \end{pmatrix}$,

   find

   (a) $\mathbf{A}^T$, (b) $\overline{\mathbf{A}}$, (c) $\mathbf{A}^*$.

5. If $\mathbf{A} = \begin{pmatrix} 3 & 2 & -1 \\ 2 & -1 & 2 \\ 1 & 2 & 1 \end{pmatrix}$ and $\mathbf{B} = \begin{pmatrix} 2 & 1 & -1 \\ -2 & 3 & 3 \\ 1 & 0 & 2 \end{pmatrix}$,

   verify that $2(\mathbf{A} + \mathbf{B}) = 2\mathbf{A} + 2\mathbf{B}$.

6. If $\mathbf{A} = \begin{pmatrix} 1 & -2 & 0 \\ 3 & 2 & -1 \\ -2 & 0 & 3 \end{pmatrix}$, $\mathbf{B} = \begin{pmatrix} 2 & 1 & -1 \\ -2 & 3 & 3 \\ 1 & 0 & 2 \end{pmatrix}$, $\mathbf{C} = \begin{pmatrix} 2 & 1 & 0 \\ 1 & 2 & 2 \\ 0 & 1 & -1 \end{pmatrix}$,

   verify that

   (a) $(\mathbf{AB})\mathbf{C} = \mathbf{A}(\mathbf{BC})$, (b) $(\mathbf{A} + \mathbf{B}) + \mathbf{C} = \mathbf{A} + (\mathbf{B} + \mathbf{C})$,
   (c) $\mathbf{A}(\mathbf{B} + \mathbf{C}) = \mathbf{AB} + \mathbf{AC}$.

7. Prove each of the following laws of matrix algebra:

   (a) $\mathbf{A} + \mathbf{B} = \mathbf{B} + \mathbf{A}$ (b) $\mathbf{A} + (\mathbf{B} + \mathbf{C}) = (\mathbf{A} + \mathbf{B}) + \mathbf{C}$
   (c) $\alpha(\mathbf{A} + \mathbf{B}) = \alpha\mathbf{A} + \alpha\mathbf{B}$ (d) $(\alpha + \beta)\mathbf{A} = \alpha\mathbf{A} + \beta\mathbf{A}$
   (e) $\mathbf{A}(\mathbf{BC}) = (\mathbf{AB})\mathbf{C}$ (f) $\mathbf{A}(\mathbf{B} + \mathbf{C}) = \mathbf{AB} + \mathbf{AC}$

8. If $\mathbf{x} = \begin{pmatrix} 2 \\ 3i \\ 1-i \end{pmatrix}$ and $\mathbf{y} = \begin{pmatrix} -1+i \\ 2 \\ 3-i \end{pmatrix}$, find

(a) $\mathbf{x}^T\mathbf{y}$, (b) $\mathbf{y}^T\mathbf{y}$, (c) $(\mathbf{x}, \mathbf{y})$, (d) $(\mathbf{y}, \mathbf{y})$.

9. If $\mathbf{x} = \begin{pmatrix} 1-2i \\ i \\ 2 \end{pmatrix}$ and $\mathbf{y} = \begin{pmatrix} 2 \\ 3-i \\ 1+2i \end{pmatrix}$, show that

(a) $\mathbf{x}^T\mathbf{y} = \mathbf{y}^T\mathbf{x}$, (b) $(\mathbf{x}, \mathbf{y}) = \overline{(\mathbf{y}, \mathbf{x})}$.

In each of Problems 10 through 19 either compute the inverse of the given matrix, or else show that it is singular.

10. $\begin{pmatrix} 1 & 4 \\ -2 & 3 \end{pmatrix}$

11. $\begin{pmatrix} 3 & -1 \\ 6 & 2 \end{pmatrix}$

12. $\begin{pmatrix} 1 & 2 & 3 \\ 2 & 4 & 5 \\ 3 & 5 & 6 \end{pmatrix}$

13. $\begin{pmatrix} 1 & 1 & -1 \\ 2 & -1 & 1 \\ 1 & 1 & 2 \end{pmatrix}$

14. $\begin{pmatrix} 1 & 2 & 1 \\ -2 & 1 & 8 \\ 1 & -2 & -7 \end{pmatrix}$

15. $\begin{pmatrix} 2 & 1 & 0 \\ 0 & 2 & 1 \\ 0 & 0 & 2 \end{pmatrix}$

16. $\begin{pmatrix} 1 & -1 & -1 \\ 2 & 1 & 0 \\ 3 & -2 & 1 \end{pmatrix}$

17. $\begin{pmatrix} 2 & 3 & 1 \\ -1 & 2 & 1 \\ 4 & -1 & -1 \end{pmatrix}$

18. $\begin{pmatrix} 1 & 0 & 0 & -1 \\ 0 & -1 & 1 & 0 \\ -1 & 0 & 1 & 0 \\ 0 & 1 & -1 & 1 \end{pmatrix}$

19. $\begin{pmatrix} 1 & -1 & 2 & 0 \\ -1 & 2 & -4 & 2 \\ 1 & 0 & 1 & 3 \\ -2 & 2 & 0 & -1 \end{pmatrix}$

20. Prove that if $\mathbf{A}$ is nonsingular, then $\mathbf{A}^{-1}$ is uniquely determined; that is, show that there cannot be two different matrices $\mathbf{B}$ and $\mathbf{C}$ such that $\mathbf{AB} = \mathbf{I}$ and $\mathbf{AC} = \mathbf{I}$.

21. Prove that if $\mathbf{A}$ is nonsingular, then $\mathbf{AA}^{-1} = \mathbf{A}^{-1}\mathbf{A}$; that is, multiplication is commutative between any nonsingular matrix and its inverse.

22. If $\mathbf{A}(t) = \begin{pmatrix} e^t & 2e^{-t} & e^{2t} \\ 2e^t & e^{-t} & -e^{2t} \\ -e^t & 3e^{-t} & 2e^{2t} \end{pmatrix}$ and $\mathbf{B}(t) = \begin{pmatrix} 2e^t & e^{-t} & 3e^{2t} \\ -e^t & 2e^{-t} & e^{2t} \\ 3e^t & -e^{-t} & -e^{2t} \end{pmatrix}$,

find

(a) $\mathbf{A} + 3\mathbf{B}$, (b) $\mathbf{AB}$, (c) $d\mathbf{A}/dt$, (d) $\int_0^1 \mathbf{A}(t)\,dt$.

In each of Problems 23 through 25 verify that the given vector satisfies the given differential equation.

23. $\mathbf{x}' = \begin{pmatrix} 3 & -2 \\ 2 & -2 \end{pmatrix}\mathbf{x}, \quad \mathbf{x} = \begin{pmatrix} 4 \\ 2 \end{pmatrix}e^{2t}$

24. $\mathbf{x}' = \begin{pmatrix} 2 & -1 \\ 3 & -2 \end{pmatrix}\mathbf{x} + \begin{pmatrix} 1 \\ -1 \end{pmatrix}e^t, \quad \mathbf{x} = \begin{pmatrix} 1 \\ 0 \end{pmatrix}e^t + 2\begin{pmatrix} 1 \\ 1 \end{pmatrix}te^t$

25. $\mathbf{x}' = \begin{pmatrix} 1 & 1 & 1 \\ 2 & 1 & -1 \\ 0 & -1 & 1 \end{pmatrix}\mathbf{x}, \quad \mathbf{x} = \begin{pmatrix} 6 \\ -8 \\ -4 \end{pmatrix}e^{-t} + 2\begin{pmatrix} 0 \\ 1 \\ -1 \end{pmatrix}e^{2t}$

In each of Problems 26 and 27 verify that the given matrix satisfies the given differential equation.

26. $\mathbf{\Psi}' = \begin{pmatrix} 1 & 1 \\ 4 & -2 \end{pmatrix} \mathbf{\Psi}, \qquad \mathbf{\Psi}(t) = \begin{pmatrix} e^{-3t} & e^{2t} \\ -4e^{-3t} & e^{2t} \end{pmatrix}$

27. $\mathbf{\Psi}' = \begin{pmatrix} 1 & -1 & 4 \\ 3 & 2 & -1 \\ 2 & 1 & -1 \end{pmatrix} \mathbf{\Psi}, \qquad \mathbf{\Psi}(t) = \begin{pmatrix} e^{t} & e^{-2t} & e^{3t} \\ -4e^{t} & -e^{-2t} & 2e^{3t} \\ -e^{t} & -e^{-2t} & e^{3t} \end{pmatrix}$

7.3 Systems of Linear Algebraic Equations; Linear Independence, Eigenvalues, Eigenvectors

In this section we review some results from linear algebra that are important for the solution of systems of linear differential equations. Some of these results are easily proved and others are not; since we are interested simply in summarizing some useful information in compact form, we give no indication of proofs in either case. All of the results in this section depend on some basic facts about the solution of systems of linear algebraic equations.

SYSTEMS OF LINEAR ALGEBRAIC EQUATIONS. A set of n simultaneous linear algebraic equations in n variables

$$\begin{aligned} a_{11}x_1 + a_{12}x_2 + \cdots + a_{1n}x_n &= b_1, \\ &\vdots \\ a_{n1}x_1 + a_{n2}x_2 + \cdots + a_{nn}x_n &= b_n, \end{aligned} \tag{1}$$

can be written as

$$\mathbf{Ax} = \mathbf{b}, \tag{2}$$

where the $n \times n$ matrix $\mathbf{A}$ and the vector $\mathbf{b}$ are given, and the components of $\mathbf{x}$ are to be determined. If $\mathbf{b} = \mathbf{0}$, the system is said to be *homogeneous*; otherwise, it is *nonhomogeneous*. Two systems having precisely the same set of solutions are said to be *equivalent systems*.

If the coefficient matrix $\mathbf{A}$ is nonsingular, that is, if $\det \mathbf{A}$ is not zero, then there is a unique solution of the system (2). Since $\mathbf{A}$ is nonsingular, $\mathbf{A}^{-1}$ exists, and the solution can be found by multiplying each side of Eq. (2) on the left by $\mathbf{A}^{-1}$; thus

$$\mathbf{x} = \mathbf{A}^{-1}\mathbf{b}. \tag{3}$$

In particular, the homogeneous problem $\mathbf{Ax} = \mathbf{0}$, corresponding to $\mathbf{b} = \mathbf{0}$ in Eq. (2), has only the trivial solution $\mathbf{x} = \mathbf{0}$.

On the other hand, if $\mathbf{A}$ is singular, that is, if $\det \mathbf{A}$ is zero, then solutions of Eq. (2) either do not exist, or do exist but are not unique. Since $\mathbf{A}$ is singular, $\mathbf{A}^{-1}$ does not exist, so Eq. (3) is no longer valid. The homogeneous system

$$\mathbf{Ax} = \mathbf{0} \tag{4}$$

has (infinitely many) nonzero solutions in addition to the trivial solution. The situation for the nonhomogeneous system (2) is more complicated. This system has no solution unless the vector **b** satisfies a certain further condition that is by no means obviously necessary. This condition is that

$$(\mathbf{b}, \mathbf{y}) = 0, \tag{5}$$

for all vectors **y** satisfying $\mathbf{A}^*\mathbf{y} = \mathbf{0}$, where $\mathbf{A}^*$ is the adjoint of **A**. If condition (5) is met, then the system (2) has (infinitely many) solutions. Each of these solutions has the form

$$\mathbf{x} = \mathbf{x}^{(0)} + \boldsymbol{\xi}, \tag{6}$$

where $\mathbf{x}^{(0)}$ is a particular solution of Eq. (2), and $\boldsymbol{\xi}$ is any solution of the homogeneous system (4). Note the resemblance between Eq. (6) and the solution of a nonhomogeneous linear differential equation, Eq. (4) of Section 3.6; $\mathbf{x}^{(0)}$ and $\boldsymbol{\xi}$ correspond, respectively, to the particular solution y_p and the complementary solution y_c. The proofs of some of the above statements are outlined in Problems 21 through 25.

The results in the preceding paragraph are important as a means of classifying the solutions of linear systems. However, for solving particular systems it is generally best to use row reduction to transform the system into a much simpler one from which the solution(s), if there are any, can be easily written down. To do this efficiently we can form the *augmented matrix*

$$\mathbf{A} \mid \mathbf{b} = \left(\begin{array}{ccc|c} a_{11} & \cdots & a_{1n} & b_1 \\ \vdots & & \vdots & \vdots \\ a_{n1} & \cdots & a_{nn} & b_n \end{array}\right) \tag{7}$$

by adjoining the vector **b** to the coefficient matrix **A** as an additional column. The dashed line replaces the equals sign and is said to partition the augmented matrix. We now perform row operations on the augmented matrix so as to transform **A** into a triangular matrix, that is, a matrix whose elements below the main diagonal are all zero. Once this is done, it is easy to see whether the system has solutions, and to find them if it does. Observe that elementary row operations on the augmented matrix (7) correspond to legitimate operations on the equations in the system (1). The following examples illustrate the process.

EXAMPLE 1

Solve the system of equations

$$\begin{aligned} x_1 - 2x_2 + 3x_3 &= 7, \\ -x_1 + x_2 - 2x_3 &= -5, \\ 2x_1 - x_2 - x_3 &= 4. \end{aligned} \tag{8}$$

The augmented matrix for the system (8) is

$$\left(\begin{array}{rrr|r} 1 & -2 & 3 & 7 \\ -1 & 1 & -2 & -5 \\ 2 & -1 & -1 & 4 \end{array}\right). \tag{9}$$

We now perform row operations on the matrix (9) with a view to introducing zeros in the lower left part of the matrix. Each step is described and the result recorded below.

(a) Add the first row to the second row and add (-2) times the first row to the third row.

$$\left(\begin{array}{rrr|r} 1 & -2 & 3 & 7 \\ 0 & -1 & 1 & 2 \\ 0 & 3 & -7 & -10 \end{array}\right)$$

(b) Multiply the second row by -1.

$$\left(\begin{array}{rrr|r} 1 & -2 & 3 & 7 \\ 0 & 1 & -1 & -2 \\ 0 & 3 & -7 & -10 \end{array}\right)$$

(c) Add (-3) times the second row to the third row.

$$\left(\begin{array}{rrr|r} 1 & -2 & 3 & 7 \\ 0 & 1 & -1 & -2 \\ 0 & 0 & -4 & -4 \end{array}\right)$$

(d) Divide the third row by -4.

$$\left(\begin{array}{rrr|r} 1 & -2 & 3 & 7 \\ 0 & 1 & -1 & -2 \\ 0 & 0 & 1 & 1 \end{array}\right)$$

The matrix obtained in this manner corresponds to the system of equations

$$\begin{aligned} x_1 - 2x_2 + 3x_3 &= 7 \\ x_2 - \ \ x_3 &= -2 \\ x_3 &= 1 \end{aligned} \tag{10}$$

which is equivalent to the original system (8). Note that the coefficients in Eqs. (10) form a triangular matrix. From the last of Eqs. (10) we have $x_3 = 1$, from the second equation $x_2 = -2 + x_3 = -1$, and from the first equation $x_1 = 7 + 2x_2 - 3x_3 = 2$. Thus we obtain

$$\mathbf{x} = \begin{pmatrix} 2 \\ -1 \\ 1 \end{pmatrix},$$

which is the solution of the given system (8). Incidentally, since the solution is unique, we conclude that the coefficient matrix is nonsingular.

EXAMPLE 2

Discuss solutions of the system

$$\begin{aligned} x_1 - 2x_2 + 3x_3 &= b_1, \\ -x_1 + x_2 - 2x_3 &= b_2, \\ 2x_1 - x_2 + 3x_3 &= b_3, \end{aligned} \tag{11}$$

for various values of b_1, b_2, and b_3.

Observe that the coefficients in the system (11) are the same as those in the system (8) except for the coefficient of x_3 in the third equation. The augmented matrix for the system (11) is

$$\left(\begin{array}{rrr|c} 1 & -2 & 3 & b_1 \\ -1 & 1 & -2 & b_2 \\ 2 & -1 & 3 & b_3 \end{array}\right). \tag{12}$$

By performing steps (a), (b), and (c) as in Example 1 we transform the matrix (12) into

$$\left(\begin{array}{rrr|r} 1 & -2 & 3 & b_1 \\ 0 & 1 & -1 & -b_1 - b_2 \\ 0 & 0 & 0 & b_1 + 3b_2 + b_3 \end{array}\right). \tag{13}$$

The equation corresponding to the third row of the matrix (13) is

$$b_1 + 3b_2 + b_3 = 0; \tag{14}$$

thus the system (11) has no solution unless the condition (14) is satisfied by b_1, b_2, and b_3. It is possible to show that this condition is just Eq. (5) for the system (11).

Let us now assume that $b_1 = 2$, $b_2 = 1$, and $b_3 = -5$, in which case Eq. (14) is satisfied. Then the first two rows of the matrix (13) correspond to the equations

$$\begin{aligned} x_1 - 2x_2 + 3x_3 &= 2, \\ x_2 - x_3 &= -3. \end{aligned} \tag{15}$$

To solve the system (15) we can choose one of the unknowns arbitrarily and then solve for the other two. Letting $x_3 = \alpha$, where α is arbitrary, it then follows that

$$\begin{aligned} x_2 &= \alpha - 3, \\ x_1 &= 2(\alpha - 3) - 3\alpha + 2 = -\alpha - 4. \end{aligned}$$

If we write the solution in vector notation, we have

$$\mathbf{x} = \begin{pmatrix} -\alpha - 4 \\ \alpha - 3 \\ \alpha \end{pmatrix} = \alpha \begin{pmatrix} -1 \\ 1 \\ 1 \end{pmatrix} + \begin{pmatrix} -4 \\ -3 \\ 0 \end{pmatrix}. \tag{16}$$

It is easy to verify that the second term on the right side of Eq. (16) is a solution of the nonhomogeneous system (11), while the first term is the most general solution of the homogeneous system corresponding to (11).

Row reduction is also useful in solving homogeneous systems, and systems in which the number of equations is different from the number of unknowns.

LINEAR INDEPENDENCE. A set of k vectors $\mathbf{x}^{(1)}, \ldots, \mathbf{x}^{(k)}$ is said to be *linearly dependent* if there exists a set of (complex) numbers $c_1, \ldots, c_k$, at least one of which is nonzero, such that

$$c_1\mathbf{x}^{(1)} + \cdots + c_k\mathbf{x}^{(k)} = \mathbf{0}. \tag{17}$$

In other words, $\mathbf{x}^{(1)}, \ldots, \mathbf{x}^{(k)}$ are linearly dependent if there is a linear relation among them. On the other hand, if the only set $c_1, \ldots, c_k$ for which Eq. (17) is satisfied is $c_1 = c_2 = \cdots = c_k = 0$, then $\mathbf{x}^{(1)}, \ldots, \mathbf{x}^{(k)}$ are said to be *linearly independent*.

Consider now a set of n vectors, each of which has n components. Let $x_{ij} = x_i^{(j)}$ be the ith component of the vector $\mathbf{x}^{(j)}$, and let $\mathbf{X} = (x_{ij})$. Then Eq. (17) can be written as

$$\begin{pmatrix} x_1^{(1)}c_1 + \cdots + x_1^{(n)}c_n \\ \vdots \qquad\qquad \vdots \\ x_n^{(1)}c_1 + \cdots + x_n^{(n)}c_n \end{pmatrix} = \begin{pmatrix} x_{11}c_1 + \cdots + x_{1n}c_n \\ \vdots \qquad\qquad \vdots \\ x_{n1}c_1 + \cdots + x_{nn}c_n \end{pmatrix} = \mathbf{Xc} = \mathbf{0}. \tag{18}$$

If $\det \mathbf{X} \neq 0$, then the only solution of Eq. (18) is $\mathbf{c} = \mathbf{0}$, but if $\det \mathbf{X} = 0$, there are nonzero solutions. Thus the set of vectors $\mathbf{x}^{(1)}, \ldots, \mathbf{x}^{(n)}$ is linearly independent if and only if $\det \mathbf{X} \neq 0$.

EXAMPLE 3

Determine whether the vectors

$$\mathbf{x}^{(1)} = \begin{pmatrix} 1 \\ 2 \\ -1 \end{pmatrix}, \qquad \mathbf{x}^{(2)} = \begin{pmatrix} 2 \\ 1 \\ 3 \end{pmatrix}, \qquad \mathbf{x}^{(3)} = \begin{pmatrix} -4 \\ 1 \\ -11 \end{pmatrix} \tag{19}$$

are linearly independent or linearly dependent. If linearly dependent, find a linear relation among them.

To determine whether $\mathbf{x}^{(1)}$, $\mathbf{x}^{(2)}$, and $\mathbf{x}^{(3)}$ are linearly dependent we compute $\det(x_{ij})$, whose columns are the components of $\mathbf{x}^{(1)}$, $\mathbf{x}^{(2)}$, and $\mathbf{x}^{(3)}$, respectively. Thus

$$\det(x_{ij}) = \begin{vmatrix} 1 & 2 & -4 \\ 2 & 1 & 1 \\ -1 & 3 & -11 \end{vmatrix},$$

and an elementary calculation shows that it is zero. Thus $\mathbf{x}^{(1)}$, $\mathbf{x}^{(2)}$, and $\mathbf{x}^{(3)}$ are linearly dependent, and there are constants c_1, c_2, and c_3 such that

$$c_1\mathbf{x}^{(1)} + c_2\mathbf{x}^{(2)} + c_3\mathbf{x}^{(3)} = \mathbf{0}. \tag{20}$$

Equation (20) can also be written in the form

$$\begin{pmatrix} 1 & 2 & -4 \\ 2 & 1 & 1 \\ -1 & 3 & -11 \end{pmatrix} \begin{pmatrix} c_1 \\ c_2 \\ c_3 \end{pmatrix} = \begin{pmatrix} 0 \\ 0 \\ 0 \end{pmatrix}, \tag{21}$$

and solved by means of elementary row operations starting from the augmented matrix

$$\left(\begin{array}{rrr|r} 1 & 2 & -4 & 0 \\ 2 & 1 & 1 & 0 \\ -1 & 3 & -11 & 0 \end{array}\right). \tag{22}$$

We proceed as in Examples 1 and 2.

(a) Add (-2) times the first row to the second row, and add the first row to the third row.

$$\left(\begin{array}{rrr|r} 1 & 2 & -4 & 0 \\ 0 & -3 & 9 & 0 \\ 0 & 5 & -15 & 0 \end{array}\right)$$

(b) Divide the second row by -3; then add (-5) times the second row to the third row.

$$\left(\begin{array}{rrr|r} 1 & 2 & -4 & 0 \\ 0 & 1 & -3 & 0 \\ 0 & 0 & 0 & 0 \end{array}\right)$$

Thus we obtain the equivalent system

$$\begin{aligned} c_1 + 2c_2 - 4c_3 &= 0, \\ c_2 - 3c_3 &= 0. \end{aligned} \tag{23}$$

From the second of Eqs. (23) we have $c_2 = 3c_3$, and from the first we obtain $c_1 = 4c_3 - 2c_2 = -2c_3$. Thus we have solved for c_1 and c_2 in terms of c_3, with the latter remaining arbitrary. If we choose $c_3 = -1$ for convenience, then $c_1 = 2$ and $c_2 = -3$. In this case the desired relation (20) becomes

$$2\mathbf{x}^{(1)} - 3\mathbf{x}^{(2)} - \mathbf{x}^{(3)} = \mathbf{0}.$$

Frequently it is useful to think of the columns (or rows) of a matrix $\mathbf{A}$ as vectors. These column (or row) vectors are linearly independent if and only if $\det \mathbf{A} \neq 0$. Further, if $\mathbf{C} = \mathbf{AB}$, then it can be shown that $\det \mathbf{C} = (\det \mathbf{A})(\det \mathbf{B})$. Therefore, if the columns (or rows) of both $\mathbf{A}$ and $\mathbf{B}$ are linearly independent, then the columns (or rows) of $\mathbf{C}$ are also linearly independent.

Now let us extend the concepts of linear dependence and independence to a set of vector functions $\mathbf{x}^{(1)}(t), \ldots, \mathbf{x}^{(k)}(t)$ defined on an interval $\alpha < t < \beta$. The vectors $\mathbf{x}^{(1)}(t), \ldots, \mathbf{x}^{(k)}(t)$ are said to be linearly dependent on $\alpha < t < \beta$ if there exists a set of constants $c_1, \ldots, c_k$, not all of which are zero, such that $c_1\mathbf{x}^{(1)}(t) + \cdots + c_k\mathbf{x}^{(k)}(t) = \mathbf{0}$ for all t in the interval. Otherwise, $\mathbf{x}^{(1)}(t), \ldots, \mathbf{x}^{(k)}(t)$ are said to be linearly independent. Note that if $\mathbf{x}^{(1)}(t), \ldots, \mathbf{x}^{(k)}(t)$ are linearly dependent on an interval, they are linearly dependent at each point in the interval. However, if $\mathbf{x}^{(1)}(t), \ldots, \mathbf{x}^{(k)}(t)$ are linearly independent on an interval, they may or may not be linearly independent at each point; they may, in fact, be linearly dependent at each point, but with different sets of constants at different points. See Problem 9 for an example.

EIGENVALUES AND EIGENVECTORS. The equation

$$\mathbf{A}\mathbf{x} = \mathbf{y} \tag{24}$$

can be viewed as a linear transformation that maps (or transforms) a given vector $\mathbf{x}$ into a new vector $\mathbf{y}$. Vectors that are transformed into multiples of themselves play an important role in many applications.[3] To find such vectors we set $\mathbf{y} = \lambda\mathbf{x}$, where λ is a scalar proportionality factor, and seek solutions of the equations

$$\mathbf{A}\mathbf{x} = \lambda\mathbf{x}, \tag{25}$$

or

$$(\mathbf{A} - \lambda\mathbf{I})\mathbf{x} = \mathbf{0}. \tag{26}$$

The latter equation has nonzero solutions if and only if λ is chosen so that

$$\Delta(\lambda) = \det(\mathbf{A} - \lambda\mathbf{I}) = 0. \tag{27}$$

Values of λ that satisfy Eq. (27) are called *eigenvalues* of the matrix $\mathbf{A}$, and the solutions of Eq. (25) or (26) that are obtained by using such a value of λ are called the *eigenvectors* corresponding to that eigenvalue. Eigenvectors are determined only up to an arbitrary multiplicative constant; if this constant is specified in some way, then the eigenvectors are said to be *normalized*. Frequently it is convenient to normalize an eigenvector $\mathbf{x}$ by requiring that $(\mathbf{x}, \mathbf{x}) = 1$. Alternatively, we may wish to set one of the components equal to one.

Since Eq. (27) is a polynomial equation of degree n in λ, there are n eigenvalues $\lambda_1, \ldots, \lambda_n$, some of which may be repeated. If a given eigenvalue appears m times as a root of Eq. (27), then that eigenvalue is said to have *multiplicity* m. Each eigenvalue has at least one associated eigenvector, and an eigenvalue of multiplicity m may have q linearly independent eigenvectors, where

$$1 \le q \le m. \tag{28}$$

That q can be less than m is illustrated by Example 4 below. If all of the eigenvalues of a matrix $\mathbf{A}$ are *simple* (have multiplicity one), then it is possible to

[3] For example, this problem is encountered in finding the principal axes of stress or strain in an elastic body, and in finding the modes of free vibration in a conservative system with a finite number of degrees of freedom.

show that the n eigenvectors of **A**, one for each eigenvalue, are linearly independent. On the other hand, if **A** has one or more repeated eigenvalues, then there may be fewer than n linearly independent eigenvectors associated with **A**, since for a repeated eigenvalue we may have $q < m$. This fact leads to complications later on in the solution of systems of differential equations.

EXAMPLE 4

Find the eigenvalues and eigenvectors of the matrix

$$\mathbf{A} = \begin{pmatrix} 1 & -1 \\ 1 & 3 \end{pmatrix}. \tag{29}$$

The eigenvalues λ and eigenvectors $\mathbf{x}$ satisfy the equation $(\mathbf{A} - \lambda\mathbf{I})\mathbf{x} = \mathbf{0}$, or

$$\begin{pmatrix} 1-\lambda & -1 \\ 1 & 3-\lambda \end{pmatrix}\begin{pmatrix} x_1 \\ x_2 \end{pmatrix} = \begin{pmatrix} 0 \\ 0 \end{pmatrix}. \tag{30}$$

The eigenvalues are the roots of the equation

$$\det(\mathbf{A} - \lambda\mathbf{I}) = \begin{vmatrix} 1-\lambda & -1 \\ 1 & 3-\lambda \end{vmatrix}$$

$$= \lambda^2 - 4\lambda + 4 = 0. \tag{31}$$

Thus the two eigenvalues are $\lambda_1 = 2$ and $\lambda_2 = 2$; that is, the eigenvalue 2 has multiplicity two.

To determine the eigenvectors we must return to Eq. (30) and use for λ the value 2. This gives

$$\begin{pmatrix} -1 & -1 \\ 1 & 1 \end{pmatrix}\begin{pmatrix} x_1 \\ x_2 \end{pmatrix} = \begin{pmatrix} 0 \\ 0 \end{pmatrix}. \tag{32}$$

Hence we obtain the single condition $x_1 + x_2 = 0$, which determines x_2 in terms of x_1, or vice versa. If $x_1 = c$, then $x_2 = -c$ and the eigenvector $\mathbf{x}^{(1)}$ is

$$\mathbf{x}^{(1)} = c\begin{pmatrix} 1 \\ -1 \end{pmatrix}. \tag{33}$$

Usually we will drop the arbitrary constant c when finding eigenvectors; thus instead of Eq. (33) we write

$$\mathbf{x}^{(1)} = \begin{pmatrix} 1 \\ -1 \end{pmatrix}, \tag{34}$$

and remember that any multiple of this vector is also an eigenvector. Observe that there is only one linearly independent eigenvector associated with the double eigenvalue.

EXAMPLE 5

Find the eigenvalues and eigenvectors of the matrix

$$\mathbf{A} = \begin{pmatrix} 0 & 1 & 1 \\ 1 & 0 & 1 \\ 1 & 1 & 0 \end{pmatrix}. \tag{35}$$

The eigenvalues λ and eigenvectors $\mathbf{x}$ satisfy the equation $(\mathbf{A} - \lambda\mathbf{I})\mathbf{x} = \mathbf{0}$, or

$$\begin{pmatrix} -\lambda & 1 & 1 \\ 1 & -\lambda & 1 \\ 1 & 1 & -\lambda \end{pmatrix} \begin{pmatrix} x_1 \\ x_2 \\ x_3 \end{pmatrix} = \begin{pmatrix} 0 \\ 0 \\ 0 \end{pmatrix}. \tag{36}$$

The eigenvalues are the roots of the equation

$$\det(\mathbf{A} - \lambda\mathbf{I}) = \begin{vmatrix} -\lambda & 1 & 1 \\ 1 & -\lambda & 1 \\ 1 & 1 & -\lambda \end{vmatrix} = -\lambda^3 + 3\lambda + 2 = 0. \tag{37}$$

Solving Eq. (37), perhaps by trial and error, we obtain three roots, namely, $\lambda_1 = 2$, $\lambda_2 = -1$, and $\lambda_3 = -1$. Thus 2 is a simple eigenvalue, and -1 is an eigenvalue of multiplicity two.

To find the eigenvector $\mathbf{x}^{(1)}$ corresponding to the eigenvalue λ_1 we substitute $\lambda = 2$ in Eq. (36); this gives the system

$$\begin{pmatrix} -2 & 1 & 1 \\ 1 & -2 & 1 \\ 1 & 1 & -2 \end{pmatrix} \begin{pmatrix} x_1 \\ x_2 \\ x_3 \end{pmatrix} = \begin{pmatrix} 0 \\ 0 \\ 0 \end{pmatrix}. \tag{38}$$

We can reduce this to the equivalent system

$$\begin{pmatrix} 2 & -1 & -1 \\ 0 & 1 & -1 \\ 0 & 0 & 0 \end{pmatrix} \begin{pmatrix} x_1 \\ x_2 \\ x_3 \end{pmatrix} = \begin{pmatrix} 0 \\ 0 \\ 0 \end{pmatrix} \tag{39}$$

by elementary row operations. Solving this system we obtain the eigenvector

$$\mathbf{x}^{(1)} = \begin{pmatrix} 1 \\ 1 \\ 1 \end{pmatrix}. \tag{40}$$

For $\lambda = -1$, Eqs. (36) reduce immediately to the single equation

$$x_1 + x_2 + x_3 = 0. \tag{41}$$

Thus values for two of the quantities x_1, x_2, x_3 can be chosen arbitrarily and the third is determined from Eq. (41). For example, if $x_1 = 1$ and $x_2 = 0$, then $x_3 = -1$ and

$$\mathbf{x}^{(2)} = \begin{pmatrix} 1 \\ 0 \\ -1 \end{pmatrix} \tag{42}$$

is an eigenvector. Any multiple of $\mathbf{x}^{(2)}$ is also an eigenvector, but a second independent eigenvector can be found by making another choice of x_1 and x_2;

for instance, $x_1 = 0$ and $x_2 = 1$. Again $x_3 = -1$ and

$$\mathbf{x}^{(3)} = \begin{pmatrix} 0 \\ 1 \\ -1 \end{pmatrix} \tag{43}$$

is an eigenvector linearly independent of $\mathbf{x}^{(2)}$. Therefore in this example two linearly independent eigenvectors are associated with the double eigenvalue.

An important special class of matrices, called *self-adjoint* or *Hermitian* matrices, are those for which $\mathbf{A}^* = \mathbf{A}$; that is, $\bar{a}_{ji} = a_{ij}$. Hermitian matrices include as a subclass real symmetric matrices; that is, matrices that have real elements and for which $\mathbf{A}^T = \mathbf{A}$. The eigenvalues and eigenvectors of Hermitian matrices always have the following useful properties.

1. All eigenvalues are real.
2. There always exists a full set of n linearly independent eigenvectors, regardless of the multiplicities of the eigenvalues.
3. If $\mathbf{x}^{(1)}$ and $\mathbf{x}^{(2)}$ are eigenvectors that correspond to different eigenvalues, then $(\mathbf{x}^{(1)}, \mathbf{x}^{(2)}) = 0$. Thus, if all eigenvalues are simple, then the associated eigenvectors form an orthogonal set of vectors.
4. Corresponding to an eigenvalue of multiplicity m, it is possible to choose m eigenvectors that are mutually orthogonal. Thus the full set of n eigenvectors can always be chosen to be orthogonal as well as linearly independent.

Example 5 above involves a real symmetric matrix and illustrates properties 1, 2, and 3, but the choice we have made for $\mathbf{x}^{(2)}$ and $\mathbf{x}^{(3)}$ does not illustrate property 4. However, it is always possible to choose an $\mathbf{x}^{(2)}$ and an $\mathbf{x}^{(3)}$ so that $(\mathbf{x}^{(2)}, \mathbf{x}^{(3)}) = 0$. For example, we could have chosen

$$\mathbf{x}^{(2)} = \begin{pmatrix} 1 \\ 0 \\ -1 \end{pmatrix}, \qquad \mathbf{x}^{(3)} = \begin{pmatrix} 1 \\ -2 \\ 1 \end{pmatrix}$$

as the eigenvectors associated with the eigenvalue $\lambda = -1$ in Example 5. Example 4 shows that if the matrix is not Hermitian, then there may be a deficiency of eigenvectors when an eigenvalue is repeated. The proofs of statements 1 and 3 are outlined in Problems 27 through 29.

In the solution of systems of algebraic equations, as well as in other situations, it is sometimes useful to transform a given matrix into a diagonal matrix, that is, one having nonzero elements only on the diagonal. Eigenvectors are useful in accomplishing such a transformation. Suppose that $\mathbf{A}$ has a full set of n linearly independent eigenvectors (whether $\mathbf{A}$ is Hermitian or not). Letting $\mathbf{x}^{(1)}, \ldots, \mathbf{x}^{(n)}$ denote these eigenvectors and $\lambda_1, \ldots, \lambda_n$ the corresponding eigenvalues, form the matrix $\mathbf{T}$ whose columns are the eigenvectors $\mathbf{x}^{(1)}, \ldots, \mathbf{x}^{(n)}$. Since the columns of $\mathbf{T}$ are linearly independent vectors, $\det \mathbf{T} \neq 0$; hence $\mathbf{T}$ is nonsingular and $\mathbf{T}^{-1}$ exists. A straightforward calculation shows that the columns of the matrix **AT** are just the vectors $\mathbf{Ax}^{(1)}, \ldots, \mathbf{Ax}^{(n)}$. Since $\mathbf{Ax}^{(k)} = \lambda_k \mathbf{x}^{(k)}$, it follows that

$$\mathbf{AT} = \begin{pmatrix} \lambda_1 x_1^{(1)} & \cdots & \lambda_n x_1^{(n)} \\ \vdots & & \vdots \\ \lambda_1 x_n^{(1)} & \cdots & \lambda_n x_n^{(n)} \end{pmatrix} = \mathbf{TD} \tag{44}$$

where

$$\mathbf{D} = \begin{pmatrix} \lambda_1 & 0 & \cdots & 0 \\ 0 & \lambda_2 & \cdots & 0 \\ \vdots & \vdots & & \vdots \\ 0 & 0 & \cdots & \lambda_n \end{pmatrix} \tag{45}$$

is a diagonal matrix whose diagonal elements are the eigenvalues of **A**. From Eq. (44) it follows that

$$\mathbf{T}^{-1}\mathbf{AT} = \mathbf{D}. \tag{46}$$

Thus, if the eigenvalues and eigenvectors of **A** are known, **A** can be transformed into a diagonal matrix by the process shown in Eq. (46). This process is known as a *similarity transformation*, and Eq. (46) is summed up in words by saying that **A** is *similar* to the diagonal matrix **D**. Alternatively, we may say that **A** is *diagonalizable*. The possibility of carrying out such a diagonalization will be important later in the chapter.

If **A** is Hermitian, then the determination of $\mathbf{T}^{-1}$ is very simple. We choose the eigenvectors $\mathbf{x}^{(1)}, \ldots, \mathbf{x}^{(n)}$ of **A** so that they are normalized by $(\mathbf{x}^{(i)}, \mathbf{x}^{(i)}) = 1$ for each i, as well as orthogonal. Then it is easy to verify that $\mathbf{T}^{-1} = \mathbf{T}^*$; in other words, the inverse of **T** is the same as its adjoint (the transpose of its complex conjugate).

Finally, we note that if **A** has fewer than n linearly independent eigenvectors, then there is no matrix **T** such that $\mathbf{T}^{-1}\mathbf{AT} = \mathbf{D}$. In this case, **A** is not similar to a diagonal matrix, or is not diagonalizable.

PROBLEMS

In each of Problems 1 through 5 either solve the given set of equations, or else show that there is no solution.

1. $\begin{aligned} x_1 - x_3 &= 0 \\ 3x_1 + x_2 + x_3 &= 1 \\ -x_1 + x_2 + 2x_3 &= 2 \end{aligned}$

2. $\begin{aligned} x_1 + 2x_2 - x_3 &= 1 \\ 2x_1 + x_2 + x_3 &= 1 \\ x_1 - x_2 + 2x_3 &= 1 \end{aligned}$

3. $\begin{aligned} x_1 + 2x_2 - x_3 &= 2 \\ 2x_1 + x_2 + x_3 &= 1 \\ x_1 - x_2 + 2x_3 &= -1 \end{aligned}$

4. $\begin{aligned} x_1 + 2x_2 - x_3 &= 0 \\ 2x_1 + x_2 + x_3 &= 0 \\ x_1 - x_2 + 2x_3 &= 0 \end{aligned}$

5. $\begin{aligned} x_1 - x_3 &= 0 \\ 3x_1 + x_2 + x_3 &= 0 \\ -x_1 + x_2 + 2x_3 &= 0 \end{aligned}$

6. Determine whether each of the following sets of vectors is linearly independent. If linearly dependent, find the linear relation among them. The vectors are written as row vectors to save space, but may be considered as column vectors; that is, the transposes of the given vectors may be used instead of the vectors themselves.

 (a) $\mathbf{x}^{(1)} = (1, 1, 0)$, $\mathbf{x}^{(2)} = (0, 1, 1)$, $\mathbf{x}^{(3)} = (1, 0, 1)$
 (b) $\mathbf{x}^{(1)} = (2, 1, 0)$, $\mathbf{x}^{(2)} = (0, 1, 0)$, $\mathbf{x}^{(3)} = (-1, 2, 0)$
 (c) $\mathbf{x}^{(1)} = (1, 2, 2, 3)$, $\mathbf{x}^{(2)} = (-1, 0, 3, 1)$, $\mathbf{x}^{(3)} = (-2, -1, 1, 0)$, $\mathbf{x}^{(4)} = (-3, 0, -1, 3)$
 (d) $\mathbf{x}^{(1)} = (1, 2, -1, 0)$, $\mathbf{x}^{(2)} = (2, 3, 1, -1)$, $\mathbf{x}^{(3)} = (-1, 0, 2, 2)$, $\mathbf{x}^{(4)} = (3, -1, 1, 3)$
 (e) $\mathbf{x}^{(1)} = (1, 2, -2)$, $\mathbf{x}^{(2)} = (3, 1, 0)$, $\mathbf{x}^{(3)} = (2, -1, 1)$, $\mathbf{x}^{(4)} = (4, 3, -2)$

7. Suppose that the vectors $\mathbf{x}^{(1)}, \ldots, \mathbf{x}^{(m)}$ each have n components where $n < m$. Show that $\mathbf{x}^{(1)}, \ldots, \mathbf{x}^{(m)}$ are linearly dependent.

8. Determine whether each of the following sets of vectors is linearly independent for $-\infty < t < \infty$. If linearly dependent, find the linear relation among them. As in Problem 6 the vectors are written as row vectors to save space.

 (a) $\mathbf{x}^{(1)}(t) = (e^{-t}, 2e^{-t})$, $\mathbf{x}^{(2)}(t) = (e^{-t}, e^{-t})$, $\mathbf{x}^{(3)}(t) = (3e^{-t}, 0)$
 (b) $\mathbf{x}^{(1)}(t) = (2\sin t, \sin t)$, $\mathbf{x}^{(2)}(t) = (\sin t, 2\sin t)$

9. Let

$$\mathbf{x}^{(1)}(t) = \begin{pmatrix} e^t \\ te^t \end{pmatrix}, \qquad \mathbf{x}^{(2)}(t) = \begin{pmatrix} 1 \\ t \end{pmatrix}.$$

Show that $\mathbf{x}^{(1)}(t)$ and $\mathbf{x}^{(2)}(t)$ are linearly dependent at each point in the interval $0 \le t \le 1$. Nevertheless, show that $\mathbf{x}^{(1)}(t)$ and $\mathbf{x}^{(2)}(t)$ are linearly independent on $0 \le t \le 1$.

In each of Problems 10 through 19 find all eigenvalues and eigenvectors of the given matrix.

10. $\begin{pmatrix} 5 & -1 \\ 3 & 1 \end{pmatrix}$
11. $\begin{pmatrix} 3 & -2 \\ 4 & -1 \end{pmatrix}$
12. $\begin{pmatrix} -2 & 1 \\ 1 & -2 \end{pmatrix}$

13. $\begin{pmatrix} 1 & i \\ -i & 1 \end{pmatrix}$
14. $\begin{pmatrix} 1 & \sqrt{3} \\ \sqrt{3} & -1 \end{pmatrix}$
15. $\begin{pmatrix} 1 & -4 \\ 4 & -7 \end{pmatrix}$

16. $\begin{pmatrix} 1 & 0 & 0 \\ 2 & 1 & -2 \\ 3 & 2 & 1 \end{pmatrix}$
17. $\begin{pmatrix} 3 & 2 & 2 \\ 1 & 4 & 1 \\ -2 & -4 & -1 \end{pmatrix}$

18. $\begin{pmatrix} 1 & 1 & 1 \\ 2 & 1 & -1 \\ -3 & 2 & 4 \end{pmatrix}$
19. $\begin{pmatrix} 3 & 2 & 4 \\ 2 & 0 & 2 \\ 4 & 2 & 3 \end{pmatrix}$

20. For each given matrix $\mathbf{A}$, find $\mathbf{T}$ such that $\mathbf{T}^{-1}\mathbf{A}\mathbf{T} = \mathbf{D}$ where $\mathbf{D}$ is a diagonal matrix. Confirm your choice of $\mathbf{T}$ by calculating $\mathbf{T}^{-1}\mathbf{A}\mathbf{T}$.

 (a) $\mathbf{A}$ is the matrix of Problem 10.
 (b) $\mathbf{A}$ is the matrix of Problem 11.
 (c) $\mathbf{A}$ is the matrix of Problem 13.

Problems 21 through 25 deal with the problem of solving $\mathbf{A}\mathbf{x} = \mathbf{b}$ when $\det \mathbf{A} = 0$.

21. Suppose that, for a given matrix $\mathbf{A}$, there is a nonzero vector $\mathbf{x}$ such that $\mathbf{Ax} = \mathbf{0}$. Show that there is also a nonzero vector $\mathbf{y}$ such that $\mathbf{A}^*\mathbf{y} = \mathbf{0}$.
22. Show that $(\mathbf{Ax}, \mathbf{y}) = (\mathbf{x}, \mathbf{A}^*\mathbf{y})$ for any vectors $\mathbf{x}$ and $\mathbf{y}$.
23. Suppose that $\det \mathbf{A} = 0$ but that $\mathbf{Ax} = \mathbf{b}$ has solutions. Show that $(\mathbf{b}, \mathbf{y}) = 0$, where $\mathbf{y}$ is any solution of $\mathbf{A}^*\mathbf{y} = \mathbf{0}$. Verify that this statement is true for the set of equations in Problem 3.
    Hint: Use the result of Problem 22.
24. Suppose that $\det \mathbf{A} = 0$ but that $\mathbf{x} = \mathbf{x}^{(0)}$ is a solution of $\mathbf{Ax} = \mathbf{b}$. Show that if $\boldsymbol{\xi}$ is a solution of $\mathbf{A}\boldsymbol{\xi} = \mathbf{0}$ and α is any constant, then $\mathbf{x} = \mathbf{x}^{(0)} + \alpha\boldsymbol{\xi}$ is also a solution of $\mathbf{Ax} = \mathbf{b}$.

*25. Suppose that $\det \mathbf{A} = 0$, and that $\mathbf{y}$ is a solution of $\mathbf{A}^*\mathbf{y} = \mathbf{0}$. Show that if $(\mathbf{b}, \mathbf{y}) = 0$ then $\mathbf{Ax} = \mathbf{b}$ has solutions. Note that this is the converse of Problem 23; the form of the solution is given by Problem 24.

26. Prove that $\lambda = 0$ is an eigenvalue of $\mathbf{A}$ if and only if $\mathbf{A}$ is singular.
27. Prove that if $\mathbf{A}$ is Hermitian, then $(\mathbf{Ax}, \mathbf{y}) = (\mathbf{x}, \mathbf{Ay})$, where $\mathbf{x}$ and $\mathbf{y}$ are any vectors.
28. In this problem we show that the eigenvalues of a Hermitian matrix $\mathbf{A}$ are real. Let $\mathbf{x}$ be an eigenvector corresponding to the eigenvalue λ.

    (a) Show that $(\mathbf{Ax}, \mathbf{x}) = (\mathbf{x}, \mathbf{Ax})$. *Hint:* See Problem 27.
    (b) Show that $\lambda(\mathbf{x}, \mathbf{x}) = \bar{\lambda}(\mathbf{x}, \mathbf{x})$. *Hint:* Recall that $\mathbf{Ax} = \lambda\mathbf{x}$.
    (c) Show that $\lambda = \bar{\lambda}$; that is, the eigenvalue λ is real.

29. Show that if λ_1 and λ_2 are eigenvalues of a Hermitian matrix $\mathbf{A}$, and if $\lambda_1 \neq \lambda_2$, then the corresponding eigenvectors $\mathbf{x}^{(1)}$ and $\mathbf{x}^{(2)}$ are orthogonal.
    Hint: Use the results of Problems 27 and 28 to show that $(\lambda_1 - \lambda_2)(\mathbf{x}^{(1)}, \mathbf{x}^{(2)}) = 0$.

7.4 Basic Theory of Systems of First Order Linear Equations

The general theory of a system of n first order linear equations

$$\begin{aligned} x_1' &= p_{11}(t)x_1 + \cdots + p_{1n}(t)x_n + g_1(t), \\ &\vdots \\ x_n' &= p_{n1}(t)x_1 + \cdots + p_{nn}(t)x_n + g_n(t), \end{aligned} \tag{1}$$

closely parallels that of a single linear equation of nth order. The discussion in this section therefore follows the same general lines as that in Sections 3.2, 3.3, and 5.2. To discuss the system (1) most effectively, we write it in matrix notation. That is, we consider $x_1 = \phi_1(t), \ldots, x_n = \phi_n(t)$ to be components of a vector $\mathbf{x} = \boldsymbol{\phi}(t)$; similarly $g_1(t), \ldots, g_n(t)$ are components of a vector $\mathbf{g}(t)$, and $p_{11}(t), \ldots, p_{nn}(t)$ are elements of an $n \times n$ matrix $\mathbf{P}(t)$. Equation (1) then takes the form

$$\mathbf{x}' = \mathbf{P}(t)\mathbf{x} + \mathbf{g}(t). \tag{2}$$

The use of vectors and matrices not only saves a great deal of space and facilitates calculations but also emphasizes the similarity between systems of equations and single (scalar) equations.

A vector $\mathbf{x} = \boldsymbol{\phi}(t)$ is said to be a solution of Eq. (2) if its components satisfy the system of equations (1). Throughout this section we assume that $\mathbf{P}$ and $\mathbf{g}$ are continuous on some interval $\alpha < t < \beta$; that is, each of the scalar functions $p_{11}, \ldots, p_{nn}, g_1, \ldots, g_n$ is continuous there. According to Theorem 7.2, this is sufficient to guarantee the existence of solutions of Eq. (2) on the interval $\alpha < t < \beta$.

It is convenient to consider first the homogeneous equation

$$\mathbf{x}' = \mathbf{P}(t)\mathbf{x} \tag{3}$$

obtained from Eq. (2) by setting $\mathbf{g}(t) = \mathbf{0}$. Once the homogeneous equation has been solved, there are several methods that can be used to solve the nonhomogeneous equation (2); this is taken up in Section 7.9. We use the notation

$$\mathbf{x}^{(1)}(t) = \begin{pmatrix} x_{11}(t) \\ x_{21}(t) \\ \vdots \\ x_{n1}(t) \end{pmatrix}, \ldots, \quad \mathbf{x}^{(k)}(t) = \begin{pmatrix} x_{1k}(t) \\ x_{2k}(t) \\ \vdots \\ x_{nk}(t) \end{pmatrix}, \ldots \tag{4}$$

to designate specific solutions of the system (3). Note that $x_{ij}(t) = x_i^{(j)}(t)$ refers to the ith component of the jth solution $\mathbf{x}^{(j)}(t)$. The main facts about the structure of solutions of the system (3) are stated in Theorems 7.3 to 7.6. They closely resemble the corresponding theorems in Sections 3.2, 3.3, and 5.2; some of the proofs are left to the reader as exercises.

Theorem 7.3. *If the vector functions* $\mathbf{x}^{(1)}$ *and* $\mathbf{x}^{(2)}$ *are solutions of the system* (3), *then the linear combination* $c_1\mathbf{x}^{(1)} + c_2\mathbf{x}^{(2)}$ *is also a solution for any constants* c_1 *and* c_2.

This is the principle of superposition; it is proved simply by differentiating $c_1\mathbf{x}^{(1)} + c_2\mathbf{x}^{(2)}$ and using the fact that $\mathbf{x}^{(1)}$ and $\mathbf{x}^{(2)}$ satisfy Eq. (3). By repeated application of Theorem 7.3 we reach the conclusion that if $\mathbf{x}^{(1)}, \ldots, \mathbf{x}^{(k)}$ are solutions of Eq. (3) then

$$\mathbf{x} = c_1\mathbf{x}^{(1)}(t) + \cdots + c_k\mathbf{x}^{(k)}(t) \tag{5}$$

is also a solution for any constants $c_1, \ldots, c_k$. As an example, it can be verified that

$$\mathbf{x}^{(1)}(t) = \begin{pmatrix} e^{3t} \\ 2e^{3t} \end{pmatrix} = \begin{pmatrix} 1 \\ 2 \end{pmatrix} e^{3t}, \quad \mathbf{x}^{(2)}(t) = \begin{pmatrix} e^{-t} \\ -2e^{-t} \end{pmatrix} = \begin{pmatrix} 1 \\ -2 \end{pmatrix} e^{-t} \tag{6}$$

satisfy the equation

$$\mathbf{x}' = \begin{pmatrix} 1 & 1 \\ 4 & 1 \end{pmatrix} \mathbf{x}. \tag{7}$$

According to Theorem 7.3

$$\begin{aligned}\mathbf{x} &= c_1\begin{pmatrix}1\\2\end{pmatrix}e^{3t} + c_2\begin{pmatrix}1\\-2\end{pmatrix}e^{-t}\\ &= c_1\mathbf{x}^{(1)}(t) + c_2\mathbf{x}^{(2)}(t)\end{aligned} \tag{8}$$

also satisfies Eq. (7).

As we indicated above, by repeatedly applying Theorem 7.3 it follows that every finite linear combination of solutions of Eq. (3) is again a solution. The question now arises as to whether all solutions of Eq. (3) can be found in this way. By analogy with previous cases it is reasonable to expect that for a system of the form (3) of nth order it is sufficient to form linear combinations of n properly chosen solutions. Therefore let $\mathbf{x}^{(1)}, \ldots, \mathbf{x}^{(n)}$ be n solutions of the nth order system (3), and consider the matrix $\mathbf{X}(t)$ whose columns are the vectors $\mathbf{x}^{(1)}(t), \ldots, \mathbf{x}^{(n)}(t)$;

$$\mathbf{X}(t) = \begin{pmatrix} x_{11}(t) & \cdots & x_{1n}(t) \\ \vdots & & \vdots \\ x_{n1}(t) & \cdots & x_{nn}(t) \end{pmatrix}. \tag{9}$$

Recall from Section 7.3 that the columns of $\mathbf{X}(t)$ are linearly independent for a given value of t if and only if $\det \mathbf{X} \neq 0$ for that value of t. This determinant is called the Wronskian of the n solutions $\mathbf{x}^{(1)}, \ldots, \mathbf{x}^{(n)}$ and is also denoted by $W[\mathbf{x}^{(1)}, \ldots, \mathbf{x}^{(n)}]$; that is

$$W[\mathbf{x}^{(1)}, \ldots, \mathbf{x}^{(n)}] = \det \mathbf{X}. \tag{10}$$

The solutions $\mathbf{x}^{(1)}, \ldots, \mathbf{x}^{(n)}$ are then linearly independent at a point if and only if $W[\mathbf{x}^{(1)}, \ldots, \mathbf{x}^{(n)}]$ is not zero there.

Theorem 7.4. *If the vector functions* $\mathbf{x}^{(1)}, \ldots, \mathbf{x}^{(n)}$ *are linearly independent solutions of the system* (3) *for each point in the interval* $\alpha < t < \beta$, *then each solution* $\mathbf{x} = \boldsymbol{\phi}(t)$ *of the system* (3) *can be expressed as a linear combination of* $\mathbf{x}^{(1)}, \ldots, \mathbf{x}^{(n)}$,

$$\boldsymbol{\phi}(t) = c_1\mathbf{x}^{(1)}(t) + \cdots + c_n\mathbf{x}^{(n)}(t), \tag{11}$$

in exactly one way.

Before proving Theorem 7.4, note that according to Theorem 7.3 all expressions of the form (11) are solutions of the system (3), while by Theorem 7.4 all solutions of Eq. (3) can be written in the form (11). If the constants $c_1, \ldots, c_n$ are thought of as arbitrary, then Eq. (11) includes all solutions of the system (3), and it is customary to call it the *general solution*. Any set of solutions $\mathbf{x}^{(1)}, \ldots, \mathbf{x}^{(n)}$ of Eq. (3), which is linearly independent at each point in the interval $\alpha < t < \beta$, is said to be a *fundamental set of solutions* for that interval.

To prove Theorem 7.4 we will show, given any solution $\boldsymbol{\phi}$ of Eq. (3), that $\boldsymbol{\phi}(t) = c_1\mathbf{x}^{(1)}(t) + \cdots + c_n\mathbf{x}^{(n)}(t)$ for suitable values of $c_1, \ldots, c_n$. Let $t = t_0$ be some point in the interval $\alpha < t < \beta$ and let $\boldsymbol{\xi} = \boldsymbol{\phi}(t_0)$. We now wish to determine whether there is any solution of the form $\mathbf{x} = c_1\mathbf{x}^{(1)}(t) + \cdots + c_n\mathbf{x}^{(n)}(t)$ that also satisfies the same initial condition $\mathbf{x}(t_0) = \boldsymbol{\xi}$. That is, we wish to know whether there are values of $c_1, \ldots, c_n$ such that

$$c_1\mathbf{x}^{(1)}(t_0) + \cdots + c_n\mathbf{x}^{(n)}(t_0) = \boldsymbol{\xi}, \tag{12}$$

or in scalar form

$$\begin{aligned} c_1 x_{11}(t_0) + \cdots + c_n x_{1n}(t_0) &= \xi_1, \\ &\vdots \\ c_1 x_{n1}(t_0) + \cdots + c_n x_{nn}(t_0) &= \xi_n. \end{aligned} \tag{13}$$

The necessary and sufficient condition that Eqs. (13) possess a unique solution $c_1, \ldots, c_n$ is precisely the nonvanishing of the determinant of coefficients, which is the Wronskian $W[\mathbf{x}^{(1)}, \ldots, \mathbf{x}^{(n)}]$ evaluated at $t = t_0$. The hypothesis that $\mathbf{x}^{(1)}, \ldots, \mathbf{x}^{(n)}$ are linearly independent throughout $\alpha < t < \beta$ guarantees that $W[\mathbf{x}^{(1)}, \ldots, \mathbf{x}^{(n)}]$ is not zero at $t = t_0$, and therefore there is a (unique) solution of Eq. (3) of the form $\mathbf{x} = c_1\mathbf{x}^{(1)}(t) + \cdots + c_n\mathbf{x}^{(n)}(t)$ that also satisfies the initial condition (12). By the uniqueness part of Theorem 7.2 this solution is identical to $\boldsymbol{\phi}(t)$, and hence $\boldsymbol{\phi}(t) = c_1\mathbf{x}^{(1)}(t) + \cdots + c_n\mathbf{x}^{(n)}(t)$, as was to be proved.

Theorem 7.5. *If $\mathbf{x}^{(1)}, \ldots, \mathbf{x}^{(n)}$ are solutions of Eq. (3) on the interval $\alpha < t < \beta$, then in this interval $W[\mathbf{x}^{(1)}, \ldots, \mathbf{x}^{(n)}]$ either is identically zero or else never vanishes.*

The significance of Theorem 7.5 lies in the fact that it relieves us of the necessity of examining $W[\mathbf{x}^{(1)}, \ldots, \mathbf{x}^{(n)}]$ at all points in the interval of interest, and enables us to determine whether $\mathbf{x}^{(1)}, \ldots, \mathbf{x}^{(n)}$ form a fundamental set of solutions merely by evaluating their Wronskian at any convenient point in the interval.

Theorem 7.5 is proved by first establishing that the Wronskian of $\mathbf{x}^{(1)}, \ldots, \mathbf{x}^{(n)}$ satisfies the differential equation (see Problem 2)

$$\frac{dW}{dt} = (p_{11} + p_{22} + \cdots + p_{nn})W. \tag{14}$$

Hence W is an exponential function, and the conclusion of the theorem follows immediately. The expression for W obtained by solving Eq. (14) is known as Abel's formula; note the analogy with Eq. (15) of Section 3.2.

Alternatively, Theorem 7.5 can be established by showing that if n solutions $\mathbf{x}^{(1)}, \ldots, \mathbf{x}^{(n)}$ of Eq. (3) are linearly dependent at one point $t = t_0$, then they must be linearly dependent at each point in $\alpha < t < \beta$ (see Problem 8). Consequently, if $\mathbf{x}^{(1)}, \ldots, \mathbf{x}^{(n)}$ are linearly independent at one point, they must be linearly independent at each point in the interval.

The next theorem states that the system (3) always has at least one fundamental set of solutions.

Theorem 7.6. *Let*

$$\mathbf{e}^{(1)} = \begin{pmatrix} 1 \\ 0 \\ 0 \\ \vdots \\ 0 \end{pmatrix}, \quad \mathbf{e}^{(2)} = \begin{pmatrix} 0 \\ 1 \\ 0 \\ \vdots \\ 0 \end{pmatrix}, \ldots, \quad \mathbf{e}^{(n)} = \begin{pmatrix} 0 \\ \vdots \\ 0 \\ 1 \end{pmatrix};$$

further let $\mathbf{x}^{(1)}, \ldots, \mathbf{x}^{(n)}$ *be the solutions of the system* (3) *satisfying the initial conditions*

$$\mathbf{x}^{(1)}(t_0) = \mathbf{e}^{(1)}, \ldots, \mathbf{x}^{(n)}(t_0) = \mathbf{e}^{(n)}, \tag{15}$$

respectively, where t_0 *is any point in* $\alpha < t < \beta$. *Then* $\mathbf{x}^{(1)}, \ldots, \mathbf{x}^{(n)}$ *form a fundamental set of solutions of the system* (3).

To prove this theorem, note that the existence and uniqueness of the solutions $\mathbf{x}^{(1)}, \ldots, \mathbf{x}^{(n)}$ mentioned in Theorem 7.6 is assured by Theorem 7.2. It is not hard to see that the Wronskian of these solutions is equal to one when $t = t_0$; therefore $\mathbf{x}^{(1)}, \ldots, \mathbf{x}^{(n)}$ are a fundamental set of solutions.

Once one fundamental set of solutions has been found, other sets can be generated by forming (independent) linear combinations of the first set. For theoretical purposes the set given by Theorem 7.6 is usually the simplest.

To summarize, any set of n linearly independent solutions of the system (3) constitutes a fundamental set of solutions. Under the conditions given in this section, such fundamental sets always exist, and every solution of the system (3) can be represented as a linear combination of any fundamental set of solutions.

PROBLEMS

1. Using matrix algebra, prove the statement following Theorem 7.3 for an arbitrary value of the integer k.
2. In this problem we outline a proof of Theorem 7.5 in the case $n = 2$. Let $\mathbf{x}^{(1)}$ and $\mathbf{x}^{(2)}$ be solutions of Eq. (3) for $\alpha < t < \beta$, and let W be the Wronskian of $\mathbf{x}^{(1)}$ and $\mathbf{x}^{(2)}$.
   (a) Show that

$$\frac{dW}{dt} = \begin{vmatrix} \dfrac{dx_1^{(1)}}{dt} & \dfrac{dx_1^{(2)}}{dt} \\ x_2^{(1)} & x_2^{(2)} \end{vmatrix} + \begin{vmatrix} x_1^{(1)} & x_1^{(2)} \\ \dfrac{dx_2^{(1)}}{dt} & \dfrac{dx_2^{(2)}}{dt} \end{vmatrix}.$$

(b) Using Eq. (3), show that

$$\frac{dW}{dt} = (p_{11} + p_{22})W.$$

(c) Find $W(t)$ by solving the differential equation obtained in part (b). Use this expression to obtain the conclusion stated in Theorem 7.5.

*(d) Generalize this procedure so as to prove Theorem 7.5 for an arbitrary value of n.

3. Show that the Wronskians of two fundamental sets of solutions of the system (3) can differ at most by a multiplicative constant.
Hint: Use Eq. (14).

4. If $x_1 = y$ and $x_2 = y'$, then the second order equation

$$y'' + p(t)y' + q(t)y = 0 \tag{i}$$

corresponds to the system

$$x_1' = x_2,$$
$$x_2' = -q(t)x_1 - p(t)x_2. \tag{ii}$$

Show that if $\mathbf{x}^{(1)}$ and $\mathbf{x}^{(2)}$ are a fundamental set of solutions of Eqs. (ii), and if $y^{(1)}$ and $y^{(2)}$ are a fundamental set of solutions of Eq. (i), then $W[y^{(1)}, y^{(2)}] = cW[\mathbf{x}^{(1)}, \mathbf{x}^{(2)}]$, where c is a nonzero constant.
Hint: $y^{(1)}(t)$ and $y^{(2)}(t)$ must be linear combinations of $x_{11}(t)$ and $x_{12}(t)$.

5. Show that the general solution of $\mathbf{x}' = \mathbf{P}(t)\mathbf{x} + \mathbf{g}(t)$ is the sum of any particular solution $\mathbf{x}^{(p)}$ of this equation and the general solution $\mathbf{x}^{(c)}$ of the corresponding homogeneous equation.

6. Consider the vectors $\mathbf{x}^{(1)}(t) = \begin{pmatrix} t \\ 1 \end{pmatrix}$ and $\mathbf{x}^{(2)}(t) = \begin{pmatrix} t^2 \\ 2t \end{pmatrix}$.

(a) Compute the Wronskian of $\mathbf{x}^{(1)}$ and $\mathbf{x}^{(2)}$.
(b) In what intervals are $\mathbf{x}^{(1)}$ and $\mathbf{x}^{(2)}$ linearly independent?
(c) What conclusion can be drawn about the coefficients in the system of homogeneous differential equations satisfied by $\mathbf{x}^{(1)}$ and $\mathbf{x}^{(2)}$?
(d) Find this system of equations and verify the conclusions of part (c).

7. Consider the vectors $\mathbf{x}^{(1)}(t) = \begin{pmatrix} t^2 \\ 2t \end{pmatrix}$ and $\mathbf{x}^{(2)}(t) = \begin{pmatrix} e^t \\ e^t \end{pmatrix}$, and answer the same questions as in Problem 6.

The following two problems indicate an alternative derivation of Theorem 7.4.

8. Let $\mathbf{x}^{(1)}, \dots, \mathbf{x}^{(m)}$ be solutions of $\mathbf{x}' = \mathbf{P}(t)\mathbf{x}$ on the interval $\alpha < t < \beta$. Assume that $\mathbf{P}$ is continuous and let t_0 be an arbitrary point in the given interval. Show that $\mathbf{x}^{(1)}, \dots, \mathbf{x}^{(m)}$ are linearly dependent for $\alpha < t < \beta$ if (and only if) $\mathbf{x}^{(1)}(t_0), \dots, \mathbf{x}^{(m)}(t_0)$ are linearly dependent. In other words, $\mathbf{x}^{(1)}, \dots, \mathbf{x}^{(m)}$ are linearly dependent on the interval (α, β) if they are linearly dependent at any point in it.
Hint: There are constants $c_1, \dots, c_m$ such that $c_1\mathbf{x}^{(1)}(t_0) + \cdots + c_m\mathbf{x}^{(m)}(t_0) = \mathbf{0}$. Let $\mathbf{z}(t) = c_1\mathbf{x}^{(1)}(t) + \cdots + c_m\mathbf{x}^{(m)}(t)$, and use the uniqueness theorem to show that $\mathbf{z}(t) = \mathbf{0}$ for each t in $\alpha < t < \beta$.

9. Let $\mathbf{x}^{(1)}, \dots, \mathbf{x}^{(n)}$ be linearly independent solutions of $\mathbf{x}' = \mathbf{P}(t)\mathbf{x}$, where $\mathbf{P}$ is continuous on $\alpha < t < \beta$.

(a) Show that any solution $\mathbf{x} = \mathbf{z}(t)$ can be written in the form

$$\mathbf{z}(t) = c_1\mathbf{x}^{(1)}(t) + \cdots + c_n\mathbf{x}^{(n)}(t)$$

for suitable constants $c_1, \dots, c_n$.
Hint: Use the result of Problem 7 of Section 7.3, and also Problem 8 above.

(b) Show that the expression for the solution $\mathbf{z}(t)$ in part (a) is unique; that is, if $\mathbf{z}(t) = k_1\mathbf{x}^{(1)}(t) + \cdots + k_n\mathbf{x}^{(n)}(t)$, then $k_1 = c_1, \ldots, k_n = c_n$.
Hint: Show that $(k_1 - c_1)\mathbf{x}^{(1)}(t) + \cdots + (k_n - c_n)\mathbf{x}^{(n)}(t) = \mathbf{0}$ for each t in $\alpha < t < \beta$ and use the linear independence of $\mathbf{x}^{(1)}, \ldots, \mathbf{x}^{(n)}$.

7.5 Homogeneous Linear Systems with Constant Coefficients

In this section we begin to show how to construct the general solution of a system of homogeneous linear equations with constant coefficients—that is, a system of the form

$$\mathbf{x}' = \mathbf{A}\mathbf{x}, \tag{1}$$

where $\mathbf{A}$ is a constant $n \times n$ matrix. By analogy with the treatment of second order linear equations in Section 3.5, we seek solutions of Eq. (1) of the form

$$\mathbf{x} = \boldsymbol{\xi} e^{rt}, \tag{2}$$

where r and the constant vector $\boldsymbol{\xi}$ are to be determined. Substituting from Eq. (2) for $\mathbf{x}$ in the system (1) gives

$$r\boldsymbol{\xi} e^{rt} = \mathbf{A}\boldsymbol{\xi} e^{rt}.$$

Upon canceling the nonzero scalar factor e^{rt} we obtain $\mathbf{A}\boldsymbol{\xi} = r\boldsymbol{\xi}$, or

$$(\mathbf{A} - r\mathbf{I})\boldsymbol{\xi} = \mathbf{0}, \tag{3}$$

where $\mathbf{I}$ is the $n \times n$ identity matrix. Thus, to solve the system of differential equations (1) we must solve the system of algebraic equations (3). This latter problem is precisely the one that determines the eigenvalues and eigenvectors of the matrix $\mathbf{A}$. Therefore the vector $\mathbf{x}$ given by Eq. (2) is a solution of Eq. (1) provided that r is an eigenvalue and $\boldsymbol{\xi}$ an associated eigenvector of the coefficient matrix $\mathbf{A}$.

EXAMPLE 1

Find the general solution of the system

$$\mathbf{x}' = \begin{pmatrix} 1 & 1 \\ 4 & 1 \end{pmatrix}\mathbf{x}. \tag{4}$$

Assuming that $\mathbf{x} = \boldsymbol{\xi} e^{rt}$, and substituting for $\mathbf{x}$ in Eq. (4), we are led to the system of algebraic equations

$$\begin{pmatrix} 1 - r & 1 \\ 4 & 1 - r \end{pmatrix}\begin{pmatrix} \xi_1 \\ \xi_2 \end{pmatrix} = \begin{pmatrix} 0 \\ 0 \end{pmatrix}. \tag{5}$$

Equations (5) have a nontrivial solution if and only if the determinant of coefficients is zero. Thus allowable values of r are found from the equation

$$\begin{vmatrix} 1-r & 1 \\ 4 & 1-r \end{vmatrix} = (1-r)^2 - 4$$
$$= r^2 - 2r - 3 = 0. \tag{6}$$

Equation (6) has the roots $r_1 = 3$ and $r_2 = -1$; these are the eigenvalues of the coefficient matrix in Eq. (4). If $r = 3$, then the system (5) reduces to the single equation

$$-2\xi_1 + \xi_2 = 0. \tag{7}$$

Thus $\xi_2 = 2\xi_1$ and the eigenvector corresponding to $r_1 = 3$ can be taken as

$$\boldsymbol{\xi}^{(1)} = \begin{pmatrix} 1 \\ 2 \end{pmatrix}. \tag{8}$$

Similarly, corresponding to $r_2 = -1$, we find that $\xi_2 = -2\xi_1$, so the eigenvector is

$$\boldsymbol{\xi}^{(2)} = \begin{pmatrix} 1 \\ -2 \end{pmatrix}. \tag{9}$$

The corresponding solutions of the differential equation are

$$\mathbf{x}^{(1)}(t) = \begin{pmatrix} 1 \\ 2 \end{pmatrix} e^{3t}, \qquad \mathbf{x}^{(2)}(t) = \begin{pmatrix} 1 \\ -2 \end{pmatrix} e^{-t}. \tag{10}$$

The Wronskian of these solutions is

$$W[\mathbf{x}^{(1)}, \mathbf{x}^{(2)}](t) = \begin{vmatrix} e^{3t} & e^{-t} \\ 2e^{3t} & -2e^{-t} \end{vmatrix} = -4e^{2t}, \tag{11}$$

which is never zero. Hence the solutions $\mathbf{x}^{(1)}$ and $\mathbf{x}^{(2)}$ form a fundamental set, and the general solution of the system (4) is

$$\mathbf{x} = c_1\mathbf{x}^{(1)}(t) + c_2\mathbf{x}^{(2)}(t)$$
$$= c_1\begin{pmatrix} 1 \\ 2 \end{pmatrix} e^{3t} + c_2\begin{pmatrix} 1 \\ -2 \end{pmatrix} e^{-t}, \tag{12}$$

where c_1 and c_2 are arbitrary constants.

To visualize the solution (12) it is helpful to consider its graph in the x_1x_2 plane for various values of the constants c_1 and c_2. We start with $\mathbf{x} = c_1\mathbf{x}^{(1)}(t)$, or in scalar form

$$x_1 = c_1 e^{3t}, \qquad x_2 = 2c_1 e^{3t}.$$

By eliminating t between these two equations, we see that this solution lies on the straight line $x_2 = 2x_1$; see Figure 7.6. This is the line through the origin in the direction of the eigenvector $\boldsymbol{\xi}^{(1)}$. If we look upon the solution as the trajectory of a moving particle, then the particle is in the first quadrant when $c_1 > 0$ and in the third quadrant when $c_1 < 0$. In either case the particle recedes from the origin as t increases. Next consider $\mathbf{x} = c_2\mathbf{x}^{(2)}(t)$, or

$$x_1 = c_2 e^{-t}, \qquad x_2 = -2c_2 e^{-t}.$$

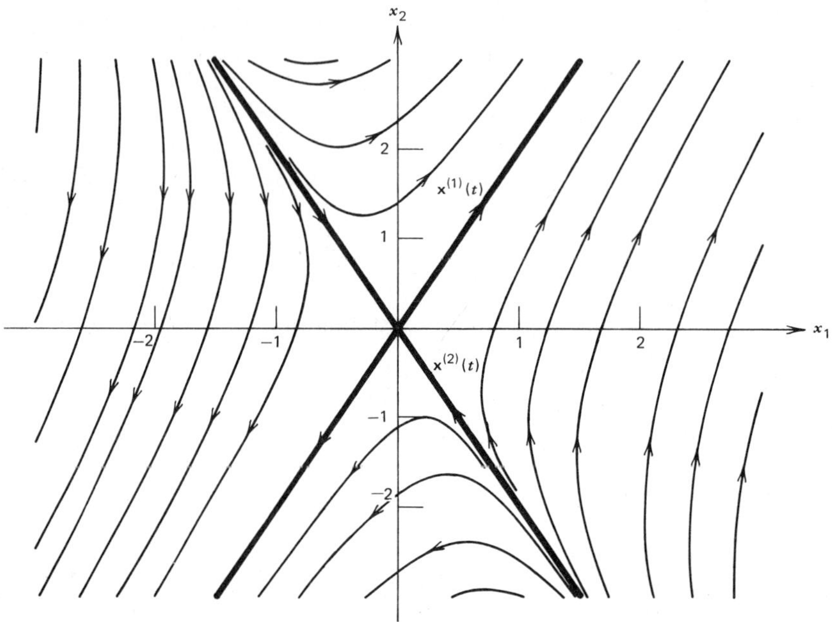

FIGURE 7.6 Trajectories of the system (4). The origin is a saddle point.

This solution lies on the line $x_2 = -2x_1$, whose direction is determined by the eigenvector $\boldsymbol{\xi}^{(2)}$. The solution is in the fourth quadrant when $c_2 > 0$ and in the second quadrant when $c_2 < 0$, as shown in Figure 7.6. In both cases the particle moves toward the origin as t increases. The solution (12) is a combination of $\mathbf{x}^{(1)}(t)$ and $\mathbf{x}^{(2)}(t)$. For large t, the term $c_1\mathbf{x}^{(1)}(t)$ is dominant and the term $c_2\mathbf{x}^{(2)}(t)$ becomes negligible. Thus all solutions for which $c_1 \neq 0$ are asymptotic to the line $x_2 = 2x_1$ as $t \to \infty$. Similarly all solutions for which $c_2 \neq 0$ are asymptotic to the line $x_2 = -2x_1$ as $t \to -\infty$. The graphs of several solutions are shown in Figure 7.6. The pattern of trajectories in this figure is typical of all second order systems $\mathbf{x}' = \mathbf{A}\mathbf{x}$ for which the eigenvalues are real and of opposite signs. The origin is often called a saddle point in this case.

Returning to the general system (1) we proceed as in the example. To find solutions of the differential equation (1) we must find the eigenvalues and eigenvectors of $\mathbf{A}$ from the associated algebraic system (3). The eigenvalues $r_1, \ldots, r_n$ (which need not all be different) are roots of the polynomial equation

$$\det(\mathbf{A} - r\mathbf{I}) = 0. \tag{13}$$

The nature of the eigenvalues and the corresponding eigenvectors determines the nature of the general solution of the system (1).

HERMITIAN SYSTEMS. The situation is simplest when **A** is a Hermitian matrix. As stated in Section 7.3, the eigenvalues $r_1, \ldots, r_n$ are all real in this case. Further, even if some of the eigenvalues are repeated, there is always a full set of n eigenvectors $\boldsymbol{\xi}^{(1)}, \ldots, \boldsymbol{\xi}^{(n)}$ that are linearly independent (in fact, orthogonal). Hence the corresponding solutions of the differential system (1) are

$$\mathbf{x}^{(1)}(t) = \boldsymbol{\xi}^{(1)} e^{r_1 t}, \ldots, \mathbf{x}^{(n)}(t) = \boldsymbol{\xi}^{(n)} e^{r_n t}. \tag{14}$$

To show that these solutions form a fundamental set, we evaluate their Wronskian:

$$\begin{aligned} W[\mathbf{x}^{(1)}, \ldots, \mathbf{x}^{(n)}](t) &= \begin{vmatrix} \xi_1^{(1)} e^{r_1 t} & \cdots & \xi_1^{(n)} e^{r_n t} \\ \vdots & & \vdots \\ \xi_n^{(1)} e^{r_1 t} & \cdots & \xi_n^{(n)} e^{r_n t} \end{vmatrix} \\ &= e^{(r_1 + \cdots + r_n)t} \begin{vmatrix} \xi_1^{(1)} & \cdots & \xi_1^{(n)} \\ \vdots & & \vdots \\ \xi_n^{(1)} & \cdots & \xi_n^{(n)} \end{vmatrix}. \end{aligned} \tag{15}$$

First we observe that the exponential function is never zero. Next, since the eigenvectors $\boldsymbol{\xi}^{(1)}, \ldots, \boldsymbol{\xi}^{(n)}$ are linearly independent, the determinant in the last term of Eq. (15) is nonzero. As a consequence, the Wronskian $W[\mathbf{x}^{(1)}, \ldots, \mathbf{x}^{(n)}](t)$ is never zero; hence $\mathbf{x}^{(1)}, \ldots, \mathbf{x}^{(n)}$ form a fundamental set of solutions. Thus when **A** is a Hermitian matrix, the general solution of Eq. (1) is

$$\mathbf{x} = c_1 \boldsymbol{\xi}^{(1)} e^{r_1 t} + \cdots + c_n \boldsymbol{\xi}^{(n)} e^{r_n t}. \tag{16}$$

An important subclass of Hermitian matrices is the class of real, symmetric matrices. If **A** is real and symmetric, then the eigenvectors $\boldsymbol{\xi}^{(1)}, \ldots, \boldsymbol{\xi}^{(n)}$ as well as the eigenvalues $r_1, \ldots, r_n$ are all real. Hence the solutions given by Eq. (14) are real valued. However, if the Hermitian matrix **A** is not real, then in general the eigenvectors have nonzero imaginary parts, and the solutions (14) are complex valued.

EXAMPLE 2

Find the general solution of

$$\mathbf{x}' = \begin{pmatrix} -3 & \sqrt{2} \\ \sqrt{2} & -2 \end{pmatrix} \mathbf{x}. \tag{17}$$

The coefficient matrix in Eq. (17) is real and symmetric, so the results just described apply to this problem. Assuming that $\mathbf{x} = \boldsymbol{\xi} e^{rt}$, we obtain the algebraic system

$$\begin{pmatrix} -3 - r & \sqrt{2} \\ \sqrt{2} & -2 - r \end{pmatrix} \begin{pmatrix} \xi_1 \\ \xi_2 \end{pmatrix} = \begin{pmatrix} 0 \\ 0 \end{pmatrix}. \tag{18}$$

The eigenvalues satisfy

$$\begin{aligned}(-3-r)(-2-r)-2 &= r^2+5r+4\\ &= (r+1)(r+4)=0,\end{aligned}\tag{19}$$

so $r_1 = -1$ and $r_2 = -4$. For $r = -1$, Eq. (18) becomes

$$\begin{pmatrix}-2 & \sqrt{2}\\ \sqrt{2} & -1\end{pmatrix}\begin{pmatrix}\xi_1\\ \xi_2\end{pmatrix} = \begin{pmatrix}0\\0\end{pmatrix}.\tag{20}$$

Hence $\xi_2 = \sqrt{2}\,\xi_1$ and the eigenvector $\boldsymbol{\xi}^{(1)}$ corresponding to the eigenvalue $r_1 = -1$ can be taken as

$$\boldsymbol{\xi}^{(1)} = \begin{pmatrix}1\\ \sqrt{2}\end{pmatrix}.\tag{21}$$

Similarly, corresponding to the eigenvalue $r_2 = -4$, we have $\xi_1 = -\sqrt{2}\,\xi_2$, so the eigenvector is

$$\boldsymbol{\xi}^{(2)} = \begin{pmatrix}-\sqrt{2}\\ 1\end{pmatrix}.\tag{22}$$

Thus a fundamental set of solutions of the system (17) is

$$\mathbf{x}^{(1)}(t) = \begin{pmatrix}1\\ \sqrt{2}\end{pmatrix}e^{-t},\qquad \mathbf{x}^{(2)}(t) = \begin{pmatrix}-\sqrt{2}\\ 1\end{pmatrix}e^{-4t},\tag{23}$$

and the general solution is

$$\mathbf{x} = c_1\mathbf{x}^{(1)}(t) + c_2\mathbf{x}^{(2)}(t) = c_1\begin{pmatrix}1\\ \sqrt{2}\end{pmatrix}e^{-t} + c_2\begin{pmatrix}-\sqrt{2}\\ 1\end{pmatrix}e^{-4t}.\tag{24}$$

Graphs of the solution (24) for several values of c_1 and c_2 are shown in Figure 7.7. The solution $\mathbf{x}^{(1)}(t)$ approaches the origin along the line $x_2 = \sqrt{2}\,x_1$, while the solution $\mathbf{x}^{(2)}(t)$ approaches the origin along the line $x_1 = -\sqrt{2}\,x_2$. The directions of these lines are determined by the eigenvectors $\boldsymbol{\xi}^{(1)}$ and $\boldsymbol{\xi}^{(2)}$, respectively. In general, we have a combination of these two fundamental solutions. As $t \to \infty$ the solution $\mathbf{x}^{(2)}(t)$ is negligible compared to $\mathbf{x}^{(1)}(t)$. Thus, unless $c_1 = 0$, the solution (24) approaches the origin tangent to the line $x_2 = \sqrt{2}\,x_1$. The pattern of trajectories shown in Figure 7.7 is typical of all second order systems $\mathbf{x}' = \mathbf{Ax}$ for which the eigenvalues are real, different, and of the same sign. The origin is often called a node for such a system. If the eigenvalues were positive rather than negative, then the trajectories would be similar but traversed in the outward direction.

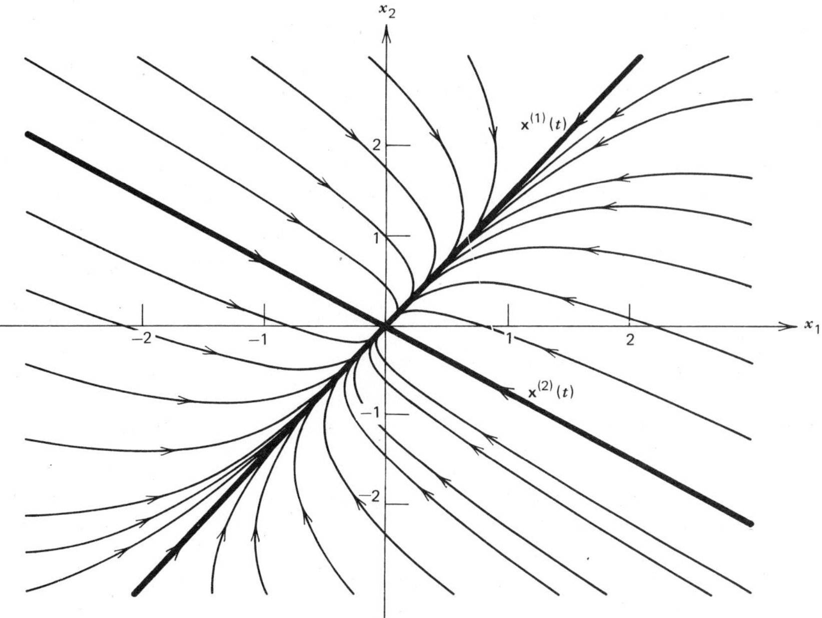

FIGURE 7.7 Trajectories of the system (17). The origin is a node.

EXAMPLE 3

Find the general solution of

$$\mathbf{x}' = \begin{pmatrix} 0 & 1 & 1 \\ 1 & 0 & 1 \\ 1 & 1 & 0 \end{pmatrix} \mathbf{x}. \tag{25}$$

Again, we observe that the coefficient matrix is real and symmetric. The eigenvalues and eigenvectors of this matrix were found in Example 5 of Section 7.3, namely,

$$r_1 = 2; \qquad \boldsymbol{\xi}^{(1)} = \begin{pmatrix} 1 \\ 1 \\ 1 \end{pmatrix} \tag{26}$$

$$r_2 = -1, \qquad r_3 = -1; \qquad \boldsymbol{\xi}^{(2)} = \begin{pmatrix} 1 \\ 0 \\ -1 \end{pmatrix}, \qquad \boldsymbol{\xi}^{(3)} = \begin{pmatrix} 0 \\ 1 \\ -1 \end{pmatrix}. \tag{27}$$

Hence a fundamental set of solutions of Eq. (25) is

$$\mathbf{x}^{(1)}(t) = \begin{pmatrix} 1 \\ 1 \\ 1 \end{pmatrix} e^{2t}, \qquad \mathbf{x}^{(2)}(t) = \begin{pmatrix} 1 \\ 0 \\ -1 \end{pmatrix} e^{-t}, \qquad \mathbf{x}^{(3)}(t) = \begin{pmatrix} 0 \\ 1 \\ -1 \end{pmatrix} e^{-t} \tag{28}$$

and the general solution is

$$\mathbf{x} = c_1 \begin{pmatrix} 1 \\ 1 \\ 1 \end{pmatrix} e^{2t} + c_2 \begin{pmatrix} 1 \\ 0 \\ -1 \end{pmatrix} e^{-t} + c_3 \begin{pmatrix} 0 \\ 1 \\ -1 \end{pmatrix} e^{-t}. \tag{29}$$

This example illustrates the fact that even though an eigenvalue ($r = -1$) has multiplicity two, it may still be possible to find two linearly independent eigenvectors $\boldsymbol{\xi}^{(2)}$ and $\boldsymbol{\xi}^{(3)}$ and, as a consequence, to construct the general solution (29).

NON-HERMITIAN SYSTEMS. If the coefficient matrix $\mathbf{A}$ in the system (1)

$$\mathbf{x}' = \mathbf{A}\mathbf{x}$$

is not Hermitian, then the situation regarding the solution is more complicated. Let us suppose first that $\mathbf{A}$ is real. Then there are three possibilities for the eigenvalues of $\mathbf{A}$:

1. All eigenvalues are real and distinct.
2. Some eigenvalues occur in complex conjugate pairs.
3. Some eigenvalues are repeated.

The first case leads to no difficulties. There is a single linearly independent real eigenvector for each eigenvalue, and consequently there are n linearly independent solutions of the form (14). Hence the general solution is still given by Eq. (16), where it is now understood that $r_1, \ldots, r_n$ are all different. This case is illustrated by Example 1 above.

If some of the eigenvalues occur in complex conjugate pairs, then there are still n linearly independent solutions of the form (14), provided all of the eigenvalues are different. Of course, the solutions arising from complex eigenvalues are complex valued. However, as in Section 3.5.1, it is possible to obtain a full set of real-valued solutions. This is discussed in Section 7.6.

More serious difficulties can occur if an eigenvalue is repeated. In this event the number of corresponding linearly independent eigenvectors may be smaller than the multiplicity of the eigenvalue, as in Example 4 of Section 7.3. If so, the number of linearly independent solutions of the form $\boldsymbol{\xi}e^{rt}$ will be smaller than n. To construct a fundamental set of solutions it is then necessary to seek additional solutions of another form. The situation is somewhat analogous to that for an nth order linear equation with constant coefficients; a repeated root of the characteristic equation gave rise to solutions of the form $e^{rx}, xe^{rx}, x^2e^{rx}, \ldots$. The case of repeated eigenvalues is treated in Section 7.7.

Finally, if $\mathbf{A}$ is complex, but not Hermitian, then complex eigenvalues need not occur in conjugate pairs, and the eigenvectors are normally complex valued even though the associated eigenvalue may be real. The solutions of the differential equation (1) are still of the form (14), provided the eigenvalues are distinct, but in general all of the solutions are complex valued.

PROBLEMS

In each of Problems 1 through 6 find the general solution of the given system of equations. Also sketch a few of the trajectories.

1. $\mathbf{x}' = \begin{pmatrix} 3 & -2 \\ 2 & -2 \end{pmatrix}\mathbf{x}$

2. $\mathbf{x}' = \begin{pmatrix} 1 & -2 \\ 3 & -4 \end{pmatrix}\mathbf{x}$

3. $\mathbf{x}' = \begin{pmatrix} 2 & -1 \\ 3 & -2 \end{pmatrix}\mathbf{x}$

4. $\mathbf{x}' = \begin{pmatrix} 1 & 1 \\ 4 & -2 \end{pmatrix}\mathbf{x}$

5. $\mathbf{x}' = \begin{pmatrix} -2 & 1 \\ 1 & -2 \end{pmatrix}\mathbf{x}$

6. $\mathbf{x}' = \begin{pmatrix} \frac{5}{4} & \frac{3}{4} \\ \frac{3}{4} & \frac{5}{4} \end{pmatrix}\mathbf{x}$

In each of Problems 7 through 9 find the general solution of the given system of equations. Also sketch a few of the trajectories. In each of these problems the coefficient matrix has a zero eigenvalue. As a result, the pattern of trajectories is different from those in the examples in the text.

7. $\mathbf{x}' = \begin{pmatrix} 4 & -3 \\ 8 & -6 \end{pmatrix}\mathbf{x}$

8. $\mathbf{x}' = \begin{pmatrix} 2 & 10 \\ -1 & -5 \end{pmatrix}\mathbf{x}$

9. $\mathbf{x}' = \begin{pmatrix} 3 & 6 \\ -1 & -2 \end{pmatrix}\mathbf{x}$

In each of Problems 10 through 15 find the general solution of the given system of equations.

10. $\mathbf{x}' = \begin{pmatrix} 1 & i \\ -i & 1 \end{pmatrix}\mathbf{x}$

11. $\mathbf{x}' = \begin{pmatrix} 2 & 2+i \\ -1 & -1-i \end{pmatrix}\mathbf{x}$

12. $\mathbf{x}' = \begin{pmatrix} 1 & 1 & 2 \\ 1 & 2 & 1 \\ 2 & 1 & 1 \end{pmatrix}\mathbf{x}$

13. $\mathbf{x}' = \begin{pmatrix} 3 & 2 & 4 \\ 2 & 0 & 2 \\ 4 & 2 & 3 \end{pmatrix}\mathbf{x}$

14. $\mathbf{x}' = \begin{pmatrix} 1 & 1 & 1 \\ 2 & 1 & -1 \\ -8 & -5 & -3 \end{pmatrix}\mathbf{x}$

15. $\mathbf{x}' = \begin{pmatrix} 1 & -1 & 4 \\ 3 & 2 & -1 \\ 2 & 1 & -1 \end{pmatrix}\mathbf{x}$

In each of Problems 16 through 19 solve the given initial value problem.

16. $\mathbf{x}' = \begin{pmatrix} 5 & -1 \\ 3 & 1 \end{pmatrix}\mathbf{x}, \quad \mathbf{x}(0) = \begin{pmatrix} 2 \\ -1 \end{pmatrix}$

17. $\mathbf{x}' = \begin{pmatrix} -2 & 1 \\ -5 & 4 \end{pmatrix}\mathbf{x}, \quad \mathbf{x}(0) = \begin{pmatrix} 1 \\ 3 \end{pmatrix}$

18. $\mathbf{x}' = \begin{pmatrix} 1 & 1 & 2 \\ 0 & 2 & 2 \\ -1 & 1 & 3 \end{pmatrix}\mathbf{x}, \quad \mathbf{x}(0) = \begin{pmatrix} 2 \\ 0 \\ 1 \end{pmatrix}$

19. $\mathbf{x}' = \begin{pmatrix} 0 & 0 & -1 \\ 2 & 0 & 0 \\ -1 & 2 & 4 \end{pmatrix}\mathbf{x}, \quad \mathbf{x}(0) = \begin{pmatrix} 7 \\ 5 \\ 5 \end{pmatrix}$

20. The system $t\mathbf{x}' = \mathbf{A}\mathbf{x}$ is analogous to the second order Euler equation (Section 4.4). Assuming that $\mathbf{x} = \boldsymbol{\xi} t^r$, where $\boldsymbol{\xi}$ is a constant vector, show that $\boldsymbol{\xi}$ and r must satisfy $(\mathbf{A} - r\mathbf{I})\boldsymbol{\xi} = \mathbf{0}$; also show that to obtain nontrivial solutions of the given differential equation, r must be a root of the characteristic equation $\det(\mathbf{A} - r\mathbf{I}) = 0$.

Referring to Problem 20, solve the given system of equations in each of Problems 21 through 24. Assume that $t > 0$.

21. $t\mathbf{x}' = \begin{pmatrix} 2 & -1 \\ 3 & -2 \end{pmatrix}\mathbf{x}$

22. $t\mathbf{x}' = \begin{pmatrix} 5 & -1 \\ 3 & 1 \end{pmatrix}\mathbf{x}$

23. $t\mathbf{x}' = \begin{pmatrix} 4 & -3 \\ 8 & -6 \end{pmatrix}\mathbf{x}$

24. $t\mathbf{x}' = \begin{pmatrix} 3 & -2 \\ 2 & -2 \end{pmatrix}\mathbf{x}$

25. Consider a second order system $\mathbf{x}' = \mathbf{A}\mathbf{x}$. Assuming that $r_1 \neq r_2$, the general solution is $\mathbf{x} = c_1\boldsymbol{\xi}^{(1)}e^{r_1 t} + c_2\boldsymbol{\xi}^{(2)}e^{r_2 t}$, provided $\boldsymbol{\xi}^{(1)}$ and $\boldsymbol{\xi}^{(2)}$ are linearly independent. In this problem we establish the linear independence of $\boldsymbol{\xi}^{(1)}$ and $\boldsymbol{\xi}^{(2)}$ by assuming that they are linearly dependent, and then showing that this leads to a contradiction.
(a) Note that $\boldsymbol{\xi}^{(1)}$ satisfies the matrix equation $(\mathbf{A} - r_1\mathbf{I})\boldsymbol{\xi}^{(1)} = \mathbf{0}$; similarly, note that $(\mathbf{A} - r_2\mathbf{I})\boldsymbol{\xi}^{(2)} = \mathbf{0}$.
(b) Show that $(\mathbf{A} - r_2\mathbf{I})\boldsymbol{\xi}^{(1)} = (r_1 - r_2)\boldsymbol{\xi}^{(1)}$.
(c) Suppose that $\boldsymbol{\xi}^{(1)}$ and $\boldsymbol{\xi}^{(2)}$ are linearly dependent. Then $c_1\boldsymbol{\xi}^{(1)} + c_2\boldsymbol{\xi}^{(2)} = \mathbf{0}$ and at least one of c_1 and c_2 is not zero; suppose $c_1 \neq 0$. Show that $(\mathbf{A} - r_2\mathbf{I})(c_1\boldsymbol{\xi}^{(1)} + c_2\boldsymbol{\xi}^{(2)}) = \mathbf{0}$, and also show that $(\mathbf{A} - r_2\mathbf{I})(c_1\boldsymbol{\xi}^{(1)} + c_2\boldsymbol{\xi}^{(2)}) = c_1(r_1 - r_2)\boldsymbol{\xi}^{(1)}$. Hence $c_1 = 0$, which is a contradiction. Therefore $\boldsymbol{\xi}^{(1)}$ and $\boldsymbol{\xi}^{(2)}$ are linearly independent.
(d) Modify the argument of part (c) in case c_1 is zero but c_2 is not.
(e) Carry out a similar argument for the case in which the order $n = 3$; note that the procedure can be extended to cover an arbitrary value of n.

26. Consider the equation

$$ay'' + by' + cy = 0 \tag{i}$$

where a, b, and c are constants. In Chapter 3 it was shown that the general solution depended upon the roots of the characteristic equation

$$ar^2 + br + c = 0. \tag{ii}$$

(a) Transform Eq. (i) into a system of first order equations by letting $x_1 = y$, $x_2 = y'$. Find the system of equations $\mathbf{x}' = \mathbf{A}\mathbf{x}$ satisfied by $\mathbf{x} = \begin{pmatrix} x_1 \\ x_2 \end{pmatrix}$.
(b) Find the equation that determines the eigenvalues of the coefficient matrix $\mathbf{A}$ in part (a). Note that this equation is just the characteristic equation (ii) of Eq. (i).

27. **Reduction of Order.** This is a method for dealing with systems that do not have a complete set of solutions of the form $\boldsymbol{\xi}e^{rt}$. Consider the system

$$\mathbf{x}' = \begin{pmatrix} 3 & -2 \\ 2 & -2 \end{pmatrix}\mathbf{x}. \tag{i}$$

(a) Verify that $\mathbf{x} = \begin{pmatrix} 2 \\ 1 \end{pmatrix}e^{2t}$ satisfies the given differential equation.
(b) Introduce a new dependent variable by the transformation

$$\mathbf{x} = \begin{pmatrix} 1 & 2e^{2t} \\ 0 & e^{2t} \end{pmatrix}\mathbf{y}. \tag{ii}$$

Observe that this transformation is obtained by replacing the second column of the identity matrix by the known solution. By substituting for $\mathbf{x}$ in Eq. (i), show that $\mathbf{y}$

satisfies the system of equations

$$\begin{pmatrix} 1 & 2e^{2t} \\ 0 & e^{2t} \end{pmatrix}\mathbf{y}' = \begin{pmatrix} 3 & 0 \\ 2 & 0 \end{pmatrix}\mathbf{y}. \tag{iii}$$

(c) Solve Eq. (iii) and show that

$$\mathbf{y} = c_1\begin{pmatrix} \frac{1}{2}e^{-t} \\ -\frac{1}{3}e^{-3t} \end{pmatrix} + c_2\begin{pmatrix} 0 \\ 1 \end{pmatrix} \tag{iv}$$

where c_1 and c_2 are arbitrary constants.
(d) Using Eq. (ii), show that

$$\mathbf{x} = -\frac{c_1}{6}\begin{pmatrix} 1 \\ 2 \end{pmatrix}e^{-t} + c_2\begin{pmatrix} 2 \\ 1 \end{pmatrix}e^{2t}; \tag{v}$$

the first term is a second independent solution of Eq. (i). This is the method of reduction of order as it applies to a second order system of equations.

In Problems 28 and 29 use the method of reduction of order (Problem 27) to solve the given system of equations.

28. $\mathbf{x}' = \begin{pmatrix} 1 & -4 \\ 4 & -7 \end{pmatrix}\mathbf{x}$

29. $\mathbf{x}' = \begin{pmatrix} 3 & -4 \\ 1 & -1 \end{pmatrix}\mathbf{x}$

Electric Circuits. Problems 30 and 31 are concerned with the electric circuit described by the system of differential equations derived in Problem 9 of Section 7.1:

$$\frac{d}{dt}\begin{pmatrix} I \\ V \end{pmatrix} = \begin{pmatrix} -\dfrac{R_1}{L} & -\dfrac{1}{L} \\ \dfrac{1}{C} & -\dfrac{1}{CR_2} \end{pmatrix}\begin{pmatrix} I \\ V \end{pmatrix}. \tag{i}$$

30. (a) Find the general solution of Eq. (i) if $R_1 = 1$ ohm, $R_2 = \frac{3}{5}$ ohm, $L = 2$ henrys, and $C = \frac{2}{3}$ farad.
(b) Show that $I(t) \to 0$ and $V(t) \to 0$ as $t \to \infty$ regardless of the initial values $I(0)$ and $V(0)$.
31. Consider the system of differential equations (i) above.
(a) Find a condition on R_1, R_2, C, and L that must be satisfied if the eigenvalues of the coefficient matrix are to be real and different.
(b) If the condition found in part (a) is satisfied, show that both eigenvalues are negative. Then show that $I(t) \to 0$ and $V(t) \to 0$ as $t \to \infty$ regardless of the initial conditions.
*(c) If the condition found in part (a) is not satisfied, then the eigenvalues are either complex or repeated. Do you think that $I(t) \to 0$ and $V(t) \to 0$ as $t \to \infty$ in these cases as well?
Hint: In part (c) one approach is to change the system (i) into a single second order equation. We also discuss complex and repeated eigenvalues in Sections 7.6 and 7.7.

7.6 Complex Eigenvalues

In this section we again consider a system of n linear homogeneous equations with constant coefficients

$$\mathbf{x}' = \mathbf{A}\mathbf{x} \tag{1}$$

where we now assume that the coefficient matrix $\mathbf{A}$ is real valued. If we seek solutions of the form $\mathbf{x} = \boldsymbol{\xi} e^{rt}$, then it follows as in Section 7.5 that r must be an eigenvalue and $\boldsymbol{\xi}$ a corresponding eigenvector of the coefficient matrix $\mathbf{A}$. Recall that the eigenvalues $r_1, \ldots, r_n$ of $\mathbf{A}$ are the roots of the equation

$$\det(\mathbf{A} - r\mathbf{I}) = 0, \tag{2}$$

and that the corresponding eigenvectors satisfy

$$(\mathbf{A} - r\mathbf{I})\boldsymbol{\xi} = \mathbf{0}. \tag{3}$$

If $\mathbf{A}$ is real, then the coefficients in the polynomial equation (2) for r are real, and any complex eigenvalues must occur in conjugate pairs. For example, if $r_1 = \lambda + i\mu$, where λ and μ are real, is an eigenvalue of $\mathbf{A}$, then so is $r_2 = \lambda - i\mu$. Further, the corresponding eigenvectors $\boldsymbol{\xi}^{(1)}$ and $\boldsymbol{\xi}^{(2)}$ are also complex conjugates. To see that this is so, suppose that r_1 and $\boldsymbol{\xi}^{(1)}$ satisfy

$$(\mathbf{A} - r_1\mathbf{I})\boldsymbol{\xi}^{(1)} = \mathbf{0}. \tag{4}$$

On taking the complex conjugate of this equation, and noting that $\mathbf{A}$ and $\mathbf{I}$ are real valued, we obtain

$$(\mathbf{A} - \bar{r}_1\mathbf{I})\bar{\boldsymbol{\xi}}^{(1)} = \mathbf{0}, \tag{5}$$

where $\bar{r}_1$ and $\bar{\boldsymbol{\xi}}^{(1)}$ are the complex conjugates of r_1 and $\boldsymbol{\xi}^{(1)}$, respectively. In other words, $r_2 = \bar{r}_1$ is also an eigenvalue, and $\boldsymbol{\xi}^{(2)} = \bar{\boldsymbol{\xi}}^{(1)}$ is the corresponding eigenvector. The corresponding solutions

$$\mathbf{x}^{(1)}(t) = \boldsymbol{\xi}^{(1)} e^{r_1 t}, \qquad \mathbf{x}^{(2)}(t) = \bar{\boldsymbol{\xi}}^{(1)} e^{\bar{r}_1 t} \tag{6}$$

of the differential equation (1) are then complex conjugates of each other. Therefore, by an extension of Theorem 3.10 to systems (which is a simple matter), we can find two real-valued solutions of Eq. (1) corresponding to the eigenvalues r_1 and r_2 by taking the real and imaginary parts of $\mathbf{x}^{(1)}(t)$ or $\mathbf{x}^{(2)}(t)$ given by Eq. (6).

Let us write $\boldsymbol{\xi}^{(1)} = \mathbf{a} + i\mathbf{b}$, where $\mathbf{a}$ and $\mathbf{b}$ are real; then we have

$$\begin{aligned} \mathbf{x}^{(1)}(t) &= (\mathbf{a} + i\mathbf{b})e^{(\lambda + i\mu)t} \\ &= (\mathbf{a} + i\mathbf{b})e^{\lambda t}(\cos\mu t + i\sin\mu t). \end{aligned} \tag{7}$$

Upon separating $\mathbf{x}^{(1)}(t)$ into its real and imaginary parts, we obtain

$$\mathbf{x}^{(1)}(t) = e^{\lambda t}(\mathbf{a}\cos\mu t - \mathbf{b}\sin\mu t) + ie^{\lambda t}(\mathbf{a}\sin\mu t + \mathbf{b}\cos\mu t). \tag{8}$$

If we write $\mathbf{x}^{(1)}(t) = \mathbf{u}(t) + i\mathbf{v}(t)$, then the vectors

$$\mathbf{u}(t) = e^{\lambda t}(\mathbf{a}\cos\mu t - \mathbf{b}\sin\mu t), \tag{9a}$$

$$\mathbf{v}(t) = e^{\lambda t}(\mathbf{a}\sin\mu t + \mathbf{b}\cos\mu t) \tag{9b}$$

are real-valued solutions of Eq. (1). It is possible to show that $\mathbf{u}$ and $\mathbf{v}$ are linearly independent solutions (see Problem 15).

For example, suppose that $r_1 = \lambda + i\mu$, $r_2 = \lambda - i\mu$, and that $r_3, \ldots, r_n$ are all real and distinct. Let the corresponding eigenvectors be $\boldsymbol{\xi}^{(1)} = \mathbf{a} + i\mathbf{b}$, $\boldsymbol{\xi}^{(2)} = \mathbf{a} - i\mathbf{b}$, $\boldsymbol{\xi}^{(3)}, \ldots, \boldsymbol{\xi}^{(n)}$. Then the general solution of Eq. (1) is

$$\mathbf{x} = c_1\mathbf{u}(t) + c_2\mathbf{v}(t) + c_3\boldsymbol{\xi}^{(3)}e^{r_3 t} + \cdots + c_n\boldsymbol{\xi}^{(n)}e^{r_n t}, \tag{10}$$

where $\mathbf{u}(t)$ and $\mathbf{v}(t)$ are given by Eqs. (9). We emphasize that this analysis applies only if the coefficient matrix $\mathbf{A}$ in Eq. (1) is real, for it is only then that complex eigenvalues and eigenvectors occur in conjugate pairs.

EXAMPLE 1

Find a fundamental set of real-valued solutions of the system

$$\mathbf{x}' = \begin{pmatrix} -\frac{1}{2} & 1 \\ -1 & -\frac{1}{2} \end{pmatrix}\mathbf{x}. \tag{11}$$

Assuming that

$$\mathbf{x} = \boldsymbol{\xi}e^{rt} \tag{12}$$

leads to the set of linear algebraic equations

$$\begin{pmatrix} -\frac{1}{2} - r & 1 \\ -1 & -\frac{1}{2} - r \end{pmatrix}\begin{pmatrix} \xi_1 \\ \xi_2 \end{pmatrix} = \begin{pmatrix} 0 \\ 0 \end{pmatrix} \tag{13}$$

for the eigenvalues and eigenvectors of $\mathbf{A}$. The characteristic equation is

$$\begin{vmatrix} -\frac{1}{2} - r & 1 \\ -1 & -\frac{1}{2} - r \end{vmatrix} = r^2 + r + \tfrac{5}{4} = 0; \tag{14}$$

therefore the eigenvalues are $r_1 = -\frac{1}{2} + i$ and $r_2 = -\frac{1}{2} - i$. From Eq. (13) a simple calculation shows that the corresponding eigenvectors are

$$\boldsymbol{\xi}^{(1)} = \begin{pmatrix} 1 \\ i \end{pmatrix}, \qquad \boldsymbol{\xi}^{(2)} = \begin{pmatrix} 1 \\ -i \end{pmatrix}. \tag{15}$$

Hence a fundamental set of solutions of the system (11) is

$$\mathbf{x}^{(1)}(t) = \begin{pmatrix} 1 \\ i \end{pmatrix}e^{(-1/2+i)t}, \qquad \mathbf{x}^{(2)}(t) = \begin{pmatrix} 1 \\ -i \end{pmatrix}e^{(-1/2-i)t}. \tag{16}$$

To obtain a set of real-valued solutions, we must find the real and imaginary parts

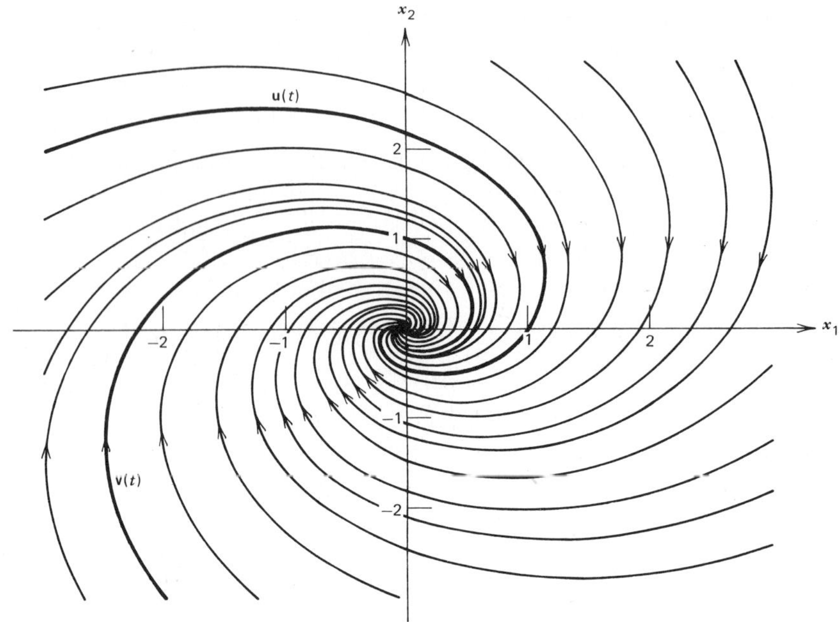

FIGURE 7.8 Trajectories of the system (11). The origin is a spiral point.

of either $\mathbf{x}^{(1)}$ or $\mathbf{x}^{(2)}$. In fact,

$$\mathbf{x}^{(1)}(t) = \begin{pmatrix} 1 \\ i \end{pmatrix} e^{-t/2}(\cos t + i \sin t) = \begin{pmatrix} e^{-t/2}\cos t \\ -e^{-t/2}\sin t \end{pmatrix} + i\begin{pmatrix} e^{-t/2}\sin t \\ e^{-t/2}\cos t \end{pmatrix}. \quad (17)$$

Hence

$$\mathbf{u}(t) = e^{-t/2}\begin{pmatrix} \cos t \\ -\sin t \end{pmatrix}, \qquad \mathbf{v}(t) = e^{-t/2}\begin{pmatrix} \sin t \\ \cos t \end{pmatrix} \quad (18)$$

is a set of real-valued solutions. To verify that $\mathbf{u}(t)$ and $\mathbf{v}(t)$ are linearly independent, we compute their Wronskian:

$$\begin{aligned} W(\mathbf{u}, \mathbf{v})(t) &= \begin{vmatrix} e^{-t/2}\cos t & e^{-t/2}\sin t \\ -e^{-t/2}\sin t & e^{-t/2}\cos t \end{vmatrix} \\ &= e^{-t}(\cos^2 t + \sin^2 t) = e^{-t}. \end{aligned}$$

Since the Wronskian is never zero, it follows that $\mathbf{u}(t)$ and $\mathbf{v}(t)$ constitute a fundamental set of (real-valued) solutions of the system (11).

The graphs of the solutions $\mathbf{u}(t)$ and $\mathbf{v}(t)$ are shown in Figure 7.8. Since

$$\mathbf{u}(0) = \begin{pmatrix} 1 \\ 0 \end{pmatrix}, \qquad \mathbf{v}(0) = \begin{pmatrix} 0 \\ 1 \end{pmatrix},$$

the graphs of $\mathbf{u}(t)$ and $\mathbf{v}(t)$ pass through the points (1, 0) and (0, 1), respectively.

Other solutions of the system (11) are linear combinations of $\mathbf{u}(t)$ and $\mathbf{v}(t)$, and graphs of a few of these solutions are also shown in Figure 7.8. In all cases the solution approaches the origin along a spiral path as $t \to \infty$; this is due to the fact that the solutions (18) are products of decaying exponential and sine or cosine factors. Figure 7.8 is typical of all second order systems $\mathbf{x}' = \mathbf{A}\mathbf{x}$ whose eigenvalues are complex with negative real part. The origin is called a spiral point. For a system whose eigenvalues have positive real part the trajectories are similar to those in Figure 7.8 but the direction of motion is away from the origin.

EXAMPLE 2

The electric circuit shown in Figure 7.9 is described by the system of differential equations

$$\frac{d}{dt}\begin{pmatrix} I \\ V \end{pmatrix} = \begin{pmatrix} -1 & -1 \\ 2 & -1 \end{pmatrix}\begin{pmatrix} I \\ V \end{pmatrix} \tag{19}$$

where I is the current through the inductance and V is the voltage drop across the capacitor. These equations were derived in Problem 8 of Section 7.1. Suppose that at time $t = 0$ the current is 1 ampere and the voltage drop is 2 volts. Find $I(t)$ and $V(t)$ at any time.

Assuming that

$$\begin{pmatrix} I \\ V \end{pmatrix} = \boldsymbol{\xi} e^{rt}, \tag{20}$$

we obtain the algebraic equations

$$\begin{pmatrix} -1 - r & -1 \\ 2 & -1 - r \end{pmatrix}\begin{pmatrix} \xi_1 \\ \xi_2 \end{pmatrix} = \begin{pmatrix} 0 \\ 0 \end{pmatrix}. \tag{21}$$

The eigenvalues are determined from the condition

$$\begin{vmatrix} -1 - r & -1 \\ 2 & -1 - r \end{vmatrix} = r^2 + 2r + 3 = 0; \tag{22}$$

thus $r_1 = -1 + \sqrt{2}\, i$ and $r_2 = -1 - \sqrt{2}\, i$. The corresponding eigenvectors are

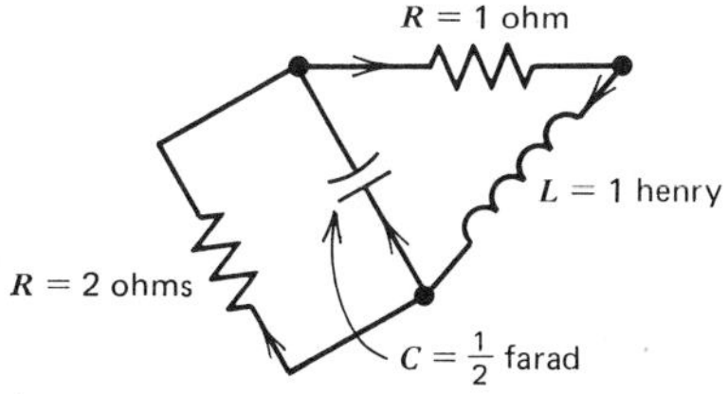

FIGURE 7.9 The circuit in Example 2.

then found from Eq. (21), namely,

$$\boldsymbol{\xi}^{(1)} = \begin{pmatrix} 1 \\ -\sqrt{2}\,i \end{pmatrix}, \qquad \boldsymbol{\xi}^{(2)} = \begin{pmatrix} 1 \\ \sqrt{2}\,i \end{pmatrix}. \tag{23}$$

The complex-valued solution corresponding to r_1 and $\boldsymbol{\xi}^{(1)}$ is

$$\begin{aligned} \boldsymbol{\xi}^{(1)} e^{r_1 t} &= \begin{pmatrix} 1 \\ -\sqrt{2}\,i \end{pmatrix} e^{(-1+\sqrt{2}\,i)t} \\ &= \begin{pmatrix} 1 \\ -\sqrt{2}\,i \end{pmatrix} e^{-t}(\cos\sqrt{2}\,t + i\sin\sqrt{2}\,t) \\ &= e^{-t}\begin{pmatrix} \cos\sqrt{2}\,t \\ \sqrt{2}\sin\sqrt{2}\,t \end{pmatrix} + ie^{-t}\begin{pmatrix} \sin\sqrt{2}\,t \\ -\sqrt{2}\cos\sqrt{2}\,t \end{pmatrix}. \end{aligned} \tag{24}$$

The real and imaginary parts of this solution form a pair of linearly independent real-valued solutions of Eq. (19):

$$\mathbf{u}(t) = e^{-t}\begin{pmatrix} \cos\sqrt{2}\,t \\ \sqrt{2}\sin\sqrt{2}\,t \end{pmatrix}, \qquad \mathbf{v}(t) = e^{-t}\begin{pmatrix} \sin\sqrt{2}\,t \\ -\sqrt{2}\cos\sqrt{2}\,t \end{pmatrix}. \tag{25}$$

Hence the general solution of Eqs. (19) is

$$\begin{pmatrix} I \\ V \end{pmatrix} = c_1 e^{-t}\begin{pmatrix} \cos\sqrt{2}\,t \\ \sqrt{2}\sin\sqrt{2}\,t \end{pmatrix} + c_2 e^{-t}\begin{pmatrix} \sin\sqrt{2}\,t \\ -\sqrt{2}\cos\sqrt{2}\,t \end{pmatrix}. \tag{26}$$

Upon imposing the initial conditions

$$\begin{pmatrix} I \\ V \end{pmatrix}(0) = \begin{pmatrix} 1 \\ 2 \end{pmatrix}, \tag{27}$$

we find that

$$c_1\begin{pmatrix} 1 \\ 0 \end{pmatrix} + c_2\begin{pmatrix} 0 \\ -\sqrt{2} \end{pmatrix} = \begin{pmatrix} 1 \\ 2 \end{pmatrix}. \tag{28}$$

Thus $c_1 = 1$ and $c_2 = -\sqrt{2}$. The solution of the stated problem is then given by Eq. (26) with these values of c_1 and c_2.

PROBLEMS

In each of Problems 1 through 8, express the general solution of the given system of equations in terms of real-valued functions.

1. $\mathbf{x}' = \begin{pmatrix} 3 & -2 \\ 4 & -1 \end{pmatrix}\mathbf{x}$

2. $\mathbf{x}' = \begin{pmatrix} -1 & -4 \\ 1 & -1 \end{pmatrix}\mathbf{x}$

3. $\mathbf{x}' = \begin{pmatrix} 2 & -5 \\ 1 & -2 \end{pmatrix}\mathbf{x}$

4. $\mathbf{x}' = \begin{pmatrix} 2 & -\frac{5}{2} \\ \frac{9}{5} & -1 \end{pmatrix}\mathbf{x}$

5. $\mathbf{x}' = \begin{pmatrix} 1 & -1 \\ 5 & -3 \end{pmatrix}\mathbf{x}$

6. $\mathbf{x}' = \begin{pmatrix} 1 & 2 \\ -5 & -1 \end{pmatrix}\mathbf{x}$

7. $\mathbf{x}' = \begin{pmatrix} 1 & 0 & 0 \\ 2 & 1 & -2 \\ 3 & 2 & 1 \end{pmatrix}\mathbf{x}$

8. $\mathbf{x}' = \begin{pmatrix} -3 & 0 & 2 \\ 1 & -1 & 0 \\ -2 & -1 & 0 \end{pmatrix}\mathbf{x}$

In each of Problems 9 and 10 find the solution of the given initial value problem.

9. $\mathbf{x}' = \begin{pmatrix} 1 & -5 \\ 1 & -3 \end{pmatrix}\mathbf{x}, \quad \mathbf{x}(0) = \begin{pmatrix} 1 \\ 1 \end{pmatrix}$

10. $\mathbf{x}' = \begin{pmatrix} -3 & 2 \\ -1 & -1 \end{pmatrix}\mathbf{x}, \quad \mathbf{x}(0) = \begin{pmatrix} 1 \\ -2 \end{pmatrix}$

In each of Problems 11 and 12 solve the given system of equations by the method of Problem 20 of Section 7.5. Assume that $t > 0$.

11. $t\mathbf{x}' = \begin{pmatrix} -1 & -1 \\ 2 & -1 \end{pmatrix}\mathbf{x}$

12. $t\mathbf{x}' = \begin{pmatrix} 2 & -5 \\ 1 & -2 \end{pmatrix}\mathbf{x}$

13. Consider the electric circuit shown in Figure 7.10. Suppose that $R_1 = R_2 = 4$ ohms, $C = \frac{1}{2}$ farad, and $L = 8$ henrys.
(a) Show that this circuit is described by the system of differential equations

$$\frac{d}{dt}\begin{pmatrix} I \\ V \end{pmatrix} = \begin{pmatrix} -\frac{1}{2} & -\frac{1}{8} \\ 2 & -\frac{1}{2} \end{pmatrix}\begin{pmatrix} I \\ V \end{pmatrix} \tag{i}$$

where I is the current through the inductance and V is the voltage drop across the capacitor.
Hint: See Problem 7 of Section 7.1.
(b) Find the general solution of Eqs. (i) in terms of real-valued functions.
(c) Find $I(t)$ and $V(t)$ if $I(0) = 2$ amperes and $V(0) = 3$ volts.
(d) Determine the limiting values of $I(t)$ and $V(t)$ as $t \to \infty$. Do these limiting values depend on the initial conditions?

FIGURE 7.10 The circuit in Problem 13.

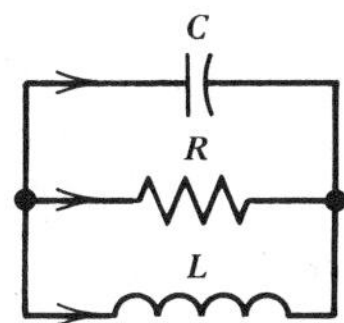

FIGURE 7.11 The circuit in Problem 14.

14. The electric circuit shown in Figure 7.11 is described by the system of differential equations

$$\frac{d}{dt}\begin{pmatrix} I \\ V \end{pmatrix} = \begin{pmatrix} 0 & \frac{1}{L} \\ -\frac{1}{C} & -\frac{1}{RC} \end{pmatrix}\begin{pmatrix} I \\ V \end{pmatrix} \tag{i}$$

where I is the current through the inductance and V is the voltage drop across the capacitor. These differential equations were derived in Problem 7 of Section 7.1.
(a) Show that the eigenvalues of the coefficient matrix are real and different if $L > 4R^2C$; show that they are complex conjugates if $L < 4R^2C$.

(b) Suppose that $R = 1$ ohm, $C = \frac{1}{2}$ farad, and $L = 1$ henry. Find the general solution of the system (i) in this case.
(c) Find $I(t)$ and $V(t)$ if $I(0) = 2$ amperes and $V(0) = 1$ volt.
(d) For the circuit of part (b) determine the limiting values of $I(t)$ and $V(t)$ as $t \to \infty$. Do these limiting values depend on the initial conditions?

15. In this problem we indicate how to show that $\mathbf{u}(t)$ and $\mathbf{v}(t)$, as given by Eqs. (9), are linearly independent.
(a) Let $r_1 = \lambda + i\mu$ and $\bar{r}_1 = \lambda - i\mu$ be a pair of conjugate eigenvalues of the coefficient matrix $\mathbf{A}$ of Eqs. (1); let $\boldsymbol{\xi}^{(1)} = \mathbf{a} + i\mathbf{b}$ and $\bar{\boldsymbol{\xi}}^{(1)} = \mathbf{a} - i\mathbf{b}$ be the corresponding eigenvectors. Recall that it was stated in Section 7.3 that if $r_1 \neq \bar{r}_1$, then $\boldsymbol{\xi}^{(1)}$ and $\bar{\boldsymbol{\xi}}^{(1)}$ are linearly independent. By computing the Wronskian of $\boldsymbol{\xi}^{(1)}$ and $\bar{\boldsymbol{\xi}}^{(1)}$, show that $\mathbf{a}$ and $\mathbf{b}$ are also linearly independent.
(b) Compute the Wronskian of $\mathbf{u}(t)$ and $\mathbf{v}(t)$ at $t = 0$, and then use the result of part (a) with Theorem 7.5 to show that $\mathbf{u}(t)$ and $\mathbf{v}(t)$ are linearly independent for all t.

7.7 Repeated Eigenvalues

We conclude our consideration of the linear homogeneous system with constant coefficients

$$\mathbf{x}' = \mathbf{A}\mathbf{x} \tag{1}$$

with a discussion of the case in which the matrix $\mathbf{A}$ has a repeated eigenvalue. Once again, the discussion in this section is valid whether $\mathbf{A}$ is real or complex. If $r = \rho$ is a k-fold root of the determinantal equation

$$\det(\mathbf{A} - r\mathbf{I}) = 0, \tag{2}$$

then ρ is said to be an eigenvalue of multiplicity k of the matrix $\mathbf{A}$. In this event, there are two possibilities: either there are k linearly independent eigenvectors corresponding to the eigenvalue ρ, or else there are fewer than k such eigenvectors. These possibilities were illustrated in Examples 4 and 5 of Section 7.3; in Example 5 a double eigenvalue was accompanied by two linearly independent eigenvectors, while in Example 4 a double eigenvalue was associated with only one linearly independent eigenvector.

In the first case, let $\boldsymbol{\xi}^{(1)}, \ldots, \boldsymbol{\xi}^{(k)}$ be k linearly independent eigenvectors associated with the eigenvalue ρ of multiplicity k. Then $\mathbf{x}^{(1)}(t) = \boldsymbol{\xi}^{(1)}e^{\rho t}, \ldots, \mathbf{x}^{(k)}(t) = \boldsymbol{\xi}^{(k)}e^{\rho t}$ are k linearly independent solutions of Eq. (1). Thus in this case it makes no difference that the eigenvalue $r = \rho$ is repeated; there is still a fundamental set of solutions of Eq. (1) of the form $\boldsymbol{\xi}e^{rt}$. This case always occurs if the coefficient matrix $\mathbf{A}$ is Hermitian.

However, if the coefficient matrix is not Hermitian, then there may be fewer than k independent eigenvectors corresponding to an eigenvalue ρ of multiplicity k, and if so, there will be fewer than k solutions of Eq. (1) of the form $\boldsymbol{\xi}e^{rt}$ associated with this eigenvalue. Therefore, to construct the general solution of Eq. (1) it is necessary to find other solutions of a different form. By analogy with previous results for linear equations of order n, it is natural to seek additional solutions involving products of polynomials and exponential functions. We first consider an example.

EXAMPLE 1

Find a fundamental set of solutions of the system

$$\mathbf{x}' = \mathbf{A}\mathbf{x} = \begin{pmatrix} 1 & -1 \\ 1 & 3 \end{pmatrix}\mathbf{x}. \tag{3}$$

Assuming that $\mathbf{x} = \boldsymbol{\xi}e^{rt}$ and substituting for $\mathbf{x}$ in Eq. (3) gives

$$\begin{pmatrix} 1-r & -1 \\ 1 & 3-r \end{pmatrix}\begin{pmatrix} \xi_1 \\ \xi_2 \end{pmatrix} = \begin{pmatrix} 0 \\ 0 \end{pmatrix}, \tag{4}$$

from which it follows that the eigenvalues of $\mathbf{A}$ are $r_1 = r_2 = 2$. For this value of r, Eqs. (4) imply that $\xi_1 + \xi_2 = 0$. Hence there is only the one eigenvector $\boldsymbol{\xi}$, given by $\boldsymbol{\xi}^T = (1, -1)$, corresponding to the double eigenvalue. Thus one solution of the system (3) is

$$\mathbf{x}^{(1)}(t) = \begin{pmatrix} 1 \\ -1 \end{pmatrix}e^{2t}, \tag{5}$$

but there is no second solution of the form $\mathbf{x} = \boldsymbol{\xi}e^{rt}$. It is natural to attempt to find a second solution of the system (3) of the form

$$\mathbf{x} = \boldsymbol{\xi}te^{2t}, \tag{6}$$

where $\boldsymbol{\xi}$ is a constant vector to be determined. Substituting for $\mathbf{x}$ in Eq. (3) gives

$$2\boldsymbol{\xi}te^{2t} + \boldsymbol{\xi}e^{2t} - \mathbf{A}\boldsymbol{\xi}te^{2t} = \mathbf{0}. \tag{7}$$

For Eq. (7) to be satisfied for all t it is necessary for the coefficients of te^{2t} and e^{2t} each to be zero. From the term in e^{2t} we find that

$$\boldsymbol{\xi} = \mathbf{0}. \tag{8}$$

Hence there is no nonzero solution of the system (3) of the form (6).

Since Eq. (7) contains terms in both te^{2t} and e^{2t}, it appears that in addition to $\boldsymbol{\xi}te^{2t}$ the second solution must contain a term of the form $\boldsymbol{\eta}e^{2t}$; in other words, we need to assume that

$$\mathbf{x} = \boldsymbol{\xi}te^{2t} + \boldsymbol{\eta}e^{2t}, \tag{9}$$

where $\boldsymbol{\xi}$ and $\boldsymbol{\eta}$ are constant vectors. Upon substituting this expression for $\mathbf{x}$ in Eq. (3) we obtain

$$2\boldsymbol{\xi}te^{2t} + (\boldsymbol{\xi} + 2\boldsymbol{\eta})e^{2t} = \mathbf{A}(\boldsymbol{\xi}te^{2t} + \boldsymbol{\eta}e^{2t}). \tag{10}$$

Equating coefficients of te^{2t} and e^{2t} on each side of Eq. (10) gives the conditions

$$(\mathbf{A} - 2\mathbf{I})\boldsymbol{\xi} = \mathbf{0}, \tag{11}$$

and

$$(\mathbf{A} - 2\mathbf{I})\boldsymbol{\eta} = \boldsymbol{\xi} \tag{12}$$

for the determination of $\boldsymbol{\xi}$ and $\boldsymbol{\eta}$. Equation (11) is satisfied if $\boldsymbol{\xi}$ is the eigenvector of $\mathbf{A}$ corresponding to the eigenvalue $r = 2$, that is, $\boldsymbol{\xi}^T = (1, -1)$. Since $\det(\mathbf{A} - 2\mathbf{I})$ is zero, we might expect that Eq. (12) cannot be solved. However, this is not necessarily true, since for some vectors $\boldsymbol{\xi}$ it is possible to solve Eq. (12). In fact, the augmented matrix for Eq. (12) is

$$\left(\begin{array}{cc|c} -1 & -1 & 1 \\ 1 & 1 & -1 \end{array}\right)$$

and it is clear that the system is solvable. We have

$$-\eta_1 - \eta_2 = 1$$

so if $\eta_1 = k$, where k is arbitrary, then $\eta_2 = -k - 1$. If we write

$$\boldsymbol{\eta} = \begin{pmatrix} k \\ -1 - k \end{pmatrix} = \begin{pmatrix} 0 \\ -1 \end{pmatrix} + k\begin{pmatrix} 1 \\ -1 \end{pmatrix}, \tag{13}$$

then by substituting for $\boldsymbol{\xi}$ and $\boldsymbol{\eta}$ in Eq. (9) we obtain

$$\mathbf{x} = \begin{pmatrix} 1 \\ -1 \end{pmatrix} te^{2t} + \begin{pmatrix} 0 \\ -1 \end{pmatrix} e^{2t} + k\begin{pmatrix} 1 \\ -1 \end{pmatrix} e^{2t}. \tag{14}$$

The last term in Eq. (14) is merely a multiple of the first solution $\mathbf{x}^{(1)}(t)$, and may

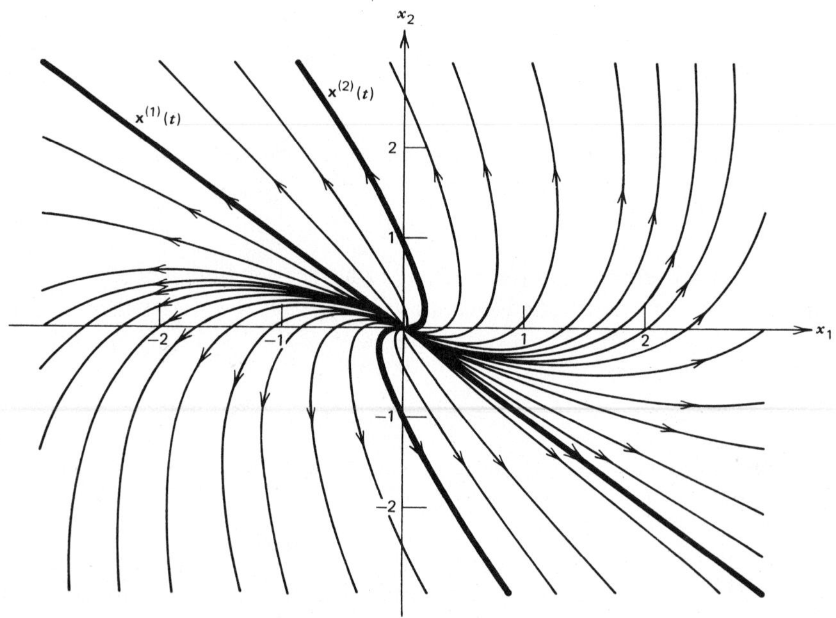

FIGURE 7.12 Trajectories of the system (3). The origin is a node.

be ignored, but the first two terms constitute a new solution

$$\mathbf{x}^{(2)}(t) = \begin{pmatrix} 1 \\ -1 \end{pmatrix} te^{2t} + \begin{pmatrix} 0 \\ -1 \end{pmatrix} e^{2t}. \tag{15}$$

An elementary calculation shows that $W[\mathbf{x}^{(1)}, \mathbf{x}^{(2)}](t) = -e^{4t}$, and therefore $\mathbf{x}^{(1)}$ and $\mathbf{x}^{(2)}$ form a fundamental set of solutions of the system (3). The general solution is

$$\begin{aligned} \mathbf{x} &= c_1\mathbf{x}^{(1)}(t) + c_2\mathbf{x}^{(2)}(t) \\ &= c_1\begin{pmatrix} 1 \\ -1 \end{pmatrix} e^{2t} + c_2\left[\begin{pmatrix} 1 \\ -1 \end{pmatrix} te^{2t} + \begin{pmatrix} 0 \\ -1 \end{pmatrix} e^{2t}\right]. \end{aligned} \tag{16}$$

The graph of the solution (16) is a little more difficult to analyze than in some of the previous examples. It is clear that $\mathbf{x}$ becomes unbounded as $t \to \infty$ and that $\mathbf{x} \to \mathbf{0}$ as $t \to -\infty$. It is possible to show that as $t \to -\infty$, all solutions approach the origin tangent to the line $x_2 = -x_1$ determined by the eigenvector. Similarly, as $t \to \infty$, each trajectory is asymptotic to a line of slope -1. The trajectories of the system (3) are shown in Figure 7.12. The pattern of trajectories in this figure is typical of second order systems $\mathbf{x}' = \mathbf{Ax}$ with equal eigenvalues and only one independent eigenvector. The origin is also called a node in this case. If the eigenvalues are negative, then the trajectories are similar, but are traversed in the inward direction.

One difference between a system of two first order equations and a single second order equation is evident from the example. Recall that for a second order linear equation with a repeated root r_1 of the characteristic equation, a term $ce^{r_1 t}$ in the second solution is not required since it is a multiple of the first solution. On the other hand, for a system of two first order equations, the term $\boldsymbol{\eta}e^{r_1 t}$ of Eq. (9) with $r_1 = 2$ is not a multiple of the first solution $\boldsymbol{\xi}e^{r_1 t}$ unless $\boldsymbol{\eta}$ is proportional to the eigenvector $\boldsymbol{\xi}$ associated with the eigenvalue r_1. Since this is not usually the case, the term $\boldsymbol{\eta}e^{r_1 t}$ must be retained.

The same procedure can be applied in the general case. Consider again the system (1), and suppose that $r = \rho$ is a double eigenvalue of $\mathbf{A}$, but that there is only one corresponding eigenvector $\boldsymbol{\xi}$. Then one solution [similar to Eq. (5)] is

$$\mathbf{x}^{(1)}(t) = \boldsymbol{\xi}e^{\rho t}, \tag{17}$$

where $\boldsymbol{\xi}$ satisfies

$$(\mathbf{A} - \rho\mathbf{I})\boldsymbol{\xi} = \mathbf{0}. \tag{18}$$

By proceeding as in the example we find that a second solution [similar to Eq. (15)] is

$$\mathbf{x}^{(2)}(t) = \boldsymbol{\xi}te^{\rho t} + \boldsymbol{\eta}e^{\rho t}, \tag{19}$$

where $\boldsymbol{\xi}$ satisfies Eq. (18) and $\boldsymbol{\eta}$ is determined from

$$(\mathbf{A} - \rho\mathbf{I})\boldsymbol{\eta} = \boldsymbol{\xi}. \tag{20}$$

Even though $\det(\mathbf{A} - \rho\mathbf{I}) = 0$, it can be shown that it is always possible to solve Eq. (20) for $\boldsymbol{\eta}$. The vector $\boldsymbol{\eta}$ is called a generalized eigenvector corresponding to the eigenvalue ρ.

If $r = \rho$ is an eigenvalue of the matrix $\mathbf{A}$ of multiplicity greater than two, then there is a greater variety of possibilities. We can illustrate this by considering an eigenvalue of multiplicity three. Suppose first that the triple eigenvalue $r = \rho$ has three linearly independent eigenvectors $\boldsymbol{\xi}^{(1)}$, $\boldsymbol{\xi}^{(2)}$, and $\boldsymbol{\xi}^{(3)}$. In this case the corresponding set of independent solutions is simply

$$\mathbf{x}^{(1)}(t) = \boldsymbol{\xi}^{(1)}e^{\rho t}, \qquad \mathbf{x}^{(2)}(t) = \boldsymbol{\xi}^{(2)}e^{\rho t}, \qquad \mathbf{x}^{(3)}(t) = \boldsymbol{\xi}^{(3)}e^{\rho t}. \tag{21}$$

Now suppose that there is a single linearly independent eigenvector associated with the triple eigenvalue $r = \rho$. Then the first solution is given by Eq. (17), a second solution by Eq. (19), and a third solution is of the form

$$\mathbf{x}^{(3)}(t) = \boldsymbol{\xi}\frac{t^2}{2!}e^{\rho t} + \boldsymbol{\eta}te^{\rho t} + \boldsymbol{\zeta}e^{\rho t}, \tag{22}$$

where $\boldsymbol{\xi}$ satisfies Eq. (18), $\boldsymbol{\eta}$ satisfies Eq. (20), and $\boldsymbol{\zeta}$ is determined from

$$(\mathbf{A} - \rho\mathbf{I})\boldsymbol{\zeta} = \boldsymbol{\eta}. \tag{23}$$

Again it is always possible to solve Eq. (23) for $\boldsymbol{\zeta}$, and the vectors $\boldsymbol{\eta}$ and $\boldsymbol{\zeta}$ are called generalized eigenvectors.

The final possibility is that there are two linearly independent eigenvectors $\boldsymbol{\xi}^{(1)}$ and $\boldsymbol{\xi}^{(2)}$ corresponding to the eigenvalue $r = \rho$. Then two solutions of the system (1) are

$$\mathbf{x}^{(1)}(t) = \boldsymbol{\xi}^{(1)}e^{\rho t}, \qquad \mathbf{x}^{(2)}(t) = \boldsymbol{\xi}^{(2)}e^{\rho t}. \tag{24}$$

A third solution is of the form

$$\mathbf{x}^{(3)}(t) = \boldsymbol{\xi}te^{\rho t} + \boldsymbol{\eta}e^{\rho t}, \tag{25}$$

where $\boldsymbol{\xi}$ satisfies Eq. (18) and $\boldsymbol{\eta}$ is a solution of Eq. (20). Here it is necessary to choose $\boldsymbol{\xi}$ as a linear combination of the eigenvectors $\boldsymbol{\xi}^{(1)}$ and $\boldsymbol{\xi}^{(2)}$ in such a way that the equation

$$(\mathbf{A} - \rho\mathbf{I})\boldsymbol{\eta} = \boldsymbol{\xi} \tag{26}$$

is solvable. If we write

$$\boldsymbol{\xi} = c_1\boldsymbol{\xi}^{(1)} + c_2\boldsymbol{\xi}^{(2)} \tag{27}$$

then we must choose c_1 and c_2 so that Eq. (26) can be solved for $\boldsymbol{\eta}$. The solutions $\mathbf{x}^{(1)}(t)$, $\mathbf{x}^{(2)}(t)$, and $\mathbf{x}^{(3)}(t)$, given by Eqs. (24) and (25), form a set of independent solutions corresponding to the eigenvalue $r = \rho$ in this case. Problem 12 illustrates this situation.

It should be clear that if there is an eigenvalue of even higher multiplicity, then the situation may become even more complicated. For example, if $r = \rho$ is an eigenvalue of multiplicity four, then we must deal with different cases according to whether there are four, three, two, or one linearly independent eigenvector. A general treatment of problems with repeated eigenvalues is possible, but it involves advanced topics from matrix theory and is beyond the scope of this book. We concentrate on the computational techniques necessary to find a fundamental set of solutions when the multiplicities of the eigenvalues do not exceed three.

EXAMPLE 2

Find the solution of the initial value problem

$$\mathbf{x}' = \begin{pmatrix} 1 & 9 \\ -1 & -5 \end{pmatrix}\mathbf{x}, \qquad \mathbf{x}(0) = \begin{pmatrix} 1 \\ -1 \end{pmatrix}. \tag{28}$$

Assuming that $\mathbf{x} = \boldsymbol{\xi}e^{rt}$, we obtain the algebraic system

$$\begin{pmatrix} 1 - r & 9 \\ -1 & -5 - r \end{pmatrix}\begin{pmatrix} \xi_1 \\ \xi_2 \end{pmatrix} = \begin{pmatrix} 0 \\ 0 \end{pmatrix}. \tag{29}$$

Thus the eigenvalues are the roots of

$$(1 - r)(-5 - r) + 9 = r^2 + 4r + 4 = (r + 2)^2 = 0,$$

so $r_1 = r_2 = -2$. From Eq. (29) the corresponding eigenvector is found to be $\boldsymbol{\xi}^T = (3, -1)$. Thus one solution of the system (28) is

$$\mathbf{x}^{(1)}(t) = \begin{pmatrix} 3 \\ -1 \end{pmatrix}e^{-2t}. \tag{30}$$

A second independent solution is of the form $\mathbf{x} = \boldsymbol{\xi}te^{-2t} + \boldsymbol{\eta}e^{-2t}$, where $\boldsymbol{\xi}$ is the same as before, and $\boldsymbol{\eta}$ satisfies $(\mathbf{A} + 2\mathbf{I})\boldsymbol{\eta} = \boldsymbol{\xi}$, or

$$\begin{pmatrix} 3 & 9 \\ -1 & -3 \end{pmatrix}\begin{pmatrix} \eta_1 \\ \eta_2 \end{pmatrix} = \begin{pmatrix} 3 \\ -1 \end{pmatrix}.$$

Hence $\eta_1 + 3\eta_2 = 1$, so that if $\eta_2 = k$, then $\eta_1 = 1 - 3k$, where k is arbitrary. Thus

$$\boldsymbol{\eta} = \begin{pmatrix} 1 - 3k \\ k \end{pmatrix} = \begin{pmatrix} 1 \\ 0 \end{pmatrix} - k\begin{pmatrix} 3 \\ -1 \end{pmatrix}.$$

The term involving k merely produces a multiple of the first solution $\mathbf{x}^{(1)}(t)$, so it can be dropped, and we obtain the second solution

$$\mathbf{x}^{(2)}(t) = \begin{pmatrix} 3 \\ -1 \end{pmatrix}te^{-2t} + \begin{pmatrix} 1 \\ 0 \end{pmatrix}e^{-2t}. \tag{31}$$

Finally, the general solution of the system (28) is

$$\mathbf{x} = c_1\begin{pmatrix} 3 \\ -1 \end{pmatrix}e^{-2t} + c_2\left[\begin{pmatrix} 3 \\ -1 \end{pmatrix}te^{-2t} + \begin{pmatrix} 1 \\ 0 \end{pmatrix}e^{-2t}\right]. \tag{32}$$

To satisfy the initial condition we set $t = 0$ in Eq. (32); this gives

$$c_1\begin{pmatrix} 3 \\ -1 \end{pmatrix} + c_2\begin{pmatrix} 1 \\ 0 \end{pmatrix} = \begin{pmatrix} 1 \\ -1 \end{pmatrix}.$$

Consequently $c_1 = 1$, $c_2 = -2$, and the solution of the given initial value problem is

$$\mathbf{x} = \begin{pmatrix} 3 \\ -1 \end{pmatrix}e^{-2t} - 2\left[\begin{pmatrix} 3 \\ -1 \end{pmatrix}te^{-2t} + \begin{pmatrix} 1 \\ 0 \end{pmatrix}e^{-2t}\right] = \begin{pmatrix} 1 \\ -1 \end{pmatrix}e^{-2t} - 2\begin{pmatrix} 3 \\ -1 \end{pmatrix}te^{-2t}. \tag{33}$$

PROBLEMS

In each of Problems 1 through 6 find the general solution of the given system of equations.

1. $\mathbf{x}' = \begin{pmatrix} 3 & -4 \\ 1 & -1 \end{pmatrix}\mathbf{x}$

2. $\mathbf{x}' = \begin{pmatrix} 4 & -2 \\ 8 & -4 \end{pmatrix}\mathbf{x}$

3. $\mathbf{x}' = \begin{pmatrix} -\frac{3}{2} & 1 \\ -\frac{1}{4} & -\frac{1}{2} \end{pmatrix}\mathbf{x}$

4. $\mathbf{x}' = \begin{pmatrix} -3 & \frac{5}{2} \\ -\frac{5}{2} & 2 \end{pmatrix}\mathbf{x}$

5. $\mathbf{x}' = \begin{pmatrix} 1 & 1 & 1 \\ 2 & 1 & -1 \\ 0 & -1 & 1 \end{pmatrix}\mathbf{x}$

6. $\mathbf{x}' = \begin{pmatrix} 0 & 1 & 1 \\ 1 & 0 & 1 \\ 1 & 1 & 0 \end{pmatrix}\mathbf{x}$

In each of Problems 7 and 8 find the solution of the given initial value problem.

7. $\mathbf{x}' = \begin{pmatrix} 1 & -4 \\ 4 & -7 \end{pmatrix}\mathbf{x}, \quad \mathbf{x}(0) = \begin{pmatrix} 3 \\ 2 \end{pmatrix}$

8. $\mathbf{x}' = \begin{pmatrix} 1 & 0 & 0 \\ -4 & 1 & 0 \\ 3 & 6 & 2 \end{pmatrix}\mathbf{x}, \quad \mathbf{x}(0) = \begin{pmatrix} -1 \\ 2 \\ -30 \end{pmatrix}$

9. Show that $r = 2$ is a triple root of the characteristic equation for the system

$$\mathbf{x}' = \begin{pmatrix} 1 & 1 & 1 \\ 2 & 1 & -1 \\ -3 & 2 & 4 \end{pmatrix}\mathbf{x},$$

and find three linearly independent solutions of this system.

In each of Problems 10 and 11 solve the given system of equations by the method of Problem 20 of Section 7.5. Assume that $t > 0$.

10. $t\mathbf{x}' = \begin{pmatrix} 3 & -4 \\ 1 & -1 \end{pmatrix}\mathbf{x}$

11. $t\mathbf{x}' = \begin{pmatrix} 1 & -4 \\ 4 & -7 \end{pmatrix}\mathbf{x}$

12. In this problem we indicate how to proceed when there is a triple eigenvalue and only two associated eigenvectors. Consider the system

$$\mathbf{x}' = \mathbf{A}\mathbf{x} = \begin{pmatrix} 5 & -3 & -2 \\ 8 & -5 & -4 \\ -4 & 3 & 3 \end{pmatrix}\mathbf{x}. \tag{i}$$

(a) Show that $r = 1$ is a triple eigenvalue of the coefficient matrix $\mathbf{A}$, and that there are only two linearly independent eigenvectors

$$\boldsymbol{\xi}^{(1)} = \begin{pmatrix} 1 \\ 0 \\ 2 \end{pmatrix}, \qquad \boldsymbol{\xi}^{(2)} = \begin{pmatrix} 0 \\ 2 \\ -3 \end{pmatrix}. \tag{ii}$$

Find two linearly independent solutions $\mathbf{x}^{(1)}(t)$ and $\mathbf{x}^{(2)}(t)$ of Eq. (i).

(b) To find a third solution assume that

$$\mathbf{x}^{(3)}(t) = \boldsymbol{\xi} te^t + \boldsymbol{\eta} e^t; \tag{iii}$$

then show that $\boldsymbol{\xi}$ and $\boldsymbol{\eta}$ must satisfy

$$(\mathbf{A} - \mathbf{I})\boldsymbol{\xi} = \mathbf{0}, \tag{iv}$$

$$(\mathbf{A} - \mathbf{I})\boldsymbol{\eta} = \boldsymbol{\xi}. \tag{v}$$

(c) Show that $\boldsymbol{\xi} = c_1\boldsymbol{\xi}^{(1)} + c_2\boldsymbol{\xi}^{(2)}$, where c_1 and c_2 are arbitrary constants, is the most general solution of Eq. (iv). Show that in order to solve Eq. (v) it is necessary that $c_1 = c_2$.

(d) It is convenient to choose $c_1 = c_2 = 2$. For this choice show that

$$\boldsymbol{\xi} = \begin{pmatrix} 2 \\ 4 \\ -2 \end{pmatrix}, \qquad \boldsymbol{\eta} = \begin{pmatrix} 0 \\ 0 \\ -1 \end{pmatrix} + k_1\begin{pmatrix} 1 \\ 0 \\ 2 \end{pmatrix} + k_2\begin{pmatrix} 0 \\ 2 \\ -3 \end{pmatrix}, \tag{vi}$$

where k_1 and k_2 are arbitrary constants. Use the results given in Eqs. (vi) to find a third linearly independent solution $\mathbf{x}^{(3)}(t)$ of Eq. (i).

13. Show that all solutions of the system

$$\mathbf{x}' = \begin{pmatrix} a & b \\ c & d \end{pmatrix}\mathbf{x}$$

approach zero as $t \to \infty$ if and only if $a + d < 0$ and $ad - bc > 0$. Compare this result with that of Problem 12 in Section 3.5.1.

14. Consider again the electric circuit in Problem 14 of Section 7.6. This circuit is described by the system of differential equations

$$\frac{d}{dt}\begin{pmatrix} I \\ V \end{pmatrix} = \begin{pmatrix} 0 & \dfrac{1}{L} \\ -\dfrac{1}{C} & -\dfrac{1}{RC} \end{pmatrix}\begin{pmatrix} I \\ V \end{pmatrix}.$$

(a) Show that the eigenvalues are real and equal if $L = 4R^2C$.

(b) Suppose that $R = 1$ ohm, $C = 1$ farad, and $L = 4$ henrys. Suppose also that $I(0) = 1$ ampere and $V(0) = 2$ volts. Find $I(t)$ and $V(t)$.

7.8 Fundamental Matrices

The theory of systems of linear differential equations can be further illuminated by introducing the idea of a fundamental matrix. This concept is particularly useful to us in the next section where we extend the method of variation of parameters to systems of first order linear equations. Suppose that

$\mathbf{x}^{(1)}, \ldots, \mathbf{x}^{(n)}$ form a fundamental set of solutions for the equation

$$\mathbf{x}' = \mathbf{P}(t)\mathbf{x} \tag{1}$$

on some interval $\alpha < t < \beta$. Then the matrix

$$\boldsymbol{\Psi}(t) = \begin{pmatrix} x_1^{(1)}(t) & \cdots & x_1^{(n)}(t) \\ \vdots & & \vdots \\ x_n^{(1)}(t) & \cdots & x_n^{(n)}(t) \end{pmatrix}, \tag{2}$$

whose columns are the vectors $\mathbf{x}^{(1)}, \ldots, \mathbf{x}^{(n)}$, is said to be a *fundamental matrix* for the system (1). Note that any fundamental matrix is nonsingular since its columns are linearly independent vectors.

EXAMPLE 1

Find a fundamental matrix for the system

$$\mathbf{x}' = \begin{pmatrix} 1 & 1 \\ 4 & 1 \end{pmatrix}\mathbf{x}. \tag{3}$$

In Example 1 of Section 7.5 we found that

$$\mathbf{x}^{(1)}(t) = \begin{pmatrix} e^{3t} \\ 2e^{3t} \end{pmatrix}, \qquad \mathbf{x}^{(2)}(t) = \begin{pmatrix} e^{-t} \\ -2e^{-t} \end{pmatrix}$$

are linearly independent solutions of Eq. (3). Thus a fundamental matrix for the system (3) is

$$\boldsymbol{\Psi}(t) = \begin{pmatrix} e^{3t} & e^{-t} \\ 2e^{3t} & -2e^{-t} \end{pmatrix}. \tag{4}$$

The solution of an initial value problem can be written very compactly in terms of a fundamental matrix. The general solution of Eq. (1) is

$$\mathbf{x} = c_1\mathbf{x}^{(1)}(t) + \cdots + c_n\mathbf{x}^{(n)}(t) \tag{5}$$

or, in terms of $\boldsymbol{\Psi}(t)$,

$$\mathbf{x} = \boldsymbol{\Psi}(t)\mathbf{c}, \tag{6}$$

where $\mathbf{c}$ is a constant vector with arbitrary components $c_1, \ldots, c_n$. For an initial value problem consisting of the differential equation (1) and the initial condition

$$\mathbf{x}(t_0) = \mathbf{x}^0, \tag{7}$$

where t_0 is a given point in $\alpha < t < \beta$ and $\mathbf{x}^0$ is a given initial vector, it is only necessary to choose the vector $\mathbf{c}$ in Eq. (6) so as to satisfy the initial condition (7). Hence $\mathbf{c}$ must satisfy

$$\boldsymbol{\Psi}(t_0)\mathbf{c} = \mathbf{x}^0. \tag{8}$$

Therefore, since $\mathbf{\Psi}(t)$ is nonsingular,

$$\mathbf{c} = \mathbf{\Psi}^{-1}(t_0)\mathbf{x}^0, \tag{9}$$

and

$$\mathbf{x} = \mathbf{\Psi}(t)\mathbf{\Psi}^{-1}(t_0)\mathbf{x}^0 \tag{10}$$

is the solution of the given initial value problem. We emphasize, however, that to solve a given initial value problem one would ordinarily solve Eq. (8) by row reduction and then substitute for $\mathbf{c}$ in Eq. (6), rather than compute $\mathbf{\Psi}^{-1}(t_0)$ and use Eq. (10).

Recall that each column of the fundamental matrix $\mathbf{\Psi}$ is a solution of Eq. (1). It follows that $\mathbf{\Psi}$ satisfies the matrix differential equation

$$\mathbf{\Psi}' = \mathbf{P}(t)\mathbf{\Psi}. \tag{11}$$

This relation is readily confirmed by comparing the two sides of Eq. (11) column by column.

Sometimes it is convenient to make use of the special fundamental matrix, denoted by $\mathbf{\Phi}(t)$, whose columns are the vectors $\mathbf{x}^{(1)}, \ldots, \mathbf{x}^{(n)}$ designated in Theorem 7.6 of Section 7.4. Besides the differential equation (1) these vectors satisfy the initial conditions

$$\mathbf{x}^{(j)}(t_0) = \mathbf{e}^{(j)}, \tag{12}$$

where $\mathbf{e}^{(j)}$ is the unit vector, defined in Theorem 7.6, with a one in the jth position and zeros elsewhere. Thus $\mathbf{\Phi}(t)$ has the property that

$$\mathbf{\Phi}(t_0) = \begin{pmatrix} 1 & 0 & \cdots & 0 \\ 0 & 1 & \cdots & 0 \\ \vdots & \vdots & & \vdots \\ 0 & 0 & \cdots & 1 \end{pmatrix} = \mathbf{I}. \tag{13}$$

We will always reserve the symbol $\mathbf{\Phi}$ to denote the fundamental matrix satisfying the initial condition (13), and use $\mathbf{\Psi}$ when an arbitrary fundamental matrix is intended. In terms of $\mathbf{\Phi}(t)$ the solution of the initial value problem (1) and (7) is even simpler in appearance; since $\mathbf{\Phi}^{-1}(t_0) = \mathbf{I}$, it follows from Eq. (11) that

$$\mathbf{x} = \mathbf{\Phi}(t)\mathbf{x}^0. \tag{14}$$

While the fundamental matrix $\mathbf{\Phi}(t)$ is often more complicated than $\mathbf{\Psi}(t)$, it is especially helpful if the same system of differential equations is to be solved repeatedly subject to many different initial conditions. This corresponds to a given physical system that can be started from many different initial states. If the fundamental matrix $\mathbf{\Phi}(t)$ has been determined, then the solution for each set of initial conditions can be found simply by matrix multiplication, as indicated by Eq. (14). The matrix $\mathbf{\Phi}(t)$ thus represents a transformation of the initial conditions $\mathbf{x}^0$ into the solution $\mathbf{x}(t)$ at an arbitrary time t. By comparing Eqs. (10) and (14) it is clear that $\mathbf{\Phi}(t) = \mathbf{\Psi}(t)\mathbf{\Psi}^{-1}(t_0)$.

EXAMPLE 2

For the system (3)

$$\mathbf{x}' = \begin{pmatrix} 1 & 1 \\ 4 & 1 \end{pmatrix}\mathbf{x}$$

in Example 1, find the fundamental matrix $\mathbf{\Phi}$ such that $\mathbf{\Phi}(0) = \mathbf{I}$.

The columns of $\mathbf{\Phi}$ are solutions of Eq. (3) which satisfy the initial conditions

$$\mathbf{x}^{(1)}(0) = \begin{pmatrix} 1 \\ 0 \end{pmatrix}, \qquad \mathbf{x}^{(2)}(0) = \begin{pmatrix} 0 \\ 1 \end{pmatrix}. \tag{15}$$

Since the general solution of Eq. (3) is

$$\mathbf{x} = c_1\begin{pmatrix} 1 \\ 2 \end{pmatrix}e^{3t} + c_2\begin{pmatrix} 1 \\ -2 \end{pmatrix}e^{-t},$$

we can find the solution satisfying the first set of these initial conditions by choosing $c_1 = c_2 = \frac{1}{2}$; similarly we obtain the solution satisfying the second set of initial conditions by choosing $c_1 = \frac{1}{4}$ and $c_2 = -\frac{1}{4}$. Hence

$$\mathbf{\Phi}(t) = \begin{pmatrix} \frac{1}{2}e^{3t} + \frac{1}{2}e^{-t} & \frac{1}{4}e^{3t} - \frac{1}{4}e^{-t} \\ e^{3t} - e^{-t} & \frac{1}{2}e^{3t} + \frac{1}{2}e^{-t} \end{pmatrix}. \tag{16}$$

Note that the elements of $\mathbf{\Phi}(t)$ are more complicated than those of the fundamental matrix $\mathbf{\Psi}(t)$ given by Eq. (4); however, it is now easy to determine the solution corresponding to any set of initial conditions.

Now let us turn again to the system

$$\mathbf{x}' = \mathbf{A}\mathbf{x}, \tag{17}$$

where $\mathbf{A}$ is a constant matrix, not necessarily real. In Sections 7.5 through 7.7 we have described how to solve such a system by starting from the assumption that $\mathbf{x} = \boldsymbol{\xi}e^{rt}$. Here we wish to provide another viewpoint. The basic reason why a system of equations presents some difficulty is that the equations are usually *coupled*; in other words, some or all of the equations involve more than one, perhaps all, of the dependent variables. This occurs whenever the coefficient matrix $\mathbf{A}$ is not a diagonal matrix. Hence the equations in the system must be solved *simultaneously*, rather than *consecutively*. This observation suggests that one way to solve a system of equations might be by transforming it into an equivalent *uncoupled* system in which each equation contains only one dependent variable. This corresponds to transforming the coefficient matrix $\mathbf{A}$ into a *diagonal* matrix.

According to the results quoted in Section 7.3, it is possible to do this whenever $\mathbf{A}$ has a full set of n linearly independent eigenvectors. Let $\boldsymbol{\xi}^{(1)}, \ldots, \boldsymbol{\xi}^{(n)}$ be eigenvectors of $\mathbf{A}$ corresponding to the eigenvalues $r_1, \ldots, r_n$, and form the transformation matrix **T** whose columns are $\boldsymbol{\xi}^{(1)}, \ldots, \boldsymbol{\xi}^{(n)}$. Then

$$\mathbf{T} = \begin{pmatrix} \xi_1^{(1)} & \cdots & \xi_1^{(n)} \\ \vdots & & \vdots \\ \xi_n^{(1)} & \cdots & \xi_n^{(n)} \end{pmatrix}. \tag{18}$$

Defining a new dependent variable **y** by the relation

$$\mathbf{x} = \mathbf{T}\mathbf{y}, \tag{19}$$

we have from Eq. (17) that

$$\mathbf{T}\mathbf{y}' = \mathbf{A}\mathbf{T}\mathbf{y}, \tag{20}$$

or

$$\mathbf{y}' = (\mathbf{T}^{-1}\mathbf{A}\mathbf{T})\mathbf{y}. \tag{21}$$

From Eq. (46) of Section 7.3,

$$\mathbf{T}^{-1}\mathbf{A}\mathbf{T} = \mathbf{D} = \begin{pmatrix} r_1 & 0 & \cdots & 0 \\ 0 & r_2 & \cdots & 0 \\ \vdots & \vdots & & \vdots \\ 0 & 0 & \cdots & r_n \end{pmatrix} \tag{22}$$

is the diagonal matrix whose diagonal elements are the eigenvalues of **A**. Thus **y** satisfies the uncoupled system of equations

$$\mathbf{y}' = \mathbf{D}\mathbf{y}, \tag{23}$$

for which a fundamental matrix is the diagonal matrix

$$\mathbf{Q}(t) = \begin{pmatrix} e^{r_1 t} & 0 & \cdots & 0 \\ 0 & e^{r_2 t} & \cdots & 0 \\ \vdots & \vdots & & \vdots \\ 0 & 0 & \cdots & e^{r_n t} \end{pmatrix}. \tag{24}$$

A fundamental matrix $\boldsymbol{\Psi}$ for the system (17) is then found from **Q** by the transformation (19),

$$\boldsymbol{\Psi} = \mathbf{T}\mathbf{Q}; \tag{25}$$

that is,

$$\boldsymbol{\Psi}(t) = \begin{pmatrix} \xi_1^{(1)} e^{r_1 t} & \cdots & \xi_1^{(n)} e^{r_n t} \\ \vdots & & \vdots \\ \xi_n^{(1)} e^{r_1 t} & \cdots & \xi_n^{(n)} e^{r_n t} \end{pmatrix}. \tag{26}$$

Equation (26) is the same result that was obtained earlier. This diagonalization procedure does not afford any computational advantage over the method of Section 7.5 since in either case it is necessary to calculate the eigenvalues and eigenvectors of the coefficient matrix in the system of differential equations. Nevertheless, it is noteworthy that the problem of solving a system of differential equations and the problem of diagonalizing a matrix are mathematically one and the same.

PROBLEMS

In each of Problems 1 through 10 find a fundamental matrix for the given system of equations. In each case also find the fundamental matrix $\mathbf{\Phi}(t)$ satisfying $\mathbf{\Phi}(0) = \mathbf{I}$.

1. $\mathbf{x}' = \begin{pmatrix} 3 & -2 \\ 2 & -2 \end{pmatrix}\mathbf{x}$

2. $\mathbf{x}' = \begin{pmatrix} 4 & -3 \\ 8 & -6 \end{pmatrix}\mathbf{x}$

3. $\mathbf{x}' = \begin{pmatrix} 2 & -1 \\ 3 & -2 \end{pmatrix}\mathbf{x}$

4. $\mathbf{x}' = \begin{pmatrix} 1 & 1 \\ 4 & -2 \end{pmatrix}\mathbf{x}$

5. $\mathbf{x}' = \begin{pmatrix} 2 & -5 \\ 1 & -2 \end{pmatrix}\mathbf{x}$

6. $\mathbf{x}' = \begin{pmatrix} -1 & -4 \\ 1 & -1 \end{pmatrix}\mathbf{x}$

7. $\mathbf{x}' = \begin{pmatrix} 5 & -1 \\ 3 & 1 \end{pmatrix}\mathbf{x}$

8. $\mathbf{x}' = \begin{pmatrix} 3 & -4 \\ 1 & -1 \end{pmatrix}\mathbf{x}$

9. $\mathbf{x}' = \begin{pmatrix} 1 & 1 & 1 \\ 2 & 1 & -1 \\ -8 & -5 & -3 \end{pmatrix}\mathbf{x}$

10. $\mathbf{x}' = \begin{pmatrix} 1 & -1 & 4 \\ 3 & 2 & -1 \\ 2 & 1 & -1 \end{pmatrix}\mathbf{x}$

11. Show that $\mathbf{\Phi}(t) = \mathbf{\Psi}(t)\mathbf{\Psi}^{-1}(t_0)$, where $\mathbf{\Phi}(t)$ and $\mathbf{\Psi}(t)$ are as defined in this section.

*12. The method of successive approximations (see Section 2.12) can also be applied to systems of equations. In this problem we indicate formally how it works for linear systems with constant coefficients. Consider the initial value problem

$$\mathbf{x}' = \mathbf{A}\mathbf{x}, \qquad \mathbf{x}(0) = \boldsymbol{\eta}, \tag{i}$$

where $\mathbf{A}$ is a constant matrix, and $\boldsymbol{\eta}$ a prescribed vector.

(a) Assuming that a solution $\mathbf{x} = \boldsymbol{\phi}(t)$ of Eq. (i) exists, show that it must satisfy the integral equation

$$\boldsymbol{\phi}(t) = \boldsymbol{\eta} + \int_0^t \mathbf{A}\boldsymbol{\phi}(s)\,ds. \tag{ii}$$

(b) Start with the initial approximation $\boldsymbol{\phi}^{(0)}(t) = \boldsymbol{\eta}$. Substitute this expression for $\boldsymbol{\phi}(s)$ in the right side of Eq. (ii), and obtain a new approximation $\boldsymbol{\phi}^{(1)}(t)$. Show that

$$\boldsymbol{\phi}^{(1)}(t) = (\mathbf{I} + \mathbf{A}t)\boldsymbol{\eta}. \tag{iii}$$

(c) Repeat this process by substituting the expression in Eq. (iii) for $\boldsymbol{\phi}(s)$ in the right side of Eq. (ii) to obtain a new approximation $\boldsymbol{\phi}^{(2)}(t)$. Show that

$$\boldsymbol{\phi}^{(2)}(t) = \left(\mathbf{I} + \mathbf{A}t + \mathbf{A}^2\frac{t^2}{2}\right)\boldsymbol{\eta}. \tag{iv}$$

(d) By continuing this process obtain a sequence of approximations $\boldsymbol{\phi}^{(0)}, \boldsymbol{\phi}^{(1)}, \boldsymbol{\phi}^{(2)}, \ldots, \boldsymbol{\phi}^{(n)}, \ldots$. Show by induction that

$$\boldsymbol{\phi}^{(n)}(t) = \left(\mathbf{I} + \mathbf{A}t + \mathbf{A}^2\frac{t^2}{2!} + \cdots + \mathbf{A}^n\frac{t^n}{n!}\right)\boldsymbol{\eta}. \tag{v}$$

(e) The sum appearing in Eq. (v) is a sum of matrices. It is possible to show that each component of this sum converges as $n \to \infty$, and therefore a new matrix $\mathbf{M}(t)$ can be defined as

$$\mathbf{M}(t) = \lim_{n\to\infty}\left(\mathbf{I} + \mathbf{A}t + \mathbf{A}^2\frac{t^2}{2!} + \cdots + \mathbf{A}^n\frac{t^n}{n!}\right) = \mathbf{I} + \sum_{k=1}^{\infty}\mathbf{A}^k\frac{t^k}{k!}. \qquad \text{(vi)}$$

Assuming that term-by-term differentiation is permissible, show that $\mathbf{M}(t)$ satisfies $\mathbf{M}'(t) = \mathbf{AM}(t)$, $\mathbf{M}(0) = \mathbf{I}$. Hence show that $\mathbf{M}(t) = \mathbf{\Phi}(t)$.

(f) The series in Eq. (vi) resembles the exponential series. Thus it is natural to adopt the notation $e^{\mathbf{A}t}$ for this sum:

$$e^{\mathbf{A}t} = \mathbf{I} + \sum_{k=1}^{\infty}\mathbf{A}^k\frac{t^k}{k!}. \qquad \text{(vii)}$$

Show that the solution of the initial value problem (i) can then be written as $\mathbf{x} = e^{\mathbf{A}t}\boldsymbol{\eta}$.

*13. Let $\mathbf{\Phi}(t)$ denote the fundamental matrix satisfying $\mathbf{\Phi}' = \mathbf{A\Phi}$, $\mathbf{\Phi}(0) = \mathbf{I}$. In Problem 12 we indicated that it was also reasonable to denote this matrix by $e^{\mathbf{A}t}$. In this problem we show that $\mathbf{\Phi}$ does indeed have the principal algebraic properties associated with the exponential function.

(a) Show that $\mathbf{\Phi}(t)\mathbf{\Phi}(s) = \mathbf{\Phi}(t+s)$; that is, $e^{\mathbf{A}t}e^{\mathbf{A}s} = e^{\mathbf{A}(t+s)}$.

Hint: Show that if s is fixed and t is variable, then both $\mathbf{\Phi}(t)\mathbf{\Phi}(s)$ and $\mathbf{\Phi}(t+s)$ satisfy the initial value problem $\mathbf{Z}' = \mathbf{AZ}$, $\mathbf{Z}(0) = \mathbf{\Phi}(s)$.

(b) Show that $\mathbf{\Phi}(t)\mathbf{\Phi}(-t) = \mathbf{I}$, or $e^{\mathbf{A}t}e^{\mathbf{A}(-t)} = \mathbf{I}$, hence that $\mathbf{\Phi}(-t) = \mathbf{\Phi}^{-1}(t)$.

(c) Show that $\mathbf{\Phi}(t-s) = \mathbf{\Phi}(t)\mathbf{\Phi}^{-1}(s)$.

7.9 Nonhomogeneous Linear Systems

In this section we turn to the nonhomogeneous system

$$\mathbf{x}' = \mathbf{P}(t)\mathbf{x} + \mathbf{g}(t), \qquad (1)$$

where the $n \times n$ matrix $\mathbf{P}(t)$ and the $n \times 1$ vector $\mathbf{g}(t)$ are continuous for $\alpha < t < \beta$. By the same argument as in Section 3.6 (see also Problem 16) the general solution of Eq. (1) can be expressed as

$$\mathbf{x} = \mathbf{x}^{(c)}(t) + \mathbf{x}^{(p)}(t), \qquad (2)$$

where $\mathbf{x}^{(c)}(t)$ is the general solution of the homogeneous system $\mathbf{x}' = \mathbf{P}(t)\mathbf{x}$, and $\mathbf{x}^{(p)}(t)$ is a particular solution of the nonhomogeneous system (1). We will briefly describe several methods for determining $\mathbf{x}^{(p)}(t)$.

We begin with systems of the form

$$\mathbf{x}' = \mathbf{Ax} + \mathbf{g}(t), \qquad (3)$$

where $\mathbf{A}$ is an $n \times n$ diagonalizable constant matrix. We proceed by the method indicated at the end of Section 7.8. Let $\mathbf{T}$ be the matrix whose columns are the eigenvectors $\boldsymbol{\xi}^{(1)}, \ldots, \boldsymbol{\xi}^{(n)}$ of $\mathbf{A}$, and define a new dependent variable $\mathbf{y}$ by

$$\mathbf{x} = \mathbf{Ty}. \qquad (4)$$

Then substituting for $\mathbf{x}$ in Eq. (3), we obtain

$$\mathbf{Ty}' = \mathbf{ATy} + \mathbf{g}(t).$$

By multiplying by $\mathbf{T}^{-1}$ it follows that

$$\mathbf{y}' = (\mathbf{T}^{-1}\mathbf{AT})\mathbf{y} + \mathbf{T}^{-1}\mathbf{g}(t) = \mathbf{Dy} + \mathbf{h}(t), \tag{5}$$

where $\mathbf{h}(t) = \mathbf{T}^{-1}\mathbf{g}(t)$, and where $\mathbf{D}$ is the diagonal matrix whose diagonal entries are the eigenvalues $r_1, \ldots, r_n$ of $\mathbf{A}$, arranged in the same order as the corresponding eigenvectors $\boldsymbol{\xi}^{(1)}, \ldots, \boldsymbol{\xi}^{(n)}$ appear as columns of $\mathbf{T}$. Equation (5) is a system of *n uncoupled* equations for $y_1(t), \ldots, y_n(t)$; as a consequence the equations can be solved separately. In scalar form Eq. (5) has the form

$$y_j'(t) = r_j y_j(t) + h_j(t), \qquad j = 1, \ldots, n, \tag{6}$$

where $h_j(t)$ is a certain linear combination of $g_1(t), \ldots, g_n(t)$. Equation (6) is a first order linear equation and can be solved by the methods of Section 2.1. In fact, we have

$$y_j(t) = e^{r_j t}\int_{t_0}^{t} e^{-r_j s} h_j(s)\,ds + c_j e^{r_j t}, \qquad j = 1, \ldots, n, \tag{7}$$

where the c_j are arbitrary constants. Finally, the solution $\mathbf{x}$ of Eq. (3) is obtained from Eq. (4). When multiplied by the transformation matrix $\mathbf{T}$, the second term on the right side of Eq. (7) produces the general solution $\mathbf{x}^{(c)}(t)$ of the homogeneous equation $\mathbf{x}' = \mathbf{Ax}$, while the first term on the right side of Eq. (7) yields a particular solution of the nonhomogeneous system (3).

EXAMPLE 1

Find the general solution of the system

$$\mathbf{x}' = \begin{pmatrix} -2 & 1 \\ 1 & -2 \end{pmatrix}\mathbf{x} + \begin{pmatrix} 2e^{-t} \\ 3t \end{pmatrix} = \mathbf{Ax} + \mathbf{g}(t). \tag{8}$$

Proceeding as in Section 7.5, we find that the eigenvalues of the coefficient matrix are $r_1 = -3$ and $r_2 = -1$, and the corresponding eigenvectors are

$$\boldsymbol{\xi}^{(1)} = \begin{pmatrix} 1 \\ -1 \end{pmatrix}, \qquad \boldsymbol{\xi}^{(2)} = \begin{pmatrix} 1 \\ 1 \end{pmatrix}. \tag{9}$$

Thus the general solution of the homogeneous system is

$$\mathbf{x} = c_1\begin{pmatrix} 1 \\ -1 \end{pmatrix}e^{-3t} + c_2\begin{pmatrix} 1 \\ 1 \end{pmatrix}e^{-t}. \tag{10}$$

Before writing down the matrix $\mathbf{T}$ of eigenvectors we recall that eventually we must find $\mathbf{T}^{-1}$. The coefficient matrix $\mathbf{A}$ is real and symmetric, so we can use the result stated at the end of Section 7.3: $\mathbf{T}^{-1}$ is simply the adjoint or (since $\mathbf{T}$ is real) the transpose of $\mathbf{T}$, provided that the eigenvectors of $\mathbf{A}$ are normalized so that $(\boldsymbol{\xi}, \boldsymbol{\xi}) = 1$. Hence, upon normalizing $\boldsymbol{\xi}^{(1)}$ and $\boldsymbol{\xi}^{(2)}$, we have

$$\mathbf{T} = \frac{1}{\sqrt{2}}\begin{pmatrix} 1 & 1 \\ -1 & 1 \end{pmatrix}, \qquad \mathbf{T}^{-1} = \frac{1}{\sqrt{2}}\begin{pmatrix} 1 & -1 \\ 1 & 1 \end{pmatrix}. \tag{11}$$

Letting $\mathbf{x} = \mathbf{T}\mathbf{y}$ and substituting for $\mathbf{x}$ in Eq. (8), we obtain the following system of equations for the new dependent variable $\mathbf{y}$:

$$\mathbf{y}' = \mathbf{D}\mathbf{y} + \mathbf{T}^{-1}\mathbf{g}(t) = \begin{pmatrix} -3 & 0 \\ 0 & -1 \end{pmatrix}\mathbf{y} + \frac{1}{\sqrt{2}}\begin{pmatrix} 2e^{-t} - 3t \\ 2e^{-t} + 3t \end{pmatrix}. \tag{12}$$

Thus

$$\begin{aligned} y_1' + 3y_1 &= \sqrt{2}\,e^{-t} - (3/\sqrt{2})t, \\ y_2' + y_2 &= \sqrt{2}\,e^{-t} + (3/\sqrt{2})t. \end{aligned} \tag{13}$$

Each of Eqs. (13) is a first order linear equation, and so can be solved by the methods of Section 2.1. In this way we obtain

$$\begin{aligned} y_1 &= (\sqrt{2}/2)e^{-t} - (3/\sqrt{2})[(t/3) - \tfrac{1}{9}] + c_1e^{-3t}, \\ y_2 &= \sqrt{2}\,te^{-t} + (3/\sqrt{2})(t-1) + c_2e^{-t}. \end{aligned} \tag{14}$$

Finally, we write the solution in terms of the original variables:

$$\begin{aligned} \mathbf{x} = \mathbf{T}\mathbf{y} &= \frac{1}{\sqrt{2}}\begin{pmatrix} y_1 + y_2 \\ -y_1 + y_2 \end{pmatrix} \\ &= \begin{pmatrix} (c_1/\sqrt{2})e^{-3t} + [(c_2/\sqrt{2}) + \tfrac{1}{2}]e^{-t} + t - \tfrac{4}{3} + te^{-t} \\ -(c_1/\sqrt{2})e^{-3t} + [(c_2/\sqrt{2}) - \tfrac{1}{2}]e^{-t} + 2t - \tfrac{5}{3} + te^{-t} \end{pmatrix} \\ &= k_1\begin{pmatrix} 1 \\ -1 \end{pmatrix}e^{-3t} + k_2\begin{pmatrix} 1 \\ 1 \end{pmatrix}e^{-t} + \frac{1}{2}\begin{pmatrix} 1 \\ -1 \end{pmatrix}e^{-t} + \begin{pmatrix} 1 \\ 1 \end{pmatrix}te^{-t} + \begin{pmatrix} 1 \\ 2 \end{pmatrix}t - \frac{1}{3}\begin{pmatrix} 4 \\ 5 \end{pmatrix}, \end{aligned} \tag{15}$$

where $k_1 = c_1/\sqrt{2}$ and $k_2 = c_2/\sqrt{2}$. The first two terms on the right side of Eq. (15) form the general solution of the homogeneous system corresponding to Eq. (8). The remaining terms are a particular solution of the nonhomogeneous system.

A second way to find a particular solution of the nonhomogeneous system (1) is the method of undetermined coefficients. To make use of this method, one assumes the form of the solution with some or all of the coefficients unspecified, and then seeks to determine these coefficients so as to satisfy the differential equation. As a practical matter, this method is applicable only if the coefficient matrix $\mathbf{P}$ is a constant matrix, and if the components of $\mathbf{g}$ are polynomial, exponential, or sinusoidal functions, or sums of products of these. Only in these cases can the correct form of the solution be predicted in a simple and systematic manner. The procedure for choosing the form of the solution is substantially the same as that given in Section 3.6.1 for linear second order equations. The main difference is illustrated by the case of a nonhomogeneous term of the form $\mathbf{v}e^{\lambda t}$, where λ is a simple root of the characteristic equation. In this situation, rather than assuming a solution of the form $\mathbf{a}te^{\lambda t}$, it is necessary to use $\mathbf{a}te^{\lambda t} + \mathbf{b}e^{\lambda t}$, where $\mathbf{a}$ and $\mathbf{b}$ are determined by substituting into the differential equation.

EXAMPLE 2

Use the method of undetermined coefficients to find a particular solution of

$$\mathbf{x}' = \begin{pmatrix} -2 & 1 \\ 1 & -2 \end{pmatrix}\mathbf{x} + \begin{pmatrix} 2e^{-t} \\ 3t \end{pmatrix} = \mathbf{A}\mathbf{x} + \mathbf{g}(t). \tag{16}$$

This is the same system of equations as in Example 1. To use the method of undetermined coefficients, we write $\mathbf{g}(t)$ in the form

$$\mathbf{g}(t) = \begin{pmatrix} 2 \\ 0 \end{pmatrix} e^{-t} + \begin{pmatrix} 0 \\ 3 \end{pmatrix} t. \tag{17}$$

Then we assume that

$$\mathbf{x}^{(p)}(t) = \mathbf{a}te^{-t} + \mathbf{b}e^{-t} + \mathbf{c}t + \mathbf{d}, \tag{18}$$

where $\mathbf{a}$, $\mathbf{b}$, $\mathbf{c}$, and $\mathbf{d}$ are vectors to be determined. Observe that $r = -1$ is an eigenvalue of the coefficient matrix, and therefore we must include both $\mathbf{a}te^{-t}$ and $\mathbf{b}e^{-t}$ in the assumed solution. By substituting Eq. (18) into Eq. (16) and collecting terms, we obtain the following algebraic equations for $\mathbf{a}$, $\mathbf{b}$, $\mathbf{c}$, and $\mathbf{d}$.

$$\begin{aligned} \mathbf{A}\mathbf{a} &= -\mathbf{a}, \\ \mathbf{A}\mathbf{b} &= \mathbf{a} - \mathbf{b} - \begin{pmatrix} 2 \\ 0 \end{pmatrix}, \\ \mathbf{A}\mathbf{c} &= -\begin{pmatrix} 0 \\ 3 \end{pmatrix}, \\ \mathbf{A}\mathbf{d} &= \mathbf{c}. \end{aligned} \tag{19}$$

From the first of Eqs. (19) we see that $\mathbf{a}$ is an eigenvector of $\mathbf{A}$ corresponding to the eigenvalue $r = -1$. Thus $\mathbf{a}^T = (1, 1)$. Then from the second of Eqs. (19) we find that

$$\mathbf{b} = k\begin{pmatrix} 1 \\ 1 \end{pmatrix} - \begin{pmatrix} 0 \\ 1 \end{pmatrix} \tag{20}$$

for any constant k. The simplest choice is $k = 0$, whence $\mathbf{b}^T = (0, -1)$. Then the third and fourth of Eqs. (19) yield $\mathbf{c}^T = (1, 2)$ and $\mathbf{d}^T = (-\frac{4}{3}, -\frac{5}{3})$, respectively. Finally, from Eq. (18) we obtain the particular solution

$$\mathbf{x}^{(p)}(t) = \begin{pmatrix} 1 \\ 1 \end{pmatrix} te^{-t} - \begin{pmatrix} 0 \\ 1 \end{pmatrix} e^{-t} + \begin{pmatrix} 1 \\ 2 \end{pmatrix} t - \frac{1}{3}\begin{pmatrix} 4 \\ 5 \end{pmatrix}. \tag{21}$$

The particular solution (21) is not identical to the one contained in Eq. (15) of

Example 1 because the term in e^{-t} is different. However, if we choose $k = \frac{1}{2}$ in Eq. (20), then $\mathbf{b}^T = (\frac{1}{2}, -\frac{1}{2})$ and the two particular solutions then agree.

Now let us turn to more general problems in which the coefficient matrix is not constant or not diagonalizable. Let

$$\mathbf{x}' = \mathbf{P}(t)\mathbf{x} + \mathbf{g}(t) \tag{22}$$

where $\mathbf{P}(t)$ and $\mathbf{g}(t)$ are continuous on $\alpha < t < \beta$. Assume that a fundamental matrix $\boldsymbol{\Psi}(t)$ for the corresponding homogeneous system

$$\mathbf{x}' = \mathbf{P}(t)\mathbf{x} \tag{23}$$

has been found. We use the method of variation of parameters to construct a particular solution, and hence the general solution, of the nonhomogeneous system (22).

Since the general solution of the homogeneous system (23) is $\boldsymbol{\Psi}(t)\mathbf{c}$, it is natural to proceed as in Section 3.6.2, and to seek a solution of the nonhomogeneous system (22) by replacing the constant vector $\mathbf{c}$ by a vector function $\mathbf{u}(t)$. Thus, we assume that

$$\mathbf{x} = \boldsymbol{\Psi}(t)\mathbf{u}(t), \tag{24}$$

where $\mathbf{u}(t)$ is a vector to be found. Upon differentiating $\mathbf{x}$ as given by Eq. (24) and requiring that Eq. (22) be satisfied, we obtain

$$\boldsymbol{\Psi}'(t)\mathbf{u}(t) + \boldsymbol{\Psi}(t)\mathbf{u}'(t) = \mathbf{P}(t)\boldsymbol{\Psi}(t)\mathbf{u}(t) + \mathbf{g}(t). \tag{25}$$

Since $\boldsymbol{\Psi}(t)$ is a fundamental matrix, $\boldsymbol{\Psi}'(t) = \mathbf{P}(t)\boldsymbol{\Psi}(t)$; hence Eq. (25) reduces to

$$\boldsymbol{\Psi}(t)\mathbf{u}'(t) = \mathbf{g}(t). \tag{26}$$

Recall that $\boldsymbol{\Psi}(t)$ is nonsingular on any interval where $\mathbf{P}$ is continuous. Hence $\boldsymbol{\Psi}^{-1}(t)$ exists, and therefore

$$\mathbf{u}'(t) = \boldsymbol{\Psi}^{-1}(t)\mathbf{g}(t). \tag{27}$$

Thus for $\mathbf{u}(t)$ we can select any vector from the class of vectors that satisfy Eq. (27); these vectors are determined only up to an arbitrary additive constant (vector); therefore we denote $\mathbf{u}(t)$ by

$$\mathbf{u}(t) = \int^t \boldsymbol{\Psi}^{-1}(s)\mathbf{g}(s)\,ds + \mathbf{c}, \tag{28}$$

where the constant vector $\mathbf{c}$ is arbitrary. Finally, substituting for $\mathbf{u}(t)$ in Eq. (24) gives the solution $\mathbf{x}$ of the system (22):

$$\mathbf{x} = \boldsymbol{\Psi}(t)\mathbf{c} + \boldsymbol{\Psi}(t)\int^t \boldsymbol{\Psi}^{-1}(s)\mathbf{g}(s)\,ds. \tag{29}$$

Since $\mathbf{c}$ is arbitrary, any initial condition at a point $t = t_0$ can be satisfied by an appropriate choice of $\mathbf{c}$. Thus, every solution of the system (22) is contained in the expression given by Eq. (29); it is therefore the general solution of Eq. (22).

Note that the first term on the right side of Eq. (29) is the general solution of the corresponding homogeneous system (23), and the second term is a particular solution of Eq. (22) itself.

Now let us consider the initial value problem consisting of the differential equation (22) and the initial condition

$$\mathbf{x}(t_0) = \mathbf{x}^0. \tag{30}$$

We can write the solution of this problem most conveniently if we choose for the particular solution in Eq. (29) the specific one that is zero when $t = t_0$. This can be done by using t_0 as the lower limit of integration in Eq. (29), so that the general solution of the differential equation takes the form

$$\mathbf{x} = \mathbf{\Psi}(t)\mathbf{c} + \mathbf{\Psi}(t)\int_{t_0}^{t}\mathbf{\Psi}^{-1}(s)\mathbf{g}(s)\,ds. \tag{31}$$

The initial condition (30) can also be satisfied provided that

$$\mathbf{c} = \mathbf{\Psi}^{-1}(t_0)\mathbf{x}^0. \tag{32}$$

Therefore

$$\mathbf{x} = \mathbf{\Psi}(t)\mathbf{\Psi}^{-1}(t_0)\mathbf{x}^0 + \mathbf{\Psi}(t)\int_{t_0}^{t}\mathbf{\Psi}^{-1}(s)\mathbf{g}(s)\,ds \tag{33}$$

is the solution of the given initial value problem. Again, while it is helpful to use $\mathbf{\Psi}^{-1}$ to write the solutions (29) and (33), it is usually better in particular cases to solve the necessary equations by row reduction rather than to calculate $\mathbf{\Psi}^{-1}$ and substitute into Eqs. (29) and (33).

The solution (33) takes a slightly simpler form if we use the fundamental matrix $\mathbf{\Phi}(t)$ satisfying $\mathbf{\Phi}(t_0) = \mathbf{I}$. In this case we have

$$\mathbf{x} = \mathbf{\Phi}(t)\mathbf{x}^0 + \mathbf{\Phi}(t)\int_{t_0}^{t}\mathbf{\Phi}^{-1}(s)\mathbf{g}(s)\,ds. \tag{34}$$

Equation (34) can be simplified further if the coefficient matrix $\mathbf{P}(t)$ is a constant matrix (see Problem 17).

EXAMPLE 3

Use the method of variation of parameters to find the general solution of the system

$$\mathbf{x}' = \begin{pmatrix} -2 & 1 \\ 1 & -2 \end{pmatrix}\mathbf{x} + \begin{pmatrix} 2e^{-t} \\ 3t \end{pmatrix}. \tag{35}$$

This is the same system of equations as in Examples 1 and 2.

The general solution of the corresponding homogeneous system was given in Eq. (10). Thus

$$\mathbf{\Psi}(t) = \begin{pmatrix} e^{-3t} & e^{-t} \\ -e^{-3t} & e^{-t} \end{pmatrix} \tag{36}$$

is a fundamental matrix. Then the solution $\mathbf{x}$ of Eq. (35) is given by $\mathbf{x} = \boldsymbol{\Psi}(t)\mathbf{u}(t)$, where $\mathbf{u}(t)$ satisfies $\boldsymbol{\Psi}(t)\mathbf{u}'(t) = \mathbf{g}(t)$, or

$$\begin{pmatrix} e^{-3t} & e^{-t} \\ -e^{-3t} & e^{-t} \end{pmatrix}\begin{pmatrix} u_1' \\ u_2' \end{pmatrix} = \begin{pmatrix} 2e^{-t} \\ 3t \end{pmatrix}. \tag{37}$$

Solving Eq. (37) by row reduction, we obtain

$$u_1' = e^{2t} - \tfrac{3}{2}te^{3t},$$

$$u_2' = 1 + \tfrac{3}{2}te^{t}.$$

Hence

$$u_1(t) = \tfrac{1}{2}e^{2t} - \tfrac{1}{2}te^{3t} + \tfrac{1}{6}e^{3t} + c_1,$$

$$u_2(t) = t + \tfrac{3}{2}te^{t} - \tfrac{3}{2}e^{t} + c_2,$$

and

$$\begin{aligned} \mathbf{x} &= \boldsymbol{\Psi}(t)\mathbf{u}(t) \\ &= c_1\begin{pmatrix} 1 \\ -1 \end{pmatrix}e^{-3t} + c_2\begin{pmatrix} 1 \\ 1 \end{pmatrix}e^{-t} + \begin{pmatrix} 1 \\ 1 \end{pmatrix}te^{-t} + \frac{1}{2}\begin{pmatrix} 1 \\ -1 \end{pmatrix}e^{-t} + \begin{pmatrix} 1 \\ 2 \end{pmatrix}t - \frac{1}{3}\begin{pmatrix} 4 \\ 5 \end{pmatrix}, \end{aligned}$$

which is the same as the solution obtained previously.

In comparing the methods for solving nonhomogeneous equations, it is clear that each has some advantages and disadvantages. The method of undetermined coefficients requires no integration, but is somewhat limited in scope and may entail the solution of several sets of algebraic equations. The method of diagonalization requires finding the inverse of the transformation matrix and the solution of a set of uncoupled first order linear equations, followed by a matrix multiplication. Its main advantage is that for Hermitian coefficient matrices the inverse of the transformation matrix can be written down without calculation, a feature that is more important for large systems. Variation of parameters is the most general method. On the other hand, it involves the solution of a set of linear algebraic equations with variable coefficients, followed by an integration and a matrix multiplication, so it may also be the most complicated from a computational viewpoint. For many small systems with constant coefficients, such as the one in the examples in this section, there may be little reason to select one of these methods over another. Keep in mind, however, that the method of diagonalization is ruled out if the coefficient matrix is not diagonalizable, and the method of undetermined coefficients is only practical for the kinds of nonhomogeneous terms mentioned earlier.

PROBLEMS

In each of Problems 1 through 12 find the general solution of the given system of equations.

1. $\mathbf{x}' = \begin{pmatrix} 2 & -1 \\ 3 & -2 \end{pmatrix}\mathbf{x} + \begin{pmatrix} e^t \\ t \end{pmatrix}$

2. $\mathbf{x}' = \begin{pmatrix} 1 & \sqrt{3} \\ \sqrt{3} & -1 \end{pmatrix}\mathbf{x} + \begin{pmatrix} e^t \\ \sqrt{3}\, e^{-t} \end{pmatrix}$

3. $\mathbf{x}' = \begin{pmatrix} 2 & -5 \\ 1 & -2 \end{pmatrix}\mathbf{x} + \begin{pmatrix} -\cos t \\ \sin t \end{pmatrix}$

4. $\mathbf{x}' = \begin{pmatrix} 1 & 1 \\ 4 & -2 \end{pmatrix}\mathbf{x} + \begin{pmatrix} e^{-2t} \\ -2e^t \end{pmatrix}$

5. $\mathbf{x}' = \begin{pmatrix} 4 & -2 \\ 8 & -4 \end{pmatrix}\mathbf{x} + \begin{pmatrix} t^{-3} \\ -t^{-2} \end{pmatrix}, \quad t > 0$

6. $\mathbf{x}' = \begin{pmatrix} -4 & 2 \\ 2 & -1 \end{pmatrix}\mathbf{x} + \begin{pmatrix} t^{-1} \\ 2t^{-1} + 4 \end{pmatrix}, \quad t > 0$

7. $\mathbf{x}' = \begin{pmatrix} 1 & 1 \\ 4 & 1 \end{pmatrix}\mathbf{x} + \begin{pmatrix} 2 \\ -1 \end{pmatrix}e^t$

8. $\mathbf{x}' = \begin{pmatrix} 2 & -1 \\ 3 & -2 \end{pmatrix}\mathbf{x} + \begin{pmatrix} 1 \\ -1 \end{pmatrix}e^t$

9. $\mathbf{x}' = \begin{pmatrix} -\frac{5}{4} & \frac{3}{4} \\ \frac{3}{4} & -\frac{5}{4} \end{pmatrix}\mathbf{x} + \begin{pmatrix} 2t \\ e^t \end{pmatrix}$

10. $\mathbf{x}' = \begin{pmatrix} -3 & \sqrt{2} \\ \sqrt{2} & -2 \end{pmatrix}\mathbf{x} + \begin{pmatrix} 1 \\ -1 \end{pmatrix}e^{-t}$

11. $\mathbf{x}' = \begin{pmatrix} 2 & -5 \\ 1 & -2 \end{pmatrix}\mathbf{x} + \begin{pmatrix} 0 \\ \cos t \end{pmatrix}, \quad 0 < t < \pi$

12. $\mathbf{x}' = \begin{pmatrix} 2 & -5 \\ 1 & -2 \end{pmatrix}\mathbf{x} + \begin{pmatrix} \csc t \\ \sec t \end{pmatrix}, \quad \frac{\pi}{2} < t < \pi$

13. The electric circuit shown in Figure 7.13 is described by the system of differential equations

$$\frac{d\mathbf{x}}{dt} = \begin{pmatrix} -\frac{1}{2} & -\frac{1}{8} \\ 2 & -\frac{1}{2} \end{pmatrix}\mathbf{x} + \begin{pmatrix} \frac{1}{2} \\ 0 \end{pmatrix} I(t), \tag{i}$$

where x_1 is the current through the inductance, x_2 is the voltage drop across the capacitance, and $I(t)$ is the current supplied by the external source.
(a) Determine a fundamental matrix $\mathbf{\Psi}(t)$ for the homogeneous system corresponding to Eqs. (i). Refer to Problem 13 of Section 7.6.
(b) If $I(t) = e^{-t/2}$, determine the solution of the system (i) that also satisfies the initial conditions $\mathbf{x}(0) = \mathbf{0}$.

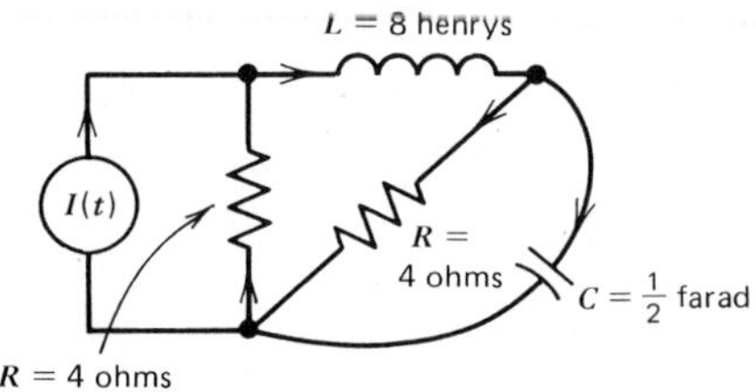

FIGURE 7.13 The circuit in Problem 13.

In each of Problems 14 and 15 verify that the given vector is the general solution of the corresponding homogeneous system, and then solve the nonhomogeneous system. Assume that $t > 0$.

14. $t\mathbf{x}' = \begin{pmatrix} 2 & -1 \\ 3 & -2 \end{pmatrix}\mathbf{x} + \begin{pmatrix} 1 - t^2 \\ 2t \end{pmatrix}, \qquad \mathbf{x}^{(c)} = c_1\begin{pmatrix} 1 \\ 1 \end{pmatrix}t + c_2\begin{pmatrix} 1 \\ 3 \end{pmatrix}t^{-1}$

15. $t\mathbf{x}' = \begin{pmatrix} 3 & -2 \\ 2 & -2 \end{pmatrix}\mathbf{x} + \begin{pmatrix} -2t \\ t^4 - 1 \end{pmatrix}, \qquad \mathbf{x}^{(c)} = c_1\begin{pmatrix} 1 \\ 2 \end{pmatrix}t^{-1} + c_2\begin{pmatrix} 2 \\ 1 \end{pmatrix}t^2$

16. Let $\mathbf{x} = \boldsymbol{\phi}(t)$ be the general solution of $\mathbf{x}' = \mathbf{P}(t)\mathbf{x} + \mathbf{g}(t)$, and let $\mathbf{x} = \mathbf{x}^{(p)}(t)$ be some particular solution of the same system. By considering the difference $\boldsymbol{\phi}(t) - \mathbf{x}^{(p)}(t)$, show that $\boldsymbol{\phi}(t) = \mathbf{x}^{(c)}(t) + \mathbf{x}^{(p)}(t)$, where $\mathbf{x}^{(c)}(t)$ is the general solution of the homogeneous system $\mathbf{x}' = \mathbf{P}(t)\mathbf{x}$.

*17. Consider the initial value problem

$$\mathbf{x}' = \mathbf{A}\mathbf{x} + \mathbf{g}(t), \qquad \mathbf{x}(0) = \mathbf{x}^0.$$

(a) By referring to Problem 13(c) in Section 7.8, show that

$$\mathbf{x} = \boldsymbol{\Phi}(t)\mathbf{x}^0 + \int_0^t \boldsymbol{\Phi}(t-s)\mathbf{g}(s)\,ds.$$

(b) Show also that

$$\mathbf{x} = e^{\mathbf{A}t}\mathbf{x}^0 + \int_0^t e^{\mathbf{A}(t-s)}\mathbf{g}(s)\,ds.$$

REFERENCES

The books listed below are representative of recent introductory books on matrices and linear algebra:

Anton, H., *Elementary Linear Algebra* (4th ed.) (New York: Wiley, 1984).

Franklin, J. N., *Matrix Theory* (Englewood Cliffs, N.J.: Prentice-Hall, 1968).

Johnson, L. W., and Riess, R. D., *Introduction to Linear Algebra* (Reading, Mass.: Addison-Wesley, 1981).

Kumpel, P. G., and Thorpe, J. A., *Linear Algebra with Applications to Differential Equations* (Philadelphia: Saunders, 1983).

Leon, S. J., *Linear Algebra with Applications* (New York: Macmillan, 1980).

Strang, G., *Linear Algebra and Its Applications* (2nd ed.) (New York: Academic Press, 1980).

8.1 Introduction

In this chapter we consider numerical methods by which we can construct, in tabular or graphical form, the solution of a given differential equation and initial conditions. For example, a table of values of e^x, which is the solution of the initial value problem

$$y' - y = 0, \qquad y(0) = 1,$$

would constitute a numerical solution of this problem. We primarily consider the first order differential equation

$$y' = f(x, y) \tag{1}$$

with the initial condition

$$y(x_0) = y_0. \tag{2}$$

We also assume that the functions f and f_y are continuous in some rectangle in the xy plane including the point (x_0, y_0) so that, according to Theorem 2.2, there exists a unique solution $y = \phi(x)$ of the given problem in some interval about x_0. As we emphasized in Section 2.3, if Eq. (1) is nonlinear, then the interval about x_0 in which the solution ϕ exists may be difficult to determine, and may have no simple relationship to the function f. In all of our discussion of Eq. (1) we assume that a unique solution satisfying the initial condition (2) exists on the interval of interest.

In some cases $f(x, y)$ is so simple that Eq. (1) can be integrated directly. However, in many engineering and scientific problems this is not the case, and it

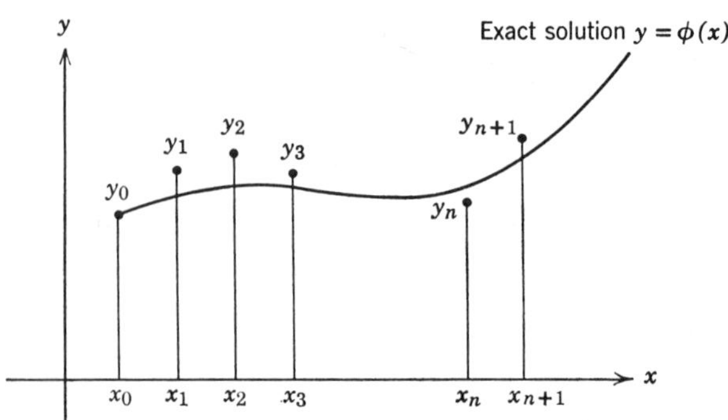

FIGURE 8.1 A numerical approximation to the solution of $y' = f(x, y)$, $y(x_0) = y_0$.

is natural to consider numerical procedures for obtaining an approximate solution of the problem. Even if Eq. (1) can be integrated directly, it may be more difficult to evaluate the analytic solution than to solve the original initial value problem numerically. For example, the analytic solution may be in the form of a complicated implicit relation $F(x, y) = 0$. Nevertheless, the importance of analytic methods should not be underestimated. They provide valuable information about the existence, uniqueness, and qualitative behavior of solutions, including their dependence on parameters in the differential equation or initial conditions.

By a numerical procedure for solving the initial value problem given by Eqs. (1) and (2) we mean a procedure for constructing approximate values $y_0, y_1, y_2, \ldots, y_n, \ldots$ of the solution ϕ at the points $x_0 < x_1 < x_2 < \cdots < x_n < \cdots$ (see Figure 8.1). More precisely, such a numerical procedure is referred to as a *discrete variable method* since we are replacing a problem involving continuous variables by one involving discrete variables.

Our first concern is to determine y_1, knowing y_0 from the initial condition and $\phi'(x_0) = f(x_0, y_0)$ from the differential equation (1). Then, knowing y_1, we determine y_2, and so on. In determining y_2 we can either use the same method as in going from x_0 to x_1, or we can use a different method that takes account of the fact that we know both y_0 at x_0 and y_1 at x_1.

Methods that require only a knowledge of y_n to determine y_{n+1} are known as *one-step*, or *stepwise*, or *starting methods*. Methods that make use of data at more than the previous point, say y_n, y_{n-1}, y_{n-2} to determine y_{n+1}, are known as *multistep* or *continuing methods*. Clearly, if a continuing method involving y_n and y_{n-1} is used, we must use a starting method (hence the name starting method) at the first step to determine y_1; the continuing method can then be used to determine y_2, and so on. On the other hand, a stepwise or starting method may be used for all of the computations. In Sections 8.2 through 8.7 we discuss several stepwise methods, and in Section 8.8 we discuss a typical multistep method. In Section 8.9 we briefly consider systems of equations.

With the advent of digital computers, the use of numerical methods to solve initial value problems has become commonplace. Accurate programs for solving general classes of initial value problems are widely available. Many problems that are hopeless to solve analytically can now be solved quickly using computers. On the other hand, not all questions about differential equations can be answered by using numerical methods. For example, it may be difficult to answer such questions as how a solution depends on y_0 (which is often of interest) or on other parameters in the problem. Such questions would be considerably simpler if the analytic solution were known. Further, analytic solutions of special or limiting cases may be extremely valuable in checking the validity of numerical solutions.

Moreover, numerical methods raise serious questions of their own. In the first place, there is the question of *convergence*. That is, as the spacing between the points $x_0, x_1, x_2, \ldots, x_n, \ldots$ approaches zero, do the values of the numerical solution $y_1, y_2, \ldots, y_n, \ldots$ approach the values of the exact solution? There is also the question of estimating the error made in computing the values $y_1, y_2, \ldots, y_n$. This error arises from two sources: first, the formula used in the numerical method is only an approximate one, which causes a *discretization error*, or *truncation error*, or *formula error*; second, it is possible to carry only a limited number of digits in any computation, which causes a *round-off error*. In Sections 8.3 and 8.7 we discuss briefly the significance of discretization and round-off errors. Another important question arises when multistep methods are used. It is possible that the numerical formula can have solutions, usually called *parasitic solutions*, that do not correspond to solutions of the original differential equation. If these extraneous solutions grow with increasing x they can cause serious difficulties. This is discussed in Section 8.8.

The notation used throughout this chapter is as follows. The exact solution of the initial value problem given by Eqs. (1) and (2) will be denoted by ϕ (or $y = \phi(x)$); then the value of the exact solution at x_n is $\phi(x_n)$. For a given numerical procedure the symbols y_n and $y_n' = f(x_n, y_n)$ will denote the approximate values of the exact solution and its derivative at the point x_n. Clearly $\phi(x_0) = y_0$, but in general $\phi(x_n) \neq y_n$ for $n \geq 1$. Similarly $\phi'(x_0) = y_0'$, but in general $\phi'(x_n) = f[x_n, \phi(x_n)]$ is not equal to $y_n' = f(x_n, y_n)$ for $n \geq 1$. Further, in all of the discussion we will use a uniform spacing or step size h on the x axis. Thus $x_1 = x_0 + h$, $x_2 = x_1 + h = x_0 + 2h$, and in general $x_n = x_0 + nh$.

Finally we would like to call the reader's attention to the following point. The first problem at the end of Sections 8.2 through 8.6 and 8.8 deals with the same initial value problem, and so do the next several problems. By using one of these problems for practice work, it is possible to compare the accuracy and amount of labor required by the various numerical procedures.

8.2 The Euler or Tangent Line Method

One of the simplest stepwise methods of solving the initial value problem

$$y' = f(x, y), \qquad y(x_0) = y_0 \tag{1}$$

numerically is the tangent line method, which was first used by Euler. Since x_0

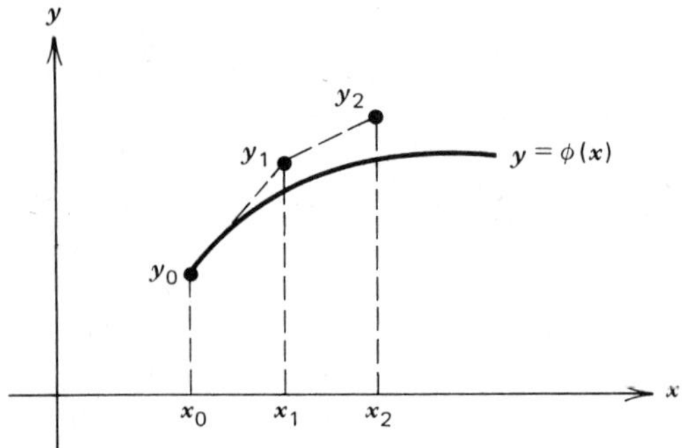

FIGURE 8.2 The Euler or tangent line approximation.

and y_0 are known, the slope of the tangent line to the solution at $x = x_0$, namely $\phi'(x_0) = f(x_0, y_0)$, is also known. Hence we can construct the tangent line to the solution at x_0, and then obtain an approximate value y_1 of $\phi(x_1)$ by moving along the tangent line to x_1 (see Figure 8.2). Thus

$$\begin{aligned} y_1 &= y_0 + \phi'(x_0)(x_1 - x_0) \\ &= y_0 + f(x_0, y_0)(x_1 - x_0). \end{aligned} \tag{2}$$

Once y_1 is determined we can compute $y_1' = f(x_1, y_1)$, which is an approximate value of $\phi'(x_1) = f[x_1, \phi(x_1)]$, the slope of the tangent line to the actual solution at x_1. Using this slope we obtain

$$\begin{aligned} y_2 &= y_1 + y_1'(x_2 - x_1) \\ &= y_1 + f(x_1, y_1)(x_2 - x_1), \end{aligned}$$

and, in general,

$$y_{n+1} = y_n + f(x_n, y_n)(x_{n+1} - x_n).$$

If we assume that there is a uniform step size h between the points $x_0, x_1, x_2, \ldots$, then $x_{n+1} = x_n + h$ and we obtain the Euler formula

$$\begin{aligned} y_{n+1} &= y_n + hf(x_n, y_n) \\ &= y_n + hy_n', \qquad n = 0, 1, 2, \ldots . \end{aligned} \tag{3}$$

The Euler formula is very simple to use and in discussing the application of the formula it is possible to illustrate many of the important ideas in numerical methods for solving initial value problems. On the other hand, it is not an accurate formula. Before discussing the Euler formula and methods of improving it, let us first illustrate how this result can be used to solve the initial value problem

$$y' = 1 - x + 4y, \tag{4}$$

$$y(0) = 1. \tag{5}$$

This example is used throughout the rest of this chapter to illustrate and compare different numerical methods. While it is a very simple problem, it illustrates the improvement that can be gained by using more accurate numerical procedures, and smaller step sizes h in the independent variable. Equation (4) is a linear first order equation, and it is easily verified that the solution satisfying the initial condition (5) is

$$y = \phi(x) = \tfrac{1}{4}x - \tfrac{3}{16} + \tfrac{19}{16}e^{4x}. \tag{6}$$

EXAMPLE

Using the Euler formula (3) and a step size $h = 0.1$, determine an approximate value of the solution $y = \phi(x)$ at $x = 0.2$ for the illustrative example $y' = 1 - x + 4y$, $y(0) = 1$.

First we compute $y_0' = f(0, 1) = 5$; hence

$$\begin{aligned} y_1 &= y_0 + hf(0, 1) \\ &= 1 + (0.1)5 = 1.5. \end{aligned}$$

Next

$$\begin{aligned} y_2 &= y_1 + hf(x_1, y_1) \\ &= 1.5 + (0.1)f(0.1, 1.5) \\ &= 1.5 + (0.1)(1 - 0.1 + 6) \\ &= 2.19. \end{aligned}$$

This result should be compared with the exact value, which is $\phi(0.2) = 2.5053299$ correct through eight digits. The error is approximately $2.51 - 2.19 = 0.32$.

Normally, an error this large (a percentage error of 12%) is not acceptable.[1] A better result can be obtained by using a smaller step size. Thus using $h = 0.05$ and taking four steps to reach $x = 0.2$ gives the approximate value 2.3249 for $\phi(0.2)$ with a percentage error of 8%. A step size $h = 0.025$ leads to a value of 2.4080117 with a percentage error of 4%. If we continue the calculations begun in Example 1, then we obtain the results given in Table 8.1. The second through the fifth columns contain approximate values of the solution, using a step size of $h = 0.1, 0.05, 0.025$, and 0.01, respectively. In the last column are the corresponding values of $\phi(x)$. Even without comparing the approximate and exact solutions, the fact that the values of y calculated with $h = 0.01$ differ significantly from

[1] Lest the student be discouraged we emphasize again that we have purposely chosen an example that will illustrate the improvement to be gained by using more accurate procedures and/or a smaller step size h. On the other hand, the present example is extremely simple compared to what may be encountered in practice. It follows that, in general, more accurate procedures are used in constructing numerical solutions of initial value problems in engineering, science, and other fields of applications.

TABLE 8.1 A COMPARISON OF RESULTS FOR THE NUMERICAL SOLUTION OF $y' = 1 - x + 4y$, $y(0) = 1$ USING THE EULER METHOD FOR DIFFERENT STEP SIZES h

x	$h = 0.1$	$h = 0.05$	$h = 0.025$	$h = 0.01$	Exact
0	1.0000000	1.0000000	1.0000000	1.0000000	1.0000000
0.1	1.5000000	1.5475000	1.5761188	1.5952901	1.6090418
0.2	2.1900000	2.3249000	2.4080117	2.4644587	2.5053299
0.3	3.1460000	3.4333560	3.6143837	3.7390345	3.8301388
0.4	4.4744000	5.0185326	5.3690304	5.6137120	5.7942260
0.5	6.3241600	7.2901870	7.9264062	8.3766865	8.7120041
0.6	8.9038240	10.550369	11.659058	12.454558	13.052522
0.7	12.505354	15.234032	17.112430	18.478797	19.515518
0.8	17.537495	21.967506	25.085110	27.384136	29.144880
0.9	24.572493	31.652708	36.746308	40.554208	43.497903
1	34.411490	45.588400	53.807866	60.037126	64.897803

those calculated with $h = 0.025$, 0.05, and 0.1 for values of x as small as 0.2 indicates that the Euler method with these step sizes is not satisfactory for this problem.

One possible way of obtaining better accuracy is to use even smaller step sizes. However, as we see in Section 8.7, this runs the risk of losing accuracy instead through round-off error. Another, and usually better, alternative is to use a more accurate procedure that will give satisfactory results without using a very small step size. For example, if the Runge-Kutta method, which is discussed in Section 8.6, is used with $h = 0.1$, the approximate value 2.5050062 for $\phi(0.2)$ is obtained. This result agrees with the exact solution through the first four figures. These points are discussed in more detail in the following sections. In this section we are primarily interested in developing a familiarity with the technique of using a numerical procedure to solve an initial value problem.

Today, when solving an initial value problem numerically, one would normally use a computer to carry out the numerical calculations. The calculations in this chapter that illustrate different numerical procedures and difficulties associated with numerical methods have been done on a computer. However, the practice problems in this chapter are sufficiently simple, although still appropriately illustrative, that the numerical calculations can easily be done with a pocket calculator or even by hand in some cases.

In any event, the essential steps required for solving the initial value problem (1) for $x_0 \le x \le x_0 + Nh$ by the Euler method are shown in Figure 8.3 in flow chart form. Notice that in the second box the variables x_n and y_n are set equal to their initial values x_0 and y_0, respectively. The third box represents the calculation of $f(x_n, y_n)$ and the updating of x_n and y_n by the Euler formula. As long as $x_n < x_0 + Nh$, the computation is repeated using the values of x_n and y_n obtained at the preceding step. When $x_n = x_0 + Nh$, the process stops.

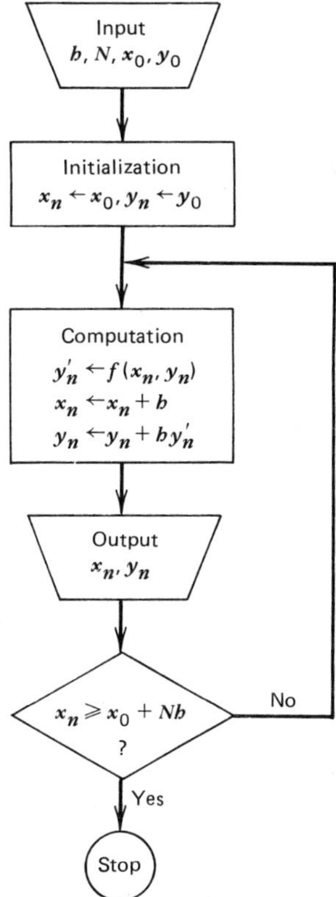

FIGURE 8.3 Flow chart for the Euler method.

In concluding this section we point out two alternative methods for deriving the Euler formula (3), which suggest ways of obtaining better formulas and also assist in our study of the error in using formula (3).

First, since $y = \phi(x)$ is a solution of the initial value problem (1), on integrating from x_n to x_{n+1} we obtain

$$\int_{x_n}^{x_{n+1}} \phi'(x)\,dx = \int_{x_n}^{x_{n+1}} f[x, \phi(x)]\,dx$$

or

$$\phi(x_{n+1}) = \phi(x_n) + \int_{x_n}^{x_{n+1}} f[x, \phi(x)]\,dx. \tag{7}$$

The integral in Eq. (7) is represented geometrically as the area under the curve in Figure 8.4 between $x = x_n$ and $x = x_{n+1}$. If we approximate the integral by replacing $f[x, \phi(x)]$ by its value $f[x_n, \phi(x_n)]$ at $x = x_n$, then we are approximating the actual area by the area of the shaded rectangle. In this way we obtain

$$\begin{aligned}\phi(x_{n+1}) &\cong \phi(x_n) + f[x_n, \phi(x_n)](x_{n+1} - x_n)\\ &= \phi(x_n) + hf[x_n, \phi(x_n)].\end{aligned} \tag{8}$$

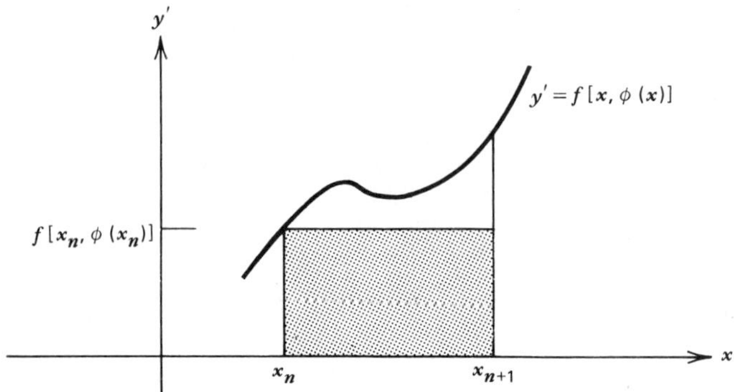

FIGURE 8.4 Integral derivation of the Euler method.

Finally, to obtain an approximation y_{n+1} for $\phi(x_{n+1})$ we make a second approximation by replacing $\phi(x_n)$ by its approximate value y_n in Eq. (8). This gives the Euler formula $y_{n+1} = y_n + hf(x_n, y_n)$. A more accurate formula can be obtained by approximating the integral more accurately. This is discussed in Section 8.4.

Second, suppose we assume that the solution $y = \phi(x)$ has a Taylor series about the point x_n. Then

$$\phi(x_n + h) = \phi(x_n) + \phi'(x_n)h + \phi''(x_n)\frac{h^2}{2!} + \cdots$$

or

$$\phi(x_{n+1}) = \phi(x_n) + f[x_n, \phi(x_n)]h + \phi''(x_n)\frac{h^2}{2!} + \cdots. \tag{9}$$

If the series is terminated after the first two terms, and $\phi(x_{n+1})$ and $\phi(x_n)$ are replaced by their approximate values y_{n+1} and y_n, we again obtain the Euler formula (3). If more terms in the series are retained, a more accurate formula is obtained. This is discussed in Section 8.5. Further, by using a Taylor series with a remainder it is possible to estimate the magnitude of the error in the formula. This is discussed in the next section.

PROBLEMS

Unless otherwise specified, round your answers to four nonzero digits. Answers in the back of the book are recorded in this way, although more digits were kept in the intermediate calculations.

In each of Problems 1 and 2

(a) Find approximate values of the solution of the given initial value problem at $x = 0.1$, 0.2, 0.3, and 0.4 using the Euler formula (3) with $h = 0.1$.

(b) Repeat part (a) with $h = 0.05$. Compare the results with those found in (a).
(c) Find the solution $y = \phi(x)$ of the given problem and evaluate $\phi(x)$ at $x = 0.1, 0.2, 0.3$, and 0.4. Compare these values with the results of (a) and (b).

1. $y' = 2y - 1, \quad y(0) = 1$
2. $y' = 0.5 - x + 2y, \quad y(0) = 1$

In each of Problems 3 through 6 find approximate values of the solution of the given initial value problem at $x = x_0 + 0.1$, $x_0 + 0.2$, $x_0 + 0.3$, and $x_0 + 0.4$

(a) using the Euler method with $h = 0.1$.
(b) using the Euler method with $h = 0.05$.

3. $y' = x^2 + y^2, \quad y(0) = 1$
4. $y' = 5x - 3\sqrt{y}, \quad y(0) = 2$
5. $y' = \sqrt{x + y}, \quad y(1) = 3$
6. $y' = 2x + e^{-xy}, \quad y(0) = 1$

*7. Consider the initial value problem

$$y' = 3x^2/(3y^2 - 4), \qquad y(1) = 0.$$

(a) Use the Euler formula (3) with $h = 0.1$ to obtain approximate values of the solution at $x = 1.2, 1.4, 1.6$, and 1.8.
(b) Repeat part (a) with $h = 0.05$.
(c) Compare the results of parts (a) and (b). Note that they are reasonably close for $x = 1.2$, 1.4, and 1.6, but are quite different for $x = 1.8$. Also note (from the differential equation) that the line tangent to the solution is parallel to the y axis when $y = \pm 2/\sqrt{3} \cong \pm 1.155$. Explain how this might cause such a difference in the calculated values.

8. Complete the calculations up to $x = 0.6$ leading to the entries in columns two and three in Table 8.1.

9. Using three terms in the Taylor series given in Eq. (9), and taking $h = 0.1$, determine approximate values of the solution of the illustrative example $y' = 1 - x + 4y$, $y(0) = 1$ at $x = 0.1$ and 0.2. Compare the results with those using the Euler method and the exact values.
Hint: If $y' = f(x, y)$ what is y''? See Section 8.5.

*10. It can be shown that under suitable conditions on f the numerical solution generated by the Euler method for the initial value problem $y' = f(x, y)$, $y(x_0) = y_0$ converges to the exact solution as the step size h decreases. This is illustrated by the following example. Consider the initial value problem

$$y' = 1 - x + y, \qquad y(x_0) = y_0.$$

(a) Show that the exact solution is $y = \phi(x) = (y_0 - x_0)e^{(x - x_0)} + x$.
(b) Show, using the Euler formula, that

$$y_k = (1 + h)\, y_{k-1} + h - hx_{k-1}, \qquad k = 1, 2, \ldots .$$

(c) Noting that $y_1 = (1 + h)(y_0 - x_0) + x_1$, show that

$$y_n = (1 + h)^n (y_0 - x_0) + x_n.$$

(d) Consider a fixed point $x > x_0$, and for a given n choose $h = (x - x_0)/n$. Then for any n, $x_n = x$. Note also that $h \to 0$ as $n \to \infty$. Substituting for h in the preceding formula and letting $n \to \infty$ gives the desired result.
Hint: $\lim\limits_{n \to \infty} (1 + a/n)^n = e^a$.

*11. Using the technique discussed in Problem 10, show that the approximate solution obtained by the Euler method converges to the exact solution at any fixed point as $h \to 0$ for each of the following problems.

(a) $y' = y, \quad y(0) = 1$
(b) $y' = 2y - 1, \quad y(0) = 1 \qquad$ *Hint:* $y_1 = (1 + 2h)/2 + 1/2$
(c) $y' = \frac{1}{2} - x + 2y, \quad y(0) = 1 \qquad$ *Hint:* $y_1 = (1 + 2h) + x_1/2$

12. An alternative procedure for constructing the solution $y = \phi(x)$ of the initial value problem $y' = f(x, y)$, $y(x_0) = y_0$ is the method of *iteration*. Integrating the differential equation from x_0 to x gives

$$\phi(x) = y_0 + \int_{x_0}^{x} f[t, \phi(t)]\, dt. \tag{i}$$

If, on the right side of Eq. (i), $\phi(t)$ is replaced by a particular function, the integral can be evaluated and a new function ϕ obtained. This suggests the iteration procedure

$$\phi_{n+1}(x) = y_0 + \int_{x_0}^{x} f[t, \phi_n(t)]\, dt. \tag{ii}$$

With an initial choice for $\phi_0(x)$, Eq. (ii) can be used to generate a sequence of functions $\phi_n(x)$ that approximates the exact solution and, under suitable conditions on $f(x, y)$, actually converges to $\phi(x)$ as $n \to \infty$. Indeed, this iteration procedure was used to establish the existence of a solution of the initial value problem in Section 2.12. For actually computing $\phi(x)$ this procedure is generally unwieldy since it may be difficult, if not impossible, to evaluate the integral of $f[t, \phi_n(t)]$, $n = 0, 1, 2, \ldots$. Taking $\phi_0(x) = 1$, determine $\phi_3(x)$ for each of the following initial value problems. Also, compute $\phi_2(0.4)$ and $\phi_3(0.4)$ and compare with the result obtained using the Euler method.

(a) $y' = 2y - 1, \quad y(0) = 1 \qquad$ (b) $y' = \frac{1}{2} - x + 2y, \quad y(0) = 1$
(c) $y' = x^2 + y^2, \quad y(0) = 1$; compute only $\phi_2(x)$

8.3 The Error

As we mentioned in the introduction to this chapter there are two fundamental sources of error in solving the initial value problem

$$y' = f(x, y), \qquad y(x_0) = y_0 \tag{1}$$

by a numerical procedure such as the Euler method:

$$y_{n+1} = y_n + hf(x_n, y_n), \qquad x_n = x_0 + nh. \tag{2}$$

Let us first assume that our computer is such that we can carry out all computations with complete accuracy; that is, we can retain an infinite number of decimal places. The difference between the exact solution $y = \phi(x)$ and the approximate solution of the initial value problem (1),

$$E_n = \phi(x_n) - y_n, \tag{3}$$

is known as the *discretization error*, or the *accumulated discretization error*. It arises from two causes: (1) at each step we use an approximate formula to determine y_{n+1}; (2) the input data at each step do not agree with the exact solution since in general $\phi(x_n)$ is not equal to y_n. If we assume that the input data are correct, the only error in going one step is that due to the use of an approximate formula. This error is known as the *local discretization error* e_n.

In practice, due to the limitations of all computers, it is in general impossible to compute y_{n+1} exactly from the given formula. Thus we have a round-off error due to a lack of computational accuracy. For example, if a computer that can carry only eight digits is used, and y_0 is 1.017325842, the last two digits must be rounded off, which immediately introduces an error in the computation of y_1. Alternatively, if $f(x, y)$ involves functions such as the logarithm or exponential we have a round-off error in carrying out these operations. Just as for the discretization error, it is possible to speak of the local round-off error and the accumulated round-off error. The *accumulated round-off error* R_n is defined as

$$R_n = y_n - Y_n, \tag{4}$$

where Y_n is the value *actually computed* by the given numerical procedure—for example, the Euler formula (2).

The absolute value of the total error in computing $\phi(x_n)$ is given by

$$|\phi(x_n) - Y_n| = |\phi(x_n) - y_n + y_n - Y_n|. \tag{5}$$

Making use of the triangle inequality, $|a + b| \le |a| + |b|$, we obtain from Eq. (5)

$$\begin{aligned} |\phi(x_n) - Y_n| &\le |\phi(x_n) - y_n| + |y_n - Y_n| \\ &\le |E_n| + |R_n|. \end{aligned} \tag{6}$$

Thus the total error is bounded by the sum of the absolute values of the discretization and round-off errors. For the numerical procedures discussed in this book it is possible to obtain useful estimates of the discretization error. However, we limit our discussion primarily to the local discretization error, which is somewhat simpler. The round-off error is clearly more random in nature. It depends on the type of computer used, the sequence in which the computations are carried out, the method of rounding off, etc. While an analysis of round-off error is beyond the scope of this introductory text, it is possible to say more about it than one would at first expect (see, for example, Henrici). Some of the dangers from round-off error are discussed in Problems 13 through 15 and in Section 8.7.

LOCAL DISCRETIZATION ERROR FOR THE EULER METHOD. Let us assume that the solution $y = \phi(x)$ of the initial value problem (1) has a continuous second derivative in the interval of interest. To assure this, we can assume that f, f_x, and f_y are continuous in the region of interest. Observe that if f has these properties and if ϕ is a solution of the initial value problem (1), then

$$\phi'(x) = f[x, \phi(x)],$$

and by the chain rule

$$\begin{aligned}\phi''(x) &= f_x[x,\phi(x)] + f_y[x,\phi(x)]\phi'(x) \\ &= f_x[x,\phi(x)] + f_y[x,\phi(x)]f[x,\phi(x)]. \qquad (7)\end{aligned}$$

Since the right side of this equation is continuous, ϕ'' is also continuous.

Then, making use of a Taylor series with a remainder to expand ϕ about x_n, we obtain

$$\phi(x_n + h) = \phi(x_n) + \phi'(x_n)h + \tfrac{1}{2}\phi''(\bar{x}_n)h^2, \qquad (8)$$

where $\bar{x}_n$ is some point in the interval $x_n < \bar{x}_n < x_n + h$. Subtracting Eq. (2) from Eq. (8), and noting that $\phi(x_n + h) = \phi(x_{n+1})$ and $\phi'(x_n) = f[x_n, \phi(x_n)]$, gives

$$\begin{aligned}\phi(x_{n+1}) - y_{n+1} = [\phi(x_n) - y_n] + h\{f[x_n, \phi(x_n)] - f(x_n, y_n)\} \\ + \tfrac{1}{2}\phi''(\bar{x}_n)h^2. \qquad (9)\end{aligned}$$

To compute the local discretization error we assume that the data at the nth step are correct, that is, $y_n = \phi(x_n)$. Then we immediately obtain from Eq. (9) that the local discretization error e_{n+1} is

$$e_{n+1} = \phi(x_{n+1}) - y_{n+1} = \tfrac{1}{2}\phi''(\bar{x}_n)h^2. \qquad (10)$$

Thus the local discretization error for the Euler method is proportional to the square of the step size h and to the second derivative of ϕ. For the fixed interval $a = x_0 \le x \le b$, the absolute value of the local discretization error for any step is bounded by $Mh^2/2$ where M is the maximum of $|\phi''(x)|$ on $a \le x \le b$. The primary difficulty in estimating the local discretization error is that of obtaining an accurate estimate of M. However, notice that M is independent of h; hence, reducing h by a factor of $\frac{1}{2}$ reduces the error bound by a factor of $\frac{1}{4}$, and a reduction by a factor of $\frac{1}{10}$ in h reduces the error bound by a factor of $\frac{1}{100}$.

More important than the local discretization error is the accumulated discretization error E_n. Nevertheless, an estimate of the local discretization error does give some understanding of the numerical procedure and a means for comparing the accuracy of different numerical procedures. The analysis for estimating E_n is more difficult than that for e_n. However, knowing the local discretization error we can make an *intuitive* estimate of the accumulated discretization error at a fixed $\bar{x} > x_0$ as follows. Suppose we take n steps in going from x_0 to $\bar{x} = x_0 + nh$. In each step the error is at most $Mh^2/2$; thus the error in n steps is at most $nMh^2/2$. Noting that $n = (\bar{x} - x_0)/h$, we find that the accumulated discretization error for the Euler method in going from x_0 to $\bar{x}$ is bounded by

$$n\frac{Mh^2}{2} = (\bar{x} - x_0)\frac{Mh}{2}. \qquad (11)$$

While this argument is not correct since it does not take account of the effect that an error at one step will have in succeeding steps, it is shown in Problem 8 that on any finite interval the accumulated discretization error using the Euler method is no greater than a constant times h. Thus in going from x_0 to a fixed point $\bar{x}$, the accumulated discretization error can be reduced by making the step size h smaller. Unfortunately, this is not the end of the story. If h is made too small, that is, too many steps are used in going from x_0 to $\bar{x}$, the accumulated round-off error may become more important than the accumulated discretization error. In practice both sources of error must be considered and an "optimum" choice of h made so that neither is too large. This is discussed further in Section 8.7. A practical procedure for improving the accuracy of the computation, which requires computation with two different step sizes, is discussed in Problems 9 and 11.

As an example of how we can use the result (10) if we have a priori information about the solution of the given initial value problem, consider the illustrative example

$$y' = 1 - x + 4y, \qquad y(0) = 1 \tag{12}$$

on the interval $0 \le x \le 1$. Let $y = \phi(x)$ be the solution of the initial value problem (12). Then $\phi'(x) = 1 - x + 4\phi(x)$, and

$$\begin{aligned} \phi''(x) &= -1 + 4\phi'(x) \\ &= -1 + 4[1 - x + 4\phi(x)] \\ &= 3 - 4x + 16\phi(x). \end{aligned}$$

From Eq. (10)

$$e_{n+1} = \frac{3 - 4\bar{x}_n + 16\phi(\bar{x}_n)}{2}h^2, \qquad x_n < \bar{x}_n < x_n + h. \tag{13}$$

Noting that $|3 - 4\bar{x}_n| \le 3$ on $0 \le x \le 1$, a rough bound for e_{n+1} is given by

$$e_{n+1} \le \left[\tfrac{3}{2} + 8 \max_{0 \le x \le 1} |\phi(x)|\right] h^2. \tag{14}$$

Since the right side of Eq. (14) does not depend on n, this provides a uniform, although not necessarily an accurate, bound for the error at any step if a bound for $|\phi(x)|$ on $0 \le x \le 1$ is known. For the present problem, $\phi(x) = (4x - 3 + 19e^{4x})/16$. Substituting for $\phi(\bar{x}_n)$ in Eq. (13) gives

$$e_{n+1} = 19e^{4\bar{x}_n}h^2/2, \qquad x_n < \bar{x}_n < x_n + h. \tag{15}$$

The appearance of the factor 19 and the rapid growth of e^{4x} (at $x = 1$, $e^{4x} = 54.598$) explain why the results in the previous section with $h = 0.1$ were not very accurate. For example, the error in the first step is

$$e_1 = \phi(x_1) - y_1 = \frac{19e^{4\bar{x}_0}(0.01)}{2}, \qquad 0 < \bar{x}_0 < 0.1.$$

It is clear that e_1 is positive and, since $e^{4\bar{x}_0} < e^{0.4}$, we have

$$e_1 \le \frac{19e^{0.4}(0.01)}{2} \cong 0.142. \tag{16}$$

Note also that $e^{4\bar{x}_0} > 1$; hence $e_1 > 19(0.01)/2 = 0.095$. The actual error is 0.1090418. It follows from Eq. (15) that the error becomes progressively worse with increasing x; this is clearly shown by the results in Table 8.1 of the last section. A similar computation for a bound for the local discretization error in going from 0.4 to 0.5 gives

$$0.47 \cong \frac{19e^{1.6}(0.01)}{2} \le e_5 \le \frac{19e^2(0.01)}{2} \cong 0.7.$$

In practice, of course, we will not know the solution ϕ of the given initial value problem. However, if we have bounds on the functions f, f_x, and f_y we can obtain a bound on $|\phi''(x)|$, although not necessarily an accurate one, from Eq. (7). Then we can use Eq. (10) to estimate the local discretization error. Even if we cannot find a useful bound for $|\phi''(x)|$, we still know that for the Euler method the local discretization error and the accumulated discretization error are bounded by a constant times h^2 and a constant times h, respectively.

We conclude this section with two practical comments based on experience.

1. One way of estimating whether the discretization error is sufficiently small is, after completing the computations with a given h, to repeat the computations using a step size $h/2$ (see Problems 9 and 11). If the changes are greater than we are willing to accept, then it is necessary to use a smaller step size or a more accurate numerical procedure or possibly both. More accurate procedures are discussed in the following sections. The results for the illustrative example using the Euler method with several different step sizes were given in Table 8.1. These results show that even with $h = 0.01$ it is not possible to obtain the first three digits correctly at $x = 0.10$. This could be expected without knowing the exact solution, since the results at $x = 0.1$ for $h = 0.025$ and $h = 0.01$ differ in the third digit.
2. It is possible to estimate the effect of the round-off error after completing the computations using single precision arithmetic by repeating the computations using double precision arithmetic. Again, if the changes are more than we are willing to accept, it is necessary to retain more digits in the computations or to use a more accurate method. Some examples of the difficulties that arise due to round-off errors are discussed in Problems 13 through 15.

PROBLEMS

In Problems 1 and 2 estimate the local discretization error in terms of the exact solution $y = \phi(x)$ if the Euler method is used. Obtain a bound for e_{n+1} in terms of x and $\phi(x)$ that is valid on the interval $0 \le x \le 1$. By using the exact solution obtain a more accurate error bound for e_{n+1}. For $h = 0.1$ compute a bound for e_1 and compare with the actual error at $x = 0.1$. Also compute a bound for the error e_4 in the fourth step.

1. $y' = 2y - 1, \qquad y(0) = 1$
2. $y' = \frac{1}{2} - x + 2y, \qquad y(0) = 1$

In Problems 3 through 6 obtain a formula for the local discretization error in terms of x and the exact solution ϕ if the Euler method is used.

3. $y' = x^2 + y^2, \quad y(0) = 1$
4. $y' = 5x - 3\sqrt{y}, \quad y(0) = 2$
5. $y' = \sqrt{x + y}, \quad y(1) = 3$
6. $y' = 2x + e^{-xy}, \quad y(0) = 1$

7. Consider the initial value problem

$$y' = \cos 5\pi x, \qquad y(0) = 1.$$

(a) Determine the exact solution $y = \phi(x)$, and draw a graph of $y = \phi(x)$ for $0 \le x \le 1$. Use a scale for the ordinate so that $1/5\pi$ is about 1 in.
(b) Determine approximate values of $\phi(x)$ at $x = 0.2$, 0.4, and 0.6 using the Euler method with $h = 0.2$. Draw a broken line graph for the approximate solution, and compare with the graph of the exact solution.
(c) Repeat the computations of part (b) for $0 \le x \le 0.4$, but take $h = 0.1$.
(d) Show by computing the local discretization error that neither of these step sizes is sufficiently small. Determine a value of h to insure that the local discretization error is less than 0.05 throughout the interval $0 \le x \le 1$. That such a small value of h is required results from the fact that $\max|\phi''(x)|$ is large or, put in rough terms, that the solution is highly oscillatory.

*8. In this problem we discuss the accumulated discretization error associated with the Euler method for the initial value problem $y' = f(x, y)$, $y(x_0) = y_0$. Assuming that the functions f and f_y are continuous in a region R of the xy plane that includes the point (x_0, y_0), it can be shown that there exists a constant L such that $|f(x, y) - f(x, \tilde{y})| < L|y - \tilde{y}|$ where (x, y) and $(x, \tilde{y})$ are any two points in R with the same x coordinate (see Problem 3 of Section 2.12). Further, we assume that f_x is continuous so that the solution ϕ has a continuous second derivative.
(a) Using Eq. (9) show that

$$|E_{n+1}| \le |E_n| + h|f[x_n, \phi(x_n)] - f(x_n, y_n)| + \tfrac{1}{2}h^2|\phi''(\bar{x}_n)| \le \alpha|E_n| + \beta h^2, \quad \text{(i)}$$

where $\alpha = 1 + hL$ and $\beta = \max|\phi''(x)|/2$ on $x_0 \le x \le x_n$.
(b) Accepting without proof that if $E_0 = 0$, and if $|E_n|$ satisfies Eq. (i), then $|E_n| \le \beta h^2(\alpha^n - 1)/(\alpha - 1)$ for $\alpha \ne 1$, show that

$$|E_n| \le \frac{(1 + hL)^n - 1}{L}\beta h. \quad \text{(ii)}$$

Equation (ii) gives a bound for $|E_n|$ in terms of h, L, n, and β. Notice that for a fixed h, this error bound increases with increasing n; that is, the error bound increases with distance from the starting point x_0.
(c) Show that $(1 + hL)^n \le e^{nhL}$, hence that

$$|E_n| \le \frac{e^{nhL} - 1}{L}\beta h$$

$$\le \frac{e^{(x_n - x_0)L} - 1}{L}\beta h.$$

For a fixed point $\bar{x} = x_0 + nh$ (that is, nh is constant and $h = (\bar{x} - x_0)/n$) this error bound is of the form constant times h and approaches zero as $h \to 0$. Also note that for $nhL = (\bar{x} - x_0)L$ small the right side of the above equation is approximately $nh^2\beta = (\bar{x} - x_0)\beta h$ which was obtained in Eq. (11) by an intuitive argument.

9. **Richardson's Extrapolation Method.** Let $y = \phi(x)$ be the solution of the initial value problem $y' = f(x, y)$, $y(x_0) = y_0$. As was discussed in the text (also see part (c) of Problem (8)), the accumulated discretization error using the Euler method to go from x_0 to a fixed point $\bar{x} = x_0 + nh$ is bounded by a constant times h. More precisely, it can be shown that $\phi(\bar{x}) = y_n(h) + Ch + M(h)$ where C is a constant that depends on the differential equation and the length of the interval but not on h, and $M(h)$ is a function that is proportional to h^2. The dependence of y_n on h is indicated by writing $y_n(h)$. The error is $Ch + M(h)$. Now suppose that the computation is repeated using a step size $h/2$. Since $2n$ steps are now required to reach $\bar{x}$, we have $\phi(\bar{x}) = y_{2n}(h/2) + Ch/2 + M(h/2)$. By solving the first equation for C and substituting in the second equation, show that

$$\phi(\bar{x}) \cong y_{2n}(h/2) + [y_{2n}(h/2) - y_n(h)] \tag{i}$$

where terms proportional to h^2 have been neglected. Thus, this new approximation to $\phi(\bar{x})$ has an error proportional to h^2 as compared to an error proportional to h for a single calculation with the Euler method. One can also use Eq. (i) to estimate the error associated with a step size of $h/2$, that is,

$$\phi(\bar{x}) - y_{2n}(h/2) \cong y_{2n}(h/2) - y_n(h). \tag{ii}$$

This procedure for computing an improved approximation to $\phi(\bar{x})$ in terms of $y_n(h)$ and $y_{2n}(h/2)$ is called Richardson's deferred approach to the limit or Richardson's extrapolation method. It was first used by L. F. Richardson in 1927.

10. Using the technique discussed in Problem 9 determine new estimates for the value of the exact solution $y = \phi(x)$ at $x = 0.2$, by making use of the results for $h = 0.2$ and $h = 0.1$ with the Euler method, for each of the following initial value problems. When possible compare the results with the values of the exact solution at $x = 0.2$.

(a) $y' = 2y, \quad y(0) = 1$ (b) $y' = 2y - 1, \quad y(0) = 1$
(c) $y' = \frac{1}{2} - x + 2y, \quad y(0) = 1$ (d) $y' = x^2 + y^2, \quad y(0) = 1$

11. A numerical method is said to have rth order accuracy if the error at $\bar{x} = x_0 + nh$ is proportional to h^r; or, more precisely, if $\phi(\bar{x}) = y_n(h) + Ch^r + M(h)$, where the constant C depends on the differential equation and $\bar{x}$ but is independent of h, and $M(h)$ is proportional to h^{r+1}. Thus the Euler method has first order accuracy. Following the procedure of Problem 9, show that an improved estimate of $\phi(\bar{x})$ is given by $\phi(\bar{x}) = y_{2n}(h/2) + [y_{2n}(h/2) - y_n(h)]/(2^r - 1)$.
12. Consider the example problem $y' = 1 - x + 4y$, $y(0) = 1$. Using Richardson's extrapolation method (Problem 9), and using the approximate values of $\phi(1)$ for $h = 0.1$, 0.05, and 0.025 given in Table 8.1, obtain new estimates for $\phi(1)$ for

(a) $h = 0.1$ and $h = 0.05$ (b) $h = 0.05$ and $h = 0.025$

Compare the results with the value of the exact solution.

13. Using a step size $h = 0.05$ and the Euler method, but retaining only three digits throughout the computations, determine approximate values of the solution at $x =$ 0.05, 0.1, 0.15, and 0.2 for each of the following initial value problems.

(a) $y' = 2y - 1, \quad y(0) = 1$ (b) $y' = \frac{1}{2} - x + 2y, \quad y(0) = 1$
(c) $y' = x^2 + y^2, \quad y(0) = 1$

Compare the results with those obtained in the previous section. The small differences between some of those results rounded to three digits and the present results are due to round-off error. The round-off error would become important if the computation required many steps.

14. The following problem illustrates a danger that occurs because of round-off error when nearly equal numbers are subtracted, and the difference then multiplied by a large number. Evaluate the quantity

$$1000 \cdot \begin{vmatrix} 6.010 & 18.04 \\ 2.004 & 6.000 \end{vmatrix}$$

as follows.

(a) First round each entry in the determinant to two digits.

(b) First round each entry in the determinant to three digits.

(c) Retain all four digits. Compare the exact value with the results in (a) and (b).

15. The distributive law $a(b - c) = ab - ac$ does not hold, in general, if the products are rounded off to a smaller number of digits. To show this in a specific case take $a = 0.22$, $b = 3.19$, and $c = 2.17$. After each multiplication round off the last digit.

8.4 An Improved Euler Method

Consider the initial value problem

$$y' = f(x, y), \qquad y(x_0) = y_0. \tag{1}$$

Letting ϕ be the solution of the initial value problem (1), and integrating from x_n to x_{n+1}, as we did in Eq. (7) of Section 8.2, we obtain

$$\phi(x_{n+1}) = \phi(x_n) + \int_{x_n}^{x_{n+1}} f[x, \phi(x)]\, dx. \tag{2}$$

The Euler formula

$$y_{n+1} = y_n + hf(x_n, y_n) \tag{3}$$

was obtained by replacing $f[x, \phi(x)]$ in Eq. (2) by its approximate value $f(x_n, y_n)$ at the left end point.

A better formula can be obtained if the integrand in Eq. (2) is approximated more accurately. One way to do this is to replace the integrand by the average of its values at the two end points, namely $\{f[x_n, \phi(x_n)] + f[x_{n+1}, \phi(x_{n+1})]\}/2$. Thus the area under the curve in Figure 8.5 between $x = x_n$ and $x = x_{n+1}$ is approximated by the area of the shaded rectangle. Further, we replace $\phi(x_n)$ and $\phi(x_{n+1})$ by their approximate values y_n and y_{n+1}. Then we obtain from Eq. (2)

$$y_{n+1} = y_n + \frac{f(x_n, y_n) + f(x_{n+1}, y_{n+1})}{2} h. \tag{4}$$

Since the unknown y_{n+1} appears as one of the arguments of f on the right side of Eq. (4), it will in general be fairly difficult to solve this equation for y_{n+1}. This difficulty can be overcome by replacing y_{n+1} on the right side of Eq. (4) by the

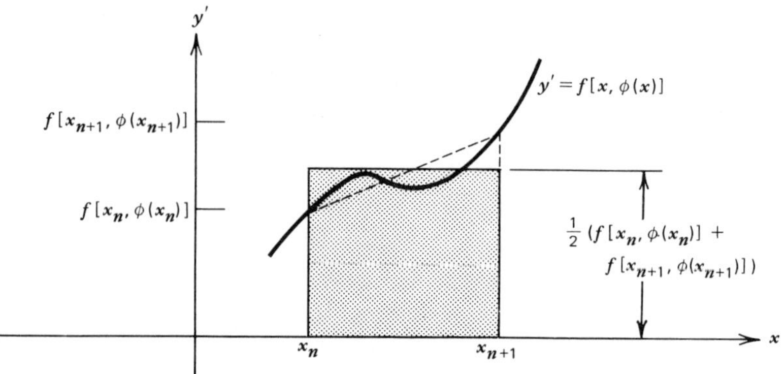

FIGURE 8.5 Derivation of the improved Euler method.

value obtained using the simple Euler formula (3). Thus

$$y_{n+1} = y_n + \frac{f(x_n, y_n) + f[x_n + h, y_n + hf(x_n, y_n)]}{2}h$$

$$= y_n + \frac{y_n' + f[x_n + h, y_n + hy_n']}{2}h, \tag{5}$$

where x_{n+1} has been replaced by $x_n + h$.

Equation (5) gives a formula for computing y_{n+1}, the approximate value of $\phi(x_{n+1})$, in terms of the data at x_n. This formula is known as the *improved Euler formula* or the *Heun formula*. That Eq. (5) does represent an improvement over the Euler formula (3) rests on the fact that the local discretization error in using Eq. (5) is proportional to h^3 while that for the Euler method is proportional to h^2. The error estimate for the improved Euler formula is established in Problem 7. It can also be shown that for a finite interval the accumulated discretization error for the improved Euler formula is bounded by a constant times h^2. Note that this greater accuracy is achieved at the expense of more computational work, since it is now necessary to evaluate $f(x, y)$ twice in order to go from x_n to x_{n+1}.

If $f(x, y)$ depends only on x and not on y, then solving the differential equation $y' = f(x, y)$ reduces to integrating $f(x)$. In this case $y_n' = f(x_n)$, $f(x_n + h, y_n + hy_n') = f(x_n + h)$, and the improved Euler formula (5) becomes

$$y_{n+1} - y_n = \tfrac{1}{2}h[f(x_n) + f(x_n + h)]. \tag{6}$$

Equation (6) is just the trapezoid rule for approximating the value of the integral $\int_{x_n}^{x_n+h} f(x)\,dx$ by finding the area of the trapezoid shown in Figure 8.6.

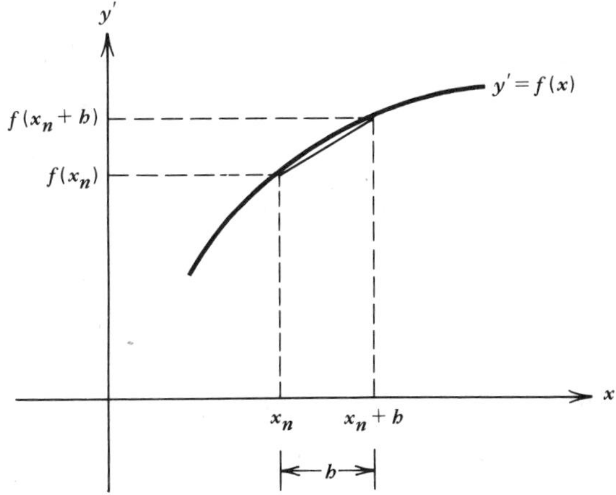

FIGURE 8.6 The trapezoid rule.

EXAMPLE

Use the improved Euler formula (5) to calculate approximate values of the solution of the initial value problem

$$y' = 1 - x + 4y, \qquad y(0) = 1. \tag{7}$$

For this problem $f(x, y) = 1 - x + 4y$; hence

$$y_n' = 1 - x_n + 4y_n$$

and

$$f(x_n + h, y_n + hy_n') = 1 - (x_n + h) + 4(y_n + hy_n').$$

A flow chart for the improved Euler method is similar to the one for the Euler method in Figure 8.3 of Section 8.2. The only difference is that in the third box y_n' is set equal to $F(x_n, y_n)$, where

$$F(x_n, y_n) = \tfrac{1}{2}\{f(x_n, y_n) + f[x_n + h, y_n + hf(x_n, y_n)]\}.$$

For the problem (7) we have $x_0 = 0$, $y_0 = 1$, and $y_0' = 1 - x_0 + 4y_0 = 5$. Further, if $h = 0.1$, then

$$f(x_0 + h, y_0 + hy_0') = 1 - 0.1 + 4[1 + (0.1)5] = 6.9.$$

Then, from Eq. (5),

$$y_1 = 1 + (0.5)(5 + 6.9)(0.1) = 1.595.$$

Further results for $0 \le x \le 1$ obtained by using the improved Euler method with $h = 0.1$ and $h = 0.05$ are given in Table 8.2. The table shows that the results using the improved Euler method with $h = 0.1$ are indeed better than those

TABLE 8.2 A COMPARISON OF RESULTS USING THE EULER AND IMPROVED EULER METHODS FOR $h = 0.1$ AND $h = 0.05$ FOR THE INITIAL VALUE PROBLEM $y' = 1 - x + 4y$, $y(0) = 1$

	Euler		Improved Euler		
x	$h = 0.1$	$h = 0.05$	$h = 0.1$	$h = 0.05$	Exact
0	1.0000000	1.0000000	1.0000000	1.0000000	1.0000000
0.1	1.5000000	1.5475000	1.5950000	1.6049750	1.6090418
0.2	2.1900000	2.3249000	2.4636000	2.4932098	2.5053299
0.3	3.1460000	3.4333560	3.7371280	3.8030484	3.8301388
0.4	4.4744000	5.0185326	5.6099494	5.7404023	5.7942260
0.5	6.3241600	7.2901870	8.3697252	8.6117498	8.7120041
0.6	8.9038240	10.550369	12.442193	12.873253	13.052522
0.7	12.505354	15.234032	18.457446	19.203865	19.515518
0.8	17.537495	21.967506	27.348020	28.614138	29.144880
0.9	24.572493	31.652708	40.494070	42.608178	43.497903
1.0	34.411490	45.588400	59.938223	63.424698	64.897803

obtained with the Euler method with $h = 0.1$, and are also better than those obtained with the Euler method with $h = 0.05$. Thus, even though the improved Euler formula requires more computations at each step, it yields substantially better results with only half as many steps. The results using the improved Euler formula with $h = 0.05$ are in fairly close agreement with the exact solution; the percentage errors at $x = 0.5$ and $x = 1.0$ are 1.15 and 2.3%, respectively.

PROBLEMS

In Problems 1 through 6 use the improved Euler formula with $h = 0.1$ to determine approximate values of the solution of the given initial value problem at $x = x_0 + 0.1$, $x_0 + 0.2$, $x_0 + 0.3$, and $x_0 + 0.4$. Compare the results with those obtained using the Euler formula and with the exact solution (if available).

1. $y' = 2y - 1, \quad y(0) = 1$
2. $y' = 0.5 - x + 2y, \quad y(0) = 1$
3. $y' = x^2 + y^2, \quad y(0) = 1$
4. $y' = 5x - 3\sqrt{y}, \quad y(0) = 2$
5. $y' = \sqrt{x + y}, \quad y(1) = 3$
6. $y' = 2x + e^{-xy}, \quad y(0) = 1$
7. In this problem we establish that the local discretization error for the improved Euler formula is proportional to h^3. Assuming that the solution ϕ of the initial value problem $y' = f(x, y)$, $y(x_0) = y_0$ has derivatives that are continuous through the third order (f has continuous second partial derivatives), it follows that

$$\phi(x_n + h) = \phi(x_n) + \phi'(x_n)h + \frac{\phi''(x_n)h^2}{2!} + \frac{\phi'''(\bar{x}_n)h^3}{3!},$$

where $x_n < \bar{x}_n < x_n + h$. Assume that $y_n = \phi(x_n)$.

(a) Show that for y_{n+1} as given by Eq. (5)

$$e_{n+1} = \phi(x_{n+1}) - y_{n+1}$$
$$= \frac{\phi''(x_n)h - \{f[x_n + h, y_n + hf(x_n, y_n)] - f(x_n, y_n)\}}{2!}h + \frac{\phi'''(\bar{x}_n)h^3}{3!}. \qquad \text{(i)}$$

(b) Making use of the fact that $\phi''(x) = f_x[x, \phi(x)] + f_y[x, \phi(x)]\phi'(x)$, and that the Taylor series with a remainder for a function of two variables (which we accept without proof) is

$$F(a+h, b+k) = F(a,b) + F_x(a,b)h + F_y(a,b)k + \frac{1}{2!}\left(h^2F_{xx} + 2hkF_{xy} + k^2F_{yy}\right)\Big|_{x=\xi,\, y=\eta}$$

where ξ lies between a and $a + h$ and η lies between b and $b + k$, show that the first term on the right side of Eq. (i) is proportional to h^3 plus higher order terms. This is the desired result.

(c) Show that if $f(x, y)$ is linear in x and y, then $e_{n+1} = \phi'''(\bar{x}_n)h^3/6$, where $x_n < \bar{x}_n < x_{n+1}$.

Hint: What are f_{xx}, f_{xy}, and f_{yy}?

8. Consider the improved Euler method for solving the initial value problem $y' = 1 - x + 4y$, $y(0) = 1$. Using the result of part (c) of Problem 7 and the exact solution of the initial value problem, determine e_{n+1} and a bound for the error at any step on $0 \le x \le 1$. Compare this error with the one obtained in the previous section, Eq. (15), using the Euler method. Also obtain a bound for e_1 for $h = 0.1$ and compare with Eq. (16) of the previous section.
9. Making use of the exact solution, determine e_{n+1} and a bound for e_{n+1} at any step on $0 \le x \le 1$ for the improved Euler method for each of the following initial value problems. Also, obtain a bound for e_1 for $h = 0.1$ and compare it with the similar estimate for the Euler method and with the actual error using the improved Euler method.

   (a) $y' = 2y - 1, \quad y(0) = 1$ (b) $y' = \frac{1}{2} - x + 2y, \quad y(0) = 1$

10. The *modified Euler formula* for the initial value problem $y' = f(x, y)$, $y(x_0) = y_0$ is given by

$$y_{n+1} = y_n + hf\left[x_n + \tfrac{1}{2}h, y_n + \tfrac{1}{2}hf(x_n, y_n)\right].$$

   Following the procedure outlined in Problem 7 show that the local discretization error in the modified Euler formula is proportional to h^3.
11. Use the modified Euler formula of Problem 10 with $h = 0.1$ to compute approximate values of the solution of each of the following initial value problems at $x = 0.1, 0.2, 0.3$, and 0.4. Compare the results with those obtained earlier.

   (a) $y' = 2y - 1, \quad y(0) = 1$ (b) $y' = 0.5 - x + 2y, \quad y(0) = 1$
   (c) $y' = x^2 + y^2, \quad y(0) = 1$ (d) $y' = 5x - 3\sqrt{y}, \quad y(0) = 2$

12. Show that the modified Euler formula of Problem 10 is identical to the improved Euler formula of Eq. (5) for $y' = f(x, y)$ if $f(x, y)$ is linear in both x and y.

8.5 The Three-Term Taylor Series Method

We saw in Section 8.2 that the Euler formula $y_{n+1} = y_n + hf(x_n, y_n)$ for solving the initial value problem

$$y' = f(x, y), \qquad y(x_0) = y_0 \tag{1}$$

can be derived by retaining the first two terms in the Taylor series for the solution $y = \phi(x)$ about the point $y = x_n$. A more accurate formula can be obtained by using the first three terms. Assuming that ϕ has at least three continuous derivatives in the interval of interest (f has continuous second partial derivatives) and using the Taylor series with a remainder, we obtain

$$\phi(x_n + h) = \phi(x_n) + \phi'(x_n)h + \phi''(x_n)\frac{h^2}{2!} + \phi'''(\bar{x}_n)\frac{h^3}{3!}, \tag{2}$$

where $\bar{x}_n$ is some point in the interval $x_n < \bar{x}_n < x_n + h$.

From Eq. (1)

$$\phi'(x_n) = f[x_n, \phi(x_n)]. \tag{3}$$

Further, $\phi''(x)$ can be computed from Eq. (1):

$$\phi''(x) = f_x[x, \phi(x)] + f_y[x, \phi(x)]\phi'(x). \tag{4}$$

Hence

$$\phi''(x_n) = f_x[x_n, \phi(x_n)] + f_y[x_n, \phi(x_n)]\phi'(x_n). \tag{5}$$

The three-term Taylor series formula is obtained by replacing $\phi(x_n)$ by its approximate value y_n in the formulas for $\phi'(x)$ and $\phi''(x)$ and then neglecting the term $\phi'''(\bar{x}_n)h^3/3!$ in Eq. (2). Thus we obtain

$$y_{n+1} = y_n + hy_n' + \frac{h^2}{2}y_n'', \tag{6}$$

where

$$y_n' = f(x_n, y_n), \tag{7}$$

and

$$y_n'' = f_x(x_n, y_n) + f_y(x_n, y_n)y_n'. \tag{8}$$

Assuming that $\phi(x_n) = y_n$, it is trivial to show that the local discretization error e_{n+1} associated with formula (6) is

$$e_{n+1} = \phi(x_{n+1}) - y_{n+1} = \tfrac{1}{6}\phi'''(\bar{x}_n)h^3, \tag{9}$$

where $x_n < \bar{x}_n < x_n + h$. Thus the local discretization error for the three-term Taylor series formula is proportional to h^3, just as for the improved Euler formula discussed in the previous section. Again, it can also be shown that for a finite interval the accumulated discretization error is no greater than a constant times h^2.

The three-term Taylor series formula requires the computation of $f_x(x, y)$ and $f_y(x, y)$, and then the evaluation of these functions as well as $f(x, y)$ at (x_n, y_n). In some problems it may be inconvenient to compute f_x and f_y. If this is the case, it is probably better to use a formula with comparable accuracy, such as the improved Euler formula, that does not require the partial derivatives f_x and f_y. In principle, four-term, or even higher, Taylor series formulas can be developed. However, such formulas involve even higher partial derivatives of f and are in general rather awkward to use.

EXAMPLE

Use the three-term Taylor series formula to calculate approximate values of the solution of the initial value problem

$$y' = 1 - x + 4y, \qquad y(0) = 1. \tag{10}$$

In this problem we have

$$f(x, y) = 1 - x + 4y, \qquad f_x(x, y) = -1, \qquad f_y(x, y) = 4.$$

Hence

$$\begin{aligned} y_n' &= 1 - x_n + 4y_n, \\ y_n'' &= -1 + 4y_n' = -1 + 4(1 - x_n + 4y_n) = 3 - 4x_n + 16y_n. \end{aligned} \tag{11}$$

It follows from the initial condition and Eqs. (11) that $y_0' = 5$ and $y_0'' = 19$. For $h = 0.1$ we then use Eq. (6) to obtain

$$y_1 = 1 + (0.1)(5) + (0.005)(19) = 1.595.$$

For this problem the results using the three-term Taylor series formula are the same, except possibly for minor differences due to round-off error, as those obtained using the improved Euler formula. The reason for this is that the two formulas are identical when $f(x, y)$ is linear in both x and y. Thus the results given in Table 8.2 for the improved Euler method are also valid for the three-term Taylor series method.

PROBLEMS

In each of Problems 1 through 6 use the three-term Taylor series formula with $h = 0.1$ to determine approximate values of the solution of the given initial value problem at $x = x_0 + 0.1$, $x_0 + 0.2$, $x_0 + 0.3$, and $x_0 + 0.4$. Compare the results with those obtained using the Euler formula, the improved Euler formula, and with the exact solution (if available).

1. $y' = 2y - 1, \quad y(0) = 1$
2. $y' = 0.5 - x + 2y, \quad y(0) = 1$
3. $y' = x^2 + y^2, \quad y(0) = 1$
4. $y' = 5x - 3\sqrt{y}, \quad y(0) = 2$
5. $y' = \sqrt{x + y}, \quad y(1) = 3$
6. $y' = 2x + e^{-xy}, \quad y(0) = 1$

7. Show that the improved Euler formula of Section 8.4 and the three-term Taylor series formula are identical for $y' = f(x, y)$ if $f(x, y)$ is linear in both x and y.
   Hint: See Problem 7(b) of Section 8.4. Note that f_{xx}, f_{xy}, and f_{yy} are each zero.
8. For the illustrative problem $y' = 1 - x + 4y$, $y(0) = 1$, determine approximate values of the solution at $x = 0.2$, 0.3, and 0.4, using the three-term Taylor series method. Take $h = 0.1$. Retain four digits. Check your results with those given in Table 8.2.

Problems 9 through 11 deal with the initial value problem $y' = f(x, y)$, $y(x_0) = y_0$. Let $y = \phi(x)$ be the exact solution.

9. Show that the local discretization error e_{n+1} using the three-term Taylor series formula is given by $\phi'''(\bar{x}_n)h^3/6$, $x_n < \bar{x}_n < x_n + h$. Remember that in computing e_{n+1} it is assumed that $y_n = \phi(x_n)$.
10. Derive a four-term Taylor series formula for y_{n+1} in terms of $f(x, y)$ and its partial derivatives evaluated at (x_n, y_n). What is the local discretization error?
11. Following the outline of Richardson's extrapolation method (Problems 9 and 11 of Section 8.3) show that an improved estimate for $\phi(\bar{x})$, $\bar{x} = x_0 + nh$, in terms of $y_n(h)$ and $y_{2n}(h/2)$ computed using the three-term Taylor series method or the improved Euler method, is given by $y_{2n}(h/2) + [y_{2n}(h/2) - y_n(h)]/(2^2 - 1)$. Consider the initial value problem $y' = 1 - x + 4y$, $y(0) = 1$. Using the approximate results for $\phi(1)$ obtained by the improved Euler method for $h = 0.1$ and $h = 0.05$ (see Table 8.2), determine a new estimate of $\phi(1)$. Compare this result with the value of the exact solution.

8.6 The Runge-Kutta Method

Again consider the initial value problem

$$y' = f(x, y), \qquad y(x_0) = y_0. \tag{1}$$

In the previous section we developed a three-term Taylor series formula that involved the functions f_x and f_y. Unfortunately, higher order Taylor series formulas rapidly become unwieldy because of the higher partial derivatives of f that must be computed.

However, it is possible to develop formulas equivalent to third, fourth, fifth, or even higher order Taylor series formulas that do not involve partial derivatives of f. By equivalent numerical procedure we mean that the local discretization errors are each proportional to the same power of h plus (possibly different) higher order terms. These formulas are known as Runge-Kutta[2] formulas; the

[2] Carl David Runge (1856–1927), German mathematician and physicist, worked for many years in spectroscopy. The analysis of data led him to consider problems in numerical computation, and the Runge-Kutta method originated in his paper on the numerical solution of differential equations in 1895. The method was extended to systems of equations in 1901 by M. Wilhelm Kutta (1867–1944). Kutta was a German mathematician and aerodynamicist who is also well-known for his important contributions to classical airfoil theory.

classical Runge-Kutta formula is equivalent to a five-term Taylor formula[3]

$$y_{n+1} = y_n + hy_n' + \frac{h^2}{2!}y_n'' + \frac{h^3}{3!}y_n''' + \frac{h^4}{4!}y_n^{\text{iv}}. \tag{2}$$

The Runge-Kutta formula involves a weighted average of values of $f(x, y)$ taken at different points in the interval $x_n \le x \le x_{n+1}$. It is given by

$$y_{n+1} = y_n + \frac{h}{6}(k_{n1} + 2k_{n2} + 2k_{n3} + k_{n4}), \tag{3}$$

where

$$k_{n1} = f(x_n, y_n), \tag{4a}$$

$$k_{n2} = f\left(x_n + \tfrac{1}{2}h, y_n + \tfrac{1}{2}hk_{n1}\right), \tag{4b}$$

$$k_{n3} = f\left(x_n + \tfrac{1}{2}h, y_n + \tfrac{1}{2}hk_{n2}\right), \tag{4c}$$

$$k_{n4} = f(x_n + h, y_n + hk_{n3}). \tag{4d}$$

The sum $(k_{n1} + 2k_{n2} + 2k_{n3} + k_{n4})/6$ can be interpreted as an average slope. Note that k_{n1} is the slope at the left end of the interval, k_{n2} is the slope at the midpoint using the Euler formula to go from x_n to $x_n + h/2$, k_{n3} is a second approximation to the slope at the midpoint, and finally k_{n4} is the slope at $x_n + h$ using the Euler formula and the slope k_{n3} to go from x_n to $x_n + h$.

While in principle it is not difficult to show that Eqs. (2) and (3) differ by terms that are proportional to h^5, the algebra is extremely lengthy. Thus we will accept the fact that the local discretization error in using Eq. (3) is proportional to h^5 and that for a finite interval the accumulated discretization error is at most a constant times h^4. A derivation of a Runge-Kutta formula that is equivalent to the three-term Taylor series formula is given in Problem 8.

Clearly the Runge-Kutta formula, Eqs. (3) and (4), is more complicated than any of the formulas discussed previously. On the other hand, it should be remembered that this is a very accurate formula (halving the step size reduces the local discretization error by the factor $\frac{1}{32}$); further, it is not necessary to compute any partial derivatives of f.

We also note that if f does not depend on y, then

$$k_{n1} = f(x_n), \qquad k_{n2} = k_{n3} = f\left(x_n + \tfrac{1}{2}h\right), \qquad k_{n4} = f(x_n + h),$$

and Eq. (3) reduces to

$$y_{n+1} - y_n = \frac{h}{6}\left[f(x_n) + 4f\left(x_n + \tfrac{1}{2}h\right) + f(x_n + h)\right]. \tag{5}$$

Equation (5) can be identified as Simpson's (1710–1761) rule (see Problem 9) for the approximate evaluation of the integral of $y' = f(x)$. The fact that Simpson's

[3]The numbers y_n'', y_n''', and y_n^{iv} are obtained by successively differentiating Eq. (1), and then evaluating at (x_n, y_n). See Eq. (8) of Section 8.5.

rule has an error proportional to h^5 is in agreement with the earlier comment concerning the error in the Runge-Kutta formula.

The Runge-Kutta formula, Eqs. (3) and (4), is one of the most widely used and most successful of all one-step formulas.

EXAMPLE

Use the Runge-Kutta method to calculate approximate values of the solution $y = \phi(x)$ of the initial value problem

$$y' = 1 - x + 4y, \qquad y(0) = 1.$$

Taking $h = 0.2$ we have

$$\begin{aligned} k_{01} &= f(0,1) = 5; & hk_{01} &= 1.0 \\ k_{02} &= f(0 + 0.1, 1 + 0.5) = 6.9; & hk_{02} &= 1.38 \\ k_{03} &= f(0 + 0.1, 1 + 0.69) = 7.66; & hk_{03} &= 1.532 \\ k_{04} &= f(0 + 0.2, 1 + 1.532) = 10.928. \end{aligned}$$

Thus

$$\begin{aligned} y_1 &= 1 + \frac{0.2}{6}[5 + 2(6.9) + 2(7.66) + 10.928] \\ &= 1 + 1.5016 = 2.5016. \end{aligned}$$

Further results using the Runge-Kutta method with $h = 0.2$ and with $h = 0.1$ are given in Table 8.3. For comparison, values obtained by using the Euler and

TABLE 8.3 COMPARISON OF RESULTS FOR THE NUMERICAL SOLUTION OF THE INITIAL VALUE PROBLEM $y' = 1 - x + 4y$, $y(0) = 1$

	Euler	Improved Euler	Runge-Kutta		
x	$h = 0.1$	$h = 0.1$	$h = 0.2$	$h = 0.1$	Exact
0	1.0000000	1.0000000	1.0000000	1.0000000	1.0000000
0.1	1.5000000	1.5950000		1.6089333	1.6090418
0.2	2.1900000	2.4636000	2.5016000	2.5050062	2.5053299
0.3	3.1460000	3.7371280		3.8294145	3.8301388
0.4	4.4774000	5.6099494	5.7776358	5.7927853	5.7942260
0.5	6.3241600	8.3697252		8.7093175	8.7120041
0.6	8.9038240	12.442193	12.997178	13.047713	13.052522
0.7	12.505354	18.457446		19.507148	19.515518
0.8	17.537495	27.348020	28.980768	29.130609	29.144880
0.9	24.572493	40.494070		43.473954	43.497903
1.0	34.411490	59.938223	64.441579	64.858107	64.897803

TABLE 8.4 COMPARISON OF RESULTS USING DIFFERENT NUMERICAL PROCEDURES AND DIFFERENT STEP SIZES FOR THE VALUE AT $x = 1$ OF THE SOLUTION OF THE INITIAL VALUE PROBLEM $y' = 1 - x + 4y$, $y(0) = 1$

h	Euler	Improved Euler	Runge-Kutta	Exact
0.2			64.441579	64.897803
0.1	34.411490	59.938223	64.858107	64.897803
0.05	45.588400	63.424698	64.894875	64.897803
0.025	53.807866	64.497931	64.897604	64.897803
0.01	60.037126	64.830722	64.897798	64.897803

improved Euler methods are also given. Note that the Runge-Kutta method yields a value at $x = 1$ that differs from the exact solution by only 0.7% if the step size is $h = 0.2$, and by only 0.06% if $h = 0.1$. The accuracy of the Runge-Kutta method for this problem is further demonstrated in Table 8.4, which contains approximate values of $\phi(1)$ obtained by using different methods and step sizes. Observe that the result for the Runge-Kutta method with $h = 0.1$ is better than those of either of the other methods with $h = 0.01$. That is, the Runge-Kutta method with 10 steps is better than the improved Euler method with 100 steps; actually, the proper comparison is 40 evaluations of $f(x, y)$ compared with 100. Further, from Table 8.4 we also see that for $h = 0.01$ the Runge-Kutta method yields essentially the exact value of $\phi(1)$.

PROBLEMS

In each of Problems 1 through 6 use the Runge-Kutta method with $h = 0.2$ to determine approximate values of the solution of the given initial value problem at $x = x_0 + 0.2$ and $x_0 + 0.4$. Compare the results with those obtained using other methods and with the exact solution (if available).

1. $y' = 2y - 1, \quad y(0) = 1$
2. $y' = 0.5 - x + 2y, \quad y(0) = 1$
3. $y' = x^2 + y^2, \quad y(0) = 1$
4. $y' = 5x - 3\sqrt{y}, \quad y(0) = 2$
5. $y' = \sqrt{x + y}, \quad y(1) = 3$
6. $y' = 2x + e^{-xy}, \quad y(0) = 1$
7. Compute an approximate value of the solution $y = \phi(x)$ at $x = 0.4$ for the illustrative example $y' = 1 - x + 4y$, $y(0) = 1$. Use $h = 0.2$ and the approximate value of $\phi(0.2)$ given in the example in the text. Compare the result with the value given in Table 8.3.

*8. Consider the initial value problem $y' = f(x, y)$, $y(x_0) = y_0$. The corresponding three-term Taylor series formula is

$$y_{n+1} = y_n + hy_n' + (h^2/2)\, y_n''. \tag{i}$$

For the moment consider the formula

$$y_{n+1} = y_n + h\{af(x_n, y_n) + bf[x_n + \alpha h, y_n + \beta hf(x_n, y_n)]\}, \qquad \text{(ii)}$$

where a, b, α, and β are arbitrary.

(a) Show by expanding the right side of formula (ii) about the point (x_n, y_n) that for $a + b = 1$, $b\alpha = \frac{1}{2}$, and $b\beta = \frac{1}{2}$ the difference between formulas (i) and (ii) is proportional to h^3.

Hint: Use the Taylor series expansion for a function of two variables given in Problem 7(b) of Section 8.4.

(b) Show that the equations for a, b, α, and β have the infinity of solutions $a = 1 - \lambda$, $b = \lambda$, and $\alpha = \beta = 1/2\lambda$, $\lambda \neq 0$.

(c) For $\lambda = \frac{1}{2}$, show that Eq. (ii) reduces to the improved Euler formula given in Section 8.4.

(d) For $\lambda = 1$, show that Eq. (ii) reduces to the modified Euler formula given in Problem 10 of Section 8.4.

9. To derive Simpson's rule for the approximate value of the integral of $f(x)$ from $x = 0$ to $x = h$, first show that

$$\int_0^h (Ax^2 + Bx + C)\,dx = \frac{Ah^3}{3} + \frac{Bh^2}{2} + Ch.$$

Next choose the constants A, B, and C so that the parabola $Ax^2 + Bx + C$ passes through the points $[0, f(0)]$, $[h/2, f(h/2)]$, and $[h, f(h)]$. Using this polynomial to represent $f(x)$ approximately on the interval $0 \le x \le h$, substitute for A, B, and C in the above formula to obtain

$$\int_0^h f(x)\,dx \cong \frac{h}{6}\left[f(0) + 4f\left(\frac{h}{2}\right) + f(h)\right].$$

It can be shown that the error in using this formula is proportional to h^5.

10. Following the outline of Richardson's extrapolation method (Problems 9 and 11 of Section 8.3), show that for the initial value problem $y' = f(x, y)$, $y(x_0) = y_0$ an improved estimate for $\phi(\bar{x})$, $\bar{x} = x_0 + nh$, in terms of $y_n(h)$ and $y_{2n}(h/2)$ for the Runge-Kutta method, is given by $y_{2n}(h/2) + [y_{2n}(h/2) - y_n(h)]/(2^4 - 1)$. Consider the illustrative initial value problem $y' = 1 - x + 4y$, $y(0) = 1$. Using the approximate values of $\phi(1)$ obtained by the Runge-Kutta method with $h = 0.2$ and $h = 0.1$ (Table 8.3) and the above formula, obtain a new estimate for $\phi(1)$. Compare the result with the exact solution.

8.7 Some Difficulties with Numerical Methods

In Section 8.3 we discussed some ideas related to the errors that can occur in a numerical solution of the initial value problem

$$y' = f(x, y), \qquad y(x_0) = y_0. \qquad (1)$$

In this section we continue that discussion and also point out several more subtle difficulties that can arise. The points that we wish to make are fairly difficult to treat in detail, and we are content to illustrate them by means of examples.

First, recall that for the Euler method we showed that the local discretization error is proportional to h^2 and that for a finite interval the accumulated discretization error is at most a constant times h. Although we do not prove it, it is true in general that if the local discretization error is proportional to h^p then for a finite interval the accumulated discretization error is bounded by a constant times h^{p-1}. To achieve high accuracy we normally use a numerical procedure for which p is large (for the Runge-Kutta method $p = 5$). As p increases, the formula used in computing y_{n+1} normally becomes more complicated, and hence more calculations are required at each step; however, this is usually not a serious problem when using a computer unless $f(x, y)$ is very complicated or if the calculation must be repeated many times (for example, for many different values of y_0 and/or different values of parameters that may appear in f).

If the step size h is decreased, the accumulated discretization error is decreased by the same factor raised to the power $p - 1$. However, as we mentioned in Section 8.3, if h is too small, too many steps will be required to cover a fixed interval, and the accumulated round-off error may be larger than the accumulated discretization error. This is illustrated by solving our example problem

$$y' = 1 - x + 4y, \qquad y(0) = 1 \tag{2}$$

on the interval $0 \leq x \leq 1$ using the Runge-Kutta method with different step sizes h. In Table 8.5 the difference between the value of the exact solution and the computed value at $x = 1$ is shown for several values of h. Notice that initially the error decreases as h is decreased to about 0.025 to 0.02 (the number of steps N is increased to about 40 to 50), but then begins to increase with decreasing h. This is the result of the accumulated round-off error. It is not possible to name an

TABLE 8.5 A COMPARISON OF THE ERROR AT $x = 1$ USING DIFFERENT STEP SIZES AND THE RUNGE-KUTTA METHOD FOR THE INITIAL VALUE PROBLEM $y' = 1 - x + 4y$, $y(0) = 1$

$N = 1/h$	h	$(-1) \times$ Error at $x = 1$
10	0.1	0.04008
20	0.05	0.00336
30	0.03333333	0.00114
40	0.025	0.00078
50	0.02	0.00089
60	0.01666667	0.00101
80	0.0125	0.00111
160	0.00625	0.00212
320	0.003125	0.00357
640	0.0015625	0.00731
1280	0.00078125	0.01567

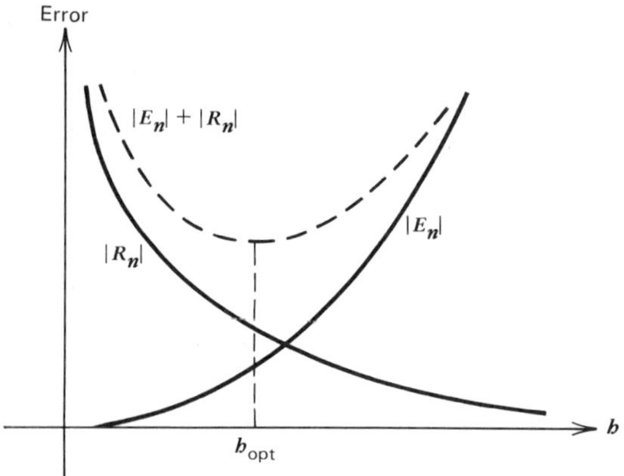

FIGURE 8.7 The dependence of discretization and round-off errors on the step size h.

optimum value of N, since for values of N from about 40 to 55 both the accumulated discretization error and the accumulated round-off error are probably affecting the last two significant digits and the small error is oscillatory.

The situation is shown schematically in Figure 8.7. We assume that the round-off error R_n is proportional to the number of computations performed and therefore is inversely proportional to the step size h. On the other hand, the discretization error E_n is proportional to a positive power of h. From Eq. (6) of Section 8.3 we know that the total error is bounded by $|E_n| + |R_n|$; hence we wish to choose h so as to minimize this quantity. As indicated in Figure 8.7, the optimum value of h is approximately the value at which the graphs of $|E_n|$ and $|R_n|$ cross. In other words, h is near the optimum when discretization and round-off errors are approximately equal in magnitude.

As a second example of the types of difficulties that can arise when we use numerical procedures carelessly, consider the problem of determining the solution $y = \phi(x)$ of

$$y' = x^2 + y^2, \qquad y(0) = 1. \tag{3}$$

This is the third problem in the problem sets for the previous sections. Since the differential equation is nonlinear, the existence and uniqueness Theorem 2.2 guarantees only that there is a solution in *some* interval about $x = 0$. Suppose we try to compute a solution of the initial value problem on the interval $0 \le x \le 1$ using different numerical procedures.

If we use the Euler method with $h = 0.1$, 0.05, and 0.01, we find the following approximate values at $x = 1$: 7.189500, 12.32054, and 90.68743, respectively. The large differences among the computed values are convincing evidence that we should use a more accurate numerical procedure—the Runge-Kutta method, for example. Using the Runge-Kutta method with $h = 0.1$ we find the approximate

TABLE 8.6 CALCULATION OF THE SOLUTION OF THE INITIAL VALUE PROBLEM $y' = x^2 + y^2$, $y(0) = 1$ USING THE RUNGE-KUTTA METHOD

h	$x = 0.90$	$x = 1.0$
0.1	14.02158	735.0004
0.05	14.27042	1.755613×10^4
0.01	14.30200	$> 10^{15}$

value 735.0004 at $x = 1$, which is quite different from those obtained using the Euler method. If we were naive and stopped now, we would make a serious mistake. Repeating the calculations using step sizes of $h = 0.05$ and $h = 0.01$ we obtain the interesting information listed in Table 8.6.

While the values at $x = 0.90$ are reasonable and we might well believe that the exact solution has a value of about 14.3 at $x = 0.90$, it is clear that something strange is happening between $x = 0.9$ and $x = 1.0$. To help determine what is happening let us turn to some analytical approximations to the solution of the initial value problem (3). Note that on $0 \le x \le 1$,

$$y^2 \le x^2 + y^2 \le 1 + y^2. \tag{4}$$

This suggests that the solution $y = \phi_1(x)$ of

$$y' = 1 + y^2, \qquad y(0) = 1 \tag{5}$$

and the solution $y = \phi_2(x)$ of

$$y' = y^2, \qquad y(0) = 1 \tag{6}$$

are upper and lower bounds, respectively, for the solution $y = \phi(x)$ of the original problem, since all of these solutions pass through the same initial point. Indeed, it can be shown (for example, by the iteration method of Section 2.12) that $\phi_2(x) \le \phi(x) \le \phi_1(x)$ as long as these functions exist. The important thing to note is that we *can solve* Eqs. (5) and (6) for ϕ_1 and ϕ_2 by separation of variables. We find that

$$\phi_1(x) = \tan\left(x + \frac{\pi}{4}\right), \qquad \phi_2(x) = \frac{1}{1 - x}. \tag{7}$$

Thus $\phi_2(x) \to \infty$ as $x \to 1$ and $\phi_1(x) \to \infty$ as $x \to \pi/4 \cong 0.785$. These calculations show that the solution of the original initial value problem must become unbounded somewhere between $x \cong 0.785$ and $x = 1$. We now know that the problem (3) has no solution on the entire interval $0 \le x \le 1$.

However, our numerical calculations suggest that we can go beyond $x = 0.785$ and probably beyond $x = 0.9$. Assuming that the solution of the initial value problem exists at $x = 0.9$ and has the value 14.3, we can obtain a more accurate appraisal of what happens for larger x by considering the initial value problems (5) and (6) with $y(0) = 1$ replaced by $y(0.9) = 14.3$. Then we obtain

$$\phi_1(x) = \tan(x + 0.6010), \qquad \phi_2(x) = 1/(0.9699 - x), \tag{8}$$

where only four decimal places have been kept. Thus $\phi_1(x) \to \infty$ as $x \to \pi/2 - 0.6010 \cong 0.9698$ and $\phi_2(x) \to \infty$ as $x \to 0.9699$. We conclude that the solution of the initial value problem (3) becomes unbounded near $x = 0.97$. We cannot be more precise than this because the initial condition $y(0.9) = 14.3$ is only approximate. This example illustrates the sort of information that can be obtained by a judicious combination of analytical and numerical work.

As a final example, consider the problem of determining two linearly independent solutions of the second order linear equation

$$y'' - 100y = 0 \tag{9}$$

for $x > 0$. The generalization of numerical techniques for first order equations to higher order equations or to systems of equations is discussed in Section 8.9, but that is not needed for the present discussion. Two linearly independent solutions of Eq. (9) are $\phi_1(x) = \cosh 10x$ and $\phi_2(x) = \sinh 10x$. The first solution $\phi_1(x) = \cosh 10x$ is generated by the initial conditions $\phi_1(0) = 1$, $\phi_1'(0) = 0$; the second solution $\phi_2(x) = \sinh 10x$ is generated by the initial conditions $\phi_2(0) = 0$, $\phi_2'(0) = 10$. While analytically we can tell the difference between $\cosh 10x$ and $\sinh 10x$, for large x we have $\cosh 10x \sim e^{10x}/2$ and $\sinh 10x \sim e^{10x}/2$; numerically these two functions look exactly the same if only a fixed number of digits are retained. For example, correct to eight significant figures,

$$\sinh 10 = \cosh 10 = 11{,}013.233.$$

If the calculations are carried out on a machine that carries only eight digits the two solutions ϕ_1 and ϕ_2 are identical at $x = 1$ and indeed for all $x > 1$! Thus, even though the solutions are linearly independent, their numerical tabulation would show that they are the same because we can retain only a finite number of digits. We call this phenomenon *numerical dependence*. It is one of the important and difficult problems always encountered when we are solving a second or higher order equation that has at least one solution that grows very rapidly.

For the present problem we can partially circumvent this difficulty by computing instead of $\sinh 10x$ and $\cosh 10x$ the linearly independent solutions $\phi_3(x) = e^{10x}$ and $\phi_4(x) = e^{-10x}$ corresponding to the initial conditions $\phi_3(0) = 1$, $\phi_3'(0) = 10$ and $\phi_4(0) = 1$, $\phi_4'(0) = -10$, respectively. The solution ϕ_3 grows exponentially while ϕ_4 decays exponentially. Even so, we encounter difficulty in calculating ϕ_4 correctly on a large interval. The reason is that at each step of the calculation for ϕ_4 we introduce discretization and round-off errors. Thus at any point x_n the data to be used in going to the next point are not precisely the values of $\phi_4(x_n)$ and $\phi_4'(x_n)$. The solution of the initial value problem with these data at x_n involves not only e^{-10x} but also e^{10x}. Because the error in the data at x_n is small, the latter function appears with a very small coefficient. Nevertheless, since e^{-10x} tends to zero and e^{10x} grows very rapidly, the latter eventually dominates, and the calculated solution is simply a multiple of $e^{10x} = \phi_3(x)$.

To be specific, suppose that we use the Runge-Kutta method to calculate the solution $y = \phi_4(x) = e^{-10x}$ of the initial value problem

$$y'' - 100y = 0, \qquad y(0) = 1, \qquad y'(0) = -10.$$

TABLE 8.7 EXACT SOLUTION OF $y'' - 100y = 0$, $y(0) = 1$, $y'(0) = -10$ AND NUMERICAL SOLUTION USING THE RUNGE-KUTTA METHOD WITH $h = 0.01$

	y	
x	Numerical	Exact
0.0	1.0	1.0
0.25	8.2085×10^{-2}	8.2085×10^{-2}
0.5	6.7395×10^{-3}	6.7379×10^{-3}
0.75	5.7167×10^{-4}	5.5308×10^{-4}
1.0	2.7181×10^{-4}	4.5400×10^{-5}
1.5	3.3602×10^{-2}	3.0590×10^{-7}
2.0	4.9870	2.0612×10^{-9}
2.5	7.4013×10^{2}	1.3888×10^{-11}
3.0	1.0984×10^{5}	9.3576×10^{-14}
3.5	1.6302×10^{7}	6.3051×10^{-16}
4.0	2.4195×10^{9}	4.2484×10^{-18}
4.5	3.5908×10^{11}	2.8625×10^{-20}
5.0	5.3292×10^{13}	1.9288×10^{-22}

Using single-precision (eight-digit) arithmetic with a step size $h = 0.01$, we obtain the results in Table 8.7. It is clearly evident from these results that the numerical solution begins to deviate significantly from the exact solution for $x > 0.5$, and soon differs from it by many orders of magnitude. The reason is the presence in the numerical solution of a small component of the exponentially growing solution $\phi_3(x) = e^{10x}$. With eight-digit arithmetic we can expect a round-off error of the order of 10^{-8} at each step. Since e^{10x} grows by a factor of 5×10^{21} from $x = 0$ to $x = 5$, an error of order 10^{-8} near $x = 0$ can produce an error of order 5×10^{13} at $x = 5$ even if no further errors are introduced in the intervening calculations. The results given in Table 8.7 demonstrate that this is exactly what happens.

Equation (9) is an example of what are called *stiff* equations. Such equations have at least one rapidly decaying solution and usually require special care when they are solved numerically. Modern computer codes for solving differential equations include provisions for dealing with stiffness. [For more discussion of stiff equations, see the books by Gear (1971) and Rice (1983) noted in the references at the end of this chapter.]

Another problem that must be investigated for each numerical procedure is the convergence of the approximate solution to the exact solution as h tends to zero. While it does not occur for the one-step methods discussed in the previous sections, there is the following danger when using multistep procedures. It can happen that the numerical formula for y_n, which will involve y_{n-1} and y_{n-2}, etc., may have extraneous solutions that do not correspond to solutions of the original

TABLE 8.8 CALCULATION OF THE SOLUTION OF THE INITIAL VALUE PROBLEM $y' = x^2 + e^y$, $y(0) = 0$ USING THE RUNGE-KUTTA METHOD

h	$x = 0.90$	$x = 1.0$
0.02	3.42975	$> 10^{38}$
0.01	3.42981	$> 10^{38}$

differential equation. If these extraneous solutions grow rather than decay they will cause difficulty. This is illustrated by a specific example in the next section.

Other practical problems, which the numerical analyst must often face, are how to compute solutions of a differential equation in the neighborhood of a regular singular point and how to calculate numerically the solution of a differential equation on an *unbounded* interval. Such problems generally require a combination of analytical and numerical work.

PROBLEMS

1. To obtain some idea of the possible dangers of small errors in the initial conditions, such as those due to round-off, consider the initial value problem

$$y' = x + y - 3, \qquad y(0) = 2.$$

(a) Show that the solution is $y = \phi_1(x) = 2 - x$.
(b) Suppose that in the initial condition a mistake is made and 2.001 is used instead of 2. Determine the solution $y = \phi_2(x)$ in this case, and compare the difference $\phi_2(x) - \phi_1(x)$ at $x = 1$ and as $x \to \infty$.

2. Consider the initial value problem

$$y' = x^2 + e^y, \qquad y(0) = 0. \tag{i}$$

Using the Runge-Kutta method with step size h, we obtain the results in Table 8.8. These results suggest that the solution has a vertical asymptote between $x = 0.9$ and $x = 1.0$.
(a) Show that for $0 \le x \le 1$ the solution $y = \phi(x)$ of the problem (i) satisfies

$$\phi_2(x) \le \phi(x) \le \phi_1(x), \tag{ii}$$

where $y = \phi_1(x)$ is the solution of

$$y' = 1 + e^y, \qquad y(0) = 0. \tag{iii}$$

and $y = \phi_2(x)$ is the solution of

$$y' = e^y, \qquad y(0) = 0. \tag{iv}$$

(b) Determine $\phi_1(x)$ and $\phi_2(x)$. Then show that $\phi(x) \to \infty$ for some x between $x = \ln 2 \cong 0.69315$ and $x = 1$.
(c) Solve the differential equations $y' = e^y$ and $y' = 1 + e^y$, respectively, with the initial condition $y(0.9) = 3.4298$. Use the results to show that $\phi(x) \to \infty$ when $x \simeq 0.932$.

In each of Problems 3 and 4

(a) Find a formula for the solution of the initial value problem, and note that it is independent of λ.

(b) Use the Runge-Kutta method with $h = 0.01$ to compute approximate values of the solution for $0 \le x \le 1$ for various values of λ such as $\lambda = 1, 10, 20$, and 50. Do not attempt this unless you have at least a programmable calculator with sufficient memory.

(c) Explain the differences, if any, between the exact solution and the numerical approximations.

3. $y' - \lambda y = 1 - \lambda x, \qquad y(0) = 0$
4. $y' - \lambda y = 2x - \lambda x^2, \qquad y(0) = 0$

8.8 A Multistep Method

In the previous sections we considered one-step or starting methods for solving numerically the initial value problem

$$y' = f(x, y), \qquad y(x_0) = y_0. \tag{1}$$

Once approximate values of the exact solution $y = \phi(x)$ have been obtained at a few points beyond x_0, it is natural to ask whether we can make use of some of this information, rather than just the value at the last point, to obtain the value of ϕ at the next point. Specifically, if y_1 at x_1, y_2 at $x_2, \ldots, y_n$ at x_n are known, how can we use this information to determine y_{n+1} at x_{n+1}?

One of the best of the multistep methods is the Adams-Moulton[4] predictor–corrector method that makes use of information at the points $x_n, x_{n-1}, x_{n-2}, x_{n-3}$. Before the method can be used, it is necessary to compute y_1, y_2, and y_3 by some starting method such as the Runge-Kutta method—the starting method should be as accurate as the multistep method. We briefly discuss the development of the Adams-Moulton predictor–corrector method.

Suppose that we integrate $\phi'(x)$ from x_n to x_{n+1}; then

$$\phi(x_{n+1}) - \phi(x_n) = \int_{x_n}^{x_{n+1}} \phi'(x)\, dx. \tag{2}$$

Assuming that $\phi(x_{n-3})$, $\phi(x_{n-2})$, $\phi(x_{n-1})$, and $\phi(x_n)$ are known, then $\phi'(x_{n-3}), \ldots, \phi'(x_n)$ can be computed from Eq. (1). Next we approximate $\phi'(x)$ by a polynomial of degree three that passes through the four points $[x_{n-3}, \phi'(x_{n-3})]$, $[x_{n-2}, \phi'(x_{n-2})]$, $[x_{n-1}, \phi'(x_{n-1})]$, and $[x_n, \phi'(x_n)]$. It can be proved that this can always be done, and further that the polynomial is unique.

[4]John Couch Adams (1819–1891), English astronomer, is most famous as codiscoverer with Joseph Leverrier of the planet Neptune in 1846. Adams was also extremely skilled at computation; his procedure for numerical integration of differential equations appeared in 1883 in a book with Francis Bashforth on capillary action. Forest Ray Moulton (1872–1952) was an American astronomer and administrator of science. While calculating ballistics trajectories during World War I, he devised substantial improvements in the Adams formula.

This polynomial is called an *interpolation polynomial* since it provides us with an approximate formula for $\phi'(x)$ at values of x other than x_{n-3}, x_{n-2}, x_{n-1}, and x_n. Substituting this interpolation polynomial for $\phi'(x)$ in Eq. (2), evaluating the integral, and replacing $\phi(x_{n-3}), \ldots, \phi(x_{n+1})$ by their approximate values gives the multistep formula

$$y_{n+1} = y_n + \frac{h}{24}\left(55y_n' - 59y_{n-1}' + 37y_{n-2}' - 9y_{n-3}'\right). \tag{3}$$

This formula is known as the Adams-Bashforth predictor formula; it predicts the approximate value y_{n+1} for $\phi(x)$ at x_{n+1}. It has a local discretization error proportional to h^5.

The details of the derivation of Eq. (3) that have been omitted are the determination of the polynomial of degree three that passes through the points (x_{n-3}, y_{n-3}'), (x_{n-2}, y_{n-2}'), (x_{n-1}, y_{n-1}'), and (x_n, y_n'), and the integration of this polynomial from x_n to x_{n+1}. The method of constructing interpolation polynomials dates back to Newton and is discussed in books on numerical methods; see, for example, Henrici (Chapter 5) or Conte and deBoor (Chapter 4). What is important at the present time is not these algebraic details, but the underlying general principle of multistep methods.

1. A polynomial of degree p is fitted to the $p + 1$ points

$$\left(x_{n-p}, y_{n-p}'\right), \ldots, \left(x_n, y_n'\right).$$

2. This interpolation polynomial is integrated over an interval in x terminating at x_{n+1}, which gives an equation for y_{n+1}.

Also note that different formulas can be obtained by integrating the same polynomial over different intervals. If the polynomial used in obtaining Eq. (3) is integrated from x_{n-3} to x_{n+1}, we obtain the Milne (1890–1971) predictor formula:

$$y_{n+1} = y_{n-3} + \frac{4h}{3}\left(2y_n' - y_{n-1}' + 2y_{n-2}'\right). \tag{4}$$

This formula also has a local discretization error proportional to h^5.

The Adams-Moulton predictor–corrector method involves, in addition to the formula (3), a *corrector formula* for improving the value of y_{n+1} predicted by Eq. (3). The corrector formula is derived by integrating $\phi'(x)$ from x_n to x_{n+1} using an interpolation polynomial that passes through the points $[x_{n-2}, \phi'(x_{n-2})]$, $[x_{n-1}, \phi'(x_{n-1})]$, $[x_n, \phi'(x_n)]$, and $[x_{n+1}, \phi'(x_{n+1})]$. In the actual calculation the approximate values y_{n-2}', y_{n-1}', y_n', and y_{n+1}' are used for $\phi'(x)$ at x_{n-2}, x_{n-1}, x_n, and x_{n+1}, and $y_{n+1}' = f(x_{n+1}, y_{n+1})$ is evaluated using the value of y_{n+1} determined by Eq. (3). The Adams-Moulton corrector formula is

$$y_{n+1} = y_n + \frac{h}{24}\left(9y_{n+1}' + 19y_n' - 5y_{n-1}' + y_{n-2}'\right). \tag{5}$$

This formula also has a local discretization error proportional to h^5. However, the proportionality constant in the error bound for the corrector formula (5) is about $\frac{1}{15}$ of the proportionality constant in the error bound for the predictor formula (3).

Once y_{n-3}, y_{n-2}, y_{n-1}, and y_n are known we can compute y'_{n-3}, y'_{n-2}, y'_{n-1} and y'_n, and then use the predictor formula (3) to obtain a first value for y_{n+1}. Then we compute y'_{n+1} and use the corrector formula (5) to obtain an improved value of y_{n+1}. We can, of course, continue to use the corrector formula (5) if the change in y_{n+1} is too large. As a general rule, however, if it is necessary to use the corrector formula more than once or twice it can be expected that the step size h is too large and should be made smaller. Note also that Eq. (5) serves as a check on the arithmetic (programming) in determining y_{n+1} from Eq. (3)—any substantial difference between the values of y_{n+1} from Eq. (3) and from Eq. (5) would indicate a definite possibility of an arithmetic error.

To summarize, we first compute y_1, y_2, and y_3 by a starting formula with a local discretization error no greater than h^5. We then switch to the predictor formula (3) to compute y_{n+1}, $n \geq 3$ and the corrector formula (5) to correct y_{n+1}. We continue to correct y_{n+1} until the relative change is less than a prescribed allowable error. A more detailed discussion of the Adams-Moulton predictor–corrector method and other multistep methods can be found in Conte and deBoor (Chapter 6) and Henrici (Chapter 5).

EXAMPLE

Using the Adams-Moulton predictor–corrector method with $h = 0.1$ determine an approximate value of the exact solution $y = \phi(x)$ at $x = 0.4$ for the initial value problem

$$y' = 1 - x + 4y, \qquad y(0) = 1. \tag{6}$$

For starting data we use the values of y_1, y_2, and y_3 determined using the Runge-Kutta method. These are tabulated in Table 8.3. Then, using Eq. (6), we obtain

$$\begin{aligned} y_0 &= 1 & y'_0 &= 5 \\ y_1 &= 1.6089333 & y'_1 &= 7.3357332 \\ y_2 &= 2.5050062 & y'_2 &= 10.820025 \\ y_3 &= 3.8294145 & y'_3 &= 16.017658. \end{aligned}$$

From Eq. (3) we find that the predicted value of y_4 is

$$y_4 = 3.8294145 + \frac{0.1(469.011853)}{24} = 5.783631.$$

Next we use the corrector formula (5) to correct y_4. Corresponding to the predicted value of y_4 we find from Eq. (6) that $y'_4 = 23.734524$. Hence, from Eq. (5), the corrected value of y_4 is

$$y_4 = 3.8294145 + \frac{0.1(471.181828)}{24} = 5.792673.$$

The exact value, correct through eight digits, is 5.7942260. Note that using the correction formula once reduces the error in y_4 to approximately one-seventh of the error before its use.

A natural question to ask is why use a predictor–corrector method with a local discretization error proportional to h^5 when the Runge-Kutta method has an error of the same order of magnitude. One reason is that for many problems the most time-consuming part of the calculation, hence the most expensive, is the evaluation of the function $f(x, y)$ at successive steps. For the Runge-Kutta method four evaluations of f are required in taking one step. On the other hand for the predictor–corrector method (3) and (5), the predictor requires only the evaluation of $f(x, y)$ at each x_n, and the use of the corrector requires an additional evaluation of $f(x, y)$ at x_{n+1} for each iteration performed. Thus the number of function evaluations can be reduced significantly. This saving can be very important in writing a general program for solving initial value problems.

STABILITY. We mentioned earlier in this chapter that numerical difficulties can arise when multistep methods are used. We illustrate what can happen by considering the initial value problem

$$y' + 2y = 0, \qquad y(0) = 1, \tag{7}$$

and the multistep method

$$y_{n+1} = y_{n-1} + 2hy_n'. \tag{8}$$

The multistep method (8) is derived in Problems 5 and 6; it has a local discretization error proportional to h^3. For later reference we note that the exact solution of the initial value problem (7) is $y = e^{-2x}$.

For this problem $y_n' = -2y_n$, so Eq. (8) becomes

$$y_{n+1} + 4hy_n - y_{n-1} = 0. \tag{9}$$

Equation (9) is a *linear difference equation*. It can be solved explicitly following a theory similar to the theory for linear constant coefficient differential equations. We look for a solution of Eq. (9) of the form $y_n = \lambda^n$ where λ is to be determined. We have

$$\lambda^{n+1} + 4h\lambda^n - \lambda^{n-1} = 0,$$

or

$$\lambda^{n-1}(\lambda^2 + 4h\lambda - 1) = 0. \tag{10}$$

The nonzero solutions of Eq. (10) are

$$\lambda_1 = -2h + \sqrt{1 + 4h^2} \quad \text{and} \quad \lambda_2 = -(2h + \sqrt{1 + 4h^2}). \tag{11}$$

The corresponding solutions of Eq. (9) are

$$y_n = \lambda_1^n \quad \text{and} \quad y_n = \lambda_2^n. \tag{12}$$

By analogy with the theory of linear second order differential equations it is reasonable to expect (and it can be shown) that the general solution of the difference equation (9) is

$$y_n = c_1\lambda_1^n + c_2\lambda_2^n, \tag{13}$$

where c_1 and c_2 are arbitrary constants.

What is important is that the difference equation (9) has two solutions, λ_1^n and λ_2^n, while the original differential equation has only the one solution e^{-2x}. This may cause trouble. We now show that as $h \to 0$ one of the solutions of the difference equation approaches the true solution e^{-2x} of the differential equation, but that the other oscillates between $\pm e^{2x}$. Let X be a fixed but arbitrary value of x, and suppose that h decreases and n increases in such a way that $nh = X$. Then, for small h and large n, we (i) approximate $(1 + 4h^2)^{1/2}$ by $1 + 2h^2$, (ii) approximate λ_1 by $1 - 2h$ and λ_2 by $-1 - 2h$, and (iii) use the fact that $(1 + \alpha/n)^n \to e^\alpha$ as $n \to \infty$. Thus we have

$$\lambda_1^n \cong (1 - 2h)^n = (1 - 2X/n)^n \to e^{-2X} \quad \text{as } n \to \infty. \tag{14}$$

A similar calculation shows that

$$(-1)^n\lambda_2^n \to e^{2X} \quad as\ n \to \infty. \tag{15}$$

Then, for small h and large n we can write

$$y_n \cong c_1e^{-2X} + c_2(-1)^ne^{2X} \tag{16}$$

Since X is arbitrary, the first term of Eq. (16) correctly gives the general solution c_1e^{-2x} of the original differential equation (7). The second term is an extraneous solution that arises from the numerical procedure.

In using the multistep formula (8) or (9) we must know in addition to the initial value $y_0 = 1$ at $x_0 = 0$ the value of y_1 at $x_1 = h$. In theory, if the correct values of y_0 and y_1 are used, then the constants c_1 and c_2 of Eq. (16) would be determined as $c_1 = 1$ and $c_2 = 0$, and the extraneous solution $(-1)^ne^{2x}$ would not appear. However, as we discussed in the previous section, there will always be small errors, and as a consequence the term $c_2(-1)^ne^{2x}$ will be present in the solution of Eq. (9). Even though c_2 may be very small corresponding to the very small error in the calculation, the term $c_2(-1)^ne^{2x}$ will soon dominate the full solution as x increases. This is vividly shown by the graph in Figure 8.8 of the numerical solution of the initial value problem (7) using the multistep method (8) with $h = 7/224 \simeq 0.03$. Clearly, beyond $x = 3$ the extraneous solution $(-1)^ne^{2x}$ dominates.

This phenomenon is known as *numerical instability*. Roughly speaking, a numerical procedure is unstable if errors introduced into the calculations grow at an exponential rate as the computation proceeds. While it is beyond the scope of this book to give a detailed account of stability of numerical methods, we can

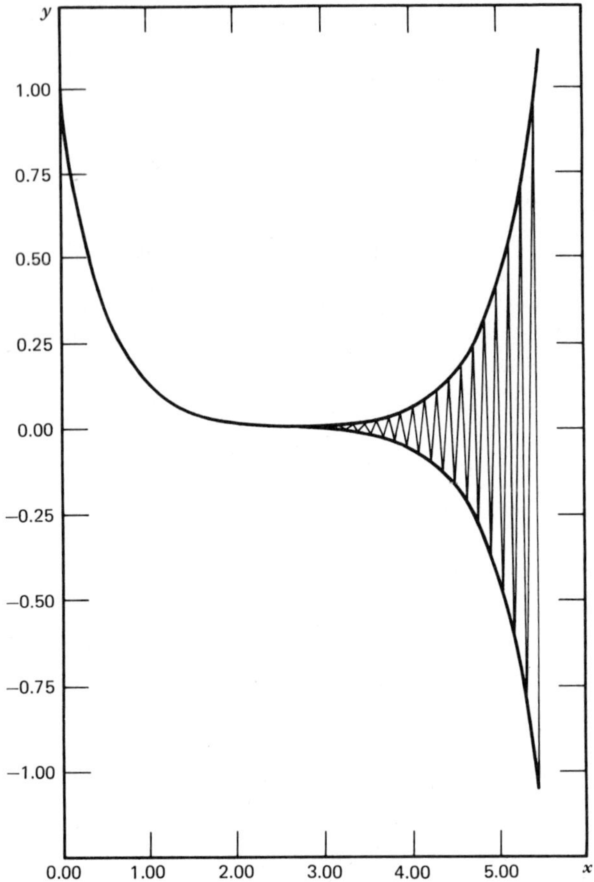

FIGURE 8.8 An example of numerical instability: $y' + 2y = 0,\ y(0) = 1;\ y_{n+1} = y_{n-1} + 2hy_n'$.

make several observations. The question of stability (or instability) of a multistep method depends not only on the method but also on the particular differential equation; the method may be stable for one problem but not for another. Also, multistep methods may be stable for some values of h but not for all values of h. To assess the stability of a multistep method when applied to an initial value problem, one must examine the solutions of the resulting difference equation. The method is said to be *strongly stable* if the extraneous solutions λ^n of the difference equation are such that $|\lambda| < 1$ as $h \to 0$. If this is the case, then $\lambda^n \to 0$ as $n \to \infty$ for h sufficiently small, and errors in the computation will not grow as n increases. Note that for the example problem just discussed the extraneous solution λ_2^n has $|\lambda_2| > 1$.

It is not possible, in general, to apply this criterion to the initial value problem $y' = f(x, y)$, $y(x_0) = y_0$. Usually, in discussing stability of a numerical method one is content to examine the *linear stability* of the method; is the procedure stable for the differential equation $y' = Ay$ where A is an arbitrary constant? It can be shown that the Adams-Moulton predictor–corrector method is strongly stable in the sense of linear stability for h sufficiently small. See Problems 8 and 9 for further discussion of these ideas.

PROBLEMS

In Problems 1 through 4 determine an approximate value of the solution at $x = 0.4$ using the Adams-Moulton predictor–corrector method. Take $x_0 = 0$ and $h = 0.1$. The values for y_1, y_2, and y_3 are those computed by the Runge-Kutta method, rounded to six digits. Use the corrector formula to compute one correction.

1. $y' = 2y - 1;$ $y_0 = 1,$ $y_1 = 1.11070,$ $y_2 = 1.24591,$ $y_3 = 1.41105$
2. $y' = \frac{1}{2} - x + 2y;$ $y_0 = 1,$ $y_1 = 1.27140,$ $y_2 = 1.59182,$ $y_3 = 1.97211$
3. $y' = x^2 + y^2;$ $y_0 = 1,$ $y_1 = 1.11146,$ $y_2 = 1.25302,$ $y_3 = 1.43967$
4. $y' = 5x - 3\sqrt{y},$ $y_0 = 2,$ $y_1 = 1.62231,$ $y_2 = 1.33253,$ $y_3 = 1.12358$
5. Given that the values of $\phi'(x)$ at x_{n-1} and x_n are y'_{n-1} and y'_n, fit a first degree polynomial $ax + b$ to $\phi'(x)$ so that it passes through these two points. By integrating $\phi'(x)$ from x_{n-1} to x_{n+1} derive the multistep formula

$$y_{n+1} = y_{n-1} + 2hy'_n.$$

   Hint: The calculations are not difficult if one sets $x_{n+1} = x_n + h$ and $x_{n-1} = x_n - h$.
6. Show that the multistep formula derived in Problem 5 has a local discretization error proportional to h^3. Note that this formula is just as easy to use (after y_1 has been computed) as the Euler formula $y_{n+1} = y_n + hy'_n$ which has a local discretization error proportional only to h^2. Unfortunately, however, as shown in the text this multistep formula is unstable.
   Hint: Evaluate $\phi(x_n \pm h)$ by expanding $\phi(x)$ in a Taylor series about $x = x_n$.
7. This problem is concerned with the Milne predictor–corrector formula. The predictor formula was given in Eq. (4) of the text. The corrector formula is derived by evaluating the integral of $\phi'(x)$ from x_{n-1} to x_{n+1} by Simpson's rule. The formula is

$$y_{n+1} = y_{n-1} + \frac{h}{3}\left(y'_{n-1} + 4y'_n + y'_{n+1}\right).$$

   Both the predictor and corrector formulas have local discretization errors proportional to h^5. The Milne method is often very effective, but the corrector can be unstable (see Problem 9). As a consequence, the Adams-Moulton predictor–corrector method is to be preferred. In each of the following problems determine an approximate value of the exact solution at $x = 0.4$ using the Milne predictor–corrector method. Use the corrector formula to compute one correction.
   (a) Problem 1 (b) Problem 2 (c) Problem 3 (d) Problem 4.

*8. In this problem we show that the Adams-Moulton corrector formula (5) is stable for the linear differential equation $y' = Ay$.

(a) Show that the solution of the differential equation is $y = ce^{Ax}$.

(b) Show that the appropriate difference equation is

$$(1 - 9\alpha)\, y_{n+1} - (1 + 19\alpha)\, y_n + 5\alpha y_{n-1} - \alpha y_{n-2} = 0 \qquad \text{(i)}$$

where $\alpha = Ah/24$.

(c) Next show that $y_n = \lambda^n$ is a solution of Eq. (i) if λ is a root of

$$(1 - 9\alpha)\lambda^3 - (1 + 19\alpha)\lambda^2 + 5\alpha\lambda - \alpha = 0. \qquad \text{(ii)}$$

(d) In the limit as $h \to 0$, which implies $\alpha \to 0$, show that the roots of Eq. (ii) are $\lambda_1 = 1$, $\lambda_2 = 0$, $\lambda_3 = 0$.

(e) We examine the root λ_1 more closely for h small. Set $\lambda_1 = 1 + \lambda_{11}h$, substitute in Eq. (ii), and neglect terms proportional to h^2 or higher. Deduce that $\lambda_{11} = A$ and show that if $y_1 = \lambda_1^n \simeq (1 + Ah)^n$ then $y_1 \to e^{Ax_n}$ as $h \to 0$, on using $x_n = nh$.

Thus the solution corresponding to λ_1 approximates the true solution of the differential equation. For small h we have $|\lambda_2| < 1$ and $|\lambda_3| < 1$; thus the corresponding extraneous solutions will not grow. As a consequence, the Adams-Moulton corrector formula is strongly stable.

*9. In this problem we show that the Milne corrector formula given in Problem 7 is unstable for the linear differential equation $y' = Ay$. The analysis parallels that of Problem 8.

(a) Show that the solution of the differential equation is $y = ce^{Ax}$.

(b) Show that the appropriate difference equation is

$$(1 - \alpha)\, y_{n+1} - 4\alpha y_n - (1 + \alpha)\, y_{n-1} = 0 \qquad \text{(i)}$$

where $\alpha = Ah/3$.

(c) Next show that $y_n = \lambda^n$ is a solution of Eq. (i) if λ is a root of

$$(1 - \alpha)\lambda^2 - 4\alpha\lambda - (1 + \alpha) = 0. \qquad \text{(ii)}$$

(d) In the limit $h \to 0$, which implies $\alpha \to 0$, show that the roots of Eq. (ii) are $\lambda_1 = 1$ and $\lambda_2 = -1$. Note that $|\lambda_1| = |\lambda_2| = 1$ and hence for small nonzero values of h one of the roots may have absolute value greater than one.

(e) Solve the quadratic equation (ii) and show that if h is small then $\lambda_1 \simeq 1 + Ah$ and $\lambda_2 \simeq -(1 - Ah/3)$. Next show that as $h \to 0$, with $x_n = nh$, the solution of the difference equation (i) is

$$y_n = c_1 e^{Ax_n} + c_2(-1)^n e^{-Ax_n/3}. \qquad \text{(iii)}$$

The first term in Eq. (iii) approximates the true solution of the differential equation. The second term, the extraneous solution corresponding to λ_2, decays if $A > 0$ but grows if $A < 0$. Thus the Milne corrector formula can be unstable. Methods whose stability (instability) depend on the sign of A are called *weakly stable*. Finally, we note that $|\lambda_2| > 1$ for $A < 0$.

8.9 Systems of First Order Equations

In the preceding sections we discussed numerical methods for solving initial value problems associated with first order differential equations. These methods can also be applied to a system of first order equations or to equations of higher order. Since a higher order equation such as $d^2x/dt^2 = f(t, x, dx/dt)$ can be reduced to the system of first order equations $dx/dt = y$, $dy/dt = f(t, x, y)$, and since such a transformation is almost always performed prior to the use of numerical methods, we consider only systems of first order equations. We use t to denote the independent variable, and $x, y, z, \ldots$ to denote the dependent variables. Primes denote differentiation with respect to t. Further, for the purpose of simplicity, we restrict our attention to a system of two first order equations

$$x' = f(t, x, y), \tag{1}$$

$$y' = g(t, x, y), \tag{2}$$

with the initial conditions

$$x(t_0) = x_0, \qquad y(t_0) = y_0. \tag{3}$$

The functions f and g are assumed to satisfy the conditions of Theorem 7.1 so that the initial value problem (1), (2), and (3) has a unique solution in some interval of the t axis containing the point t_0. We wish to determine approximate values $x_1, x_2, \ldots, x_n, \ldots$ and $y_1, y_2, \ldots, y_n, \ldots$ of the solution $x = \phi(t)$, $y = \psi(t)$ at the points $t_n = t_0 + nh$.

As an example of how the methods of the previous sections can be generalized, consider the Euler method of Section 8.2. In this case we use the initial conditions to determine $\phi'(t_0)$ and $\psi'(t_0)$, and then move along the tangent lines to the curves $x = \phi(t)$ and $y = \psi(t)$. Thus

$$x_1 = x_0 + hf(t_0, x_0, y_0), \qquad y_1 = y_0 + hg(t_0, x_0, y_0).$$

In general

$$\begin{aligned} x_{n+1} &= x_n + hf(t_n, x_n, y_n) & \qquad y_{n+1} &= y_n + hg(t_n, x_n, y_n) \\ &= x_n + hx'_n & &= y_n + hy'_n. \end{aligned} \tag{4}$$

The other numerical methods can be generalized in a similar manner. The formulas are given in Problems 2, 4, and 5.

EXAMPLE

Determine approximate values of $x = \phi(t)$, $y = \psi(t)$ at the points $t = 0.1$ and $t = 0.2$ using the Euler formula (4) for the initial value problem

$$x' = x - 4y, \qquad y' = -x + y, \tag{5}$$

with

$$x(0) = 1, \qquad y(0) = 0. \tag{6}$$

Take $h = 0.1$.

For this problem $x'_n = x_n - 4y_n$ and $y'_n = -x_n + y_n$; hence

$$x'_0 = 1 - 4(0) \qquad y'_0 = -1 + 0$$
$$= 1 \qquad\qquad = -1.$$

Thus

$$x_1 = 1 + 0.1(1) \qquad y_1 = 0 + 0.1(-1)$$
$$= 1.1 \qquad\qquad = -0.1.$$

Similarly

$$x'_1 = 1.1 - 4(-0.1) \qquad y'_1 = -1.1 + (-0.1)$$
$$= 1.5 \qquad\qquad = -1.2;$$

and

$$x_2 = 1.1 + 0.1(1.5) \qquad y_2 = -0.1 + 0.1(-1.2)$$
$$= 1.25 \qquad\qquad = -0.22.$$

It is easily verified that the exact solution of the initial value problem (5) and (6) is

$$\phi(t) = \frac{e^{-t} + e^{3t}}{2}, \qquad \psi(t) = \frac{e^{-t} - e^{3t}}{4}. \tag{7}$$

The values of the exact solution, rounded to eight digits, at $t = 0.2$ are $\phi(0.2) = 1.3204248$ and $\psi(0.2) = -0.25084701$.

PROBLEMS

Some of the problems for this section deal with the following initial value problems.

(a) $x' = x + y + t, \quad y' = 4x - 2y; \qquad x(0) = 1, \quad y(0) = 0$
(b) $x' = 2x + ty, \quad y' = xy; \qquad x(0) = 1, \quad y(0) = 1$
(c) $x' = -tx - y - 1, \quad y' = x; \qquad x(0) = 1, \quad y(0) = 1$

1. For each of initial value problems (a), (b), and (c), determine approximate values of the solution at $t = 0.1$ and $t = 0.2$ using the Euler method with $h = 0.1$.
2. Show that the generalization of the three-term Taylor formula of Section 8.5 for the initial value problem $x' = f(t, x, y)$, $y' = g(t, x, y)$, with $x(t_0) = x_0$, $y(x_0) = y_0$ is

$$x_{n+1} = x_n + hx'_n + \tfrac{1}{2}h^2x''_n, \qquad y_{n+1} = y_n + hy'_n + \tfrac{1}{2}h^2y''_n$$

where

$$x''_n = f_t(t_n, x_n, y_n) + f_x(t_n, x_n, y_n)x'_n + f_y(t_n, x_n, y_n)y'_n,$$
$$y''_n = g_t(t_n, x_n, y_n) + g_x(t_n, x_n, y_n)x'_n + g_y(t_n, x_n, y_n)y'_n.$$

Determine an approximate value of the solution at $t = 0.2$ for the illustrative initial value problem in the text. Take $h = 0.2$ and compare the results with the exact values.

3. Using the three-term Taylor formula derived in Problem 2 determine an approximate value of the solution at $x = 0.2$ for each of the initial value problems (a), (b), and (c). Take $h = 0.2$.
4. The generalization of the Runge-Kutta method of Section 8.6 for the initial value problem $x' = f(t, x, y)$, $y' = g(t, x, y)$, with $x(t_0) = x_0$, $y(t_0) = y_0$ is

$$x_{n+1} = x_n + \tfrac{1}{6}h(k_{n1} + 2k_{n2} + 2k_{n3} + k_{n4}),$$
$$y_{n+1} = y_n + \tfrac{1}{6}h(l_{n1} + 2l_{n2} + 2l_{n3} + l_{n4}),$$

where

$$k_{n1} = f(t_n, x_n, y_n), \qquad l_{n1} = g(t_n, x_n, y_n),$$
$$k_{n2} = f\left(t_n + \tfrac{1}{2}h, x_n + \tfrac{1}{2}hk_{n1}, y_n + \tfrac{1}{2}hl_{n1}\right), \quad l_{n2} = g\left(t_n + \tfrac{1}{2}h, x_n + \tfrac{1}{2}hk_{n1}, y_n + \tfrac{1}{2}hl_{n1}\right),$$
$$k_{n3} = f\left(t_n + \tfrac{1}{2}h, x_n + \tfrac{1}{2}hk_{n2}, y_n + \tfrac{1}{2}hl_{n2}\right), \quad l_{n3} = g\left(t_n + \tfrac{1}{2}h, x_n + \tfrac{1}{2}hk_{n2}, y_n + \tfrac{1}{2}hl_{n2}\right),$$
$$k_{n4} = f(t_n + h, x_n + hk_{n3}, y_n + hl_{n3}), \qquad l_{n4} = g(t_n + h, x_n + hk_{n3}, y_n + hl_{n3}).$$

Determine an approximate value of the solution at $t = 0.2$ for the illustrative initial value problem in the text. Take $h = 0.2$ and compare the results with the exact values.
5. Consider the initial value problem $x' = f(t, x, y)$ and $y' = g(t, x, y)$ with $x(t_0) = x_0$ and $y(t_0) = y_0$. The generalization of the Adams-Moulton predictor–corrector method of Section 8.8 is

$$x_{n+1} = x_n + \tfrac{1}{24}h\left(55x'_n - 59x'_{n-1} + 37x'_{n-2} - 9x'_{n-3}\right),$$
$$y_{n+1} = y_n + \tfrac{1}{24}h\left(55y'_n - 59y'_{n-1} + 37y'_{n-2} - 9y'_{n-3}\right),$$

and

$$x_{n+1} = x_n + \tfrac{1}{24}h\left(9x'_{n+1} + 19x'_n - 5x'_{n-1} + x'_{n-2}\right),$$
$$y_{n+1} = y_n + \tfrac{1}{24}h\left(9y'_{n+1} + 19y'_n - 5y'_{n-1} + y'_{n-2}\right),$$

Determine an approximate value of the solution at $t = 0.4$ for the illustrative initial value problem $x' = x - 4y$, $y' = -x + y$ with $x(0) = 1$, $y(0) = 0$. Take $h = 0.1$. Correct the predicted value once. For the values of $x_1, \ldots, y_3$ use the values of the exact solution rounded to six digits: $x_1 = 1.12883$, $x_2 = 1.32042$, $x_3 = 1.60021$, $y_1 = -0.110527$, $y_2 = -0.250847$, and $y_3 = -0.429696$.
6. Consider the initial value problem

$$x'' + t^2x' + 3x = t, \qquad x(0) = 1, \quad x'(0) = 2.$$

Change this problem to a system of two equations and determine approximate values of the solution at $t = 0.1$ and $t = 0.2$ using the Euler method with a step size $h = 0.1$.

REFERENCES

There are many books of varying degrees of sophistication dealing with numerical analysis in general, programming, and the numerical solution of ordinary differential equations and partial differential equations. The following books were mentioned in the text:

Conte, S. D., and deBoor, Carl, *Elementary Numerical Analysis: An Algorithmic Approach* (3rd ed.) (New York: McGraw-Hill, 1980).

Henrici, Peter, *Discrete Variable Methods in Ordinary Differential Equations* (New York: Wiley, 1962).

The text by Henrici provides a wealth of theoretical information about numerical methods for solving ordinary differential equations; however, it is fairly advanced. Two additional books on numerical methods for ordinary differential equations are:

Gear, C. William, *Numerical Initial Value Problems in Ordinary Differential Equations* (Englewood Cliffs, N.J.: Prentice-Hall, 1971).

Lambert, J. D., *Computational Methods in Ordinary Differential Equations* (New York: Wiley, 1973).

A detailed exposition of Adams predictor–corrector methods, including practical guidelines for implementation, may be found in:

Shampine, L. F., and Gordon, M. K., *Computer Solution of Ordinary Differential Equations: The Initial Value Problem* (San Francisco: Freeman, 1975).

Many books on numerical analysis have chapters on differential equations. For example, at an elementary level:

Burden, R. L., Faires, J. D., and Reynolds, A. C., *Numerical Analysis* (3rd ed.) (Boston: Prindle, Weber, and Schmidt, 1985).

At a more advanced level, see:

Rice, J. R., *Numerical Methods, Software and Analysis* (New York: McGraw-Hill, 1983).

The latter book also includes a discussion of some of the subroutines available in the International Mathematics and Statistics Library (IMSL).

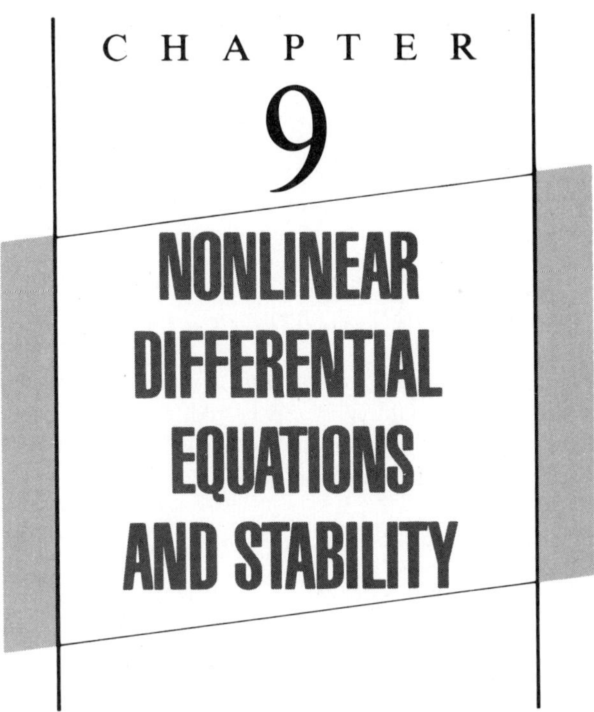

CHAPTER 9

NONLINEAR DIFFERENTIAL EQUATIONS AND STABILITY

9.1 Autonomous Systems and Critical Points

Up to now we have been concerned primarily with methods for solving differential equations. For the most part we have considered linear equations, for which a number of highly developed methods are available. However, we must face the fact that it is often very difficult, if not impossible, to find a solution for a given differential equation in a reasonably convenient form. This is especially true of nonlinear equations. Therefore, it is important to consider what qualitative[1] information can be obtained about the solutions of differential equations, particularly nonlinear ones, without actually solving the equations.

The questions that we consider in this chapter are mainly associated with the idea of stability of a solution, and the methods that we employ are largely

[1] The qualitative theory of differential equations, the subject of this chapter, is largely the creation of Henri Poincaré (1854–1912) in several major papers between 1880 and 1886. Poincaré was professor at the University of Paris and is generally considered the leading mathematician of his time. He made fundamental discoveries in several different areas of mathematics, including complex function theory, partial differential equations, and celestial mechanics. In a series of papers beginning in 1894 he initiated the use of modern methods in topology. In differential equations he was a pioneer in the use of asymptotic series, one of the most powerful tools of contemporary applied mathematics. Among other things, he used asymptotic expansions to obtain solutions about irregular singular points, thereby extending the work of Fuchs and Frobenius discussed in Chapter 4.

geometrical in nature. Both the concept of stability and the use of geometric analysis were introduced in Section 2.6 for a single first order equation. In this chapter we refine the ideas and extend the discussion to systems of equations.

We are concerned with systems of two simultaneous differential equations of the form

$$dx/dt = F(x, y), \qquad dy/dt = G(x, y). \tag{1}$$

We assume that the functions F and G are continuous and have continuous partial derivatives in some domain D in the xy plane. By Theorem 7.1, if (x_0, y_0) is a point in this domain, then there exists a unique solution $x = \phi(t)$, $y = \psi(t)$ of the system (1) satisfying the initial conditions

$$x(t_0) = x_0, \qquad y(t_0) = y_0. \tag{2}$$

The solution is defined in some interval $\alpha < t < \beta$ which contains the point t_0.

Sometimes we will write the initial value problem (1), (2) in the vector form

$$d\mathbf{x}/dt = \mathbf{f}(\mathbf{x}), \qquad \mathbf{x}(t_0) = \mathbf{x}^0, \tag{3}$$

where $\mathbf{x} = x\mathbf{i} + y\mathbf{j}$, $\mathbf{f}(\mathbf{x}) = F(x, y)\mathbf{i} + G(x, y)\mathbf{j}$, and $\mathbf{x}^0 = x_0\mathbf{i} + y_0\mathbf{j}$. In this case the solution is expressed as $\mathbf{x} = \boldsymbol{\phi}(t)$, where $\boldsymbol{\phi}(t) = \phi(t)\mathbf{i} + \psi(t)\mathbf{j}$.

It is helpful to interpret a solution $\mathbf{x} = \boldsymbol{\phi}(t)$ as a curve traced by a moving point in the xy plane. The equations $x = \phi(t)$, $y = \psi(t)$ are a parametric representation of this curve. The xy plane is referred to as the *phase plane*. A curve described parametrically by a solution of the system (1) is often called a *trajectory* of the system.

In Section 2.6 we found that the constant, or equilibrium, solutions were of particular significance. For the two-dimensional system (1) the equilibrium solutions are found by setting dx/dt and dy/dt equal to zero and solving the algebraic system

$$F(x, y) = 0, \qquad G(x, y) = 0. \tag{4}$$

Points that satisfy Eqs. (4) are called *equilibrium points* or *critical points* of Eqs. (1). One of our main purposes in this chapter is to analyze the nature and importance of critical points for two-dimensional systems.

Among the questions that we wish to discuss are the following:

1. What is the behavior of the solution $\mathbf{x} = \boldsymbol{\phi}(t)$ for large t? In particular, does the solution approach a limit as $t \to \infty$ that is independent of the initial conditions?
2. Is the solution stable or unstable with respect to the initial conditions? That is, if the initial conditions are altered slightly, does the solution change only slightly, or does it experience large modifications?

3. If the system (1) is replaced by a simpler one, for instance, by a linear system, does the solution of the linear system constitute a good approximation to the solution of the original nonlinear system (1)?

Observe that the system (1) does not explicitly contain the independent variable t. A system with this property is said to be *autonomous*. The geometric reasoning introduced in Section 2.6 can be effectively extended to autonomous systems of two equations, but is not nearly as useful for nonautonomous systems. To clarify this important difference between autonomous and nonautonomous systems, we will consider two simple examples.

EXAMPLE 1

Describe the trajectories through the point $(1, 2)$ for the system

$$dx/dt = x, \qquad dy/dt = y. \tag{5}$$

We can visualize this problem by supposing that particles are continuously emitted at the point $(1, 2)$, and then move in the xy plane according to the differential equations (5). The particle emitted at the time $t = s$ is specified by the initial conditions $x(s) = 1$, $y(s) = 2$. What is the path of the particle for $t \geq s$? The solution of Eqs. (5) satisfying these initial conditions is

$$x = \phi(t; s) = e^{t-s}, \qquad y = \psi(t; s) = 2e^{t-s}, \qquad t \geq s. \tag{6}$$

The paths followed by the particles emitted at different times s can be most readily visualized by eliminating t from Eqs. (6). This gives

$$y = 2x, \tag{7}$$

where it is clear from Eqs. (6) that $x \geq 1$, $y \geq 2$. The portion of the straight line

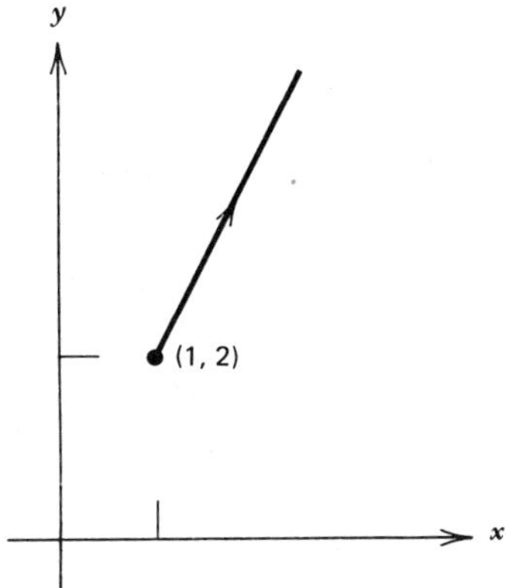

FIGURE 9.1 A trajectory of an autonomous system.

$y = 2x$ corresponding to Eqs. (6), with the direction of motion shown by an arrow, is sketched in Figure 9.1. What is important is that the initial time s does not appear in Eq. (7). Therefore, no matter when a particle is emitted from the point $(1, 2)$, it always moves along the same curve; there is only one trajectory passing through the point $(1, 2)$.

EXAMPLE 2

Describe the trajectories through the point $(1, 2)$ for the system

$$dx/dt = x/t, \qquad dy/dt = y. \tag{8}$$

Observe that the first of Eqs. (8) involves t explicitly on the right side of the equation; hence the system (8) is not autonomous. The solution of Eqs. (8) satisfying the initial conditions $x(s) = 1$, $y(s) = 2$ is

$$x = \phi(t; s) = t/s, \qquad y = \psi(t; s) = 2e^{t-s}, \qquad t \geq s. \tag{9}$$

Solving the first of Eqs. (9) gives $t = sx$. Then substituting in the second of Eqs. (9) we obtain

$$y = 2e^{s(x-1)}, \tag{10}$$

where again $x \geq 1$. Since s appears in Eq. (10), the path that a particle follows depends on the time at which it is emitted, that is, the time at which the initial condition is applied. The paths followed by particles emitted at several different times are sketched in Figure 9.2.

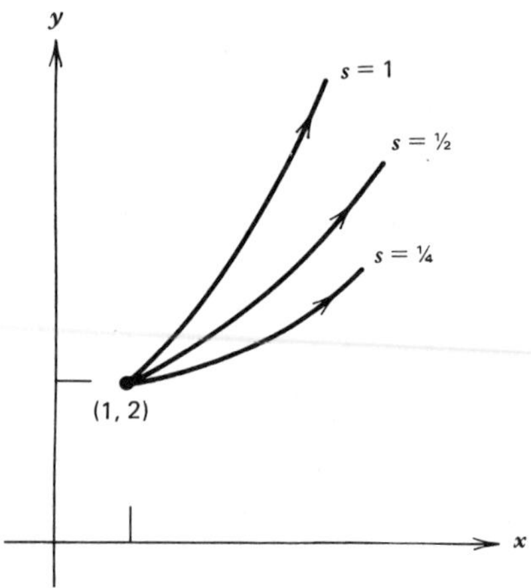

FIGURE 9.2 Trajectories of a nonautonomous system.

In the first example, when we eliminated the parameter t we automatically eliminated s, since t and s appeared only in the combination $t - s$. In the second example this was not the case, and s remains in the final result (10). These examples illustrate the special property of autonomous systems, not shared by nonautonomous systems, that all particles passing through a given point follow the same curve in the phase plane. An equivalent, and the more usual, statement is the following. Suppose that $x = \phi(t)$, $y = \psi(t)$ is a solution of Eqs. (1) for $-\infty < t < \infty$. If s is any constant then $x = \phi(t - s)$, $y = \psi(t - s)$ is also a solution of Eqs. (1) for $-\infty < t < \infty$. This is readily verified by direct substitution. Thus the same curve (trajectory) in the phase plane is represented parametrically by many different solutions differing from one another by a translation of the parameter t. This is illustrated in Figure 9.1 where every solution of the system (5) passing through the point $(1, 2)$ lies on the straight line $y = 2x$.

The trajectories of two-dimensional autonomous systems can be found, as in Example 1, by solving the system and then eliminating the parameter t. In many cases, they can also be found by solving a related first order differential equation. From Eqs. (1) we have

$$\frac{dy}{dx} = \frac{dy/dt}{dx/dt} = \frac{G(x, y)}{F(x, y)}, \tag{11}$$

which is a first order equation in the variables x and y. If Eq. (11) can be solved by any of the methods of Chapter 2, then its solution provides an equation for the trajectories of the system (1).

For example, for the system

$$dx/dt = x, \qquad dy/dt = y$$

in Example 1, we have

$$dy/dx = y/x,$$

whose general solution is $y = cx$. If the initial point is $(1, 2)$, then we must choose $c = 2$, so $y = 2x$, with $x \geq 1$, is the equation of the trajectory passing through $(1, 2)$.

Autonomous systems occur frequently in practice. Physically, an autonomous system is one in which the parameters of the system are independent of time. In this chapter we draw examples mainly from the fields of ecology and mechanics. To begin to illustrate the importance of critical points we turn to a preliminary geometrical discussion of problems in each of these areas.

TWO COMPETING SPECIES. Suppose that we have two similar species competing for a limited food supply; for example, two species of fish in a pond that do not prey on each other, but do compete for the available food. Let x and y be the populations of the two species at time t. As discussed in Section 2.6, we assume that the population of each of the species, in the absence of the other, is governed by a logistic equation. Thus

$$dx/dt = x(\epsilon_1 - \sigma_1 x), \tag{12a}$$

$$dy/dt = y(\epsilon_2 - \sigma_2 y), \tag{12b}$$

respectively, where ϵ_1 and ϵ_2 are the growth rates of the two populations, and ϵ_1/σ_1 and ϵ_2/σ_2 are their saturation levels. However, when both species are present each will impinge on the available food supply for the other. In effect, they reduce the growth rates and saturation populations of each other. The simplest expression for reducing the growth rate of species x due to the presence of species y is to replace the growth rate factor $\epsilon_1 - \sigma_1 x$ of Eq. (12a) by $\epsilon_1 - \sigma_1 x - \alpha_1 y$ where α_1 is a measure of the degree to which species y interferes with species x. Similarly, in Eq. (12b) we replace $\epsilon_2 - \sigma_2 y$ by $\epsilon_2 - \sigma_2 y - \alpha_2 x$. Thus we have the system of equations

$$dx/dt = x(\epsilon_1 - \sigma_1 x - \alpha_1 y), \tag{13a}$$

$$dy/dt = y(\epsilon_2 - \sigma_2 y - \alpha_2 x). \tag{13b}$$

The values of the positive constants $\epsilon_1, \sigma_1, \alpha_1, \epsilon_2, \sigma_2, \alpha_2$ depend on the particular species under consideration. We are interested in solutions of Eqs. (13) for which x and y are nonnegative.

The nature of the solutions of Eqs. (13) depends on the relative magnitudes of the parameters in the equations, and we give a more complete discussion in Section 9.4. For the present, we simplify matters in the following example by assuming that the parameters in Eqs. (13) have certain numerical values.

EXAMPLE 3

Discuss the qualitative behavior of solutions of the system

$$dx/dt = x(1 - x - y), \tag{14a}$$

$$dy/dt = y(0.75 - y - 0.5x). \tag{14b}$$

We find the critical points by solving the system of algebraic equations

$$x(1 - x - y) = 0, \qquad y(0.75 - y - 0.5x) = 0. \tag{15}$$

There are four points that satisfy Eqs. (15), namely, $(0, 0)$, $(0, 0.75)$, $(1, 0)$, and $(0.5, 0.5)$; they correspond to equilibrium solutions of the system (14). As we have mentioned, other solutions correspond to curves, or trajectories, in the xy plane. It is significant that we can obtain considerable qualitative information about the trajectories, and hence about the solutions themselves, without actually solving the differential equations (14).

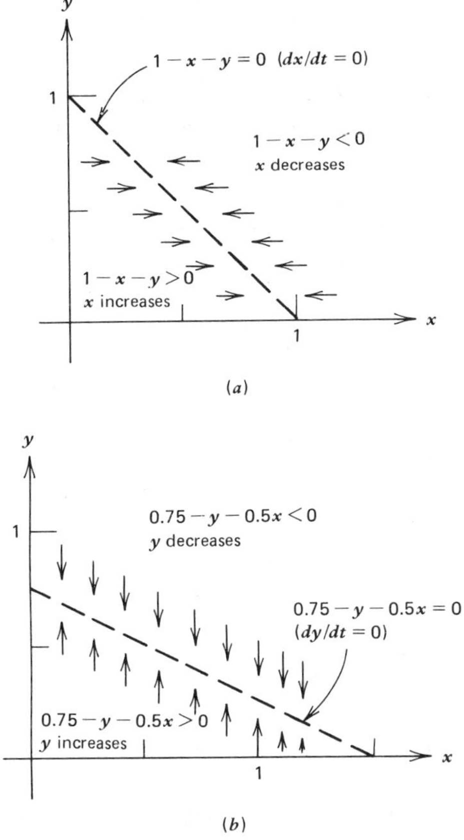

FIGURE 9.3 Directions of trajectories for the competing species system (14).

Since x is nonnegative, it follows from Eq. (14a) that x increases or decreases according to whether $1 - x - y > 0$ or $1 - x - y < 0$; similarly, from Eq. (14b) we see that y increases or decreases as $0.75 - y - 0.5x > 0$ or $0.75 - y - 0.5x < 0$. This situation is shown geometrically in Figure 9.3. To see what is happening to the two populations simultaneously we must superimpose the two diagrams of Figure 9.3. This is done in Figure 9.4, where the critical points are also indicated by the heavy dots.

The two intersecting straight lines create four zones in the first quadrant of the xy plane. In zone I we note that $x' > 0$ and $y' > 0$; therefore, whenever the point (x, y) is in zone I it must be moving up and to the right. Similarly, the point (x, y) moves up and to the left in zone II, down and to the left in zone III, and down and to the right in zone IV. Thus in all cases, the point (x, y) moves toward the critical point $(0.5, 0.5)$. Consequently, if $x_0 > 0$ and $y_0 > 0$, then, after a long time has passed, we should expect to see the point (x, y) close to the critical point

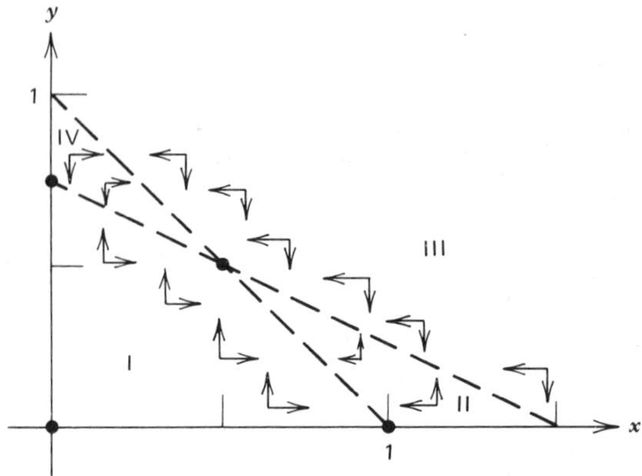

FIGURE 9.4 Critical points and directions of trajectories for the competing species system (14).

(0.5, 0.5), which represents the population levels of the two species that can coexist in equilibrium with each other and with the available food supply. This equilibrium configuration does not depend on the initial populations x_0 and y_0, so long as they are positive.

If $x_0 = 0$, then only the species y is present. In this case, the point (x, y) remains on the y axis and approaches the critical point (0, 0.75), which represents the saturation level of the species y. The situation is similar if $y_0 = 0$.

More can be said about the qualitative behavior of solutions of Eqs. (14), or of the more general system (13). We continue the discussion in Section 9.4.

OSCILLATING PENDULUM. An undamped pendulum of length l is governed by the differential equation

$$\frac{d^2\theta}{dt^2} + \frac{g}{l}\sin\theta = 0. \tag{16}$$

Letting $x = \theta$ and $y = d\theta/dt$, we can rewrite Eq. (16) as an autonomous system of two equations:

$$dx/dt = y, \qquad dy/dt = -(g/l)\sin x. \tag{17}$$

For small x (or θ) the system (17) can be simplified by replacing $\sin x$ by x; this yields the linear system

$$dx/dt = y, \qquad dy/dt = -(g/l)x. \tag{18}$$

Since Eqs. (18) can be easily solved, as in Chapter 7, an important question is whether the solutions of Eqs. (18) are an adequate approximation to the solutions of Eqs. (17). We return to this question in Section 9.3.

EXAMPLE 4

Find the trajectories of Eqs. (18), the linearized pendulum equations.

We can find the trajectories by writing Eq. (11) for this system. However, we will proceed in a different way. Multiply the first and second of Eqs. (18) by $2x$ and $2y/\omega^2$, respectively, where $\omega^2 = g/l$. Thus

$$2xx' = 2xy, \qquad 2yy'/\omega^2 = -2xy.$$

By adding these two equations, we obtain

$$2xx' + 2yy'/\omega^2 = 0,$$

or

$$(d/dt)(x^2 + y^2/\omega^2) = 0. \tag{19}$$

Integrating Eq. (19) yields

$$x^2 + y^2/\omega^2 = c^2, \tag{20}$$

where c^2 is an arbitrary constant. Equation (20) describes the trajectories of the system (18); they are a family of ellipses centered at the origin, as shown in Figure 9.5. Note that the origin is the only critical point of Eqs. (18). In Example 3 the trajectories approached or receded from the critical points; however, in this problem they are closed curves that surround the critical point of the system.

The direction of motion on the trajectories can be determined directly from Eqs. (18). For instance, for points in the first quadrant ($x > 0,\ y > 0$), Eqs. (18) imply that $dx/dt > 0$ and $dy/dt < 0$; therefore the motion is clockwise.

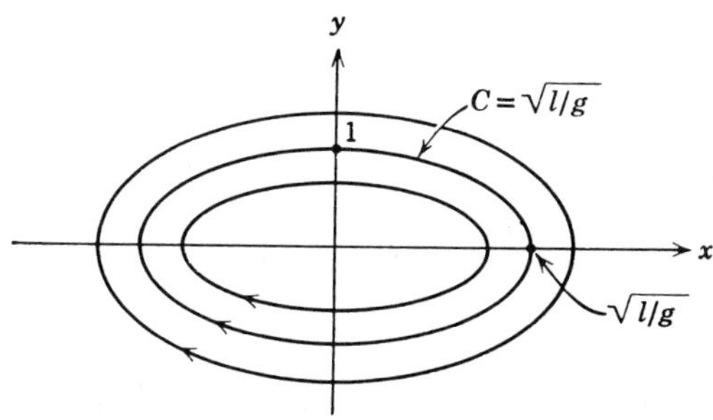

FIGURE 9.5 Trajectories of the linearized undamped pendulum equations.

Equations (18) can also be readily solved by the methods of Section 7.6, which lead to the general solution

$$\begin{pmatrix} x \\ y \end{pmatrix} = c_1 \begin{pmatrix} \cos \omega t \\ -\omega \sin \omega t \end{pmatrix} + c_2 \begin{pmatrix} \sin \omega t \\ \omega \cos \omega t \end{pmatrix}. \tag{21}$$

By eliminating the parameter t from Eqs. (21) we can again obtain Eq. (20) for the trajectories.

Example 4 illustrates the possible occurrence of periodic solutions of an autonomous system (1). In this example it is clear from Eqs. (21) that every solution of the linearized pendulum equations (18) is periodic with period $2\pi/\sqrt{g/l}$. More generally, a solution $x = \phi(t)$, $y = \psi(t)$ of Eqs. (1) is said to be *periodic* with period T if it exists for all t, and if $\phi(t + T) = \phi(t)$, $\psi(t + T) = \psi(t)$ for all t. The trajectory of a periodic solution is always a closed curve, as illustrated by the elliptical trajectories in Figure 9.5.

We have already seen the important role that critical points played in Examples 3 and 4. The nature of a critical point of an autonomous system can be further illustrated by an appeal to mechanics. The equation

$$m\frac{d^2x}{dt^2} = f\left(x, \frac{dx}{dt}\right) \tag{22}$$

can be interpreted as describing the motion of a particle of mass m acted on by a force that depends on the position x and velocity dx/dt. If we let $y = dx/dt$, then Eq. (22) can be rewritten as

$$\frac{dx}{dt} = y, \qquad \frac{dy}{dt} = \frac{1}{m}f(x, y). \tag{23}$$

At a critical point of this system, the velocity $y = dx/dt$ and acceleration dy/dt vanish simultaneously. Thus, when the mass is at a critical point it is at rest (zero velocity) with zero force acting on it.

A question of interest is whether a particle slightly displaced from an equilibrium point

(a) moves back toward the equilibrium point;
(b) moves away from the equilibrium point; or
(c) moves about in the neighborhood of the equilibrium point, but does not approach it.

In case (a) we say that the equilibrium (critical) point is asymptotically stable, in case (b) it is unstable, and in case (c) it is stable but not asymptotically stable.

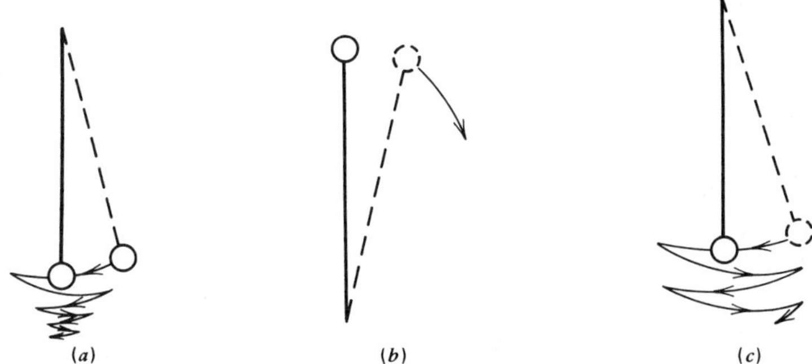

FIGURE 9.6 Qualitative motion of a pendulum: (a) with air resistance; (b) with or without air resistance; (c) without air resistance.

These possibilities are illustrated by the motion of a pendulum bob after it is slightly displaced from (a) a vertical downward position with air resistance, (b) a vertical upward position with or without air resistance, and (c) a vertical downward position without air resistance; see Figure 9.6.

Indeed, in Example 4 we have shown, for the linearized pendulum without air resistance, that the origin is a stable critical point. Further, for the problem of competing species in Example 3, our analysis suggests that the critical point $(\frac{1}{2}, \frac{1}{2})$ is asymptotically stable, and that the other critical points are unstable. In these specific problems the intuitive understanding of stability is clear, but we postpone precise definitions until Section 9.3.

The significance of critical points in the analysis of the autonomous system (1) is due to the fact that they are constant solutions of the system, and that the qualitative behavior of all trajectories in the phase plane is determined to a very large extent by the location of the critical points and the behavior of trajectories near them.

The following statements can be made about the trajectories of autonomous systems. The proofs follow from the existence and uniqueness theorem and are outlined in the problems.

1. Through any point (x_0, y_0) in the phase plane there is at most one trajectory of the system (1) (see Problem 10).
2. A particle starting at a point that is not a critical point cannot reach a critical point in a finite time. Thus if (x_0, y_0) is a critical point, and a trajectory corresponding to a solution approaches (x_0, y_0), then necessarily $t \to \pm\infty$ (see Problem 11).
3. A trajectory passing through at least one point that is not a critical point cannot cross itself unless it is a closed curve. In this case the trajectory corresponds to a periodic solution of the system (1) (see Problem 12).

The substance of these three results is the following. If a particle (solution) starts at a point that is not a critical point, then it moves on the same trajectory no matter at what time it starts, it can never come back to its initial point unless the motion is periodic, it can never cross another trajectory, and it can only "reach" a critical point in the limit as $t \to \pm\infty$. This suggests that such a particle (solution) either approaches a critical point (x_0, y_0), moves on a closed trajectory or approaches a closed trajectory as $t \to \pm\infty$, or else goes off to infinity. Thus for autonomous systems a study of the critical points and periodic solutions is of fundamental importance. These facts will be developed in greater detail in the next two sections.

In conclusion, we note that many results for autonomous systems can be generalized to nonautonomous systems though the analysis becomes more complicated. Formally, a nonautonomous system of n first order equations for $x_1(t), x_2(t), \ldots, x_n(t)$ can be written as an autonomous system of $n + 1$ equations by letting $t = x_{n+1}$ wherever t appears explicitly and by appending the equation $dx_{n+1}/dt = 1$. For $n = 2$, this would require an analysis of the trajectories in three-dimensional space, which is considerably more difficult than in two dimensions.

PROBLEMS

1. Sketch the trajectory corresponding to the solution satisfying the specified initial conditions, and indicate the direction of motion for increasing t.

   (a) $dx/dt = -x, \quad dy/dy = -2y; \quad x(0) = 4, \quad y(0) = 2$
   (b) $dx/dt = -x, \quad dy/dt = 2y;$
   $x(0) = 4, \quad y(0) = 2$ and $x(0) = 4, \quad y(0) = 0$
   (c) $dx/dt = -y, \quad dy/dt = x;$
   $x(0) = 4, \quad y(0) = 0$ and $x(0) = 0, \quad y(0) = 4$
   (d) $dx/dt = ay, \quad dy/dt = -bx, \quad a > 0,\ b > 0; \quad x(0) = \sqrt{a}, \quad y(0) = 0$

2. Determine the critical points for each of the following systems.

   (a) $dx/dt = 2x - 3y, \quad dy/dt = 2x - 2y$
   (b) $dx/dt = x - xy, \quad dy/dt = y + 2xy$
   (c) $dx/dt = x - x^2 - xy, \quad dy/dt = \frac{1}{2}y - \frac{1}{4}y^2 - \frac{3}{4}xy$
   (d) $dx/dt = y, \quad dy/dt = -(g/l)x - (c/ml)y; \quad g, l, c, m > 0$
   (e) $dx/dt = y, \quad dy/dt = -(g/l)\sin x - (c/ml)y; \quad g, l, c, m > 0$
   (f) The van der Pol equation: $dx/dt = y, \quad dy/dt = \mu(1 - x^2)y - x, \quad \mu > 0$

3. Sketch the trajectories of the following systems either by solving the system and eliminating t to obtain a one-parameter family of curves, or by integrating the differential equation [Eq. (11)], which gives the slope of the tangent of a trajectory. Indicate the direction of motion with increasing t.

   (a) $dx/dt = -x, \quad dy/dt = -2y$ (b) $dx/dt = -x, \quad dy/dt = 2y$
   (c) $dx/dt = x, \quad dy/dt = -2y$ (d) $dx/dt = -x, \quad dy/dt = 3x + 2y$

4. Given that $x = \phi(t)$, $y = \psi(t)$ is a solution of the autonomous system

$$dx/dt = F(x, y), \qquad dy/dt = G(x, y)$$

for $\alpha < t < \beta$, show that $x = \Phi(t) = \phi(t - s)$, $y = \Psi(t) = \psi(t - s)$ is a solution for $\alpha + s < t < \beta + s$.

5. Show by direct integration that even though the right side of the system

$$\frac{dx}{dt} = \frac{x}{1+t}, \qquad \frac{dy}{dt} = \frac{y}{1+t}$$

depends on t, the paths followed by particles emitted at (x_0, y_0) at $t = s$ are the same regardless of the value of s. Why is this so?
Hint: Consider Eq. (11) for this case.

6. Consider the system

$$dx/dt = F(x, y, t), \qquad dy/dt = G(x, y, t).$$

Show that if the functions $F/(F^2 + G^2)^{1/2}$ and $G/(F^2 + G^2)^{1/2}$ are independent of t, then the solutions corresponding to the initial conditions $x(s) = x_0$, $y(s) = y_0$ give the same trajectories regardless of the value of s.
Hint: Use Eq. (11).

7. Show that the trajectories of the nonlinear undamped pendulum equation $d^2\theta/dt^2 + (g/l)\sin\theta = 0$ are given by

$$(g/l)(1 - \cos x) + (y^2/2) = c,$$

where $x = \theta$ and $y = d\theta/dt$.

8. Find an equation for the trajectories of

$$d^2\theta/dt^2 + \theta - \alpha\theta^3 = 0,$$

where α is a constant. Let $x = \theta$ and $y = d\theta/dt$.

9. Show that the trajectories of

$$d^2\theta/dt^2 + f(\theta) = 0$$

are given by

$$F(x) + (y^2/2) = c,$$

where F is an antiderivative of f, $x = \theta$, and $y = d\theta/dt$.

10. Prove that for the system

$$dx/dt = F(x, y), \qquad dy/dt = G(x, y)$$

there is at most one trajectory passing through a given point (x_0, y_0).
Hint: Let C_0 be the trajectory generated by the solution $x = \phi_0(t)$, $y = \psi_0(t)$, with $\phi_0(t_0) = x_0$, $\psi_0(t_0) = y_0$, and let C_1 be the trajectory generated by the solution $x = \phi_1(t)$, $y = \psi_1(t)$, with $\phi_1(t_1) = x_0$, $\psi_1(t_1) = y_0$. Use the fact that the system is autonomous and the existence and uniqueness theorem to show that C_0 and C_1 are the same.

11. Prove that if a trajectory starts at a noncritical point of the system

$$dx/dt = F(x, y), \qquad dy/dt = G(x, y)$$

then it cannot reach a critical point (x_0, y_0) in a finite length of time.
Hint: Assume the contrary; that is, assume that the solution $x = \phi(t)$, $y = \psi(t)$

satisfies $\phi(a) = x_0$, $\psi(a) = y_0$. Then use the fact that $x = x_0$, $y = y_0$ is a solution of the given system satisfying the initial condition $x = x_0$, $y = y_0$, at $t = a$.

12. Assuming that the trajectory corresponding to a solution $x = \phi(t)$, $y = \psi(t)$, $-\infty < t < \infty$, of an autonomous system is closed, show that the solution is periodic.
Hint: Since the trajectory is closed there exists at least one point (x_0, y_0) such that $\phi(t_0) = x_0$, $\psi(t_0) = y_0$ and a number $T > 0$ such that $\phi(t_0 + T) = x_0$, $\psi(t_0 + T) = y_0$. Show that $x = \Phi(t) = \phi(t + T)$ and $y = \Psi(t) = \psi(t + T)$ is a solution and then use the existence and uniqueness theorem to show that $\Phi(t) = \phi(t)$ and $\Psi(t) = \psi(t)$ for all t.[2]

9.2 The Phase Plane: Linear Systems

In this section and the following one we discuss the behavior of the trajectories of the two-dimensional autonomous system

$$d\mathbf{x}/dt = \mathbf{f}(\mathbf{x}) \tag{1}$$

in the neighborhood of a critical point $\mathbf{x}^0$. First (in this section) we consider the linear system

$$d\mathbf{x}/dt = \mathbf{A}\mathbf{x}, \tag{2}$$

where $\mathbf{A}$ is a 2×2 constant matrix. Later (in Section 9.3) we relate the results for linear systems to an important class of nonlinear autonomous systems.

We assume that the coefficient matrix $\mathbf{A}$ in Eq. (2) is nonsingular, or that $\det \mathbf{A} \neq 0$. It follows that $\mathbf{x} = \mathbf{0}$ is the only solution of the equation $\mathbf{A}\mathbf{x} = \mathbf{0}$, and consequently $\mathbf{x} = \mathbf{0}$ is the only critical point of the system (2). On the other hand, if $\mathbf{A}$ were singular, then the system (2) would have infinitely many critical points, lying on a straight line through the origin.

Linear systems of the form (2) were solved in Sections 7.5 through 7.7. Recall that if we seek solutions of the form $\mathbf{x} = \boldsymbol{\xi} e^{rt}$, then by substituting for $\mathbf{x}$ in Eq. (2) we find that

$$(\mathbf{A} - r\mathbf{I})\boldsymbol{\xi} = \mathbf{0}. \tag{3}$$

Thus r must be an eigenvalue and $\boldsymbol{\xi}$ a corresponding eigenvector of the coefficient matrix $\mathbf{A}$. The eigenvalues are the roots of the polynomial equation

$$\det(\mathbf{A} - r\mathbf{I}) = 0, \tag{4}$$

[2] If it is originally known only that the functions ϕ and ψ are defined on an interval containing t_0 and $t_0 + T$, the proof is more difficult. However, it can be shown that the interval of definition must be $-\infty < t < \infty$ and that the solution is periodic with period T.

and the eigenvectors are determined from Eq. (3) up to an arbitrary multiplicative constant.

Observe that $r = 0$ is not an eigenvalue of the matrix $\mathbf{A}$. If it were, then it would follow at once from Eq. (4) that $\det \mathbf{A} = 0$, whereas we have assumed that $\det \mathbf{A} \neq 0$.

A number of different cases must be considered, depending on the nature of the eigenvalues of $\mathbf{A}$. These cases also occurred in Sections 7.5 through 7.7, where we were primarily interested in finding a convenient formula for the general solution. Now our main interest is in classifying the critical point $\mathbf{x} = \mathbf{0}$ for the system (2) on the basis of the pattern of trajectories that it generates. In each case we discuss the behavior of the trajectories in general and illustrate it with an example. It is important that the student become familiar with the type of behavior that the trajectories have for each case because these are the basic building blocks of the qualitative theory of differential equations.

CASE 1. REAL UNEQUAL EIGENVALUES OF THE SAME SIGN. The general solution of Eq. (2) is

$$\mathbf{x} = c_1\boldsymbol{\xi}^{(1)}e^{r_1t} + c_2\boldsymbol{\xi}^{(2)}e^{r_2t}, \tag{5}$$

where r_1 and r_2 are either both positive or both negative. Suppose first that $r_1 < r_2 < 0$, and that the eigenvectors $\boldsymbol{\xi}^{(1)}$ and $\boldsymbol{\xi}^{(2)}$ are as shown in Figure 9.7. It is clear from Eq. (5) that $\mathbf{x} \to \mathbf{0}$ as $t \to \infty$ regardless of the values of c_1 and c_2; in other words, all solutions approach the critical point at the origin as $t \to \infty$. If the solution starts at an initial point on the line through $\boldsymbol{\xi}^{(1)}$, then it follows that

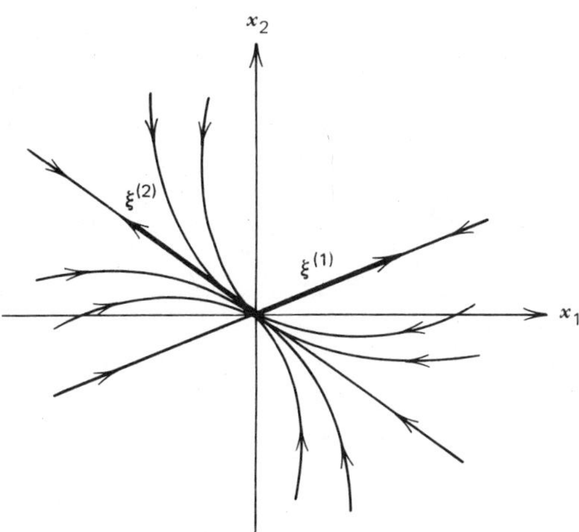

FIGURE 9.7 An improper node; $r_1 < r_2 < 0$.

$c_2 = 0$. Consequently, the solution remains on the line through $\boldsymbol{\xi}^{(1)}$ for all t, and approaches the origin as $t \to \infty$. Similarly, if the initial point is on the line through $\boldsymbol{\xi}^{(2)}$, then the solution approaches the origin along that line. In the general situation, it is helpful to rewrite Eq. (5) in the form

$$\mathbf{x} = e^{r_2 t}\left[c_1\boldsymbol{\xi}^{(1)}e^{(r_1 - r_2)t} + c_2\boldsymbol{\xi}^{(2)}\right]. \tag{6}$$

As long as $c_2 \neq 0$, the term $c_1\boldsymbol{\xi}^{(1)}\exp[(r_1 - r_2)t]$ is negligible compared to $c_2\boldsymbol{\xi}^{(2)}$ for t sufficiently large. Thus, as $t \to \infty$, the trajectory not only approaches the origin, it also tends toward the line through $\boldsymbol{\xi}^{(2)}$. Hence all solutions approach the critical point tangent to $\boldsymbol{\xi}^{(2)}$ except for those solutions that start exactly on the line through $\boldsymbol{\xi}^{(1)}$. Several trajectories are sketched in Figure 9.7. This type of critical point is called a *node*, or sometimes an *improper node*, to distinguish it from another type of node to be mentioned later.

If r_1 and r_2 are both positive, and $0 < r_1 < r_2$, then the trajectories have the same pattern as in Figure 9.7, but the direction of motion is away from, rather than toward, the critical point at the origin.

An example of an improper node occurs in Example 2 of Section 7.5, and its trajectories are shown in Figure 7.7.

CASE 2. REAL EIGENVALUES OF OPPOSITE SIGN. The general solution of Eq. (2) is

$$\mathbf{x} = c_1\boldsymbol{\xi}^{(1)}e^{r_1 t} + c_2\boldsymbol{\xi}^{(2)}e^{r_2 t}, \tag{7}$$

where $r_1 > 0$ and $r_2 < 0$. Suppose that the eigenvectors $\boldsymbol{\xi}^{(1)}$ and $\boldsymbol{\xi}^{(2)}$ are as shown in Figure 9.8. If the solution starts at an initial point on the line through $\boldsymbol{\xi}^{(1)}$, then

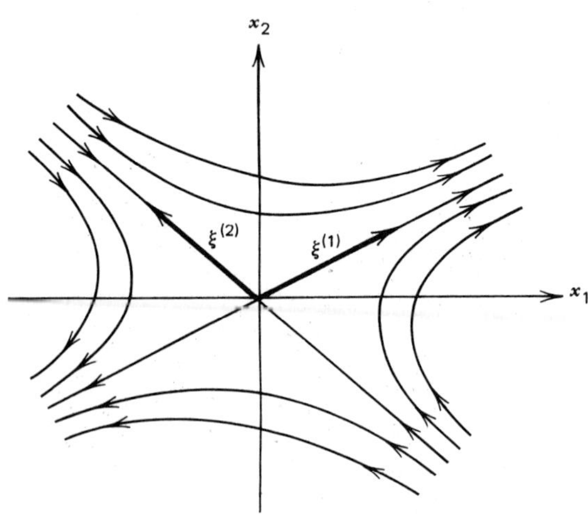

FIGURE 9.8 A saddle point; $r_1 > 0$, $r_2 < 0$.

it follows that $c_2 = 0$. Consequently, the solution remains on the line through $\boldsymbol{\xi}^{(1)}$ for all t, and since $r_1 > 0$, $\|\mathbf{x}\| \to \infty$ as $t \to \infty$. If the solution starts at an initial point on the line through $\boldsymbol{\xi}^{(2)}$, then the situation is similar except that $\|\mathbf{x}\| \to 0$ as $t \to \infty$ because $r_2 < 0$. Solutions starting at other initial points follow trajectories such as those shown in Figure 9.8. The positive exponential is the dominant term in Eq. (7) for large t, so eventually all these solutions approach infinity asymptotic to the line determined by the eigenvector $\boldsymbol{\xi}^{(1)}$ corresponding to the positive eigenvalue r_1. The only solutions that approach the critical point at the origin are those that start precisely on the line determined by $\boldsymbol{\xi}^{(2)}$. The origin is called a *saddle point*.

A specific example of a saddle point is in Example 1 of Section 7.5 whose trajectories are shown in Figure 7.6.

CASE 3. EQUAL EIGENVALUES. We now suppose that $r_1 = r_2 = r$. We consider the case in which the eigenvalues are negative; if they are positive, the trajectories are similar but the direction of motion is reversed. There are two subcases, depending on whether the repeated eigenvalue has two independent eigenvectors or only one.

(a) Two independent eigenvectors. The general solution of Eq. (2) is

$$\mathbf{x} = c_1\boldsymbol{\xi}^{(1)}e^{rt} + c_2\boldsymbol{\xi}^{(2)}e^{rt}, \tag{8}$$

where $\boldsymbol{\xi}^{(1)}$ and $\boldsymbol{\xi}^{(2)}$ are the independent eigenvectors. The ratio x_2/x_1 is independent of t, but depends on the components of $\boldsymbol{\xi}^{(1)}$ and $\boldsymbol{\xi}^{(2)}$, and on the arbitrary constants c_1 and c_2. Thus every trajectory lies on a straight line through the origin, as shown in Figure 9.9. The critical point is called a *proper node*.

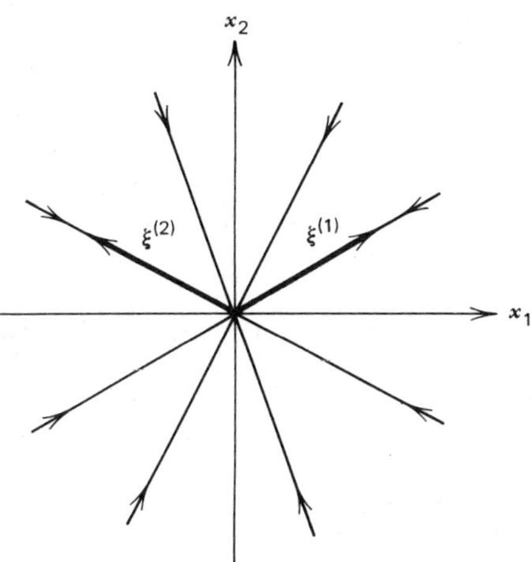

FIGURE 9.9 A proper node, two independent eigenvectors; $r_1 = r_2 < 0$.

(b) One independent eigenvector. As shown in Section 7.7, the general solution of Eq. (2) in this case is

$$\mathbf{x} = c_1\boldsymbol{\xi}e^{rt} + c_2(\boldsymbol{\xi}te^{rt} + \boldsymbol{\eta}e^{rt}), \tag{9}$$

where $\boldsymbol{\xi}$ is the eigenvector and $\boldsymbol{\eta}$ is the generalized eigenvector associated with the repeated eigenvalue. For large t the dominant term in Eq. (9) is $c_2\boldsymbol{\xi}te^{rt}$. Thus, as $t \to \infty$, every trajectory approaches the origin tangent to the line through the eigenvector. This is true even if $c_2 = 0$, for then the solution $\mathbf{x} = c_1\boldsymbol{\xi}e^{rt}$ lies on this line. Similarly, for large negative t the term $c_2\boldsymbol{\xi}te^{rt}$ is again the dominant one, so as $t \to -\infty$, each trajectory is asymptotic to a line parallel to $\boldsymbol{\xi}$.

The orientation of the trajectories depends on the relative positions of $\boldsymbol{\xi}$ and $\boldsymbol{\eta}$. One possible situation is shown in Figure 9.10*a*. To locate the trajectories it is helpful to write the solution (9) in the form

$$\mathbf{x} = [(c_1\boldsymbol{\xi} + c_2\boldsymbol{\eta}) + c_2\boldsymbol{\xi}t]e^{rt} = \mathbf{y}e^{rt}, \tag{10}$$

where $\mathbf{y} = (c_1\boldsymbol{\xi} + c_2\boldsymbol{\eta}) + c_2\boldsymbol{\xi}t$. Observe that the vector $\mathbf{y}$ determines the direction of $\mathbf{x}$ whereas the scalar quantity e^{rt} affects only the magnitude of $\mathbf{x}$. Also note that, for fixed values of c_1 and c_2, the expression for $\mathbf{y}$ is a vector equation of the straight line through the point $c_1\boldsymbol{\xi} + c_2\boldsymbol{\eta}$ and parallel to $\boldsymbol{\xi}$.

To sketch the trajectory corresponding to a given pair of values of c_1 and c_2, one can proceed in the following way. First, draw the line given by $(c_1\boldsymbol{\xi} + c_2\boldsymbol{\eta}) + c_2\boldsymbol{\xi}t$ and note the direction of increasing t on this line. Two such lines are shown in Figure 9.10*a*, one for $c_2 > 0$ and the other for $c_2 < 0$. Next, note that the given trajectory passes through the point $c_1\boldsymbol{\xi} + c_2\boldsymbol{\eta}$ when $t = 0$. Further, as t increases, the direction of the vector $\mathbf{x}$ given by Eq. (10) follows the direction of increasing t on the line, but the magnitude of $\mathbf{x}$ rapidly decreases and approaches zero because of the decaying exponential factor e^{rt}. Finally, as t decreases toward $-\infty$ the direction of $\mathbf{x}$ is determined by points on the corresponding part of the line and the magnitude of $\mathbf{x}$ approaches infinity. In this way we obtain the heavy trajectories in Figure 9.10*a*. A few other trajectories are lightly sketched as well to help complete the diagram.

The other possible situation is shown in Figure 9.10*b*, where the relative orientation of $\boldsymbol{\xi}$ and $\boldsymbol{\eta}$ is reversed. As indicated in the figure, this results in a reversal in the orientation of the trajectories.

If $r_1 = r_2 > 0$, one can sketch the trajectories by following the same procedure. In this event the trajectories are traversed in the outward direction, and the orientation of the trajectories with respect to that of $\boldsymbol{\xi}$ and $\boldsymbol{\eta}$ is also reversed.

When a double eigenvalue has only a single independent eigenvector, the critical point is again called an *improper node*. A specific example of this case is Example 1 in Section 7.7; the trajectories are shown in Figure 7.12.

CASE 4. COMPLEX EIGENVALUES. Suppose that the eigenvalues are $\lambda \pm i\mu$, where λ and μ are real, and $\mu > 0$. It is possible to write down the general

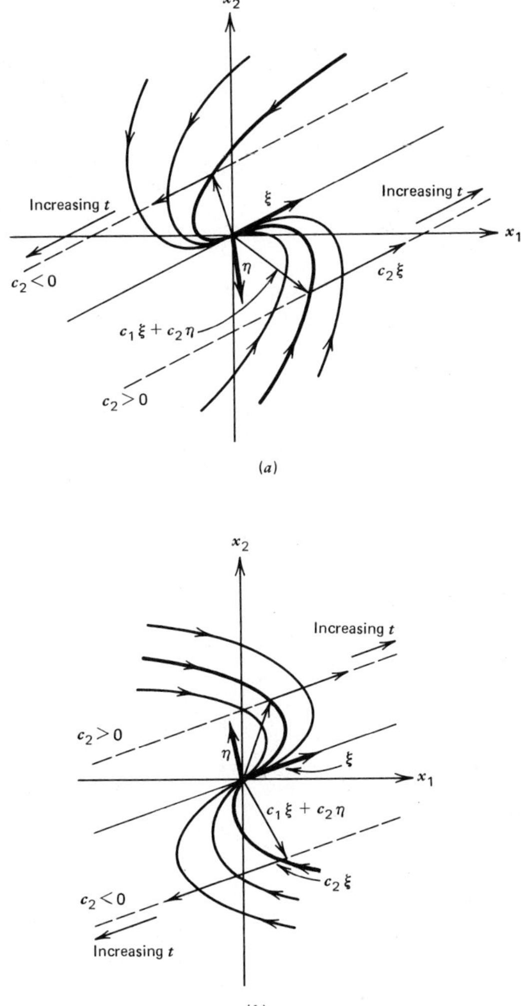

FIGURE 9.10 An improper node, one independent eigenvector; $r_1 = r_2 < 0$.

solution in terms of the eigenvalues and eigenvectors, as shown in Section 7.6. However, we proceed in a different way.

Systems having the eigenvalues $\lambda \pm i\mu$ are typified by

$$\mathbf{x}' = \begin{pmatrix} \lambda & \mu \\ -\mu & \lambda \end{pmatrix} \mathbf{x} \tag{11}$$

or, in scalar form,

$$x_1' = \lambda x_1 + \mu x_2, \qquad x_2' = -\mu x_1 + \lambda x_2. \tag{12}$$

We introduce the polar coordinates r, θ given by

$$r^2 = x_1^2 + x_2^2, \qquad \tan\theta = x_2/x_1.$$

By differentiating these equations we obtain

$$rr' = x_1x_1' + x_2x_2', \qquad (\sec^2\theta)\theta' = (x_1x_2' - x_2x_1')/x_1^2. \tag{13}$$

Substituting from Eqs. (12) in the first of Eqs. (13), we find that

$$r' = \lambda r, \tag{14}$$

and hence

$$r = ce^{\lambda t}, \tag{15}$$

where c is a constant. Similarly, substituting from Eqs. (12) in the second of Eqs.

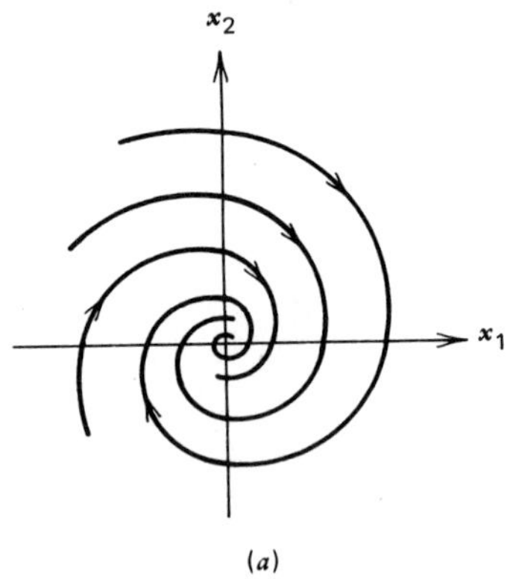

(a)

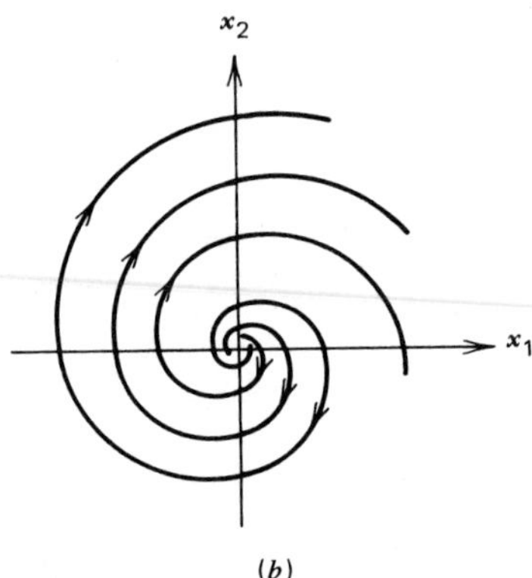

(b)

FIGURE 9.11 A spiral point; $r_1 = \lambda + i\mu$, $r_2 = \lambda - i\mu$. (a) $\lambda < 0$; (b) $\lambda > 0$.

(13), and using the fact that $\sec^2\theta = r^2/x_1^2$, we have

$$\theta' = -\mu. \tag{16}$$

Hence

$$\theta = -\mu t + \theta_0, \tag{17}$$

where θ_0 is the value of θ when $t = 0$.

Equations (15) and (17) are parametric equations in polar coordinates of the trajectories of the system (11). Since $\mu > 0$, it follows from Eq. (17) that θ decreases as t increases, so the direction of motion on a trajectory is clockwise. As $t \to \infty$ we see from Eq. (15) that $r \to 0$ if $\lambda < 0$ and $r \to \infty$ if $\lambda > 0$. Thus the trajectories are spirals, which approach or recede from the origin depending on the sign of λ. Both possibilities are shown in Figure 9.11. The critical point is called a *spiral point* in this case.

More generally, it is possible to show that for any system with complex eigenvalues $\lambda \pm i\mu$, the trajectories are always spirals. They are directed inward or outward, respectively, depending on whether λ is negative or positive. They may be elongated and skewed with respect to the coordinate axes, and the direction of motion may be either clockwise or counterclockwise. While a detailed analysis is moderately difficult, it is easy to obtain a general idea of the orientation of the trajectories directly from the differential equations. Consider the system

$$\begin{pmatrix} dx/dt \\ dy/dt \end{pmatrix} = \begin{pmatrix} a & b \\ c & d \end{pmatrix}\begin{pmatrix} x \\ y \end{pmatrix} \tag{18}$$

and look at the point $(0, 1)$ on the positive y axis. At this point it follows from Eqs. (18) that $dx/dt = b$ and $dy/dt = d$. Depending on the signs of b and d, one can infer the direction of motion and the approximate orientation of the trajecto-

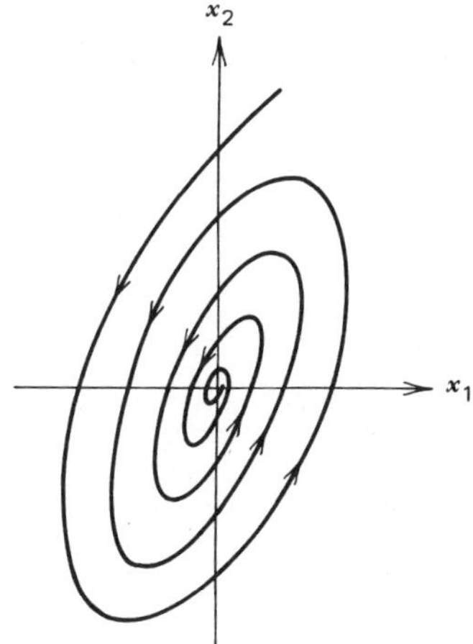

FIGURE 9.12 A spiral point; $r = \lambda \pm i\mu$ with $\lambda < 0$.

ries. For instance, if b and d are both negative, then the trajectories cross the positive y axis so as to move down and into the second quadrant. If $\lambda < 0$ also, then the trajectories must be inward-pointing spirals resembling the one in Figure 9.12. Another case was given in Example 1 of Section 7.6, whose trajectories are shown in Figure 7.8.

CASE 5. PURE IMAGINARY EIGENVALUES. This is actually a subcase of Case 4, but it is usually considered separately. If $\lambda = 0$, then the system (11) reduces to

$$\mathbf{x}' = \begin{pmatrix} 0 & \mu \\ -\mu & 0 \end{pmatrix} \mathbf{x} \tag{19}$$

with eigenvalues $\pm i\mu$. Using the same argument as in Case 4, we find that

$$r' = 0, \qquad \theta' = -\mu, \tag{20}$$

and consequently

$$r = c, \qquad \theta = -\mu t + \theta_0, \tag{21}$$

where c and θ_0 are constants. Thus the trajectories are circles, with center at the origin, that are traversed clockwise if $\mu > 0$ and counterclockwise if $\mu < 0$. A complete circuit about the origin is made in a time interval of length 2π, so all solutions are periodic with period 2π. The critical point is called a *center*.

In general, when the eigenvalues are pure imaginary, it is possible to show (see Problem 21) that the trajectories are ellipses centered at the origin. A typical situation is shown in Figure 9.13. The linearized undamped pendulum equations in Example 4 of Section 9.1 also belong to this case; their trajectories are shown in Figure 9.5.

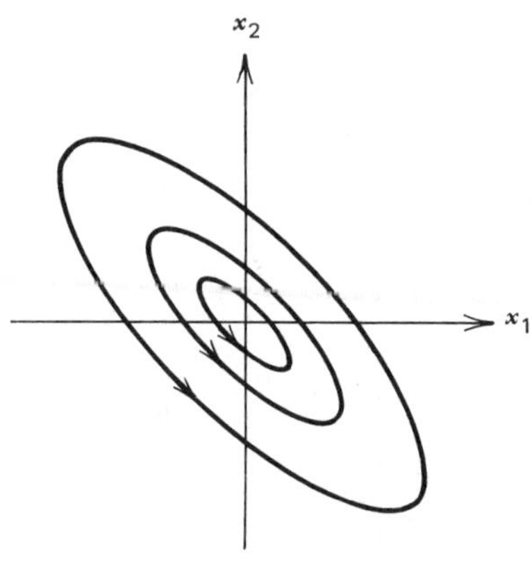

FIGURE 9.13 A center; $r_1 = i\mu$, $r_2 = -i\mu$.

TABLE 9.1 STABILITY PROPERTIES OF LINEAR SYSTEMS

$\mathbf{x}' = \mathbf{A}\mathbf{x}$	$\det(\mathbf{A} - r\mathbf{I}) = 0$ $\det \mathbf{A} \neq 0$	
Eigenvalues	Type of critical point	Stability
$r_1 > r_2 > 0$	Improper node	Unstable
$r_1 < r_2 < 0$	Improper node	Asymptotically stable
$r_2 < 0 < r_1$	Saddle point	Unstable
$r_1 = r_2 > 0$	Proper or improper node	Unstable
$r_1 = r_2 < 0$	Proper or improper node	Asymptotically stable
$r_1, r_2 = \lambda \pm i\mu$	Spiral point	
$\lambda > 0$		Unstable
$\lambda < 0$		Asymptotically stable
$r_1 = i\mu,\ r_2 = -i\mu$	Center	Stable

In each of the cases we have discussed, the trajectories of the linear system (2) exhibited one of the following three types of behavior.

1. All of the trajectories approach the critical point as $t \to \infty$. This is the case if the eigenvalues are real and negative or complex with negative real part.
2. The trajectories neither approach the critical point nor tend to infinity as $t \to \infty$. This is the case if the eigenvalues are pure imaginary.
3. At least one (possibly all) of the trajectories tends to infinity as $t \to \infty$. This is the case if at least one of the eigenvalues is positive or if the eigenvalues have positive real part.

These three possibilities illustrate the concepts of asymptotic stability, stability, and instability, respectively, of the critical point at the origin of the system (2). The precise definitions of these terms are given in Section 9.3, but their basic meaning should be clear from the geometrical discussion above and the discussion following Eq. (23) of Section 9.1. Our analysis of the trajectories of the linear system (2) is summarized in Table 9.1. Also see Problems 22 and 23 and Figure 9.17, which accompanies those problems.

It is clear from our discussion and Table 9.1 that the eigenvalues r_1, r_2 of the coefficient matrix $\mathbf{A}$ determine the type of critical point and its stability characteristics. In turn, the values of r_1, r_2 depend on the coefficients in the system of differential equations (2). When such a system (2) arises in some applied field, the coefficients usually result from the measurements of certain physical quantities. Such measurements are often subject to small uncertainties, so it is of interest to investigate whether small changes (perturbations) in the coefficients can affect the stability or instability of a critical point and/or significantly alter the pattern of trajectories.

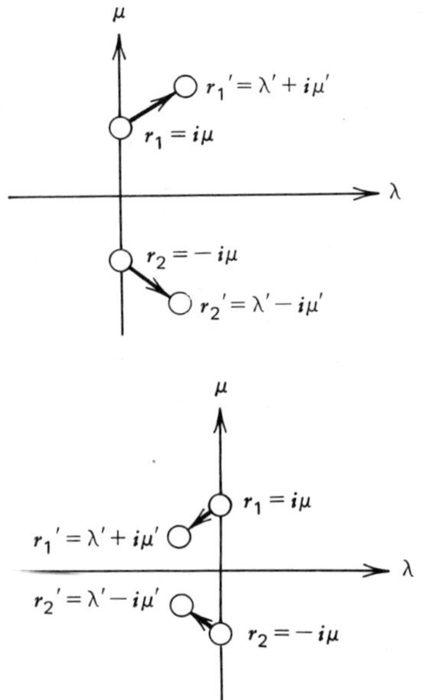

FIGURE 9.14 Schematic perturbation of $r_1 = i\mu$, $r_2 = -i\mu$.

It is possible to show from Eq. (4) that *small* perturbations in some or all of the coefficients are reflected in *small* perturbations in the eigenvalues r_1 and r_2. The most sensitive situation occurs for $r_1 = i\mu$ and $r_2 = -i\mu$; that is, when the critical point is a center and the trajectories are closed curves. If a slight change is made in the coefficients, then the eigenvalues r_1 and r_2 will take on a new values, $r_1' = \lambda' + i\mu'$ and $r_2' = \lambda' - i\mu'$, where λ' is small in magnitude and $\mu' \cong \mu$ (see Figure 9.14). If $\lambda' \neq 0$, which almost always occurs, then the trajectories of the perturbed system are spirals, rather than closed curves. The system is asymptotically stable if $\lambda' < 0$, but unstable if $\lambda' > 0$. Thus in the case of a center, small perturbations in the coefficients may well change a stable system into an unstable one, and in any case may be expected to alter radically the pattern of trajectories in the phase plane (see Problem 19).

Another slightly less sensitive case occurs if the eigenvalues r_1 and r_2 are equal; in this case the critical point is a node. Small perturbations in the coefficients will normally cause the two equal roots to separate (bifurcate). If the separated roots are real, then the critical point of the perturbed system remains a node, but if the separated roots are complex conjugates, then the critical point becomes a spiral point. These two possibilities are shown schematically in Figure 9.15. In this case the stability or instability of the system is not affected by small perturbations in the coefficients, but the trajectories may be altered considerably (see Problem 20).

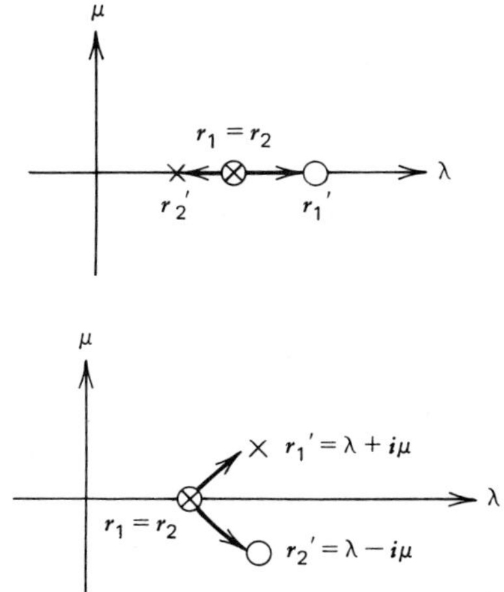

FIGURE 9.15 Schematic perturbation of $r_1 = r_2$.

In all other cases the stability or instability of the system is not changed, nor is the type of critical point altered, by sufficiently small perturbations in the coefficients of the system. For example, if r_1 and r_2 are real, negative, and unequal, then a *small* change in the coefficients will not change the sign of r_1 and r_2 nor allow them to coalesce. Thus the critical point remains an asymptotically stable improper node.

In Section 9.3 we discuss how the results of this section may be used to investigate almost linear systems.

PROBLEMS

In each of the Problems 1 through 12 determine the eigenvalues, classify the critical point (0, 0) as to type, and determine whether it is stable, asymptotically stable, or unstable.

1. $\dfrac{d\mathbf{x}}{dt} = \begin{pmatrix} 3 & -2 \\ 2 & -2 \end{pmatrix}\mathbf{x}$

2. $\dfrac{d\mathbf{x}}{dt} = \begin{pmatrix} 5 & -1 \\ 3 & 1 \end{pmatrix}\mathbf{x}$

3. $\dfrac{d\mathbf{x}}{dt} = \begin{pmatrix} 2 & -1 \\ 3 & -2 \end{pmatrix}\mathbf{x}$

4. $\dfrac{d\mathbf{x}}{dt} = \begin{pmatrix} 1 & -4 \\ 4 & -7 \end{pmatrix}\mathbf{x}$

5. $\dfrac{d\mathbf{x}}{dt} = \begin{pmatrix} 1 & -5 \\ 1 & -3 \end{pmatrix}\mathbf{x}$

6. $\dfrac{d\mathbf{x}}{dt} = \begin{pmatrix} 2 & -5 \\ 1 & -2 \end{pmatrix}\mathbf{x}$

7. $\dfrac{d\mathbf{x}}{dt} = \begin{pmatrix} 3 & -2 \\ 4 & -1 \end{pmatrix}\mathbf{x}$

8. $\dfrac{d\mathbf{x}}{dt} = \begin{pmatrix} -1 & -1 \\ 0 & -0.25 \end{pmatrix}\mathbf{x}$

9. $\dfrac{d\mathbf{x}}{dt} = \begin{pmatrix} 3 & -4 \\ 1 & -1 \end{pmatrix}\mathbf{x}$

10. $\dfrac{d\mathbf{x}}{dt} = \begin{pmatrix} 1 & 2 \\ -5 & -1 \end{pmatrix}\mathbf{x}$

11. $\dfrac{d\mathbf{x}}{dt} = \begin{pmatrix} -1 & 0 \\ 0 & -1 \end{pmatrix}\mathbf{x}$

12. $\dfrac{d\mathbf{x}}{dt} = \begin{pmatrix} 2 & -\frac{5}{2} \\ \frac{9}{5} & -1 \end{pmatrix}\mathbf{x}$

In each of Problems 13 through 16 determine the critical point $\mathbf{x} = \mathbf{x}^0$, and then classify its type and examine its stability by making the transformation $\mathbf{x} = \mathbf{x}^0 + \mathbf{u}$.

13. $\dfrac{d\mathbf{x}}{dt} = \begin{pmatrix} 1 & 1 \\ 1 & -1 \end{pmatrix}\mathbf{x} - \begin{pmatrix} 2 \\ 0 \end{pmatrix}$

14. $\dfrac{d\mathbf{x}}{dt} = \begin{pmatrix} -2 & 1 \\ 1 & -2 \end{pmatrix}\mathbf{x} + \begin{pmatrix} -2 \\ 1 \end{pmatrix}$

15. $\dfrac{d\mathbf{x}}{dt} = \begin{pmatrix} -1 & -1 \\ 2 & -1 \end{pmatrix}\mathbf{x} + \begin{pmatrix} -1 \\ 5 \end{pmatrix}$

16. $\dfrac{d\mathbf{x}}{dt} = \begin{pmatrix} 0 & -\beta \\ \delta & 0 \end{pmatrix}\mathbf{x} + \begin{pmatrix} \alpha \\ -\gamma \end{pmatrix}$; $\alpha, \beta, \gamma, \delta > 0$

17. Thc equation of motion of a spring–mass system with damping (see Section 3.7) is

$$m\frac{d^2u}{dt^2} + c\frac{du}{dt} + ku = 0,$$

where m, c, and k are positive. Write this second order equation as a system of two first order equations for $x = u$, $y = du/dt$. Show that $x = 0$, $y = 0$ is a critical point, and analyze the nature and stability of the critical point as a function of the parameters m, c, and k. A similar analysis can be applied to the electric circuit equation (see Section 3.8)

$$L\frac{d^2I}{dt^2} + R\frac{dI}{dt} + \frac{1}{C}I = 0.$$

18. Consider the system $\mathbf{x}' = \mathbf{A}\mathbf{x}$, and suppose that $\mathbf{A}$ has one zero eigenvalue.
(a) Show that $\mathbf{x} = \mathbf{0}$ is a critical point, but not an isolated one.
(b) Let $r_1 = 0$ and $r_2 \neq 0$, and let $\boldsymbol{\xi}^{(1)}$ and $\boldsymbol{\xi}^{(2)}$ be corresponding eigenvectors. Show that the trajectories are as indicated in Figure 9.16. What is the direction of motion on the trajectories?

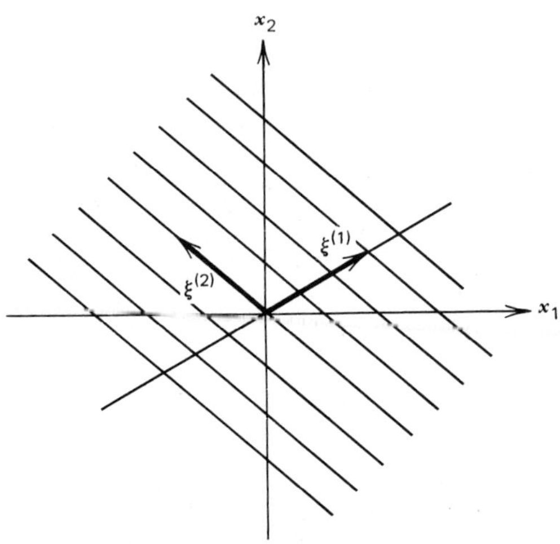

FIGURE 9.16 A nonisolated critical point; $r_1 = 0$, $r_2 \neq 0$.

19. In this problem we show how small changes in the coefficients of a system of linear equations can affect a critical point that is a center. Consider the system

$$\mathbf{x}' = \begin{pmatrix} 0 & 1 \\ -1 & 0 \end{pmatrix} \mathbf{x}.$$

Show that the eigenvalues are $\pm i$ so that $(0, 0)$ is a center. Now consider the system

$$\mathbf{x}' = \begin{pmatrix} \epsilon & 1 \\ -1 & \epsilon \end{pmatrix} \mathbf{x}$$

where $|\epsilon|$ is arbitrarily small. Show that the eigenvalues are $\epsilon \pm i$. Thus no matter how small $|\epsilon| \neq 0$ is, the center becomes a spiral point. If $\epsilon < 0$ then the spiral point is asymptotically stable; if $\epsilon > 0$ the spiral point is unstable.

20. In this problem we show how small changes in the coefficients of a system of linear equations can affect the nature of a critical point when the eigenvalues are equal. Consider the system

$$\mathbf{x}' = \begin{pmatrix} -1 & 1 \\ 0 & -1 \end{pmatrix} \mathbf{x}.$$

Show that the eigenvalues are $r_1 = -1$, $r_2 = -1$ so that the critical point $(0, 0)$ is an asymptotically stable node. Now consider the system

$$\mathbf{x}' = \begin{pmatrix} -1 & 1 \\ -\epsilon & -1 \end{pmatrix} \mathbf{x}$$

where $|\epsilon|$ is arbitrarily small. Show that if $\epsilon > 0$, then the eigenvalues are $-1 \pm i\sqrt{\epsilon}$, so that the asymptotically stable node becomes an asymptotically stable spiral point. If $\epsilon < 0$ then the roots are $-1 \pm \sqrt{|\epsilon|}$, and the critical point remains an asymptotically stable node.

21. In this problem we indicate how to show that the trajectories are ellipses when the eigenvalues are pure imaginary. Consider the system

$$\begin{pmatrix} x \\ y \end{pmatrix}' = \begin{pmatrix} a & b \\ c & d \end{pmatrix} \begin{pmatrix} x \\ y \end{pmatrix}. \tag{i}$$

(a) Show that the eigenvalues of the coefficient matrix are pure imaginary if and only if

$$a + d = 0, \qquad ad - bc > 0. \tag{ii}$$

(b) The trajectories of the system (i) can be found by converting Eqs. (i) into the single equation

$$\frac{dy}{dx} = \frac{dy/dt}{dx/dt} = \frac{cx + dy}{ax + by}. \tag{iii}$$

Use the first of Eqs. (ii) to show that Eq. (iii) is exact.

(c) By integrating Eq. (iii) show that

$$cx^2 + 2dxy - by^2 = k, \tag{iv}$$

where k is a constant. Use Eqs. (ii) to conclude that the graph of Eq. (iv) is always an ellipse. *Hint:* What is the discriminant of the quadratic form in Eq. (iv)?

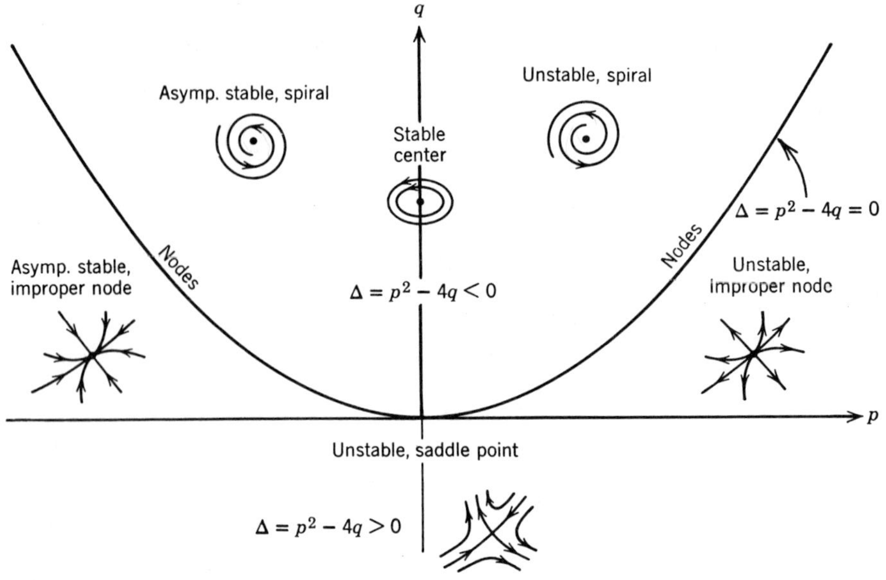

FIGURE 9.17 Stability diagram.

22. Consider the linear autonomous system

$$dx/dt = ax + by, \qquad dy/dt = cx + dy,$$

where a, b, c, and d are real constants. Let $p = a + d$, $q = ad - bc$, and $\Delta = p^2 - 4q$. Show that the critical point $(0, 0)$ is a
(a) node if $q > 0$ and $\Delta \geq 0$ (b) saddle point if $q < 0$
(c) spiral point if $p \neq 0$ and $\Delta < 0$ (d) center if $p = 0$ and $q > 0$.
Hint: These conclusions can be obtained by studying the eigenvalues r_1 and r_2. It may also be helpful to show, and then to use, the relations $r_1 r_2 = q$ and $r_1 + r_2 = p$.

23. Continuing Problem 22, show that the critical point $(0, 0)$ is
(a) asymptotically stable if $q > 0$ and $p < 0$
(b) stable if $q > 0$ and $p = 0$
(c) unstable if $q < 0$ or $p > 0$.
Notice that the results (a), (b), and (c), together with the fact that $q \neq 0$, show that the critical point is asymptotically stable if, and only if, $q > 0$ and $p < 0$.
The results of Problems 22 and 23 are conveniently summarized in Figure 9.17.

9.3 Stability: Almost Linear Systems

In the previous sections of this chapter, we have on several occasions referred to the concepts of stability, asymptotic stability, and instability of a solution of the autonomous system

$$\mathbf{x}' = \mathbf{f}(\mathbf{x}) \tag{1}$$

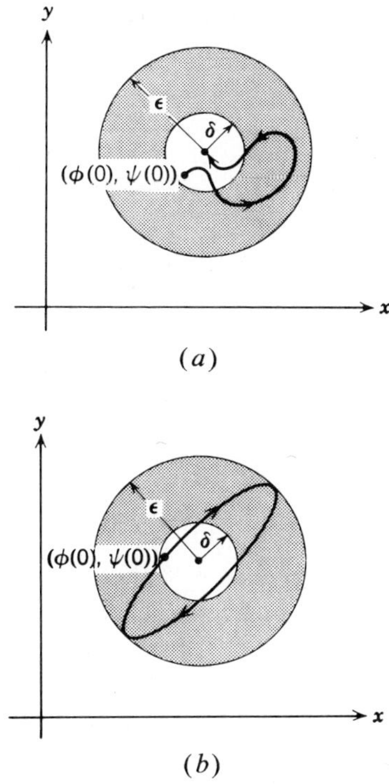

FIGURE 9.18 (*a*) *Asymptotic stability*. (*b*) *Stability*.

or, in scalar form,

$$x' = F(x, y), \qquad y' = G(x, y). \tag{2}$$

Now we give a precise mathematical meaning to these concepts. In the following definitions, and elsewhere, we use the notation $\|\mathbf{x}\|$ to designate the length, or magnitude, of the vector $\mathbf{x}$.

A critical point $\mathbf{x}^0$, corresponding to an equilibrium solution $\mathbf{x} = \mathbf{x}^0$, of the autonomous system (1) is said to be a *stable critical point* if, given any $\epsilon > 0$, there is a $\delta > 0$ such that every solution $\mathbf{x} = \boldsymbol{\phi}(t)$ of the system (1), which at $t = 0$ satisfies

$$\|\boldsymbol{\phi}(0) - \mathbf{x}^0\| < \delta, \tag{3}$$

exists and satisfies

$$\|\boldsymbol{\phi}(t) - \mathbf{x}^0\| < \epsilon \tag{4}$$

for all $t \geq 0$. This is illustrated geometrically in Figures 9.18*a*, *b*. These mathematical statements say that all solutions that start "sufficiently close" (that is, within the distance δ) to $\mathbf{x}^0$ stay "close" (within the distance ϵ) to $\mathbf{x}^0$. Note that in Figure 9.18*a* the trajectory is within the circle $\|\mathbf{x} - \mathbf{x}^0\| = \delta$ at $t = 0$ and, while it soon passes outside of this circle, it remains within the circle $\|\mathbf{x} - \mathbf{x}^0\| = \epsilon$ for all $t \geq 0$. However, the trajectory of the solution does not have to approach the critical point $\mathbf{x}^0$ as $t \to \infty$, as is illustrated in Figure 9.18*b*.

A critical point $\mathbf{x}^0$ is said to be *asymptotically stable* if it is stable and if there exists a δ_0, with $0 < \delta_0 < \delta$, such that if a solution $\mathbf{x} = \boldsymbol{\phi}(t)$ satisfies

$$\|\boldsymbol{\phi}(0) - \mathbf{x}^0\| < \delta_0, \tag{5}$$

then

$$\lim_{t \to \infty} \boldsymbol{\phi}(t) = \mathbf{x}^0. \tag{6}$$

Thus trajectories that start "sufficiently close" to $\mathbf{x}^0$ must not only stay "close" but must eventually approach $\mathbf{x}^0$ as $t \to \infty$. This is the case for the trajectory in Figure 9.18*a* but not for the one in Figure 9.18*b*. Note that asymptotic stability is a stronger requirement than stability, since a critical point must be stable before we can even talk about whether it is asymptotically stable. On the other hand, the limit condition (6), which is an essential feature of asymptotic stability, does not by itself imply even ordinary stability. Indeed, examples can be constructed in which all of the trajectories approach $\mathbf{x}^0$ as $t \to \infty$, but for which $\mathbf{x}^0$ is not a stable critical point.[3] Geometrically, all that is needed is a family of trajectories having members that start arbitrarily close to $\mathbf{x}^0$, then recede an arbitrarily large distance before eventually approaching $\mathbf{x}^0$ as $t \to \infty$. Finally, a critical point that is not stable is said to be *unstable*.

Our main interest is in examining the behavior of the trajectories of the system (1) in the neighborhood of a critical point $\mathbf{x}^0$. It is convenient to choose the critical point to be the origin. This involves no loss of generality, since if $\mathbf{x}^0 \neq \mathbf{0}$, it is always possible to make the substitution $\mathbf{u} = \mathbf{x} - \mathbf{x}^0$ in Eq. (1). Then $\mathbf{u}$ will satisfy an autonomous system with a critical point at the origin.

We now suppose that the system (1) has the form

$$\mathbf{x}' = \mathbf{A}\mathbf{x} + \mathbf{g}(\mathbf{x}), \tag{7}$$

and that $\mathbf{x} = \mathbf{0}$ is an *isolated critical point* of the system (7). This means that there is some circle about the origin within which there are no other critical points. In addition, we assume that $\det \mathbf{A} \neq 0$, so that $\mathbf{x} = \mathbf{0}$ is also an isolated critical point of the linear system

$$\mathbf{x}' = \mathbf{A}\mathbf{x}. \tag{8}$$

[3]Such examples are fairly complicated (see Cesari, p. 96).

Finally, we assume that the components of **g** have continuous first partial derivatives and are small near the origin in the sense that

$$\|\mathbf{g}(\mathbf{x})\|/\|\mathbf{x}\| \to 0 \quad \text{as} \quad \mathbf{x} \to \mathbf{0}. \tag{9}$$

Such a system is called an *almost linear system* in the neighborhood of the critical point $\mathbf{x} = \mathbf{0}$.

It may be helpful to express the condition (9) in scalar form. If we let $\mathbf{x}^T = (x, y)$, then $\|\mathbf{x}\| = (x^2 + y^2)^{1/2} = r$. Similarly, if $\mathbf{g}^T(\mathbf{x}) = (g_1(x, y), g_2(x, y))$, then $\|\mathbf{g}(\mathbf{x})\| = [g_1^2(x, y) + g_2^2(x, y)]^{1/2}$. Then it follows that condition (9) is satisfied if and only if

$$g_1(x, y)/r \to 0, \qquad g_2(x, y)/r \to 0 \quad \text{as} \quad r \to 0. \tag{10}$$

EXAMPLE 1

Determine whether the system

$$\begin{pmatrix} x \\ y \end{pmatrix}' = \begin{pmatrix} 1 & 0 \\ 0 & 0.5 \end{pmatrix}\begin{pmatrix} x \\ y \end{pmatrix} + \begin{pmatrix} -x^2 - xy \\ -0.75xy - 0.25y^2 \end{pmatrix} \tag{11}$$

is almost linear in the neighborhood of the origin.

Observe that the system (11) is of the form (7), that $(0, 0)$ is a critical point, and that $\det \mathbf{A} \neq 0$. It is not hard to show that the other critical points of Eqs. (11) are $(0, 2)$, $(1, 0)$, and $(0.5, 0.5)$; consequently, the origin is an isolated critical point. In checking the condition (10) it is convenient to introduce polar coordinates by letting $x = r\cos\theta$, $y = r\sin\theta$. Then

$$\begin{aligned}\frac{g_1(x, y)}{r} &= \frac{-x^2 - xy}{r} = \frac{-r^2\cos^2\theta - r^2\sin\theta\cos\theta}{r} \\ &= -r(\cos^2\theta + \sin\theta\cos\theta) \to 0\end{aligned}$$

as $r \to 0$. In a similar way one can show that $g_2(x, y)/r \to 0$ as $r \to 0$. Hence the system (11) is almost linear near the origin.

EXAMPLE 2

The undamped motion of a pendulum is described by the system [see Eq. (17) of Section 9.1]

$$dx/dt = y, \qquad dy/dt = -(g/l)\sin x. \tag{12a}$$

The origin is an isolated critical point of this system. Show that the system is almost linear near the origin.

To compare Eqs. (12a) with Eq. (7) we must rewrite the former so that the linear and nonlinear terms are clearly identified. If we write $\sin x = x + (\sin x - x)$ and substitute this expression in the second of Eqs. (12a), we obtain the equivalent system

$$\begin{pmatrix} x \\ y \end{pmatrix}' = \begin{pmatrix} 0 & 1 \\ -g/l & 0 \end{pmatrix}\begin{pmatrix} x \\ y \end{pmatrix} - \frac{g}{l}\begin{pmatrix} 0 \\ \sin x - x \end{pmatrix}. \tag{12b}$$

On comparing Eq. (12b) with Eq. (7) we see that $g_1(x, y) = 0$ and $g_2(x, y) = -(g/l)(\sin x - x)$. From the Taylor series for $\sin x$ we know that $\sin x - x$ behaves like $-x^3/3! = -(r^3\cos^3\theta)/3!$ when x is small. Consequently, $(\sin x - x)/r \to 0$ as $r \to 0$. Thus the conditions (10) are satisfied and the system (12b) is almost linear near the origin.

More generally, one can show that the system (2) is almost linear in the neighborhood of a critical point (x_0, y_0) whenever the functions F and G have continuous partial derivatives up to order two. In that case, using Taylor expansions about the point (x_0, y_0), we can write

$$F(x, y) = F(x_0, y_0) + F_x(x_0, y_0)(x - x_0) + F_y(x_0, y_0)(y - y_0) + \eta_1(x, y),$$

$$G(x, y) = G(x_0, y_0) + G_x(x_0, y_0)(x - x_0) + G_y(x_0, y_0)(y - y_0) + \eta_2(x, y),$$

where $\eta_1(x, y)/[(x - x_0)^2 + (y - y_0)^2]^{1/2} \to 0$ as $(x, y) \to (x_0, y_0)$, and similarly for η_2. Note that $F(x_0, y_0) = G(x_0, y_0) = 0$, and that $dx/dt = d(x - x_0)/dt$ and $dy/dt = d(y - y_0)/dt$. Then the system (2) reduces to

$$\frac{d}{dt}\begin{pmatrix} x - x_0 \\ y - y_0 \end{pmatrix} = \begin{pmatrix} F_x(x_0, y_0) & F_y(x_0, y_0) \\ G_x(x_0, y_0) & G_y(x_0, y_0) \end{pmatrix}\begin{pmatrix} x - x_0 \\ y - y_0 \end{pmatrix} + \begin{pmatrix} \eta_1(x, y) \\ \eta_2(x, y) \end{pmatrix}, \tag{13}$$

which is of the form (7) for the vector $\mathbf{u}^T = (x - x_0, y - y_0)$. Observe that the linear part of the system (13) has the coefficient matrix consisting of the partial derivatives of F and G evaluated at (x_0, y_0). Equation (13) provides a general method for finding the linear system corresponding to an almost linear system near a given point (x_0, y_0).

Since the nonlinear term $\mathbf{g}(\mathbf{x})$ in the system (7) is small compared to the linear term $\mathbf{Ax}$ when $\mathbf{x}$ is near the origin, it is reasonable to hope that the trajectories of the linear system (8) are satisfactory approximations to those of the nonlinear system (7), at least near the origin. This turns out to be true in many (but not all) cases.

For the linear system $\mathbf{x}' = \mathbf{A}\mathbf{x}$, with $\det \mathbf{A} \neq 0$, the type and stability of the critical point $\mathbf{x} = \mathbf{0}$ as a function of the eigenvalues r_1 and r_2 of the coefficient matrix $\mathbf{A}$ were listed in Table 9.1 of the preceding section. The stability characteristics are summarized in the following theorem.

Theorem 9.1. *The critical point* $(0,0)$ *of the linear system* (8) *is*
(1) *asymptotically stable if the eigenvalues* r_1, r_2 *are real and negative or have negative real part*;
(2) *stable, but not asymptotically stable, if* r_1 *and* r_2 *are pure imaginary*;
(3) *unstable if* r_1 *and* r_2 *are real and either is positive, or if they have positive real part.*

This theorem was established by intuitive arguments in the preceding section. For a rigorous proof, it is necessary to show how to compute δ for a given ϵ so that Eq. (4) is satisfied and, for asymptotic stability, how to compute δ_0 so that Eq. (6) is true. We will not take up this type of detailed analysis.

Next we turn to the almost linear system (7). In most cases the type and stability of a critical point of the almost linear system are closely related to the type and stability of the critical point of the corresponding linear system (8).

Theorem 9.2. *Let* r_1 *and* r_2 *be the eigenvalues of the linear system* (8) *corresponding to the almost linear system* (7). *Then the type and stability of the critical point* $(0,0)$ *of the linear system* (8) *and the almost linear system* (7) *are as shown in Table* 9.2.

At this stage, the proof of this theorem is too difficult to give, and we will accept the results of Theorem 9.2 without proof. The results for asymptotic stability and for instability follow as a consequence of a result discussed in Section 9.5, and a proof is sketched in Problems 7 to 9 of Section 9.5. Essentially, Theorem 9.2 says that for $\mathbf{x}$ small the nonlinear terms are also small and do not affect the stability and type of critical point as determined by the linear terms except in two sensitive cases: r_1 and r_2 pure imaginary, and r_1 and r_2 real and equal. Recall that at the end of Section 9.2 we stated that small perturbations in the coefficients of the linear system (8), and hence in the eigenvalues r_1 and r_2, can alter the type and stability of the critical point only in these two sensitive cases. When r_1 and r_2 are pure imaginary, a small perturbation can change the stable center into either an asymptotically stable or an unstable spiral point or

TABLE 9.2 STABILITY AND INSTABILITY PROPERTIES OF LINEAR AND ALMOST LINEAR SYSTEMS

r_1, r_2	Linear system		Almost linear system	
	Type[a]	Stability	Type[a]	Stability
$r_1 > r_2 > 0$	IN	Unstable	IN	Unstable
$r_1 < r_2 < 0$	IN	Asymptotically stable	IN	Asymptotically stable
$r_2 < 0 < r_1$	SP	Unstable	SP	Unstable
$r_1 = r_2 > 0$	PN or IN	Unstable	PN, IN, or SpP	Unstable
$r_1 = r_2 < 0$	PN or IN	Asymptotically stable	PN, IN, or SpP	Asymptotically stable
$r_1, r_2 = \lambda \pm i\mu$				
$\lambda > 0$	SpP	Unstable	SpP	Unstable
$\lambda < 0$	SpP	Asymptotically stable	SpP	Asymptotically stable
$r_1 = i\mu,\ r_2 = -i\mu$	C	Stable	C or SpP	Indeterminate

[a] IN, improper node; PN, proper node; SP, saddle point; SpP, spiral point; C, center.

even leave it as a center. When $r_1 = r_2$ a small perturbation does not affect the stability of the critical point, but may change the node into a spiral point. It is reasonable to expect that the small nonlinear terms in Eqs. (7) might have a similarly substantial effect, at least in these two sensitive cases. This is so, but the main significance of Theorem 9.2 is that in *all other cases* the small nonlinear terms do not alter the type or stability of the critical point. Thus, except in the two sensitive cases, the type and stability of the critical point of the nonlinear system (7) can be determined from a study of the much simpler linear system (8).

Even if the critical point is of the same type as that of the linear system, the trajectories of the almost linear system may be considerably different in appearance from those of the corresponding linear system. This is illustrated in Problems 12 and 13. However, it can be shown that the slope at which trajectories "enter" or "leave" the critical point is given correctly by the linear equations.

Further, if a critical point of the linear system (8) is asymptotically stable, then not only do trajectories that start close to the critical point approach the critical point, but, in fact, since all solutions are linear combinations of $e^{r_1 t}$ and $e^{r_2 t}$, *every* trajectory approaches the critical point. In this case the critical point is said to be *globally asymptotically stable*. This property of linear systems is not, in general, true for nonlinear systems. For a nonlinear system an important practical problem is to estimate the set of initial conditions for which a critical point is

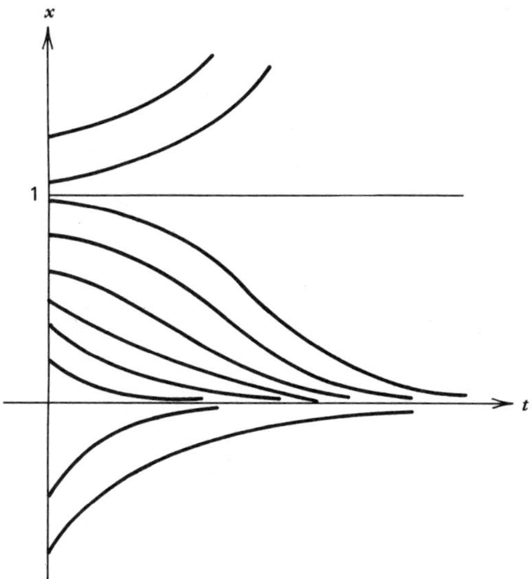

FIGURE 9.19 Trajectories of $x' = -(1 - x)x$.

asymptotically stable. This set of initial points is called the *region of asymptotic stability* for the critical point.

EXAMPLE 3

Consider the single first order equation

$$dx/dt = -(1 - x)x. \tag{14}$$

The trajectories of this equation are sketched in Figure 9.19. A similar equation was discussed in Section 2.6; see Figure 2.18. The critical points of Eq. (14) are $x = 0$ and $x = 1$; the former is asymptotically stable and the latter is unstable. The region of asymptotic stability for the critical point $x = 0$ is the set of points $(-\infty, 1)$. Trajectories that start in this interval approach the equilibrium solution $x = 0$ as $t \to \infty$. On the other hand, trajectories with initial points in $[1, \infty)$ do not approach the solution $x = 0$ as $t \to \infty$. Since not all trajectories approach the equilibrium solution $x = 0$ as $t \to \infty$, this solution is not globally asymptotically stable.

We illustrate the relation between linear and almost linear systems by considering the motion of a damped pendulum and several problems in ecology. The pendulum problem is discussed here; the ecological problems are in Section 9.4.

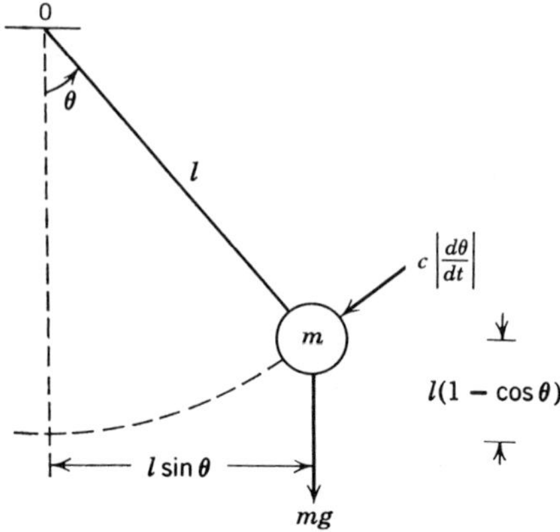

FIGURE 9.20 An oscillating pendulum.

DAMPED PENDULUM. We consider the motion of a damped pendulum, as shown in Figure 9.20. We assume that the magnitude of the damping force is given by $c|d\theta/dt|$, where the damping constant c is positive. The principle of angular momentum states that the rate of change of angular momentum about any fixed point is equal to the moment of the resultant force about the point. Since the angular momentum about the origin is $ml^2(d\theta/dt)$, we immediately obtain the governing equation

$$ml^2\frac{d^2\theta}{dt^2} = -cl\frac{d\theta}{dt} - mgl\sin\theta,$$

or

$$\frac{d^2\theta}{dt^2} + \frac{c}{ml}\frac{d\theta}{dt} + \frac{g}{l}\sin\theta = 0. \tag{15}$$

Letting $x = \theta$ and $y = d\theta/dt$ gives the system

$$\frac{dx}{dt} = y, \qquad \frac{dy}{dt} = -\frac{g}{l}\sin x - \frac{c}{ml}y. \tag{16}$$

The point $x = 0$, $y = 0$ is a critical point of the system (16). Because of the damping mechanism, we expect any small motion about $\theta = 0$ to decay in amplitude. Thus intuitively the equilibrium point $(0, 0)$ should be asymptotically stable. To show this we rewrite the system (16) as

$$\frac{dx}{dt} = y, \qquad \frac{dy}{dt} = -\frac{g}{l}x - \frac{c}{ml}y - \frac{g}{l}(\sin x - x). \tag{17}$$

It was shown in Example 2 that $(\sin x - x)/r \to 0$ as $r \to 0$, so the system (17) is an almost linear system; hence Theorem 9.2 is applicable. The eigenvalues of the corresponding linear system

$$\frac{dx}{dt} = y, \qquad \frac{dy}{dt} = -\frac{g}{l}x - \frac{c}{ml}y \tag{18}$$

are

$$r_1, r_2 = \frac{-c/ml \pm \sqrt{(c/ml)^2 - 4g/l}}{2}. \tag{19}$$

1. If $(c/ml)^2 - 4g/l > 0$ the eigenvalues are real, unequal, and negative. The critical point $(0, 0)$ is an asymptotically stable improper node of the linear system (18) and of the almost linear system (17).
2. If $(c/ml)^2 - 4g/l = 0$ the eigenvalues are real, equal, and negative. The critical point $(0, 0)$ is an asymptotically stable node of the linear system (18). It may be either an asymptotically stable node or an asymptotically stable spiral point of the almost linear system (17).
3. If $(c/ml)^2 - 4g/l < 0$ the eigenvalues are complex with negative real part. The critical point $(0, 0)$ is an asymptotically stable spiral point of the linear system (18) and of the almost linear system (17).

Thus for small damping the critical point $(0, 0)$ is a spiral point. In addition to $(0, 0)$, the almost linear system (16) has the critical points $x = n\pi$, $y = 0$, $n = \pm 1, \pm 2, \pm 3, \ldots$ corresponding to $\theta = \pm\pi, \pm 2\pi, \pm 3\pi, \ldots, d\theta/dt = 0$. Again we expect (from Figure 9.20) that the points corresponding to $\theta = \pm 2\pi, \pm 4\pi, \ldots$ are asymptotically stable spiral points, and the points corresponding to $\theta = \pm\pi, \pm 3\pi, \ldots$ are unstable saddle points. Consider the critical point $x = \pi$, $y = 0$. To examine the stability of this point, we can find the corresponding linear system from Eq. (13). Alternatively, we can introduce the change of variables

$$x = \pi + u, \qquad y = 0 + v, \tag{20}$$

and investigate the resulting system for small u and v. Substituting for x and y in Eqs. (16), and making use of the fact that $\sin(\pi + u) = -\sin u$, we obtain

$$\frac{du}{dt} = v, \qquad \frac{dv}{dt} = -\frac{c}{ml}v + \frac{g}{l}\sin u. \tag{21}$$

We are interested in studying the critical point $u = v = 0$ of the system (21). We can rewrite the second of Eqs. (21) as

$$\frac{dv}{dt} = \frac{g}{l}u - \frac{c}{ml}v + \frac{g}{l}(\sin u - u). \tag{22}$$

It is clear that the first of Eqs. (21) and Eq. (22) are the same as the system (17), except that $-g/l$ is replaced by g/l. Thus the system is almost linear, and the eigenvalues of the corresponding linear system are given by

$$r_1, r_2 = \frac{-c/ml \pm \sqrt{(c/ml)^2 + 4g/l}}{2}. \tag{23}$$

One eigenvalue (r_1) is positive and the other (r_2) is negative. Therefore, the critical point $x = \pi$, $y = 0$ is an unstable saddle point of both the linear system and the almost linear system, as expected.

Let us consider the case $(c/ml)^2 - 4g/l < 0$ in more detail. We will try to make a schematic sketch of the trajectories in the xy (phase) plane. The critical points are at $y = 0$, $x = n\pi$, $n = 0, \pm 1, \pm 2, \ldots$.

We have seen that the critical point $(0, 0)$ is an asymptotically stable spiral point. This point corresponds to the pendulum hanging vertically downward with zero velocity. It is not difficult to show that the critical points $(2n\pi, 0)$, $n = \pm 1, \pm 2, \ldots$, which also correspond to the pendulum hanging vertically downward with zero velocity, are also asymptotically stable spiral points. (Let $x = 2n\pi + u$, $y = 0 + v$ and proceed as above.) The direction of motion on the spirals near $(0, 0)$ can be obtained from Eqs. (16). Consider the point at which a spiral intersects the positive y axis ($x = 0$, $y > 0$). At such a point it follows from Eqs. (16) that $dx/dt > 0$. Thus the point (x, y) on the trajectory is moving to the right so the direction of motion on the spirals is clockwise. The situation is the same for the spirals near the critical points $(2n\pi, 0)$, $n = \pm 1, \pm 2, \ldots$. The spirals are sketched in Figure 9.21.

We have shown above that the critical point $(\pi, 0)$ is an unstable saddle point. This point corresponds to the pendulum in a vertically upward position with zero velocity. Again, we can establish that the critical points $((2n + 1)\pi, 0)$, $n = \pm 1, \pm 2, \ldots$, are unstable saddle points. Thus in the phase plane the asymptotically stable spiral points are interspersed with unstable saddle points. Recall from Section 9.2 that only one pair of trajectories "enters" a saddle point. To determine the direction of the entering trajectories at the saddle point $(\pi, 0)$,

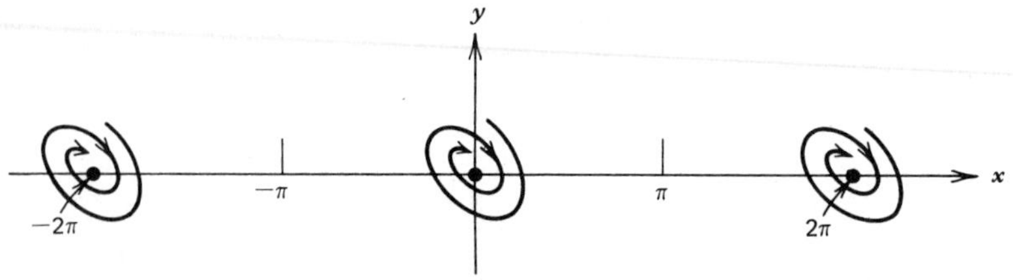

FIGURE 9.21 Asymptotically stable spiral points for the damped pendulum.

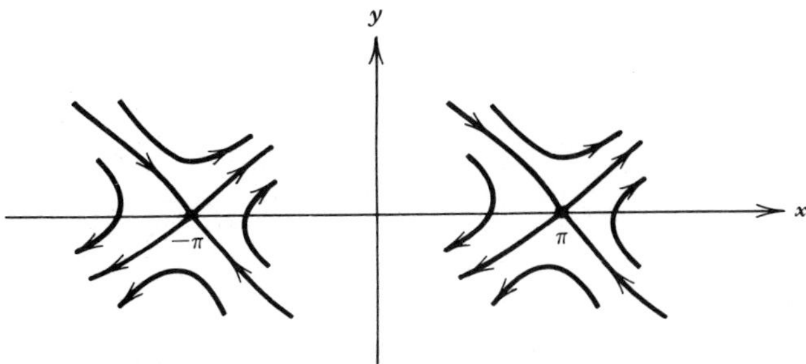

FIGURE 9.22 Unstable saddle points for the damped pendulum.

we consider the linear system corresponding to Eqs. (21):

$$\frac{du}{dt} = v, \qquad \frac{dv}{dt} = \frac{g}{l}u - \frac{c}{ml}v. \tag{24}$$

On making use of Eq. (23) and the first of Eqs. (24), we can write the general solution of Eq. (24) in the form

$$\begin{pmatrix} u \\ v \end{pmatrix} = C_1\begin{pmatrix} 1 \\ r_1 \end{pmatrix}e^{r_1 t} + C_2\begin{pmatrix} 1 \\ r_2 \end{pmatrix}e^{r_2 t},$$

where C_1 and C_2 are arbitrary constants. Since $r_1 > 0$ and $r_2 < 0$, it follows that the solution that approaches zero as $t \to \infty$ corresponds to $C_1 = 0$. For this solution $v/u = r_2$ so the slope of the entering trajectories is negative; one lies in the second quadrant ($C_2 < 0$) and the other lies in the fourth quadrant ($C_2 > 0$). For $C_2 = 0$ we obtain the pair of trajectories "exiting" from the saddle point. These trajectories have slope $r_1 > 0$; one lies in the first quadrant ($C_1 > 0$) and the other lies in the third quadrant ($C_1 < 0$). The situation is the same for the other unstable saddle points $((2n + 1)\pi, 0)$, $n = \pm 1, \pm 2, \ldots$. A possible picture of the trajectories in the neighborhood of the unstable saddle points is shown in Figure 9.22.

If we put Figures 9.21 and 9.22 together, we obtain the schematic picture of the trajectories in the phase plane shown in Figure 9.23. Note that the trajectories entering the saddle points separate the phase plane into sectors. Such a trajectory is called a *separatrix*. The initial conditions on θ and $d\theta/dt$ determine the position of an initial point (x, y) in the phase plane. The subsequent motion of the pendulum is represented by the trajectory passing through the initial point as it spirals toward the asymptotically stable point of that sector. Note that it is mathematically possible (but physically unrealizable) to choose initial conditions on a separatrix so that the resulting motion leads to a balanced pendulum in a vertically upward position of unstable equilibrium.

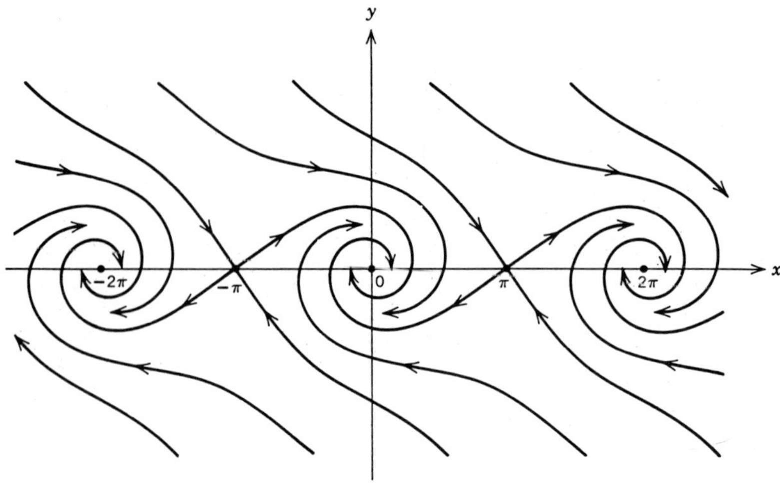

FIGURE 9.23 Phase plane for the damped pendulum.

PROBLEMS

In each of Problems 1 through 10, verify that (0, 0) is a critical point, show that the system is almost linear, and discuss the type and stability of the critical point (0, 0).

1. $dx/dt = x - y + xy$
   $dy/dt = 3x - 2y - xy$
2. $dx/dt = x + x^2 + y^2$
   $dy/dt = y - xy$
3. $dx/dt = -2x - y - x(x^2 + y^2)$
   $dy/dt = x - y + y(x^2 + y^2)$
4. $dx/dt = y + x(1 - x^2 - y^2)$
   $dy/dt = -x + y(1 - x^2 - y^2)$
5. $dx/dt = 2x + y + xy^3$
   $dy/dt = x - 2y - xy$
6. $dx/dt = x + 2x^2 - y^2$
   $dy/dt = x - 2y + x^3$
7. $dx/dt = y$
   $dy/dt = -x + \mu y(1 - x^2), \ \mu > 0$
8. $dx/dt = 1 + y - e^{-x}$
   $dy/dt = y - \sin x$
9. $dx/dt = (1 + x)\sin y$
   $dy/dt = 1 - x - \cos y$
10. $dx/dt = e^{-x+y} - \cos x$
    $dy/dt = \sin(x - 3y)$

11. Determine all real critical points of each of the following systems of equations and discuss their type and stability.

    (a) $dx/dy = x + y^2$
    $dy/dt = x + y$

    (b) $dx/dt = 1 - xy$
    $dy/dt = x - y^3$

    (c) $dx/dt = x - x^2 - xy$
    $dy/dt = 3y - xy - 2y^2$

    (d) $dx/dt = 1 - y$
    $dy/dt = x^2 - y^2$

12. Consider the autonomous system

$$dx/dt = y, \qquad dy/dt = x + 2x^3.$$

(a) Show that the critical point $(0, 0)$ is a saddle point.
(b) Sketch the trajectories for the corresponding linear system by integrating the equation for dy/dx. Show from the parametric form of the solution that the only trajectory on which $x \to 0$, $y \to 0$ as $t \to \infty$ is $y = -x$.
(c) Determine the trajectories for the nonlinear system by integrating the equation for dy/dx. Sketch the trajectories for the nonlinear system that correspond to $y = -x$ and $y = x$ for the linear system.

13. Consider the autonomous system

$$dx/dt = x, \qquad dy/dt = -2y + x^3.$$

(a) Show that the critical point $(0, 0)$ is a saddle point.
(b) Sketch the trajectories for the corresponding linear system [Problem 3(c) of Section 9.1], and show that the trajectory for which $x \to 0$, $y \to 0$ as $t \to \infty$ is given by $x = 0$.
(c) Determine the trajectories for the nonlinear system for $x \neq 0$ by integrating the equation for dy/dx. Show that the trajectory corresponding to $x = 0$ for the linear system is unaltered, but that the one corresponding to $y = 0$ is $y = x^3/5$. Sketch several of the trajectories for the nonlinear system.

14. Theorem 9.2 provides no information about the stability of a critical point of an almost linear system if that point is a center of the corresponding linear system. That this must be the case is illustrated by the following two systems

$$\text{(i)} \begin{cases} dx/dt = y + x(x^2 + y^2) \\ dy/dt = -x + y(x^2 + y^2) \end{cases} \qquad \text{(ii)} \begin{cases} dx/dt = y - x(x^2 + y^2) \\ dy/dt = -x - y(x^2 + y^2). \end{cases}$$

(a) Show that $(0, 0)$ is a critical point of each system and, furthermore, is a center of the corresponding linear system.
(b) Show that each system is almost linear.
(c) Let $r^2 = x^2 + y^2$, and note that $x\,dx/dt + y\,dy/dt = r\,dr/dt$. For system (ii) show that $dr/dt < 0$ and that $r \to 0$ as $t \to \infty$, and hence the critical point is asymptotically stable. For system (i) show that the solution of the initial value problem for r with $r = r_0$ at $t = 0$ becomes unbounded as $t \to 1/2r_0^2$, and hence the critical point is unstable.

15. The equation of motion of an undamped pendulum is $d^2\theta/dt^2 + (g/l)\sin\theta = 0$. Let $x = \theta$, $y = d\theta/dt$, and $k^2 = g/l$ to obtain the system of equations

$$dx/dt = y, \qquad dy/dt = -k^2 \sin x.$$

(a) Show that the critical points are $(\pm n\pi, 0)$, $n = 0, 1, 2, \ldots$, and that the system is almost linear in the neighborhood of each critical point.
(b) Show that the critical point $(0, 0)$ is a (stable) center of the corresponding linear system. Using Theorem 9.2 what can be said about the almost linear problem? The situation is similar at the critical points $(\pm 2n\pi, 0)$, $n = 1, 2, 3, \ldots$. What is the physical interpretation of these critical points?
(c) Show that the critical point $(\pi, 0)$ is an (unstable) saddle point. The situation is similar at the critical points $[\pm(2n - 1)\pi, 0]$, $n = 1, 2, 3, \ldots$. What is the physical interpretation of these critical points?

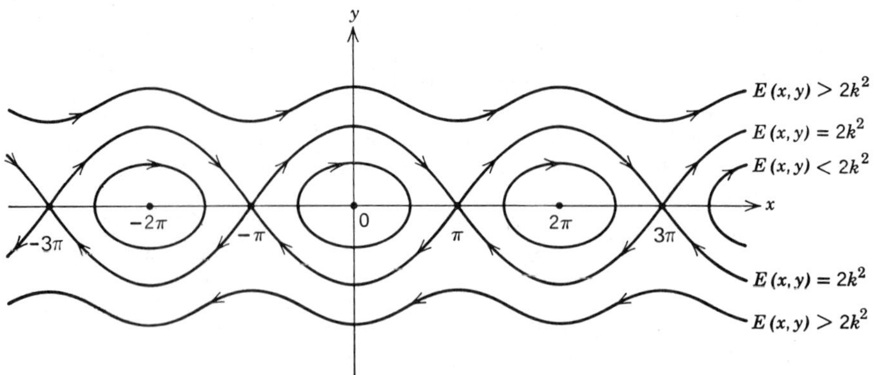

FIGURE 9.24 Phase plane for the undamped pendulum.

16. In this problem we give some of the details of the analysis for sketching the trajectories of the undamped pendulum of Problem 15. Show by eliminating t that the equation of the trajectories can be written as

$$\tfrac{1}{2}y^2 + k^2(1 - \cos x) = E.$$

To discover the significance of the constant E, observe that $\frac{1}{2}y^2 = \frac{1}{2}(dx/dt)^2 = \frac{1}{2}(d\theta/dt)^2$ is proportional to the kinetic energy of the pendulum. Also, $k^2(1 - \cos x) = \int_0^x k^2 \sin x\, dx$ is proportional to the potential energy of the pendulum due to the gravity force. Thus the constant E is the "energy" of the motion. It is constant along a trajectory (during the course of the motion), and is determined by the initial values of x and y.

In sketching the trajectories it is only necessary to consider the interval $-\pi \le x \le \pi$ since the equation is periodic in x with period 2π. For $E = 2k^2$ show that $y = \pm 2k \cos x/2$ and sketch these trajectories. Observe that these trajectories enter or leave the unstable saddle points at $(\pm\pi, 0)$. Determine the direction of motion on each trajectory by using the differential equations given in Problem 15.

It can be shown that the trajectories are closed curves for $E < 2k^2$ and are not closed curves for $E > 2k^2$. We will not pursue these details here, but a schematic sketch of the trajectories for an undamped pendulum is given in Figure 9.24. The trajectories for $E < 2k^2$ correspond to periodic motions about a center; the trajectories for $E > 2k^2$ correspond to whirling motions.

*17. In this problem we derive a formula for the natural period of an undamped nonlinear pendulum [$c = 0$ in Eq. (15)]. Suppose that the bob is pulled through a positive angle α and then released with zero velocity. Assume that the relation θ as a function of t can be solved for t as a function of θ so that $d\theta/dt$ can be considered a function of

θ. Derive the following sequence of equations:

$$\tfrac{1}{2}ml^2\frac{d}{d\theta}\left[\left(\frac{d\theta}{dt}\right)^2\right] = -mgl\sin\theta,$$

$$\tfrac{1}{2}m\left(l\frac{d\theta}{dt}\right)^2 = mgl(\cos\theta - \cos\alpha),$$

$$dt = -\sqrt{\frac{l}{2g}}\frac{d\theta}{\sqrt{\cos\theta - \cos\alpha}}.$$

Why was the negative square root chosen in the last equation?

If T is the natural period of oscillation, derive the formula

$$\frac{T}{4} = -\sqrt{\frac{l}{2g}}\int_{\alpha}^{0}\frac{d\theta}{\sqrt{\cos\theta - \cos\alpha}}.$$

By making the change of variables $\cos\theta = 1 - 2\sin^2\theta/2$, $\cos\alpha = 1 - 2\sin^2\alpha/2$ followed by $\sin\theta/2 = k\sin\phi$ with $k = \sin\alpha/2$, show that

$$T = 4\sqrt{\frac{l}{g}}\int_0^{\pi/2}\frac{d\phi}{\sqrt{1 - k^2\sin^2\phi}} = 4\sqrt{\frac{l}{g}}\,F(k, \pi/2).$$

The function F is called the *elliptic integral of the first kind*. Note that the period depends on the ratio l/g and also the initial displacement α through $k = \sin\alpha/2$. The corresponding period for the linearized pendulum is $2\pi(l/g)^{1/2}$ and is independent of the initial displacement. To obtain this special result from the general formula we must consider the limiting case of small α (small angular displacement) in which case k is small. In the limit $k \to 0$ the above formula gives $T = 4(l/g)^{1/2}F(0, \pi/2) = 2\pi(l/g)^{1/2}$.

18. A generalization of the damped pendulum equation discussed in the text, or a damped spring–mass system, is the Liénard equation

$$\frac{d^2x}{dt^2} + c(x)\frac{dx}{dt} + g(x) = 0.$$

If $c(x)$ is a constant and $g(x) = kx$ then this equation has the form of the linear pendulum equation [replace $\sin\theta$ with θ in Eq. (15)]; otherwise the damping force $c(x)\,dx/dt$ and restoring force $g(x)$ are nonlinear. Assume that c is continuously differentiable, g is twice continuously differentiable and $g(0) = 0$.

(a) Write the Liénard equation as a system of two first order equations by introducing the variable $y = dx/dt$.

(b) Show that $(0, 0)$ is a critical point and that the system is almost linear in the neighborhood of $(0, 0)$.

(c) Show that if $c(0) > 0$ and $g'(0) > 0$ then the critical point is asymptotically stable, and that if $c(0) < 0$ or $g'(0) < 0$ then the critical point is unstable.

Hint: Use Taylor series to approximate c and g in the neighborhood of $x = 0$.

19. Consider the equations for two competing species derived in Section 9.1:

$$dx/dt = x(\epsilon_1 - \sigma_1 x - \alpha_1 y), \qquad dy/dt = y(\epsilon_2 - \sigma_2 y - \alpha_2 x).$$

Suppose that $\epsilon_1/\sigma_1 < \epsilon_2/\alpha_2$ and $\epsilon_2/\sigma_2 < \epsilon_1/\alpha_1$.
(a) Find the critical point (X, Y) for which both species can coexist.
(b) By making the change of variables $x = X + u$, $y = Y + v$ transform the system of equations to one with a critical point at $u = 0$, $v = 0$. Observe that the system is almost linear.
(c) Classify the critical point as to type and stability.

20. Carry out the calculations of Problem 19 for the case $\epsilon_1/\sigma_1 > \epsilon_2/\alpha_2$ and $\epsilon_2/\sigma_2 > \epsilon_1/\alpha_1$.

9.4 Competing Species and Predator–Prey Problems

In this section we consider two problems in ecology: competing species and predator–prey.

COMPETING SPECIES. In Section 9.1 we showed that a model for the competition between two species with population densities x and y leads to the differential equations

$$dx/dt = x(\epsilon_1 - \sigma_1 x - \alpha_1 y), \tag{1a}$$

$$dy/dt = y(\epsilon_2 - \sigma_2 y - \alpha_2 x), \tag{1b}$$

where the parameters $\epsilon_1, \sigma_1, \ldots, \alpha_2$ are positive. As we saw then, we can analyze these equations by dividing the phase plane into regions according to the sign of dx/dt and dy/dt and then drawing typical trajectories. Let us now see how we can use the theory of almost linear systems to obtain a more precise understanding of what happens.

We start by considering the following specific example:

$$\begin{aligned} dx/dt &= x(1 - x - y), \\ dy/dt &= y(0.5 - 0.75x - 0.25y), \end{aligned} \tag{2}$$

or

$$\frac{d}{dt}\begin{pmatrix} x \\ y \end{pmatrix} = \begin{pmatrix} 1 & 0 \\ 0 & 0.5 \end{pmatrix}\begin{pmatrix} x \\ y \end{pmatrix} - \begin{pmatrix} x^2 + xy \\ 0.75xy + 0.25y^2 \end{pmatrix}. \tag{3}$$

For convenience we think of x and y as the population densities of two species of bacteria competing with each other for the same supply of food. We ask whether there are equilibrium states that might be reached, or whether a periodic growth and decay will be observed, and how such possibilities depend on the initial state of the two cultures.

The critical points of the system (2) are the solutions of the nonlinear algebraic equations

$$\begin{aligned} x(1 - x - y) &= 0, \\ y(0.5 - 0.75x - 0.25y) &= 0. \end{aligned} \tag{4}$$

There are four critical points, namely, $(0, 0)$, $(1, 0)$, $(0, 2)$, and $(0.5, 0.5)$. The system (2) is almost linear in the neighborhood of each critical point, so we investigate the trajectories near each of these points by considering the corresponding linear system.

$\boldsymbol{x = 0,\ y = 0.}$ This corresponds to a state in which both bacteria die as a result of their competition. From Eq. (3) the corresponding linear system is

$$\frac{d}{dt}\begin{pmatrix} x \\ y \end{pmatrix} = \begin{pmatrix} 1 & 0 \\ 0 & 0.5 \end{pmatrix}\begin{pmatrix} x \\ y \end{pmatrix}, \tag{5}$$

whose eigenvalues and eigenvectors are

$$r_1 = 1, \quad \boldsymbol{\xi}^{(1)} = \begin{pmatrix} 1 \\ 0 \end{pmatrix}; \qquad r_2 = 0.5, \quad \boldsymbol{\xi}^{(2)} = \begin{pmatrix} 0 \\ 1 \end{pmatrix}. \tag{6}$$

Thus the general solution of the system (5) is

$$\begin{pmatrix} x \\ y \end{pmatrix} = c_1\begin{pmatrix} 1 \\ 0 \end{pmatrix}e^{t} + c_2\begin{pmatrix} 0 \\ 1 \end{pmatrix}e^{t/2}. \tag{7}$$

The origin is an unstable improper node of both the linear system (5) and the nonlinear system (3). In the neighborhood of the origin all the trajectories are tangent to the y axis except for one pair of trajectories that lies along the x axis. Since the origin is an unstable critical point, this equilibrium solution will not occur in practice.

$\boldsymbol{x = 1,\ y = 0.}$ This corresponds to a state in which bacteria x survives the competition but bacteria y does not. To examine this critical point let $x = 1 + u$, $y = 0 + v$. Substituting for x and y in Eqs. (2) or (3) and simplifying, we obtain

$$\frac{d}{dt}\begin{pmatrix} u \\ v \end{pmatrix} = \begin{pmatrix} -1 & -1 \\ 0 & -0.25 \end{pmatrix}\begin{pmatrix} u \\ v \end{pmatrix} - \begin{pmatrix} u^2 + uv \\ 0.75uv + 0.25v^2 \end{pmatrix}. \tag{8}$$

The corresponding linear system is

$$\frac{d}{dt}\begin{pmatrix} u \\ v \end{pmatrix} = \begin{pmatrix} -1 & -1 \\ 0 & -0.25 \end{pmatrix}\begin{pmatrix} u \\ v \end{pmatrix}. \tag{9}$$

Its eigenvalues and eigenvectors are

$$r_1 = -1, \quad \boldsymbol{\xi}^{(1)} = \begin{pmatrix} 1 \\ 0 \end{pmatrix}; \qquad r_2 = -0.25, \quad \boldsymbol{\xi}^{(2)} = \begin{pmatrix} 4 \\ -3 \end{pmatrix}, \tag{10}$$

and its general solution is

$$\begin{pmatrix} u \\ v \end{pmatrix} = c_1\begin{pmatrix} 1 \\ 0 \end{pmatrix}e^{-t} + c_2\begin{pmatrix} 4 \\ -3 \end{pmatrix}e^{-t/4}. \tag{11}$$

The point $(1, 0)$ is an asymptotically stable improper node of the linear system (9) and hence of the nonlinear systems (8) and (3) as well. If the initial values of x and y are sufficiently close to $(1, 0)$, then the interaction process will lead ultimately to that state.

The behavior of the trajectories near the point $(1, 0)$ can be seen from Eq. (11). If $c_2 = 0$, then there is one pair of trajectories that approaches the critical point along the x axis. All other trajectories approach $(1, 0)$ tangent to the line (with slope $-\frac{3}{4}$) that is determined by the eigenvector $\boldsymbol{\xi}^{(2)}$.

$\boldsymbol{x = 0,\ y = 2.}$ In this case bacteria y survives but bacteria x does not. The analysis is similar to that for the point $(1, 0)$. If we let $x = u$, $y = 2 + v$, then we obtain

$$\frac{d}{dt}\begin{pmatrix} u \\ v \end{pmatrix} = \begin{pmatrix} -1 & 0 \\ -1.5 & -0.5 \end{pmatrix}\begin{pmatrix} u \\ v \end{pmatrix} - \begin{pmatrix} u^2 + uv \\ 0.75uv + 0.25v^2 \end{pmatrix}. \tag{12}$$

The eigenvalues and eigenvectors of the corresponding linear system are

$$r_1 = -1, \quad \boldsymbol{\xi}^{(1)} = \begin{pmatrix} 1 \\ 3 \end{pmatrix}; \qquad r_2 = -0.5, \quad \boldsymbol{\xi}^{(2)} = \begin{pmatrix} 0 \\ 1 \end{pmatrix}, \tag{13}$$

and its general solution is

$$\begin{pmatrix} u \\ v \end{pmatrix} = c_1\begin{pmatrix} 1 \\ 3 \end{pmatrix}e^{-t} + c_2\begin{pmatrix} 0 \\ 1 \end{pmatrix}e^{-t/2}. \tag{14}$$

The critical point $(0, 2)$ is an asymptotically stable improper node. All trajectories approach the critical point along the y axis except for one pair that approaches along the line with slope 3.

$\boldsymbol{x = 0.5,\ y = 0.5.}$ This critical point corresponds to a mixed equilibrium state or coexistence; a standoff, so to speak, in the competition between the two bacteria cultures. To examine the nature of this critical point we let $x = 0.5 + u$, $y = 0.5 + v$. Substituting for x and y in Eqs. (2) or (3), we obtain

$$\frac{d}{dt}\begin{pmatrix} u \\ v \end{pmatrix} = \begin{pmatrix} -0.5 & -0.5 \\ -0.375 & -0.125 \end{pmatrix}\begin{pmatrix} u \\ v \end{pmatrix} - \begin{pmatrix} u^2 + uv \\ 0.75uv + 0.25v^2 \end{pmatrix}. \tag{15}$$

The eigenvalues and eigenvectors of the corresponding linear system are

$$r_1, r_2 = (-5 \pm \sqrt{57})/16; \qquad \boldsymbol{\xi}^{(1)}, \boldsymbol{\xi}^{(2)} = \begin{pmatrix} 1 \\ (-3 \mp \sqrt{57})/8 \end{pmatrix}. \tag{16}$$

Since the eigenvalues are of opposite sign, the critical point (0.5, 0.5) is a saddle point, and hence is unstable. One pair of trajectories approaches the critical point as $t \to \infty$; the others recede from it. The entering trajectories have the slope $(\sqrt{57} - 3)/8 \simeq 0.57$ determined from the eigenvector associated with the negative eigenvalue.

A diagram of what the trajectories look like in the neighborhood of each critical point is shown in Figure 9.25*a*. With a little additional work it is possible to extend the local pictures and obtain a global picture of the trajectories in the phase plane. First, we are only interested in x and y positive. Since trajectories cannot cross other trajectories, and since the x and y axes are trajectories, it follows that a trajectory that starts in the first quadrant must stay in the first quadrant, and a trajectory that starts in any other quadrant cannot enter the first quadrant. Second, we accept without proof two facts that follow from more advanced theory: (i) the system (2) does not have any periodic solutions, that is, trajectories that are closed curves; and (ii) a trajectory that is not a closed curve must either enter a critical point or recede to infinity as $t \to \infty$. For x and y large the nonlinear terms $-(x^2 + xy)$ and $-\frac{1}{4}(y^2 + 3xy)$ in the first and second of Eqs. (2), respectively, outweigh the linear terms. Since they are negative, dx/dt and dy/dt are negative for x and y large. Thus for x and y large the direction of motion on every trajectory is inward. The trajectories cannot escape to infinity! Eventually they must head toward one or the other of the two stable nodes. The *schematic* sketch shown in Figure 9.25*b* is not an unreasonable representation[4] of what must be happening in the first quadrant. If the initial values of x and y are in region I of Figure 9.25*b*, then x wins the competition; if the initial values are in region II, then y wins. "Peaceful coexistence" is not possible unless the initial point lies exactly on the dividing trajectory (separatrix). Of particular interest would be the determination of the dividing trajectories that enter the saddle point (0.5, 0.5) which separate regions I and II.

Now let us return to the general system (1). Recall from Section 9.1 that four cases must be considered depending on the relative orientation of the lines

$$\epsilon_1 - \sigma_1 x - \alpha_1 y = 0 \quad \text{and} \quad \epsilon_2 - \sigma_2 y - \alpha_2 x = 0, \tag{17}$$

as shown in Figure 9.26. Let (X, Y) denote any critical point in any one of the four cases. To study the system (1) in the neighborhood of this critical point we let

$$x = X + u, \qquad y = Y + v, \tag{18}$$

[4] The dotted trajectories that appear to go, for example, from (0, 0) to (1, 0) actually correspond to trajectories through particular initial points. As $t \to \infty$ they approach the stable node (1, 0); as $t \to -\infty$ they approach the unstable node (0, 0).

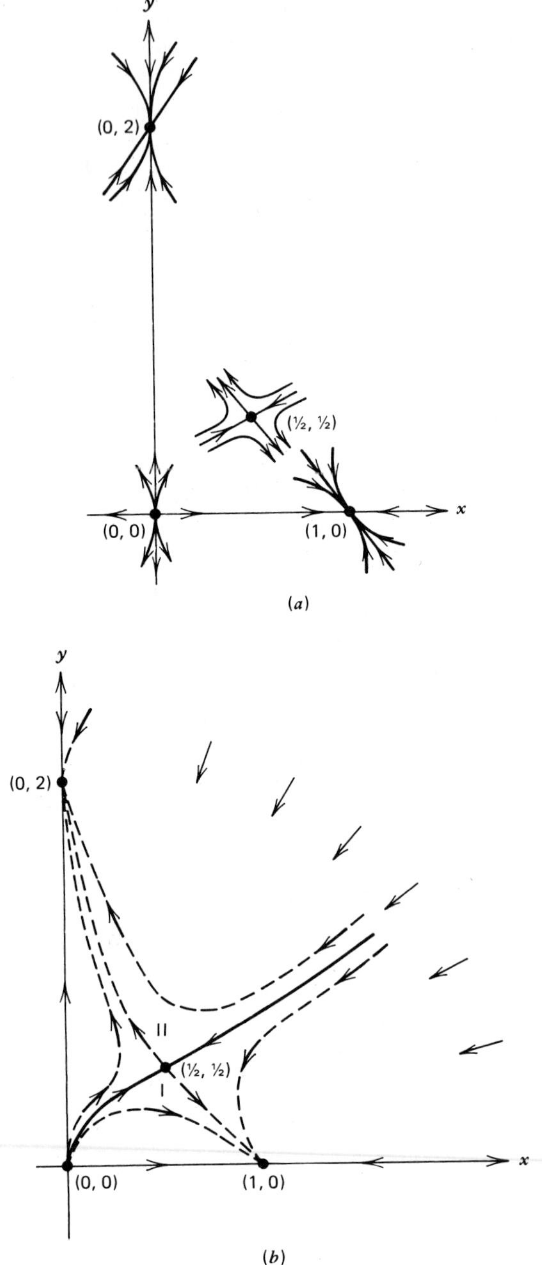

FIGURE 9.25 (*a*) *Trajectories in the neighborhood of each critical point of the system* (2). (*b*) *Overall pattern of trajectories for the system* (2).

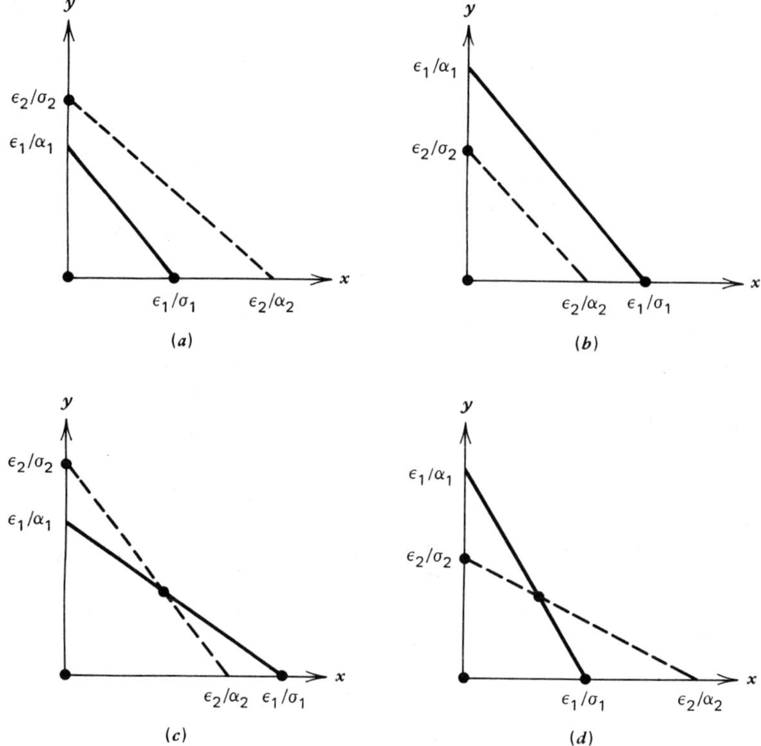

FIGURE 9.26 The various cases for the competing species system (1).

and substitute in Eqs. (1):

$$\begin{aligned}\frac{d}{dt}(X+u) &= (X+u)[\epsilon_1 - \sigma_1(X+u) - \alpha_1(Y+v)],\\ \frac{d}{dt}(Y+v) &= (Y+v)[\epsilon_2 - \sigma_2(Y+v) - \alpha_2(X+u)].\end{aligned} \tag{19}$$

Since $dX/dt = dY/dt = 0$, we have

$$\frac{du}{dt} = (X+u)[(\epsilon_1 - \sigma_1 X - \alpha_1 Y) - \sigma_1 u - \alpha_1 v], \tag{20a}$$

$$\frac{dv}{dt} = (Y+v)[(\epsilon_2 - \sigma_2 Y - \alpha_2 X) - \sigma_2 v - \alpha_2 u]. \tag{20b}$$

The right side of Eq. (20a) has the form $X(\epsilon_1 - \sigma_1 X - \alpha_1 Y) + (\ )u + (\ )v + (\ )u^2 + (\ )uv$. The constant term is zero since either $X = 0$ or $\epsilon_1 - \sigma_1 X - \alpha_1 Y = 0$. Similarly, the right side of Eq. (20b) reduces to the form $(\ )u + (\ )v + (\ )uv + (\ )v^2$. Next, it is clear that Eqs. (20) are almost linear, so we consider the corresponding linear system

$$\begin{aligned} du/dt &= \left[(\epsilon_1 - \sigma_1 X - \alpha_1 Y) - \sigma_1 X\right]u - \alpha_1 X v, \\ dv/dt &= -\alpha_2 Y u + \left[(\epsilon_2 - \sigma_2 Y - \alpha_2 X) - \sigma_2 Y\right]v. \end{aligned} \tag{21}$$

Equations (21), along with Theorem 9.2 of Section 9.3, can be used to determine the type and stability of any critical point (X, Y) of the original system (1). These linear equations are referred to as the linearized equations for small perturbations in the neighborhood of the critical point (X, Y). The process of deriving them is called linearization.

We use Eqs. (21) to determine whether the model given by the system (1) can ever lead to coexistence for the two species x and y, and if so, under what conditions on the parameters $\epsilon_1, \sigma_1, \ldots, \alpha_2$. The four possible situations are shown in Figure 9.26; coexistence is possible only in cases (c) and (d). The values of $X \neq 0$ and $Y \neq 0$ are obtained by solving the system of simultaneous linear algebraic equations (17). We readily obtain

$$X = \frac{\epsilon_1\sigma_2 - \epsilon_2\alpha_1}{\sigma_1\sigma_2 - \alpha_1\alpha_2}, \qquad Y = \frac{\epsilon_2\sigma_1 - \epsilon_1\alpha_2}{\sigma_1\sigma_2 - \alpha_1\alpha_2}. \tag{22}$$

Moreover, since $\epsilon_1 - \sigma_1 X - \alpha_1 Y = 0$ and $\epsilon_2 - \sigma_2 Y - \alpha_2 X = 0$, Eqs. (21) immediately reduce to

$$\frac{d}{dt}\begin{pmatrix} u \\ v \end{pmatrix} = \begin{pmatrix} -\sigma_1 X & -\alpha_1 X \\ -\alpha_2 Y & -\sigma_2 Y \end{pmatrix}\begin{pmatrix} u \\ v \end{pmatrix}. \tag{23}$$

The eigenvalues of the system (23) are found from the equation

$$r^2 + (\sigma_1 X + \sigma_2 Y)r + (\sigma_1\sigma_2 XY - \alpha_1\alpha_2 XY) = 0. \tag{24}$$

Thus

$$r = \frac{-(\sigma_1 X + \sigma_2 Y) \pm \sqrt{(\sigma_1 X + \sigma_2 Y)^2 - 4(\sigma_1\sigma_2 - \alpha_1\alpha_2)XY}}{2}. \tag{25}$$

If $\sigma_1\sigma_2 - \alpha_1\alpha_2 < 0$ then the radicand of Eq. (25) is positive and greater than $(\sigma_1 X + \sigma_2 Y)^2$. Thus the eigenvalues are real and of opposite sign. Consequently, the critical point (X, Y) is an unstable saddle point, and coexistence is not possible. This is case in the specific example given by Eqs. (2); there $\sigma_1 = 1$, $\alpha_1 = 1$, $\sigma_2 = \frac{1}{4}$, $\alpha_2 = \frac{3}{4}$ and $\sigma_1\sigma_2 - \alpha_1\alpha_2 = -\frac{1}{2}$.

On the other hand, if $\sigma_1\sigma_2 - \alpha_1\alpha_2 > 0$, then the radicand of Eq. (25) is less than $(\sigma_1 X + \sigma_2 Y)^2$. Thus the eigenvalues are real, negative, and unequal, or complex with negative real part. A simple analysis of the radicand of Eq. (25) shows that the eigenvalues cannot be complex (see Problem 6). Thus the critical point is an asymptotically stable improper node. Coexistence is possible if $\sigma_1\sigma_2 > \alpha_1\alpha_2$.

Let us relate this result to Figures 9.26c and 9.26d. In Figure 9.26c we have

$$\frac{\epsilon_1}{\sigma_1} > \frac{\epsilon_2}{\alpha_2} \quad \text{or} \quad \epsilon_1\alpha_2 > \epsilon_2\sigma_1 \quad \text{and} \quad \frac{\epsilon_2}{\sigma_2} > \frac{\epsilon_1}{\alpha_1} \quad \text{or} \quad \epsilon_2\alpha_1 > \epsilon_1\sigma_2. \tag{26}$$

These inequalities coupled with the condition that X and Y given by Eqs. (22) be positive yield the inequality $\sigma_1\sigma_2 < \alpha_1\alpha_2$. Hence in this case the mixed state is an unstable saddle point. Corresponding to Figure 9.26d, we have

$$\frac{\epsilon_2}{\alpha_2} > \frac{\epsilon_1}{\sigma_1} \quad \text{or} \quad \epsilon_2\sigma_1 > \epsilon_1\alpha_2 \quad \text{and} \quad \frac{\epsilon_1}{\alpha_1} > \frac{\epsilon_2}{\sigma_2} \quad \text{or} \quad \epsilon_1\sigma_2 > \epsilon_2\alpha_1. \tag{27}$$

Now, the condition X and Y positive yields $\sigma_1\sigma_2 > \alpha_1\alpha_2$. Hence this mixed state is asymptotically stable. For this case we can also show that the critical points $(0, 0)$, $(\epsilon_1/\sigma_1, 0)$ and $(0, \epsilon_2/\sigma_2)$ are unstable. Thus no matter what the initial values of $x \neq 0$ and $y \neq 0$ are, the two species approach an equilibrium state of coexistence given by Eqs. (22).

Equations (1) provide the biological interpretation of the result that $\sigma_1\sigma_2 > \alpha_1\alpha_2$ leads to coexistence and $\alpha_1\alpha_2 > \sigma_1\sigma_2$ does not allow coexistence. The σ's are a measure of the inhibitory effect the growth of each species has on its own growth rate, while the α's are a measure of the inhibiting effect the growth of each species has on the other species (interaction). Thus when $\sigma_1\sigma_2 > \alpha_1\alpha_2$ interaction is "small" and the species can coexist; when $\alpha_1\alpha_2 > \sigma_1\sigma_2$ interaction (competition) is "large" and the species cannot coexist—one must die out.

PREDATOR–PREY. As a second example we consider the classical predator–prey problem. We study an ecological situation involving two species, one of which preys on the other (does not compete with it for food but actually preys on it) while the other lives on a different source of food. An example is foxes and rabbits in a closed forest; the foxes prey on the rabbits, the rabbits live on the vegetation in the forest. Other examples are bass in a lake as predators and redear (sunfish) as prey, and lady bugs as predators and aphids (insects that suck the juice of plants) as prey. Let $H(t)$ and $P(t)$ be the populations of prey and predator, respectively, at time t.

We build as simple a model of the interaction as possible. We make the following assumptions:

1. In the absence of the predator the prey grows without bound; thus $dH/dt = aH$, $a > 0$, for $P = 0$.
2. In the absence of the prey the predator dies out, thus $dP/dt = -cP$, $c > 0$, for $H = 0$.
3. The increase in the number of predators is wholly dependent on the food supply (the prey) and the prey are consumed at a rate proportional to the

number of encounters between predators and prey. Thus, for example, if the number of prey is doubled the number of encounters is doubled. Encounters decrease the number of prey and increase the number of predators. A fixed proportion of prey is killed in each encounter, and the rate of population growth of the predator is enhanced by a factor proportional to the amount of prey consumed.

As a consequence, we have the equations

$$
\begin{aligned}
dH/dt &= aH - \alpha HP = H(a - \alpha P), \\
dP/dt &= -cP + \gamma HP = P(-c + \gamma H).
\end{aligned}
\tag{28}
$$

The constants a, c, α, and γ are positive; a and c are the growth rate of the prey and the death rate of the predator, respectively, and α and γ are measures of the effect of the interaction between the two species. Equations (28) are known as the Lotka-Volterra equations. They were developed in papers by Lotka[5] in 1925 and Volterra[6] in 1926. Although these equations are simple, they do characterize a wide class of problems. Ways of making them more realistic are discussed at the end of this section and in the problems.

What happens for given initial values of $P > 0$ and $H > 0$? Will the predators eat all of their prey and in turn die out, will the predators die out because of a too low level of prey and then the prey grow without bound, will an equilibrium state be reached, or will a cyclic fluctuation of prey and predator occur?

The critical points of Eqs. (28) are the solutions of

$$
H(a - \alpha P) = 0, \qquad P(-c + \gamma H) = 0. \tag{29}
$$

These solutions are $H = 0$, $P = 0$ and $H = c/\gamma$, $P = a/\alpha$. It is easy to show that the critical point $(0, 0)$ is a saddle point, and hence unstable. Entrance to the saddle point is along the line $H = 0$; all other trajectories recede from the critical point.

[5]Alfred J. Lotka (1880–1949), an American biophysicist, was born in what is now the Ukraine, and was educated mainly in Europe. He is remembered chiefly for his formulation of the Lotka-Volterra equations. He was also the author, in 1924, of the first book on mathematical biology; it is now available as *Elements of Mathematical Biology* (New York: Dover, 1956).

[6]Vito Volterra (1860–1940), a distinguished Italian mathematician, held professorships at Pisa, Turin, and Rome. He is particularly famous for his work in integral equations and functional analysis. Indeed, one of the major classes of integral equations is named for him; see Problem 14 of Section 6.5. His theory of interacting species was motivated by data collected by a friend, D'Ancona, concerning fish catches in the Adriatic Sea. A translation of his 1926 paper can be found in an appendix of R. N. Chapman, *Animal Ecology with Special Reference to Insects* (New York: McGraw-Hill, 1931).

To study the critical point $(c/\gamma, a/\alpha)$ we let

$$H = (c/\gamma) + u, \qquad P = (a/\alpha) + v. \tag{30}$$

Substituting for H and P in Eqs. (28), we obtain

$$\frac{d}{dt}\begin{pmatrix} u \\ v \end{pmatrix} = \begin{pmatrix} 0 & -\alpha c/\gamma \\ \gamma a/\alpha & 0 \end{pmatrix}\begin{pmatrix} u \\ v \end{pmatrix} + \begin{pmatrix} -\alpha uv \\ \gamma uv \end{pmatrix}. \tag{31}$$

The system (31) is almost linear, and the corresponding linear system is

$$\frac{d}{dt}\begin{pmatrix} u \\ v \end{pmatrix} = \begin{pmatrix} 0 & -\alpha c/\gamma \\ \gamma a/\alpha & 0 \end{pmatrix}\begin{pmatrix} u \\ v \end{pmatrix}. \tag{32}$$

The eigenvalues of the system (32) are $r = \pm i\sqrt{ac}$, so the critical point is a (stable) center of the linear system. To find the trajectories of the system (32) we can divide the second equation by the first:

$$\frac{dv}{du} = \frac{dv/dt}{du/dt} = -\frac{(\gamma a/\alpha)u}{(\alpha c/\gamma)v}, \tag{33}$$

or

$$(\gamma a/\alpha)u\,du + (\alpha c/\gamma)v\,dv = 0. \tag{34}$$

Consequently,

$$(\gamma a/\alpha)u^2 + (\alpha c/\gamma)v^2 = k, \tag{35}$$

where k is an arbitrary nonnegative constant of integration. Thus the trajectories are ellipses, a few of which are sketched in Figure 9.27.

While the critical point is a stable center of the linear system (32), we need to assess its character for the almost linear system (31). Here, as we know, our theory for almost linear systems fails. The effect of the nonlinear terms may be to

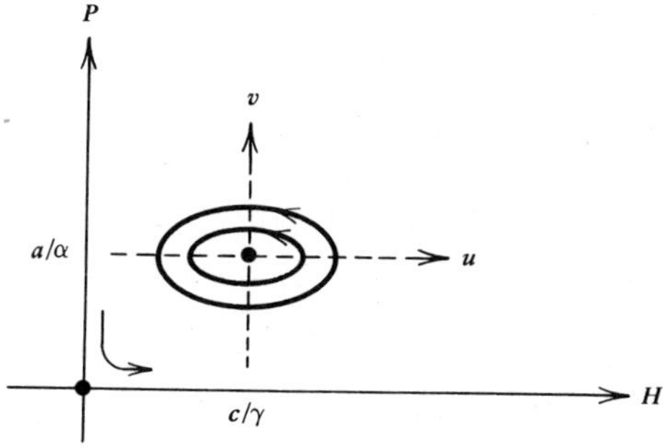

FIGURE 9.27 Trajectories of the linearized predator–prey equation.

change the center into a stable spiral point, or into an unstable spiral point, or it may remain as a stable center. Fortunately, for the predator–prey problem we can actually solve the nonlinear equations (28) and determine what happens. Dividing the second of Eqs. (28) by the first equation, we obtain

$$\frac{dP}{dH} = \frac{P(-c + \gamma H)}{H(a - \alpha P)}. \tag{36}$$

Upon separating the variables in Eq. (36), we have

$$\frac{a - \alpha P}{P}\, dP = \frac{-c + \gamma H}{H}\, dH,$$

from which it follows that

$$a \ln P - \alpha P = -c \ln H + \gamma H + \ln C, \tag{37}$$

where C is a constant of integration. We cannot solve Eq. (37) explicitly for P in terms of H or for H in terms of P, but it can be shown that the graph of this equation for a fixed value of C is a closed curve (not an ellipse, of course) enclosing the critical point $(c/\gamma, a/\alpha)$.[7] Thus the predator and prey have a cyclic variation about the critical point and the critical point is also a center of the nonlinear system.

We can analyze this cyclic variation in more detail when the deviation from the point $(c/\gamma, a/\alpha)$ is small; that is, when it is permissible to linearize the perturbation equations (31) for u and v. As we have noted, the trajectories are the family of ellipses given by Eq. (35). We can also verify, either by solving Eqs. (32) or by direct substitution, that the solution of Eqs. (32) is

$$u(t) = \frac{c}{\gamma} K \cos\left(\sqrt{ac}\, t + \phi\right), \qquad v(t) = \frac{a}{\alpha}\sqrt{\frac{c}{a}}\, K \sin\left(\sqrt{ac}\, t + \phi\right), \tag{38}$$

where the constants K and ϕ are determined by the initial conditions. Thus

$$\begin{aligned} H(t) &= \frac{c}{\gamma} + \frac{c}{\gamma} K \cos\left(\sqrt{ac}\, t + \phi\right), \\ P(t) &= \frac{a}{\alpha} + \frac{a}{\alpha}\sqrt{\frac{c}{a}}\, K \sin\left(\sqrt{ac}\, t + \phi\right). \end{aligned} \tag{39}$$

These equations are valid for the elliptical trajectories close to the critical point $(c/\gamma, a/\alpha)$. We can use them to draw several conclusions about the cyclic variation of the predator and prey on such trajectories.

[7]Volterra gave a clever elementary geometric proof of this result. Shorter (but mathematically more advanced) proofs have also been discovered.

1. The size of the predator and prey populations varies sinusoidally with period $2\pi/\sqrt{ac}$. This period of oscillation is independent of the initial conditions.
2. The predator and prey populations are out of phase by one-quarter of a cycle. The prey leads and the predator lags as one might expect. This is discussed in Problem 9.
3. The amplitudes of the oscillations are Kc/γ for the prey and $a\sqrt{c}\,K/\alpha\sqrt{a}$ for the predator and hence depend on the initial conditions as well as the parameters of the problem.
4. The average numbers of predators and prey over one complete cycle are c/γ and a/α, respectively. These are the same as the equilibrium populations (see Problem 10).

Cyclic variations of predator and prey as predicted by Eqs. (28) have been observed in nature. One striking example is described by Odum (pp. 191–192): based on the records of the Hudson Bay Company of Canada, the abundance of lynx and snowshoe hare as indicated by the number of pelts turned in over the period 1845–1935 shows a distinct periodic variation with a period of nine to ten years. The peaks of abundance are followed by very rapid declines, and the peaks of abundance of the lynx and hare are out of phase, with that of the hare preceding that of the lynx by a year or more.

The Volterra-Lotka model of the predator–prey problem has revealed a cyclic variation that was perhaps intuitively expected. On the other hand, the use of the Volterra-Lotka model in other situations can lead to conclusions that are not intuitively obvious. An example revealing a possible danger in using insecticides is given in Problem 12.

One criticism of the Volterra-Lotka predator–prey model is that in the absence of the predator the prey will grow without bound. This can be corrected by allowing for the natural inhibiting effect that an increasing population has on the growth rate of the population; for example, by modifying the first of Eqs. (28) so that when $P = 0$ it reduces to a logistic equation for H (see Problem 13). The models of two competing species and predator–prey discussed here can be modified to allow for the effect of time delays; probabilistic and statistical effects can also be included. Finally, we mention that there are *discrete* analogs of each of the problems we have discussed corresponding to species that breed only at certain times. The mathematics of the discrete problems are often interesting and some of the results are unexpected. These generalizations are discussed in the references given at the end of this chapter as well as in other books on mathematical biology and ecology.

We conclude with a final warning. Using only elementary phase plane theory for one and two nonlinear ordinary differential equations, we have been able to illustrate several of the fundamental principles of simple biological systems. But one should not be misled—ecology is not this simple.

PROBLEMS

Each of Problems 1 through 5 can be interpreted as describing the interaction of two species with population densities x and y. In each of these problems
(a) Find the critical points.
(b) For each critical point find the corresponding linear system. Find the eigenvalues of the linear system, and determine the type and stability of each critical point.
(c) Sketch the trajectories in the neighborhood of each critical point. Determine the limiting behavior of x and y as $t \to \infty$.

1. $dx/dt = x(1 - x + 0.5y)$
   $dy/dt = y(2.5 - 1.5y + 0.25x)$
2. $dx/dt = x(1.5 - x - 0.5y)$
   $dy/dt = y(2 - y - 0.75x)$
3. $dx/dt = x(1 - 0.5x - 0.5y)$
   $dy/dt = y(-0.25 + 0.5x)$
4. $dx/dt = x(1.5 - x - 0.5y)$
   $dy/dt = y(2 - 0.5y - 1.5x)$
5. $dx/dt = x(1.125 - x - 0.5y)$
   $dy/dt = y(-1 + x)$
6. Show that

$$(\sigma_1 X + \sigma_2 Y)^2 - 4(\sigma_1\sigma_2 - \alpha_1\alpha_2)XY = (\sigma_1 X - \sigma_2 Y)^2 + 4\alpha_1\alpha_2 XY.$$

   Hence conclude that the eigenvalues given by Eq. (25) can never be complex.
7. Two species of fish that compete with each other for food, but do not prey on each other, are bluegill and redear. Suppose that a pond is stocked with bluegill and redear and let x and y be the populations of bluegill and redear, respectively, at time t. Suppose further that the competition is modeled by the equations

$$dx/dt = x(\epsilon_1 - \sigma_1 x - \alpha_1 y),$$

$$dy/dt = y(\epsilon_2 - \sigma_2 y - \alpha_2 x).$$

   (a) If $\epsilon_2/\alpha_2 > \epsilon_1/\sigma_1$ and $\epsilon_2/\sigma_2 > \epsilon_1/\alpha_1$, show that the only equilibrium populations in the pond are no fish, no redear, or no bluegill. What will happen?
   (b) If $\epsilon_1/\sigma_1 > \epsilon_2/\alpha_2$ and $\epsilon_1/\alpha_1 > \epsilon_2/\sigma_2$, show that the only equilibrium populations in the pond are no fish, no redear, or no bluegill. What will happen?
8. Consider the competition between bluegill and redear mentioned in Problem 7. Suppose that $\epsilon_2/\alpha_2 > \epsilon_1/\sigma_1$ and $\epsilon_1/\alpha_1 > \epsilon_2/\sigma_2$, so, as shown in the text, there is a stable equilibrium point at which both species can coexist. It is convenient to rewrite the equations of Problem 7 in terms of the carrying capacities of the pond for bluegill ($B = \epsilon_1/\sigma_1$) in the absence of redear and for redear ($R = \epsilon_2/\sigma_2$) in the absence of bluegill.
   (a) Show that the equations of Problem 7 take the form

$$\frac{dx}{dt} = \epsilon_1 x\left(1 - \frac{1}{B}x - \frac{\gamma_1}{B}y\right), \qquad \frac{dy}{dt} = \epsilon_2 y\left(1 - \frac{1}{R}y - \frac{\gamma_2}{R}x\right),$$

   where $\gamma_1 = \alpha_1/\sigma_1$ and $\gamma_2 = \alpha_2/\sigma_2$. Determine the coexistence equilibrium point (X, Y) in terms of B, R, γ_1, and γ_2.

(b) Now suppose that a fisherman fishes only for bluegill with the effect that B is reduced. What effect does this have on the equilibrium populations? Is it possible, by fishing, to reduce the population of bluegill to such a level that they will die out?

9. In this problem we examine the phase difference between the cyclic variations of the predator and prey populations as given by Eqs. (39) of the text. Suppose we assume that $K > 0$ and that T is measured from the time that the prey population (H) is a maximum; then $\phi = 0$. Show that the predator population (P) is a maximum at $t = \pi/2\sqrt{ac} = T/4$, where T is the period of the oscillation. When is the prey population increasing most rapidly, decreasing most rapidly, a minimum? Answer the same questions for the predator population. Draw a typical elliptic trajectory enclosing the point $(c/\gamma, a/\alpha)$, and mark these points on it.

10. The average sizes of the prey and predator populations are defined as
$$\overline{H} = \frac{1}{T}\int_A^{A+T} H(t)\,dt, \qquad \overline{P} = \frac{1}{T}\int_A^{A+T} P(t)\,dt,$$
respectively, where T is the period of a full cycle and A is any nonnegative constant. Show for trajectories near the critical point that $\overline{H} = c/\gamma$, $\overline{P} = a/\alpha$.

11. Suppose that the predator–prey equations (28) of the text govern foxes (P) and rabbits (H) in a forest. A trapping company is engaging in trapping foxes and rabbits for their pelts. Explain why it is reasonable for the company to conduct its operation in such a way as to move the population of each species closer to the center $(c/\gamma, a/\alpha)$. When is it best to trap foxes? Rabbits? Rabbits and foxes? Neither? *Hint:* See Problem 9. A mathematical argument is not required.

12. Suppose that an insect population (H) is controlled by a natural predator population (P) according to the model (28), so that there are small cyclic variations of the populations about the critical point $(c/\gamma, a/\alpha)$. Show that it is self-defeating to employ an insecticide if the insecticide also kills the predator. Assume that the insecticide kills both prey and predator at rates proportional to each population, respectively. To ban insecticides on the basis of this very simple model would certainly be ill-advised. On the other hand, it is also rash to ignore the possible genuine existence of a phenomenon suggested by a simple model.

13. As was mentioned in the text, one improvement in the predator–prey model is to modify the equation for the prey so that it has the form of a logistic equation in the absence of the predator. Thus in place of Eqs. (28) we consider the model system
$$dH/dt = H(a - \sigma H - \alpha P),$$
$$dP/dt = P(-c + \gamma H),$$
where a, σ, α, c, and γ are positive constants. Determine all critical points and discuss their nature and stability. Assume that $a/\sigma \gg c/\gamma$. What happens for initial data $H \neq 0$, $P \neq 0$?

9.5 Liapunov's Second Method

In Section 9.3 we showed how the stability of a critical point of an almost linear system can usually be determined from a study of the corresponding linear system. However, no conclusion can be drawn when the critical point is a center of the corresponding linear system. Examples of this situation are the undamped pendulum, Eqs. (1) and (2) below, and the predator–prey problem discussed in Section 9.4. Also, for an asymptotically stable critical point it may be important to investigate the region of asymptotic stability; that is, the domain such that all solutions starting within that domain approach the critical point. Since the theory of almost linear systems is a local theory, it provides no information about this question.

In this section we discuss another approach, known as *Liapunov's*[8] *second method* or *direct method*. The method is referred to as a direct method because no knowledge of the solution of the system of differential equations is required. Rather, conclusions about the stability or instability of a critical point are obtained by constructing a suitable auxiliary function. The technique is a very powerful one that provides a more global type of information; for example, an estimate of the extent of the region of asymptotic stability of a critical point. In addition, Liapunov's second method can also be used to study systems of equations that are not almost linear; however, we will not discuss such problems.

Basically, Liapunov's second method is a generalization of the physical principles that for a conservative system, (i) a rest position is stable if the potential energy is a local minimum, otherwise it is unstable, and (ii) the total energy is a constant during any motion. To illustrate these concepts, again consider the undamped pendulum (a conservative mechanical system), which is governed by the equation

$$\frac{d^2\theta}{dt^2} + \frac{g}{l}\sin\theta = 0. \tag{1}$$

The corresponding system of first order equations is

$$\frac{dx}{dt} = y, \qquad \frac{dy}{dt} = -\frac{g}{l}\sin x, \tag{2}$$

where $x = \theta$ and $y = d\theta/dt$. Omitting an arbitrary constant, the potential energy U is the work done in lifting the pendulum above its lowest position, namely,

$$U(x, y) = mgl(1 - \cos x); \tag{3}$$

see Figure 9.20 in Section 9.3. The critical points of the system (2) are $x = \pm n\pi$, $y = 0$, $n = 0, 1, 2, 3, \ldots$, corresponding to $\theta = \pm n\pi$, $d\theta/dt = 0$. Physically, we

[8]Aleksandr M. Liapunov (1857–1918) was a Russian mathematician who lived mainly in Kharkov and St. Petersburg. His so-called second method was first presented in his doctoral dissertation in 1892, but did not become widely known until some years later.

expect that the points $x = 0$, $y = 0$; $x = \pm 2\pi$, $y = 0$; ... corresponding to $\theta = 0, \pm 2\pi, \ldots$ for which the pendulum bob is vertical with the weight down are stable; and that the points $x = \pm\pi$, $y = 0$; $x = \pm 3\pi$, $y = 0$; ... corresponding to $\theta = \pm\pi, \pm 3\pi, \ldots$ for which the pendulum bob is vertical with the weight up are unstable. This agrees with statement (i), for at the former points U is a minimum equal to zero, and at the latter points U is a maximum equal to $2mgl$.

Next consider the total energy V, which is the sum of the potential energy U and the kinetic energy $\frac{1}{2}ml^2(d\theta/dt)^2$. In terms of x and y

$$V(x, y) = mgl(1 - \cos x) + \tfrac{1}{2}ml^2y^2. \tag{4}$$

On a trajectory corresponding to a solution $x = \phi(t)$, $y = \psi(t)$ of Eqs. (2), V can be considered as a function of t. The derivative of $V[\phi(t), \psi(t)]$ is called the rate of change of V following the trajectory. By the chain rule

$$\begin{aligned}\frac{dV[\phi(t), \psi(t)]}{dt} &= V_x[\phi(t), \psi(t)]\frac{d\phi(t)}{dt} + V_y[\phi(t), \psi(t)]\frac{d\psi(t)}{dt} \\ &= (mgl\sin x)\frac{dx}{dt} + ml^2y\frac{dy}{dt},\end{aligned} \tag{5}$$

where it is understood that $x = \phi(t)$, $y = \psi(t)$. Finally, substituting in Eqs. (5) for dx/dt and dy/dt from Eqs. (2) we find that $dV/dt = 0$. Hence V is a constant along any trajectory of the system (2), which is in agreement with our earlier remark (ii) that the total energy is constant during any motion of a conservative system.

It is important to note that at any point (x, y) the rate of change of V along the trajectory through that point was computed *without actually solving* the system (2). It is precisely this fact that allows us to use Liapunov's second method for systems whose solution we do not know, and hence makes it such an important technique.

At the stable critical points, $x = \pm 2n\pi$, $y = 0$, $n = 0, 1, 2, \ldots$, the energy V is zero. If the initial state, say (x_1, y_1), of the pendulum is sufficiently near a stable critical point, then the energy $V(x_1, y_1)$ is small, and the motion (trajectory) associated with this energy stays close to the critical point. It can be shown that if $V(x_1, y_1)$ is sufficiently small, then the trajectory is closed and contains the critical point. For example, suppose that (x_1, y_1) is near $(0, 0)$ and that $V(x_1, y_1)$ is very small. The equation of the trajectory with energy $V(x_1, y_1)$ is

$$V(x, y) = mgl(1 - \cos x) + \tfrac{1}{2}ml^2y^2 = V(x_1, y_1).$$

For x small we have $1 - \cos x = 1 - (1 - x^2/2! + \cdots) \simeq x^2/2$. Thus the equation of the trajectory is approximately

$$\tfrac{1}{2}mglx^2 + \tfrac{1}{2}ml^2y^2 = V(x_1, y_1)$$

or

$$\frac{x^2}{2V(x_1, y_1)/mgl} + \frac{y^2}{2V(x_1, y_1)/ml^2} = 1.$$

This is an ellipse enclosing the critical point (0, 0); the smaller $V(x_1, y_1)$ is, the smaller are the major and minor axes of the ellipse. Physically, the closed trajectory corresponds to a solution that is periodic in time—the motion is a small oscillation about the equilibrium point.

If damping is present, however, it is natural to expect that the amplitude of the motion decays in time and that the stable critical point (center) becomes an asymptotically stable critical point (spiral point). See the sketch of the trajectories for the damped pendulum in Figure 9.23 of Section 9.3. This can almost be argued from a consideration of dV/dt. For the damped pendulum, the total energy is still given by Eq. (4), but now from Eqs. (16) of Section 9.3 $dx/dt = y$ and $dy/dt = -(g/l)\sin x - (c/lm)y$. Substituting for dx/dt and dy/dt in Eq. (5) gives $dV/dt = -cly^2 \le 0$. Thus the energy is nonincreasing along any trajectory and, except for the line $y = 0$, the motion is such that the energy decreases, and hence each trajectory must approach a point of minimum energy—a stable equilibrium point. If $dV/dt < 0$ instead of $dV/dt \le 0$, it is reasonable to expect that this would be true for all trajectories that start sufficiently close to the origin.

To pursue these ideas further, consider the autonomous system

$$dx/dt = F(x, y), \qquad dy/dt = G(x, y), \tag{6}$$

and suppose that the point $x = 0$, $y = 0$ is an asymptotically stable critical point. Then there exists some domain D containing (0, 0) such that every trajectory that starts in D must approach the origin as $t \to \infty$. Suppose that there exists an "energy" function V such that $V(x, y) \ge 0$ for (x, y) in D with $V = 0$ only at the origin. Since each trajectory in D approaches the origin as $t \to \infty$, then following any particular trajectory, V decreases to zero as t approaches infinity. The type of result we want to prove is essentially the converse: if, on every trajectory, V decreases to zero as t increases, then the trajectories must approach the origin as $t \to \infty$, and hence the origin as asymptotically stable. First, however, it is necessary to make several definitions.

Let V be defined on some domain D containing the origin. Then V is said to be *positive definite* on D if $V(0,0) = 0$ and $V(x, y) > 0$ for all other points in D. Similarly, V is said to be *negative definite* on D if $V(0,0) = 0$ and $V(x, y) < 0$ for all other points in D. If the inequalities $>$ and $<$ are replaced by $\ge$ and $\le$, then V is said to be *positive semidefinite* and *negative semidefinite*, respectively. We emphasize that in speaking of a positive definite (negative definite, . . .) function on a domain D containing the origin, the function must be zero at the origin in addition to satisfying the proper inequality at all other points in D.

EXAMPLE 1

The function

$$V(x, y) = \sin(x^2 + y^2)$$

is positive definite on $x^2 + y^2 < \pi/2$ since $V(0,0) = 0$ and $V(x, y) > 0$ for $0 < x^2 + y^2 < \pi/2$. However, the function

$$V(x, y) = (x + y)^2$$

is only positive semidefinite since $V(x, y) = 0$ on the line $y = -x$.

We also want to consider the function

$$\dot{V}(x, y) = V_x(x, y)F(x, y) + V_y(x, y)G(x, y), \tag{7}$$

where F and G are the same functions as in Eqs. (6). We choose this notation because $\dot{V}(x, y)$ can be identified as the rate of change of V along the trajectory of the system (6) that passes through the point (x, y). That is, if $x = \phi(t)$, $y = \psi(t)$ is a solution of the system (6), then

$$\begin{aligned} \frac{dV[\phi(t), \psi(t)]}{dt} &= V_x[\phi(t), \psi(t)]\frac{d\phi(t)}{dt} + V_y[\phi(t), \psi(t)]\frac{d\psi(t)}{dt} \\ &= V_x(x, y)F(x, y) + V_y(x, y)G(x, y) \\ &= \dot{V}(x, y). \end{aligned} \tag{8}$$

The function $\dot{V}$ is sometimes referred to as the derivative of V with respect to the system (6).

We now state two Liapunov theorems, the first dealing with stability, the second with instability.

Theorem 9.3. *Suppose that the autonomous system* (6) *has an isolated critical point at the origin. If there exists a function V that is continuous and has continuous first partial derivatives, is positive definite, and for which the function $\dot{V}$ given by Eq.* (7) *is negative definite on some domain D in the xy plane containing* (0, 0), *then the origin is an asymptotically stable critical point. If $\dot{V}$ is negative semidefinite, then the origin is a stable critical point.*

Theorem 9.4. *Let the origin be an isolated critical point of the autonomous system* (6). *Let V be a function that is continuous and has continuous first partial derivatives. Suppose that* $V(0,0) = 0$ *and that in every neighborhood of the origin there is at least one point at which V is positive (negative). Then if there exists a domain D containing the origin such that the function $\dot{V}$ as given by Eq.* (7) *is positive definite (negative definite) on D, then the origin is an unstable critical point.*

The function V is called a *Liapunov function*. Before sketching geometric proofs of Theorems 9.3 and 9.4, we note that the difficulty in using these theorems is that they tell us nothing about how to construct a Liapunov function, assuming that one exists. In cases where the autonomous system (6) represents a physical problem, it is natural first to consider the actual total energy function of the system as a possible Liapunov function. However, we emphasize that Theorems 9.3 and 9.4 are applicable in cases where the concept of physical energy is not pertinent. In such cases a judicious trial-and-error approach may be necessary.

Now consider the second part of Theorem 9.3; that is, the case $\dot{V}(x, y) \leq 0$. Let $c \geq 0$ be a constant and consider the curve in the xy plane given by the equation $V(x, y) = c$. For $c = 0$ the curve reduces to the single point $x = 0$, $y = 0$. However, for $c > 0$ and sufficiently small, it can be shown by using the continuity of V that the curve is a closed curve containing the origin as illustrated in Figure 9.28*a*. There may, of course, be other curves in the xy plane corresponding to the same value of c, but these are not of interest. Further, again by continuity, as c gets smaller and smaller the closed curves $V(x, y) = c$ enclosing the origin shrink to the origin. We show that a trajectory starting inside of a closed curve $V(x, y) = c$ cannot cross to the outside. Thus, given a circle of radius ϵ about the origin, by taking c sufficiently small we can insure that every trajectory starting inside of the closed curve $V(x, y) = c$ stays within the circle of radius ϵ; indeed, it stays within the closed curve $V(x, y) = c$ itself. Thus the origin is a stable critical point.

It is shown in calculus that the vector

$$\nabla V(x, y) = V_x(x, y)\mathbf{i} + V_y(x, y)\mathbf{j}, \tag{9}$$

known as the gradient of V, is normal to the curve $V(x, y) = c$ and points in the direction of increasing V. In the present case, V increases outward from the origin, so ∇V points away from the origin, as indicated in Figure 9.28*b*. Next, consider a trajectory $x = \phi(t)$, $y = \psi(t)$ of the system (6), and recall that the vector $\mathbf{T}(t) = \phi'(t)\mathbf{i} + \psi'(t)\mathbf{j}$ is tangent to the trajectory at each point; see Figure 9.28*b*. Let $x_1 = \phi(t_1)$, $y_1 = \psi(t_1)$ be a point of intersection of the trajectory and a closed curve $V(x, y) = c$. At this point $\phi'(t_1) = F(x_1, y_1)$, $\psi'(t_1) = G(x_1, y_1)$,

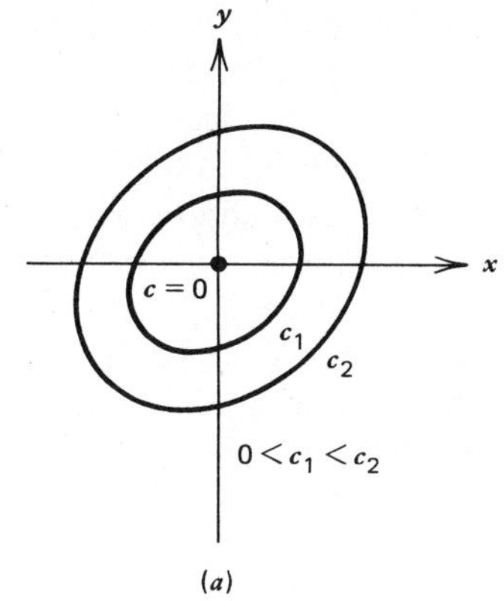

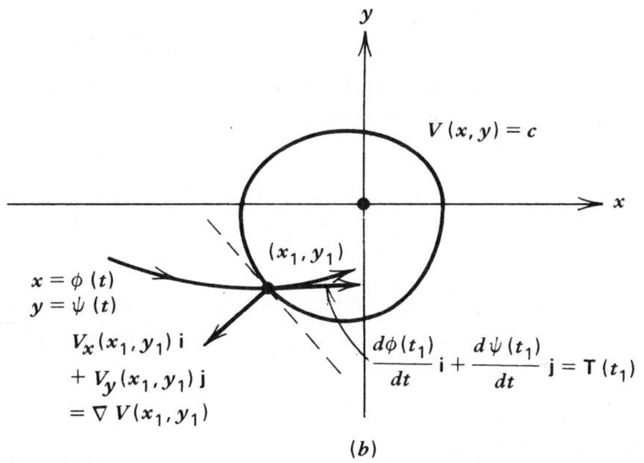

FIGURE 9.28 Geometrical interpretation of Liapunov's method.

so from Eq. (7) we obtain

$$\begin{aligned} \dot{V}(x_1, y_1) &= V_x(x_1, y_1)\phi'(t_1) + V_y(x_1, y_1)\psi'(t_1) \\ &= \left[V_x(x_1, y_1)\mathbf{i} + V_y(x_1, y_1)\mathbf{j}\right] \cdot \left[\phi'(t_1)\mathbf{i} + \psi'(t_1)\mathbf{j}\right] \\ &= \nabla V(x_1, y_1) \cdot \mathbf{T}(t_1). \end{aligned} \tag{10}$$

Thus $\dot{V}(x_1, y_1)$ is the scalar product of the vector $\nabla V(x_1, y_1)$ and the vector $\mathbf{T}(t_1)$. Since $\dot{V}(x_1, y_1) \le 0$, it follows that the cosine of the angle between $\nabla V(x_1, y_1)$ and $\mathbf{T}(t_1)$ is also less than or equal to zero; hence the angle itself is in the range $[\pi/2, 3\pi/2]$. Thus the direction of motion on the trajectory is inward with respect to $V(x, y) = c$ or, at worst, tangent to this curve. Trajectories starting inside a closed curve $V(x, y) = c$ (no matter how small c is) cannot escape, so the origin is a stable point. If $\dot{V}(x, y) < 0$, then the trajectories passing through points on the curve are actually pointed inward. As a consequence, it can be shown that trajectories starting sufficiently close to the origin must approach the origin; hence the origin is asymptotically stable.

A geometric proof of Theorem 9.4 follows by somewhat similar arguments. Briefly, suppose that $\dot{V}$ is positive definite, and suppose that given any circle about the origin there is an interior point (x_1, y_1) at which $V(x_1, y_1) > 0$. Consider a trajectory that starts at (x_1, y_1). Along this trajectory it follows from Eq. (8) that V must increase, since $\dot{V}(x, y) > 0$; furthermore, since $V(x_1, y_1) > 0$ the trajectory cannot approach the origin because $V(0,0) = 0$. This shows that the origin cannot be asymptotically stable. By further exploiting the fact that $\dot{V}(x, y) > 0$, it is possible to show that the origin is an unstable point; however, we will not pursue this argument.

To illustrate the use of Theorem 9.3 we consider the question of the stability of the critical point $(0,0)$ of the undamped pendulum equations (2). While the system (2) is almost linear, the point $(0,0)$ is a center of the corresponding linear system, so no conclusion can be drawn from Theorem 9.2. Since the mechanical system is conservative, it is natural to suspect that the total energy function V given by Eq. (4) is a Liapunov function. For example, if we take D to be the domain $-\pi/2 < x < \pi/2$, $-\infty < y < \infty$, then V is positive definite. As we have seen, $\dot{V}(x, y) = 0$, so it follows from the second part of Theorem 9.3 that the critical point $(0,0)$ of Eqs. (2) is a stable critical point. The damped pendulum is discussed in Problem 4.

From a practical point of view one is usually interested more in the region of asymptotic stability. One of the simplest results dealing with this question is given by the following theorem.

Theorem 9.5. *Let the origin be an isolated critical point of the autonomous system* (6). *Let the function V be continuous and have continuous first partial derivatives. If there is a bounded domain D_K containing the origin where $V(x, y) < K$, V is positive definite, and $\dot{V}$ is negative definite, then every solution of Eqs.* (6) *that starts at a point in D_K approaches the origin as t approaches infinity.*

In other words, Theorem 9.5 says that if $x = \phi(t)$, $y = \psi(t)$ is the solution of Eqs. (6) for initial data lying in D_K then (x, y) approaches the critical point $(0,0)$ as $t \to \infty$. Thus D_K gives a region of asymptotic stability; of course, it may not be the entire region of asymptotic stability. This theorem is proved by showing that (i) there are no periodic solutions of the system (6) in D_K, and (ii) there are no other critical points in D_K. It then follows that trajectories starting in D_K cannot escape and, therefore, must tend to the origin as t tends to infinity.

Theorems 9.3 and 9.4 give sufficient conditions for stability and instability, respectively. However, these conditions are not necessary, nor does our failure to determine a suitable Liapunov function mean that there is not one. Unfortunately, there are no general methods for the construction of Liapunov functions; however, there has been extensive work on the construction of Liapunov functions for special classes of equations. One simple result from elementary algebra, which is often useful in constructing positive definite or negative definite functions, is stated without proof in the following theorem.

Theorem 9.6. *The function*

$$V(x, y) = ax^2 + bxy + cy^2 \tag{11}$$

is positive definite if, and only if,

$$a > 0 \quad \text{and} \quad 4ac - b^2 > 0, \tag{12}$$

and is negative definite if, and only if,

$$a < 0 \quad \text{and} \quad 4ac - b^2 > 0. \tag{13}$$

The use of Theorem 9.6 is illustrated in the following example.

EXAMPLE 2

Show that the critical point (0, 0) of the autonomous system

$$\frac{dx}{dt} = -x - xy^2, \qquad \frac{dy}{dt} = -y - x^2y \tag{14}$$

is asymptotically stable.

We try to construct a Liapunov function of the form (11). Then $V_x(x, y) = 2ax + by$, $V_y(x, y) = bx + 2cy$ so

$$\begin{aligned} \dot{V}(x, y) &= (2ax + by)(-x - xy^2) + (bx + 2cy)(-y - x^2y) \\ &= -\left[2a(x^2 + x^2y^2) + b(2xy + xy^3 + yx^3) + 2c(y^2 + x^2y^2)\right]. \end{aligned}$$

If we choose $b = 0$, and a and c to be any positive numbers, then $\dot{V}$ is negative definite and V is positive definite by Theorem 9.6. Thus by Theorem 9.3 the origin is an asymptotically stable critical point.

EXAMPLE 3

Consider the system

$$\begin{aligned} dx/dt &= x(1 - x - y), \\ dy/dt &= y(0.75 - y - 0.5x). \end{aligned} \tag{15}$$

In Example 3 of Section 9.1 we found that this system models a certain pair of competing species, and that the critical point (0.5, 0.5) appears to be asymptotically stable. Confirm this conclusion by finding a suitable Liapunov function.

It is helpful to transform the point (0.5, 0.5) to the origin. To this end let

$$x = 0.5 + u, \quad y = 0.5 + v. \tag{16}$$

Then, substituting for x and y in Eqs. (15), we obtain the new system

$$\begin{aligned} du/dt &= -0.5u - 0.5v - u^2 - uv, \\ dv/dt &= -0.25u - 0.5v - 0.5uv - v^2. \end{aligned} \tag{17}$$

To keep the calculations relatively simple, consider the function $V(u, v) = u^2 + v^2$ as a possible Liapunov function. This function is clearly positive definite, so the only question is whether there is a region containing the origin in the uv plane where the derivative $\dot{V}$ with respect to the system (17) is negative definite. We compute $\dot{V}(u, v)$ and find that

$$\begin{aligned} \dot{V}(u,v) &= V_u \frac{du}{dt} + V_v \frac{dv}{dt} \\ &= 2u(-0.5u - 0.5v - u^2 - uv) + 2v(-0.25u - 0.5v - 0.5uv - v^2), \end{aligned}$$

or

$$\dot{V}(u,v) = -\left[(u^2 + 1.5uv + v^2) + (2u^3 + 2u^2v + uv^2 + 2v^3)\right], \tag{18}$$

where we have collected the quadratic and cubic terms together. We want to show that the expression in square brackets in Eq. (18) is positive definite, at least for u and v sufficiently small. Observe that the quadratic terms can be written as

$$u^2 + 1.5uv + v^2 = 0.25(u^2 + v^2) + 0.75(u + v)^2, \tag{19}$$

so these terms are positive definite. On the other hand, the cubic terms in Eq. (18) may be of either sign. Thus we must show that, in some neighborhood of $u = 0, v = 0$, the cubic terms are smaller in magnitude than the quadratic terms; that is,

$$|2u^3 + 2u^2v + uv^2 + 2v^3| \le 0.25(u^2 + v^2) + 0.75(u + v)^2. \tag{20}$$

To estimate the left side of Eq. (20) we introduce polar coordinates $u = r\cos\theta$, $v = r\sin\theta$. Then

$$\begin{aligned} &|2u^3 + 2u^2v + uv^2 + 2v^3| \\ &\qquad = r^3|2\cos^3\theta + 2\cos^2\theta\sin\theta + \cos\theta\sin^2\theta + 2\sin^3\theta| \\ &\qquad \le r^3\left[2|\cos^3\theta| + 2\cos^2\theta|\sin\theta| + |\cos\theta|\sin^2\theta + 2|\sin^3\theta|\right] \\ &\qquad \le 7r^3, \end{aligned}$$

since $|\sin\theta|, |\cos\theta| \le 1$. To satisfy Eq. (20) it is now certainly sufficient to satisfy the more stringent requirement

$$7r^3 \le 0.25(u^2 + v^2) = 0.25r^2,$$

which yields $r \le 1/28$. Thus, at least in this circle, the hypotheses of Theorem 9.3

are satisfied, so the origin is an asymptotically stable critical point of the system (17). The same is then true of the critical point $(0.5, 0.5)$ of the orignal system (15).

Referring to Theorem 9.5, the preceding argument also shows that the circle with center $(0.5, 0.5)$ and radius $1/28$ is a region of asymptotic stability for the system (15). This is a severe underestimate of the full region of asymptotic stability, as the discussion in Sections 9.1 and 9.3 shows. To obtain a better estimate of the actual region of asymptotic stability from Theorem 9.5, one would have to estimate the terms in Eq. (20) more accurately, use a better (and presumably more complicated) Liapunov function, or both.

PROBLEMS

1. By constructing suitable Liapunov functions of the form $ax^2 + cy^2$, where a and c are to be determined, show that for each of the following systems the critical point at the origin is of the indicated type.

   (a) $dx/dt = -x^3 + xy^2$, $dy/dt = -2x^2y - y^3$; asymptotically stable
   (b) $dx/dt = -\frac{1}{2}x^3 + 2xy^2$, $dy/dt = -y^3$; asymptotically stable
   (c) $dx/dt = -x^3 + 2y^3$, $dy/dt = -2xy^2$; stable (at least)
   (d) $dx/dt = x^3 - y^3$, $dy/dt = 2xy^2 + 4x^2y + 2y^3$; unstable

2. Consider the system of equations

$$\frac{dx}{dt} = y - xf(x, y), \qquad \frac{dy}{dt} = -x - yf(x, y),$$

   where f is continuous and has continuous first partial derivatives. Show that if $f(x, y) > 0$ in some neighborhood of the origin, then the origin is an asymptotically stable critical point, and if $f(x, y) < 0$ in some neighborhood of the origin, then the origin is an unstable critical point.
   Hint: Construct a Liapunov function of the form $c(x^2 + y^2)$.
3. A generalization of the undamped pendulum equation is

$$\frac{d^2u}{dt^2} + g(u) = 0, \tag{i}$$

   where $g(0) = 0$, $g(u) > 0$ for $0 < u < k$, and $g(u) < 0$ for $-k < u < 0$; that is, $ug(u) > 0$ for $u \neq 0$, $-k < u < k$. Notice that $g(u) = \sin u$ has this property on $(-\pi/2, \pi/2)$.
   (a) Letting $x = u$, $y = du/dt$, write Eq. (i) as a system of two equations, and show that $x = 0$, $y = 0$ is a critical point.
   (b) Show that

$$V(x, y) = \tfrac{1}{2}y^2 + \int_0^x g(s)\, ds, \qquad -k < x < k, \tag{ii}$$

   is positive definite, and use this result to show that the critical point $(0, 0)$ is stable. Note that the Liapunov function V given by Eq. (ii) corresponds to the energy function $V(x, y) = \frac{1}{2}y^2 + (1 - \cos x)$ for the case $g(u) = \sin u$.

4. By introducing suitable dimensionless variables, the system of nonlinear equations for the damped pendulum [Eqs. (16) of Section 9.3] can be written as

$$dx/dt = y, \qquad dy/dt = -y - \sin x.$$

(a) Show that the origin is a critical point.
*(b) Show that while $V(x, y) = x^2 + y^2$ is positive definite, $\dot{V}(x, y)$ takes on both positive and negative values in any domain containing the origin, so that V is not a Liapunov function.
Hint: $x - \sin x > 0$ for $x > 0$ and $x - \sin x < 0$ for $x < 0$. Consider these cases with y positive but y so small that y^2 can be ignored compared to y.
(c) Using the energy function $V(x, y) = \frac{1}{2}y^2 + (1 - \cos x)$ mentioned Problem 3(b), show that the origin is a stable critical point. Note, however, that even though there is damping and we can expect that the origin is asymptotically stable, it is not possible to draw this conclusion using the present Liapunov function.
*(d) To show asymptotic stability it is necessary to construct a better Liapunov function than the one used in part (c). Show that $V(x, y) = \frac{1}{2}(x + y)^2 + x^2 + \frac{1}{2}y^2$ is such a Liapunov function, and conclude that the origin is an asymptotically stable critical point.
Hint: From Taylor's formula with a remainder it follows that $\sin x = x - \alpha x^3/3!$, where α depends on x but $0 < \alpha < 1$ for $-\pi/2 < x < \pi/2$. Then letting $x = r\cos\theta$, $y = r\sin\theta$ show that $\dot{V}(r\cos\theta, r\sin\theta) = -r^2[1 + h(r, \theta)]$ where $|h(r, \theta)| < 1$ if r is sufficiently small.

5. The Liénard equation (Problem 18 of Section 9.3) is

$$\frac{d^2u}{dt^2} + c(u)\frac{du}{dt} + g(u) = 0,$$

where g satisfies the conditions of Problem 3 and $c(u) \geq 0$. Show that the point $u = 0$, $du/dt = 0$ is a stable critical point.

6. (a) A special case of the Liénard equation of Problem 5 is

$$\frac{d^2u}{dt^2} + \frac{du}{dt} + g(u) = 0,$$

where g satisfies the conditions of Problem 3. Letting $x = u$, $y = du/dt$, show that the origin is a critical point of the resulting system. This equation can be interpreted as giving the motion of a spring–mass system with damping proportional to the velocity and a nonlinear restoring force. Using the Liapunov function of Problem 3, show that the origin is a stable critical point, but note that even with damping we cannot conclude asymptotic stability using this Liapunov function.
*(b) Asymptotic stability of the critical point (0, 0) can be shown by constructing a better Liapunov function as was done in part (d) of Problem 4. However, the analysis for a general function g is somewhat sophisticated and we only mention that V will have the form

$$V(x, y) = \tfrac{1}{2}y^2 + Ayg(x) + \int_0^x g(s)\,ds,$$

where A is a positive constant to be chosen so that V is positive definite and $\dot{V}$ is negative definite. For the pendulum problem $[g(x) = \sin x]$ use V as given by the above equation with $A = \frac{1}{2}$ to show that the origin is asymptotically stable.
Hint: Use $\sin x = x - \alpha x^3/3!$ and $\cos x = 1 - \beta x^2/2!$ where α and β depend on x, but $0 < \alpha < 1$ and $0 < \beta < 1$ for $-\pi/2 < x < \pi/2$; let $x = r\cos\theta$, $y = r\sin\theta$, and show that $\dot{V}(r\cos\theta, r\sin\theta) = -\frac{1}{2}r^2[1 + \frac{1}{2}\sin 2\theta + h(r,\theta)]$ where $|h(r,\theta)| < \frac{1}{2}$ if r is sufficiently small. To show that V is positive definite use $\cos x = 1 - x^2/2 + \gamma x^4/4!$ where γ depends on x, but $0 < \gamma < 1$ for $-\pi/2 < x < \pi/2$.

In Problems 7 and 8 we will prove part of Theorem 9.2: if the critical point $(0,0)$ of the almost linear system

$$dx/dt = ax + by + F_1(x,y), \qquad dy/dt = cx + dy + G_1(x,y) \tag{i}$$

is an asymptotically stable critical point of the corresponding linear system

$$dx/dt = ax + by, \qquad dy/dt = cx + dy, \tag{ii}$$

then it is an asymptotically stable critical point of the almost linear system (i). Problem 9 deals with the corresponding result for instability.

7. Consider the linear system (ii).
(a) Since $(0,0)$ is an asymptotically stable critical point, show that $a + d < 0$ and $ad - bc > 0$. (See Problem 23 of Section 9.2.)
(b) Construct a Liapunov function $V(x,y) = Ax^2 + Bxy + Cy^2$, such that V is positive definite and $\dot{V}$ is negative definite. One way to insure that $\dot{V}$ is negative definite is to choose A, B, and C so that $\dot{V}(x,y) = -x^2 - y^2$. Show that this leads to the result

$$A = -\frac{c^2 + d^2 + (ad - bc)}{2\Delta}, \qquad B = \frac{bd + ac}{\Delta}, \qquad C = -\frac{a^2 + b^2 + (ad - bc)}{2\Delta},$$

where $\Delta = (a + d)(ad - bc)$.
(c) Using the result of part (a) show that $A > 0$, and then show (several steps of algebra are required) that

$$4AC - B^2 = \frac{(a^2 + b^2 + c^2 + d^2)(ad - bc) + 2(ad - bc)^2}{\Delta^2} > 0.$$

Thus by Theorem 9.6, V is positive definite.
8. In this problem we show that the Liapunov function constructed in the previous problem is also a Liapunov function for the almost linear system (i). We must show that there is some region containing the origin for which $\dot{V}$ is negative definite.
(a) Show that

$$\dot{V}(x,y) = -(x^2 + y^2) + (2Ax + By)F_1(x,y) + (Bx + 2Cy)G_1(x,y).$$

(b) Recall that $F_1(x,y)/r \to 0$ and $G_1(x,y)/r \to 0$ as $r = (x^2 + y^2)^{1/2} \to 0$. This means that given any $\epsilon > 0$ there exists a circle $r = R$ about the origin such that for $0 \le r < R$, $|F_1(x,y)| < \epsilon r$, and $|G_1(x,y)| < \epsilon r$. Letting M be the maximum of $|2A|$,

$|B|$, and $|2C|$, show by introducing polar coordinates that R can be chosen so that $\dot{V}(x,y) < 0$ for $r < R$.

Hint: Choose ϵ sufficiently small in terms of M.

9. In this problem we prove a part of Theorem 9.2 relating to instability.

(a) Show that if $p = (a + d) > 0$ and $q = ad - bc > 0$ then the critical point $(0, 0)$ of the linear system (ii) is unstable.

(b) The same result holds for the almost linear system (i). Following Problems 7 and 8 construct a positive definite function V such that $\dot{V}(x, y) = x^2 + y^2$ and hence is positive definite, and then invoke Theorem 9.4.

REFERENCES

Several books on nonlinear differential equations present a number of applications in addition to theory. A few of these are:

LaSalle, J., and Lefschetz, S., *Stability by Liapunov's Direct Method with Applications* (New York: Academic Press, 1961).

Minorsky, N., *Nonlinear Oscillations* (Princeton: van Nostrand, 1962).

Stoker, J. J., *Nonlinear Vibrations* (New York: Wiley-Interscience, 1950).

A few of the more theoretically oriented books are:

Birkhoff, G., and Rota, G. C., *Ordinary Differential Equations* (3rd ed.) (New York: Wiley, 1978).

Brauer, F., and Nohel, J., *Qualitative Theory of Ordinary Differential Equations* (New York: W. A. Benjamin, 1969).

Cesari, L., *Asymptotic Behavior and Stability Problems in Ordinary Differential Equations* (2nd ed.) (Berlin: Springer, 1963).

Guckenheimer, J. C., and Holmes, P., *Nonlinear Oscillations, Dynamical Systems, and Bifurcations of Vector Fields*, (New York: Springer, 1983).

Hirsch, W. H., and Smale, S., *Differential Equations, Dynamical Systems and Linear Algebra* (New York: Academic Press, 1974).

Struble, R. A., *Nonlinear Differential Equations* (New York: McGraw-Hill, 1962).

Of these books, the ones by Struble and Brauer and Nohel are the more elementary.

A classic reference on ecology is:

Odum, E. P., *Fundamentals of Ecology* (3rd ed.) (Philadelphia: W. B. Saunders, 1971).

Two books dealing with ecology and population dynamics on a more mathematical level are:

May, R. M., *Stability and Complexity in Model Ecosystems* (Princeton, N.J.: Princeton Univ. Press, 1973).

Pielou, E. C., *Mathematical Ecology* (New York: Wiley, 1977).

10.1 Introduction

In this chapter we consider a number of problems involving partial differential equations. No effort is made to present a general theory of such equations; rather, we limit our discussion to the basic partial differential equations of heat conduction, wave propagation, and potential theory. These equations are associated with three distinct types of physical phenomena: diffusive processes, oscillatory processes, and time-independent or steady processes. Consequently, they are of fundamental importance in many branches of physics.

They are also of considerable significance from a mathematical point of view. The partial differential equations whose theory is best developed and whose applications are most significant and varied are the linear equations of second order. All such equations can be classified into one of three categories; the heat conduction equation, the wave equation, and the potential equation, respectively, are prototypes of each of these categories. Thus a study of these three equations yields much information about more general second order linear partial differential equations.

During the last two centuries several methods have been advanced for solving partial differential equations. Among these, we consider only one, a technique known as the method of separation of variables. This is perhaps the oldest systematic method for solving partial differential equations, having been used by D'Alembert, D. Bernoulli, and Euler in about 1750 in their treatments of the wave equation. It has been considerably refined and generalized in the meantime, and remains a method of great importance today. Its essential feature is the replacement of the partial differential equation by a set of ordinary differential equations. The required solution of the partial differential equation is then expressed as a sum, usually an infinite series, involving solutions of the set of ordinary differential equations. In many cases these solutions are sine or cosine functions. Experience shows that many significant problems in partial differential equations can be solved by the method of separation of variables; this method is one of the classical techniques of mathematical physics.

In Section 10.2 we show how the method of separation of variables can be used to solve a basic problem arising in the study of heat conduction and other diffusive processes. In solving this problem it is eventually necessary to consider the question of whether an essentially arbitrary function can be expressed as an infinite series of sine and cosine functions. Series of this kind are called Fourier series, and Sections 10.3 to 10.5 are devoted to the development of their elementary properties. Then, in Sections 10.6 to 10.8 we return to some typical problems in heat conduction, wave propagation, and potential theory, solving them by means of separation of variables and Fourier series. Some further extensions and applications of this method appear also in Sections 11.4 and 11.6.

10.2 Heat Conduction and Separation of Variables

One of the classic partial differential equations of mathematical physics is the equation describing the conduction of heat in a solid body. The study of this equation originated[1] about 1800, and it continues to command the attention of modern scientists. For example, the analysis of the dissipation and transfer of heat away from its sources in high-speed machinery is frequently an important technological problem.

Let us now consider a heat conduction problem for a straight bar of uniform cross section and homogeneous material. Let the x axis be chosen to lie along the axis of the bar, and let $x = 0$ and $x = l$ denote the ends of the bar (see Figure 10.1). Suppose further that the sides of the bar are perfectly insulated so that no heat passes through them. We also assume that the cross-sectional dimensions are so small that the temperature u can be considered as constant on any given cross section. Then u is a function only of the axial coordinate x and the time t.

[1] The first important investigation of heat conduction was carried out by Joseph Fourier (1768–1830) in his spare time while he was serving as prefect of the department of Isère (Grenoble) from 1801 to 1815. He presented basic papers on the subject to the Academy of Sciences of Paris in 1807 and 1811. However, these papers were criticized by the referees (principally Lagrange) for a lack of rigor and so were not published. Fourier continued to develop his ideas and eventually wrote one of the classics of applied mathematics, *Théorie analytique de la chaleur*, published in 1822.

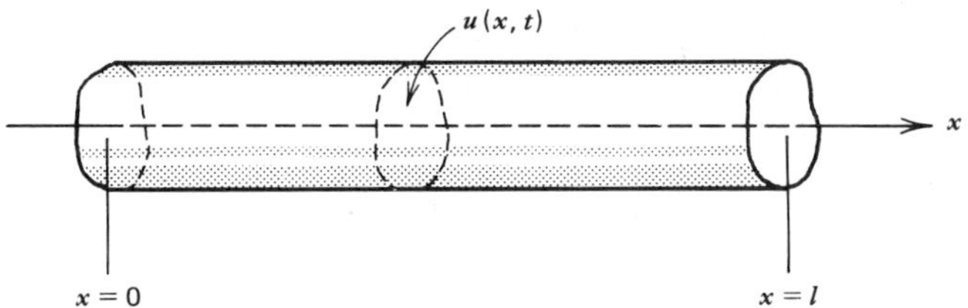

FIGURE 10.1 A heat-conducting solid bar

The variation of temperature in the bar is governed by a partial differential equation whose derivation appears in Appendix A at the end of the chapter. This equation is called the *heat conduction equation*, and has the form

$$\alpha^2 u_{xx} = u_t, \qquad 0 < x < l, \quad t > 0, \tag{1}$$

where α^2 is a constant known as the *thermal diffusivity*. The parameter α^2 depends only on the material from which the bar is made, and is defined by

$$\alpha^2 = \kappa/\rho s, \tag{2}$$

where κ is the thermal conductivity, ρ is the density, and s is the specific heat of the material in the bar. The dimensions of α^2 are (length)^2/time. Typical values of α^2 are given in Table 10.1.

In addition, we assume that the initial temperature distribution in the bar is given; thus

$$u(x,0) = f(x), \qquad 0 \leq x \leq l, \tag{3}$$

where f is a given function. Finally, we assume that the ends of the bar are held at fixed temperatures: the temperature T_1 at $x = 0$ and the temperature T_2 at $x = l$. However, it turns out that we need only consider the case where $T_1 = T_2 = 0$. We show in Section 10.6 how to reduce the more general problem to this special case. Thus in this section we will assume that u is always zero when $x = 0$

TABLE 10.1 VALUES OF THE THERMAL DIFFUSIVITY FOR SOME COMMON MATERIALS

Material	α^2 (cm^2/sec)
Silver	1.71
Copper	1.14
Aluminum	0.86
Cast iron	0.12
Granite	0.011
Brick	0.0038
Water	0.00144

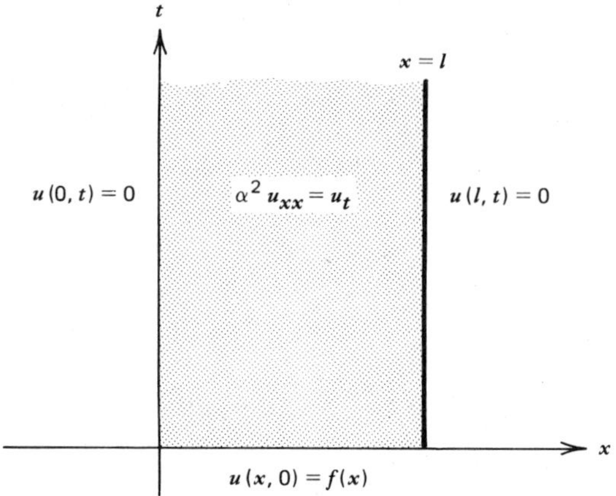

FIGURE 10.2 Boundary value problem for the heat conduction equation.

or $x = l$:

$$u(0,t) = 0, \qquad u(l,t) = 0, \qquad t > 0. \tag{4}$$

The fundamental problem of heat conduction is to find $u(x,t)$, which satisfies the differential equation (1), the initial condition (3), and the boundary conditions (4).

The problem described by Eqs. (1), (3), and (4) is an initial value problem in the time variable t; an initial condition is given and the differential equation governs what happens later. However, with respect to the space variable x, the problem is of a different type known as a boundary value problem. The solution of the differential equation is desired in a certain interval, and conditions (called boundary conditions) are imposed at the ends of the interval. Alternatively, we can consider the problem as a boundary value problem in the xt plane (see Figure 10.2). The solution $u(x,t)$ of Eq. (1) is sought in the semi-infinite strip $0 < x < l$, $t > 0$ subject to the requirement that $u(x,t)$ must assume a prescribed value at each point on the boundary of this strip.

The heat conduction problem (1), (3), (4) is *linear* since u appears only to the first power throughout. The differential equation and boundary conditions are also *homogeneous*. This suggests that we might approach the problem by seeking solutions of the differential equation and boundary conditions, and then superposing them to satisfy the initial condition. To find the solutions we need, we use a technique known as the method of separation of variables.

This method is based on the idea of finding certain solutions of the differential equation (1) having the form of a product of a function of x only and a function of t only; thus we assume that

$$u(x,t) = X(x)T(t). \tag{5}$$

Substituting Eq. (5) for u in the differential equation (1) yields

$$\alpha^2 X''T = XT', \tag{6}$$

where primes refer to ordinary differentiation with respect to the independent variable, whether x or t. Equation (6) is equivalent to

$$\frac{X''}{X} = \frac{1}{\alpha^2}\frac{T'}{T}, \tag{7}$$

in which the variables are separated; that is, the left side depends only on x and the right side only on t. For Eq. (7) to be valid for $0 < x < l$, $t > 0$, it is necessary that both sides of Eq. (7) be equal to the same constant. Otherwise, by keeping one independent variable (say x) fixed and varying the other, one side (the left in this case) of Eq. (7) would remain unchanged while the other varied, thus violating the equality. If we call this separation constant σ, then Eq. (7) becomes

$$\frac{X''}{X} = \frac{1}{\alpha^2}\frac{T'}{T} = \sigma. \tag{8}$$

Hence we obtain the following two ordinary differential equations for $X(x)$ and $T(t)$:

$$X'' - \sigma X = 0, \tag{9}$$

$$T' - \alpha^2\sigma T = 0. \tag{10}$$

The partial differential equation (1) has thus been replaced by two ordinary differential equations. Each of these equations can be readily solved for *any* value of the separation constant σ. The product of two solutions of Eq. (9) and (10), respectively, for *any* value of σ provides a solution of the partial differential equation (1). However, we are interested only in those solutions of Eq. (1) also satisfying the boundary conditions (4). As we now show, this severely restricts the possible values of σ.

Substituting for $u(x,t)$ from Eq. (5) in the boundary condition at $x = 0$, we obtain

$$u(0,t) = X(0)T(t) = 0. \tag{11}$$

If Eq. (11) is satisfied by choosing $T(t)$ to be zero for all t, then $u(x,t)$ is zero for all x and t. This is unacceptable, since in this case $u(x,t)$ would not satisfy the initial condition (3) unless the initial temperature distribution $f(x)$ is zero for all x. Therefore Eq. (11) must be satisfied by requiring that

$$X(0) = 0. \tag{12}$$

Similarly, the boundary condition at $x = l$ requires that

$$X(l) = 0. \tag{13}$$

We now want to consider Eq. (9) subject to the boundary conditions (12) and (13). Two cases must be distinguished, depending on whether $\sigma = 0$ or $\sigma \neq 0$.

If $\sigma = 0$, then the general solution of Eq. (9) is

$$X(x) = k_1 x + k_2. \tag{14}$$

To satisfy the boundary condition (12) we must choose $k_2 = 0$ and, to satisfy the second boundary condition, we must have $k_1 = 0$. Hence $X(x)$ is identically zero, and therefore $u(x, t)$ is also identically zero. As before, this is unacceptable, and we conclude that σ must be nonzero.

If $\sigma \neq 0$, it is convenient to replace it by $-\lambda^2$, where λ is a new parameter, not necessarily real. Then Eq. (9) becomes

$$X'' + \lambda^2 X = 0, \tag{15}$$

and its general solution is

$$X(x) = k_1 e^{i\lambda x} + k_2 e^{-i\lambda x}. \tag{16}$$

Applying the boundary conditions (12) and (13), we obtain

$$\begin{aligned} k_1 + k_2 &= 0, \\ k_1 e^{i\lambda l} + k_2 e^{-i\lambda l} &= 0. \end{aligned} \tag{17}$$

The system of equations (17) always has the trivial solution $k_1 = 0$, $k_2 = 0$; as before, this leads to the unacceptable conclusion that $u(x, t)$ is identically zero. Nontrivial solutions of Eqs. (17) exist if and only if the determinant of coefficients is zero; that is,

$$\begin{vmatrix} 1 & 1 \\ e^{i\lambda l} & e^{-i\lambda l} \end{vmatrix} = e^{-i\lambda l} - e^{i\lambda l} = 0. \tag{18}$$

If we now write $\lambda = \mu + i\nu$, where μ and ν are real, Eq. (18) reduces to

$$e^{-i\mu l} e^{\nu l} - e^{i\mu l} e^{-\nu l} = 0. \tag{19}$$

Using Euler's relation

$$e^{i\mu l} = \cos \mu l + i \sin \mu l, \tag{20}$$

and separating Eq. (19) into real and imaginary parts, we find that

$$(e^{\nu l} - e^{-\nu l}) \cos \mu l - i(e^{\nu l} + e^{-\nu l}) \sin \mu l = 0. \tag{21}$$

Equation (21) is satisfied only if the real and imaginary parts of the left side are separately zero; thus

$$(e^{\nu l} - e^{-\nu l}) \cos \mu l = 0, \tag{22a}$$

$$(e^{\nu l} + e^{-\nu l}) \sin \mu l = 0. \tag{22b}$$

Since $e^{\nu l} + e^{-\nu l} > 0$ for all values of ν and l, Eq. (22b) requires that $\sin \mu l = 0$. Hence μ must be chosen so that μl is a multiple of π, that is, $\mu = n\pi/l$, where n is an integer. For this choice of μ, $\cos \mu l \neq 0$, and Eq. (22a) reduces to

$$e^{\nu l} - e^{-\nu l} = 2 \sinh \nu l = 0;$$

consequently $\nu = 0$. Since $\nu = 0$, we cannot also have $\mu = 0$; otherwise $\sigma = 0$, which we have already rejected. Hence

$$\lambda = \mu = n\pi/l, \tag{23}$$

where n is a *nonzero* integer.

Returning to Eqs. (17) we see that $k_2 = -k_1$; hence from Eq. (16)

$$X(x) = k_1(e^{i\mu x} - e^{-i\mu x}). \tag{24}$$

Using Euler's relation (20) again we find that $X(x)$ is proportional to $\sin \mu x$.

Summarizing our results to this point, we have shown that we can satisfy the boundary conditions (4) only if the separation constant σ has certain real negative values, given by

$$\sigma = -\lambda^2 = -n^2\pi^2/l^2, \tag{25}$$

where $n \neq 0$ is an integer. The corresponding functions X are proportional to $\sin(n\pi x/l)$. Upon introducing the values of σ given by Eq. (25) into Eq. (10), we find that $T(t)$ is proportional to $\exp(-n^2\pi^2\alpha^2 t/l^2)$. Hence, neglecting arbitrary constants of proportionality, we conclude that the functions

$$u_n(x,t) = e^{-n^2\pi^2\alpha^2 t/l^2} \sin(n\pi x/l), \qquad n = 1, 2, 3, \ldots \tag{26}$$

satisfy the boundary conditions (4) as well as the differential equation (1). It is sufficient to consider only positive values of n because if n assumes negative values, the same functions as given by Eq. (26) are generated a second time. The functions u_n are sometimes called fundamental solutions of the heat conduction problem (1), (3), and (4).

It remains only to satisfy the initial condition (3),

$$u(x,0) = f(x), \qquad 0 \le x \le l.$$

We emphasize that the differential equation (1) and the boundary conditions (4) are linear and homogeneous, and that they are satisfied by $u_n(x,t)$ for $n = 1, 2, \ldots$. From the principle of superposition we know that any linear combination of the $u_n(x,t)$ also satisfies the differential equation and boundary conditions. Consequently, we assume that

$$u(x,t) = \sum_{n=1}^{m} c_n u_n(x,t) = \sum_{n=1}^{m} c_n e^{-n^2\pi^2\alpha^2 t/l^2} \sin\frac{n\pi x}{l}, \tag{27}$$

where the coefficients c_n are as yet undetermined and where m is some positive integer. Since $u(x,t)$ as given by Eq. (27) satisfies the differential equation (1) and boundary conditions (4) for any choice of c_n, we wish to investigate whether the c_n can be chosen so as to satisfy the initial condition (3) as well. Let us first consider two simple examples.

EXAMPLE 1

Find the solution of the heat conduction problem (1), (3), and (4) if

$$f(x) = 3\sin(4\pi x/l). \tag{28}$$

In this case we need to use only one of the fundamental solutions (26), namely, the one corresponding to $n = 4$. We assume that

$$u(x,t) = c_4 e^{-16\pi^2\alpha^2 t/l^2} \sin(4\pi x/l), \tag{29}$$

which satisfies the differential equation (1) and the boundary conditions (4) for any value of c_4. Setting $t = 0$, we obtain

$$u(x,0) = c_4 \sin(4\pi x/l);$$

hence we must choose $c_4 = 3$ to satisfy the initial condition. Substituting this value c_4 into Eq. (29) gives the solution of the complete boundary value problem.

EXAMPLE 2

Find the solution of the heat conduction problem (1), (3), and (4) if

$$f(x) = b_1 \sin(\pi x/l) + \cdots + b_m \sin(m\pi x/l), \tag{30}$$

where $b_1, \ldots, b_m$ are given constants.

If we assume that $u(x,t)$ is given by Eq. (27), then the differential equation (1) and boundary conditions (4) are satisfied for any choice of $c_1, \ldots, c_m$. At $t - 0$ we have

$$\begin{aligned} u(x,0) &= c_1 \sin(\pi x/l) + \cdots + c_m \sin(m\pi x/l) \\ &= b_1 \sin(\pi x/l) + \cdots + b_m \sin(m\pi x/l). \end{aligned}$$

Thus the initial condition is satisfied if we choose $c_1 = b_1, \ldots, c_m = b_m$. The solution of the complete problem is

$$u(x,t) = \sum_{n=1}^{m} b_n e^{-n^2\pi^2\alpha^2 t/l^2} \sin\frac{n\pi x}{l}. \tag{31}$$

Let us return to the general problem of Eqs. (1), (3), and (4) with f an arbitrary function. Unless $f(x)$ takes the form given in Eq. (30), it is not possible to satisfy the initial condition (3) by means of a finite sum of the form (27). This suggests that we formally extend the principle of superposition to include *infinite series*; that is, we assume that

$$u(x,t) = \sum_{n=1}^{\infty} c_n e^{-n^2\pi^2\alpha^2 t/l^2} \sin\frac{n\pi x}{l}. \tag{32}$$

We have seen that the individual terms in Eq. (32) satisfy the partial differential equation (1) and boundary conditions (4) and that any finite sum of such terms does so as well. We will assume that the infinite series of Eq. (32) converges and also satisfies Eqs. (1) and (4). To satisfy the initial condition (3) we must have

$$u(x,0) = \sum_{n=1}^{\infty} c_n \sin\frac{n\pi x}{l} = f(x). \tag{33}$$

Now let us suppose that it is possible to express $f(x)$ by means of an infinite series of sine terms,

$$f(x) = \sum_{n=1}^{\infty} b_n \sin\frac{n\pi x}{l}, \tag{34}$$

and that we know how to compute the coefficients b_n in this infinite series. Then, just as in Example 2, we can satisfy Eq. (33) by choosing $c_n = b_n$ for each n. With the coefficients c_n selected in this manner, Eq. (32) gives the solution of the boundary value problem (1), (3), and (4).

Thus, to solve the given heat conduction problem for a fairly arbitrary initial temperature distribution by the method of separation of variables, it is necessary to express the initial temperature distribution $f(x)$ as a series of the form (34). This raises the following questions:

1. How can we identify functions that can be written in the form (34)?
2. If f is such a function, how can we determine the coefficients $b_1, b_2, \ldots, b_n, \ldots$?

In the next three sections we show how a large class of functions can be represented by series similar to that in Eq. (34); furthermore, we show that the coefficients in these series can be determined in a very simple manner.

PROBLEMS

1. State exactly the boundary value problem determining the temperature in a copper bar 1 m in length if the entire bar is originally at 20°C, and one end is then suddenly heated to 60°C and held at that temperature while the other end is held at 20°C.
2. State exactly the boundary value problem determining the temperature in a silver rod 2 m long if the ends are held at the temperatures 30°C and 50°C, respectively. Assume that the initial temperature in the bar is given by a quadratic function of distance along the bar consistent with the above boundary conditions, and with the condition that the temperature at the center of the rod is 60°C.
3. Suppose that $u_n(x,t)$ is given by Eq. (26). Show that $c_1u_1(x,t) + \cdots + c_mu_m(x,t)$, where $c_1, \ldots, c_m$ are constants, satisfies the differential equation (1) and the boundary conditions (4).
4. Find the solution of the heat conduction problem
$$100u_{xx} = u_t, \qquad 0 < x < 1, \quad t > 0$$
$$u(0,t) = 0, \qquad u(1,t) = 0, \quad t > 0$$
$$u(x,0) = \sin 2\pi x - 2\sin 5\pi x, \qquad 0 \le x \le 1.$$
5. Find the solution of the heat conduction problem
$$u_{xx} = 4u_t, \qquad 0 < x < 2, \quad t > 0$$
$$u(0,t) = 0, \qquad u(2,t) = 0, \quad t > 0$$
$$u(x,0) = 2\sin(\pi x/2) - \sin \pi x + 4\sin 2\pi x, \qquad 0 \le x \le 2.$$
6. Use Eq. (8) to determine the physical dimensions of the separation constant σ.
7. Determine whether the method of separation of variables can be used to replace each of the following partial differential equations by pairs of ordinary differential equations. If so, find the equations.

(a) $xu_{xx} + u_t = 0$ (b) $tu_{xx} + xu_t = 0$
(c) $u_{xx} + u_{xt} + u_t = 0$ (d) $[p(x)u_x]_x - r(x)u_{tt} = 0$
(e) $u_{xx} + (x+y)u_{yy} = 0$

8. In solving differential equations the computations can almost always be simplified by the use of *dimensionless variables*. Show that if the dimensionless variable ξ defined by $\xi = x/l$ is introduced, the heat conduction equation becomes

$$\frac{\partial^2 u}{\partial \xi^2} = \frac{l^2}{\alpha^2}\frac{\partial u}{\partial t}, \qquad 0 < \xi < 1, \quad t > 0.$$

Since l^2/α^2 has the dimension of time, it is convenient to use this quantity as the unit on the time scale. Defining a dimensionless time variable $\tau = (\alpha^2/l^2)t$, show that the heat conduction equation reduces to

$$\frac{\partial^2 u}{\partial \xi^2} = \frac{\partial u}{\partial \tau}, \qquad 0 < \xi < 1, \quad \tau > 0.$$

9. Consider the equation

$$au_{xx} - bu_t + cu = 0 \tag{i}$$

where a, b, and c are constants.
(a) Let $u(x,t) = e^{\delta t}w(x,t)$, where δ is a constant, and find the corresponding partial differential equation for w.
(b) If $b \neq 0$, show that δ can be chosen so that the partial differential equation found in part (a) has no term in w. Thus, by a change of dependent variable, it is possible to reduce Eq. (i) to the heat conduction equation.

*10. The heat conduction equation in two space dimensions is

$$\alpha^2(u_{xx} + u_{yy}) = u_t.$$

Assuming that $u(x,y,t) = X(x)Y(y)T(t)$, find ordinary differential equations satisfied by $X(x)$, $Y(y)$, and $T(t)$.

*11. The heat conduction equation in two space dimensions may be expressed in terms of polar coordinates as

$$\alpha^2\left[u_{rr} + (1/r)u_r + (1/r^2)u_{\theta\theta}\right] = u_t.$$

Assuming that $u(r,\theta,t) = R(r)\Theta(\theta)T(t)$, find ordinary differential equations satisfied by $R(r)$, $\Theta(\theta)$, and $T(t)$.

10.3 Fourier Series

In the last section we showed how to solve the fundamental boundary value problem of heat conduction provided that it is possible to express a given function defined on $0 \le x \le l$ as a series of sines, $\Sigma b_m \sin(m\pi x/l)$. We now begin to consider somewhat more general series of the form

$$\frac{a_0}{2} + \sum_{m=1}^{\infty}\left(a_m \cos\frac{m\pi x}{l} + b_m \sin\frac{m\pi x}{l}\right). \tag{1}$$

On the set of points where the series (1) converges, it defines a function f, whose value at each point is the sum of the series for that value of x. In this case the series (1) is said to be the *Fourier series*[2] for f. Our immediate objects are to

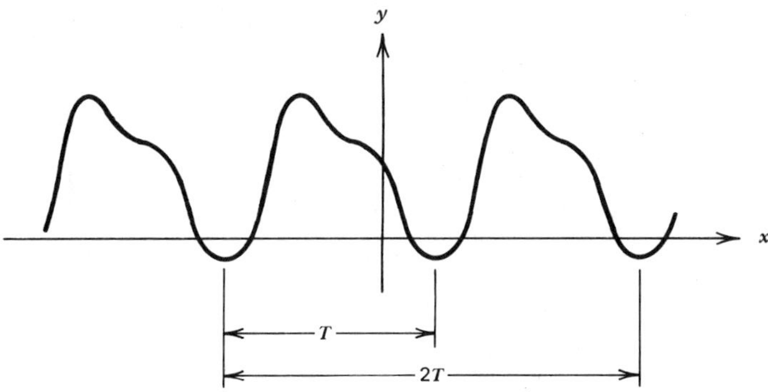

FIGURE 10.3 A periodic function.

determine what functions can be represented as a sum of a Fourier series and to find some means of computing the coefficients in the series. We note in passing that the first term in the series (1) is written as $a_0/2$ rather than as a_0 to simplify a formula for the coefficients that we derive below. Besides their association with the method of separation of variables, Fourier series are also useful in various other ways, such as in the analysis of mechanical or electrical systems acted upon by periodic external forces.

PERIODICITY OF THE SINE AND COSINE FUNCTIONS. To discuss Fourier series it is necessary to develop certain properties of the trigonometric functions $\sin(m\pi x/l)$ and $\cos(n\pi x/l)$, where m and n are positive integers. The first is their periodic character. A function f is said to be *periodic with period* $T > 0$ if the domain of f contains $x + T$ whenever it contains x, and if

$$f(x + T) = f(x) \tag{2}$$

for every value of x. An example of a periodic function is shown in Figure 10.3. It follows immediately from the definition that if T is a period of f, then $2T$ is also a period, and so indeed is any integral multiple of T.

The smallest value of T for which Eq. (2) holds is called the *fundamental period* of f. In this connection it should be noted that a constant may be thought of as a periodic function with an arbitrary period, but no fundamental period.

[2] Fourier series are named for Joseph Fourier, who made the first systematic use, although not a completely rigorous investigation, of them in 1807 and 1811 in his papers on heat conduction. According to Riemann, when Fourier presented his first paper to the Paris Academy in 1807, stating that a completely arbitrary function could be expressed as a series of the form (1), the mathematician Lagrange was so surprised that he denied the possibility in the most definite terms. Although it turned out that Fourier's claim of complete generality was somewhat too strong, his results inspired a flood of important research that has continued to the present day. See Grattan-Guinness or Carslaw [Historical Introduction] for a detailed history of Fourier series.

For any nonconstant periodic function the fundamental period is uniquely defined and all other periods are multiples of it.

If f and g are any two periodic functions with common period T, then their product fg and any linear combination $c_1 f + c_2 g$ are also periodic with period T. To prove the latter statement, let $F(x) = c_1 f(x) + c_2 g(x)$; then for any x

$$F(x+T) = c_1 f(x+T) + c_2 g(x+T) = c_1 f(x) + c_2 g(x) = F(x).$$

Moreover, it can be shown that the sum of any finite number, or even the sum of a convergent infinite series, of functions of period T is also periodic with period T.

In particular, the functions $\sin(m\pi x/l)$ and $\cos(m\pi x/l)$, $m = 1, 2, 3, \ldots$, are periodic with the fundamental period $T = 2l/m$. To show that $\sin(m\pi x/l)$ has this fundamental period, we must show that there are values of T for which

$$\sin \frac{m\pi(x+T)}{l} = \sin \frac{m\pi x}{l} \tag{3}$$

for all x, and that the minimum such value of T is $2l/m$. Expanding the left side of Eq. (3) gives

$$\sin \frac{m\pi x}{l} \cos \frac{m\pi T}{l} + \cos \frac{m\pi x}{l} \sin \frac{m\pi T}{l} = \sin \frac{m\pi x}{l}. \tag{4}$$

Equation (4) is satisfied for all x if we can choose T so that $\cos(m\pi T/l) = 1$ and $\sin(m\pi T/l) = 0$. These requirements are met if we take $m\pi T/l = 2\pi, 4\pi, \ldots$. To show that there are no other values of T for which Eq. (3) is satisfied, assume that there is such a T, and choose x so that $\cos(m\pi x/l) = 0$. Then Eq. (4) becomes

$$\sin \frac{m\pi x}{l} \cos \frac{m\pi T}{l} = \sin \frac{m\pi x}{l},$$

which is true only for the values of $m\pi T/l$ given above. The fundamental period of $\sin(m\pi x/l)$ is then

$$T = \frac{2\pi}{m\pi/l} = \frac{2l}{m}. \tag{5}$$

A similar argument applies to $\cos(m\pi x/l)$.

Note also that, since every positive integral multiple of a period is also a period, each of the functions $\sin(m\pi x/l)$ and $\cos(m\pi x/l)$ has the common period $2l$.

ORTHOGONALITY OF THE SINE AND COSINE FUNCTIONS. To describe a second essential property of the functions $\sin(m\pi x/l)$ and $\cos(m\pi x/l)$ we generalize the concept of orthogonality of vectors (see Section 7.2). The standard *inner product* (u, v) of two real-valued functions u and v on the interval $\alpha \le x \le \beta$ is defined by

$$(u, v) = \int_\alpha^\beta u(x) v(x)\, dx.$$

The functions u and v are said to be *orthogonal* on $\alpha \le x \le \beta$ if their inner product is zero; that is, if

$$\int_\alpha^\beta u(x)v(x)\,dx = 0.$$

A set of functions is said to be *mutually orthogonal* if each distinct pair of functions in the set is orthogonal.

The functions $\sin(m\pi x/l)$ and $\cos(m\pi x/l)$, $m = 1, 2, \ldots$, form a mutually orthogonal set of functions on the interval $-l \le x \le l$. In fact, they satisfy the following orthogonality relations:

$$\int_{-l}^{l} \cos\frac{m\pi x}{l}\cos\frac{n\pi x}{l}\,dx = \begin{cases} 0, & m \ne n, \\ l, & m = n; \end{cases} \tag{6}$$

$$\int_{-l}^{l} \cos\frac{m\pi x}{l}\sin\frac{n\pi x}{l}\,dx = 0, \qquad \text{all } m, n; \tag{7}$$

$$\int_{-l}^{l} \sin\frac{m\pi x}{l}\sin\frac{n\pi x}{l}\,dx = \begin{cases} 0, & m \ne n, \\ l, & m = n. \end{cases} \tag{8}$$

These results can be obtained by direct integration. For example, to derive Eq. (8), note that

$$\begin{aligned}
&\int_{-l}^{l} \sin\frac{m\pi x}{l}\sin\frac{n\pi x}{l}\,dx \\
&\quad = \frac{1}{2}\int_{-l}^{l}\left[\cos\frac{(m-n)\pi x}{l} - \cos\frac{(m+n)\pi x}{l}\right]dx \\
&\quad = \frac{1}{2}\frac{l}{\pi}\left\{\frac{\sin[(m-n)\pi x/l]}{m-n} - \frac{\sin[(m+n)\pi x/l]}{m+n}\right\}\Bigg|_{-l}^{l} \\
&\quad = 0,
\end{aligned}$$

as long as $m + n$ and $m - n$ are not zero. Since m and n are positive, $m + n \ne 0$. On the other hand, if $m - n = 0$, then $m = n$, and the integral must be evaluated in a different way. In this case

$$\begin{aligned}
\int_{-l}^{l} \sin\frac{m\pi x}{l}\sin\frac{n\pi x}{l}\,dx &= \int_{-l}^{l}\left(\sin\frac{m\pi x}{l}\right)^2 dx \\
&= \frac{1}{2}\int_{-l}^{l}\left[1 - \cos\frac{2m\pi x}{l}\right]dx \\
&= \frac{1}{2}\left\{x - \frac{\sin(2m\pi x/l)}{2m\pi/l}\right\}\Bigg|_{-l}^{l} \\
&= l.
\end{aligned}$$

This establishes Eq. (8); Eqs. (6) and (7) can be verified by similar computations.

THE EULER-FOURIER FORMULAS. Now let us suppose that a series of the form (1) converges, and let us call its sum $f(x)$:

$$f(x) = \frac{a_0}{2} + \sum_{m=1}^{\infty} \left(a_m \cos \frac{m\pi x}{l} + b_m \sin \frac{m\pi x}{l} \right). \tag{9}$$

The coefficients a_m and b_m can be related very simply to $f(x)$ as a consequence of the orthogonality conditions (6), (7), and (8). First multiply Eq. (9) by $\cos(n\pi x/l)$, where n is a *fixed* positive integer ($n > 0$), and integrate with respect to x from $-l$ to l. Assuming that the integration can be legitimately carried out term by term,[3] we obtain

$$\int_{-l}^{l} f(x) \cos \frac{n\pi x}{l}\, dx = \frac{a_0}{2} \int_{-l}^{l} \cos \frac{n\pi x}{l}\, dx + \sum_{m=1}^{\infty} a_m \int_{-l}^{l} \cos \frac{m\pi x}{l} \cos \frac{n\pi x}{l}\, dx$$

$$+ \sum_{m=1}^{\infty} b_m \int_{-l}^{l} \sin \frac{m\pi x}{l} \cos \frac{n\pi x}{l}\, dx. \tag{10}$$

Keeping in mind that n is fixed whereas m ranges over the positive integers, it follows from the orthogonality relations (6) and (7) that the only nonvanishing term on the right side of Eq. (10) is the one for which $m = n$ in the first summation. Hence

$$\int_{-l}^{l} f(x) \cos \frac{n\pi x}{l}\, dx = la_n, \qquad n = 1, 2, \ldots. \tag{11}$$

To determine a_0 we can integrate Eq. (9) from $-l$ to l, obtaining

$$\int_{-l}^{l} f(x)\, dx = \frac{a_0}{2} \int_{-l}^{l} dx + \sum_{m=1}^{\infty} a_m \int_{-l}^{l} \cos \frac{m\pi x}{l}\, dx + \sum_{m=1}^{\infty} b_m \int_{-l}^{l} \sin \frac{m\pi x}{l}\, dx$$

$$= la_0, \tag{12}$$

since all of the integrals involving trigonometric functions vanish. Thus

$$a_n = \frac{1}{l} \int_{-l}^{l} f(x) \cos \frac{n\pi x}{l}\, dx, \qquad n = 0, 1, 2, \ldots. \tag{13}$$

By writing the constant term in Eq. (9) as $a_0/2$, it is possible to compute all of the a_n from Eq. (13). Otherwise, a separate formula would have to be used for a_0.

A similar expression for b_n may be obtained by multiplying Eq. (9) by $\sin(n\pi x/l)$, integrating termwise from $-l$ to l, and using the orthogonality relations (7) and (8); thus

$$b_n = \frac{1}{l} \int_{-l}^{l} f(x) \sin \frac{n\pi x}{l}\, dx, \qquad n = 1, 2, 3, \ldots. \tag{14}$$

[3] This is a nontrivial assumption, since not all convergent series with variable terms can be so integrated. For the special case of Fourier series, however, term-by-term integration can always be justified.

Equations (13) and (14) are known as the Euler-Fourier formulas for the coefficients in a Fourier series. Hence, if the series (9) converges to $f(x)$, and if the series can be integrated term by term, then the coefficients *must be given* by Eqs. (13) and (14).

Note that Eqs. (13) and (14) are explicit formulas for a_n and b_n in terms of f, and that the determination of any particular coefficient is independent of all of the other coefficients. Of course, the difficulty in evaluating the integrals in Eqs. (13) and (14) depends very much on the particular function f involved.

Note also that the formulas (13) and (14) depend only on the values of $f(x)$ in the interval $-l \leq x \leq l$. Since each of the terms in the Fourier series (9) is periodic with period $2l$, the series converges for all x whenever it converges in $-l \leq x \leq l$, and its sum is also a periodic function with period $2l$. Hence $f(x)$ is determined for all x by its values in the interval $-l \leq x \leq l$.

It is possible to show (see Problem 2) that if g is periodic with period T, then every integral of g over an interval of length T has the same value. Applying this result to the Euler-Fourier formulas (13) and (14), it follows that the interval of integration, $-l \leq x \leq l$, can be replaced, if it is more convenient to do so, by any other interval of length $2l$.

EXAMPLE 1

Assume that there is a Fourier series converging to the function f defined by

$$f(x) = \begin{cases} -x, & -l \leq x < 0, \\ x, & 0 \leq x < l; \end{cases} \tag{15}$$

$$f(x + 2l) = f(x).$$

Determine the coefficients in this Fourier series.

This function represents a triangular wave (see Figure 10.4) and is periodic with period $2l$. Thus the Fourier series is of the form

$$f(x) = \frac{a_0}{2} + \sum_{m=1}^{\infty} \left(a_m \cos \frac{m\pi x}{l} + b_m \sin \frac{m\pi x}{l} \right), \tag{16}$$

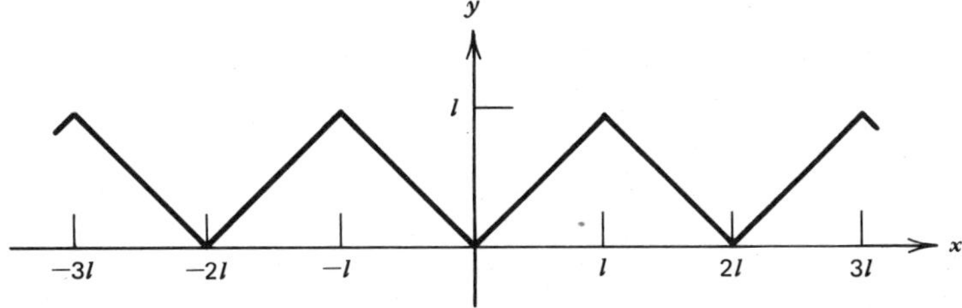

FIGURE 10.4 Triangular wave.

where the coefficients are computed from Eqs. (13) and (14). Substituting for $f(x)$ in Eq. (13) with $m = 0$, we have

$$a_0 = \frac{1}{l}\int_{-l}^{0}(-x)\,dx + \frac{1}{l}\int_{0}^{l} x\,dx$$

$$= \frac{1}{l}\frac{l^2}{2} + \frac{1}{l}\frac{l^2}{2} = l. \tag{17}$$

For $m > 0$, Eq. (13) yields

$$a_m = \frac{1}{l}\int_{-l}^{0}(-x)\cos\frac{m\pi x}{l}\,dx + \frac{1}{l}\int_{0}^{l} x\cos\frac{m\pi x}{l}\,dx.$$

These integrals can be evaluated through integration by parts, with the result that

$$a_m = \frac{1}{l}\left[-\frac{l}{m\pi}x\sin\frac{m\pi x}{l} - \left(\frac{l}{m\pi}\right)^2\cos\frac{m\pi x}{l}\right]\Bigg|_{-l}^{0}$$

$$+ \frac{1}{l}\left[\frac{l}{m\pi}x\sin\frac{m\pi x}{l} + \left(\frac{l}{m\pi}\right)^2\cos\frac{m\pi x}{l}\right]\Bigg|_{0}^{l}$$

$$= \frac{1}{l}\left[-\left(\frac{l}{m\pi}\right)^2 + \left(\frac{l}{m\pi}\right)^2\cos m\pi + \left(\frac{l}{m\pi}\right)^2\cos m\pi - \left(\frac{l}{m\pi}\right)^2\right]$$

$$= \frac{2l}{(m\pi)^2}(\cos m\pi - 1), \qquad m = 1, 2, \ldots$$

$$= \begin{cases} -4l/(m\pi)^2, & m \text{ odd}, \\ 0, & m \text{ even}. \end{cases} \tag{18}$$

Finally, from Eq. (14) it follows in a similar way that

$$b_m = 0, \qquad m = 1, 2, \ldots. \tag{19}$$

By substituting the coefficients from Eqs. (17), (18), and (19) in the series (16) we obtain the Fourier series for f:

$$f(x) = \frac{l}{2} - \frac{4l}{\pi^2}\left(\cos\frac{\pi x}{l} + \frac{1}{3^2}\cos\frac{3\pi x}{l} + \frac{1}{5^2}\cos\frac{5\pi x}{l} + \cdots\right)$$

$$= \frac{l}{2} - \frac{4l}{\pi^2}\sum_{m=1,3,5,\ldots}^{\infty}\frac{\cos(m\pi x/l)}{m^2}$$

$$= \frac{l}{2} - \frac{4l}{\pi^2}\sum_{n=1}^{\infty}\frac{\cos(2n-1)\pi x/l}{(2n-1)^2}. \tag{20}$$

The partial sums of the series (20) corresponding to $n = 1$ and $n = 2$ are shown in Figure 10.5. The series converges fairly rapidly, since the coefficients diminish as $(2n-1)^{-2}$, and the figure bears this out.

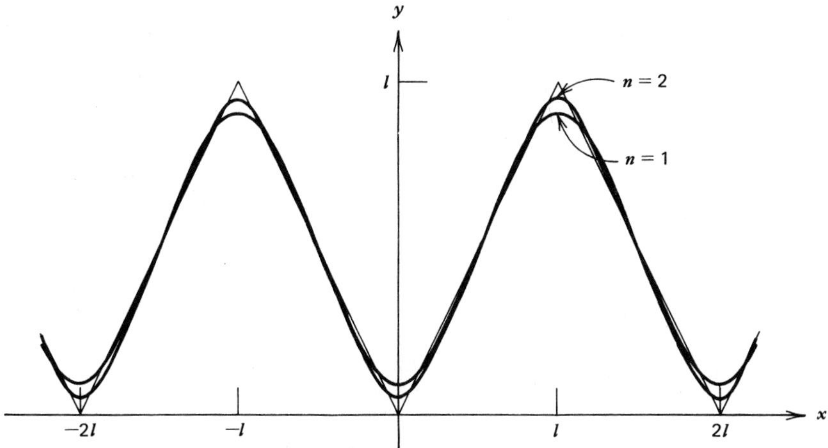

FIGURE 10.5 Partial sums in the Fourier series (Eq. (20)) for the triangular wave.

EXAMPLE 2

Let

$$f(x) = \begin{cases} 0, & -3 < x < -1 \\ 1, & -1 < x < 1 \\ 0, & 1 < x < 3 \end{cases} \tag{21}$$

and suppose that $f(x + 6) = f(x)$; see Figure 10.6. Find the coefficients in the Fourier series for f.

Since f has period 6 it follows that $l = 3$ in this problem. Consequently, the Fourier series for f has the form

$$f(x) = \frac{a_0}{2} + \sum_{n=1}^{\infty} \left(a_n \cos \frac{n\pi x}{3} + b_n \sin \frac{n\pi x}{3} \right), \tag{22}$$

where the coefficients a_n and b_n are given by Eqs. (13) and (14) with $l = 3$. We have

$$a_0 = \tfrac{1}{3} \int_{-3}^{3} f(x)\, dx = \tfrac{1}{3} \int_{-1}^{1} dx = \tfrac{2}{3}. \tag{23}$$

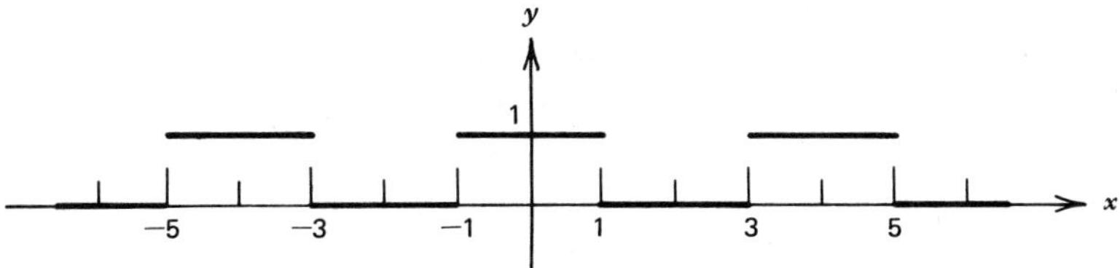

FIGURE 10.6 Graph of $f(x)$ in Example 2.

Similarly,

$$a_n = \frac{1}{3}\int_{-1}^{1} \cos\frac{n\pi x}{3}\,dx = \frac{1}{n\pi}\sin\frac{n\pi x}{3}\bigg|_{-1}^{1} = \frac{2}{n\pi}\sin\frac{n\pi}{3},\ n = 1, 2, \ldots \tag{24}$$

and

$$b_n = \frac{1}{3}\int_{-1}^{1} \sin\frac{n\pi x}{3}\,dx = -\frac{1}{n\pi}\cos\frac{n\pi x}{3}\bigg|_{-1}^{1} = 0, \qquad n = 1, 2, \ldots. \tag{25}$$

Thus the Fourier series for f is

$$\begin{aligned} f(x) &= \frac{1}{3} + \sum_{n=1}^{\infty} \frac{2}{n\pi}\sin\frac{n\pi}{3}\cos\frac{n\pi x}{3} \\ &= \frac{1}{3} + \frac{\sqrt{3}}{\pi}\left[\cos(\pi x/3) + \frac{\cos(2\pi x/3)}{2}\right. \\ &\qquad \left. - \frac{\cos(4\pi x/3)}{4} - \frac{\cos(5\pi x/3)}{5} + \cdots\right]. \end{aligned} \tag{26}$$

In this text Fourier series appear mainly as a means of solving certain problems in partial differential equations. However, such series have much wider application in science and engineering and in general are valuable tools in the investigation of periodic phenomena. Frequently the problem takes the form of resolving an incoming signal into its harmonic components, which amounts to constructing its Fourier series representation. In some frequency ranges the separate terms correspond to different colors or to different audible tones. The magnitude of the coefficient determines the amplitude of each component. If, on the one hand, the signal is given analytically (as in Examples 1 and 2), then one can use the Euler-Fourier formulas to determine the coefficients. On the other hand, in practice one often receives the signal only as a sequence of numbers at discrete times. In such cases it is common to use a numerical procedure known as the *fast Fourier transform*[4] to compute a finite number of the Fourier coefficients. The orthogonality and symmetry of the sine and cosine functions can be exploited to make this procedure extremely efficient.

PROBLEMS

1. Determine whether each of the following functions is periodic. If so, find its fundamental period.

(a) $\sin \pi x/l$ (b) $\cos 2\pi x$ (c) $\sinh 2x$ (d) $\tan \pi x$
(e) x^2 (f) $\sin 5x$ (g) $\sin mx$ (h) e^x

[4]A useful introduction to the fast Fourier transform may be found in Chapter 6 of Hamming (1962) listed in the references at the end of this chapter.

(i) $f(x) = \begin{cases} 0, & 2n - 1 \le x < 2n, \\ 1, & 2n \le x < 2n + 1; \end{cases} \quad n = 0, \pm 1, \pm 2, \ldots$

(j) $f(x) = \begin{cases} (-1)^n, & 2n - 1 \le x < 2n, \\ 1, & 2n \le x < 2n + 1; \end{cases} \quad n = 0, \pm 1, \pm 2, \ldots$

2. Suppose that g is an integrable periodic function with period T.
(a) If $0 \le a \le T$, show that

$$\int_0^T g(x)\,dx = \int_a^{a+T} g(x)\,dx.$$

Hint: Show first that $\int_0^a g(x)\,dx = \int_T^{a+T} g(x)\,dx$. Consider the change of variable $s = x - T$ in the second integral.
(b) Show that for any value of a, not necessarily in $0 \le a \le T$,

$$\int_0^T g(x)\,dx = \int_a^{a+T} g(x)\,dx.$$

(c) Show that for any values of a and b,

$$\int_a^{a+T} g(x)\,dx = \int_b^{b+T} g(x)\,dx.$$

3. If f is differentiable and is periodic with period T, show that f' is also periodic with period T. Determine whether

$$F(x) = \int_0^x f(t)\,dt$$

is always periodic.
4. Verify Eqs. (6) and (7) of the text by direct integration.

In each of Problems 5 through 14 find the Fourier series corresponding to the given function.

5. $f(x) = -x, \quad -l \le x < l; \quad f(x + 2l) = f(x)$

6. $f(x) = \begin{cases} 1, & -l \le x < 0, \\ 0, & 0 \le x < l; \end{cases} \quad f(x + 2l) = f(x)$

7. $f(x) = \begin{cases} -l - x, & -l \le x < 0, \\ l - x, & 0 \le x < l; \end{cases} \quad f(x + 2l) = f(x)$

8. $f(x) = x, \quad -1 \le x < 1; \quad f(x + 2) = f(x)$

9. $f(x) = \begin{cases} x + 1, & -1 \le x < 0, \\ x, & 0 \le x < 1; \end{cases} \quad f(x + 2) = f(x)$

10. $f(x) = \begin{cases} -1, & -2 \le x < 0, \\ 1, & 0 \le x < 2; \end{cases} \quad f(x + 4) = f(x)$

11. $f(x) = \begin{cases} x, & -\pi \le x < 0, \\ 0, & 0 \le x < \pi; \end{cases} \quad f(x + 2\pi) = f(x)$

12. $f(x) = \begin{cases} x + 1, & -1 \le x < 0, \\ 1 - x, & 0 \le x < 1; \end{cases} \quad f(x + 2) = f(x)$

13. $f(x) = \begin{cases} x + l, & -l \le x \le 0, \\ l, & 0 < x < l; \end{cases} \quad f(x + 2l) = f(x)$

14. $f(x) = \begin{cases} 0, & -2 \le x \le -1, \\ x, & -1 < x < 1, \\ 0, & 1 \le x < 2; \end{cases} \qquad f(x+4) = f(x)$

15. If $f(x) = -x$ for $-l < x < l$, and $f(x + 2l) = f(x)$, find a formula for $f(x)$ in the interval $l < x < 2l$; in the interval $-3l < x < -2l$.

16. If $f(x) = \begin{cases} x+1, & -1 < x < 0, \\ x, & 0 < x < 1, \end{cases}$
and $f(x+2) = f(x)$, find a formula for $f(x)$ in the interval $1 < x < 2$; in the interval $8 < x < 9$.

17. If $f(x) = l - x$ for $0 < x < 2l$, and $f(x + 2l) = f(x)$, find a formula for $f(x)$ in the interval $-l < x < 0$.

18. If $f(x) = \begin{cases} 1, & -l \le x < 0, \\ 0, & 0 \le x < l, \end{cases}$
and $f(x + 2l) = f(x)$, find the coefficients in the Fourier series corresponding to f by integrating over the interval $0 \le x \le 2l$. Compare the result with that of Problem 6.

19. If $f(x) = x$ for $-1 \le x < 1$, and $f(x + 2) = f(x)$, find the coefficients in the Fourier series corresponding to f by integrating over the interval $0 \le x \le 2$. Compare the result with that of Problem 8.

*20. In this problem we indicate certain similarities between three-dimensional geometric vectors and Fourier series.

(a) Let $\mathbf{v}_1$, $\mathbf{v}_2$, and $\mathbf{v}_3$ be a set of mutually orthogonal vectors in three dimensions and let $\mathbf{u}$ be any three-dimensional vector. Show that

$$\mathbf{u} = a_1\mathbf{v}_1 + a_2\mathbf{v}_2 + a_3\mathbf{v}_3, \tag{i}$$

where

$$a_i = \frac{\mathbf{u} \cdot \mathbf{v}_i}{\mathbf{v}_i \cdot \mathbf{v}_i}, \qquad i = 1, 2, 3. \tag{ii}$$

Show that a_i can be interpreted as the projection of $\mathbf{u}$ in the direction of $\mathbf{v}_i$ divided by the length of $\mathbf{v}_i$.

(b) Define the inner product (u, v) by

$$(u, v) = \int_{-l}^{l} u(x)v(x)\,dx. \tag{iii}$$

Also let

$$\begin{aligned} \phi_n(x) &= \cos(n\pi x/l), \qquad n = 0, 1, 2, \ldots; \\ \psi_n(x) &= \sin(n\pi x/l), \qquad n = 1, 2, \ldots. \end{aligned} \tag{iv}$$

Show that Eq. (10) can be written in the form

$$(f, \phi_n) = \frac{a_0}{2}(\phi_0, \phi_n) + \sum_{m=1}^{\infty} a_m(\phi_m, \phi_n) + \sum_{m=1}^{\infty} b_m(\psi_m, \phi_n). \tag{v}$$

(c) Use Eq. (v) and the orthogonality relations to show that

$$a_n = \frac{(f, \phi_n)}{(\phi_n, \phi_n)}, \qquad n = 0, 1, 2, \ldots; \qquad b_n = \frac{(f, \psi_n)}{(\psi_n, \psi_n)}, \qquad n = 1, 2, \ldots. \tag{vi}$$

Note the resemblance between Eqs. (vi) and Eq. (ii). The functions ϕ_n and ψ_n play a role for functions similar to that of the orthogonal vectors $\mathbf{v}_1$, $\mathbf{v}_2$, and $\mathbf{v}_3$ in three-dimensional space. The coefficients a_n and b_n can be interpreted as projections of the function f onto the base functions ϕ_n and ψ_n.

Observe also that any vector in three dimensions can be expressed as a linear combination of three mutually orthogonal vectors. In a somewhat similar way any sufficiently smooth function defined on $-l \leq x \leq l$ can be expressed as a linear combination of the mutually orthogonal functions $\cos(n\pi x/l)$ and $\sin(n\pi x/l)$; that is, as a Fourier series. In general, infinitely many cosines and sines are required for this purpose. This is perhaps not surprising in view of the great variety of functions defined on $-l \leq x \leq l$. What may be surprising (and was to Lagrange and others) is that only sines and cosines are needed.

10.4 The Fourier Theorem

In the last section we showed that if the Fourier series

$$\frac{a_0}{2} + \sum_{m=1}^{\infty} \left(a_m \cos \frac{m\pi x}{l} + b_m \sin \frac{m\pi x}{l} \right) \tag{1}$$

converges and thereby defines a function f, then f is periodic with period $2l$, and the coefficients a_m and b_m are related to $f(x)$ by the Euler-Fourier formulas:

$$a_m = \frac{1}{l} \int_{-l}^{l} f(x) \cos \frac{m\pi x}{l}\, dx, \qquad m = 0, 1, 2, \ldots; \tag{2}$$

$$b_m = \frac{1}{l} \int_{-l}^{l} f(x) \sin \frac{m\pi x}{l}\, dx, \qquad m = 1, 2, \ldots. \tag{3}$$

In this section we adopt a somewhat different point of view. Suppose that a function f is given. If this function is periodic with period $2l$ and integrable on the interval $[-l, l]$, then a set of coefficients a_m and b_m can be computed from Eqs. (2) and (3), and a series of the form (1) can be formally constructed. The question is whether this series converges for each value of x and, if so, whether its sum is $f(x)$. Examples have been discovered that show that the Fourier series corresponding to a function f may not converge to $f(x)$, or may even diverge. Functions whose Fourier series do not converge to the value of the function at isolated points are easily constructed, and examples will appear later in this section. Functions whose Fourier series diverge at one or more points are more pathological, and we will not consider them in this book.

To guarantee convergence of a Fourier series to the function from which its coefficients were computed it is essential to place additional conditions on the function. From a practical point of view, such conditions should be broad enough to cover all situations of interest, yet simple enough to be easily checked for particular functions. Through the years several sets of conditions have been devised to serve this purpose.

Before stating a convergence theorem for Fourier series, we define a term that appears in the theorem. A function f is said to be *piecewise continuous* on an

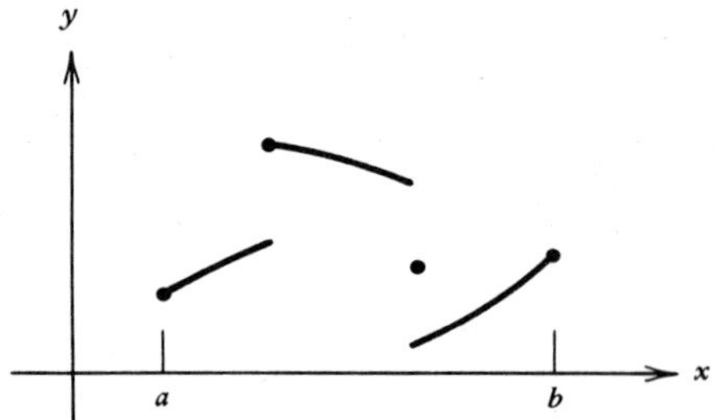

FIGURE 10.7 A piecewise continuous function.

interval $a \leq x \leq b$ if the interval can be partitioned by a finite number of points $a = x_0 < x_1 < \cdots < x_n = b$ so that

1. f is continuous on each open subinterval $x_{i-1} < x < x_i$.
2. f approaches a finite limit as the end points of each subinterval are approached from within the subinterval.

The graph of a piecewise continuous function is shown in Figure 10.7.

The notation $f(c+)$ is used to denote the limit of $f(x)$ as $x \to c$ from the right; that is,

$$f(c+) = \lim_{x \to c+} f(x).$$

Similarly, $f(c-) = \lim_{x \to c-} f(x)$ denotes the limit of $f(x)$ as x approaches c from the left.

Note that it is not essential that the function even be defined at the partition points x_i. For example, in the following theorem we assume that f' is piecewise continuous; but certainly f' does not exist at those points where f itself is discontinuous. It is also not essential that the interval be closed; it may also be open, or open at one end and closed at the other.

Theorem 10.1. *Suppose that f and f' are piecewise continuous on the interval $-l \leq x < l$. Further, suppose that f is defined outside the interval $-l \leq x < l$ so that it is periodic with period $2l$. Then f has a Fourier series*

$$f(x) = \frac{a_0}{2} + \sum_{m=1}^{\infty} \left(a_m \cos \frac{m\pi x}{l} + b_m \sin \frac{m\pi x}{l} \right), \tag{4}$$

whose coefficients are given by Eqs. (2) *and* (3). *The Fourier series converges to $f(x)$ at all points where f is continuous, and to $[f(x+) + f(x-)]/2$ at all points where f is discontinuous.*

Note that $[f(x+)+f(x-)]/2$ is the mean value of the right- and left-hand limits at the point x. At any point where f is continuous, $f(x+)=f(x-)=f(x)$. Thus it is correct to say that the Fourier series converges to $[f(x+)+f(x-)]/2$ at all points. Whenever we say that a Fourier series converges to a function f, we always mean that is converges in this sense.

It should be emphasized that the conditions given in this theorem are only sufficient for the convergence of a Fourier series; they are by no means necessary. Neither are they the most general sufficient conditions that have been discovered. In spite of this the proof of the theorem is fairly difficult and is not be given here.[5]

To obtain a better understanding of the content of the theorem it is helpful to consider some classes of functions that fail to satisfy the assumed conditions. Functions that are not included in the theorem are primarily those with infinite discontinuities in the interval $[-l, l]$, such as $1/x^2$ as $x \to 0$, or $\ln|x-l|$ as $x \to l$. Functions having an infinite number of jump discontinuities in this interval are also excluded; however, such functions are rarely encountered.

It is noteworthy that a Fourier series may converge to a sum that is not differentiable, or even continuous, in spite of the fact that each term in the series (4) is continuous, and even differentiable infinitely many times. The example below is an illustration of this, as is Example 2 in Section 10.3.

EXAMPLE

Let

$$f(x)=\begin{cases} 0, & -l<x<0, \\ l, & 0<x<l, \end{cases} \tag{5}$$

and let f be defined outside this interval so that $f(x+2l)=f(x)$ for all x. We will temporarily leave open the definition of f at the points $x=0, \pm l$, except that its values must be finite. Find the Fourier series for this function and determine where it converges.

The equation $y=f(x)$ has the graph shown in Figure 10.8, extended to infinity in both directions. It can be thought of as representing a square wave. The interval $[-l, l]$ can be partitioned to give the two open subintervals $(-l, 0)$ and $(0, l)$. In $(0, l)$, $f(x)=l$ and $f'(x)=0$. Clearly, both f and f' are continuous and furthermore have limits as $x \to 0$ from the right and as $x \to l$ from the left. The situation in $(-l, 0)$ is similar. Consequently, both f and f' are piecewise continuous on $[-l, l)$, so f satisfies the conditions of Theorem 10.1. If the coefficients a_m and b_m are computed from Eqs. (2) and (3) the convergence of the resulting Fourier series to $f(x)$ is assured at all points where f is continuous. Note that the values of a_m and b_m are the same regardless of the definition of f at its points of discontinuity. This is true because the value of an integral is

[5] Proofs of the convergence of a Fourier series can be found in most books on advanced calculus. See, for example, Kaplan (Chap. 7) or Buck (Chap. 6).

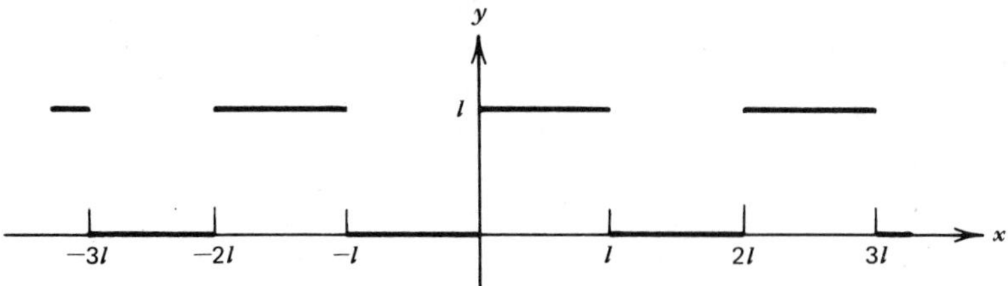

FIGURE 10.8 Square wave.

unaffected by changing the value of the integrand at a finite number of points. From Eq. (2)

$$a_0 = \frac{1}{l}\int_{-l}^{l} f(x)\,dx = \int_0^l dx = l;$$

$$a_m = \frac{1}{l}\int_{-l}^{l} f(x)\cos\frac{m\pi x}{l}\,dx = \int_0^l \cos\frac{m\pi x}{l}\,dx$$

$$= 0, \qquad m \neq 0.$$

Similarly, from Eq. (3),

$$b_m = \frac{1}{l}\int_{-l}^{l} f(x)\sin\frac{m\pi x}{l}\,dx = \int_0^l \sin\frac{m\pi x}{l}\,dx$$

$$= \frac{l}{m\pi}(1 - \cos m\pi)$$

$$= \begin{cases} 0, & m \text{ even;} \\ \dfrac{2l}{m\pi}, & m \text{ odd.} \end{cases}$$

Hence

$$f(x) = \frac{l}{2} + \frac{2l}{\pi}\left(\sin\frac{\pi x}{l} + \frac{1}{3}\sin\frac{3\pi x}{l} + \frac{1}{5}\sin\frac{5\pi x}{l} + \cdots\right)$$

$$= \frac{l}{2} + \frac{2l}{\pi}\sum_{m=1,3,5,\ldots}^{\infty}\frac{\sin(m\pi x/l)}{m}$$

$$= \frac{l}{2} + \frac{2l}{\pi}\sum_{n=1}^{\infty}\frac{\sin(2n-1)\pi x/l}{2n-1}. \tag{6}$$

At the points $x = 0, \pm nl$, where the function f in the example is not continuous, all terms in the series after the first vanish and the sum is $l/2$. This is the mean value of the limits from the right and left, as it should be. Thus we might as well define f at these points to have the value $l/2$. If we choose to define it otherwise the series still gives the value $l/2$ at these points since none of the preceding calculations is altered in any detail; it simply does not converge to

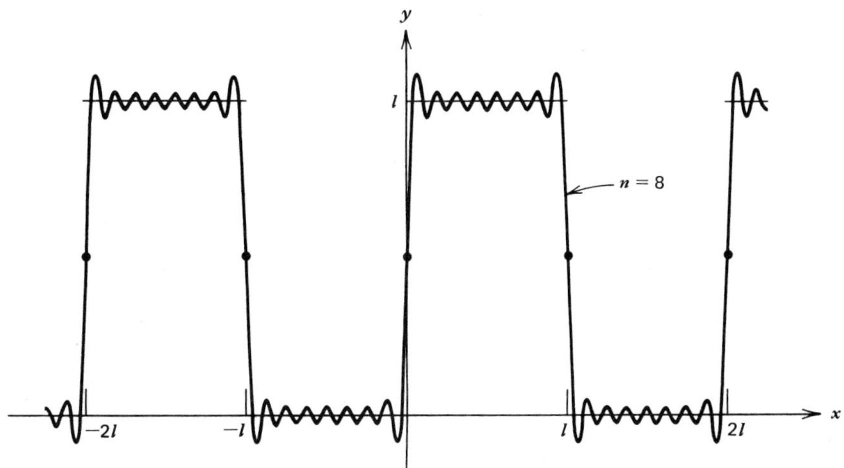

FIGURE 10.9 A partial sum in the Fourier series (Eq. (6)) for the square wave.

the function at those points unless f is defined to have this value. This illustrates the possibility that the Fourier series corresponding to a function may not converge to it at points of discontinuity unless the function is properly defined at such points.

The manner in which the partial sums

$$S_n(x) = \frac{l}{2} + \frac{2l}{\pi}\left(\sin\frac{\pi x}{l} + \cdots + \frac{1}{2n-1}\sin\frac{(2n-1)\pi x}{l}\right),$$
$$n = 0, 1, 2, \ldots .$$

of the Fourier series (6) converge to $f(x)$ is indicated in Figure 10.9, where the graph of $S_8(x)$ is sketched. The figure shows that at points where f is continuous the partial sums do approach $f(x)$ as n increases. However, in the neighborhood of points of discontinuity, such as $x = 0$ and $x = l$, the partial sums do not converge smoothly to the mean value. Instead they tend to overshoot the mark at each end of the jump, as though they cannot quite accommodate themselves to the sharp turn required at this point. This behavior is typical of Fourier series at points of discontinuity, and is known as the Gibbs[6] phenomenon.

Figure 10.9 also shows that the series in this example converges more slowly than the one in Example 1 in Section 10.3. This is due to the fact that the coefficients in the series (6) are proportional only to $1/(2n-1)$.

---

[6] The Gibbs phenomenon is named after Josiah Willard Gibbs (1839–1903), who is better known for his work on vector analysis and statistical mechanics. Gibbs was professor of mathematical physics at Yale, and one of the first American scientists to achieve an international reputation. Gibbs' phenomenon is discussed in more detail by Guillemin (Chap. 7) and Carslaw (Chap. 9).

PROBLEMS

Find the Fourier series for each of the functions in Problems 1 through 8. Assume that the functions are periodically extended outside the original interval. Sketch the function to which each series converges for several periods.

1. $f(x) = \begin{cases} -1, & -1 \le x < 0 \\ 1, & 0 \le x < 1 \end{cases}$

2. $f(x) = \begin{cases} 0, & -\pi \le x < 0 \\ x, & 0 \le x < \pi \end{cases}$

3. $f(x) = \sin^2 x, \quad -\pi \le x \le \pi$

4. $f(x) = \begin{cases} 1, & 0 \le x < s \\ 0, & s \le x < 2 - s \\ 1, & 2 - s \le x < 2 \end{cases}$

5. $f(x) = \begin{cases} l + x, & -l \le x < 0 \\ l - x, & 0 \le x < l \end{cases}$

6. $f(x) = 1 - x^2, \quad -1 \le x \le 1$

7. $f(x) = \begin{cases} 0, & -\pi \le x < -\pi/2 \\ 1, & -\pi/2 \le x < \pi/2 \\ 0, & \pi/2 \le x < \pi \end{cases}$

8. $f(x) = \begin{cases} 0, & -1 \le x < 0 \\ x^2, & 0 \le x < 1 \end{cases}$

Periodic Forcing Terms. In this chapter we are mainly concerned with the use of Fourier series to solve boundary value problems for certain partial differential equations. However, Fourier series are also useful in many other situations where periodic phenomena occur. Problems 9 through 12 indicate how they can be employed to solve initial value problems with periodic forcing terms.

9. Find the solution of the initial value problem

$$y'' + \omega^2 y = \sin nt, \qquad y(0) = 0, \quad y'(0) = 0,$$

where n is a positive integer, and $\omega^2 \neq n^2$. What happens if $\omega^2 = n^2$?

10. Find the formal solution of the initial value problem

$$y'' + \omega^2 y = \sum_{n=1}^{\infty} b_n \sin nt, \qquad y(0) = 0, \quad y'(0) = 0,$$

where $\omega > 0$ is not equal to a positive integer. How is the solution altered if $\omega = m$, where m is a positive integer?

11. Find the formal solution of the initial value problem

$$y'' + \omega^2 y = f(t), \qquad y(0) = 0, \quad y'(0) = 0,$$

where f is periodic with period 2π and

$$f(t) = \begin{cases} 1, & 0 < t < \pi; \\ 0, & t = 0, \pi, 2\pi; \\ -1, & \pi < t < 2\pi. \end{cases}$$

12. Find the formal solution of the initial value problem

$$y'' + \omega^2 y = f(t), \qquad y(0) = 1, \quad y'(0) = 0,$$

where f is periodic with period 2 and

$$f(t) = \begin{cases} 1 - t, & 0 \le t < 1; \\ -1 + t, & 1 \le t < 2. \end{cases}$$

13. Assuming that

$$f(x) = \frac{a_0}{2} + \sum_{n=1}^{\infty} \left(a_n \cos \frac{n\pi x}{l} + b_n \sin \frac{n\pi x}{l} \right), \tag{i}$$

show formally that

$$\frac{1}{l} \int_{-l}^{l} [f(x)]^2 \, dx = \frac{a_0^2}{2} + \sum_{n=1}^{\infty} (a_n^2 + b_n^2).$$

This relation between a function f and its Fourier coefficients is known as Parseval's (1755–1836) equation. Parseval's equation is very important in the theory of Fourier series and is discussed further in Section 11.7.
Hint: Multiply Eq. (i) by $f(x)$, integrate from $-l$ to l, and use the Euler-Fourier formulas.

*14. (a) If f and f' are piecewise continuous on $-l \le x < l$, and if f is periodic with period $2l$, show that na_n and nb_n are bounded as $n \to \infty$.
Hint: Use integration by parts.
(b) If f is continuous on $-l \le x \le l$ and periodic with period $2l$, and if f' and f'' are piecewise continuous on $-l \le x < l$, show that $n^2 a_n$ and $n^2 b_n$ are bounded as $n \to \infty$. Use this fact to show that the Fourier series for f converges at each point in $-l \le x \le l$. Why must f be continuous on the *closed* interval?
Hint: Again, use integration by parts.

Acceleration of convergence. In the next problem we show how it is sometimes possible to improve the speed of convergence of a Fourier, or other infinite, series.

15. Suppose that we wish to calculate values of the function g, where

$$g(x) = \sum_{n=1}^{\infty} \frac{(2n-1)}{1+(2n-1)^2} \sin(2n-1)\pi x. \tag{i}$$

It is possible to show that this series converges, albeit rather slowly. However, observe that for large n the terms in the series (i) are approximately equal to $[\sin(2n-1)\pi x]/(2n-1)$ and that the latter terms are similar to those in the example in the text, Eq. (6).
(a) Show that

$$\sum_{n=1}^{\infty} [\sin(2n-1)\pi x]/(2n-1) = (\pi/2)[f(x) - \tfrac{1}{2}], \tag{ii}$$

where f is the square wave in the example with $l = 1$.
(b) Subtract Eq. (ii) from Eq. (i) and show that

$$g(x) = \frac{\pi}{2}[f(x) - \tfrac{1}{2}] - \sum_{n=1}^{\infty} \frac{\sin(2n-1)\pi x}{(2n-1)[1+(2n-1)^2]}. \tag{iii}$$

The series (iii) converges much faster than the series (i) and thus provides a better way to calculate values of $g(x)$.

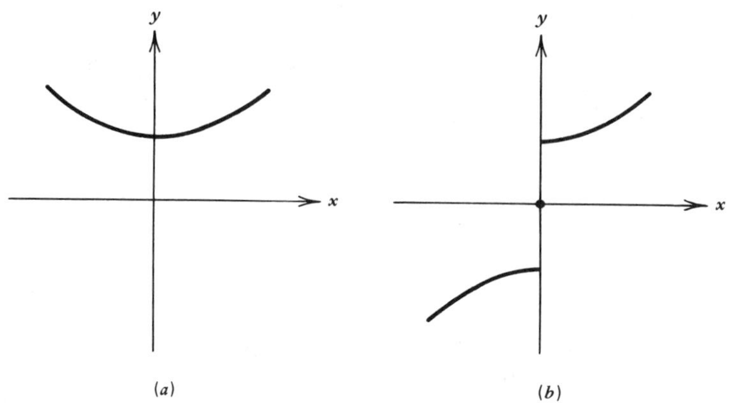

FIGURE 10.10 Even (a) and odd (b) functions.

10.5 Even and Odd Functions

Before looking at further examples of Fourier series it is useful to distinguish two classes of functions for which the Euler-Fourier formulas can be simplified. These are even and odd functions, which are characterized geometrically by the property of symmetry with respect to the y axis and the origin, respectively (see Figure 10.10).

Analytically, f is an *even function* if its domain contains the point $-x$ whenever it contains the point x, and if

$$f(-x) = f(x) \tag{1}$$

for each x in the domain of f. Similarly, f is an *odd function* if its domain contains $-x$ whenever it contains x, and if

$$f(-x) = -f(x) \tag{2}$$

for each x in the domain of f. Examples of even functions are 1, x^2, $\cos nx$, $|x|$, and x^{2n}. The functions x, x^3, $\sin nx$, and x^{2n+1} are examples of odd functions. Note that according to Eq. (2), $f(0)$ must be zero if f is an odd function whose domain contains the origin. Most functions are neither even nor odd; for instance, e^x. Only one function, f identically zero, is both even and odd.

Elementary properties of even and odd functions include the following:

1. The sum (difference) and product (quotient) of two even functions are even.
2. The sum (difference) of two odd functions is odd; the product (quotient) of two odd functions is even.
3. The sum (difference) of an odd function and an even function is neither even nor odd; the product (quotient) of two such functions is odd.[7]

[7]These statements may need to be modified if either function vanishes identically.

The proofs of all these assertions are simple and follow directly from the definitions. For example, if both f_1 and f_2 are odd, and if $g(x) = f_1(x) + f_2(x)$, then

$$\begin{aligned} g(-x) = f_1(-x) + f_2(-x) &= -f_1(x) - f_2(x) \\ &= -[f_1(x) + f_2(x)] = -g(x), \end{aligned} \tag{3}$$

so $f_1 + f_2$ is an odd function also. Similarly, if $h(x) = f_1(x)f_2(x)$, then

$$h(-x) = f_1(-x)f_2(-x) = [-f_1(x)][-f_2(x)] = f_1(x)f_2(x) = h(x), \tag{4}$$

so that $f_1 f_2$ is even.

Also of importance are the following two integral properties of even and odd functions:

4. If f is an even function, then

$$\int_{-l}^{l} f(x)\,dx = 2\int_0^l f(x)\,dx. \tag{5}$$

5. If f is an odd function, then

$$\int_{-l}^{l} f(x)\,dx = 0. \tag{6}$$

These properties are intuitively clear from the interpretation of an integral in terms of area under a curve, and also follow immediately from the definitions. For example, if f is even, then

$$\int_{-l}^{l} f(x)\,dx = \int_{-l}^{0} f(x)\,dx + \int_0^l f(x)\,dx.$$

Letting $x = -s$ in the first term on the right side, and using Eq. (1), gives

$$\int_{-l}^{l} f(x)\,dx = -\int_l^0 f(s)\,ds + \int_0^l f(x)\,dx = 2\int_0^l f(x)\,dx.$$

The proof of the corresponding property for odd functions is similar.

Even and odd functions are particularly important in applications of Fourier series since their Fourier series have special forms, which occur frequently in physical problems.

COSINE SERIES. Suppose that f and f' are piecewise continuous on $-l \le x < l$, and that f is an even periodic function with period $2l$. Then it follows from properties 1 and 3 that $f(x)\cos(n\pi x/l)$ is even and $f(x)\sin(n\pi x/l)$ is odd. As a consequence of Eqs. (5) and (6), the Fourier coefficients of f are then given by

$$\begin{aligned} a_n &= \frac{2}{l}\int_0^l f(x)\cos\frac{n\pi x}{l}\,dx, && n = 0, 1, 2, \ldots; \\ b_n &= 0, && n = 1, 2, \ldots. \end{aligned} \tag{7}$$

Thus f has the Fourier series

$$f(x) = \frac{a_0}{2} + \sum_{n=1}^{\infty} a_n \cos \frac{n\pi x}{l}.$$

In other words, the Fourier series of any even function consists only of the even trigonometric functions $\cos(n\pi x/l)$ and the constant term; it is natural to call such a series a *Fourier cosine series*. From a computational point of view, observe that only the coefficients a_n, for $n = 0, 1, 2, \ldots$, need to be calculated from the integral formula (7). Each of the b_n, for $n = 1, 2, \ldots$, is automatically zero for any even function, and so does not need to be calculated by integration.

SINE SERIES. Suppose that f and f' are piecewise continuous on $-l \le x < l$, and that f is an odd periodic function of period $2l$. Then it follows from properties 2 and 3 that $f(x)\cos(n\pi x/l)$ is odd and $f(x)\sin(n\pi x/l)$ is even. In this case the Fourier coefficients of f are

$$\begin{aligned} a_n &= 0, && n = 0, 1, 2, \ldots, \\ b_n &= \frac{2}{l}\int_0^l f(x) \sin \frac{n\pi x}{l}\, dx, && n = 1, 2, \ldots, \end{aligned} \tag{8}$$

and the Fourier series for f is of the form

$$f(x) = \sum_{n=1}^{\infty} b_n \sin \frac{n\pi x}{l}.$$

Thus the Fourier series for any odd function consists only of the odd trigonometric functions $\sin(n\pi x/l)$; such a series is called a *Fourier sine series*. Again observe that only half of the coefficients need to be calculated by integration, since each a_n, for $n = 0, 1, 2, \ldots$, is zero for any odd function.

EXAMPLE 1

Let $f(x) = x$, $-l < x < l$, and let $f(-l) = f(l) = 0$. Let f be defined elsewhere so that it is periodic of period $2l$ (see Figure 10.11). The function defined in this manner is known as a sawtooth wave. Find the Fourier series for this function.

Since f is an odd function, its Fourier coefficients are, according to Eq. (8):

$$a_n = 0, \qquad n = 0, 1, 2, \ldots;$$

$$\begin{aligned} b_n &= \frac{2}{l}\int_0^l x \sin \frac{n\pi x}{l}\, dx \\ &= \frac{2}{l}\left(\frac{l}{n\pi}\right)^2 \left\{\sin \frac{n\pi x}{l} - \frac{n\pi x}{l} \cos \frac{n\pi x}{l}\right\}\Bigg|_0^l \\ &= \frac{2l}{n\pi}(-1)^{n+1}, \qquad n = 1, 2, \ldots . \end{aligned}$$

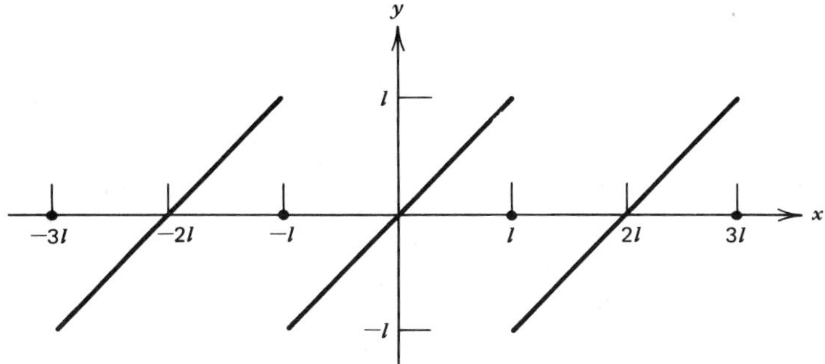

FIGURE 10.11 *Sawtooth wave.*

Hence the Fourier series for f, the sawtooth wave, is

$$f(x) = \frac{2l}{\pi} \sum_{n=1}^{\infty} \frac{(-1)^{n+1}}{n} \sin \frac{n\pi x}{l}. \tag{9}$$

Observe that the periodic function f is discontinuous at the points $\pm l, \pm 3l, \ldots,$ as shown in Figure 10.11. At these points the series (9) converges to the mean value of the left and right limits, namely zero. The partial sum of the series (9) for $n = 9$ is shown in Figure 10.12. The Gibbs phenomenon (mentioned in Section 10.4) again occurs near the points of discontinuity.

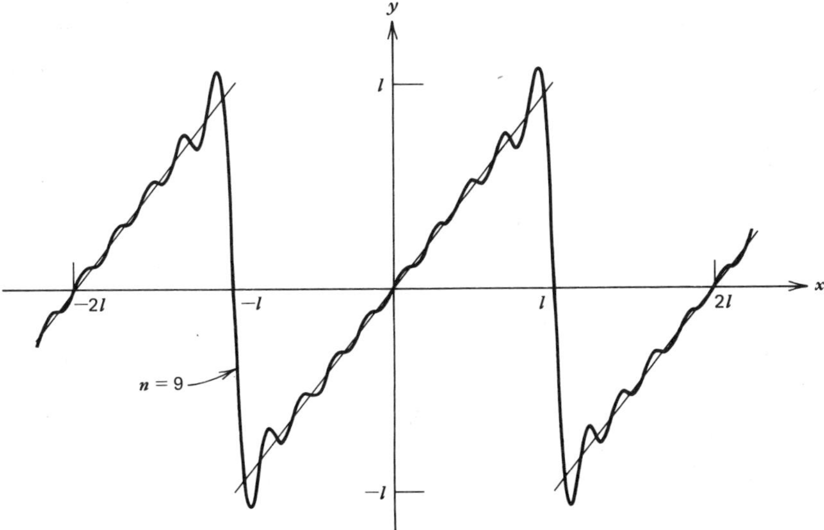

FIGURE 10.12 *A partial sum in the Fourier series (Eq. (9)) for the sawtooth wave.*

Note that in this example $f(-l) = f(l) = 0$, as well as $f(0) = 0$. This is required if the function f is to be both odd and periodic with period $2l$. When we speak of constructing a sine series for a function defined on $0 \le x \le l$, it is understood that, if necessary, we must first redefine the function to be zero at $x = 0$ and $x = l$.

It is worthwhile to observe that the triangular wave function (Example 1 of Section 10.3) and the sawtooth wave function just considered are identical on the interval $0 \le x < l$. Therefore, their Fourier series converge to the same function, $f(x) = x$, on this interval. Thus, if it is required to represent the function $f(x) = x$ on $0 \le x < l$ by a Fourier series, it is possible to do this by *either a cosine series or a sine series*. In the former case f is extended as an *even* function into the interval $-l < x < 0$ and elsewhere periodically (the triangular wave). In the latter case f is extended into $-l < x < 0$ as an *odd* function, and elsewhere periodically (the sawtooth wave). If f is extended in any other way, the resulting Fourier series will still converge to x in $0 \le x < l$ but will involve both sine and cosine terms.

In solving problems in differential equations it is often useful to expand in a Fourier series a function f originally defined only on the interval $[0, l]$. As indicated above for the function $f(x) = x$ several alternatives are available. Explicitly, we can:

1. Define a function g of period $2l$ so that

$$g(x) = \begin{cases} f(x), & 0 \le x \le l, \\ f(-x), & -l < x < 0. \end{cases} \tag{10}$$

The function g is thus the even periodic extension of f. Its Fourier series, which is a cosine series, represents f on $[0, l]$.

2. Define a function h of period $2l$ so that

$$h(x) = \begin{cases} f(x), & 0 < x < l, \\ 0, & x = 0, l, \\ -f(-x), & -l < x < 0. \end{cases} \tag{11}$$

The function h is thus the odd periodic extension of f. Its Fourier series, which is a sine series, also represents f on $(0, l)$.

3. Define a function k of period $2l$ so that

$$k(x) = f(x), \qquad 0 \le x \le l, \tag{12}$$

and let $k(x)$ be defined for $(-l, 0)$ in any way consistent with the conditions of Theorem 10.1. Sometimes it is convenient to define $k(x)$ to be zero for $-l < x < 0$. The Fourier series for k, which involves both sine and cosine terms, also represents f on $[0, l]$, regardless of the manner in which $k(x)$ is defined in $(-l, 0)$. Thus there are infinitely many such series, all of which converge to $f(x)$ in the original interval.

Usually the form of the expansion to be used will be dictated (or at least suggested) by the purpose for which it is needed. However, if there is a choice as

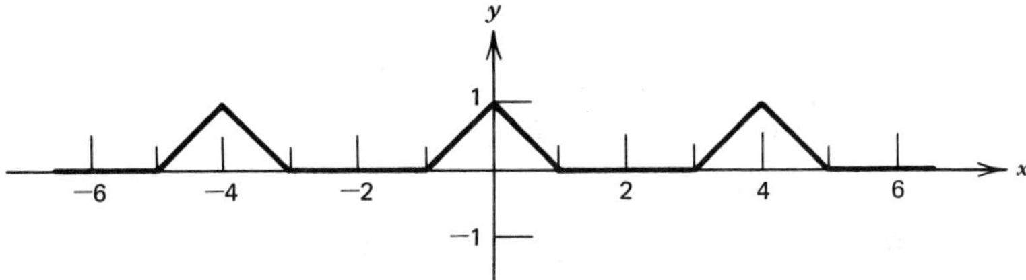

FIGURE 10.13 Even periodic extension of $f(x)$ given by Eq. (13).

to the kind of Fourier series to be used, the selection can sometimes be based on the rapidity of convergence. For example, the cosine series for the triangular wave (Eq. (20) of Section 10.3) converges more rapidly than the sine series for the sawtooth wave (Eq. (9) of Section 10.5), although both converge to the same function for $0 \leq x < l$. This is due to the fact that the triangular wave is a smoother function than the sawtooth wave and is therefore easier to approximate. In general, the more continuous derivatives possessed by a function over the entire interval $-\infty < x < \infty$, the faster its Fourier series will converge. See Problem 14 of Section 10.4.

EXAMPLE 2

Suppose that

$$f(x) = \begin{cases} 1 - x, & 0 < x \leq 1, \\ 0, & 1 < x \leq 2. \end{cases} \tag{13}$$

As indicated above, we can represent f either by a cosine series or by a sine series. Sketch the graph of the sum of each of these series for $-6 \leq x \leq 6$.

In this example $l = 2$, so the cosine series for f converges to the even periodic extension of f of period 4, whose graph is sketched in Figure 10.13.

Similarly, the sine series for f converges to the odd periodic extension of f of period 4. The graph of this function is shown in Figure 10.14.

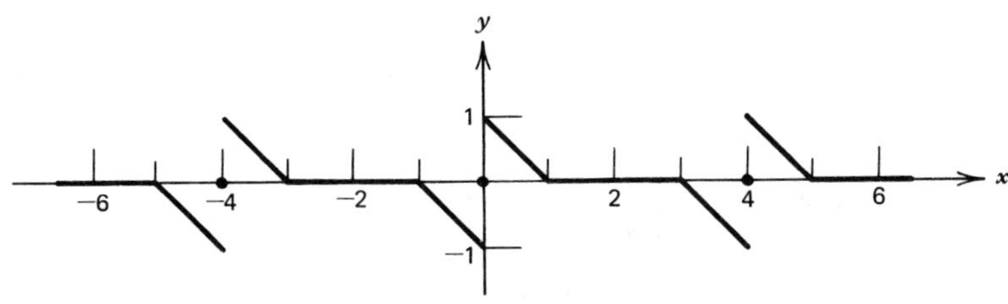

FIGURE 10.14 Odd periodic extension of $f(x)$ given by Eq. (13).

PROBLEMS

1. Determine whether the following functions are even, odd, or neither.

   (a) x^3 (b) $x^3 - 2x$ (c) $x^3 - 2x + 1$
   (d) $\tan 2x$ (e) $\sec x$ (f) $|x|^3$
   (g) e^{-x} (h) $\ln|\sin x|$ (i) $\ln|\cos x|$
   (j) $(2x - x^3)^4$
   (k) the function of Problem 1(i), Section 10.3
   (l) the function of Problem 1(j), Section 10.3

2. Using the properties of even and odd functions, evaluate the following integrals.

   (a) $\int_{-1}^{1} x\,dx$ (b) $\int_{-1}^{1} x^4\,dx$ (c) $\int_{-\pi}^{\pi} x \sin nx\,dx$
   (d) $\int_{-l}^{l} \sin \frac{n\pi x}{l} \cos \frac{n\pi x}{l}\,dx$ (e) $\int_{-\pi}^{\pi} x^4 \sin nx\,dx$ (f) $\int_{-\pi}^{\pi} x \cos nx\,dx$

3. Prove that any function can be expressed as the sum of two other functions, one of which is even and the other odd. That is, for any function f, whose domain contains $-x$ whenever it contains x, show that there is an even function g and an odd function h such that $f(x) = g(x) + h(x)$.
   Hint: Assuming $f(x) = g(x) + h(x)$, what is $f(-x)$?
4. Find the coefficients in the cosine and sine series described in Example 2.

In each of Problems 5 through 14 find the required Fourier series for the given function; sketch the graph of the function to which the series converges over two periods.

5. $f(x) = \begin{cases} 1, & 0 < x < 1, \\ 0, & 1 < x < 2; \end{cases}$ cosine series, period 4.
   Compare with the example and Problems 4 and 7 of Section 10.4.

6. $f(x) = \begin{cases} x, & 0 \le x < 1, \\ 1, & 1 \le x < 2; \end{cases}$ sine series, period 4.

7. $f(x) = 1, \quad 0 \le x \le \pi;$ cosine series, period 2π.
8. $f(x) = 1, \quad 0 < x < \pi;$ sine series, period 2π.

9. $f(x) = \begin{cases} 0, & 0 < x < \pi, \\ 1, & \pi < x < 2\pi, \\ 2, & 2\pi < x < 3\pi. \end{cases}$ sine series, period 6π.

10. $f(x) = x, \quad 0 \le x < 1;$ series of period 1.
11. $f(x) = l - x, \quad 0 \le x \le l;$ cosine series, period $2l$.
    Compare with Example 1 of Section 10.3.

12. $f(x) = l - x, \quad 0 < x < l;$ sine series, period $2l$.

13. $f(x) = \begin{cases} x, & 0 < x < \pi, \\ 0, & \pi < x < 2\pi; \end{cases}$ cosine series, period 4π

14. $f(x) = -x, \quad -\pi < x < 0;$ sine series, period 2π

15. Prove that if f is an odd function, then

$$\int_{-l}^{l} f(x)\,dx = 0.$$

16. Prove properties 2 and 3 of even and odd functions, as stated in the text.

17. Prove that the derivative of an even function is odd, and that the derivative of an odd function is even.
18. Let $F(x) = \int_0^x f(t)\,dt$. Show that if f is even, then F is odd, and that if f is odd, then F is even.
19. From the Fourier series for the square wave in the example in Section 10.4, show that

$$\frac{\pi}{4} = 1 - \frac{1}{3} + \frac{1}{5} - \frac{1}{7} + \cdots = \sum_{n=0}^{\infty} \frac{(-1)^n}{2n+1}.$$

This relation between π and the odd positive integers was discovered by Leibniz in 1673. Compute an approximate value of π by using the first four terms of this series.
20. (a) From the Fourier series for the triangular wave (Example 1 of Section 10.3), show that

$$\frac{\pi^2}{8} = 1 + \frac{1}{3^2} + \frac{1}{5^2} + \cdots = \sum_{n=0}^{\infty} \frac{1}{(2n+1)^2}.$$

(b) Compute an approximate value of π by using the first four terms of the series in part (a).
21. Assume that f has a Fourier sine series

$$f(x) = \sum_{n=1}^{\infty} b_n \sin(n\pi x/l)$$

for $0 \le x \le l$.
(a) Show formally that

$$\frac{2}{l}\int_0^l [f(x)]^2\,dx = \sum_{n=1}^{\infty} b_n^2.$$

Compare this result with that of Problem 13 of Section 10.4. What is the corresponding result if f has a cosine series?
(b) Apply the result of part (a) to the series for the sawtooth wave given in Eq. (9), and thereby show that

$$\frac{\pi^2}{6} = 1 + \frac{1}{2^2} + \frac{1}{3^2} + \cdots = \sum_{n=1}^{\infty} \frac{1}{n^2}.$$

Special Fourier Series. Let f be a function originally defined on $0 \le x \le l$. In this section we have shown that it is possible to represent f either by a sine series or by a cosine series by constructing odd or even periodic extensions of f, respectively. Problems 22 to 24 concern some other specialized Fourier series that converge to the given function f on $(0, l)$.

22. Let f be extended into $(l, 2l]$ in an arbitrary manner. Then extend the resulting function into $(-2l, 0)$ as an odd function and elsewhere as a periodic function of period $4l$ (see Figure 10.15). Show that this function has a Fourier sine series in terms of the functions $\sin(n\pi x/2l)$, $n = 1, 2, 3, \ldots$; that is,

$$f(x) = \sum_{n=1}^{\infty} b_n \sin(n\pi x/2l),$$

where

$$b_n = \frac{1}{l}\int_0^{2l} f(x)\sin(n\pi x/2l)\,dx.$$

This series converges to the original function on $(0, l)$.

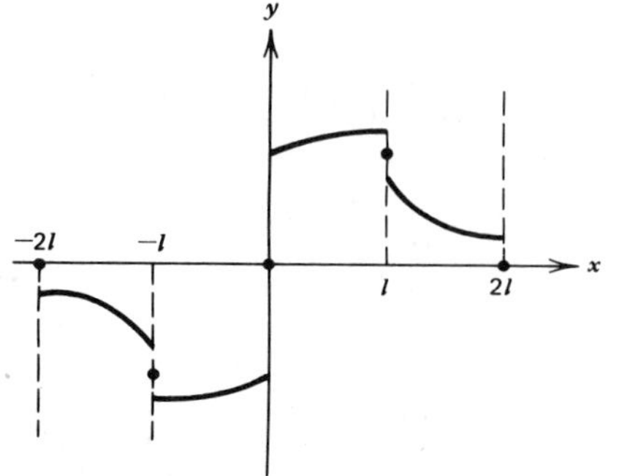

FIGURE 10.15 Graph of the function in Problem 22.

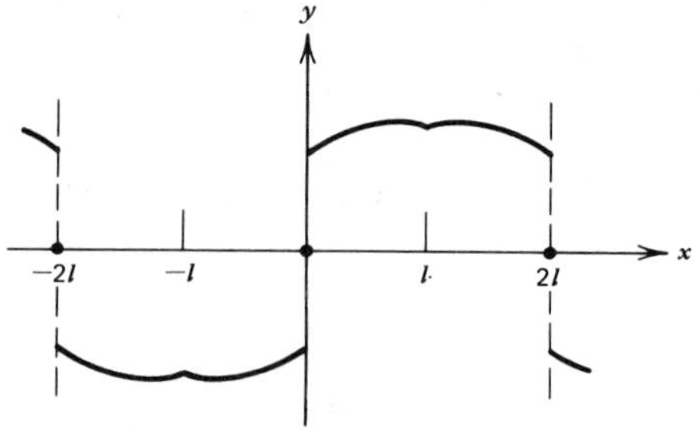

FIGURE 10.16 Graph of the function in Problem 23.

23. Let f first be extended into $(l, 2l]$ so that it is symmetric about $x = l$; that is, so as to satisfy $f(2l - x) = f(x)$ for $0 \leq x < l$. Let the resulting function be extended into $(-2l, 0)$ as an odd function and elsewhere (see Figure 10.16) as a periodic function of period $4l$. Show that this function has a Fourier series in terms of the functions $\sin(\pi x/2l), \sin(3\pi x/2l), \sin(5\pi x/2l), \ldots$; that is,

$$f(x) = \sum_{n=1}^{\infty} b_n \sin \frac{(2n-1)\pi x}{2l},$$

where

$$b_n = \frac{2}{l} \int_0^l f(x) \sin \frac{(2n-1)\pi x}{2l}\, dx.$$

This series converges to the original function on $(0, l]$.

24. How should f, originally defined on $[0, l]$, be extended so as to obtain a Fourier series involving the functions $\cos(\pi x/2l), \cos(3\pi x/2l), \cos(5\pi x/2l), \ldots$? Refer to Problems 22 and 23. If $f(x) = x$ for $0 \leq x \leq l$, sketch the function to which the Fourier series converges for $-4l \leq x \leq 4l$.

10.6 Solution of Other Heat Conduction Problems

In Section 10.2 we considered the problem consisting of the heat conduction equation

$$\alpha^2 u_{xx} = u_t, \qquad 0 < x < l, \qquad t > 0, \tag{1}$$

the boundary conditions

$$u(0,t) = 0, \qquad u(l,t) = 0, \qquad t > 0, \tag{2}$$

and the initial condition

$$u(x,0) = f(x), \qquad 0 \leq x \leq l. \tag{3}$$

We found the solution to be

$$u(x,t) = \sum_{n=1}^{\infty} b_n e^{-n^2\pi^2\alpha^2 t/l^2} \sin \frac{n\pi x}{l}, \tag{4}$$

where the coefficients b_n are the same as in the series

$$f(x) = \sum_{n=1}^{\infty} b_n \sin \frac{n\pi x}{l}. \tag{5}$$

The series in Eq. (5) is just the Fourier sine series for f; according to Section 10.5 its coefficients are given by

$$b_n = \frac{2}{l}\int_0^l f(x) \sin \frac{n\pi x}{l}\, dx. \tag{6}$$

Hence the solution of the heat conduction problem, Eqs. (1) to (3), is given by the series in Eq. (4) with the coefficients computed from Eq. (6).

EXAMPLE 1

Find the temperature $u(x,t)$ at any time in a metal rod of unit length, insulated on the sides, which initially has a uniform temperature of 10°C throughout, and whose ends are maintained at 0°C for all $t > 0$.

The temperature in the rod satisfies the heat conduction problem (1), (2), (3) with $l = 1$ and $f(x) = 10$ for $0 < x < 1$. Thus, from Eq. (4), the solution is

$$u(x,t) = \sum_{n=1}^{\infty} b_n e^{-n^2\pi^2\alpha^2 t} \sin n\pi x \tag{7}$$

where, from Eq. (6),

$$\begin{aligned} b_n &= 20\int_0^1 \sin n\pi x\, dx \\ &= \frac{20}{n\pi}(1 - \cos n\pi) \\ &= \begin{cases} 40/n\pi, & n \text{ odd}; \\ 0, & n \text{ even}. \end{cases} \end{aligned} \tag{8}$$

Finally, by substituting for b_n in Eq. (7) we obtain

$$u(x,t) = \frac{40}{\pi} \sum_{n=1,3,5,\ldots}^{\infty} \frac{1}{n} e^{-n^2\pi^2\alpha^2 t} \sin n\pi x. \tag{9}$$

The expression (9) for the temperature is moderately complicated, but the behavior of the solution can be easily seen from the graphs in Figures 10.17 and 10.18, where we have chosen $\alpha^2 = 1$ for convenience. In Figure 10.17 we show the temperature distribution in the bar at several different times. Observe that the temperature diminishes steadily as heat in the bar is lost through the end points. The way in which the temperature decays at a given point in the bar is indicated in Figure 10.18, where temperature is plotted against time for a few selected points in the bar.

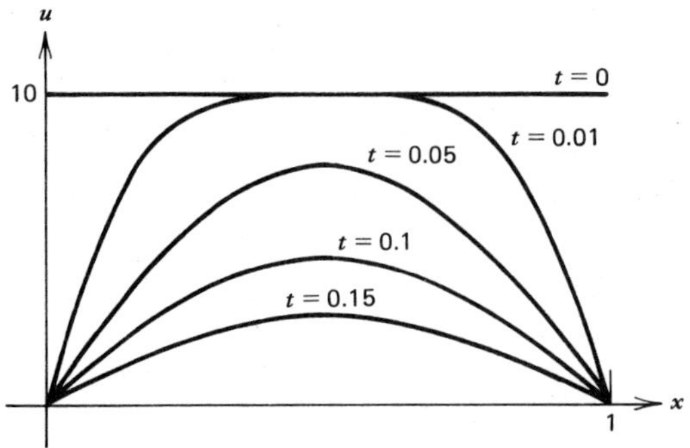

FIGURE 10.17 Temperature distributions at several times for the heat conduction problem of Example 1.

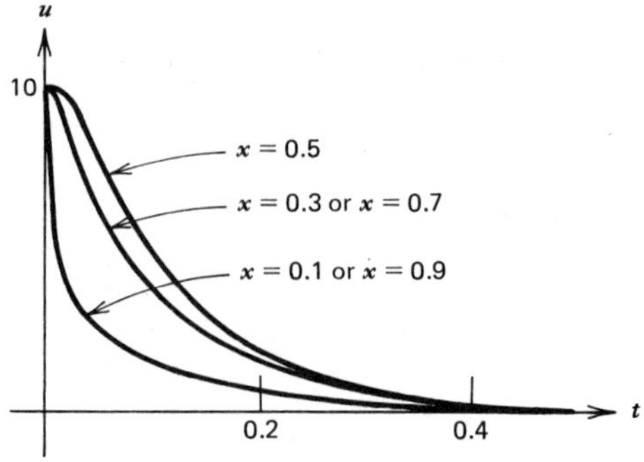

FIGURE 10.18 Dependence of temperature on time at several locations for the heat conduction problem of Example 1.

The negative exponential factor in each term of the series solution (9) causes the series to converge quite rapidly, except for very small values of t or α^2. For example, none of the graphs in Figure 10.17 requires that more than four terms in the series (9) be used. If additional terms are calculated, the resulting graphs are indistinguishable.

We emphasize that at this stage the solution (4) must be regarded as a *formal* solution; that is, we obtained it without rigorous justification of the limiting processes involved. Such a justification is beyond the scope of this book. However, once the series (4) has been obtained, it is possible to show that in $0 < x < l, t > 0$ it converges to a continuous function, that the derivatives u_{xx} and u_t can be computed by differentiating the series (4) term by term, and that the heat equation (1) is indeed satisfied. The argument rests heavily upon the fact that each term of the series (4) contains a negative exponential factor, and this results in relatively rapid convergence of the series. A further argument establishes that the function u given by Eq. (4) also satisfies the boundary and initial conditions; this completes the justification of the formal solution.

It is interesting to note that although f must satisfy the conditions of the Fourier convergence theorem (Section 10.4), it may have points of discontinuity. In this case the initial temperature distribution $u(x,0) = f(x)$ is discontinuous at one or more points. Nevertheless, the solution $u(x,t)$ is continuous for arbitrarily small values of $t > 0$. This illustrates the fact that heat conduction is a diffusive process that instantly smooths out any discontinuities that may be present in the initial temperature distribution. Finally, since f is bounded it follows from Eq. (6) that the coefficients b_n are also bounded. Consequently, the presence of the negative exponential factor in each term of the series (4) guarantees that

$$\lim_{t \to \infty} u(x,t) = 0 \tag{10}$$

for all x regardless of the initial condition. This is in accord with the result expected from physical intuition.

We now consider two other problems of one-dimensional heat conduction, which can be handled by the method developed in Section 10.2.

NONHOMOGENEOUS BOUNDARY CONDITIONS. Suppose now that one end of the bar is held at a constant temperature T_1 and the other is maintained at a constant temperature T_2. Then the boundary conditions are

$$u(0, t) = T_1, \qquad u(l, t) = T_2, \qquad t > 0. \tag{11}$$

The differential equation (1) and the initial condition (3) remain unchanged.

This problem is somewhat more difficult, because of the nonhomogeneous boundary conditions, than the one in Section 10.2. Nevertheless, if appropriate modifications are made, the problem can be solved directly by the method of separation of variables. In particular, the solution corresponding to the zero value of the separation constant (Eq. (14) of Section 10.2) must be retained and used to satisfy the boundary conditions (11). We can reach the same result in a slightly different way by *reducing the present problem to one having homogeneous boundary conditions*, which can then be solved as in Section 10.2. The technique for doing this is suggested by the following physical argument.

After a long time, that is, as $t \to \infty$, we anticipate that a steady-state temperature distribution $v(x)$ will be reached, which is independent of the time t and the initial conditions. Since $v(x)$ must satisfy the equation of heat conduction (1), we have

$$v''(x) = 0, \qquad 0 < x < l. \tag{12}$$

Further, $v(x)$ must satisfy the boundary conditions

$$v(0) = T_1, \qquad v(l) = T_2, \tag{13}$$

which apply even as $t \to \infty$. The solution of Eq. (12) satisfying Eqs. (13) is

$$v(x) = (T_2 - T_1)\frac{x}{l} + T_1. \tag{14}$$

Hence the steady-state temperature distribution is a linear function of x.

Returning to the original problem, Eqs. (1), (3), and (11), we will try to express $u(x, t)$ as the sum of the steady-state temperature distribution $v(x)$ and another (transient) temperature distribution $w(x, t)$; thus we write

$$u(x, t) = v(x) + w(x, t). \tag{15}$$

Since $v(x)$ is given by Eq (14), the problem will be solved provided we can determine $w(x, t)$. The boundary value problem for $w(x, t)$ is found by substituting the expression in Eq. (15) for $u(x, t)$ in Eqs. (1), (3), and (11).

From Eq. (1) we have

$$\alpha^2(v + w)_{xx} = (v + w)_t;$$

it follows that

$$\alpha^2 w_{xx} = w_t, \tag{16}$$

since $v_{xx} = 0$ and $v_t = 0$. Similarly, from Eqs. (15), (11), and (13),

$$\begin{aligned} w(0,t) &= u(0,t) - v(0) = T_1 - T_1 = 0, \\ w(l,t) &= u(l,t) - v(l) = T_2 - T_2 = 0. \end{aligned} \tag{17}$$

Finally, from Eqs. (15) and (3),

$$w(x,0) = u(x,0) - v(x) = f(x) - v(x), \tag{18}$$

where $v(x)$ is given by Eq. (14). Thus the transient part of the solution to the original problem is found by solving the problem consisting of Eqs. (16), (17), and (18). This latter problem is precisely the one solved in Section 10.2 provided that $f(x) - v(x)$ is now regarded as the initial temperature distribution. Hence

$$u(x,t) = (T_2 - T_1)\frac{x}{l} + T_1 + \sum_{n=1}^{\infty} b_n e^{-n^2\pi^2\alpha^2 t/l^2} \sin\frac{n\pi x}{l}, \tag{19}$$

where

$$b_n = \frac{2}{l}\int_0^l \left[f(x) - (T_2 - T_1)\frac{x}{l} - T_1 \right] \sin\frac{n\pi x}{l}\, dx. \tag{20}$$

This is another case in which a more difficult problem is solved by reducing it to a simpler problem that has already been solved. The technique of reducing a problem with nonhomogeneous boundary conditions to one with homogeneous boundary conditions by subtracting the steady-state solution is capable of wide application.

BAR WITH INSULATED ENDS. A slightly different problem occurs if the ends of the bar are insulated so that there is no passage of heat through them. According to Newton's law of cooling, as discussed in Appendix A, the rate of flow of heat across a cross section is proportional to the rate of change of temperature in the x direction. Thus in the case of no heat flow the boundary conditions are

$$u_x(0,t) = 0, \qquad u_x(l,t) = 0, \qquad t > 0. \tag{21}$$

The problem posed by Eqs. (1), (3), and (21) can also be solved by the method of separation of variables. If we let

$$u(x,t) = X(x)T(t), \tag{22}$$

and substitute for u in Eq. (1), then it follows as in Section 10.2 that

$$\frac{X''}{X} = \frac{1}{\alpha^2}\frac{T'}{T} = \sigma, \tag{23}$$

where σ is a constant. Thus we obtain again the two ordinary differential equations

$$X'' - \sigma X = 0, \tag{24}$$

$$T' - \alpha^2 \sigma T = 0. \tag{25}$$

For any value of σ a product of solutions of Eqs. (24) and (25) is a solution of the partial differential equation (1). However, we are interested only in those solutions that also satisfy the boundary conditions (21).

If we substitute for $u(x, t)$ from Eq. (22) in the boundary condition at $x = 0$, we obtain $X'(0)T(t) = 0$. We cannot permit $T(t)$ to be zero for all t, since then $u(x, t)$ would also be zero for all t. Hence we must have

$$X'(0) = 0. \tag{26}$$

Proceeding in the same way with the boundary condition at $x = l$, we find that

$$X'(l) = 0. \tag{27}$$

Thus we wish to solve Eq. (24) subject to the boundary conditions (26) and (27). It is possible to show that nontrivial solutions of this problem can exist only if σ is real. One way to show this is by an argument very similar to that in Section 10.2 (see Problem 7); alternatively, one can appeal to a more general theory to be discussed later in Section 11.3. We will assume that σ is real and consider in turn the three cases $\sigma > 0$, $\sigma = 0$, and $\sigma < 0$.

If $\sigma > 0$, it is convenient to let $\sigma = \lambda^2$, where λ is real and positive. It is sufficient to consider $\lambda > 0$ since every positive value of σ can be obtained by using only positive values of λ. Then Eq. (24) becomes $X'' - \lambda^2 X = 0$ and its general solution is

$$X(x) = k_1 \sinh \lambda x + k_2 \cosh \lambda x. \tag{28}$$

In this case the boundary conditions can be satisfied only by choosing $k_1 = k_2 = 0$. Since this is unacceptable, the separation constant σ cannot be positive.

If $\sigma = 0$, then Eq. (24) is $X'' = 0$, and therefore

$$X(x) = k_1 x + k_2. \tag{29}$$

The boundary conditions (26) and (27) require that $k_1 = 0$ but do not determine k_2. For $\sigma = 0$ it follows from Eq. (25) that $T(t)$ is also a constant, which can be combined with k_2. Hence, for $\sigma = 0$, we obtain the constant solution $u(x, t) = k_2$.

Finally, if $\sigma < 0$, let $\sigma = -\lambda^2$, where λ is real and positive. It is again sufficient to consider $\lambda > 0$ only, since all negative values of σ are obtained from positive values of λ. Then Eq. (24) becomes $X'' + \lambda^2 X = 0$, and consequently

$$X(x) = k_1 \sin \lambda x + k_2 \cos \lambda x. \tag{30}$$

The boundary condition (26) requires that $k_1 = 0$, and the boundary condition (27) requires that $\lambda = n\pi/l$ for $n = 1, 2, 3, \ldots$, but leaves k_2 arbitrary. Thus $X(x)$ is proportional to $\cos(n\pi x/l)$, where n is a positive integer, corresponding to $\sigma = -(n\pi/l)^2$. For these values of σ the solutions $T(t)$ of Eq. (25) are proportional to $\exp(-n^2\pi^2\alpha^2 t/l^2)$.

Combining all of these results, we have the following fundamental solutions for the problem (1), (3), and (21):

$$u_0(x,t) = 1,$$

$$u_n(x,t) = e^{-n^2\pi^2\alpha^2 t/l^2} \cos \frac{n\pi x}{l}, \qquad n = 1, 2, \ldots, \tag{31}$$

where arbitrary constants of proportionality have been dropped. Each of these functions satisfies the differential equation (1) and the boundary conditions (21). Because both the differential equation and boundary conditions are linear and homogeneous, any finite linear combination of the fundamental solutions satisfies them. We will assume that this is true for convergent infinite linear combinations of fundamental solutions as well. Thus, to satisfy the initial condition (3), we assume that $u(x,t)$ has the form

$$\begin{aligned} u(x,t) &= \frac{c_0}{2} u_0(x,t) + \sum_{n=1}^{\infty} c_n u_n(x,t) \\ &= \frac{c_0}{2} + \sum_{n=1}^{\infty} c_n e^{-n^2\pi^2\alpha^2 t/l^2} \cos \frac{n\pi x}{l}. \end{aligned} \tag{32}$$

The coefficients c_n are determined by the requirement that

$$u(x,0) = \frac{c_0}{2} + \sum_{n=1}^{\infty} c_n \cos \frac{n\pi x}{l} = f(x). \tag{33}$$

Thus the unknown coefficients in Eq. (32) must be the coefficients in the Fourier cosine series of period $2l$ for f. Hence

$$c_n = \frac{2}{l} \int_0^l f(x) \cos \frac{n\pi x}{l}\, dx, \qquad n = 0, 1, 2, \ldots. \tag{34}$$

With this choice of the coefficients $c_0, c_1, c_2, \ldots,$ the series (32) provides the solution to the heat conduction problem for a rod with insulated ends, Eqs. (1), (3), and (21).

It is worth observing that the solution (32) can also be thought of as the sum of a steady-state temperature distribution (given by the constant $c_0/2$), which is independent of time t, and a transient distribution (given by the rest of the infinite series) that vanishes in the limit as t approaches infinity. That the steady state is a constant is consistent with the expectation that the process of heat conduction will gradually smooth out the temperature distribution in the bar as long as no heat is allowed to escape to the outside. The physical interpretation of the term

$$\frac{c_0}{2} = \frac{1}{l} \int_0^l f(x)\, dx \tag{35}$$

is that it is the mean value of the original temperature distribution.

EXAMPLE 2

Find the temperature $u(x, t)$ at any time in a metal rod of unit length that is insulated on the ends as well as on the sides and whose initial temperature distribution is $u(x, 0) = x$ for $0 < x < 1$.

The temperature in the rod satisfies the heat conduction problem (1), (3), (21) with $l = 1$. Thus, from Eq. (32), the solution is

$$u(x, t) = \frac{c_0}{2} + \sum_{n=1}^{\infty} c_n e^{-n^2\pi^2\alpha^2 t} \cos n\pi x, \tag{36}$$

where the coefficients are determined from Eq. (34). We have

$$c_0 = 2\int_0^1 x\,dx = 1 \tag{37}$$

and, for $n \geq 1$,

$$\begin{aligned} c_n &= 2\int_0^1 x \cos n\pi x\,dx \\ &= 2(\cos n\pi - 1)/(n\pi)^2 \\ &= \begin{cases} -4/(n\pi)^2, & n \text{ odd;} \\ 0, & n \text{ even.} \end{cases} \end{aligned} \tag{38}$$

Thus

$$u(x, t) = \frac{1}{2} - \frac{4}{\pi^2} \sum_{n=1,3,5,\ldots}^{\infty} \frac{1}{n^2} e^{-n^2\pi^2\alpha^2 t} \cos n\pi x \tag{39}$$

is the solution of the given problem.

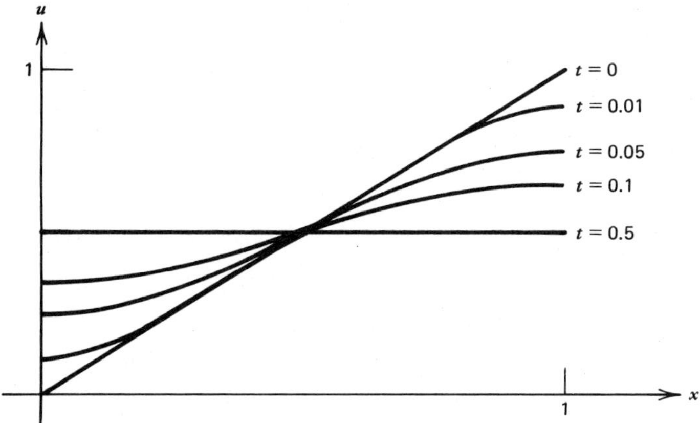

FIGURE 10.19 Temperature distributions at several times for the heat conduction problem of Example 2.

For $\alpha^2 = 1$ plots of the temperature distribution in the bar at several times are shown in Figure 10.19. Observe that the constant steady-state distribution has essentially been reached by $t = 0.5$. Again the convergence of the series is rapid so that only a few terms are needed to generate the graphs in Figure 10.19.

MORE GENERAL PROBLEMS. The method of separation of variables can also be used to solve heat conduction problems with other boundary conditions than those given by Eqs. (11) and Eqs. (21). For example, the left end of the bar might be held at a fixed temperature T while the other end is insulated. In this case the boundary conditions are

$$u(0, t) = T, \qquad u_x(l, t) = 0, \qquad t > 0. \tag{40}$$

The first step in solving this problem is to reduce the given boundary conditions to homogeneous ones by subtracting the steady-state solution. The resulting problem is solved by essentially the same procedure as in the problems considered above. However, the extension of the initial function f outside of the interval $[0, l]$ is somewhat different from that in any case considered so far (see Problems 8 through 11).

A more general type of boundary condition, called a radiation condition, occurs when the rate of heat flow through the end of the bar is proportional to the temperature. It is shown in Appendix A that the boundary conditions in this case are of the form

$$u_x(0, t) - h_1 u(0, t) = 0, \qquad u_x(l, t) + h_2 u(l, t) = 0, \qquad t > 0, \tag{41}$$

where h_1 and h_2 are nonnegative constants. If we apply the method of separation of variables to the problem consisting of Eqs. (1), (3), and (41), we find that $X(x)$ must be a solution of

$$X'' - \sigma X = 0, \qquad X'(0) - h_1 X(0) = 0, \qquad X'(l) + h_2 X(l) = 0, \tag{42}$$

where σ is the separation constant. Once again it is possible to show that nontrivial solutions can exist only for certain nonpositive real values of σ, but these values are not given by a simple formula (see Problem 14). It is also possible to show that the corresponding solutions of Eqs. (42) satisfy an orthogonality relation, and that one can satisfy the initial condition (3) by superposing solutions of Eqs. (42). However, the resulting series is not, in general, a Fourier series, and so is not included in the discussion of this chapter. There is a more general theory that covers such problems, and it is outlined in Chapter 11.

PROBLEMS

1. Consider the conduction of heat in a copper rod 100 cm in length whose ends are maintained at 0°C for all $t > 0$. Find an expression for the temperature $u(x, t)$ if the

initial temperature distribution in the rod is given by

(a) $u(x,0) = 50, \qquad 0 \le x \le 100$

(b) $u(x,0) = \begin{cases} x, & 0 \le x < 50 \\ 100 - x, & 50 \le x \le 100 \end{cases}$

(c) $u(x,0) = \begin{cases} 0, & 0 \le x < 25 \\ 50, & 25 \le x \le 75 \\ 0, & 75 < x \le 100 \end{cases}$

2. Let a metallic rod 20 cm long be heated to a uniform temperature of 100°C. Suppose that at $t = 0$ the ends of the bar are plunged into an ice bath at 0°C, and thereafter maintained at this temperature, but that no heat is allowed to escape through the lateral surface. Find an expression for the temperature at any point in the bar at any later time. Use two terms in the series expansion for the temperature to determine approximately the temperature at the center of the bar at time $t = 30$ sec if the bar is made of (a) silver, (b) aluminum, or (c) cast iron.
3. For the situation of Problem 2 find the time that will elapse before the center of the bar cools to a temperature of 25°C if the bar is made of (a) silver, (b) aluminum, or (c) cast iron. Use only one term in the series expansion for $u(x,t)$.
4. Let an aluminum rod of length l be initially at the uniform temperature of 25°C. Suppose that at time $t = 0$ the end $x = 0$ is cooled to 0°C while the end $x = l$ is heated to 60°C, and both are thereafter maintained at those temperatures.
   (a) Find the temperature distribution in the rod at any time t.
   Assume in the remaining parts of this problem that $l = 20$ cm.
   (b) Use only the first term in the series for the temperature $u(x,t)$ to find the approximate temperature at $x = 5$ cm when $t = 30$ sec; when $t = 60$ sec.
   (c) Use the first two terms in the series for $u(x,t)$ to find an approximate value of $u(5,30)$. What is the percentage difference between the one- and two-term approximations? Does the third term in the series have any appreciable effect for this value of t?
   (d) Use the first term in the series for $u(x,t)$ to estimate the time interval that must elapse before the temperature at $x = 5$ cm comes within 1 percent of its steady-state value.
   *(e) Observe that $u(5,30)$ from part (c) is less than the steady state temperature (15°C) at $x = 5$, even though the initial temperature (25°C) at $x = 5$ is higher than the steady state value. Can you explain[8] why this is so?
5. (a) Let the ends of a copper rod 100 cm long be maintained at 0°C. Suppose that the center of the bar is heated to 100°C by an external heat source and that this situation is maintained until a steady state results. Find this steady-state temperature distribution.
   (b) At a time $t = 0$ (after the steady state of part (a) has been reached) let the heat source be removed. At the same instant let the end $x = 0$ be placed in thermal contact with a reservoir at 20°C while the other end remains at 0°C. Find the temperature as a function of position and time.
6. Consider a uniform rod of length l with an initial temperature given by $\sin(\pi x/l)$, $0 \le x \le l$. Assume that both ends of the bar are insulated. Find a formal series

[8]A full discussion of this problem appears in Knop, L., and Robbins, D., "Intermediate Time Behavior of a Heat Equation Problem," *Mathematics Magazine*, *57* (1984), 217–219.

expansion for the temperature $u(x,t)$. What is the steady-state temperature as $t \to \infty$?

7. Consider the problem

$$X'' - \sigma X = 0, \qquad X'(0) = 0, \qquad X'(l) = 0. \tag{i}$$

Let $\sigma = -\lambda^2$, where $\lambda = \mu + i\nu$ with μ and ν real. Show that if $\nu \neq 0$ then the only solution of Eqs. (i) is the trivial solution $X(x) \equiv 0$.
Hint: Use an argument similar to that in Section 10.2.

Problems 8 through 11 deal with the heat conduction problem in a bar that is insulated at one end and held at a constant temperature at the other end.

8. Find the steady-state temperature in a bar that is insulated at the end $x = 0$ and held at the constant temperature 0°C at the end $x = l$.
9. Find the steady-state temperature in a bar that is insulated at the end $x = 0$ and held at the constant temperature T at the end $x = l$.
10. Consider a uniform bar of length l having an initial temperature distribution given by $f(x), 0 \le x \le l$. Assume that the temperature at the end $x = 0$ is held at 0°C, while the end $x = l$ is insulated so that no heat passes through it.
(a) Show that the fundamental solutions of the partial differential equation and boundary conditions are

$$u_n(x,t) = e^{-(2n-1)^2\pi^2\alpha^2 t/4l^2} \sin[(2n-1)\pi x/2l], \qquad n = 1,2,3,\ldots.$$

(b) Find a formal series expansion for the temperature $u(x,t)$,

$$u(x,t) = \sum_{n=1}^{\infty} c_n u_n(x,t),$$

that also satisfies the initial condition $u(x,0) = f(x)$.
Hint: Even though the fundamental solutions involve only the odd sines, it is still possible to represent f by a Fourier series involving only these functions. See Problem 23 of Section 10.5.
11. Consider the bar of Problem 10, but assume that the temperature of the end $x = 0$ is held at the constant value T. Find a formal series expansion for the temperature $u(x,t)$ at any point in the bar at any time.

12. Find the steady-state temperature in a bar that satisfies the radiation condition $u_x(0,t) - u(0,t) = 0$ at the end $x = 0$ and is held at the constant temperature T at the end $x = l$.
13. The right end of a bar of length a with thermal conductivity κ_1 and cross-sectional area A_1 is joined to the left end of a bar of thermal conductivity κ_2 and cross-sectional area A_2. The composite bar has a total length l. Suppose that the end $x = 0$ is held at temperature zero, while the end $x = l$ is held at temperature T. Find the steady-state temperature in the composite bar, assuming that the temperature and rate of heat flow are continuous at $x = a$.
Hint: See Eq. (2) of Appendix A.

14. Consider the problem

$$\alpha^2 u_{xx} = u_t, \qquad 0 < x < l, \quad t > 0$$

$$u(0,t) = 0, \qquad u_x(l,t) + \gamma u(l,t) = 0, \qquad t > 0 \tag{i}$$

$$u(x,0) = f(x), \qquad 0 \le x \le l.$$

(a) Let $u(x,t) = X(x)T(t)$ and show that

$$X'' - \sigma X = 0, \qquad X(0) = 0, \qquad X'(l) + \gamma X(l) = 0, \tag{ii}$$

and

$$T' - \sigma\alpha^2 T = 0,$$

where σ is the separation constant.

(b) Assume that σ is real, and show that problem (ii) has no nontrivial solutions if $\sigma \ge 0$.

(c) If $\sigma < 0$, let $\sigma = -\lambda^2$ with $\lambda > 0$. Show that problem (ii) has nontrivial solutions only if λ is a solution of the equation

$$\lambda \cos \lambda l + \gamma \sin \lambda l = 0. \tag{iii}$$

*(d) By drawing the graphs of $y = \tan \lambda l$ and $y = -\lambda l / \gamma l$ for $\lambda > 0$ on the same set of axes, show that Eq. (iii) is satisfied by infinitely many positive values of λ; denote these by $\lambda_1, \lambda_2, \ldots, \lambda_n, \ldots,$ ordered in increasing size.

*(e) Determine the set of fundamental solutions $u_n(x,t)$ corresponding to the values λ_n found in part (d).

10.7 The Wave Equation: Vibrations of an Elastic String

A second partial differential equation occurring frequently in applied mathematics is the wave[9] equation. Some form of this equation, or a generalization of it, almost inevitably arises in any mathematical analysis of phenomena involving the propagation of waves in a continuous medium. For example, the studies of acoustic waves, water waves, and electromagnetic waves are all based on this equation. Perhaps the easiest situation to visualize occurs in the investigation of mechanical vibrations. Suppose that an elastic string of length l is tightly stretched between two supports at the same horizontal level, so that the x axis lies along the string (see Figure 10.20). The elastic string may be thought of as a violin

[9] The solution of the wave equation was one of the major mathematical problems of the mid-eighteenth century. The wave equation was first derived and studied by D'Alembert in 1746. It also attracted the attention of Euler (1748), Daniel Bernoulli (1753), and Lagrange (1759). Solutions were obtained in several different forms, and the merits of, and relations among, these solutions were argued, sometimes heatedly, in a series of papers extending over more than 25 years. The major points at issue concerned the nature of a function, and the kinds of functions that can be represented by trigonometric series. These questions were not resolved until the nineteenth century.

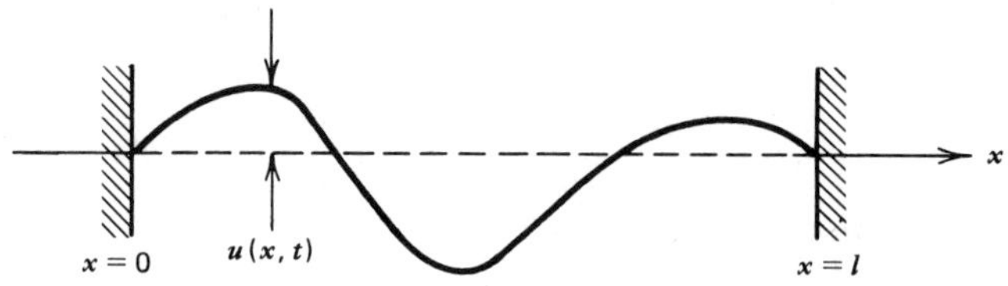

FIGURE 10.20 A vibrating string.

string, a guy wire, or possibly an electric power line. Suppose that the string is set in motion (by plucking, for example) so that it vibrates in a vertical plane, and let $u(x,t)$ denote the vertical displacement experienced by the string at the point x at time t. If damping effects, such as air resistance, are neglected, and if the amplitude of the motion is not too large, then $u(x,t)$ satisfies the partial differential equation

$$a^2 u_{xx} = u_{tt} \tag{1}$$

in the domain $0 < x < l, t > 0$. Equation (1) is known as the *wave equation* and is derived in Appendix B at the end of the chapter. The constant coefficient a^2 appearing in Eq. (1) is given by

$$a^2 = T/\rho, \tag{2}$$

where T is the tension (force) in the string, and ρ is the mass per unit length of the string material. Thus a has the physical dimensions of length/time, that is, of velocity. In Problem 9 it is shown that a is the velocity of propagation of waves along the string.

To describe the motion of the string completely it is necessary also to specify suitable initial and boundary conditions for the displacement $u(x,t)$. The ends are assumed to remain fixed, and therefore the boundary conditions are

$$u(0,t) = 0, \qquad u(l,t) = 0, \qquad t \geq 0. \tag{3}$$

Since the differential equation (1) is of second order with respect to t, it is plausible that two initial conditions must be prescribed. These are the initial position of the string

$$u(x,0) = f(x), \qquad 0 \leq x \leq l, \tag{4}$$

and its initial velocity

$$u_t(x,0) = g(x), \qquad 0 \leq x \leq l, \tag{5}$$

where f and g are given functions. In order that Eqs. (3), (4), and (5) be consistent it is also necessary to require that

$$f(0) = f(l) = 0, \qquad g(0) = g(l) = 0. \tag{6}$$

The mathematical problem then is to determine the solution of the wave equation (1) that also satisfies the boundary conditions (3) and the initial

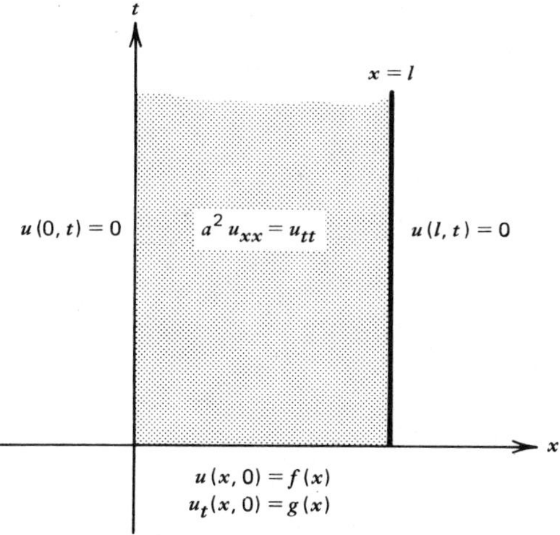

FIGURE 10.21 Boundary value problem for the wave equation.

conditions (4) and (5). Like the heat conduction problem of Sections 10.2 and 10.6, this problem is an initial value problem in the time variable t, and a boundary value problem in the space variable x. Alternatively, it can be considered as a boundary value problem in the semi-infinite strip $0 < x < l, t > 0$ of the xt plane (see Figure 10.21). One condition is imposed at each point on the semi-infinite sides, and two are imposed at each point on the finite base. In contrast, the heat conduction problem, being first order in time, required only one condition on the initial line $t = 0$.

It is important to realize that Eq. (1) governs a large number of other wave problems besides the transverse vibrations of an elastic string. For example, it is only necessary to interpret the function u and the constant a appropriately to have problems dealing with water waves in an ocean, acoustic or electromagnetic waves in the atmosphere, or elastic waves in a solid body. If more than one space dimension is significant then Eq. (1) must be slightly generalized. The two-dimensional wave equation is

$$a^2(u_{xx} + u_{yy}) = u_{tt}. \tag{7}$$

This equation would arise, for example, if we considered the motion of a thin elastic sheet, such as a drumhead. Similarly, in three dimensions the wave equation is

$$a^2(u_{xx} + u_{yy} + u_{zz}) = u_{tt}. \tag{8}$$

In connection with the latter two equations, the boundary and initial conditions must also be suitably generalized.

We now solve some typical boundary value problems involving the one-dimensional wave equation.

ELASTIC STRING WITH NONZERO INITIAL DISPLACEMENT. First suppose that the string is disturbed from its equilibrium position, and then released with zero velocity at time $t = 0$ to vibrate freely. Then the vertical displacement $u(x,t)$ must satisfy the wave equation (1)

$$a^2 u_{xx} = u_{tt}, \qquad 0 < x < l, \quad t > 0;$$

the boundary conditions (3)

$$u(0,t) = 0, \qquad u(l,t) = 0, \qquad t \geq 0;$$

and the initial conditions

$$u(x,0) = f(x), \qquad u_t(x,0) = 0, \qquad 0 \leq x \leq l; \tag{9}$$

where f is a given function describing the configuration of the string at $t = 0$.

The method of separation of variables can be used to obtain the solution of Eqs. (1), (3), and (9). Assuming that

$$u(x,t) = X(x)T(t) \tag{10}$$

and substituting for u in Eq. (1), we obtain

$$\frac{X''}{X} = \frac{1}{a^2}\frac{T''}{T} = \sigma, \tag{11}$$

where σ is a constant. Thus we find that $X(x)$ and $T(t)$ satisfy the ordinary differential equations

$$X'' - \sigma X = 0, \tag{12}$$

$$T'' - a^2\sigma T = 0. \tag{13}$$

As in the case of the heat conduction problems considered previously, we will determine permissible values of the separation constant σ from the boundary conditions. By substituting from Eq. (10) for $u(x,t)$ in the boundary conditions (3) we find that

$$X(0) = 0, \qquad X(l) = 0. \tag{14}$$

The problem of solving the differential equation (12) subject to the boundary conditions (14) is *precisely the same problem* that arose in Section 10.2 in connection with the heat conduction problem. Thus we can use the results obtained there: the problem (12) and (14) has nontrivial solutions if and only if

$$\sigma = -n^2\pi^2/l^2, \qquad n = 1, 2, \ldots, \tag{15}$$

and the corresponding solutions for $X(x)$ are proportional to $\sin(n\pi x/l)$. Using the values of σ given by Eq. (15) in Eq. (13), we find that $T(t)$ is a linear combination of $\sin(n\pi at/l)$ and $\cos(n\pi at/l)$. Hence functions of the form

$$u_n(x,t) = \sin\frac{n\pi x}{l}\sin\frac{n\pi at}{l}, \qquad n = 1, 2, \ldots \tag{16}$$

$$v_n(x,t) = \sin\frac{n\pi x}{l}\cos\frac{n\pi at}{l}, \qquad n = 1, 2, \ldots \tag{17}$$

satisfy the partial differential equation (1) and the boundary conditions (3). These functions are the fundamental solutions of the given problem.

We now seek a superposition of the fundamental solutions (16), (17) that also satisfies the initial conditions (9). Thus we assume that $u(x, t)$ is given by

$$u(x,t) = \sum_{n=1}^{\infty} \left[c_n u_n(x,t) + k_n v_n(x,t)\right]$$

$$= \sum_{n=1}^{\infty} \sin\frac{n\pi x}{l}\left(c_n \sin\frac{n\pi at}{l} + k_n \cos\frac{n\pi at}{l}\right), \tag{18}$$

where c_n and k_n are constants that will be chosen to satisfy the initial conditions.

The initial condition $u(x, 0) = f(x)$ gives

$$u(x,0) = \sum_{n=1}^{\infty} k_n \sin\frac{n\pi x}{l} = f(x). \tag{19}$$

Consequently, the k_n must be the coefficients in the Fourier sine series of period $2l$ for f, and are given by

$$k_n = \frac{2}{l}\int_0^l f(x)\sin\frac{n\pi x}{l}\,dx, \qquad n = 1, 2, \ldots. \tag{20}$$

Next we further assume that the series (18) can be differentiated term by term with respect to t. Then the initial condition $u_t(x, 0) = 0$ yields

$$u_t(x,0) = \sum_{n=1}^{\infty} c_n \frac{n\pi a}{l}\sin\frac{n\pi x}{l} = 0. \tag{21}$$

Hence the quantities $c_n(n\pi a/l)$ must be the coefficients in the Fourier sine series of period $2l$ for the function that is identically zero. From the Euler-Fourier formulas it follows that $c_n = 0$ for all n. Thus the formal solution[10] of the problem of Eqs. (1), (3), and (9) is

$$u(x,t) = \sum_{n=1}^{\infty} k_n \sin\frac{n\pi x}{l}\cos\frac{n\pi at}{l}, \tag{22}$$

with the coefficients k_n given by Eq. (20).

For a fixed value of n the expression $\sin(n\pi x/l)\cos(n\pi at/l)$ in Eq. (22) is periodic in time t with the period $2l/na$; it therefore represents a vibratory motion of the string having this period, or having the frequency $n\pi a/l$. The quantities $\lambda a = n\pi a/l$ for $n = 1, 2, \ldots$ are the *natural frequencies* of the string, that is, the frequencies at which the string will freely vibrate. The factor $k_n \sin(n\pi x/l)$ represents the displacement pattern occurring in the string when it is executing vibrations of the given frequency. Each displacement pattern is called a *natural mode* of vibration and is periodic in the space variable x; the spatial

[10] Solutions of the form (22) were given by Euler in 1749, and in greater detail by Daniel Bernoulli in 1753.

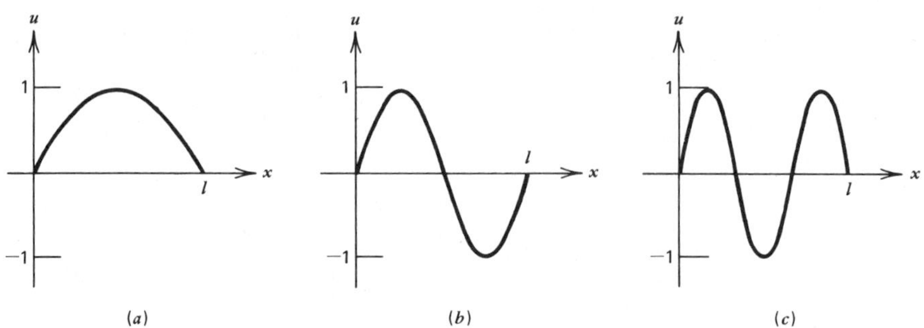

FIGURE 10.22 First three fundamental modes of vibration of an elastic string. (a) frequency $= \pi a/l$, *wavelength* $= 2l$; *(b) frequency* $= 2\pi a/l$, *wavelength* $= l$; *(c) frequency* $= 3\pi a/l$, *wavelength* $= 2l/3$.

period $2l/n$ is called the *wavelength* of the mode of frequency $n\pi a/l$. The first three natural modes are sketched in Figure 10.22. The total motion of the string, given by the function $u(x, t)$ of Eq. (22), is thus a combination of the natural modes of vibration, and is also a periodic function of time with period $2l/a$.

JUSTIFICATION OF THE SOLUTION. As in the heat conduction problem considered earlier, Eq. (22) with the coefficients k_n given by Eq. (20) is only a *formal* solution of Eqs. (1), (3), and (9). To ascertain whether Eq. (22) *actually* represents the solution of the given problem requires some further investigation. As in the heat conduction problem, it is tempting to try to show this directly by substituting Eq. (22) for $u(x, t)$ in Eqs. (1), (3), and (9). However, on formally computing u_{xx}, for example, we obtain

$$u_{xx}(x, t) = -\sum_{n=1}^{\infty} k_n \left(\frac{n\pi}{l}\right)^2 \sin\frac{n\pi x}{l} \cos\frac{n\pi a t}{l};$$

due to the presence of the n^2 factor in the numerator this series may not converge. This would not necessarily mean that the series (22) for $u(x, t)$ is incorrect, but only that the series (22) cannot be used to calculate u_{xx} and u_{tt}. The basic difference between solutions of the wave equation and those of the heat conduction equation is that the latter contain negative exponential terms that approach zero very rapidly with increasing n, which insures the convergence of the series solution and its derivatives. In contrast, series solutions of the wave equation contain only oscillatory terms which do not decay with increasing n.

However, there is an alternative way to establish the validity of Eq. (22) indirectly. At the same time we will gain additional information about the structure of the solution. First we will show that Eq. (22) is equivalent to

$$u(x, t) = \tfrac{1}{2}[h(x - at) + h(x + at)], \tag{23}$$

where h is the function obtained by extending the initial data f into $(-l, 0)$ as an odd function, and to other values of x as a periodic function of period $2l$. That is,

$$h(x) = \begin{cases} f(x), & 0 \le x \le l, \\ -f(-x), & -l < x < 0; \end{cases} \tag{24}$$

$$h(x + 2l) = h(x).$$

To establish Eq. (23) note that h has the Fourier series

$$h(x) = \sum_{n=1}^{\infty} k_n \sin \frac{n\pi x}{l}. \tag{25}$$

Then, using the trigonometric identities for the sine of a sum or difference, we obtain

$$h(x - at) = \sum_{n=1}^{\infty} k_n \left(\sin \frac{n\pi x}{l} \cos \frac{n\pi at}{l} - \cos \frac{n\pi x}{l} \sin \frac{n\pi at}{l} \right),$$

$$h(x + at) = \sum_{n=1}^{\infty} k_n \left(\sin \frac{n\pi x}{l} \cos \frac{n\pi at}{l} + \cos \frac{n\pi x}{l} \sin \frac{n\pi at}{l} \right),$$

and Eq. (23) follows immediately upon adding the last two equations. From Eq. (23) we see that $u(x, t)$ is continuous for $0 < x < l, t > 0$ provided that h is continuous on the interval $(-\infty, \infty)$. This requires that f be continuous on the original interval $[0, l]$ and, since h is the odd periodic extension of f, that f be zero at $x = 0$ and $x = l$. Similarly, u is twice continuously differentiable with respect to either variable in $0 < x < l, t > 0$, provided that h is twice continuously differentiable on $(-\infty, \infty)$. This requires that f' and f'' be continuous on $[0, l]$. Furthermore, since h'' is the odd extension of f'', we must also have $f''(0) = f''(l) = 0$. However, since h' is the even extension of f', no further conditions are required on f'. Provided that these conditions are met, u_{xx} and u_{tt} can be computed from Eq. (23), and it is an elementary exercise to show that these derivatives satisfy the wave equation. Some of the details of the argument indicated above are given in Problems 15 and 16.

If some of the continuity requirements stated in the last paragraph are not met, then u is not differentiable at some points in the semi-infinite strip $0 < x < l, t > 0$, and thus is a solution of the wave equation only in a somewhat restricted sense. An important physical consequence of this observation is that if there are any discontinuities present in the initial data f, then they will be preserved in the solution $u(x, t)$ for all time. In contrast, in heat conduction problems initial discontinuities are instantly damped out (Section 10.6). Suppose that the initial displacement f has a jump discontinuity at $x = x_0, 0 \le x_0 \le l$. Since h is a periodic extension of f, the same discontinuity is present in $h(\xi)$ at $\xi = x_0 + 2nl$ and at $\xi = -x_0 + 2nl$, where n is any integer. Thus $h(x - at)$ is discontinuous when $x - at = x_0 + 2nl$, or when $x - at = -x_0 + 2nl$. For a fixed x in $[0, l]$ the discontinuity that was originally at x_0 will reappear in $h(x - at)$ at the times $t = (x \pm x_0 - 2nl)/a$. Similarly, $h(x + at)$ is discontinuous at the point x at the times $t = (-x \pm x_0 + 2ml)/a$, where m is any integer. Referring to Eq. (23), it then follows that the solution $u(x, t)$ is also discontinuous at the given point x at these times. Since the physical problem is posed for $t > 0$, only those values of m and n yielding positive values of t are of interest.

GENERAL PROBLEM FOR THE ELASTIC STRING. Let us suppose that the string is set in motion from a specified initial position with a given velocity. Then the vertical displacement $u(x, t)$ must satisfy the wave equation (1),

$$a^2 u_{xx} = u_{tt}, \qquad 0 < x < l, \quad t > 0;$$

the boundary conditions (3),

$$u(0, t) = 0, \qquad u(l, t) = 0, \qquad t \geq 0;$$

and the initial conditions

$$u(x, 0) = f(x), \qquad u_t(x, 0) = g(x), \qquad 0 \leq x \leq l; \tag{26}$$

where f and g are given functions defining the initial position and velocity of the string, respectively.

We recall that the fundamental solutions given by Eqs. (16) and (17) satisfy the differential equation (1) and boundary conditions (3), and we again assume that $u(x, t)$ is given by Eq. (18). The coefficients c_n and k_n must now be determined from the initial conditions (26). Applying the condition $u(x, 0) = f(x)$ yields

$$u(x, 0) = \sum_{n=1}^{\infty} k_n \sin \frac{n\pi x}{l} = f(x). \tag{27}$$

Hence the coefficients k_n are again the coefficients in the Fourier sine series of period $2l$ for f, and are thus given by Eq. (20),

$$k_n = \frac{2}{l} \int_0^l f(x) \sin \frac{n\pi x}{l}\, dx, \qquad n = 1, 2, \ldots.$$

Differentiating Eq. (18) with respect to t and applying the second initial condition $u_t(x, 0) = g(x)$ now yields

$$u_t(x, 0) = \sum_{n=1}^{\infty} \frac{n\pi a}{l} c_n \sin \frac{n\pi x}{l} = g(x). \tag{28}$$

Hence the quantities $(n\pi a/l)c_n$ are the coefficients in the Fourier sine series of period $2l$ for g; therefore

$$\frac{n\pi a}{l} c_n = \frac{2}{l} \int_0^l g(x) \sin \frac{n\pi x}{l}\, dx, \qquad n = 1, 2, \ldots. \tag{29}$$

Thus Eq. (18), with the coefficients given by Eqs. (20) and (29), constitutes a formal solution to the problem of Eqs. (1), (3), and (26). The validity of this formal solution can be established by arguments similar to those previously indicated for the solution of Eqs. (1), (3), and (9).

PROBLEMS

1. Find the displacement $u(x,t)$ in an elastic string that is fixed at its ends and is set in motion by plucking it at its center. In this case $u(x,t)$ satisfies Eqs. (1), (3), and (9) with $f(x)$ defined by

$$f(x) = \begin{cases} Ax, & 0 \le x \le l/2, \\ A(l-x), & l/2 < x \le l. \end{cases}$$

2. Find the displacement $u(x,t)$ in an elastic string, fixed at both ends, that is set in motion with no initial velocity from the initial position $u(x,0) = f(x)$, where

$$f(x) = \begin{cases} Ax, & 0 \le x \le l/4, \\ Al/4, & l/4 < x < 3l/4, \\ A(l-x), & 3l/4 \le x \le l. \end{cases}$$

3. Find the displacement $u(x,t)$ in an elastic string, held fixed at the ends, that is set in motion from its equilibrium position, $u(x,0) = 0$, with an initial velocity $u_t(x,0) = g(x)$, where g is a given function.
4. Find the displacement $u(x,t)$ in an elastic string of length l, fixed at both ends, that is set in motion from its straight equilibrium position with the initial velocity g defined by

$$g(x) = \begin{cases} Ax, & 0 \le x \le l/2, \\ A(l-x), & l/2 < x \le l. \end{cases}$$

5. Show that the solution $u(x,t)$ of the problem

$$a^2 u_{xx} = u_{tt},$$

$$u(0,t) = 0, \qquad u(l,t) = 0, \qquad u(x,0) = f(x), \qquad u_t(x,0) = g(x),$$

can be written as

$$u(x,t) = v(x,t) + w(x,t),$$

where $v(x,t)$ is the solution of the same problem with $g(x) = 0$ and $w(x,t)$ is the solution of the same problem with $f(x) = 0$. Thus $v(x,t)$ represents the motion of a string started from rest with initial displacement $f(x)$, and $w(x,t)$ represents the motion of a string set in motion from its equilibrium position with initial velocity $g(x)$. The solution of the general problem can therefore be obtained by solving two subsidiary problems.
6. If an elastic string is free at one end, the boundary condition to be satisfied there is that $u_x = 0$. Find the displacement $u(x,t)$ in an elastic string of length l, fixed at $x = 0$ and free at $x = l$, set in motion with no initial velocity from the initial position $u(x,0) = f(x)$, where f is a given function.
Hint: Show that the fundamental solutions for this problem, satisfying all conditions except the nonhomogeneous initial condition, are

$$u_n(x,t) = \sin \lambda_n x \cos \lambda_n at,$$

where $\lambda_n = (2n-1)\pi/2l$, $n = 1, 2, \ldots$. Compare this problem with Problem 10 of Section 10.6; pay particular attention to the extension of the initial data out of the original interval $[0, l]$.

7. Dimensionless variables can be introduced into the wave equation (1) in the following manner. Letting $s = x/l$, show that the wave equation becomes

$$a^2 u_{ss} = l^2 u_{tt}.$$

Then show that l/a has the dimension of time, and thus can be used as the unit on the time scale. Finally, let $\tau = at/l$ and show that the wave equation then reduces to

$$u_{ss} = u_{\tau\tau}.$$

Problems 8 and 9 indicate the form of the general solution of the wave equation and the physical significance of the constant a.

8. Show that the wave equation

$$a^2 u_{xx} = u_{tt}$$

can be reduced to the form $u_{\xi\eta} = 0$ by the change of variables $\xi = x - at$, $\eta = x + at$. Show that $u(x,t)$ must be of the form

$$u(x,t) = \phi(x - at) + \psi(x + at),$$

where ϕ and ψ are arbitrary functions.

9. Plot the value of $\phi(x - at)$ for $t = 0, 1/a, 2/a, t_0/a$ if $\phi(s) = \sin s$. Note that for any $t \neq 0$ the graph of $y = \phi(x - at)$ is the same as that of $y = \phi(x)$ when $t = 0$, but displaced a distance at in the positive x direction. Thus a represents the velocity at which a disturbance moves along the string. What is the interpretation of $\phi(x + at)$?

10. A steel wire 5 ft in length is stretched by a tensile force of 50 lb. The wire has a weight per unit length of 0.026 lb/ft.
(a) Find the velocity of propagation of transverse waves in the wire.
(b) Find the natural frequencies of vibration.
(c) If the tension in the wire is increased, how are the natural frequencies changed? Are the natural modes also changed?

11. A vibrating string moving in an elastic medium satisfies the equation

$$a^2 u_{xx} - \alpha^2 u = u_{tt},$$

where α^2 is proportional to the coefficient of elasticity of the medium. Suppose that the string is fixed at the ends, and is released with no initial velocity from the initial position $u(x,0) = f(x), 0 < x < l$. Find the displacement $u(x,t)$.

12. Consider the wave equation

$$a^2 u_{xx} = u_{tt}$$

in an infinite one-dimensional medium subject to the initial conditions

$$u(x,0) = f(x), \qquad u_t(x,0) = 0, \qquad -\infty < x < \infty.$$

(a) Using the form of the solution obtained in Problem 8, show that ϕ and ψ must satisfy

$$\phi(x) + \psi(x) = f(x),$$
$$-\phi'(x) + \psi'(x) = 0.$$

(b) Solve the equations of part (a) for ϕ and ψ, and thereby show that

$$u(x,t) = \tfrac{1}{2}[f(x - at) + f(x + at)].$$

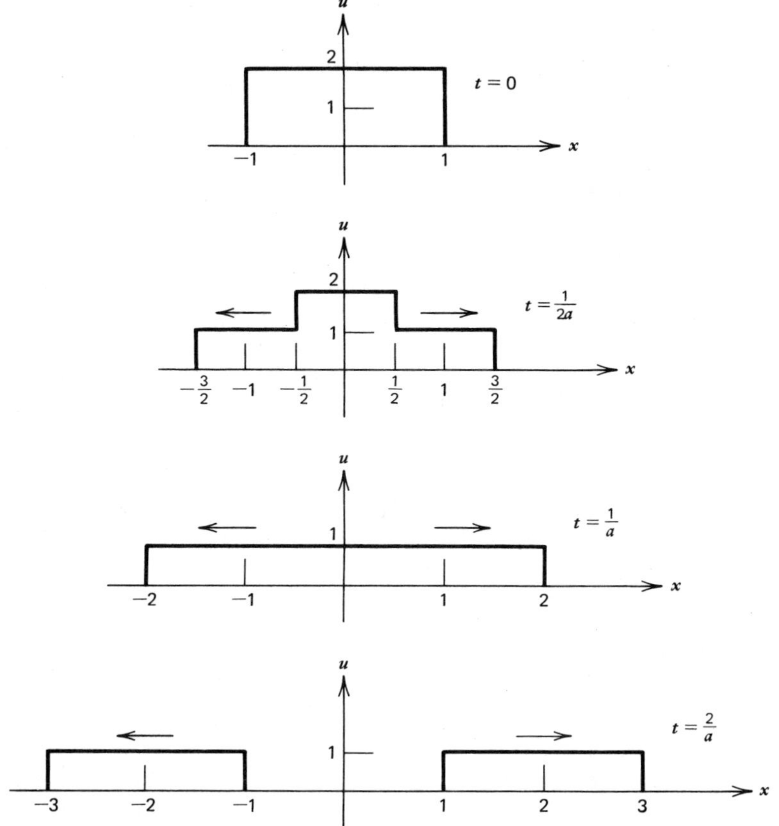

FIGURE 10.23 Propagation of initial disturbance in an infinite one-dimensional medium.

This form of the solution was obtained by D'Alembert in 1746.
Hint: Note that the equation $\psi'(x) = \phi'(x)$ is solved by choosing $\psi(x) = \phi(x) + c$.
(c) Let

$$f(x) = \begin{cases} 2, & -1 < x < 1 \\ 0, & \text{otherwise.} \end{cases}$$

Show that

$$f(x - at) = \begin{cases} 2, & -1 + at < x < 1 + at \\ 0, & \text{otherwise.} \end{cases}$$

Also determine $f(x + at)$.
(d) Sketch the solution found in part (b) at $t = 0$, $t = 1/2a$, $t = 1/a$, and $t = 2/a$, obtaining the results shown in Figure 10.23. Observe that an initial displacement produces two waves moving in opposite directions away from the original location; each wave consists of one-half of the initial displacement.

13. Consider the wave equation

$$a^2 u_{xx} = u_{tt}$$

in an infinite one-dimensional medium subject to the initial conditions

$$u(x,0) = 0, \qquad u_t(x,0) = g(x), \qquad -\infty < x < \infty.$$

(a) Using the form of the solution obtained in Problem 8, show that

$$\phi(x) + \psi(x) = 0,$$
$$-a\phi'(x) + a\psi'(x) = g(x).$$

(b) Use the first equation of part (a) to show that $\psi'(x) = -\phi'(x)$. Then use the second equation to show that $-2a\phi'(x) = g(x)$, and therefore that

$$\phi(x) = -\frac{1}{2a}\int_{x_0}^{x} g(\xi)\,d\xi + \phi(x_0),$$

where x_0 is arbitrary. Finally, determine $\psi(x)$.
(c) Show that

$$u(x,t) = \frac{1}{2a}\int_{x-at}^{x+at} g(\xi)\,d\xi.$$

14. By combining the results of Problems 12 and 13 show that the solution of the problem

$$a^2 u_{xx} = u_{tt}$$
$$u(x,0) = f(x), \qquad u_t(x,0) = g(x), \qquad -\infty < x < \infty$$

is given by

$$u(x,t) = \tfrac{1}{2}[f(x-at) + f(x+at)] + \frac{1}{2a}\int_{x-at}^{x+at} g(\xi)\,d\xi.$$

Problems 15 and 16 indicate how the formal solution (22) of Eqs. (1), (3), and (9) can be shown to constitute the actual solution of that problem.

15. By using the trigonometric identity $\sin A \cos B = \frac{1}{2}[\sin(A+B) + \sin(A-B)]$ show that the solution (22) of the problem of Eqs. (1), (3), and (9) can be written in the form (23).

*16. Let $h(\xi)$ represent the initial displacement in $[0, l]$, extended into $(-l, 0)$ as an odd function and extended elsewhere as a periodic function of period $2l$. Assuming that h, h', and h'' are all continuous, show by direct differentiation that $u(x,t)$ as given in Eq. (23) satisfies the wave equation (1) and also the initial conditions (9). Note also that since Eq. (22) clearly satisfies the boundary conditions (3), the same is true of Eq. (23). Comparing Eq. (23) with the solution of the corresponding problem for the infinite string (Problem 12), we see that they have the same form provided that the initial data for the finite string, defined originally only on the interval $0 \le x \le l$, are extended in the given manner over the entire x axis. If this is done, the solution for the infinite string is also applicable to the finite one.

17. The motion of a circular elastic membrane, such as a drum head, is governed by the two-dimensional wave equation in polar coordinates

$$u_{rr} + (1/r)u_r + (1/r^2)u_{\theta\theta} = a^{-2}u_{tt}.$$

Assuming that $u(r,\theta,t) = R(r)\Theta(\theta)T(t)$, find ordinary differential equations satisfied by $R(r)$, $\Theta(\theta)$, and $T(t)$.

*18. The total energy $E(t)$ of the vibrating string is given as a function of time by

$$E(t) = \int_0^l \left[\tfrac{1}{2}\rho u_t^2(x,t) + \tfrac{1}{2}Hu_x^2(x,t)\right] dx; \tag{i}$$

the first term is the kinetic energy due to the motion of the string, and the second term is the potential energy created by the displacement of the string away from its equilibrium position.

For the displacement $u(x,t)$ given by Eq. (22), that is, for the solution of the string problem with zero initial velocity, show that

$$E(t) = \frac{\pi^2 H}{4l} \sum_{n=1}^{\infty} n^2 k_n^2. \tag{ii}$$

Note that the right side of Eq. (ii) does not depend on t. Thus the total energy E is a constant, and therefore is *conserved* during the motion of the string.

Hint: Use Parseval's equation (Problem 21 of Section 10.5 and Problem 13 of Section 10.4), and recall that $a^2 = H/\rho$.

10.8 Laplace's Equation

One of the most important of all the partial differential equations occurring in applied mathematics is that associated with the name of Laplace[11]: in two dimensions

$$u_{xx} + u_{yy} = 0, \tag{1}$$

and in three dimensions

$$u_{xx} + u_{yy} + u_{zz} = 0. \tag{2}$$

For example, in a two-dimensional heat conduction problem, the temperature $u(x, y, t)$ must satisfy the differential equation

$$\alpha^2(u_{xx} + u_{yy}) = u_t, \tag{3}$$

where α^2 is the thermal diffusivity. If a steady state exists, u is a function of x and y only, and the time derivative vanishes; in this case Eq. (3) reduces to Eq. (1). Similarly, for the steady-state heat conduction problem in three dimensions, the temperature must satisfy the three-dimensional form of Laplace's equation. Equations (1) and (2) also occur in other branches of mathematical physics. In the consideration of electrostatic fields the electric potential function in a dielectric medium containing no electric charges must satisfy either Eq. (1) or Eq. (2),

[11] Laplace's equation is named for Pierre-Simon de Laplace, who, beginning in 1782, studied its solutions extensively while investigating the gravitational attraction of arbitrary bodies in space. However, the equation first appeared in a paper by Euler in 1752 on hydrodynamics.

depending on the number of space dimensions involved. Similarly, the potential energy function of a particle in free space acted only by gravitational forces satisfies the same equations. Consequently, Laplace's equation is often referred to as the *potential equation*. Another example arises in the study of the steady (time-independent), two-dimensional, inviscid, irrotational motion of an incompressible fluid, which centers about two functions, known as the velocity potential function and the stream function, both of which satisfy Eq. (1). In elasticity the displacements that occur when a perfectly elastic bar is twisted are described in terms of the so-called warping function, which also satisfies Eq. (1).

Since there is no time dependence in any of the problems mentioned above, there are no initial conditions to be satisfied by the solutions of Eq. (1) or (2). They must, however, satisfy certain boundary conditions on the bounding curve or surface of the region in which the differential equation is to be solved. Since Laplace's equation is of second order it might be plausible to expect that two boundary conditions would be required to determine the solution completely. This, however, is not the case. Recall that in the heat conduction problem for the finite bar (Sections 10.2 and 10.6) it was necessary to prescribe one condition at each end of the bar, that is, *one condition at each point of the boundary*. Generalizing this observation to multidimensional problems, it is then natural to prescribe one condition on the function u at each point on the boundary of the region in which a solution of Eq. (1) or (2) is sought. The most common boundary condition occurs when the value of u is specified at each boundary point; in terms of the heat conduction problem this corresponds to prescribing the temperature on the boundary. In some problems the value of the derivative, or rate of change, of u in the direction normal to the boundary is specified instead; the condition on the boundary of a thermally insulated body, for example, is of this type. It is entirely possible for more complicated boundary conditions to occur; for example, u might be prescribed on part of the boundary, and its normal derivative specified on the remainder. The problem of finding a solution of Laplace's equation that takes on given boundary values is known as a *Dirichlet problem*, in honor of P. G. L. Dirichlet[12]. In contrast, if the values of the normal derivative are prescribed on the boundary, the problem is said to be a *Neumann problem*, in honor of K. G. Neumann[13]. The Dirichlet and Neumann problems are also known as the first and second boundary value problems of potential theory, respectively.

Physically, it is plausible to expect that the types of boundary conditions just mentioned will be sufficient to determine the solution completely. Indeed, it is

[12] Peter Gustav Lejeune Dirichlet (1805–1859) was a professor at Berlin and, after the death of Gauss, at Göttingen. In 1829 he gave the first set of conditions sufficient to guarantee the convergence of a Fourier series. The definition of function usually used today in elementary calculus is essentially the one given by Dirichlet in 1837. While he is best known today for his work in analysis and differential equations, Dirichlet was also one of the leading number theorists of the nineteenth century.

[13] Karl Gottfried Neumann (1832–1925), professor at Leipzig, made contributions to differential equations, integral equations, and complex variables.

possible to establish the existence and uniqueness of the solution of Laplace's equation under the boundary conditions mentioned, provided the shape of the boundary and the functions appearing in the boundary conditions satisfy certain very mild requirements. However, the proofs of these theorems, and even their accurate statement, are beyond the scope of the present book. Our only concern will be solving some typical problems by means of separation of variables and Fourier series.

While the problems chosen as examples are capable of interesting physical interpretations (in terms of electrostatic potentials or steady-state temperature distributions, for instance), our purpose here is primarily to point out some of the features that may occur during their mathematical solution. It is also worth noting again that more complicated problems can sometimes by solved by expressing the solution as the sum of solutions of several simpler problems (see Problems 3 and 4).

DIRICHLET PROBLEM FOR A RECTANGLE. Consider the mathematical problem of finding the function u satisfying Laplace's equation (1),

$$u_{xx} + u_{yy} = 0$$

in the rectangle $0 < x < a, 0 < y < b$, and also satisfying the boundary conditions

$$\begin{aligned} u(x,0) &= 0, \qquad u(x,b) = 0, \qquad 0 < x < a, \\ u(0,y) &= 0, \qquad u(a,y) = f(y), \qquad 0 \le y \le b, \end{aligned} \tag{4}$$

where f is a given function on $0 \le y \le b$ (see Figure 10.24).

To solve this problem we wish to construct a fundamental set of solutions satisfying the partial differential equation and the homogeneous boundary conditions; then we will superpose these solutions so as to satisfy the remaining

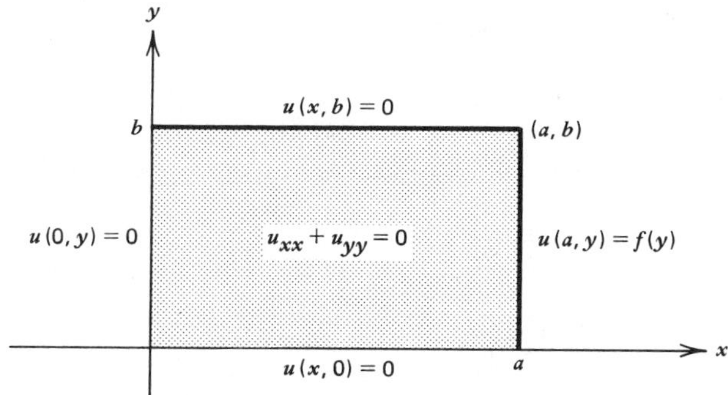

FIGURE 10.24 Dirichlet problem for a rectangle.

boundary condition. Let us assume that

$$u(x, y) = X(x)Y(y) \tag{5}$$

and substitute for u in Eq. (1). This yields

$$\frac{X''}{X} = -\frac{Y''}{Y} = \sigma,$$

where σ is the separation constant. Thus we obtain the two ordinary differential equations

$$X'' - \sigma X = 0, \tag{6}$$

$$Y'' + \sigma Y = 0. \tag{7}$$

If we now substitute for u from Eq. (5) in each of the homogeneous boundary conditions, we find that

$$X(0) = 0 \tag{8}$$

and

$$Y(0) = 0, \qquad Y(b) = 0. \tag{9}$$

We will first determine the solution of the differential equation (7) subject to the boundary conditions (9). However, this problem is essentially identical to the one considered at length in Section 10.2. We conclude that there are nontrivial solutions if and only if σ takes on certain real values, namely

$$\sigma = (n\pi/b)^2, \qquad n = 1, 2, \ldots; \tag{10}$$

the corresponding solutions for $Y(y)$ are proportional to $\sin(n\pi y/b)$. Next, we substitute from Eq. (10) for σ in Eq. (6), and solve this equation subject to the boundary condition (8). It follows at once that $X(x)$ must be proportional to $\sinh(n\pi x/b)$. Thus we obtain the fundamental solutions

$$u_n(x, y) = \sinh\frac{n\pi x}{b}\sin\frac{n\pi y}{b}, \qquad n = 1, 2, \ldots. \tag{11}$$

These solutions satisfy the differential equation (1) and all of the homogeneous boundary conditions for each value of n.

To satisfy the remaining nonhomogeneous boundary condition at $x = a$ we assume, as usual, that we can represent the solution $u(x, y)$ in the form

$$u(x, y) = \sum_{n=1}^{\infty} c_n u_n(x, y) = \sum_{n=1}^{\infty} c_n \sinh\frac{n\pi x}{b}\sin\frac{n\pi y}{b}. \tag{12}$$

The coefficients c_n are determined by the boundary condition

$$u(a, y) = \sum_{n=1}^{\infty} c_n \sinh\frac{n\pi a}{b}\sin\frac{n\pi y}{b} = f(y). \tag{13}$$

Therefore the quantities $c_n \sinh(n\pi a/b)$ must be the coefficients in the Fourier sine series of period $2b$ for f and are given by

$$c_n \sinh\frac{n\pi a}{b} = \frac{2}{b}\int_0^b f(y)\sin\frac{n\pi y}{b}\,dy. \tag{14}$$

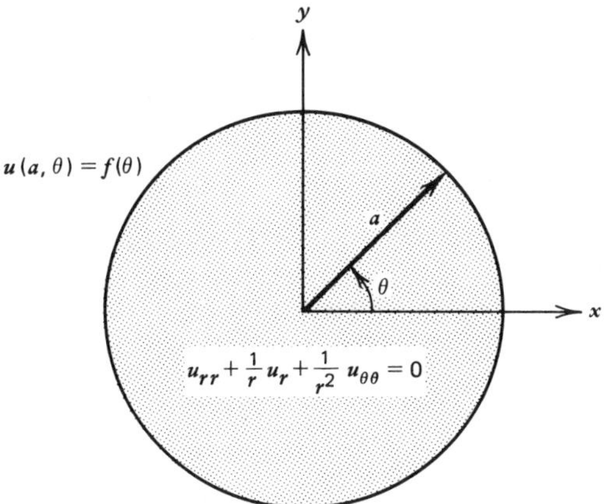

FIGURE 10.25 Dirichlet problem for a circle.

Thus the solution of the partial differential equation (1) satisfying the boundary conditions (4) is given by Eq. (12) with the coefficients c_n computed from Eq. (14).

From Eqs. (12) and (14) we see that the solution contains the factor $\sinh(n\pi x/b)/\sinh(n\pi a/b)$. To estimate this quantity for large n we can use the approximation $\sinh \xi \simeq e^{\xi}/2$, and thereby obtain

$$\frac{\sinh(n\pi x/b)}{\sinh(n\pi a/b)} \cong \frac{\frac{1}{2}\exp(n\pi x/b)}{\frac{1}{2}\exp(n\pi a/b)} = \exp[-n\pi(a-x)/b].$$

Thus this factor has the character of a negative exponential; consequently the series (12) converges fairly rapidly unless $a - x$ is quite small.

DIRICHLET PROBLEM FOR A CIRCLE. Consider the problem of solving Laplace's equation in the circular region $r < a$ subject to the boundary condition

$$u(a, \theta) = f(\theta), \tag{15}$$

where f is a given function on $0 \leq \theta < 2\pi$ (see Figure 10.25). In polar coordinates Laplace's equation takes the form

$$u_{rr} + \frac{1}{r}u_r + \frac{1}{r^2}u_{\theta\theta} = 0. \tag{16}$$

To complete the statement of the problem we note that for $u(r, \theta)$ to be single valued, it is necessary that u be periodic in θ with period 2π. Moreover, we state explicitly that $u(r, \theta)$ must be bounded for $r \leq a$, since this will become important later.

To apply the method of separation of variables to this problem we assume that

$$u(r, \theta) = R(r)\Theta(\theta), \tag{17}$$

and substitute for u in the differential equation (16). This yields

$$R''\Theta + \frac{1}{r}R'\Theta + \frac{1}{r^2}R\Theta'' = 0,$$

or

$$r^2\frac{R''}{R} + r\frac{R'}{R} = -\frac{\Theta''}{\Theta} = \sigma, \tag{18}$$

where σ is the separation constant. Thus we obtain the two ordinary differential equations

$$r^2R'' + rR' - \sigma R = 0, \tag{19}$$

$$\Theta'' + \sigma\Theta = 0. \tag{20}$$

In this problem there are no homogeneous boundary conditions; recall, however, that solutions must be bounded and also periodic in θ with period 2π. It is possible to show (Problem 9) that the periodicity condition requires that σ must be real. We will consider in turn the cases in which σ is negative, zero, and positive.

If $\sigma < 0$, let $\sigma = -\lambda^2$, where $\lambda > 0$. Then Eq. (20) becomes $\Theta'' - \lambda^2\Theta = 0$, and consequently

$$\Theta(\theta) = c_1e^{\lambda\theta} + c_2e^{-\lambda\theta}. \tag{21}$$

Thus $\Theta(\theta)$ can be periodic only if $c_1 = c_2 = 0$, and we conclude that σ cannot be negative.

If $\sigma = 0$, then Eq. (20) becomes $\Theta'' = 0$, and thus

$$\Theta(\theta) = c_1 + c_2\theta. \tag{22}$$

For $\Theta(\theta)$ to be periodic we must have $c_2 = 0$, so that $\Theta(\theta)$ is a constant. Further, for $\sigma = 0$, Eq. (19) becomes

$$r^2R'' + rR' = 0. \tag{23}$$

This equation is of the Euler type, and has the solution

$$R(r) = k_1 + k_2\ln r. \tag{24}$$

The logarithmic term cannot be accepted if $u(r,\theta)$ is to remain bounded as $r \to 0$; hence $k_2 = 0$. Thus, corresponding to $\sigma = 0$, we obtain the solution

$$u_0(r,\theta) = 1. \tag{25}$$

Finally, if $\sigma > 0$, we let $\sigma = \lambda^2$ where $\lambda > 0$. Then Eqs. (19) and (20) become

$$r^2R'' + rR' - \lambda^2R = 0 \tag{26}$$

and

$$\Theta'' + \lambda^2\Theta = 0, \tag{27}$$

respectively. Equation (26) is an Euler equation and has the solution

$$R(r) = k_1r^{\lambda} + k_2r^{-\lambda}, \tag{28}$$

while Eq. (27) has the solution

$$\Theta(\theta) = c_1 \sin \lambda\theta + c_2 \cos \lambda\theta. \tag{29}$$

In order that Θ be periodic with period 2π it is necessary that λ be a positive integer n. With $\lambda = n$ it follows that the solution $r^{-\lambda}$ in Eq. (28) must be discarded since it becomes unbounded as $r \to 0$. Consequently, $k_2 = 0$ and the appropriate solutions of Eq. (16) are

$$u_n(r,\theta) = r^n \cos n\theta, \qquad v_n(r,\theta) = r^n \sin n\theta, \qquad n = 1, 2, \dots. \tag{30}$$

These functions, together with $u_0(r,\theta) = 1$, form a set of fundamental solutions for the present problem.

In the usual way we now assume that u can be expressed as a linear combination of the fundamental solutions; that is,

$$u(r,\theta) = \frac{c_0}{2} + \sum_{n=1}^{\infty} r^n (c_n \cos n\theta + k_n \sin n\theta). \tag{31}$$

The boundary condition (15) then requires that

$$u(a,\theta) = \frac{c_0}{2} + \sum_{n=1}^{\infty} a^n (c_n \cos n\theta + k_n \sin n\theta) = f(\theta) \tag{32}$$

for $0 \le \theta < 2\pi$. The function f may be extended outside of this interval so that it is periodic of period 2π, and therefore has a Fourier series of the form (32). Since the extended function has period 2π, we may compute its Fourier coefficients by integrating over any period of the function. In particular, it is convenient to use the original interval $(0, 2\pi)$; then

$$a^n c_n = \frac{1}{\pi} \int_0^{2\pi} f(\theta) \cos n\theta \, d\theta, \qquad n = 0, 1, 2, \dots; \tag{33}$$

$$a^n k_n = \frac{1}{\pi} \int_0^{2\pi} f(\theta) \sin n\theta \, d\theta, \qquad n = 1, 2, \dots. \tag{34}$$

With this choice of the coefficients, Eq. (31) represents the solution of the boundary value problem of Eqs. (15) and (16). Note that in this problem we needed both sine and cosine terms in the solution. This is because the boundary data were given on $0 \le \theta < 2\pi$ and have period 2π. As a consequence, the full Fourier series is required, rather than sine or cosine terms alone.

PROBLEMS

1. (a) Find the solution $u(x, y)$ of Laplace's equation in the rectangle $0 < x < a$, $0 < y < b$, also satisfying the boundary conditions

$$u(0, y) = 0, \qquad u(a, y) = 0, \qquad 0 < y < b$$
$$u(x, 0) = 0, \qquad u(x, b) = g(x), \qquad 0 \le x \le a.$$

(b) Find the solution if $g(x)$ is given by the formula

$$g(x) = \begin{cases} x, & 0 \le x < a/2, \\ a - x, & a/2 \le x \le a. \end{cases}$$

2. Find the solution $u(x, y)$ of Laplace's equation in the rectangle $0 < x < a, 0 < y < b$, also satisfying the boundary conditions

$$u(0, y) = 0, \qquad u(a, y) = 0, \qquad 0 < y < b,$$
$$u(x,0) = h(x), \qquad u(x, b) = 0, \qquad 0 \le x \le a.$$

3. Find the solution $u(x, y)$ of Laplace's equation in the rectangle $0 < x < a, 0 < y < b$, also satisfying the boundary conditions

$$u(0, y) = 0, \qquad u(a, y) = f(y), \qquad 0 < y < b,$$
$$u(x,0) = h(x), \qquad u(x, b) = 0, \qquad 0 \le x \le a.$$

Hint: Consider the possibility of adding the solutions of two problems, one with homogeneous boundary conditions except for $u(a, y) = f(y)$, and the other with homogeneous boundary conditions except for $u(x, 0) = h(x)$.

4. Show how to find the solution $u(x, y)$ of Laplace's equation in the rectangle $0 < x < a, 0 < y < b$, also satisfying the boundary conditions

$$u(0, y) = k(y), \qquad u(a, y) = f(y), \qquad 0 < y < b,$$
$$u(x,0) = h(x), \qquad u(x, b) = g(x), \qquad 0 \le x \le a.$$

Hint: See Problem 3.

5. Find the solution $u(r, \theta)$ of Laplace's equation

$$u_{rr} + (1/r)u_r + (1/r^2)u_{\theta\theta} = 0$$

outside the circle $r = a$, also satisfying the boundary condition

$$u(a, \theta) = f(\theta), \qquad 0 \le \theta < 2\pi,$$

on the circle. Assume that $u(r, \theta)$ is single valued and bounded for $r > a$.

6. Find the solution $u(r, \theta)$ of Laplace's equation in the semicircular region $r < a$, $0 < \theta < \pi$, also satisfying the boundary conditions

$$u(r,0) = 0, \qquad u(r, \pi) = 0, \qquad 0 \le r < a,$$
$$u(a, \theta) = f(\theta), \qquad 0 \le \theta \le \pi.$$

Assume that u is single valued and bounded in the given region.

7. Find the solution $u(r, \theta)$ of Laplace's equation in the circular sector $0 < r < a$, $0 < \theta < \alpha$, also satisfying the boundary conditions

$$u(r,0) = 0, \qquad u(r, \alpha) = 0, \qquad 0 \le r < a$$
$$u(a, \theta) = f(\theta), \qquad 0 \le \theta \le \alpha.$$

Assume that u is single valued and bounded in the sector.

8. Find the solution $u(x, y)$ of Laplace's equation in the semi-infinite strip $y > 0$, $0 < x < a$, also satisfying the boundary conditions

$$u(0, y) = 0, \qquad u(a, y) = 0, \qquad y > 0$$
$$u(x,0) = f(x), \qquad 0 \le x \le a$$

and the additional condition that $u(x, y) \to 0$ as $y \to \infty$.

9. Show that Eq. (20) has periodic solutions only if σ is real.
Hint: Let $\sigma = -\lambda^2$ where $\lambda = \mu + i\nu$ with μ and ν real.

10. Consider the problem of finding a solution $u(x, y)$ of Laplace's equation in the rectangle $0 < x < a, 0 < y < b$, also satisfying the boundary conditions

$$u_x(0, y) = 0, \qquad u_x(a, y) = f(y), \qquad 0 < y < b,$$
$$u_y(x, 0) = 0, \qquad u_y(x, b) = 0, \qquad 0 \le x \le a.$$

This is an example of a Neumann problem.
(a) Show that Laplace's equation and the homogeneous boundary conditions determine the fundamental set of solutions

$$u_0(x, y) = c_0,$$
$$u_n(x, y) = c_n \cosh(n\pi x/b)\cos(n\pi y/b); \qquad n = 1, 2, 3, \ldots.$$

(b) By superposing the fundamental solutions of part (a), formally determine a function u also satisfying the nonhomogeneous boundary condition $u_x(a, y) = f(y)$. Note that when $u_x(a, y)$ is calculated, the constant term in $u(x, y)$ is eliminated, and there is no condition from which to determine c_0. Furthermore, it must be possible to express f by means of a Fourier cosine series of period $2b$, which does not have a constant term. This means that

$$\int_0^b f(y)\,dy = 0$$

is a necessary condition for the given problem to be solvable. Finally, note that c_0 remains arbitrary, and hence the solution is only determined up to this additive constant. This is a property of all Neumann problems.

11. Find a solution $u(r, \theta)$ of Laplace's equation inside the circle $r = a$, also satisfying the boundary condition on the circle

$$u_r(a, \theta) = g(\theta), \qquad 0 \le \theta < 2\pi.$$

Note that this is a Neumann problem, and that its solution is determined only up to an arbitrary additive constant. State a necessary condition on $g(\theta)$ for this problem to be solvable by the method of separation of variables (see Problem 10).

12. Find the solution $u(x, y)$ of Laplace's equation in the rectangle $0 < x < a, 0 < y < b$, also satisfying the boundary conditions

$$u_x(0, y) = 0, \qquad u(a, y) = f(y), \qquad 0 < y < b,$$
$$u(x, 0) = 0, \qquad u(x, b) = 0, \qquad 0 \le x \le a.$$

Note that this is neither a Dirichlet nor a Neumann problem, but a mixed problem in which u is prescribed on part of the boundary and its normal derivative on the rest.

13. Find the solution $u(x, y)$ of Laplace's equation in the rectangle $0 < x < a, 0 < y < b$, also satisfying the boundary conditions

$$u(0, y) = 0, \qquad u(a, y) = f(y), \qquad 0 < y < b,$$
$$u(x, 0) = 0, \qquad u_y(x, b) = 0, \qquad 0 \le x \le a.$$

Hint: Eventually it will be necessary to expand $f(y)$ in a series making use of the functions $\sin(\pi y/2b), \sin(3\pi y/2b), \sin(5\pi y/2b), \ldots$ (see Problem 23 of Section 10.5).

14. Find the solution $u(x, y)$ of Laplace's equation in the rectangle $0 < x < a, 0 < y < b$, also satisfying the boundary conditions

$$u_x(0, y) = 0, \qquad u_x(a, y) = 0, \qquad 0 < y < b,$$

$$u(x,0) = 0, \qquad u(x, b) = f(x), \qquad 0 \le x \le a.$$

15. By writing Laplace's equation in cylindrical coordinates (r, θ, z) and then assuming that the solution is axially symmetric (no dependence on θ), we obtain the equation

$$u_{rr} + (1/r)u_r + u_{zz} = 0.$$

Assuming that $u(r, z) = R(r)Z(z)$, show that R and Z satisfy the equations

$$rR'' + R' + \lambda^2 rR = 0, \qquad Z'' - \lambda^2 Z = 0.$$

The equation for R is Bessel's equation of order zero with independent variable λr.

Appendix A

DERIVATION OF THE HEAT CONDUCTION EQUATION. In this section we derive the differential equation which, to a first approximation at least, governs the conduction of heat in solids. It is important to understand that the mathematical analysis of a physical situation or process such as this ultimately rests on a foundation of empirical knowledge of the phenomenon involved. The mathematician must have a place to start, so to speak, and this place is furnished by experience. In the case of heat conduction it has been demonstrated many times that if two parallel plates of the same area A and different constant temperatures T_1 and T_2, respectively, are separated by a small distance d, an amount of heat per unit time will pass from the warmer to the cooler. Moreover, to a high degree of approximation, this amount of heat is proportional to the area A, the temperature difference $|T_2 - T_1|$, and inversely proportional to the separation distance d. Thus

$$\text{amount of heat per unit time} = \kappa A|T_2 - T_1|/d, \tag{1}$$

where the positive proportionality factor κ is called the thermal conductivity and depends only on the material[14] between the two plates. The physical law expressed by Eq. (1) is known as Newton's law of cooling. We repeat that Eq. (1) is an empirical, not a theoretical, result, and that it can be, and has often been, verified by careful experiment. It is the basis of the mathematical theory of heat conduction.

Now consider a straight rod of uniform cross section and homogeneous material, oriented so that the x axis lies along the axis of the rod (see Figure 10.26). Let $x = 0$ and $x = l$ designate the ends of the bar.

[14]Actually, κ also depends on the temperature, but if the temperature range is not too great, it is satisfactory to assume that κ is independent of temperature.

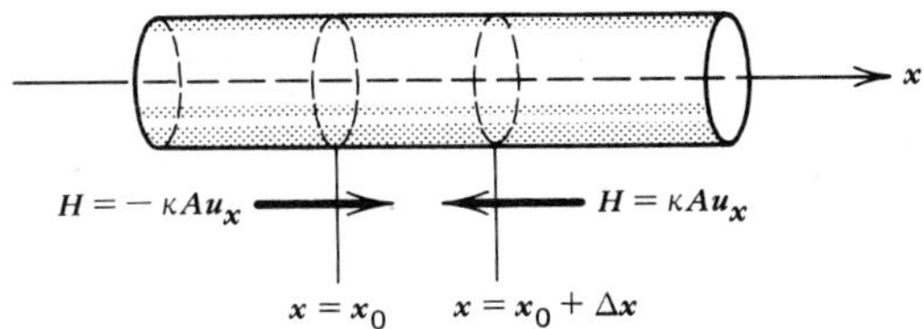

FIGURE 10.26 Conduction of heat in an element of a rod.

We will assume that the sides of the bar are perfectly insulated so that there is no passage of heat through them. We will also assume that the temperature u depends only on the axial position x and the time t, and not on the lateral coordinates y and z. In other words, we assume that the temperature remains constant on any cross section of the bar. This assumption is usually satisfactory when the lateral dimensions of the rod are small compared to its length.

The differential equation governing the temperature in the bar is an expression of a fundamental physical balance; the rate at which heat flows into any portion of the bar is equal to the rate at which heat is absorbed in that portion of the bar. The terms in the equation are called the flux (flow) term and the absorption term, respectively.

We will first calculate the flux term. Consider an element of the bar lying between the cross sections $x = x_0$ and $x = x_0 + \Delta x$, where x_0 is arbitrary and Δx is small. The instantaneous rate of heat transfer $H(x_0, t)$ from left to right across the cross section $x = x_0$ is given by

$$\begin{aligned} H(x_0, t) &= -\lim_{d \to 0} \kappa A \frac{u(x_0 + d/2, t) - u(x_0 - d/2, t)}{d} \\ &= -\kappa A u_x(x_0, t). \end{aligned} \tag{2}$$

The minus sign appears in this equation since there will be a positive flow of heat from left to right only if the temperature is greater to the left of $x = x_0$ than to the right; in this case $u_x(x_0, t)$ is negative. In a similar manner, the rate at which heat passes from left to right through the cross section $x = x_0 + \Delta x$ is given by

$$H(x_0 + \Delta x, t) = -\kappa A u_x(x_0 + \Delta x, t). \tag{3}$$

The net rate at which heat flows into the segment of the bar between $x = x_0$ and $x = x_0 + \Delta x$ is thus given by

$$Q = H(x_0, t) - H(x_0 + \Delta x, t) = \kappa A\left[u_x(x_0 + \Delta x, t) - u_x(x_0, t)\right]. \tag{4}$$

and the amount of heat entering this bar element in time Δt is

$$Q\,\Delta t = \kappa A\left[u_x(x_0 + \Delta x, t) - u_x(x_0, t)\right]\Delta t. \tag{5}$$

Let us now calculate the absorption term. The average change in temperature Δu, in the time interval Δt, is proportional to the amount of heat $Q\,\Delta t$ introduced and inversely proportional to the mass Δm of the element. Thus

$$\Delta u = \frac{1}{s}\frac{Q\,\Delta t}{\Delta m} = \frac{Q\,\Delta t}{s\rho A\,\Delta x}, \tag{6}$$

where the constant of proportionality s is known as the specific heat of the material of the bar, and ρ is its density.[15] The average temperature change Δu in the bar element under consideration is the actual temperature change at some intermediate point $x = x_0 + \theta\Delta x$, where $0 < \theta < 1$. Thus Eq. (6) can be written as

$$u(x_0 + \theta\Delta x, t + \Delta t) - u(x_0 + \theta\Delta x, t) = \frac{Q\,\Delta t}{s\rho A\,\Delta x}, \tag{7}$$

or as

$$Q\,\Delta t = \left[u(x_0 + \theta\Delta x, t + \Delta t) - u(x_0 + \theta\Delta x, t)\right] s\rho A\,\Delta x. \tag{8}$$

To balance the flux and absorption terms, we equate the two expressions for $Q\,\Delta t$:

$$\begin{aligned} &\kappa A\left[u_x(x_0 + \Delta x, t) - u_x(x_0, t)\right]\Delta t \\ &\qquad = s\rho A\left[u(x_0 + \theta\Delta x, t + \Delta t) - u(x_0 + \theta\Delta x, t)\right]\Delta x. \end{aligned} \tag{9}$$

On dividing Eq. (9) by $\Delta x\,\Delta t$ and then letting $\Delta x \to 0$ and $\Delta t \to 0$, we obtain the *heat conduction* or *diffusion* equation

$$\alpha^2 u_{xx} = u_t. \tag{10}$$

The quantity α^2 defined by

$$\alpha^2 = \kappa/\rho s \tag{11}$$

is called the *thermal diffusivity*, and is a parameter depending only on the material of the bar. The dimensions of α^2 are (length)^2/time. Typical values of α^2 are given in Table 10.1 in Section 10.2.

Several relatively simple conditions may be imposed at the ends of the bar. For example, the temperature at an end may be maintained at some constant value T. This might be accomplished by placing the end of the bar in thermal contact with some reservoir of sufficient size so that any heat that may flow between the bar and reservoir does not appreciably alter the temperature of the reservoir. At an end where this is done the boundary condition is

$$u = T. \tag{12}$$

Another simple boundary condition occurs if the end is insulated so that no heat passes through it. Recalling the expression (2) for the amount of heat crossing any cross section of the bar, the condition for insulation is clearly that

[15] The dependence of the density and specific heat on temperature is relatively small and will be neglected. Thus both ρ and s will be considered as constants.

this quantity vanish. Thus

$$u_x = 0 \tag{13}$$

is the boundary condition at an insulated end.

A more general type of boundary condition occurs if the rate of flow of heat through an end of the bar is proportional to the temperature there; this type of boundary condition is called a radiation condition. Let us consider the end $x = 0$, where the rate of flow of heat from left to right is given by $-\kappa A u_x(0, t)$; see Eq. (2). Hence the rate of heat flow out of the bar (from right to left) at $x = 0$ is $\kappa A u_x(0, t)$. If this quantity is proportional to the temperature $u(0, t)$, then we obtain the boundary condition

$$u_x(0, t) - h_1 u(0, t) = 0, \qquad t > 0, \tag{14}$$

where h_1 is a nonnegative constant of proportionality. Note that $h_1 = 0$ corresponds to an insulated end, while $h_1 \to \infty$ corresponds to an end held at zero temperature.

If radiation is taking place at the right end of the bar ($x = l$), then in a similar way we obtain the boundary condition

$$u_x(l, t) + h_2 u(l, t) = 0, \qquad t > 0, \tag{15}$$

where again h_2 is a nonnegative constant of proportionality.

Finally, to determine completely the flow of heat in the bar it is necessary to state the temperature distribution at one fixed instant, usually taken as the initial time $t = 0$. This initial condition is of the form

$$u(x, 0) = f(x), \qquad 0 \le x \le l. \tag{16}$$

The problem then is to determine the solution of the differential equation (10) subject to one of the boundary conditions (12) to (15) at each end, and to the initial condition (16) at $t = 0$.

Several generalizations of the heat equation (10) also occur in practice. First, the bar material may be nonuniform and the cross section may not be constant along the length of the bar. In this case, the parameters κ, ρ, s, and A may depend on the axial variable x. Going back to Eq. (2) we see that the rate of heat transfer from left to right across the cross section at $x = x_0$ is now given by

$$H(x_0, t) = -\kappa(x_0) A(x_0) u_x(x_0, t) \tag{17}$$

with a similar expression for $H(x_0 + \Delta x, t)$. If we introduce these quantities into Eq. (4) and eventually into Eq. (9), and proceed as before, we obtain the partial differential equation

$$(\kappa A u_x)_x = s\rho A\, u_t. \tag{18}$$

We will usually write Eq. (18) in the form

$$r(x) u_t = [p(x) u_x]_x, \tag{19}$$

where $p(x) = \kappa(x) A(x)$, and $r(x) = s(x)\rho(x) A(x)$. Note that both of these quantities are intrinsically positive.

A second generalization occurs if there are other ways in which heat enters or leaves the bar. Suppose that there is a *source* that adds heat to the bar at a rate $G(x, t, u)$ per unit time per unit length, where $G(x, t, u) > 0$. In this case we must add the term $G(x, t, u)\,\Delta x\,\Delta t$ to the left side of Eq. (9), and this leads to the differential equation

$$r(x)u_t = [p(x)u_x]_x + G(x, t, u). \tag{20}$$

If $G(x, t, u) < 0$, then we speak of a *sink* that removes heat from the bar at the rate $G(x, t, u)$ per unit time per unit length. To make the problem tractable we must restrict the form of the function G. In particular, we assume that G is linear in u and that the coefficient of u does not depend on t. Thus we write

$$G(x, t, u) = F(x, t) - q(x)u. \tag{21}$$

The minus sign in Eq. (21) has been introduced so that certain equations that appear later will have their customary forms. Substituting from Eq. (21) into Eq. (20), we obtain

$$r(x)u_t = [p(x)u_x]_x - q(x)u + F(x, t). \tag{22}$$

This equation is sometimes called the generalized heat conduction equation. Boundary value problems for Eq. (22) will be discussed to some extent in Chapter 11.

Finally, if instead of a one-dimensional bar, we consider a body with more than one significant space dimension, then the temperature is a function of two or three space coordinates rather than of x alone. Considerations similar to those leading to Eq. (10) can be employed to derive the heat conduction equation in two dimensions

$$\alpha^2(u_{xx} + u_{yy}) = u_t, \tag{23}$$

or in three dimensions

$$\alpha^2(u_{xx} + u_{yy} + u_{zz}) = u_t. \tag{24}$$

The boundary conditions corresponding to Eqs. (12) and (13) for multidimensional problems correspond to a prescribed temperature distribution on the boundary, or to an insulated boundary. Similarly the initial temperature distribution will in general be a function of x and y for Eq. (23), and a function of x, y, and z for Eq. (24).

Appendix B

DERIVATION OF THE WAVE EQUATION. In this section we derive the wave equation in one space dimension as it applies to the transverse vibrations of an elastic string, or cable; the elastic string may be thought of as a violin string, a guy wire, or possibly an electric power line. The same equation, however, with the

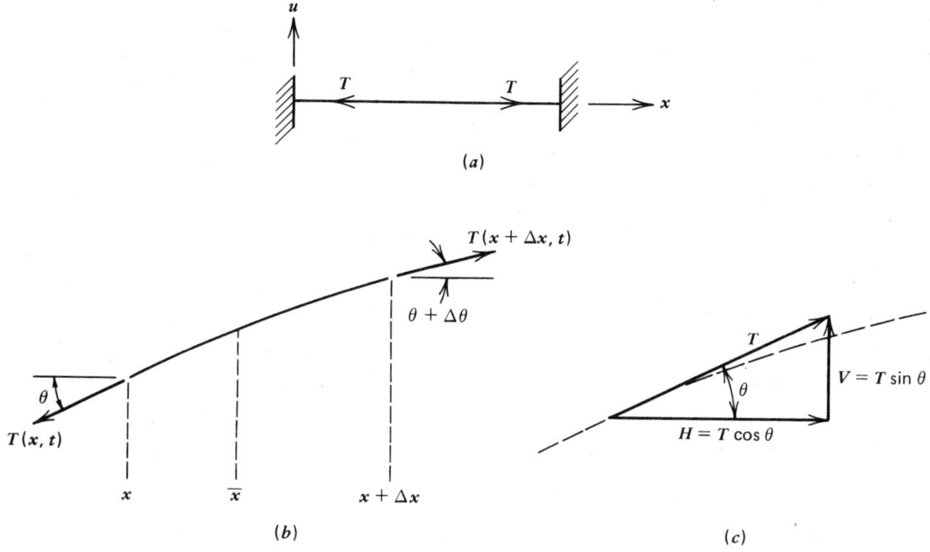

FIGURE 10.27 (a) An elastic string under tension. (b) An element of the displaced string. (c) Resolution of the tension T into components.

variables properly interpreted, occurs in many other wave problems having only one significant space variable.

Consider a perfectly flexible elastic string stretched tightly between supports fixed at the same horizontal level (see Figure 10.27*a*). Let the x axis lie along the string with the end points located at $x = 0$ and $x = l$. If the string is set in motion at some initial time $t = 0$ (by plucking, for example) and is thereafter left undisturbed, it will vibrate freely in a vertical plane provided damping effects, such as air resistance, are neglected. To determine the differential equation governing this motion we will consider the forces acting on a small element of the string of length Δx lying between the points x and $x + \Delta x$ (see Figure 10.27*b*). We assume that the motion of the string is small, and as a consequence, each point on the string moves solely on a vertical line. We denote by $u(x, t)$ the vertical displacement of the point x at the time t. Let the tension in the string, which always acts in the tangential direction, be denoted by $T(x, t)$, and let ρ denote the mass per unit length of the string.

Newton's law, as it applies to the element Δx of the string, states that the net external force, due to the tension at the ends of the element, must be equal to the product of the mass of the element and the acceleration of its mass center. Since there is no horizontal acceleration, the horizontal components must satisfy

$$T(x + \Delta x, t)\cos(\theta + \Delta\theta) - T(x, t)\cos\theta = 0. \tag{1}$$

If we denote the horizontal component of the tension (see Figure 10.27*c*) by H, then Eq. (1) states that H is independent of x.

On the other hand, the vertical components satisfy

$$T(x + \Delta x, t)\sin(\theta + \Delta\theta) - T(x, t)\sin\theta = \rho\Delta x\, u_{tt}(\bar{x}, t), \tag{2}$$

where $\bar{x}$ is the coordinate of the center of mass of the element of the string under consideration. Clearly $\bar{x}$ lies in the interval $x < \bar{x} < x + \Delta x$. The weight of the string, which acts vertically downward, is assumed to be negligible, and has been neglected in Eq. (2).

If the vertical component of T is denoted by V, then Eq. (2) can be written as

$$\frac{V(x + \Delta x, t) - V(x, t)}{\Delta x} = \rho u_{tt}(\bar{x}, t).$$

Passing to the limit as $\Delta x \to 0$ gives

$$V_x(x, t) = \rho u_{tt}(x, t). \tag{3}$$

To express Eq. (3) entirely in terms of u we note that

$$V(x, t) = H(t)\tan\theta = H(t)u_x(x, t).$$

Hence Eq. (3) becomes

$$(Hu_x)_x = \rho u_{tt},$$

or, since H is independent of x,

$$Hu_{xx} = \rho u_{tt}. \tag{4}$$

For small motions of the string it is permissible to replace $H = T\cos\theta$ by T. Then Eq. (4) takes its customary form

$$a^2 u_{xx} = u_{tt}, \tag{5}$$

where

$$a^2 = T/\rho. \tag{6}$$

We will assume further that a^2 is a constant, although this is not required in our derivation, even for small motions. Equation (5) is called the wave equation for one space dimension. Since T has the dimension of force, and ρ that of mass/length, it follows that the constant a has the dimension of velocity. It is possible to identify a as the velocity with which a small disturbance (wave) moves along the string. According to Eq. (6) the wave velocity a varies directly with the tension in the string, but inversely with the density of the string material. These facts are in agreement with experience.

As in the case of the heat conduction equation, there are various generalizations of the wave equation (5). One important equation is known as the *telegraph equation* and has the form

$$u_{tt} + cu_t + ku = a^2 u_{xx} + F(x, t), \tag{7}$$

where c and k are nonnegative constants. The terms cu_t, ku, and $F(x, t)$ arise from a viscous damping force, an elastic restoring force, and an external force, respectively. Note the similarity of Eq. (7), except for the term a^2u_{xx}, with the equation for the spring–mass system derived in Section 3.7; the additional term a^2u_{xx} arises from a consideration of internal elastic forces.

For a vibrating system with more than one significant space coordinate, it may be necessary to consider the wave equation in two dimensions

$$a^2(u_{xx} + u_{yy}) = u_{tt}, \tag{8}$$

or in three dimensions

$$a^2(u_{xx} + u_{yy} + u_{zz}) = u_{tt}. \tag{9}$$

REFERENCES

The following books contain additional information on Fourier series:

Buck, R. C., and Buck, E. F., *Advanced Calculus*, (3rd ed.) (New York: McGraw-Hill, 1978).

Carslaw, H. S., *Introduction to the Theory of Fourier's Series and Integrals* (3rd ed.) (Cambridge: Cambridge University Press, 1930); reprinted by Dover, New York, 1952.

Courant, R., and John, F., *Introduction to Calculus and Analysis* (New York: Wiley-Interscience, 1965).

Guillemin, E. A., *The Mathematics of Circuit Analysis* (New York: Wiley, 1949).

Hamming, R. W., *Numerical Methods for Scientists and Engineers* (New York: McGraw-Hill, 1962).

Kaplan, W., *Advanced Calculus* (3rd ed.) (Reading, Mass.: Addison-Wesley, 1984).

A brief biography of Fourier and an annotated copy of his 1807 paper are contained in:

Grattan-Guinness, I., *Joseph Fourier 1768–1830* (Cambridge, Mass.: MIT Press, 1972).

Useful references on partial differential equations and the method of separation of variables include the following:

Berg, P. W., and McGregor, J. L., *Elementary Partial Differential Equations* (San Francisco: Holden-Day, 1966).

Churchill, R. V., and Brown, J. W., *Fourier Series and Boundary Value Problems* (3rd ed.) (New York: McGraw-Hill, 1978).

Pinsky, M. A., *Introduction to Partial Differential Equations with Applications* (New York: McGraw-Hill, 1984).

Powers, D. L., *Boundary Value Problems* (2nd ed.) (New York: Academic Press, 1979).

Sagan, H., *Boundary and Eigenvalue Problems in Mathematical Physics*, (New York: Wiley, 1961).

Weinberger, H. F., *A First Course in Partial Differential Equations*, (New York: Wiley, 1965).

CHAPTER 11

BOUNDARY VALUE PROBLEMS AND STURM-LIOUVILLE THEORY

11.1 Introduction

In Chapter 10 we used the method of separation of variables to solve several problems of physical importance involving partial differential equations. For example, in Section 10.2 we discussed the heat conduction problem consisting of the partial differential equation

$$\alpha^2 u_{xx} = u_t, \qquad 0 < x < l, \quad t > 0 \tag{1}$$

subject to the boundary conditions

$$u(0,t) = 0, \qquad u(l,t) = 0, \quad t > 0 \tag{2}$$

and the initial condition

$$u(x,0) = f(x), \qquad 0 \le x \le l. \tag{3}$$

The method of separation of variables, as used in Chapter 10, consists of two principal steps: (1) the reduction of the partial differential equation to two (or more) ordinary differential equations, and (2) the expansion of certain functions in Fourier series. The sine or cosine functions used in forming the Fourier series usually occurred as solutions of the differential equation

$$X'' + \lambda X = 0, \qquad 0 < x < l \tag{4}$$

satisfying boundary conditions such as

$$X(0) = 0, \qquad X(l) = 0 \tag{5}$$

or

$$X'(0) = 0, \qquad X'(l) = 0. \tag{6}$$

In this chapter we extend and generalize the results of Chapter 10. Our main goal is to show how the method of separation of variables can be used to solve problems somewhat more general than that of Eqs. (1), (2), and (3). We are interested in three types of generalizations.

First, we wish to consider more general partial differential equations; for example, the equation

$$r(x)u_t = [p(x)u_x]_x - q(x)u + F(x,t). \tag{7}$$

This equation can arise, as indicated in Appendix A of Chapter 10, in the study of heat conduction in a bar of variable material properties in the presence of heat sources. If p and r are constants, and if the source terms qu and F are zero, then Eq. (7) reduces to Eq. (1). The partial differential equation (7) also occurs in the investigation of other phenomena of a diffusive character.

A second generalization is to allow more general boundary conditions. In particular, we wish to consider boundary conditions of the form

$$u_x(0,t) - h_1 u(0,t) = 0, \qquad u_x(l,t) + h_2 u(l,t) = 0. \tag{8}$$

Such conditions occur when the rate of heat flow through an end of the bar is proportional to the temperature there. Usually h_1 and h_2 are nonnegative constants, but in some cases they may be negative or depend on t. The boundary conditions (2) are obtained in the limit as $h_1 \to \infty$ and $h_2 \to \infty$. The other important limiting case, $h_1 = h_2 = 0$, gives the boundary conditions for insulated ends.

The final generalization that we discuss in this chapter concerns the geometry of the region in which the problem is posed. The results of Chapter 10 are adequate only for a rather restricted class of problems, chiefly those in which the region of interest is rectangular or, in a few cases, circular. Later in this chapter we consider certain problems posed in other geometrical regions.

Let us consider the equation

$$r(x)u_t = [p(x)u_x]_x - q(x)u \tag{9}$$

obtained by setting the term $F(x,t)$ in Eq. (7) equal to zero. To separate the variables we assume that

$$u(x,t) = X(x)T(t), \tag{10}$$

and substitute for u in Eq. (9). We obtain

$$r(x)XT' = [p(x)X']'T - q(x)XT \tag{11}$$

or, upon dividing by $r(x)XT$,

$$\frac{T'}{T} = \frac{[p(x)X']'}{r(x)X} - \frac{q(x)}{r(x)} = -\lambda. \tag{12}$$

We have denoted the separation constant by $-\lambda$ in anticipation of the fact that usually it will turn out to be real and negative. From Eq. (12) we obtain the following two ordinary differential equations for X and T:

$$[p(x)X']' - q(x)X + \lambda r(x)X = 0, \tag{13}$$

$$T' + \lambda T = 0. \tag{14}$$

If we substitute from Eq. (10) for u in Eqs. (8) and assume that h_1 and h_2 are constants, then we obtain the boundary conditions

$$X'(0) - h_1 X(0) = 0, \qquad X'(l) + h_2 X(l) = 0. \tag{15}$$

To proceed further we need to solve Eq. (13) subject to the boundary conditions (15). This is a generalization of the problem consisting of the differential equation (4) and the boundary conditions (5) or (6). In general, the problem of solving an ordinary differential equation in an interval, subject to boundary conditions at each end, is called a *two-point boundary value problem*. In contrast, in an initial value problem values of the unknown functions and its derivatives are prescribed at a single initial point, and the solution of the differential equation is usually sought in the semi-infinite interval beginning there.

It is useful now to distinguish between two types of linear boundary value problems, namely homogeneous and nonhomogeneous problems. The most general second order linear differential equation is

$$P(x)y'' + Q(x)y' + R(x)y = G(x). \tag{16}$$

Such equations were classified in Chapter 3 as homogeneous if $G(x)$ is identically zero, and nonhomogeneous otherwise. Similarly, the most general linear boundary condition at a point $x = x_0$ is

$$a_1 y(x_0) + a_2 y'(x_0) = c, \tag{17}$$

where a_1, a_2, and c are constants. This boundary condition is said to be homogeneous if c is zero, and nonhomogeneous if it is not. A boundary value problem is homogeneous if both the differential equation and all boundary conditions are homogeneous; if not, then the problem is nonhomogeneous. A typical second order linear homogeneous boundary value problem is of the form

$$P(x)y'' + Q(x)y' + R(x)y = 0, \qquad 0 < x < 1, \tag{18}$$

$$a_1 y(0) + a_2 y'(0) = 0, \tag{19}$$

$$b_1 y(1) + b_2 y'(1) = 0. \tag{20}$$

The problem given by Eqs. (13) and (15) is also linear and homogeneous. Nonhomogeneous boundary value problems are easily identified by the presence in the differential equation or a boundary condition of at least one nonzero term that is independent of y and its derivatives.

As indicated by Eqs. (18) to (20), we usually consider problems posed on the interval $0 \le x \le 1$. Corresponding results hold for problems posed on an arbitrary finite interval. Indeed, if the interval is originally $\alpha \le s \le \beta$, it can be transformed into $0 \le x \le 1$ by the change of variable $x = (s - \alpha)/(\beta - \alpha)$.

Boundary value problems with higher order differential equations can also occur; in them the number of boundary conditions must equal the order of the differential equation. As a rule, the order of the differential equation is even, and half of the boundary conditions are given at each end of the basic interval. It is also possible for a single boundary condition to involve values of the solution and/or its derivatives at both boundary points; for example,

$$y(0) - y(1) = 0. \tag{21}$$

In this chapter we discuss some of the properties of solutions of two-point boundary value problems for second order linear equations. Where convenient, we consider the general linear equation (16), or the equation

$$y'' + p(x)y' + q(x)y = g(x), \tag{22}$$

obtained by dividing Eq. (16) by $P(x)$. However, for most purposes it is better to discuss equations in which the first and second derivative terms are related as in Eq. (13). We note that it is always possible to transform the general equation (16) so that the derivative terms appear as in Eq. (13) (see Problem 2). We will find that solutions of the problem (13), (15) are in many important respects very similar to the solutions $\sin(n\pi x/l)$ of the simple problem (4), (5). In particular, solutions of Eqs. (13) and (15) can be used to construct series solutions of a variety of problems in partial differential equations. For illustrative purposes we will use the heat conduction problem composed of the partial differential equation (7), the boundary conditions (8), and the initial condition (3).

PROBLEMS

1. State whether each of the following boundary value problems is homogeneous or nonhomogeneous.

   (a) $y'' + 4y = 0, \quad y(-1) = 0, \quad y(1) = 0$
   (b) $[(1 + x^2)y']' + 4y = 0, \quad y(0) = 0, \quad y(1) = 1$
   (c) $y'' + 4y = \sin x, \quad y(0) = 0, \quad y(1) = 0$
   (d) $-y'' + x^2y = \lambda y, \quad y'(0) - y(0) = 0, \quad y'(1) + y(1) = 0$
   (e) $-[(1 + x^2)y']' = \lambda y + 1, \quad y(-1) = 0, \quad y(1) = 0$
   (f) $y'' = \lambda(1 + x^2)y, \quad y(0) = 0, \quad y'(1) + 3y(1) = 0$

2. Consider the general linear homogeneous second order equation

$$P(x)y'' + Q(x)y' + R(x)y = 0. \tag{i}$$

   We seek an integrating factor $\mu(x)$ such that, upon multiplying Eq. (i) by $\mu(x)$, the resulting equation can be written in the form

$$[\mu(x)P(x)y']' + \mu(x)R(x)y = 0. \tag{ii}$$

(a) By equating coefficients of y', show that μ must be a solution of

$$P\mu' = (Q - P')\mu. \tag{iii}$$

(b) Solve Eq. (iii) and thereby show that

$$\mu(x) = \frac{1}{P(x)} \exp \int_{x_0}^{x} \frac{Q(s)}{P(s)}\, ds. \tag{iv}$$

Compare this result with the result of Problem 18 in Section 3.2.

3. Use the method of Problem 2 to transform each of the following equations into the form $[p(x)y']' + q(x)y = 0$.
(a) $y'' - 2xy' + \lambda y = 0,$ Hermite equation
(b) $x^2y'' + xy' + (x^2 - \nu^2)y = 0,$ Bessel equation
(c) $xy'' + (1 - x)y' + \lambda y = 0,$ Laguerre equation
(d) $(1 - x^2)y'' - xy' + \alpha^2 y = 0,$ Tchebycheff equation

4. The equation

$$u_{tt} + cu_t + ku = a^2 u_{xx} + F(x,t), \tag{i}$$

where $a^2 > 0$, $c \geq 0$, and $k \geq 0$ are constants, is known as the *telegraph equation*. It arises in the study of an elastic string under tension (see Appendix B of Chapter 10). Equation (i) also occurs in other applications.

Assuming that $F(x,t) = 0$, let $u(x,t) = X(x)T(t)$, separate the variables in Eq. (i), and derive ordinary differential equations for X and T.

11.2 Linear Homogeneous Boundary Value Problems: Eigenvalues and Eigenfunctions

Consider the boundary value problem consisting of the differential equation

$$y'' + p(x)y' + q(x)y = 0, \qquad 0 < x < 1, \tag{1}$$

and the boundary conditions

$$a_1 y(0) + a_2 y'(0) = 0, \qquad b_1 y(1) + b_2 y'(1) = 0. \tag{2}$$

According to the definitions in Section 11.1, the boundary value problem (1), (2) is linear and homogeneous. It was stated in Section 3.1 that an initial value problem for the differential equation (1) has a unique solution on any interval about the initial point in which the functions p and q are continuous. No such sweeping statement can be made about the boundary value problem (1), (2). In the first place, all homogeneous problems have the solution $y = 0$. This trivial solution is usually of no interest, and the significant question is whether there are other, nontrivial, solutions. Assuming that $y = \phi(x)$ is such a solution, then it follows from the linear homogeneous nature of the problem that $y = k\phi(x)$ is also a solution for any value of the constant k. Thus, corresponding to a nontrivial solution ϕ there is an infinite family of solutions, each member of which is proportional to ϕ. For problems of the form (1), (2) it is possible to show that there is at most one such family of nontrivial solutions (see Section 11.3). However, for more general second order boundary value problems there may be two families of solutions corresponding to two linearly independent solutions ϕ_1 and ϕ_2. The three possibilities (no nontrivial solutions, a single infinite family of solutions, and a doubly infinite family of solutions) are illustrated by the following three examples.

EXAMPLE 1

Find $y = \phi(x)$ satisfying

$$\begin{gathered} y'' + y = 0, \\ y(0) = 0, \quad y(1) = 0. \end{gathered} \tag{3}$$

The general solution of the differential equation is

$$y = c_1 \sin x + c_2 \cos x.$$

Since $y(0) = c_2$, the first boundary condition requires that $c_2 = 0$. Then we have $y = c_1 \sin x$, and $y(1) = c_1 \sin 1$. Since $\sin 1 \neq 0$, the second boundary condition requires that $c_1 = 0$. Hence the only solution is $\phi(x) = 0$, the trivial solution.

EXAMPLE 2

Find $y = \phi(x)$ satisfying

$$\begin{gathered} y'' + \pi^2 y = 0, \\ y(0) = 0, \quad y(1) = 0. \end{gathered} \tag{4}$$

The general solution of the differential equation is now

$$y = c_1 \sin \pi x + c_2 \cos \pi x,$$

and the first boundary condition is satisfied if $c_2 = 0$. The function $y = c_1 \sin \pi x$ also satisfies the second boundary condition for any constant c_1 inasmuch as $\sin \pi = 0$. Hence $\phi(x) = c_1 \sin \pi x$, where c_1 is arbitrary. The given problem thus has a single infinite family of solutions, each of which is a multiple of $\sin \pi x$.

EXAMPLE 3

Find $y = \phi(x)$ satisfying

$$y'' + \pi^2 y = 0,$$

$$y(0) + y(1) = 0, \quad y'(0) + y'(1) = 0. \tag{5}$$

Again, the general solution of the differential equation is

$$y = c_1 \sin \pi x + c_2 \cos \pi x. \tag{6}$$

Applying the first boundary condition, we obtain

$$\begin{aligned} y(0) + y(1) &= c_1 \sin 0 + c_2 \cos 0 + c_1 \sin \pi + c_2 \cos \pi \\ &= 0 + c_2 + 0 - c_2 = 0. \end{aligned}$$

Thus the first boundary condition imposes no restriction on the arbitrary constants c_1 and c_2. In a similar way the second boundary condition is also satisfied for all values of c_1 and c_2. Thus the function $y = \phi(x)$ given by Eq. (6) is actually the most general solution of the boundary value problem. In this case there are two infinite families of solutions, one proportional to $\phi_1(x) = \sin \pi x$, the other proportional to $\phi_2(x) = \cos \pi x$, and ϕ itself is a linear combination of them. Note also that the boundary conditions are not of the form (2), since each involves both boundary points.

More generally we often wish to consider a boundary value problem in which the differential equation contains an arbitrary parameter λ; for example,

$$y'' + \lambda y = 0, \tag{7}$$

$$y(0) = 0, \quad y(1) = 0. \tag{8}$$

If $\lambda = 1$ or $\lambda = \pi^2$, we obtain the problems occurring in Examples 1 and 2, respectively. In these examples we found that if $\lambda = 1$, the problem (7), (8) has only the trivial solution, whereas if $\lambda = \pi^2$, then the problem has nontrivial solutions. This raises the question of what happens when λ takes on some other value.

The boundary value problem (7), (8) was considered in detail in Section 10.2, and it also appeared in Sections 10.7 and 10.8. In all of those places the parameter λ occurred as a separation constant. It was shown in Section 10.2 that nontrivial solutions of this problem exist only for certain real values of λ, namely,

$$\lambda = \pi^2, 4\pi^2, 9\pi^2, \ldots, n^2\pi^2, \ldots. \tag{9}$$

For other values of λ there is only the trivial solution. The values given by Eq. (9) are called *eigenvalues* of the boundary value problem (7), (8). The corresponding nontrivial solutions $\phi_n(x) = \sin n\pi x$ are called *eigenfunctions*. The existence of such values and functions is typical of linear homogeneous boundary value problems containing a parameter.

In problems arising from physics, the eigenvalues and eigenfunctions usually have important physical interpretations. For example, problem (7), (8) occurs in the study of the vibrating elastic string (Section 10.7). There the eigenvalues are proportional to the squares of the natural frequencies of vibration of the string, and the eigenfunctions describe the configuration of the string when it is vibrating in one of its natural modes.

Now let us take up the question of existence of eigenvalues and eigenfunctions for a more general problem. Consider the differential equation

$$y'' + p(x, \lambda)y' + q(x, \lambda)y = 0, \qquad 0 < x < 1, \tag{10}$$

whose coefficients depend on a parameter λ, and the boundary conditions (2). For which values of λ, if any, do there exist nontrivial solutions of the given boundary value problem? We will assume that the functions p and q are continuous for $0 \leq x \leq 1$ and for all values λ.

Clearly the general solution of Eq. (10) must depend on both x and λ, and can be written in the form

$$y = c_1 y_1(x, \lambda) + c_2 y_2(x, \lambda), \tag{11}$$

where y_1 and y_2 are a fundamental set of solutions of Eq. (10). Substituting for y in the boundary conditions (2) yields the equations

$$\begin{aligned} c_1\left[a_1 y_1(0,\lambda) + a_2 y_1'(0,\lambda)\right] + c_2\left[a_1 y_2(0,\lambda) + a_2 y_2'(0,\lambda)\right] &= 0, \\ c_1\left[b_1 y_1(1,\lambda) + b_2 y_1'(1,\lambda)\right] + c_2\left[b_1 y_2(1,\lambda) + b_2 y_2'(1,\lambda)\right] &= 0 \end{aligned} \tag{12}$$

for c_1 and c_2. This set of linear homogeneous equations has solutions (other than $c_1 - c_2 = 0$) if and only if the determinant of coefficients $D(\lambda)$ is zero, that is, if and only if

$$D(\lambda) = \begin{vmatrix} a_1 y_1(0,\lambda) + a_2 y_1'(0,\lambda) & a_1 y_2(0,\lambda) + a_2 y_2'(0,\lambda) \\ b_1 y_1(1,\lambda) + b_2 y_1'(1,\lambda) & b_1 y_2(1,\lambda) + b_2 y_2'(1,\lambda) \end{vmatrix} = 0. \tag{13}$$

Values of λ, if any, satisfying this determinantal equation are the eigenvalues of the boundary value problem (10), (2). Corresponding to each eigenvalue there is at least one nontrivial solution, an eigenfunction, which is determined at most to within an arbitrary multiplicative constant.

Because of the arbitrary way in which λ appears in the differential equation (10) it is difficult to say more about the eigenvalues and eigenfunctions of the boundary value problem (10), (2). Therefore, we now restrict ourselves to problems in which the coefficient p in Eq. (10) is independent of λ and the coefficient q is a linear function of λ. For a certain large class of problems of this kind there is a well-developed theory; this is the subject of later sections in this chapter. It is fortunate that this class of two-point boundary value problems also includes many of practical significance, such as those arising from the study of the heat conduction equation, the wave equation, or the potential equation in Chapter 10, or their generalizations mentioned in Section 11.1. The following example illustrates the calculation of eigenvalues and eigenfunctions.

EXAMPLE 4

Find the eigenvalues and the corresponding eigenfunctions of the boundary value problem

$$y'' + \lambda y = 0, \tag{14}$$

$$y(0) = 0, \quad y'(1) + y(1) = 0. \tag{15}$$

One place where this problem occurs is in the heat conduction problem in a bar of unit length. The boundary condition at $x = 0$ corresponds to a zero temperature there. The boundary condition at $x = 1$ corresponds to a radiation boundary condition; the rate of heat flow is proportional to the temperature there, and units are chosen so that the constant of proportionality is one (see Appendix A of Chapter 10).

The solution of the differential equation may have one of several forms, depending on λ, so it is necessary to consider several cases. First, if $\lambda = 0$, the general solution of the differential equation is

$$y = c_1 x + c_2. \tag{16}$$

The two boundary conditions require that

$$c_2 = 0, \qquad 2c_1 + c_2 = 0, \tag{17}$$

respectively. The only solution of Eqs. (17) is $c_1 = c_2 = 0$, so the boundary value problem has no nontrivial solution in this case. Hence $\lambda = 0$ is not an eigenvalue.

If $\lambda > 0$, then the general solution of the differential equation (14) is

$$y = c_1 \sin\sqrt{\lambda}\, x + c_2 \cos\sqrt{\lambda}\, x, \tag{18}$$

where $\sqrt{\lambda} > 0$. The boundary condition at $x = 0$ requires that $c_2 = 0$; from the boundary condition at $x = 1$ we then obtain the equation

$$c_1(\sin\sqrt{\lambda} + \sqrt{\lambda}\cos\sqrt{\lambda}) = 0.$$

For a nontrivial solution y we must have $c_1 \neq 0$, and thus λ must satisfy

$$\sin\sqrt{\lambda} + \sqrt{\lambda}\cos\sqrt{\lambda} = 0. \tag{19}$$

Note that if λ is such that $\cos\sqrt{\lambda} = 0$, then $\sin\sqrt{\lambda} \neq 0$, and Eq. (19) is not satisfied. Hence we may assume that $\cos\sqrt{\lambda} \neq 0$; dividing Eq. (19) by $\cos\sqrt{\lambda}$, we obtain

$$\sqrt{\lambda} = -\tan\sqrt{\lambda}. \tag{20}$$

The solutions of Eq. (20) can be determined numerically. They can also be found approximately by sketching the graphs of $f(\sqrt{\lambda}) = \sqrt{\lambda}$ and $g(\sqrt{\lambda}) = -\tan\sqrt{\lambda}$ for $\sqrt{\lambda} > 0$ on the same set of axes, and identifying the points of intersection of the two curves (see Figure 11.1). The point $\sqrt{\lambda} = 0$ is specifically excluded from this argument because the solution (18) is valid only for $\sqrt{\lambda} \neq 0$. Despite the fact that the curves intersect there, $\lambda = 0$ is not an eigenvalue, as we have already shown. The first positive solution of Eq. (20) is $\sqrt{\lambda_1} \cong 2.029$. As can be seen from Figure 11.1, the other roots are given with reasonable accuracy by $\sqrt{\lambda_n} \cong (2n - 1)\pi/2$ for $n = 2, 3, \ldots,$ the precision of this estimate improving as n increases. Hence the eigenvalues are

$$\lambda_1 \cong 4.116, \qquad \lambda_n \cong (2n - 1)^2\pi^2/4 \quad \text{for } n = 2, 3, \ldots. \tag{21}$$

Finally, since $c_2 = 0$, the eigenfunction corresponding to the eigenvalue λ_n is

$$\phi_n(x, \lambda_n) = k_n \sin\sqrt{\lambda_n}\, x; \qquad n = 1, 2, \ldots, \tag{22}$$

where the constant k_n remains arbitrary.

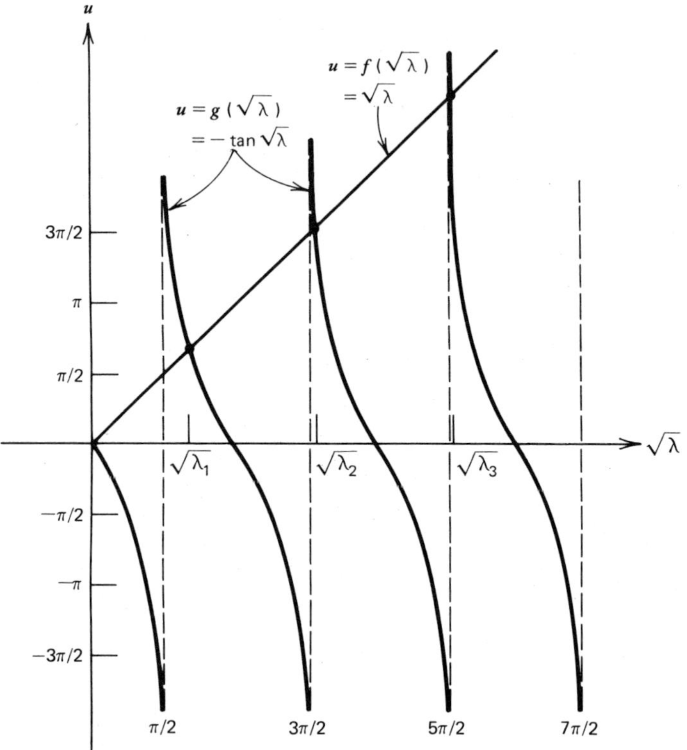

FIGURE 11.1 Graphical solution of $\sqrt{\lambda} = -\tan\sqrt{\lambda}$.

Next consider $\lambda < 0$. In this case it is convenient to let $\lambda = -\mu$ so that $\mu > 0$. Then Eq. (14) becomes

$$y'' - \mu y = 0, \tag{23}$$

and its general solution is

$$y = c_1 \sinh\sqrt{\mu}\,x + c_2 \cosh\sqrt{\mu}\,x, \tag{24}$$

where $\sqrt{\mu} > 0$. Proceeding as in the previous case, we find that μ must satisfy the equation

$$\sqrt{\mu} = -\tanh\sqrt{\mu}. \tag{25}$$

From Figure 11.2 it is clear that the graphs of $f(\sqrt{\mu}) = \sqrt{\mu}$ and $g(\sqrt{\mu}) = -\tanh\sqrt{\mu}$ intersect only at the origin. Hence there are no positive values of μ that satisfy Eq. (25), and hence the boundary value problem (14), (15) has no negative eigenvalues.

Finally, it is necessary to consider the possibility that λ may be complex. It is possible to show by direct calculation that the problem (14), (15) has no complex

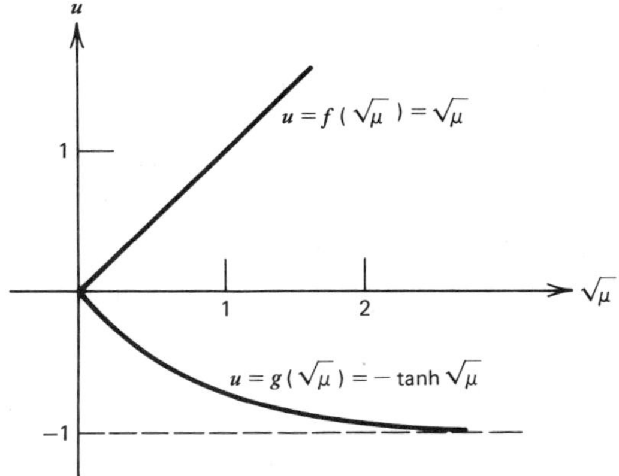

FIGURE 11.2 Graphical solution of $\sqrt{\mu} = -\tanh\sqrt{\mu}$.

eigenvalues. However, in Section 11.3 we consider in more detail a large class of problems that includes this example. One of the things we show there is that every problem in this class has only real eigenvalues. Therefore we omit the discussion of the nonexistence of complex eigenvalues here. Thus we conclude that all of the eigenvalues and eigenfunctions of the problem (14), (15) are given by Eqs. (21) and (22).

PROBLEMS

In each of Problems 1 through 4 find the eigenvalues and eigenfunctions of the given boundary value problem. Assume that all eigenvalues are real.

1. $y'' + \lambda y = 0$
   $y(0) = 0, \quad y'(1) = 0$
2. $y'' + \lambda y = 0$
   $y'(0) = 0, \quad y(1) = 0$
3. $y'' + \lambda y = 0$
   $y'(0) = 0, \quad y'(1) = 0$
4. $y'' - \lambda y = 0$
   $y(0) = 0, \quad y'(1) = 0$

In each of Problems 5 through 8 determine the form of the eigenfunctions and the determinantal equation satisfied by the nonzero eigenvalues. Determine whether $\lambda = 0$ is an eigenvalue, and find an approximate value for λ_1, the nonzero eigenvalue of smallest absolute value. Estimate λ_n for large values of n. Assume that all eigenvalues are real.

5. $y'' - \lambda y = 0$
   $y(0) + y'(0) = 0, \quad y(1) = 0$
6. $y'' + \lambda y = 0$
   $y(0) = 0, \quad y(\pi) + y'(\pi) = 0$
7. $y'' + \lambda y = 0$
   $y'(0) = 0, \quad y(1) + y'(1) = 0$
8. $y'' + \lambda y = 0$
   $y(0) - y'(0) = 0, \quad y(1) + y'(1) = 0$
9. Consider the boundary value problem

$$y'' - 2y' + (1 + \lambda)y = 0, \qquad y(0) = 0, \quad y(1) = 0.$$

(a) Introduce a new dependent variable u by the relation $y = s(x)u$. Determine $s(x)$ so that the differential equation for u has no u' term.
(b) Solve the boundary value problem for u and thereby determine the eigenvalues and eigenfunctions of the original problem. Assume that all eigenvalues are real.
(c) Also solve the given problem directly (without introducing u).

*10. Consider the boundary value problem

$$y'' + 4y' + (4 + 9\lambda)y = 0, \qquad y(0) = 0, \quad y'(l) = 0.$$

(a) Determine, at least approximately, the real eigenvalues and the corresponding eigenfunctions by proceeding as in Problem 9(a, b).
(b) Also solve the given problem directly (without introducing a new variable).
Hint: In part (a) be sure to pay attention to the boundary conditions as well as the differential equation.

The differential equations in Problems 11 and 12 differ from those in the previous problems in that the parameter λ multiplies the y' term as well as the y term. In each of these problems determine the real eigenvalues and the corresponding eigenfunctions.

11. $y'' + y' + \lambda(y' + y) = 0$
$y'(0) = 0, \quad y(1) = 0$

*12. $x^2y'' - \lambda(xy' - y) = 0$
$y(1) = 0, \quad y(2) - y'(2) = 0$

13. Consider the problem

$$y'' + \lambda y = 0, \qquad 2y(0) + y'(0) = 0, \quad y(1) = 0.$$

(a) Find the determinantal equation satisfied by the positive eigenvalues. Show that there is an infinite sequence of such eigenvalues and that $\lambda_n \cong (2n + 1)\pi/2$ for large n.
(b) Find the determinantal equation satisfied by the negative eigenvalues. Show that there is exactly one negative eigenvalue.

*14. Consider the problem

$$y'' + \lambda y = 0, \qquad \alpha y(0) + y'(0) = 0, \quad y(1) = 0,$$

where α is a given real constant.
(a) Show that for all values of α there is an infinite sequence of positive eigenvalues.
(b) If $\alpha < 1$, show that all (real) eigenvalues are positive. Show that the smallest eigenvalue approaches zero as α approaches one from below.
(c) Show that $\lambda = 0$ is an eigenvalue only if $\alpha = 1$.
(d) If $\alpha > 1$, show that there is exactly one negative eigenvalue and that this eigenvalue decreases as α increases.

15. Consider the problem

$$y'' + \lambda y = 0, \qquad y(0) = 0, \quad y'(l) = 0.$$

Show that if ϕ_m and ϕ_n are eigenfunctions, corresponding to the eigenvalues λ_m and λ_n, respectively, with $\lambda_m \neq \lambda_n$, then

$$\int_0^l \phi_m(x)\phi_n(x)\,dx = 0.$$

Hint: Note that

$$\phi_m'' + \lambda_m\phi_m = 0, \qquad \phi_n'' + \lambda_n\phi_n = 0.$$

Multiply the first of these equations by ϕ_n, the second by ϕ_m, and integrate from 0 to l, using integration by parts.

*16. In this problem we consider a higher order eigenvalue problem. In the study of transverse vibrations of a uniform elastic bar one is led to the differential equation

$$y^{\text{iv}} - \lambda y = 0,$$

where y is the transverse displacement and $\lambda = m\omega^2/EI$; m is the mass per unit length of the rod, E is Young's modulus, I is the moment of inertia of the cross section about an axis through the centroid perpendicular to the plane of vibration, and ω is the frequency of vibration. Thus for a bar whose material and geometric properties are given, the eigenvalues determine the natural frequencies of vibration. Boundary conditions at each end are usually one of the following types:

$$\begin{aligned} y = y' &= 0, && \text{clamped end,} \\ y = y'' &= 0, && \text{simply supported or hinged end,} \\ y'' = y''' &= 0, && \text{free end.} \end{aligned}$$

For each of the following three cases find the form of the eigenfunctions and the equation satisfied by the eigenvalues of this fourth order boundary value problem. Assume that the eigenvalues are real and positive.

(a) $y(0) = y''(0) = 0, \qquad y(l) = y''(l) = 0$
(b) $y(0) = y''(0) = 0, \qquad y(l) = y'(l) = 0$
(c) $y(0) = y'(0) = 0, \qquad y''(l) = y'''(l) = 0$ (cantilevered bar)

*17. This problem illustrates that the eigenvalue parameter sometimes appears in the boundary conditions as well as in the differential equation. Consider the longitudinal vibrations of a uniform straight elastic bar of length l. It can be shown that the axial displacement $u(x, t)$ satisfies the partial differential equation

$$(E/\rho)u_{xx} = u_{tt}; \qquad 0 < x < l, \quad t > 0 \tag{i}$$

where E is Young's modulus and ρ is the mass per unit volume. If the end $x = 0$ is fixed, then the boundary condition there is

$$u(0, t) = 0, \qquad t > 0. \tag{ii}$$

Suppose that the end $x = l$ is rigidly attached to a mass m but is otherwise unrestrained. We can obtain the boundary condition here by writing Newton's law for the mass. From the theory of elasticity it can be shown that the force exerted by the bar on the mass is given by $-EAu_x(l, t)$. Hence the boundary condition is

$$EAu_x(l, t) + mu_{tt}(l, t) = 0, \qquad t > 0. \tag{iii}$$

(a) Assume that $u(x, t) = X(x)T(t)$, and show that $X(x)$ and $T(t)$ satisfy the differential equations

$$X'' + \lambda X = 0, \tag{iv}$$

$$T'' + \lambda(E/\rho)T = 0. \tag{v}$$

(b) Show that the boundary conditions are

$$X(0) = 0, \qquad X'(l) - \gamma\lambda l X(l) = 0, \tag{vi}$$

where $\gamma = m/\rho A l$ is a dimensionless parameter that gives the ratio of the end mass to the mass of the rod.

Hint: Use the differential equation for $T(t)$ in simplifying the boundary condition at $x = l$.

(c) Determine the form of the eigenfunctions and the equation satisfied by the real eigenvalues of Eqs. (iv) and (vi).

11.3 Sturm-Liouville Boundary Value Problems

We now consider two-point boundary value problems of the type obtained in Section 11.1 by separating the variables in a heat conduction problem for a bar of variable material properties and with a source term proportional to the temperature. This kind of problem also occurs in many other applications.

These boundary value problems are commonly associated with the names of Sturm and Liouville.[1] They consist of a differential equation of the form

$$[p(x)y']' - q(x)y + \lambda r(x)y = 0 \tag{1}$$

on the interval $0 < x < 1$, together with the boundary conditions

$$a_1 y(0) + a_2 y'(0) = 0, \qquad b_1 y(1) + b_2 y'(1) = 0 \tag{2}$$

at the end points. It is often convenient to introduce the linear homogeneous differential operator L defined by

$$L[y] = -[p(x)y']' + q(x)y. \tag{3}$$

Then the differential equation (1) can be written as

$$L[y] = \lambda r(x)y. \tag{4}$$

We assume that the functions p, p', q, and r are continuous on the interval $0 \le x \le 1$ and, further, that $p(x) > 0$ and $r(x) > 0$ at all points in $0 \le x \le 1$. These assumptions are necessary to render the theory as simple as possible while retaining considerable generality. It turns out that these conditions are satisfied in many significant problems in mathematical physics. For example, the equation $y'' + \lambda y = 0$, which arose repeatedly in the previous chapter, is of the form (1) with $p(x) = 1$, $q(x) = 0$, and $r(x) = 1$. The boundary conditions (2) are said to be *separated*; that is, each involves only one of the boundary points. These are the most general separated boundary conditions that are possible for a second order differential equation.

Before proceeding to establish some of the properties of the Sturm-Liouville problem (1), (2), it is necessary to derive an identity, known as *Lagrange's identity*, which is basic to the study of linear boundary value problems. Let u and

[1] Charles Sturm (1803–1855) and Joseph Liouville (1809–1882), in a series of papers in 1836 and 1837, discovered many properties of the class of boundary value problems associated with their names, including the results stated in Theorems 11.1 to 11.4. Liouville is also known for his work in other areas of analysis and differential equations. In addition, he was the first to show (in 1844) the existence of transcendental numbers.

v be functions having continuous second derivatives on the interval $0 \le x \le 1$. Then[2]

$$\int_0^1 L[u]v\,dx = \int_0^1 \left\{ -(pu')'v + quv \right\} dx.$$

Integrating the first term on the right side twice by parts, we obtain

$$\begin{aligned}\int_0^1 L[u]v\,dx &= -p(x)u'(x)v(x)\Big|_0^1 + p(x)u(x)v'(x)\Big|_0^1 \\ &\quad + \int_0^1 \left\{ -u(pv')' + uqv \right\} dx \\ &= -p(x)\left[u'(x)v(x) - u(x)v'(x)\right]\Big|_0^1 + \int_0^1 uL[v]\,dx.\end{aligned}$$

Hence, on transposing the integral on the right side, we have

$$\int_0^1 \{L[u]v - uL[v]\}\,dx = -p(x)\left[u'(x)v(x) - u(x)v'(x)\right]\Big|_0^1, \tag{5}$$

which is Lagrange's identity.

Now let us suppose that the functions u and v in Eq. (5) also satisfy the boundary conditions (2). Then, assuming that $a_2 \ne 0$ and $b_2 \ne 0$, the right side of Eq. (5) becomes

$$\begin{aligned}&-p(x)\left[u'(x)v(x) - u(x)v'(x)\right]\Big|_0^1 \\ &\quad = -p(1)\left[u'(1)v(1) - u(1)v'(1)\right] + p(0)\left[u'(0)v(0) - u(0)v'(0)\right] \\ &\quad = -p(1)\left[-\frac{b_1}{b_2}u(1)v(1) + \frac{b_1}{b_2}u(1)v(1)\right] \\ &\qquad + p(0)\left[-\frac{a_1}{a_2}u(0)v(0) + \frac{a_1}{a_2}u(0)v(0)\right] \\ &\quad = 0.\end{aligned}$$

The same result holds if either a_2 or b_2 is zero; the proof in this case is even simpler, and is left to the reader. Thus, for the differential operator L defined by Eq. (3), and if the functions u and v satisfy the boundary conditions (2), Lagrange's identity reduces to

$$\int_0^1 \{L[u]v - uL[v]\}\,dx = 0. \tag{6}$$

Let us now write Eq. (6) in a slightly different way. In Section 10.3 we introduced the inner product (u, v) of two real-valued functions u and v on a given interval; using the interval $0 \le x \le 1$, we have

$$(u, v) = \int_0^1 u(x)v(x)\,dx. \tag{7}$$

[2] For brevity we sometimes use the notation $\int_0^1 f\,dx$ rather than $\int_0^1 f(x)\,dx$ in this chapter.

In this notation Eq. (6) becomes

$$(L[u], v) - (u, L[v]) = 0. \tag{8}$$

In proving Theorem 11.1 below it is necessary to deal with complex-valued functions. By analogy with the definition in Section 7.2 for vectors, we define the inner product of two complex-valued functions on $0 \le x \le 1$ as

$$(u, v) = \int_0^1 u(x)\bar{v}(x)\,dx, \tag{9}$$

where $\bar{v}$ is the complex conjugate of v. Clearly Eq. (9) coincides with Eq. (7) if $u(x)$ and $v(x)$ are real. It is important to know that Eq. (8) remains valid under the stated conditions if u and v are complex-valued functions and if the inner product (9) is used. To see this, one can start with the quantity $\int_0^1 L[u]\bar{v}\,dx$ and retrace the steps leading to Eq. (6), making use of the fact that $p(x)$, $q(x)$, a_1, a_2, b_1, and b_2 are all real quantities (see Problem 8).

We now consider some of the implications of Eq. (8) for the Sturm-Liouville boundary value problem (1), (2). We assume without proof[3] that this problem actually has eigenvalues and eigenfunctions. In Theorems 11.1 to 11.4 below, we state several of their important, but relatively elementary, properties. Each of these properties is illustrated by the very simple Sturm-Liouville problem

$$y'' + \lambda y = 0, \qquad y(0) = 0, \quad y(1) = 0, \tag{10}$$

whose eigenvalues are $\lambda_n = n^2\pi^2$, with the corresponding eigenfunctions $\phi_n(x) = \sin n\pi x$.

> ***Theorem 11.1.*** *All of the eigenvalues of the Sturm-Liouville problem* (1), (2) *are real.*

To prove this theorem let us suppose that λ is a (possibly complex) eigenvalue of the problem (1), (2) and that ϕ is a corresponding eigenfunction, also possibly complex valued. Let us write $\lambda = \mu + i\nu$ and $\phi(x) = U(x) + iV(x)$, where μ, ν, $U(x)$, and $V(x)$ are real. Then, if we let $u = \phi$ and also $v = \phi$ in Eq. (8), we have

$$(L[\phi], \phi) = (\phi, L[\phi]). \tag{11}$$

However, we know that $L[\phi] = \lambda r\phi$, so Eq. (11) becomes

$$(\lambda r\phi, \phi) = (\phi, \lambda r\phi). \tag{12}$$

Writing out Eq. (12) in full, using the definition (9) of the inner product, we obtain

$$\int_0^1 \lambda r(x)\phi(x)\bar{\phi}(x)\,dx = \int_0^1 \phi(x)\bar{\lambda}\bar{r}(x)\bar{\phi}(x)\,dx. \tag{13}$$

[3] The proof may be found, for example, in the references by Sagan (Chap. 5) or Birkhoff and Rota (Chap. 10).

Since $r(x)$ is real, Eq. (13) reduces to

$$(\lambda - \bar{\lambda})\int_0^1 r(x)\phi(x)\bar{\phi}(x)\,dx = 0,$$

or

$$(\lambda - \bar{\lambda})\int_0^1 r(x)\left[U^2(x) + V^2(x)\right]dx = 0. \tag{14}$$

The integrand in Eq. (14) is nonnegative and not identically zero. Since the integrand is also continuous, it follows that the integral is positive. Therefore, the factor $\lambda - \bar{\lambda} = 2i\nu$ must be zero. Hence $\nu = 0$ and λ is real, so the theorem is proved.

An important consequence of Theorem 11.1 is that in finding eigenvalues and eigenfunctions of a Sturm-Liouville boundary value problem, one need only look for real eigenvalues. It is also possible to show that the eigenfunctions of the boundary value problem (1), (2) are real. A proof is sketched in Problem 9.

Theorem 11.2. *If* ϕ_1 *and* ϕ_2 *are two eigenfunctions of the Sturm-Liouville problem* (1), (2) *corresponding to eigenvalues* λ_1 *and* λ_2, *respectively, and if* $\lambda_1 \neq \lambda_2$, *then*

$$\int_0^1 r(x)\phi_1(x)\phi_2(x)\,dx = 0. \tag{15}$$

This theorem expresses the property of *orthogonality* of the eigenfunctions with respect to the weight function r. To prove the theorem we note that ϕ_1 and ϕ_2 satisfy the differential equations

$$L[\phi_1] = \lambda_1 r\phi_1 \tag{16}$$

and

$$L[\phi_2] = \lambda_2 r\phi_2, \tag{17}$$

respectively. If we let $u = \phi_1$, $v = \phi_2$, and substitute for $L[u]$ and $L[v]$ in Eq. (8), we obtain

$$(\lambda_1 r\phi_1, \phi_2) - (\phi_1, \lambda_2 r\phi_2) = 0,$$

or, using Eq. (9),

$$\lambda_1\int_0^1 r(x)\phi_1(x)\bar{\phi}_2(x)\,dx - \bar{\lambda}_2\int_0^1 \phi_1(x)\bar{r}(x)\bar{\phi}_2(x)\,dx = 0.$$

Because λ_2, $r(x)$, and $\phi_2(x)$ are all real, this equation becomes

$$(\lambda_1 - \lambda_2)\int_0^1 r(x)\phi_1(x)\phi_2(x)\,dx = 0. \tag{18}$$

Since by hypothesis $\lambda_1 \neq \lambda_2$, it follows that ϕ_1 and ϕ_2 must satisfy Eq. (15), and the theorem is proved.

Theorem 11.3. *The eigenvalues of the Sturm-Liouville problem* (1), (2) *are all simple; that is, to each eigenvalue there corresponds only one linearly independent eigenfunction. Further, the eigenvalues form an infinite sequence, and can be ordered according to increasing magnitude so that*

$$\lambda_1 < \lambda_2 < \lambda_3 < \cdots < \lambda_n < \cdots$$

Moreover, $\lambda_n \to \infty$ *as* $n \to \infty$.

The proof of this theorem is somewhat more advanced than those of the two previous theorems, and will be omitted. However, a proof that the eigenvalues are simple is indicated in Problem 6.

Again we note that all of the properties stated in Theorems 11.1 to 11.3 are exemplified by the eigenvalues $\lambda_n = n^2\pi^2$ and eigenfunctions $\phi_n(x) = \sin n\pi x$ of the example problem (10). Clearly, the eigenvalues are real. The eigenfunctions satisfy the orthogonality relation

$$\int_0^1 \phi_m(x)\phi_n(x)\,dx = \int_0^1 \sin m\pi x \sin n\pi x\,dx = 0, \qquad m \neq n, \tag{19}$$

which was established in Section 10.3 by another method. Further, the eigenvalues can be ordered so that $\lambda_1 < \lambda_2 < \cdots$, and $\lambda_n \to \infty$ as $n \to \infty$. Finally, to each eigenvalue there corresponds a single linearly independent eigenfunction.

We will now assume that the eigenvalues of the Sturm-Liouville problem (1), (2) are ordered as indicated in Theorem 11.3. Associated with the eigenvalue λ_n is a corresponding eigenfunction ϕ_n, determined up to a multiplicative constant. It is often convenient to choose the arbitrary constant multiplying each eigenfunction so as to satisfy the condition

$$\int_0^1 r(x)\phi_n^2(x)\,dx = 1, \qquad n = 1, 2, \ldots. \tag{20}$$

Equation (20) is called a normalization condition, and eigenfunctions satisfying this condition are said to be *normalized*. Indeed, in this case, the eigenfunctions are said to form an *orthonormal set* (with respect to the weight function r) since they automatically satisfy the orthogonality relation (15). It is sometimes useful to combine Eqs. (15) and (20) into a single equation. To this end we introduce the symbol δ_{mn}, known as the Kronecker (1823–1891) delta and defined by

$$\delta_{mn} = \begin{cases} 0, & \text{if } m \neq n, \\ 1, & \text{if } m = n. \end{cases} \tag{21}$$

Making use of the Kronecker delta, we can write Eqs. (15) and (20) as

$$\int_0^1 r(x)\phi_m(x)\phi_n(x)\,dx = \delta_{mn}. \tag{22}$$

EXAMPLE 1

Determine the normalized eigenfunctions of the problem (10):

$$y'' + \lambda y = 0, \qquad y(0) = 0, \quad y(1) = 0.$$

The eigenvalues of this problem are $\lambda_1 = \pi^2$, $\lambda_2 = 4\pi^2, \ldots, \lambda_n = n^2\pi^2, \ldots$, and the corresponding eigenfunctions are $k_1 \sin \pi x$, $k_2 \sin 2\pi x, \ldots$, $k_n \sin n\pi x, \ldots$, respectively. In this case the weight function is $r(x) = 1$. To satisfy Eq. (20) we must choose k_n so that

$$\int_0^1 (k_n \sin n\pi x)^2\, dx = 1 \tag{23}$$

for each value of n. Since

$$k_n^2 \int_0^1 \sin^2 n\pi x\, dx = k_n^2 \int_0^1 \left(\tfrac{1}{2} - \tfrac{1}{2}\cos 2n\pi x\right) dx = \tfrac{1}{2}k_n^2,$$

Eq. (23) will be satisfied if k_n is chosen to be $\sqrt{2}$ for each value of n. Hence the normalized eigenfunctions of the given boundary value problem are

$$\phi_n(x) = \sqrt{2}\, \sin n\pi x, \qquad n = 1, 2, 3, \ldots. \tag{24}$$

EXAMPLE 2

Determine the normalized eigenfunctions of the problem

$$y'' + \lambda y = 0, \qquad y(0) = 0, \quad y'(1) + y(1) = 0. \tag{25}$$

In Example 4 of Section 11.2 we found that the eigenvalues λ_n satisfy the equation

$$\sin\sqrt{\lambda_n} + \sqrt{\lambda_n}\cos\sqrt{\lambda_n} = 0, \tag{26}$$

and that the corresponding eigenfunctions are

$$\phi_n(x) = k_n \sin\sqrt{\lambda_n}\, x, \tag{27}$$

where k_n is arbitrary. We can determine k_n from the normalization condition (20). Since $r(x) = 1$ in this problem, we have

$$\begin{aligned}
\int_0^1 \phi_n^2(x)\, dx &= k_n^2 \int_0^1 \sin^2\sqrt{\lambda_n}\, x\, dx \\
&= k_n^2 \int_0^1 \left(\frac{1}{2} - \frac{1}{2}\cos 2\sqrt{\lambda_n}\, x\right) dx = k_n^2 \left(\frac{x}{2} - \frac{\sin 2\sqrt{\lambda_n}\, x}{4\sqrt{\lambda_n}}\right)\Bigg|_0^1 \\
&= k_n^2 \frac{2\sqrt{\lambda_n} - \sin 2\sqrt{\lambda_n}}{4\sqrt{\lambda_n}} = k_n^2 \frac{\sqrt{\lambda_n} - \sin\sqrt{\lambda_n}\cos\sqrt{\lambda_n}}{2\sqrt{\lambda_n}} \\
&= k_n^2 \frac{1 + \cos^2\sqrt{\lambda_n}}{2},
\end{aligned}$$

where in the last step we have used Eq. (26). Hence, to normalize the eigenfunctions ϕ_n we must choose

$$k_n = \left(\frac{2}{1+\cos^2\sqrt{\lambda_n}}\right)^{1/2}. \tag{28}$$

The normalized eigenfunctions of the given problem are

$$\phi_n(x) = \frac{\sqrt{2}\sin\sqrt{\lambda_n}\,x}{\left(1+\cos^2\sqrt{\lambda_n}\right)^{1/2}}; \qquad n = 1, 2, \ldots. \tag{29}$$

We now turn to the question of expressing a given function f as a series of eigenfunctions of the Sturm-Liouville problem (1), (2). We have already seen examples of such expansions in Sections 10.3 to 10.5. For example, it was shown there that if f is continuous and has a piecewise continuous derivative on $0 \le x \le 1$, and satisfies the boundary conditions $f(0) = f(1) = 0$, then f can be expanded in a Fourier sine series of the form

$$f(x) = \sum_{n=1}^{\infty} b_n \sin n\pi x. \tag{30}$$

The functions $\sin n\pi x$, $n = 1, 2, \ldots$, are precisely the eigenfunctions of the boundary value problem (10). The coefficients b_n are given by

$$b_n = 2\int_0^1 f(x)\sin n\pi x\,dx \tag{31}$$

and the series (30) converges for each x in $0 \le x \le 1$. In a similar way f can be expanded in a Fourier cosine series using the eigenfunctions $\cos n\pi x$, $n = 0, 1, 2, \ldots$, of the boundary value problem $y'' + \lambda y = 0$, $y'(0) = 0$, $y'(1) = 0$.

Now suppose that a given function f, satisfying suitable conditions, can be expanded in an infinite series of eigenfunctions of the more general Sturm-Liouville problem (1), (2). If this can be done, then we have

$$f(x) = \sum_{n=1}^{\infty} c_n\phi_n(x), \tag{32}$$

where the $\phi_n(x)$ satisfy Eqs. (1), (2), and also the orthonormality condition (22). To compute the coefficients c_n in the series (32) we multiply Eq. (32) by $r(x)\phi_m(x)$ and integrate from $x = 0$ to $x = 1$. Assuming that the series can be integrated term by term we obtain

$$\int_0^1 r(x)f(x)\phi_m(x)\,dx = \sum_{n=1}^{\infty} c_n\int_0^1 r(x)\phi_m(x)\phi_n(x)\,dx = \sum_{n=1}^{\infty} c_n\delta_{mn}. \tag{33}$$

Hence, using the definition of δ_{mn},

$$c_m = \int_0^1 r(x)f(x)\phi_m(x)\,dx = (f, r\phi_m), \qquad m = 1, 2, \ldots. \tag{34}$$

The coefficients in the series (32) have thus been formally determined. Equation (34) has the same structure as the Euler-Fourier formulas for the coefficients in a Fourier series, and the eigenfunction series (32) also has convergence properties similar to those of Fourier series. The following theorem is analogous to Theorem 10.1 in Section 10.4.

Theorem 11.4. *Let $\phi_1, \phi_2, \ldots, \phi_n, \ldots$ be the normalized eigenfunctions of the Sturm-Liouville problem* (1), (2):

$$[p(x)y']' - q(x)y + \lambda r(x)y = 0,$$

$$a_1 y(0) + a_2 y'(0) = 0, \qquad b_1 y(1) + b_2 y'(1) = 0.$$

Let f and f' be piecewise continuous on $0 \le x \le 1$. Then the series (32) *whose coefficients c_m are given by Eq.* (34) *converges to $[f(x+) + f(x-)]/2$ at each point in the open interval $0 < x < 1$.*

If f satisfies further conditions, then a stronger conclusion can be established. Suppose that, in addition to the hypotheses of Theorem 11.4, the function f is continuous on $0 \le x \le 1$. If $a_2 = 0$ in Eq. (2a) [so that $\phi_n(0) = 0$], then let $f(0) = 0$. Similarly, if $b_2 = 0$ in Eq. (2b), let $f(1) = 0$. Otherwise no boundary conditions need be prescribed for f. Then the series (32) converges to $f(x)$ at each point in the closed interval $0 \le x \le 1$.

EXAMPLE 3

Expand the function

$$f(x) = x, \qquad 0 \le x \le 1 \tag{35}$$

in terms of the normalized eigenfunctions $\phi_n(x)$ of the problem (25).

In Example 2 we found the normalized eigenfunctions to be

$$\phi_n(x) = k_n \sin\sqrt{\lambda_n}\, x \tag{36}$$

where k_n is given by Eq. (28) and λ_n satisfies Eq. (26). To find the expansion for f in terms of the ϕ_n we write

$$f(x) = \sum_{n=1}^{\infty} c_n \phi_n(x), \tag{37}$$

where the coefficients are given by Eq. (34). Thus

$$c_n = \int_0^1 f(x)\phi_n(x)\,dx = k_n \int_0^1 x \sin\sqrt{\lambda_n}\, x\,dx.$$

Integrating by parts, we obtain

$$c_n = k_n\left(\frac{\sin\sqrt{\lambda_n}}{\lambda_n} - \frac{\cos\sqrt{\lambda_n}}{\sqrt{\lambda_n}}\right) = k_n \frac{2\sin\sqrt{\lambda_n}}{\lambda_n},$$

where we have used Eq. (26) in the last step. Upon substituting for k_n from Eq. (28) we obtain

$$c_n = \frac{2\sqrt{2}\sin\sqrt{\lambda_n}}{\lambda_n\left(1+\cos^2\sqrt{\lambda_n}\right)^{1/2}}. \tag{38}$$

Thus

$$f(x) = 4\sum_{n=1}^{\infty}\frac{\sin\sqrt{\lambda_n}\,\sin\sqrt{\lambda_n}\,x}{\lambda_n\left(1+\cos^2\sqrt{\lambda_n}\right)}. \tag{39}$$

Observe that although the right side of Eq. (39) is a series of sines, it is not a Fourier sine series, as discussed in Section 10.5.

SELF-ADJOINT PROBLEMS. Sturm-Liouville boundary value problems are of great importance in their own right, but they can also be viewed as belonging to a much more extensive class of problems that have many of the same properties. For example, there are many similarities between Sturm-Liouville problems and the algebraic system

$$\mathbf{Ax} = \lambda\mathbf{x}, \tag{40}$$

where the $n \times n$ matrix $\mathbf{A}$ is real symmetric or Hermitian. Comparing the results mentioned in Section 7.3 with those of this section, we note that in both cases the eigenvalues are real and the eigenfunctions or eigenvectors form an orthogonal set. Further, the eigenfunctions or eigenvectors can be used as the basis for expressing an essentially arbitrary function or vector, respectively, as a sum. The most important difference is that a matrix has only a finite number of eigenvalues and eigenvectors, while a Sturm-Liouville problem has infinitely many. It is interesting and of fundamental importance in mathematics that these seemingly different problems—the matrix problem (40) and the Sturm-Liouville problem (1), (2)—which arise in different ways, are actually part of a single underlying theory. This theory is usually referred to as linear operator theory, and is part of the subject of functional analysis.

We now point out some ways in which Sturm-Liouville problems can be generalized, while still preserving the main results of Theorems 11.1 to 11.4—the existence of a sequence of real eigenvalues tending to infinity, the orthogonality of the eigenfunctions, and the possibility of expressing an arbitrary function as a series of eigenfunctions. In making these generalizations it is essential that the crucial relation (8) remain valid.

Let us consider the boundary value problem consisting of the differential equation

$$L[y] = \lambda r(x)y, \qquad 0 < x < 1 \tag{41}$$

where

$$L[y] = P_n(x)\frac{d^n y}{dx^n} + \cdots + P_1(x)\frac{dy}{dx} + P_0(x)y, \tag{42}$$

and n linear homogeneous boundary conditions at the end points. If Eq. (8) is valid for every pair of sufficiently differentiable functions that satisfy the boundary conditions, then the given problem is said to be *self-adjoint*. It is important to observe that Eq. (8) involves restrictions both on the differential equation and also on the boundary conditions. The differential operator L must be such that the same operator appears in both terms of Eq. (8). This requires that L be of even order. Further, a second order operator must have the form (3), a fourth order operator must have the form

$$L[y] = [p(x)y'']'' - [q(x)y']' + s(x)y, \tag{43}$$

and higher order operators must have an analogous structure. In addition, the boundary conditions must be such as to eliminate the boundary terms that arise during the integration by parts used in deriving Eq. (8). For example, in a second order problem this is true for the separated boundary conditions (2) and also in certain other cases, one of which is given in Example 4 below.

Let us suppose that we have a self-adjoint boundary value problem for Eq. (41), where $L[y]$ is given now by Eq. (43). We assume that p, q, r, and s are continuous on $0 \le x \le 1$, and that the derivatives of p and q indicated in Eq. (43) are also continuous. If in addition $p(x) > 0$ and $r(x) > 0$ for $0 \le x \le 1$, then there is an infinite sequence of real eigenvalues tending to $+\infty$, the eigenfunctions are orthogonal with respect to the weight function r, and an arbitrary function can be expressed as a series of eigenfunctions. However, the eigenfunctions may not be simple in these more general problems.

We turn now to the relation between Sturm-Liouville problems and Fourier series. We have noted previously that Fourier sine and cosine series can be obtained by using the eigenfunctions of certain Sturm-Liouville problems involving the differential equation $y'' + \lambda y = 0$. This raises the question of whether we can obtain a full Fourier series, including both sine and cosine terms, by choosing a suitable set of boundary conditions. The answer is provided by the following example, which also serves to illustrate the occurrence of nonseparated boundary conditions.

EXAMPLE 4

Find the eigenvalues and eigenfunctions of the boundary value problem

$$y'' + \lambda y = 0, \tag{44}$$

$$y(-l) - y(l) = 0, \qquad y'(-l) - y'(l) = 0. \tag{45}$$

This is not a Sturm-Liouville problem because the boundary conditions are not separated. The boundary conditions (45) are called *periodic boundary conditions* since they require that y and y' assume the same values at $x = l$ as at $x = -l$. Nevertheless, it is straightforward to show that the problem (44), (45) is self-adjoint. A simple calculation establishes that $\lambda_0 = 0$ is an eigenvalue and that the corresponding eigenfunction is $\phi_0(x) = 1$. Further, there are additional eigenvalues $\lambda_1 = (\pi/l)^2$, $\lambda_2 = (2\pi/l)^2, \ldots, \lambda_n = (n\pi/l)^2, \ldots$. To each of these nonzero eigenvalues there correspond *two* linearly independent eigenfunctions; for example, corresponding to λ_n are the two eigenfunctions $\phi_n(x) = \cos(n\pi x/l)$ and $\psi_n(x) = \sin(n\pi x/l)$. This illustrates that the eigenvalues may not be simple unless the boundary conditions are separated. Further, if we seek to expand a given function f of period $2l$ in a series of eigenfunctions of the problem (44), (45), we obtain the series

$$f(x) = \frac{a_0}{2} + \sum_{n=1}^{\infty}\left(a_n \cos\frac{n\pi x}{l} + b_n \sin\frac{n\pi x}{l}\right),$$

which is just the Fourier series for f.

We will not give further consideration to problems that have nonseparated boundary conditions, nor will we deal with problems of higher than second order, except in a few problems. There is, however, one other kind of generalization that we do wish to discuss. That is the case in which the coefficients p, q, and r in Eq. (1) do not quite satisfy the rather strict continuity and positivity requirements laid down at the beginning of this section. Such problems are called singular Sturm-Liouville problems, and are the subject of Section 11.5.

PROBLEMS

1. Determine the normalized eigenfunctions of each of the following problems.

   (a) $y'' + \lambda y = 0, \quad y(0) = 0, \quad y'(1) = 0;$ see Section 11.2, Problem 1.
   (b) $y'' + \lambda y = 0, \quad y'(0) = 0, \quad y(1) = 0;$ see Section 11.2, Problem 2.
   (c) $y'' + \lambda y = 0, \quad y'(0) = 0, \quad y'(1) = 0;$ see Section 11.2, Problem 3.
   (d) $y'' + \lambda y = 0, \quad y'(0) = 0, \quad y'(1) + y(1) = 0;$ see Section 11.2, Problem 7.
   (e) $y'' - 2y' + (1 + \lambda)y = 0, \quad y(0) = 0, \quad y(1) = 0;$ see Section 11.2, Problem 9.

2. Find the eigenfunction expansion $\sum_{n=1}^{\infty} a_n\phi_n(x)$ of each of the following functions, using the normalized eigenfunctions of Problem 1(a).

   (a) $f(x) = 1, \quad 0 \le x \le 1$
   (b) $f(x) = x, \quad 0 \le x \le 1$
   (c) $f(x) = \begin{cases} 1, & 0 \le x < \frac{1}{2} \\ 0, & \frac{1}{2} \le x \le 1 \end{cases}$
   (d) $f(x) = \begin{cases} 2x, & 0 \le x < \frac{1}{2} \\ 1, & \frac{1}{2} \le x \le 1 \end{cases}$

3. Find the eigenfunction expansion $\sum_{n=1}^{\infty} a_n\phi_n(x)$ of each of the following functions, using the normalized eigenfunctions of Problem 1(d).

   (a) $f(x) = 1, \quad 0 \le x \le 1$
   (b) $f(x) = x, \quad 0 \le x \le 1$
   (c) $f(x) = 1 - x, \quad 0 \le x \le 1$
   (d) $f(x) = \begin{cases} 1, & 0 \le x < \frac{1}{2} \\ 0, & \frac{1}{2} \le x \le 1 \end{cases}$

4. Determine whether each of the following boundary value problems is self-adjoint.

(a) $y'' + y' + 2y = 0, \quad y(0) = 0, \quad y(1) = 0$
(b) $(1 + x^2)y'' + 2xy' + y = 0, \quad y'(0) = 0, \quad y(1) + 2y'(1) = 0$
(c) $y'' + y = \lambda y, \quad y(0) - y'(1) = 0, \quad y'(0) - y(1) = 0$
(d) $(1 + x^2)y'' + 2xy' + y = \lambda(1 + x^2)y, \quad y(0) - y'(1) = 0,$
$y'(0) + 2y(1) = 0$
(e) $y'' + \lambda y = 0, \quad y(0) = 0, \quad y(\pi) + y'(\pi) = 0$

5. Show that if the functions u and v satisfy Eqs. (2), and either $a_2 = 0$ or $b_2 = 0$, or both, then

$$p(x)[u'(x)v(x) - u(x)v'(x)]\Big|_0^1 = 0.$$

6. In this problem we outline a proof of the first part of Theorem 11.3: that the eigenvalues of the Sturm-Liouville problem (1), (2) are simple.

For a given λ suppose that ϕ_1 and ϕ_2 are two linearly independent eigenfunctions. Compute the Wronskian $W(\phi_1, \phi_2)(x)$ and use the boundary conditions (2) to show that $W(\phi_1, \phi_2)(0) = 0$. Then use Theorems 3.5 and 3.9 to conclude that ϕ_1 and ϕ_2 cannot be linearly independent as assumed.

7. Consider the Sturm-Liouville problem

$$-[p(x)y']' + q(x)y = \lambda r(x)y,$$

$$a_1 y(0) + a_2 y'(0) = 0, \qquad b_1 y(1) + b_2 y'(1) = 0,$$

where p, q, and r satisfy the conditions stated in the text.

(a) Show that if λ is an eigenvalue and ϕ a corresponding eigenfunctior, then

$$\lambda\int_0^1 r\phi^2\,dx = \int_0^1 (p\phi'^2 + q\phi^2)\,dx + \frac{b_1}{b_2}p(1)\phi^2(1) - \frac{a_1}{a_2}p(0)\phi^2(0)$$

provided that $a_2 \neq 0$ and $b_2 \neq 0$. How must this result be modified if $a_2 = 0$ or $b_2 = 0$?

(b) Show that if $q(x) \geq 0$ and if b_1/b_2 and $-a_1/a_2$ are nonnegative, then the eigenvalue λ is nonnegative.

(c) Under the conditions of part (b) show that the eigenvalue λ is strictly positive unless $q(x) = 0$ for each x in $0 \leq x \leq 1$ and also $a_1 = b_1 = 0$.

8. Derive Eq. (8), using the inner product (9) and assuming that u and v are complex-valued functions.

Hint: Consider the quantity $\int_0^1 L[u]\bar{v}\,dx$, split u and v into real and imaginary parts, and proceed as in the text.

9. In this problem we indicate a proof that the eigenfunctions of the Sturm-Liouville problem (1), (2) are real.

(a) Let λ be an eigenvalue and ϕ a corresponding eigenfunction. Let $\phi(x) = u(x) + iv(x)$, and show that u and v are also eigenfunctions corresponding to λ.

(b) Using Theorem 11.3, or the result of Problem 6, show that u and v are linearly dependent.

(c) Show that ϕ must be real, apart from an arbitrary multiplicative constant that may be complex.

10. Consider the problem (Problem 9 of Section 11.2)

$$y'' - 2y' + (1 + \lambda)y = 0, \qquad y(0) = 0, \quad y(1) = 0.$$

(a) Show that this problem is not self-adjoint.
(b) Show that, nevertheless, all eigenvalues are real. Thus the conclusion of Theorem 11.1 is also true for certain nonself-adjoint problems.
(c) Show that the eigenfunctions are not orthogonal (with respect to the weight function arising from the coefficient of λ in the differential equation).

*11. Consider the problem

$$x^2y'' = \lambda(xy' - y), \qquad y(1) = 0, \quad y(2) = 0.$$

Note that λ appears as a coefficient of y' as well as of y itself. It is possible to extend the definition of self-adjointness to this type of problem, and to show that this particular problem is not self-adjoint. Show that this problem has eigenvalues, but that none of them is real. This shows that in general nonself-adjoint problems may have eigenvalues that are not real.

Buckling of an Elastic Column. In an investigation of the buckling of a uniform elastic column of length l by an axial load P (Figure 11.3a) one is led to the differential equation

$$y^{\text{iv}} + \lambda y'' = 0, \qquad 0 < x < l. \tag{i}$$

The parameter $\lambda = P/EI$, where E is Young's modulus and I is the moment of inertia of the cross section about an axis through the centroid perpendicular to the xy plane. The boundary conditions at $x = 0$ and $x = l$ depend on how the ends of the column are supported. Typical boundary conditions are

$$y = y' = 0; \qquad \text{clamped end}$$

$$y = y'' = 0; \qquad \text{simply supported (hinged) end.}$$

The bar is shown in Figure 11.3a is simply supported at $x = 0$ and clamped at $x = l$. It is desired to determine the eigenvalues and eigenfunctions of Eq. (i) subject to suitable boundary conditions. In particular, the smallest eigenvalue λ_1 gives the load at which the column buckles, or can assume a curved equilibrium position, as shown in Figure 11.3b. The corresponding eigenfunction describes the configuration

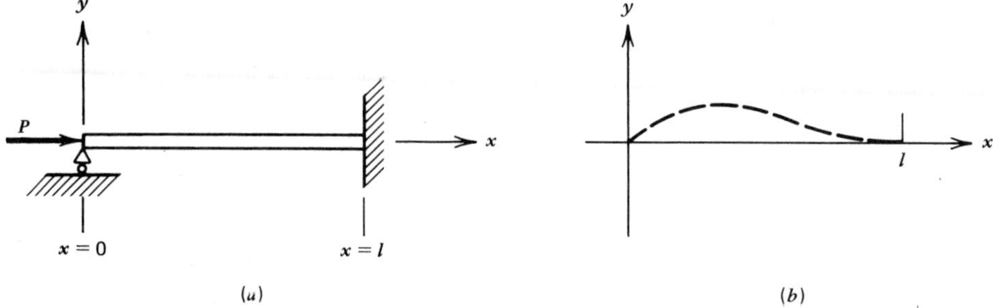

FIGURE 11.3 (a) A column under compression; (b) shape of the buckled column.

of the buckled column. Note that the differential equation (i) does not fall within the theory discussed in this section. It is possible to show, however, that in each of the cases given here the eigenvalues are all real and positive. Problems 12 and 13 deal with column buckling problems.

12. For each of the following boundary conditions find the smallest eigenvalue (the buckling load) of $y^{\text{iv}} + \lambda y'' = 0$, and also find the corresponding eigenfunction (the shape of the buckled column).
(a) $y(0) = y''(0) = 0, \quad y(l) = y''(l) = 0$
(b) $y(0) = y''(0) = 0, \quad y(l) = y'(l) = 0$
(c) $y(0) = y'(0) = 0, \quad y(l) = y'(l) = 0.$

*13. In some buckling problems the eigenvalue parameter appears in the boundary conditions as well as in the differential equation. One such case occurs when one end of the column is clamped and the other end is free. In this case the differential equation $y^{\text{iv}} + \lambda y'' = 0$ must be solved subject to the boundary conditions
$$y(0) = 0, \qquad y'(0) = 0, \qquad y''(l) = 0, \qquad y'''(l) + \lambda y'(l) = 0.$$
Find the smallest eigenvalue and the corresponding eigenfunction.

11.4 Nonhomogeneous Boundary Value Problems

In this section we discuss how to solve nonhomogeneous boundary value problems, both for ordinary and partial differential equations. Most of our attention is directed toward problems in which the differential equation alone is nonhomogeneous, while the boundary conditions are homogeneous. We assume that the solution can be expanded in a series of eigenfunctions of a related homogeneous problem, and then determine the coefficients in this series so that the nonhomogeneous problem is satisfied. We first describe this method as it applies to boundary value problems for second order linear ordinary differential equations. Later we illustrate its use for partial differential equations by solving a heat conduction problem in a bar with variable material properties and in the presence of source terms.

NONHOMOGENEOUS STURM-LIOUVILLE PROBLEMS. Consider the boundary value problem consisting of the nonhomogeneous differential equation
$$L[y] = -[p(x)y']' + q(x)y = \mu r(x)y + f(x), \tag{1}$$
where μ is a given constant and f is a given function on $0 \le x \le 1$, and the boundary conditions
$$a_1 y(0) + a_2 y'(0) = 0, \qquad b_1 y(1) + b_2 y'(1) = 0. \tag{2}$$
As in Section 11.3 we assume that p, p', q, and r are continuous on $0 \le x \le 1$ and that $p(x) > 0$ and $r(x) > 0$ there. We will solve the problem (1), (2) by making use of the eigenfunctions of the corresponding homogeneous problem consisting of the differential equation
$$L[y] = \lambda r(x)y \tag{3}$$

and the boundary conditions (2). Let $\lambda_1 < \lambda_2 < \cdots < \lambda_n < \cdots$ be the eigenvalues of this problem, and let $\phi_1, \phi_2, \ldots, \phi_n, \ldots$ be the corresponding normalized eigenfunctions.

We now assume that the solution $y = \phi(x)$ of the nonhomogeneous problem (1), (2) can be expressed as a series of the form

$$\phi(x) = \sum_{n=1}^{\infty} b_n \phi_n(x). \tag{4}$$

From Eq. (34) of Section 11.3 we know that

$$b_n = \int_0^1 r(x)\phi(x)\phi_n(x)\,dx, \qquad n = 1, 2, \ldots. \tag{5}$$

However, since we do not know $\phi(x)$, we cannot use Eq. (5) to calculate b_n. Instead, we will try to determine b_n so that the problem (1), (2) is satisfied, and then use Eq. (4) to find $\phi(x)$. Note first that ϕ as given by Eq. (4) always satisfies the boundary conditions (2) since each ϕ_n does.

Now consider the differential equation that ϕ must satisfy. This is just Eq. (1) with y replaced by ϕ:

$$L[\phi](x) = \mu r(x)\phi(x) + f(x). \tag{6}$$

We substitute the series (4) into the differential equation (6) and attempt to determine b_n so that the differential equation is satisfied. The term on the left side of Eq. (6) becomes

$$\begin{aligned} L[\phi](x) &= L\left[\sum_{n=1}^{\infty} b_n \phi_n\right](x) = \sum_{n=1}^{\infty} b_n L[\phi_n](x) \\ &= \sum_{n=1}^{\infty} b_n \lambda_n r(x)\phi_n(x), \end{aligned} \tag{7}$$

where we have assumed that we can interchange the operations of summation and differentiation.

Note that the function r appears in Eq. (7) and also in the term $\mu r(x)\phi(x)$ in Eq. (6). This suggests that we rewrite the nonhomogeneous term in Eq. (6) as $r(x)[f(x)/r(x)]$ so that $r(x)$ also appears as a multiplier in this term. If the function f/r satisfies the conditions of Theorem 11.4, then

$$\frac{f(x)}{r(x)} = \sum_{n=1}^{\infty} c_n \phi_n(x), \tag{8}$$

where, using Eq. (5) with ϕ replaced by f/r,

$$c_n = \int_0^1 r(x)\frac{f(x)}{r(x)}\phi_n(x)\,dx = \int_0^1 f(x)\phi_n(x)\,dx, \qquad n = 1, 2, \ldots. \tag{9}$$

Upon substituting for $\phi(x)$, $L[\phi](x)$, and $f(x)$ in Eq. (6) from Eqs. (4), (7), and (8), respectively, we find that

$$\sum_{n=1}^{\infty} b_n \lambda_n r(x)\phi_n(x) = \mu r(x) \sum_{n=1}^{\infty} b_n \phi_n(x) + r(x) \sum_{n=1}^{\infty} c_n \phi_n(x).$$

After collecting terms and canceling the common nonzero factor $r(x)$ we have

$$\sum_{n=1}^{\infty} [(\lambda_n - \mu)b_n - c_n]\phi_n(x) = 0. \tag{10}$$

If Eq. (10) is to hold for each x in the interval $0 \le x \le 1$, then each coefficient in the series must separately be zero; see Problem 4 for a proof of this fact. Hence

$$(\lambda_n - \mu)b_n - c_n = 0; \qquad n = 1, 2, \ldots. \tag{11}$$

We must now distinguish two main cases, one of which also has two subcases.

First suppose that $\mu \neq \lambda_n$ for $n = 1, 2, 3, \ldots$; that is, μ is not equal to any eigenvalue of the corresponding homogeneous problem. Then

$$b_n = \frac{c_n}{\lambda_n - \mu}, \qquad n = 1, 2, 3, \ldots, \tag{12}$$

and

$$y = \phi(x) = \sum_{n=1}^{\infty} \frac{c_n}{\lambda_n - \mu}\phi_n(x). \tag{13}$$

Equation (13), with c_n given by Eq. (9), is a formal solution of the nonhomogeneous boundary value problem (1), (2). Our argument does not prove that the series (13) converges. However, any solution of the boundary value problem (1), (2) clearly satisfies the conditions of Theorem 11.4; indeed, it satisfies the more stringent conditions given in the paragraph following that theorem. Thus it is reasonable to expect that the series (13) does converge at each point, and this fact can be established provided, for example, that f is continuous.

Now suppose that μ is equal to one of the eigenvalues of the corresponding homogeneous problem, say $\mu = \lambda_m$; then the situation is quite different. In this event, for $n = m$ Eq. (11) has the form $0 \cdot b_m - c_m = 0$. Again we must consider two cases.

If $\mu = \lambda_m$ and $c_m \neq 0$, then it is impossible to solve Eq. (11) for b_m, and the nonhomogeneous problem (1), (2) has no solution.

If $\mu = \lambda_m$ and $c_m = 0$, then Eq. (11) is satisfied regardless of the value of b_m; in other words, b_m remains arbitrary. In this case the boundary value problem (1), (2) does have a solution, but it is not unique, since it contains an arbitrary multiple of the eigenfunction ϕ_m.

Whether or not $c_m = 0$ depends on the nonhomogeneous term f. In particular,

$$c_m = \int_0^1 f(x)\phi_m(x)\,dx. \tag{14}$$

Thus, if $\mu = \lambda_m$ the nonhomogeneous boundary value problem (1), (2) can be solved only if f is orthogonal to the eigenfunction corresponding to the eigenvalue λ_m.

The results we have formally obtained are summarized in the following theorem.

> ***Theorem 11.5.*** *The nonhomogeneous boundary value problem* (1), (2) *has a unique solution for each continuous f whenever μ is different from all of the eigenvalues of the corresponding homogeneous problem; the solution is given by Eq.* (13), *and the series converges for each x in $0 \le x \le 1$. If μ is equal to an eigenvalue λ_m of the corresponding homogeneous problem, then the nonhomogeneous boundary value problem has no solution unless f is orthogonal to ϕ_m; that is, unless the condition* (14) *holds; in that case, the solution is not unique and contains an arbitrary multiple of $\phi_m(x)$.*

The main part of Theorem 11.5 is sometimes stated in the following way.

> ***Theorem 11.6.*** *For a given value of μ, either the nonhomogeneous problem* (1), (2) *has a unique solution for each continuous f (if μ is not equal to any eigenvalue λ_m of the corresponding homogeneous problem), or else the homogeneous problem* (3), (2) *has a nontrivial solution (the eigenfunction corresponding to λ_m).*

This latter form of the theorem is known as the Fredholm[4] alternative theorem. This is one of the basic theorems of mathematical analysis and occurs in many different contexts. The reader is probably familiar with it in connection with sets of linear algebraic equations, where the vanishing or nonvanishing of the determinant of coefficients replaces the statements about μ and λ_m. See the discussion in Section 7.3.

EXAMPLE 1

Solve the boundary value problem

$$y'' + 2y = -x, \tag{15}$$

$$y(0) = 0, \quad y(1) + y'(1) = 0. \tag{16}$$

This particular problem can be solved directly in an elementary way and has the solution

$$y = \frac{\sin\sqrt{2}\,x}{\sin\sqrt{2} + \sqrt{2}\cos\sqrt{2}} - \frac{x}{2}. \tag{17}$$

[4]The Swedish mathematician Erik Ivar Fredholm (1866–1927), professor at the University of Stockholm, established the modern theory of integral equations in a fundamental paper in 1903. Fredholm's work emphasized the similarities between integral equations and systems of linear algebraic equations. There are also many interrelations between differential and integral equations; for example, see Section 2.12 and Problem 15 of Section 6.5.

The method of solution described below illustrates the use of eigenfunction expansions, a method that can be employed in many problems not accessible by elementary procedures. To identify Eq. (15) with Eq. (1) it is helpful to write the former as

$$-y'' = 2y + x. \tag{18}$$

We seek the solution of the given problem as a series of normalized eigenfunctions ϕ_n of the corresponding homogeneous problem

$$y'' + \lambda y = 0, \qquad y(0) = 0, \quad y(1) + y'(1) = 0. \tag{19}$$

These eigenfunctions were found in Example 2 of Section 11.3, and are

$$\phi_n(x) = k_n \sin\sqrt{\lambda_n}\, x, \tag{20}$$

where

$$k_n = \left(\frac{2}{1 + \cos^2\sqrt{\lambda_n}}\right)^{1/2} \tag{21}$$

and λ_n satisfies

$$\sin\sqrt{\lambda_n} + \sqrt{\lambda_n}\cos\sqrt{\lambda_n} = 0. \tag{22}$$

We assume that y is given by Eq. (4),

$$y = \sum_{n=1}^{\infty} b_n\phi_n(x),$$

and it follows that the coefficients b_n are found from Eq. (12),

$$b_n = \frac{c_n}{\lambda_n - 2},$$

where the c_n are the expansion coefficients of the nonhomogeneous term $f(x) = x$ in Eq. (18) in terms of the eigenfunctions ϕ_n. These coefficients were found in Example 3 of Section 11.3, and are

$$c_n = \frac{2\sqrt{2}\sin\sqrt{\lambda_n}}{\lambda_n\left(1 + \cos^2\sqrt{\lambda_n}\right)^{1/2}}. \tag{23}$$

Putting everything together, we finally obtain the solution

$$y = 4\sum_{n=1}^{\infty} \frac{\sin\sqrt{\lambda_n}}{\lambda_n(\lambda_n - 2)\left(1 + \cos^2\sqrt{\lambda_n}\right)} \sin\sqrt{\lambda_n}\, x. \tag{24}$$

While Eqs. (17) and (24) are quite different in appearance, they are actually two different expressions for the same function. This follows from the uniqueness part of Theorems 11.5 or 11.6 since $\lambda = 2$ is not an eigenvalue of the homogeneous problem (19). Alternatively, one can show the equivalence of Eqs. (17) and

(24) by expanding the right side of Eq. (17) in terms of the eigenfunctions $\phi_n(x)$. For this problem it is fairly obvious that the formula (17) is more convenient than the series formula (24). However, we emphasize again that in other problems we may not be able to obtain the solution except by series (or numerical) methods.

NONHOMOGENEOUS HEAT CONDUCTION PROBLEMS. To show how eigenfunction expansions can be used to solve nonhomogeneous problems for partial differential equations, let us consider the generalized heat conduction equation

$$r(x)u_t = [p(x)u_x]_x - q(x)u + F(x,t) \tag{25}$$

with the boundary conditions

$$u_x(0,t) - h_1u(0,t) = 0, \qquad u_x(1,t) + h_2u(1,t) = 0 \tag{26}$$

and the initial condition

$$u(x,0) = f(x). \tag{27}$$

This problem was previously discussed in Appendix A of Chapter 10 and in Section 11.1. In the latter section we let $u(x,t) = X(x)T(t)$ in the homogeneous equation obtained by setting $F(x,t) = 0$, and showed that $X(x)$ must be a solution of the boundary value problem

$$-[p(x)X']' + q(x)X = \lambda r(x)X, \tag{28}$$

$$X'(0) - h_1X(0) = 0, \qquad X'(1) + h_2X(1) = 0. \tag{29}$$

Assuming that p, q, and r satisfy the proper continuity requirements and that $p(x)$ and $r(x)$ are always positive, the problem (28), (29) is a Sturm-Liouville problem as discussed in Section 11.3. Thus we obtain a sequence of eigenvalues $\lambda_1 < \lambda_2 < \cdots < \lambda_n < \cdots$ and corresponding normalized eigenfunctions $\phi_1(x), \phi_2(x), \ldots, \phi_n(x), \ldots$.

We will solve the given nonhomogeneous boundary value problem (25) to (27) by assuming that $u(x,t)$ can be expressed as a series of eigenfunctions,

$$u(x,t) = \sum_{n=1}^{\infty} b_n(t)\phi_n(x), \tag{30}$$

and then showing how to determine the coefficients $b_n(t)$. The procedure is basically the same as that used in the problem considered earlier in this section, although it is more complicated in certain respects. For instance, the coefficients b_n must now depend on t, because otherwise u would be a function of x only. Note that the boundary conditions (26) are automatically satisfied by an expression of the form (30) because each $\phi_n(x)$ satisfies the boundary conditions (29).

Next we substitute from Eq. (30) for u in Eq. (25). From the first two terms on the right side of Eq. (25) we formally obtain

$$[p(x)u_x]_x - q(x)u = \frac{\partial}{\partial x}\left[p(x)\sum_{n=1}^{\infty} b_n(t)\phi_n'(x)\right] - q(x)\sum_{n=1}^{\infty} b_n(t)\phi_n(x)$$

$$= \sum_{n=1}^{\infty} b_n(t)\left\{[p(x)\phi_n'(x)]' - q(x)\phi_n(x)\right\}. \tag{31}$$

Since $[p(x)\phi_n'(x)]' - q(x)\phi_n(x) = -\lambda_n r(x)\phi_n(x)$, we obtain finally

$$[p(x)u_x]_x - q(x)u = -r(x)\sum_{n=1}^{\infty} b_n(t)\lambda_n\phi_n(x). \tag{32}$$

Now consider the term on the left side of Eq. (25). We have

$$r(x)u_t = r(x)\frac{\partial}{\partial t}\sum_{n=1}^{\infty} b_n(t)\phi_n(x)$$

$$= r(x)\sum_{n=1}^{\infty} b_n'(t)\phi_n(x). \tag{33}$$

We must also express the nonhomogeneous term in Eq. (25) as a series of eigenfunctions. Once again, it is convenient to look at the ratio $F(x,t)/r(x)$ and to write

$$\frac{F(x,t)}{r(x)} = \sum_{n=1}^{\infty} \gamma_n(t)\phi_n(x), \tag{34}$$

where the coefficients are given by

$$\gamma_n(t) = \int_0^1 r(x)\frac{F(x,t)}{r(x)}\phi_n(x)\,dx$$

$$= \int_0^1 F(x,t)\phi_n(x)\,dx, \qquad n = 1, 2, \ldots. \tag{35}$$

Since $F(x,t)$ is given, we can consider the functions $\gamma_n(t)$ to be known.

Gathering all of these results together, we substitute from Eqs. (32), (33), and (34) in Eq. (25), and find that

$$r(x)\sum_{n=1}^{\infty} b_n'(t)\phi_n(x) = -r(x)\sum_{n=1}^{\infty} b_n(t)\lambda_n\phi_n(x) + r(x)\sum_{n=1}^{\infty}\gamma_n(t)\phi_n(x). \tag{36}$$

To simplify Eq. (36) we cancel the common nonzero factor $r(x)$ from all terms, and write everything in one summation:

$$\sum_{n=1}^{\infty}\left[b_n'(t) + \lambda_n b_n(t) - \gamma_n(t)\right]\phi_n(x) = 0. \tag{37}$$

Once again, if Eq. (37) is to hold for all x in $0 < x < 1$, it is necessary that the quantity in square brackets be zero for each n (again see Problem 4). Hence $b_n(t)$ is a solution of the first order linear ordinary differential equation

$$b_n'(t) + \lambda_n b_n(t) = \gamma_n(t), \qquad n = 1, 2, \ldots, \tag{38}$$

where $\gamma_n(t)$ is given by Eq. (35). To determine $b_n(t)$ completely we must have an initial condition

$$b_n(0) = \alpha_n, \qquad n = 1, 2, \ldots \tag{39}$$

for Eq. (38). This we obtain from the initial condition (27). Setting $t = 0$ in Eq. (30) and using Eq. (27), we have

$$u(x,0) = \sum_{n=1}^{\infty} b_n(0)\phi_n(x) = \sum_{n=1}^{\infty} \alpha_n\phi_n(x) = f(x). \tag{40}$$

Thus the initial values α_n are the coefficients in the eigenfunction expansion for $f(x)$. Therefore,

$$\alpha_n = \int_0^1 r(x)f(x)\phi_n(x)\,dx, \qquad n = 1, 2, \ldots. \tag{41}$$

Note that everything on the right side of Eq. (41) is known, so we can consider α_n as known.

The initial value problem (38), (39) is solved by the methods of Section 2.1. The integrating factor is $\mu(t) = \exp(\lambda_n t)$, and it follows that

$$b_n(t) = \alpha_n e^{-\lambda_n t} + \int_0^t e^{-\lambda_n(t-s)}\gamma_n(s)\,ds, \qquad n = 1, 2, \ldots. \tag{42}$$

The details of this calculation are left for the reader. Note that the first term on the right side of Eq. (42) depends on the function f through the coefficients α_n, while the second depends on the nonhomogeneous term F through the coefficients $\gamma_n(s)$.

Thus an explicit solution of the boundary value problem (25) to (27)

$$r(x)u_t = [p(x)u_x]_x - q(x)u + F(x,t),$$
$$u_x(0,t) - h_1 u(0,t) = 0, \quad u_x(1,t) + h_2 u(1,t) = 0,$$
$$u(x,0) = f(x),$$

is given by Eq. (30),

$$u(x,t) = \sum_{n=1}^{\infty} b_n(t)\phi_n(x),$$

where the coefficients $b_n(t)$ are determined from Eq. (42). The quantities α_n and $\gamma_n(s)$ in Eq. (42) are found in turn from Eqs. (41) and (35), respectively.

In summary, to use this method to solve a boundary value problem such as that given by Eqs. (25) to (27) we must

1. Find the eigenvalues λ_n and eigenfunctions ϕ_n of the appropriate homogeneous problem (28), (29).

2. Calculate the coefficients α_n and $\gamma_n(t)$ from Eqs. (41) and (35), respectively.
3. Evaluate the integral in Eq. (42) to determine $b_n(t)$.
4. Sum the infinite series (30).

Since any or all of these steps may be difficult, the entire process can be quite formidable. One redeeming feature is that often the series (30) converges rapidly, in which case only a very few terms may be needed to obtain an adequate approximation to the solution.

EXAMPLE 2

Find the solution of the heat conduction problem

$$u_t = u_{xx} + xe^{-t}, \tag{43}$$

$$u(0,t) = 0, \quad u_x(1,t) + u(1,t) = 0, \tag{44}$$

$$u(x,0) = 0. \tag{45}$$

Again we use the normalized eigenfunctions ϕ_n of the problem (19), and assume that u is given by Eq. (30),

$$u(x,t) = \sum_{n=1}^{\infty} b_n(t)\phi_n(x).$$

The coefficients b_n are determined from the differential equation

$$b_n' + \lambda_n b_n = \gamma_n(t), \tag{46}$$

where λ_n is the nth eigenvalue of the problem (19) and the $\gamma_n(t)$ are the expansion coefficients of the nonhomogeneous term xe^{-t} in terms of the eigenfunctions ϕ_n. Thus we have

$$\gamma_n(t) = \int_0^1 xe^{-t}\phi_n(x)\,dx = e^{-t}\int_0^1 x\phi_n(x)\,dx$$

$$= c_n e^{-t}, \tag{47}$$

where $c_n = \int_0^1 x\phi_n(x)\,dx$ is given by Eq. (23). The initial condition for Eq. (46) is

$$b_n(0) = 0 \tag{48}$$

since the initial temperature distribution (45) is zero everywhere. The solution of the initial value problem (46), (48) is

$$b_n(t) = e^{-\lambda_n t}\int_0^t e^{\lambda_n s} c_n e^{-s}\,ds = c_n e^{-\lambda_n t}\frac{e^{(\lambda_n - 1)t} - 1}{\lambda_n - 1}$$

$$= \frac{c_n}{\lambda_n - 1}\left(e^{-t} - e^{-\lambda_n t}\right). \tag{49}$$

Thus the solution of the heat conduction problem (43) to (45) is given by

$$u(x,t) = 4\sum_{n=1}^{\infty} \frac{\sin\sqrt{\lambda_n}\left(e^{-t} - e^{-\lambda_n t}\right)\sin\sqrt{\lambda_n}\,x}{\lambda_n(\lambda_n - 1)\left(1 + \cos^2\sqrt{\lambda_n}\right)}. \tag{50}$$

The solution given by Eq. (50) is exact, but complicated. To judge whether a satisfactory approximation to the solution can be obtained by using only a few terms in this series, we must estimate its speed of convergence. First we split the right side of Eq. (50) into two parts:

$$u(x,t) = 4e^{-t}\sum_{n=1}^{\infty}\frac{\sin\sqrt{\lambda_n}\,\sin\sqrt{\lambda_n}\,x}{\lambda_n(\lambda_n - 1)\left(1 + \cos^2\sqrt{\lambda_n}\right)}$$
$$-4\sum_{n=1}^{\infty}\frac{e^{-\lambda_n t}\sin\sqrt{\lambda_n}\,\sin\sqrt{\lambda_n}\,x}{\lambda_n(\lambda_n - 1)\left(1 + \cos^2\sqrt{\lambda_n}\right)}. \tag{51}$$

Recall from Example 4 in Section 11.2 that the eigenvalues λ_n are very nearly proportional to n^2. In the first series on the right side of Eq. (51) the trigonometric factors are all bounded as $n \to \infty$; therefore this series converges similarly to the series $\sum_{n=1}^{\infty}\lambda_n^{-2}$ or $\sum_{n=1}^{\infty}n^{-4}$. Hence at most two or three terms are required to obtain an excellent approximation to this part of the solution. The second series contains the additional factor $e^{-\lambda_n t}$, so its convergence is even more rapid for $t > 0$; all terms after the first are almost surely negligible.

FURTHER DISCUSSION. Eigenfunction expansions can be used to solve a much greater variety of problems than the preceding discussion and examples may suggest. The following brief remarks are intended to indicate to some extent the scope of this method.

1. **Nonhomogeneous Time-Independent Boundary Conditions.** Suppose that the boundary conditions (26) in the heat conduction problem are replaced by
$$u(0,t) = T_1, \quad u(1,t) = T_2. \tag{52}$$
Then we can proceed much as in Section 10.6 to reduce the problem to one with homogeneous boundary conditions; that is, we subtract from u a function v that is chosen to satisfy Eqs. (52). Then the difference $w = u - v$ satisfies a problem with homogeneous boundary conditions, but with a modified forcing term and initial condition. This problem can be solved by the procedure described in this section.
2. **Nonhomogeneous Time-Dependent Boundary Conditions.** Suppose that the boundary conditions (26) are replaced by
$$u(0,t) = T_1(t), \quad u(1,t) = T_2(t). \tag{53}$$
If T_1 and T_2 are differentiable functions, one can proceed just as in paragraph 1. In the same way it is possible to deal with more general boundary conditions of the radiation type. If T_1 and T_2 are not differentiable, however, then the method fails. A further refinement of the method of eigenfunction expansions can be used to cope with this type of problem, but the procedure is more complicated and we will not discuss it.

3. **Use of Functions Other Than Eigenfunctions.** One potential difficulty in using eigenfunction expansions is that the normalized eigenfunctions of the corresponding homogeneous problem must be found. For a differential equation with variable coefficients this may be difficult, if not impossible. In such a case it is sometimes possible to use other functions, such as eigenfunctions of a simpler problem, that satisfy the same boundary conditions. For instance, if the boundary conditions are

$$u(0,t) = 0, \quad u(1,t) = 0, \tag{54}$$

then it may be convenient to replace the functions $\phi_n(x)$ in Eq. (30) by $\sin n\pi x$. These functions at least satisfy the correct boundary conditions, although in general they are not solutions of the corresponding homogeneous differential equation. Next we expand the nonhomogeneous term $F(x,t)$ in a series of the form (34), again with $\phi_n(x)$ replaced by $\sin n\pi x$, and then substitute for both u and F in Eq. (25). Upon collecting the coefficients of $\sin n\pi x$ for each n, we have an infinite set of linear first order differential equations from which to determine $b_1(t), b_2(t), \ldots$. The essential difference between this case and the one considered earlier is that now the equations for the $b_n(t)$ are *coupled*. Thus they cannot be solved one by one, as before, but must be dealt with simultaneously. In practice, the infinite system is replaced by an approximating finite system, from which approximations to a finite number of coefficients are calculated. Despite its complexity, this procedure has proved very useful in solving certain types of difficult problems.
4. **Higher Order Problems.** Boundary value problems for equations of higher than second order can often be solved by eigenfunction expansions. In some cases the procedure parallels almost exactly that for second order problems. However, a variety of complications can also arise, and we will not discuss such problems in this book.

Finally, we emphasize that the discussion in this section has been purely formal. Separate and sometimes elaborate arguments must be used to establish convergence of eigenfunction expansions, or to justify some of the steps used, such as term-by-term differentiation of eigenfunction series.

There are also other altogether different methods for solving nonhomogeneous boundary value problems. One of these leads to a solution expressed as a definite integral rather than as an infinite series. This approach involves certain functions known as Green's functions, and for ordinary differential equations is the subject of Problems 14 through 19.

PROBLEMS

1. Solve each of the following problems by means of an eigenfunction expansion.

(a) $y'' + 2y = -x, \qquad y(0) = 0, \quad y(1) = 0$

(b) $y'' + 2y = -x$, $y(0) = 0$, $y'(1) = 0$; see Section 11.3, Problem 2(b).
(c) $y'' + 2y = -x$, $y'(0) = 0$, $y'(1) = 0$; see Section 11.3, Problem 1(c).
(d) $y'' + 2y = -x$, $y'(0) = 0$, $y'(1) + y(1) = 0$; see Section 11.3, Problem 3(b).
(e) $y'' + 2y = -1 + |1 - 2x|$, $y(0) = 0$, $y(1) = 0$

2. Determine a formal eigenfunction series expansion for the solution of each of the following problems. Assume that f satisfies the conditions of Theorem 11.5. State the values of μ for which the solution exists.

(a) $y'' + \mu y = -f(x)$, $y(0) = 0$, $y'(1) = 0$
(b) $y'' + \mu y = -f(x)$, $y'(0) = 0$, $y(1) = 0$
(c) $y'' + \mu y = -f(x)$, $y'(0) = 0$, $y'(1) = 0$
(d) $y'' + \mu y = -f(x)$, $y'(0) = 0$, $y'(1) + y(1) = 0$

3. For each of the following problems, determine whether there is any value of the constant a for which the problem has a solution. Find the solution for each such value.

(a) $y'' + \pi^2 y = a + x$, $y(0) = 0$, $y(1) = 0$
(b) $y'' + 4\pi^2 y = a + x$, $y(0) = 0$, $y(1) = 0$
(c) $y'' + \pi^2 y = a$, $y'(0) = 0$, $y'(1) = 0$
(d) $y'' + \pi^2 y = a - \cos \pi x$, $y(0) = 0$, $y(1) = 0$

4. Let $\phi_1, \ldots, \phi_n, \ldots$ be the normalized eigenfunctions of the differential equation (3) subject to the boundary conditions (2). If $\sum_{n=1}^{\infty} c_n\phi_n(x)$ converges to $f(x)$, where $f(x) = 0$ for each x in $0 \le x \le 1$, show that $c_n = 0$ for each n.
Hint: Multiply by $r(x)\phi_m(x)$, integrate, and use the orthogonality property of the eigenfunctions.
5. Let L be a second order linear differential operator. Show that the solution $y = \phi(x)$ of the problem

$$L[y] = f(x),$$

$$a_1 y(0) + a_2 y'(0) = \alpha, \quad b_1 y(1) + b_2 y'(1) = \beta,$$

can be written as $y = u + v$ *provided* that $u = \phi_1(x)$ and $v = \phi_2(x)$ are solutions of the problems

$$L[u] = 0,$$

$$a_1 u(0) + a_2 u'(0) = \alpha, \quad b_1 u(1) + b_2 u'(1) = \beta,$$

and

$$L[v] = f(x),$$

$$a_1 v(0) + a_2 v'(0) = 0, \quad b_1 v(1) + b_2 v'(1) = 0,$$

respectively.
6. Show that the problem

$$y'' + \pi^2 y = \pi^2 x, \qquad y(0) = 1, \quad y(1) = 0$$

has the solution

$$y = c_1 \sin \pi x + \cos \pi x + x.$$

Also show that this solution cannot be obtained by splitting the problem as suggested in Problem 5, since neither of the two subsidiary problems can be solved in this case.

7. Consider the problem

$$y'' + p(x)y' + q(x)y = 0, \qquad y(0) = a, \quad y(1) = b.$$

Let $y = u + v$, where v is any twice differentiable function satisfying the boundary conditions (but not necessarily the differential equation). Show that u is a solution of the problem

$$u'' + p(x)u' + q(x)u = g(x), \qquad u(0) = 0, \quad u(1) = 0,$$

where $g(x) = -[v'' + p(x)v' + q(x)v]$, and is known once v is chosen. Thus nonhomogeneities can be transferred from the boundary conditions to the differential equation. Find a function v for this problem.

8. Using the method of Problem 7, transform the problem

$$y'' + 2y = 2 - 4x, \qquad y(0) = 1, \quad y(1) + y'(1) = -2$$

into a new problem in which the boundary conditions are homogeneous. Solve the latter problem by reference to Example 1 of the text.

9. Use eigenfunction expansions to find the solution of each of the following boundary value problems.

(a) $u_t = u_{xx} - x, \qquad u(0, t) = 0, \quad u_x(1, t) = 0,$
$u(x, 0) = \sin(\pi x/2)$; see Problem 1(b).

(b) $u_t = u_{xx} + e^{-t}, \qquad u_x(0, t) = 0, \quad u_x(1, t) + u(1, t) = 0,$
$u(x, 0) = 1 - x$; see Section 11.3, Problem 3(a, c).

(c) $u_t = u_{xx} + 1 - |1 - 2x|, \qquad u(0, t) = 0, \quad u(1, t) = 0,$
$u(x, 0) = 0$; see Problem 1(e).

(d) $u_t = u_{xx} + e^{-t}(1 - x), \qquad u(0, t) = 0, \quad u_x(1, t) = 0,$
$u(x, 0) = 0$; see Section 11.3, Problem 2(a, b).

10. Consider the boundary value problem

$$r(x)u_t = [p(x)u_x]_x - q(x)u + F(x),$$

$$u(0, t) = T_1, \quad u(1, t) = T_2, \quad u(x, 0) = f(x).$$

(a) Let $v(x)$ be a solution of the problem

$$[p(x)v']' - q(x)v = -F(x), \qquad v(0) = T_1, \quad v(1) = T_2.$$

If $w(x, t) = u(x, t) - v(x)$, find the boundary value problem satisfied by w. Note that this problem can be solved by the method of this section.

(b) Generalize the procedure of part (a) to the case where u satisfies the boundary conditions

$$u_x(0, t) - h_1 u(0, t) = T_1, \quad u_x(1, t) + h_2 u(1, t) = T_2.$$

11. Use the method indicated in Problem 10 to solve each of the following boundary value problems.

(a) $u_t = u_{xx} - 2,$
$u(0,t) = 1, \quad u(1,t) = 0,$
$u(x,0) = x^2 - 2x + 2$

(b) $u_t = u_{xx} - \pi^2 \cos \pi x$
$u_x(0,t) = 0, \quad u(1,t) = 1,$
$u(x,0) = \cos(3\pi x/2) - \cos \pi x.$

12. The method of eigenfunction expansions is often useful for nonhomogeneous problems related to the wave equation or its generalizations. Consider the problem

$$r(x)u_{tt} = [p(x)u_x]_x - q(x)u + F(x,t), \tag{i}$$

$$u_x(0,t) - h_1 u(0,t) = 0, \quad u_x(1,t) + h_2 u(1,t) = 0, \tag{ii}$$

$$u(x,0) = f(x), \quad u_t(x,0) = g(x). \tag{iii}$$

This problem can arise in connection with generalizations of the telegraph equation (Problem 4 in Section 11.1) or the longitudinal vibrations of an elastic bar (Problem 17 in Section 11.2).

(a) Let $u(x,t) = X(x)T(t)$ in the homogeneous equation corresponding to Eq. (i) and show that $X(x)$ satisfies Eqs. (28) and (29) of the text. Let λ_n and $\phi_n(x)$ denote the eigenvalues and normalized eigenfunctions of this problem.

(b) Assume that $u(x,t) = \sum_{n=1}^{\infty} b_n(t)\phi_n(x)$, and show that $b_n(t)$ must satisfy the initial value problem

$$b_n''(t) + \lambda_n b_n(t) = \gamma_n(t), \qquad b_n(0) = \alpha_n, \quad b_n'(0) = \beta_n,$$

where α_n, β_n, and $\gamma_n(t)$ are the expansion coefficients for $f(x)$, $g(x)$, and $F(x,t)$ in terms of the eigenfunctions $\phi_1(x), \ldots, \phi_n(x), \ldots$.

13. In this problem we explore a little further the analogy between Sturm-Liouville boundary value problems and Hermitian matrices. Let $\mathbf{A}$ be an $n \times n$ Hermitian matrix with eigenvalues $\lambda_1, \ldots, \lambda_n$ and corresponding orthonormal eigenvectors $\boldsymbol{\xi}^{(1)}, \ldots, \boldsymbol{\xi}^{(n)}$.

Consider the nonhomogeneous system of equations

$$\mathbf{Ax} - \mu\mathbf{x} = \mathbf{b}, \tag{i}$$

where μ is a given real number and $\mathbf{b}$ is a given vector. We will point out a way of solving Eq. (i) that is analogous to the method presented in the text for solving Eqs. (1) and (2).

(a) Show that $\mathbf{b} = \sum_{i=1}^{n} b_i \boldsymbol{\xi}^{(i)}$, where $b_i = (\mathbf{b}, \boldsymbol{\xi}^{(i)})$.

(b) Assume that $\mathbf{x} = \sum_{i=1}^{n} a_i \boldsymbol{\xi}^{(i)}$ and show that for Eq. (i) to be satisfied, it is necessary that $a_i = b_i/(\lambda_i - \mu)$. Thus

$$\mathbf{x} = \sum_{i=1}^{n} \frac{(\mathbf{b}, \boldsymbol{\xi}^{(i)})}{\lambda_i - \mu} \boldsymbol{\xi}^{(i)}, \tag{ii}$$

provided that μ is not one of the eigenvalues of $\mathbf{A}$, $\mu \neq \lambda_i$ for $i = 1, \ldots, n$. Compare this result with Eq. (13).

Green's[5] Functions. Consider the nonhomogeneous system of algebraic equations

$$\mathbf{Ax} - \mu\mathbf{x} = \mathbf{b}, \tag{i}$$

where $\mathbf{A}$ is an $n \times n$ Hermitian matrix, μ is a given real number, and $\mathbf{b}$ is a given vector. Instead of using an eigenvector expansion as in Problem 13, we can solve Eq. (i) by computing the inverse matrix $(\mathbf{A} - \mu\mathbf{I})^{-1}$, which exists if μ is not an eigenvalue of $\mathbf{A}$. Then

$$\mathbf{x} = (\mathbf{A} - \mu\mathbf{I})^{-1}\mathbf{b}. \tag{ii}$$

Problems 14 through 19 indicate a way of solving nonhomogeneous boundary value problems that is analogous to using the inverse matrix for a system of linear algebraic equations. The Green's function plays a part similar to the inverse of the matrix of coefficients. This method leads to solutions expressed as definite integrals rather than as infinite series. Except in Problem 18 we will assume that $\mu = 0$ for simplicity.

14. (a) Show by the method of variation of parameters that the general solution of the differential equation

$$-y'' = f(x)$$

can be written in the form

$$y = \phi(x) = c_1 + c_2 x - \int_0^x (x - s) f(s)\, ds,$$

where c_1 and c_2 are arbitrary constants.

(b) Let $y = \phi(x)$ also be required to satisfy the boundary conditions $y(0) = 0$, $y(1) = 0$. Show that in this case

$$c_1 = 0, \qquad c_2 = \int_0^1 (1 - s) f(s)\, ds.$$

(c) Show that, under the conditions of parts (a) and (b), $\phi(x)$ can be written in the form

$$\phi(x) = \int_0^x s(1 - x) f(s)\, ds + \int_x^1 x(1 - s) f(s)\, ds.$$

(d) Defining

$$G(x, s) = \begin{cases} s(1 - x), & 0 \le s \le x, \\ x(1 - s), & x \le s \le 1, \end{cases}$$

show that the solution takes the form

$$\phi(x) = \int_0^1 G(x, s) f(s)\, ds.$$

[5] Green's functions are named after George Green (1793–1841) of England. He was almost entirely self-taught in mathematics but made significant contributions to electricity and magnetism, fluid mechanics, and partial differential equations. His most important work was an essay on electricity and magnetism that was published privately in 1828. In this paper Green was the first to recognize the importance of potential functions. He introduced the functions now known as Green's functions as a means of solving boundary value problems, and developed the integral transformation theorems of which Green's theorem in the plane is a particular case. However, these results did not become widely known until Green's essay was republished in the 1850s through the efforts of William Thomson (Lord Kelvin).

The function $G(x, s)$ appearing under the integral sign is a Green's function. The usefulness of a Green's function solution rests on the fact that the Green's function is independent of the nonhomogeneous term in the differential equation. Thus, once the Green's function is determined, the solution of the boundary value problem for different nonhomogeneous terms $f(x)$ is obtained by a single integration. Note further that no determination of arbitrary constants is required, since $\phi(x)$ as given by the Green's function integral formula automatically satisfies the boundary conditions.

15. By a procedure similar to that in the previous problem show that the solution of the boundary value problem

$$-(y'' + y) = f(x), \qquad y(0) = 0, \quad y(1) = 0,$$

is

$$y = \phi(x) = \int_0^1 G(x, s) f(s)\, ds,$$

where

$$G(x, s) = \begin{cases} \dfrac{\sin s \sin(1 - x)}{\sin 1}, & 0 \le s \le x, \\ \dfrac{\sin x \sin(1 - s)}{\sin 1}, & x \le s \le 1. \end{cases}$$

16. It is possible to show that the Sturm-Liouville problem

$$L[y] = -[p(x) y']' + q(x) y = f(x), \tag{i}$$

$$a_1 y(0) + a_2 y'(0) = 0, \qquad b_1 y(1) + b_2 y'(1) = 0, \tag{ii}$$

has a Green's function solution

$$y = \phi(x) = \int_0^1 G(x, s) f(s)\, ds, \tag{iii}$$

provided that $\lambda = 0$ is not an eigenvalue of $L[y] = \lambda y$ subject to the boundary conditions (ii). Further, $G(x, s)$ is given by

$$G(x, s) = \begin{cases} -y_1(s) y_2(x)/p(x) W(y_1, y_2)(x), & 0 \le s \le x, \\ -y_1(x) y_2(s)/p(x) W(y_1, y_2)(x), & x \le s \le 1, \end{cases} \tag{iv}$$

where y_1 is a solution of $L[y] = 0$ satisfying the boundary condition at $x = 0$, y_2 is a solution of $L[y] = 0$ satisfying the boundary condition at $x = 1$, and $W(y_1, y_2)$ is the Wronskian of y_1 and y_2.
(a) Verify that the Green's function obtained in Problem 14 is given by formula (iv).
(b) Verify that the Green's function obtained in Problem 15 is given by formula (iv).
(c) Show that $p(x)W(y_1, y_2)(x)$ is a constant, by showing that its derivative is zero.
(d) Using Eq. (iv) and the result of part (c), show that $G(x, s) = G(s, x)$.

17. Solve each of the following boundary value problems by determining the appropriate Green's function and expressing the solution as a definite integral. Use Eqs. (i) to (iv) of Problem 16.

(a) $-y'' = f(x), \qquad y'(0) = 0, \quad y(1) = 0$
(b) $-y'' = f(x), \qquad y(0) = 0, \quad y(1) + y'(1) = 0$

(c) $-(y'' + y) = f(x), \quad y'(0) = 0, \quad y(1) = 0$
(d) $-y'' = f(x), \quad y(0) = 0, \quad y'(1) = 0$

*18. Consider the boundary value problem

$$L[y] = -[p(x)y']' + q(x)y = \mu r(x)y + f(x), \tag{i}$$

$$a_1 y(0) + a_2 y'(0) = 0, \qquad b_1 y(1) + b_2 y'(1) = 0. \tag{ii}$$

According to the text, the solution $y = \phi(x)$ is given by Eq. (13), where c_n is defined by Eq. (9), provided that μ is not an eigenvalue of the corresponding homogeneous problem. In this case it can also be shown that the solution is given by a Green's function integral of the form

$$y = \phi(x) = \int_0^1 G(x, s, \mu) f(s)\, ds. \tag{iii}$$

Note that in this problem the Green's function also depends on the parameter μ.
(a) Show that if these two expressions for $\phi(x)$ are to be equivalent, then

$$G(x, s, \mu) = \sum_{i=1}^{\infty} \frac{\phi_i(x)\phi_i(s)}{\lambda_i - \mu}, \tag{iv}$$

where λ_i and ϕ_i are the eigenvalues and eigenfunctions, respectively, of Eqs. (3), (2) of the text. Again we see from Eq. (iv) that μ cannot be equal to any eigenvalue λ_i.
(b) Derive Eq. (iv) directly by assuming that $G(x, s, \mu)$ has the eigenfunction expansion

$$G(x, s, \mu) = \sum_{i=1}^{\infty} a_i(x, \mu)\phi_i(s); \tag{v}$$

determine $a_i(x, \mu)$ by multiplying Eq. (v) by $r(s)\phi_j(s)$ and integrating with respect to s from $s = 0$ to $s = 1$.
Hint: Show first that λ_i and ϕ_i satisfy the equation

$$\phi_i(x) = (\lambda_i - \mu)\int_0^1 G(x, s, \mu) r(s)\phi_i(s)\, ds. \tag{vi}$$

*19. Consider the boundary value problem

$$-d^2y/ds^2 = \delta(s - x), \qquad y(0) = 0, \quad y(1) = 0,$$

where s is the independent variable, $s = x$ is a definite point in the interval $0 < s < 1$, and δ is the Dirac delta function (see Section 6.4). Show that the solution of this problem is the Green's function $G(x, s)$ obtained in Problem 14.

In solving the given problem, note that $\delta(s - x) = 0$ in the intervals $0 \le s < x$ and $x < s \le 1$. Note further that $-dy/ds$ experiences a jump of magnitude 1 as s passes through the value x.

This problem illustrates a general property, namely, that the Green's function $G(x, s)$ can be identified as the response at the point s to a unit impulse at the point x. A more general nonhomogeneous term f on $0 \le x \le 1$ can be regarded as a set of impulses with $f(x)$ giving the magnitude of the impulse at the point x. The solution of a nonhomogeneous boundary value problem in terms of a Green's function integral can then be interpreted as the result of superposing the responses to the set of impulses represented by the nonhomogeneous term $f(x)$.

*11.5 Singular Sturm-Liouville Problems

In the preceding sections of this chapter we considered Sturm-Liouville boundary value problems: the differential equation

$$L[y] = -[p(x)y']' + q(x)y = \lambda r(x)y, \qquad 0 < x < 1, \tag{1}$$

together with boundary conditions of the form

$$a_1 y(0) + a_2 y'(0) = 0, \tag{2a}$$

$$b_1 y(1) + b_2 y'(1) = 0. \tag{2b}$$

Until now, we have always assumed that the problem is regular: that is, p is differentiable, q and r are continuous, and $p(x) > 0$ and $r(x) > 0$ at all points in the *closed* interval. However, there are also equations of physical interest in which some of these conditions are not satisfied.

For example, suppose we wish to study Bessel's equation of order ν on the interval $0 < x < 1$. This equation is sometimes written in the form[6]

$$-(xy')' + \frac{\nu^2}{x}y = \lambda xy, \tag{3}$$

so that $p(x) = x$, $q(x) = \nu^2/x$, and $r(x) = x$. Thus $p(0) = 0$, $r(0) = 0$, and $q(x)$ is unbounded and hence discontinuous as $x \to 0$. However, the conditions imposed on regular Sturm-Liouville problems are met elsewhere in the interval.

Similarly, for Legendre's equation we have

$$-[(1 - x^2)y']' = \lambda y, \qquad -1 < x < 1 \tag{4}$$

where $\lambda = \alpha(\alpha + 1)$, $p(x) = 1 - x^2$, $q(x) = 0$, and $r(x) = 1$. Here the required conditions on p, q, and r are satisfied in the interval $0 \le x \le 1$ except at $x = 1$ where p is zero.

We use the term singular Sturm-Liouville problem to refer to a certain class of boundary value problems for the differential equation (1) in which the functions p, q, and r satisfy the conditions previously laid down on the open interval $0 < x < 1$, but at least one fails to satisfy them at one or both of the boundary points. We also prescribe suitable separated boundary conditions, of a kind to be described in more detail later in this section. Singular problems also occur if the basic interval is unbounded; for example, $0 \le x < \infty$. We do not consider this latter kind of singular problem in this book.

As an example of a singular problem on a finite interval, consider the equation

$$xy'' + y' + \lambda xy = 0, \tag{5a}$$

or

$$-(xy')' = \lambda xy, \tag{5b}$$

[6]The substitution $t = \sqrt{\lambda}\, x$ reduces Eq. (3) to the standard form $t^2y'' + ty' + (t^2 - \nu^2)y = 0$.

on the interval $0 < x < 1$, and suppose that $\lambda > 0$. This equation arises in the study of free vibrations of a circular elastic membrane, and is discussed further in Section 11.6. If we introduce the new independent variable t defined by $t = \sqrt{\lambda}\, x$, then

$$\frac{dy}{dx} = \sqrt{\lambda}\,\frac{dy}{dt}, \qquad \frac{d^2y}{dx^2} = \lambda\frac{d^2y}{dt^2}.$$

Hence Eq. (5a) becomes

$$\frac{t}{\sqrt{\lambda}}\lambda\frac{d^2y}{dt^2} + \sqrt{\lambda}\,\frac{dy}{dt} + \lambda\frac{t}{\sqrt{\lambda}}y = 0$$

or, canceling the common factor $\sqrt{\lambda}$ in each term,

$$t\frac{d^2y}{dt^2} + \frac{dy}{dt} + ty = 0. \tag{6}$$

Equation (6) is Bessel's equation of order zero (see Section 4.7). The general solution of Eq. (6) for $t > 0$ is

$$y = c_1J_0(t) + c_2Y_0(t);$$

hence the general solution of Eq. (5) for $x > 0$ is

$$y = c_1J_0(\sqrt{\lambda}\,x) + c_2Y_0(\sqrt{\lambda}\,x), \tag{7}$$

where J_0 and Y_0 denote the Bessel functions of the first and second kinds of order zero. From Eqs. (7) and (13) of Section 4.7 we have

$$J_0(\sqrt{\lambda}\,x) = 1 + \sum_{m=1}^{\infty}\frac{(-1)^m\lambda^m x^{2m}}{2^{2m}(m!)^2}, \qquad x > 0 \tag{8}$$

$$Y_0(\sqrt{\lambda}\,x) = \frac{2}{\pi}\left[\left(\gamma + \ln\frac{\sqrt{\lambda}\,x}{2}\right)J_0(\sqrt{\lambda}\,x) + \sum_{m=1}^{\infty}\frac{(-1)^{m+1}H_m\lambda^m x^{2m}}{2^{2m}(m!)^2}\right], \qquad x > 0, \tag{9}$$

where $H_m = 1 + (1/2) + \cdots + (1/m)$, and $\gamma = \lim_{m\to\infty}(H_m - \ln m)$. The graphs of $y = J_0(x)$ and $y = Y_0(x)$ are given in Figure 4.7.

Suppose that we seek a solution of Eq. (5) that also satisfies the boundary conditions

$$y(0) = 0, \tag{10a}$$

$$y(1) = 0, \tag{10b}$$

which are typical of those we have met in other problems in this chapter. Since $J_0(0) = 1$ and $Y_0(x) \to -\infty$ as $x \to 0$, the condition $y(0) = 0$ can be satisfied only by choosing $c_1 = c_2 = 0$ in Eq. (7). Thus the boundary value problem (5), (10) has only the trivial solution.

One interpretation of this result is that the boundary condition (10a) at $x = 0$ is too restrictive for the differential equation (5). This illustrates the general situation; namely, that at a singular boundary point it is necessary to consider a modified type of boundary condition. In the present problem, suppose that we require only that the solution (7) of Eq. (5) and its derivative remain bounded. In other words, we take as the boundary condition at $x = 0$ the requirement

$$y, y' \text{ bounded as } x \to 0. \tag{11}$$

This condition can be satisfied by choosing $c_2 = 0$, so as to eliminate the unbounded solution Y_0. The second boundary condition $y(1) = 0$ then yields

$$J_0(\sqrt{\lambda}) = 0. \tag{12}$$

It is possible to show[7] that Eq. (12) has an infinite set of discrete positive roots, which yield the eigenvalues $0 < \lambda_1 < \lambda_2 < \cdots < \lambda_n < \cdots$ of the given problem. The corresponding eigenfunctions are

$$\phi_n(x) = J_0\left(\sqrt{\lambda_n}\,x\right), \tag{13}$$

determined only up to a multiplicative constant. The boundary value problem (5), (10b), and (11) is an example of a singular Sturm-Liouville problem. This example illustrates that if the boundary conditions are relaxed in an appropriate way, then a singular Sturm-Liouville problem may have an infinite sequence of eigenvalues and eigenfunctions, just as a regular Sturm-Liouville problem does.

Because of their importance in applications, it is worthwhile to investigate singular boundary value problems a little further. There are two main questions that are of concern.

1. Precisely what type of boundary conditions can be allowed in a singular Sturm-Liouville problem?
2. To what extent do the eigenvalues and eigenfunctions of a singular problem share the properties of eigenvalues and eigenfunctions of regular Sturm-Liouville problems? In particular, are the eigenvalues real, are the eigenfunctions orthogonal, and can a given function be expanded as a series of eigenfunctions?

Both of these questions can be answered by a study of the identity

$$\int_0^1 \{L[u]v - uL[v]\}\,dx = 0, \tag{14}$$

which played an essential part in the development of the theory of regular Sturm-Liouville problems. We therefore investigate the conditions under which this relation holds for singular problems, where the integral in Eq. (14) may now have to be examined as an improper integral. To be definite we consider the

[7]The function J_0 is well tabulated; the roots of Eq. (12) can be found in various tables, for example, Jahnke and Emde or Abramowitz and Stegun. The first three roots of Eq. (12) are $\sqrt{\lambda} = 2.405$, 5.520, and 8.654, respectively, to four significant figures; $\sqrt{\lambda_n} \cong (n - 1/4)\pi$ for large n.

differential equation (1) and assume that $x = 0$ is a singular boundary point, but that $x = 1$ is not. The boundary condition (2b) is imposed at the nonsingular boundary point $x = 1$, but we leave unspecified, for the moment, the boundary condition at $x = 0$. Indeed, our principal object is to determine what kinds of boundary conditions are allowable at a singular boundary point if Eq. (14) is to hold.

Since the boundary value problem under investigation is singular at $x = 0$, we consider the integral $\int_\epsilon^1 L[u]v\,dx$, instead of $\int_0^1 L[u]v\,dx$, as in Section 11.3, and afterwards let ϵ approach zero. Assuming that u and v have at least two continuous derivatives on $\epsilon \le x \le 1$, and integrating twice by parts, we find that

$$\int_\epsilon^1 \{L[u]v - uL[v]\}\,dx = -p(x)\big[u'(x)v(x) - u(x)v'(x)\big]\Big|_\epsilon^1. \tag{15}$$

The boundary term at $x = 1$ is again eliminated if both u and v satisfy the boundary condition (2b), and thus,

$$\int_\epsilon^1 \{L[u]v - uL[v]\}\,dx = p(\epsilon)\big[u'(\epsilon)v(\epsilon) - u(\epsilon)v'(\epsilon)\big]. \tag{16}$$

Taking the limit as $\epsilon \to 0$ yields

$$\int_0^1 \{L[u]v - uL[v]\}\,dx = \lim_{\epsilon\to 0} p(\epsilon)\big[u'(\epsilon)v(\epsilon) - u(\epsilon)v'(\epsilon)\big]. \tag{17}$$

Hence Eq. (14) holds if and only if, in addition to the assumptions stated previously,

$$\lim_{\epsilon\to 0} p(\epsilon)\big[u'(\epsilon)v(\epsilon) - u(\epsilon)v'(\epsilon)\big] = 0 \tag{18}$$

for every pair of functions u and v in the class under consideration. Equation (18) is therefore the criterion that determines what boundary conditions are allowable at $x = 0$ if that point is a singular boundary point. A similar condition applies at $x = 1$ if that boundary point is singular; namely

$$\lim_{\epsilon\to 0} p(1-\epsilon)\big[u'(1-\epsilon)v(1-\epsilon) - u(1-\epsilon)v'(1-\epsilon)\big] = 0. \tag{19}$$

In summary, as in Section 11.3, a singular boundary value problem for Eq. (1) is said to be *self-adjoint* if Eq. (14) is valid, possibly as an improper integral, for each pair of functions u and v with the following properties: they are twice continuously differentiable on the open interval $0 < x < 1$, they satisfy a boundary condition of the form (2) at each regular boundary point, and they satisfy a boundary condition sufficient to insure Eq. (18) if $x = 0$ is a singular boundary point, or Eq. (19) if $x = 1$ is a singular boundary point. If at least one boundary point is singular, then the differential equation (1), together with two boundary conditions of the type just described, are said to form a *singular Sturm-Liouville problem.*

For example, for Eq. (5) we have $p(x) = x$. If both u and v satisfy the boundary condition (11) at $x = 0$ it is clear that Eq. (18) will hold. Hence the singular boundary value problem, consisting of the differential equation (5), the boundary condition (11) at $x = 0$, and any boundary condition of the form (2b) at $x = 1$, is self-adjoint.

The most striking difference between regular and singular Sturm-Liouville problems is that in a singular problem the eigenvalues may not be discrete. That is, the problem may have nontrivial solutions for every value of λ, or for every value of λ in some interval. In such a case the problem is said to have a *continuous spectrum*. It may happen that a singular problem has a mixture of discrete eigenvalues and also a continuous spectrum. Finally, it is possible that only a discrete set of eigenvalues exists, just as in the regular case discussed in Section 11.3. For example, this is true of the problem consisting of Eqs. (5), (10b), and (11). In general, it may be difficult to determine which case actually occurs in a given problem.

A systematic discussion of singular Sturm-Liouville problems is quite sophisticated[8] indeed, requiring a substantial extension of the methods presented in this book. We restrict ourselves to some examples related to physical applications; in each of these examples it is known that there is an infinite set of discrete eigenvalues.

If a singular Sturm-Liouville problem does have only a discrete set of eigenvalues and eigenfunctions, then Eq. (14) can be used, just as in Section 11.3, to prove that the eigenvalues of such a problem are real, and that the eigenfunctions are orthogonal with respect to the weight function r. The expansion of a given function in terms of a series of eigenfunctions then follows as in Section 11.3.

Such expansions are useful, as in the regular case, for solving nonhomogeneous boundary value problems. The procedure is very similar to that described in Section 11.4. Some examples for ordinary differential equations are indicated in Problems 1 to 4, and some problems for partial differential equations appear in Section 11.6.

For instance, the eigenfunctions $\phi_n(x) = J_0(\sqrt{\lambda_n}\, x)$ of the singular Sturm-Liouville problem

$$-(xy')' = \lambda xy, \qquad 0 < x < 1$$

$$y,\ y' \text{ bounded as } x \to 0, \qquad y(1) = 0,$$

satisfy the orthogonality relation

$$\int_0^1 x\phi_m(x)\phi_n(x)\, dx = 0, \qquad m \neq n \tag{20}$$

with respect to the weight function $r(x) = x$. Then, if f is a given function, we assume that

$$f(x) = \sum_{n=1}^{\infty} c_n J_0\left(\sqrt{\lambda_n}\, x\right). \tag{21}$$

[8] See, for example, Chapter 5 of the book by Yosida.

Multiplying Eq. (21) by $xJ_0(\sqrt{\lambda_m}x)$ and integrating term by term from $x = 0$ to $x = 1$ yields

$$\int_0^1 xf(x)J_0\left(\sqrt{\lambda_m}x\right)dx = \sum_{n=1}^{\infty} c_n \int_0^1 xJ_0\left(\sqrt{\lambda_m}x\right)J_0\left(\sqrt{\lambda_n}x\right)dx. \tag{22}$$

Because of the orthogonality condition (20), the right side of Eq. (22) collapses to a single term; hence

$$c_m = \frac{\int_0^1 xf(x)J_0\left(\sqrt{\lambda_m}x\right)dx}{\int_0^1 xJ_0^2\left(\sqrt{\lambda_m}x\right)dx}, \tag{23}$$

which determines the coefficients in the series (21).

The convergence of the series (21) is established by an extension of Theorem 11.4 to cover this case. This theorem can also be shown to hold for other sets of Bessel functions, which are solutions of appropriate boundary value problems, for Legendre polynomials, and for solutions of a number of other singular Sturm-Liouville problems of considerable interest.

It must be emphasized that the singular problems mentioned here are not necessarily typical. In general, singular boundary value problems are characterized by continuous spectra, rather than by discrete sets of eigenvalues. The corresponding sets of eigenfunctions are therefore not denumerable, and series expansions of the type described in Theorem 11.4 do not exist. They are replaced by appropriate integral representations.

PROBLEMS

1. Find a formal solution of the nonhomogeneous boundary value problem

$$-(xy')' = \mu xy + f(x),$$

$$y, y' \text{ bounded as } x \to 0, \qquad y(1) = 0,$$

where f is a given continuous function on $0 \le x \le 1$, and μ is not an eigenvalue of the corresponding homogeneous problem.
Hint: Use a series expansion similar to those in Section 11.4.

2. Consider the boundary value problem

$$-(xy')' = \lambda xy,$$

$$y, y' \text{ bounded as } x \to 0, \qquad y'(1) = 0.$$

(a) Show that $\lambda_0 = 0$ is an eigenvalue of this problem corresponding to the eigenfunction $\phi_0(x) = 1$. If $\lambda > 0$, show formally that the eigenfunctions are given by $\phi_n(x) = J_0(\sqrt{\lambda_n}x)$, where $\sqrt{\lambda_n}$ is the nth positive root (in increasing order) of the equation $J_0'(\sqrt{\lambda}) = 0$. It is possible to show that there is an infinite sequence of such roots.

(b) Show that if $m, n = 0, 1, 2, \ldots,$ then

$$\int_0^1 x\phi_m(x)\phi_n(x)\,dx = 0, \qquad m \neq n.$$

(c) Find a formal solution of the nonhomogeneous problem

$$-(xy')' = \mu xy + f(x),$$

$$y, y' \text{ bounded as } x \to 0, \qquad y'(1) = 0,$$

where f is a given continuous function on $0 \le x \le 1$, and μ is not an eigenvalue of the corresponding homogeneous problem.

3. Consider the problem

$$-(xy')' + (k^2/x)y = \lambda xy,$$

$$y, y' \text{ bounded as } x \to 0, \qquad y(1) = 0,$$

where k is a positive integer.
(a) Using the substitution $t = \sqrt{\lambda}\, x$, show that the above differential equation reduces to Bessel's equation of order k (see Problem 6 of Section 4.7). One solution is $J_k(t)$; a second linearly independent solution, denoted by $Y_k(t)$, is unbounded as $t \to 0$.
(b) Show formally that the eigenvalues $\lambda_1, \lambda_2, \ldots$ of the given problem are the squares of the positive zeros of $J_k(\sqrt{\lambda})$, and that the corresponding eigenfunctions are $\phi_n(x) = J_k(\sqrt{\lambda_n}\, x)$. It is possible to show that there is an infinite sequence of such zeros.
(c) Show that the eigenfunctions $\phi_n(x)$ satisfy the orthogonality relation

$$\int_0^1 x\phi_m(x)\phi_n(x)\,dx = 0, \qquad m \neq n.$$

(d) Determine the coefficients in the formal series expansion

$$f(x) = \sum_{n=1}^{\infty} a_n\phi_n(x).$$

(e) Find a formal solution of the nonhomogeneous problem

$$-(xy')' + (k^2/x)y = \mu xy + f(x),$$

$$y, y' \text{ as bounded as } x \to 0, \qquad y(1) = 0,$$

where f is a given continuous function on $0 \le x \le 1$, and μ is not an eigenvalue of the corresponding homogeneous problem.

4. Consider Legendre's equation (see Problems 10 and 11 of Section 4.2.1)

$$-\left[(1 - x^2)y'\right]' = \lambda y$$

subject to the boundary conditions

$$y(0) = 0, \qquad y, y' \text{ bounded as } x \to 1.$$

The eigenfunctions of this problem are the odd Legendre polynomials $\phi_1(x) = P_1(x) = x$, $\phi_2(x) = P_3(x) = (5x^3 - 3x)/2, \ldots, \phi_n(x) = P_{2n-1}(x), \ldots$ corresponding to the eigenvalues $\lambda_1 = 2$, $\lambda_2 = 4 \cdot 3, \ldots, \lambda_n = 2n(2n-1), \ldots$.

(a) Show that

$$\int_0^1 \phi_m(x)\phi_n(x)\,dx = 0, \qquad m \neq n.$$

(b) Find a formal solution of the nonhomogeneous problem

$$-[(1-x^2)y']' = \mu y + f(x),$$

$$y(0) = 0, \qquad y, y' \text{ bounded as } x \to 1,$$

where f is a given continuous function on $0 \le x \le 1$, and μ is not an eigenvalue of the corresponding homogeneous problem.

5. The equation

$$(1-x^2)y'' - xy' + \lambda y = 0 \tag{i}$$

is Tchebycheff's equation; see Problem 4 of Section 4.2.1.

(a) Show that Eq. (i) can be written in the form

$$-\left[(1-x^2)^{1/2}y'\right]' = \lambda(1-x^2)^{-1/2}y, \qquad -1 < x < 1. \tag{ii}$$

(b) Consider the boundary conditions

$$y, y' \text{ bounded as } x \to -1, \qquad y, y' \text{ bounded as } x \to 1. \tag{iii}$$

Show that the boundary value problem (ii), (iii) is self-adjoint.

(c) It can be shown that the boundary value problem (ii), (iii) has the eigenvalues $\lambda_0 = 0$, $\lambda_1 = 1$, $\lambda_2 = 4, \ldots, \lambda_n = n^2, \ldots$. The corresponding eigenfunctions are the Tchebycheff polynomials $T_n(x)$: $T_0(x) = 1$, $T_1(x) = x$, $T_2(x) = 1 - 2x^2, \ldots$. Show that

$$\int_{-1}^{1} \frac{T_m(x)T_n(x)}{(1-x^2)^{1/2}}\,dx = 0, \qquad m \neq n. \tag{iv}$$

Note that this is a convergent improper integral.

*11.6 Further Remarks on the Method of Separation of Variables: A Bessel Series Expansion

In this chapter we are interested in extending the method of separation of variables developed in Chapter 10 to a larger class of problems—to problems involving more general differential equations, more general boundary conditions, or different geometrical regions. We indicated in Section 11.4 how to deal with a class of more general differential equations or boundary conditions. Here we concentrate on problems posed in various geometrical regions, with emphasis on those leading to singular Sturm-Liouville problems when the variables are separated.

Because of its relative simplicity, as well as the considerable physical significance of many problems to which it is applicable, the method of separation of variables merits its important place in the theory and application of partial differential equations. However, this method does have certain limitations that should not be forgotten. In the first place, the problem must be linear so that the principle of superposition can be invoked to construct additional solutions by forming linear combinations of the fundamental solutions of an appropriate homogeneous problem.

As a practical matter, we must also be able to solve the ordinary differential equations, obtained after separating the variables, in a reasonably convenient manner. In some problems to which the method of separation of variables can be applied in principle, it is of very limited practical value due to lack of information about the solutions of the ordinary differential equations that appear.

Furthermore, the geometry of the region involved in the problem is subject to rather severe restrictions. On one hand, a coordinate system must be employed in which the variables can be separated, and the partial differential equation replaced by a set of ordinary differential equations. For Laplace's equation there are about a dozen such coordinate systems; only rectangular, circular cylindrical, and spherical coordinates are likely to be familiar to most readers of this book. On the other hand, the boundary of the region of interest must consist of coordinate curves or surfaces; that is, of curves or surfaces on which one variable remains constant. Thus, at an elementary level, one is limited to regions bounded by straight lines or circular arcs in two dimensions, or by planes, circular cylinders, circular cones, or spheres in three dimensions.

In three-dimensional problems the separation of variables in Laplace's operator $u_{xx} + u_{yy} + u_{zz}$ leads to the equation $X'' + \lambda X = 0$ in rectangular coordinates, to Bessel's equation in cylindrical coordinates, and to Legendre's equation in spherical coordinates. It is this fact that is largely responsible for the intensive study that has been made of these equations and the functions defined by them. It is also noteworthy that two of the three most important situations lead to singular, rather than regular, Sturm-Liouville problems. Thus, singular problems are by no means exceptional and may be of even greater interest than regular ones. The remainder of this section is devoted to an example involving an expansion of a given function as a series of Bessel functions.

THE VIBRATIONS OF A CIRCULAR ELASTIC MEMBRANE. In Section 10.7 [Eq. (7)] it was noted that the transverse vibrations of a thin elastic membrane are governed by the two-dimensional wave equation

$$a^2(u_{xx} + u_{yy}) = u_{tt}. \tag{1}$$

To study the motion of a circular membrane it is convenient to write Eq. (1) in polar coordinates:

$$a^2\left(u_{rr} + \frac{1}{r}u_r + \frac{1}{r^2}u_{\theta\theta}\right) = u_{tt}. \tag{2}$$

We will assume that the membrane has unit radius, that it is fixed rigidly around its circumference, and that initially it occupies a displaced position independent of the angular variable θ, from which it is released at time $t = 0$. Because of the circular symmetry of the initial and boundary conditions, it is natural to assume also that u is independent of θ; that is, u is a function of r and t only. In this event the differential equation (2) becomes

$$a^2\left(u_{rr} + \frac{1}{r}u_r\right) = u_{tt}, \qquad 0 < r < 1, \quad t > 0. \tag{3}$$

The boundary condition at $r = 1$ is

$$u(1,t) = 0, \qquad t \geq 0, \tag{4}$$

and the initial conditions are

$$u(r,0) = f(r), \qquad 0 \leq r \leq 1, \tag{5}$$

$$u_t(r,0) = 0, \qquad 0 \leq r \leq 1, \tag{6}$$

where $f(r)$ describes the initial configuration of the membrane. For consistency we also require that $f(1) = 0$. Finally, we state explicitly the requirement that $u(r,t)$ is to be bounded for $0 \leq r \leq 1$.

Assuming that $u(r,t) = R(r)T(t)$, and substituting for $u(r,t)$ in Eq. (3), we obtain

$$\frac{R'' + (1/r)R'}{R} = \frac{1}{a^2}\frac{T''}{T} = \sigma. \tag{7}$$

As in Chapter 10 we consider all possible cases for the separation constant σ. We find that to obtain nontrivial bounded solutions satisfying the homogeneous boundary and initial conditions we must have $\sigma < 0$. Hence we let $\sigma = -\lambda^2$ with $\lambda > 0$. Then Eq. (7) yields

$$r^2R'' + rR' + \lambda^2 r^2 R = 0, \tag{8}$$

$$T'' + \lambda^2 a^2 T = 0. \tag{9}$$

Thus

$$T(t) = k_1 \sin \lambda at + k_2 \cos \lambda at. \tag{10}$$

Introducing the new independent variable $\xi = \lambda r$ into Eq. (8), we obtain

$$\xi^2 \frac{d^2R}{d\xi^2} + \xi\frac{dR}{d\xi} + \xi^2 R = 0, \tag{11}$$

which is Bessel's equation of order zero. Thus,

$$R = c_1 J_0(\xi) + c_2 Y_0(\xi), \tag{12}$$

where J_0 and Y_0 are Bessel functions of the first and second kinds, respectively, of order zero (see Section 11.5). In terms of r we have

$$R = c_1 J_0(\lambda r) + c_2 Y_0(\lambda r). \tag{13}$$

The boundedness condition on $u(r,t)$ requires that R remain bounded as $r \to 0$. Since $Y_0(\lambda r) \to -\infty$ as $r \to 0$, we must choose $c_2 = 0$. The boundary condition (4) then requires that

$$J_0(\lambda) = 0. \tag{14}$$

Consequently, the allowable values of the separation constant are obtained from the roots of the transcendental equation (14). Recall from Section 11.5 that $J_0(\lambda)$ has an infinite set of discrete positive zeros, which we denote by $\lambda_1, \lambda_2, \lambda_3, \ldots, \lambda_n, \ldots$, ordered in increasing magnitude. Further, the functions $J_0(\lambda_n r)$ are the eigenfunctions of a singular Sturm-Liouville problem, and can be used as the basis of a series expansion for the given function f. The fundamental solutions of this problem, satisfying the partial differential equation (3), the boundary condition (4), and the boundedness condition, are

$$u_n(r,t) = J_0(\lambda_n r)\sin\lambda_n at, \qquad n = 1, 2, \ldots \tag{15}$$

$$v_n(r,t) = J_0(\lambda_n r)\cos\lambda_n at, \qquad n = 1, 2, \ldots. \tag{16}$$

Next we assume that $u(r,t)$ can be expressed as an infinite linear combination of the fundamental solutions (15), (16):

$$\begin{aligned} u(r,t) &= \sum_{n=1}^{\infty} [k_n u_n(r,t) + c_n v_n(r,t)] \\ &= \sum_{n=1}^{\infty} [k_n J_0(\lambda_n r)\sin\lambda_n at + c_n J_0(\lambda_n r)\cos\lambda_n at]. \end{aligned} \tag{17}$$

The initial conditions require that

$$u(r,0) = \sum_{n=1}^{\infty} c_n J_0(\lambda_n r) = f(r) \tag{18}$$

and

$$u_t(r,0) = \sum_{n=1}^{\infty} \lambda_n a k_n J_0(\lambda_n r) = 0. \tag{19}$$

From Eq. (23) of Section 11.5 we obtain

$$k_n = 0, \qquad c_n = \frac{\int_0^1 rf(r)J_0(\lambda_n r)\,dr}{\int_0^1 r[J_0(\lambda_n r)]^2\,dr}; \qquad n = 1, 2, \ldots. \tag{20}$$

Thus, the solution of the partial differential equation (3) satisfying the boundary conditions (4) and the initial conditions (5) and (6), is given by

$$u(r,t) = \sum_{n=1}^{\infty} c_n J_0(\lambda_n r)\cos\lambda_n at \tag{21}$$

with the coefficients c_n defined by Eq. (20).

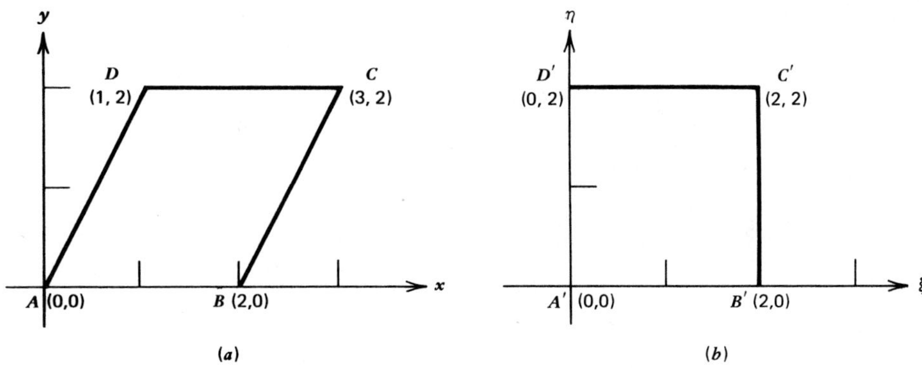

FIGURE 11.4 The region in Problem 1.

PROBLEMS

1. Consider Laplace's equation $u_{xx} + u_{yy} = 0$ in the parallelogram whose vertices are $(0,0)$, $(2,0)$, $(3,2)$, and $(1,2)$. Suppose that on the side $y = 2$ the boundary condition is $u(x,2) = f(x)$ for $1 \le x \le 3$, and that on the other three sides $u = 0$ (see Figure 11.4).
(a) Show that there are no nontrivial solutions of the partial differential equation of the form $u(x, y) = X(x)Y(y)$ that also satisfy the homogeneous boundary conditions.
(b) Let $\xi = x - \frac{1}{2}y$, $\eta = y$. Show that the given parallelogram in the xy plane transforms into the square $0 \le \xi \le 2$, $0 \le \eta \le 2$ in the $\xi\eta$ plane. Show that the differential equation transforms into
$$\tfrac{5}{4}u_{\xi\xi} - u_{\xi\eta} + u_{\eta\eta} = 0.$$
How are the boundary conditions transformed?
(c) Show that in the $\xi\eta$ plane the differential equation possesses no solutions of the form
$$u(\xi,\eta) = U(\xi)V(\eta).$$
Thus in the xy plane the shape of the boundary precludes a solution by the method of separation of variables, while in the $\xi\eta$ plane the region is acceptable but the variables in the differential equation can no longer be separated.
2. Find the displacement $u(r,t)$ in a vibrating circular elastic membrane of radius one that satisfies the boundary condition
$$u(1,t) = 0, \qquad t \ge 0,$$
and the initial conditions
$$u(r,0) = 0, \qquad u_t(r,0) = g(r), \qquad 0 \le r \le 1,$$
where $g(1) = 0$.
Hint: The differential equation to be satisfied is Eq. (3) of the text.

3. Find the displacement $u(r,t)$ in a vibrating circular elastic membrane of radius one that satisfies the boundary condition

$$u(1,t) = 0, \qquad t \ge 0,$$

and the initial conditions

$$u(r,0) = f(r), \qquad u_t(r,0) = g(r), \qquad 0 \le r \le 1,$$

where $f(1) = g(1) = 0$.

4. The wave equation in polar coordinates is

$$u_{rr} + (1/r)\,u_r + (1/r^2)\,u_{\theta\theta} = a^{-2}u_{tt}.$$

Show that if $u(r,\theta,t) = R(r)\Theta(\theta)T(t)$, then R, Θ, and T satisfy the ordinary differential equations

$$r^2R'' + rR' + (\lambda^2r^2 - n^2)\,R = 0,$$

$$\Theta'' + n^2\Theta = 0,$$

$$T'' + \lambda^2a^2T = 0.$$

5. In the circular cylindrical coordinates r, θ, z defined by

$$x = r\cos\theta, \qquad y = r\sin\theta, \qquad z = z,$$

Laplace's equation takes the form

$$u_{rr} + (1/r)\,u_r + (1/r^2)\,u_{\theta\theta} + u_{zz} = 0.$$

(a) Show that if $u(r,\theta,z) = R(r)\Theta(\theta)Z(z)$, then R, Θ, and Z satisfy the ordinary differential equations

$$r^2R'' + rR' + (\lambda^2r^2 - n^2)\,R = 0,$$

$$\Theta'' + n^2\Theta = 0,$$

$$Z'' - \lambda^2Z = 0.$$

(b) Show that if $u(r,\theta,z)$ is independent of θ, then the first equation in part (a) becomes

$$r^2R'' + rR' + \lambda^2r^2R = 0,$$

the second is omitted altogether, and the third is unchanged.

6. Find the steady-state temperature in a semi-infinite rod $0 < z < \infty$, $0 \le r < 1$, if the temperature is independent of θ and approaches zero as $z \to \infty$. Assume that the temperature $u(r,z)$ satisfies the boundary conditions

$$u(1,z) = 0, \qquad z > 0,$$

$$u(r,0) = f(r), \qquad 0 \le r \le 1.$$

Hint: Refer to Problem 5.

7. The equation

$$v_{xx} + v_{yy} + k^2v = 0$$

is a generalization of Laplace's equation, and is sometimes called the Helmholtz (1821–1894) equation.

(a) In polar coordinates the Helmholtz equation becomes

$$v_{rr} + (1/r)\,v_r + (1/r^2)\,v_{\theta\theta} + k^2 v = 0.$$

If $v(r,\theta) = R(r)\Theta(\theta)$ show that R and Θ satisfy the ordinary differential equations

$$r^2R'' + rR' + (k^2r^2 - \lambda^2)\,R = 0, \qquad \Theta'' + \lambda^2\Theta = 0.$$

(b) Find the solution of the Helmholtz equation in the disk $r < c$, which remains bounded at all points in the disk, which is periodic in θ with period 2π, and satisfies the boundary condition $v(c,\theta) = f(\theta)$, where f is a given function on $0 \le \theta < 2\pi$. *Hint:* The equation for R is a Bessel equation. See Problem 3 of Section 11.5.

8. Consider the flow of heat in an infinitely long cylinder of radius one: $0 \le r < 1$, $0 \le \theta < 2\pi$, $-\infty < z < \infty$. Let the surface of the cylinder be held at temperature zero, and the initial temperature distribution be a function of the radial variable r only. Then the temperature u is a function of r and t only, and satisfies the heat conduction equation

$$\alpha^2[\,u_{rr} + (1/r)\,u_r\,] = u_t, \qquad 0 < r < 1, \quad t > 0,$$

and the following initial and boundary conditions:

$$u(r,0) = f(r), \qquad 0 \le r \le 1,$$
$$u(1,t) = 0, \qquad t > 0.$$

Show that

$$u(r,t) = \sum_{n=1}^{\infty} c_n J_0(\lambda_n r)\,e^{-\alpha^2\lambda_n^2 t}$$

where $J_0(\lambda_n) = 0$. Find a formula for c_n.

9. In the spherical coordinates ρ, θ, ϕ $(\rho > 0, 0 \le \theta < 2\pi, 0 \le \phi \le \pi)$ defined by the equations

$$x = \rho\cos\theta\sin\phi, \qquad y = \rho\sin\theta\sin\phi, \qquad z = \rho\cos\phi,$$

Laplace's equation takes the form

$$\rho^2 u_{\rho\rho} + 2\rho u_\rho + (\csc^2\phi)\,u_{\theta\theta} + u_{\phi\phi} + (\cot\phi)\,u_\phi = 0.$$

(a) Show that if $u(\rho,\theta,\phi) = P(\rho)\Theta(\theta)\Phi(\phi)$, then P, Θ, and Φ satisfy ordinary differential equations of the form

$$\rho^2P'' + 2\rho P' - \mu^2 P = 0,$$
$$\Theta'' + \lambda^2\Theta = 0,$$
$$(\sin^2\phi)\,\Phi'' + (\sin\phi\cos\phi)\,\Phi' + (\mu^2\sin^2\phi - \lambda^2)\,\Phi = 0.$$

The first of these equations is of the Euler type, while the third is related to Legendre's equation.

(b) Show that if $u(\rho,\theta,\phi)$ is independent of θ, then the first equation in part (a) is unchanged, the second is omitted, and the third becomes

$$(\sin^2\phi)\,\Phi'' + (\sin\phi\cos\phi)\,\Phi' + (\mu^2\sin^2\phi)\,\Phi = 0.$$

(c) Show that if a new independent variable s is defined by $s = \cos\phi$, then the equation for Φ in part (b) becomes

$$(1 - s^2)\frac{d^2\Phi}{ds^2} - 2s\frac{d\Phi}{ds} + \mu^2\Phi = 0, \qquad -1 \le s \le 1.$$

Note that this is Legendre's equation.

10. Find the steady-state temperature $u(\rho, \phi)$ in a sphere of unit radius if the temperature is independent of θ and satisfies the boundary condition

$$u(1, \phi) = f(\phi), \qquad 0 \le \phi \le \pi.$$

Hint: Refer to Problem 9 and to Problems 10 through 16 of Section 4.2.1. Use the fact that the only solutions of Legendre's equation that are finite at both ± 1 are the Legendre polynomials.

*11.7 Series of Orthogonal Functions: Mean Convergence

In Section 11.3 we stated that under certain restrictions a given function f can be expanded in a series of eigenfunctions of a Sturm-Liouville boundary value problem, the series converging to $[f(x+) + f(x-)]/2$ at each point in the open interval. Under somewhat more restrictive conditions the series converges to $f(x)$ at each point in the closed interval. This type of convergence is referred to as pointwise convergence. In this section we describe a different kind of convergence that is especially useful for series of orthogonal functions, such as eigenfunctions.

Suppose that we are given the set of functions $\phi_1, \phi_2, \ldots, \phi_n$, which are continuous on the interval $0 \le x \le 1$ and satisfy the orthonormality condition

$$\int_0^1 r(x)\phi_i(x)\phi_j(x)\,dx = \begin{cases} 0, & i \ne j, \\ 1, & i = j, \end{cases} \tag{1}$$

where r is a nonnegative weight function. Suppose also that we wish to approximate a given function f, defined on $0 \le x \le 1$, by a linear combination of $\phi_1, \ldots, \phi_n$. That is, if

$$S_n(x) = \sum_{i=1}^{n} a_i \phi_i(x), \tag{2}$$

we wish to choose the coefficients $a_1, \ldots, a_n$ so that the function S_n will best approximate f on $0 \le x \le 1$. The first problem that we must face in doing this is to state precisely what we mean by "best approximate f on $0 \le x \le 1$." There are several reasonable meanings that can be attached to this phrase.

1. We can choose n points $x_1, \ldots, x_n$ in the interval $0 \le x \le 1$, and require that $S_n(x)$ have the same value as $f(x)$ at each of these points. The coefficients $a_1, \ldots, a_n$ are found by solving the set of linear algebraic equations

$$\sum_{i=1}^{n} a_i \phi_i(x_j) = f(x_j), \qquad j = 1, \ldots, n. \tag{3}$$

This procedure is known as the *method of collocation*. It has the advantage that it is very easy to write down Eqs. (3); one needs only to evaluate the functions involved at the points $x_1, \ldots, x_n$. If these points are well chosen, and if n is fairly large, then presumably $S_n(x)$ will not only be equal to $f(x)$ at the chosen points but will be reasonably close to it at other points as well. However, collocation has several deficiencies. One is that if one more base function ϕ_{n+1} is added, then one more point x_{n+1} is required, and *all* of the coefficients must be recomputed. Thus, it is inconvenient to improve the accuracy of a collocation approximation by including additional terms. Further, the a_i depend on the location of the points $x_1, \ldots, x_n$, and it is not clear how best to select these points.

2. Alternatively, we can consider the difference $|f(x) - S_n(x)|$ and try to make it as small as possible. The trouble here is that $|f(x) - S_n(x)|$ is a function of x as well as of the coefficients $a_1, \ldots, a_n$, and it is not obvious what criterion should be used in selecting the a_i. The choice of a_i that makes $|f(x) - S_n(x)|$ small at one point may make it large at another. One way to proceed is to consider instead the least upper bound[9] of $|f(x) - S_n(x)|$ for x in $0 \le x \le 1$, and then to choose $a_1, \ldots, a_n$ so as to make this quantity as small as possible. That is, if

$$E_n(a_1, \ldots, a_n) = \underset{0 \le x \le 1}{\text{lub}} |f(x) - S_n(x)|, \tag{4}$$

then choose $a_1, \ldots, a_n$ so as to minimize E_n. This approach is intuitively appealing, and is often used in theoretical calculations. However, in practice, it is usually very hard, if not impossible, to write down an explicit formula for $E_n(a_1, \ldots, a_n)$. Further, this procedure also shares one of the disadvantages of collocation; namely, on adding an additional term to $S_n(x)$, one must recompute all of the preceding coefficients. Thus, it is not often useful in practical problems.

3. Another way to proceed is to consider

$$I_n(a_1, \ldots, a_n) = \int_0^1 r(x)|f(x) - S_n(x)|\,dx. \tag{5}$$

If $r(x) = 1$, then I_n is the area between the graphs of $y = f(x)$ and $y = S_n(x)$ (see Figure 11.5). We can then determine the coefficients a_i so as to minimize I_n. To avoid the complications resulting from calculations with absolute values, it is more convenient to consider instead

$$R_n(a_1, \ldots, a_n) = \int_0^1 r(x)[f(x) - S_n(x)]^2\,dx \tag{6}$$

as our measure of the quality of approximation of the linear combination $S_n(x)$ to $f(x)$. While R_n is clearly similar in some ways to I_n, it lacks the simple geometric interpretation of the latter. Nevertheless, it is much easier mathematically to deal with R_n than with I_n. The quantity R_n is called the

[9] The least upper bound (lub) is an upper bound that is smaller than any other upper bound. The lub of a bounded function always exists, and is equal to the maximum if the function has one. Recall, however, that a discontinuous function may not have a maximum. See Problem 2 for examples.

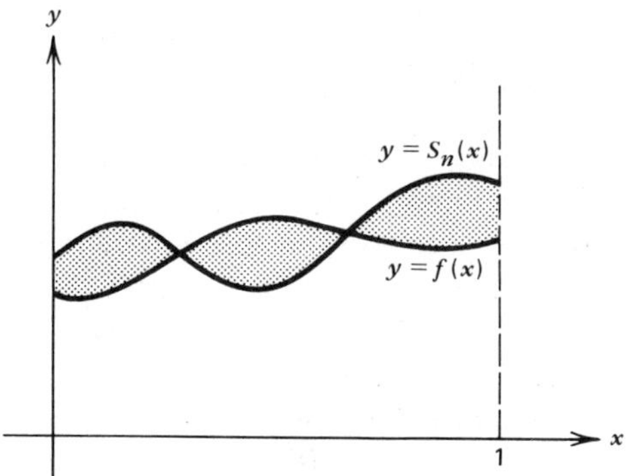

FIGURE 11.5 Approximation of f(x) by $S_n(x)$.

mean square error of the approximation S_n to f. If $a_1, \ldots, a_n$ are chosen so as to minimize R_n, then S_n is said to approximate f in the mean square sense.

To choose $a_1, \ldots, a_n$ so as to minimize R_n, we must satisfy the necessary conditions

$$\partial R_n/\partial a_i = 0, \qquad i = 1, \ldots, n. \tag{7}$$

Writing out Eq. (7), and noting that $\partial S_n(x; a_1, \ldots, a_n)/\partial a_i$ is equal to $\phi_i(x)$, we obtain

$$-\frac{\partial R_n}{\partial a_i} = 2\int_0^1 r(x)[f(x) - S_n(x)]\phi_i(x)\,dx = 0. \tag{8}$$

Substituting for $S_n(x)$ from Eq. (2) and making use of the orthogonality relation (1) leads to

$$a_i = \int_0^1 r(x)f(x)\phi_i(x)\,dx, \qquad i = 1, \ldots, n. \tag{9}$$

The coefficients defined by Eq. (9) are called the Fourier coefficients of f with respect to the orthonormal set $\phi_1, \phi_2, \ldots, \phi_n$ and the weight function r. Since the conditions (7) are only necessary and not sufficient for R_n to be a minimum, a separate argument is required to show that R_n is actually minimized if the a_i are chosen by Eq. (9). This argument is outlined in Problem 3.

Note that the coefficients (9) are the same as those in the eigenfunction series whose convergence, under certain conditions, was stated in Theorem 11.4. Thus, $S_n(x)$ is the nth partial sum in this series, and constitutes the best mean square approximation to $f(x)$ that is possible with the functions $\phi_1, \ldots, \phi_n$. We will assume hereafter that the coefficients a_i in $S_n(x)$ are given by Eq. (9).

Equation (9) is noteworthy in two other important respects. In the first place, it gives a formula for each a_i *separately*, rather than a set of linear algebraic equations for $a_1, \ldots, a_n$, as in the method of collocation, for example. This is due to the orthogonality of the base functions $\phi_1, \ldots, \phi_n$. Further, the formula for a_i is *independent* of n, the number of terms in $S_n(x)$. The practical significance of this is as follows. Suppose that, to obtain a better approximation to f, we desire to use an approximation with more terms, say k terms, where $k > n$. It is then unnecessary to recompute the first n coefficients in $S_k(x)$. All that is required is to compute from Eq. (9) the coefficients $a_{n+1}, \ldots, a_k$ arising from the additional base functions $\phi_{n+1}, \ldots, \phi_k$. Of course, if f, r, and the ϕ_n are complicated functions it may be necessary to evaluate the integrals by routine numerical procedures.

Now let us suppose that there is an infinite sequence of functions $\phi_1, \ldots, \phi_n, \ldots,$ which are continuous and orthonormal on the interval $0 \le x \le 1$. Suppose further that as n increases without bound the mean square error R_n approaches zero. In this event the infinite series

$$\sum_{i=1}^{\infty} a_i \phi_i(x)$$

is said to *converge in the mean square sense* (or, more simply, *in the mean*) to $f(x)$. Mean convergence is an essentially different type of convergence than the pointwise convergence considered up to now. A series may converge in the mean without converging at each point. This is plausible geometrically because the area between two curves, which behaves in the same way as the mean square error, may be zero even though the functions are not the same at every point. They may differ on any finite set of points, for example, without affecting the mean square error. It is less obvious, but also true, that even if an infinite series converges at every point, it may not converge in the mean. Indeed, the mean square error may even become unbounded. An example of this phenomenon is given in Problem 1.

Now suppose that we wish to know what class of functions, defined on $0 \le x \le 1$, can be represented as an infinite series of the orthonormal set ϕ_i, $i = 1, 2, \ldots$. The answer depends on what kind of convergence we require. We say that the set $\phi_1, \ldots, \phi_n, \ldots$ is *complete* with respect to mean square convergence for a set of functions $\mathscr{F}$, if for each function f in $\mathscr{F}$, the series

$$f(x) = \sum_{i=1}^{\infty} a_i \phi_i(x), \tag{10}$$

with coefficients given by Eq. (9), converges in the mean. There is a similar definition for completeness with respect to pointwise convergence.

Theorems having to do with the convergence of series such as that in Eq. (10), can now be restated in terms of the idea of completeness. For example, Theorem 11.4 can be restated as follows: The eigenfunctions of the Sturm-Liouville problem

$$-[p(x)y']' + q(x)y = \lambda r(x)y, \qquad 0 < x < 1, \tag{11}$$

$$a_1 y(0) + a_2 y'(0) = 0, \quad b_1 y(1) + b_2 y'(1) = 0, \tag{12}$$

are complete with respect to ordinary pointwise convergence for the set of functions which are continuous on $0 \leq x \leq 1$, and have a piecewise continuous derivative there.

If pointwise convergence is replaced by mean convergence, Theorem 11.4 can be considerably generalized. Before we state such a companion theorem to Theorem 11.4, we first define what is meant by a square integrable function. A function f is said to be *square integrable* on the interval $0 \leq x \leq 1$ if both f and f^2 are integrable[10] on that interval. The following theorem is similar to Theorem 11.4 except that it involves mean convergence.

> ***Theorem 11.7.*** *The eigenfunctions ϕ_i of the Sturm-Liouville problem* (11), (12) *are complete with respect to mean convergence for the set of functions which are square integrable on $0 \leq x \leq 1$. In other words, given any square integrable function f, the series* (10) *whose coefficients are given by Eq.* (9), *converges to $f(x)$ in the mean square sense.*

It is significant that the class of functions specified in Theorem 11.7 is very large indeed. The class of square integrable functions contains some functions with many discontinuities, including some kinds of infinite discontinuities, as well as some functions that are not differentiable at any point. All of these functions have mean convergent expansions in the eigenfunctions of the boundary value problem (11), (12). However, in many cases these series do not converge pointwise, at least not at every point. Thus mean convergence is more naturally associated with series of orthogonal functions, such as eigenfunctions, than ordinary pointwise convergence.

The theory of Fourier series discussed in Chapter 10 is just a special case of the general theory of Sturm-Liouville problems. For instance, the functions $\phi_n(x) = \sqrt{2}\sin n\pi x$ are the normalized eigenfunctions of the Sturm-Liouville problem $y'' + \lambda y = 0$, $y(0) = 0$, $y(1) = 0$. Thus, if f is a given square integrable function on $0 \leq x \leq 1$, then according to Theorem 11.7, the series

$$f(x) = \sum_{n=1}^{\infty} b_n\phi_n(x) = \sqrt{2}\sum_{n=1}^{\infty} b_n \sin n\pi x, \tag{13}$$

where

$$b_n = \int_0^1 f(x)\phi_n(x)\,dx = \sqrt{2}\int_0^1 f(x)\sin n\pi x\,dx, \tag{14}$$

converges in the mean. The series (13) is precisely the Fourier sine series discussed

[10] For the Riemann integral used in elementary calculus the hypotheses that f and f^2 are integrable are independent; that is, there are functions such that f is integrable but f^2 is not, and conversely (see Problem 4). A generalized integral, known as the Lebesgue integral, can be defined and has the property (among others) that if f^2 is integrable, then f is also necessarily integrable. The term *square integrable* came into common use in connection with this type of integration.

in Section 10.5. If f satisfies the further conditions stated in Theorem 11.4, then this series converges pointwise as well as in the mean. Similarly, the Fourier cosine series corresponds to the Sturm-Liouville problem $y'' + \lambda y = 0$, $y'(0) = 0$, $y'(1) = 0$.

Theorem 11.7 can be extended to cover self-adjoint boundary value problems having periodic boundary conditions, such as the problem

$$y'' + \lambda y = 0, \tag{15}$$

$$y(-l) - y(l) = 0, \quad y'(-l) - y'(l) = 0, \tag{16}$$

considered in Example 4 of Section 11.3. The eigenfunctions of the problem (15), (16) are $\phi_n(x) = \cos(n\pi x/l)$ for $n = 0, 1, 2\ldots$ and $\psi_n(x) = \sin(n\pi x/l)$ for $n = 1, 2, \ldots$. If f is a given square integrable function on $-l \le x \le l$, then its expansion in terms of the eigenfunctions ϕ_n and ψ_n is of the form

$$f(x) = \frac{a_0}{2} + \sum_{n=1}^{\infty} \left(a_n \cos\frac{n\pi x}{l} + b_n \sin\frac{n\pi x}{l} \right), \tag{17}$$

where

$$a_n = \frac{1}{l}\int_{-l}^{l} f(x)\cos\frac{n\pi x}{l}\,dx, \qquad n = 0, 1, 2, \ldots, \tag{18}$$

$$b_n = \frac{1}{l}\int_{-l}^{l} f(x)\sin\frac{n\pi x}{l}\,dx, \qquad n = 1, 2, \ldots. \tag{19}$$

This expansion is exactly the Fourier series for f discussed in Sections 10.3 and 10.4. According to the generalization of Theorem 11.7, the series (17) converges in the mean for any square integrable function f, even though f may not satisfy the conditions of Theorem 10.1, which assure pointwise convergence.

PROBLEMS

1. In this problem we show that pointwise convergence of a sequence $S_n(x)$ does not imply mean convergence, and conversely.

   (a) Let $S_n(x) = n\sqrt{x}\,e^{-nx^2/2}$, $0 \le x \le 1$. Show that $S_n(x) \to 0$ as $n \to \infty$ for each x in $0 \le x \le 1$. Show also that

   $$R_n = \int_0^1 [0 - S_n(x)]^2\,dx = \frac{n}{2}(1 - e^{-n});$$

   hence that $R_n \to \infty$ as $n \to \infty$. Thus pointwise convergence does not imply mean convergence.

   (b) Let $S_n(x) = x^n$ for $0 \le x \le 1$ and let $f(x) = 0$ for $0 \le x \le 1$. Show that

   $$R_n = \int_0^1 [f(x) - S_n(x)]^2\,dx = \frac{1}{2n+1},$$

   and hence $S_n(x)$ converges to $f(x)$ in the mean. Also show that $S_n(x)$ does not converge to $f(x)$ pointwise throughout $0 \le x \le 1$. Thus mean convergence does not imply pointwise convergence.

2. Find the least upper bound of each of the following functions on the given interval. Also determine whether the function has a maximum on the interval.

(a) $f(x) = \begin{cases} x, & 0 \le x < 1 \\ 0, & x = 1 \end{cases}$ (b) $f(x) = x, \quad 0 < x < 1$

(c) $f(x) = \dfrac{x^2}{1 + x^2}, \quad 0 \le x < \infty$

(d) $f(x) = \begin{cases} 1 - x, & 0 \le x < 1 \\ x - 2, & 1 \le x \le 2 \end{cases}$ (e) $f(x) = \begin{cases} 1/x, & 0 < x \le 1 \\ 0, & x = 0 \end{cases}$

3. Suppose that the functions $\phi_1, \dots, \phi_n$ satisfy the orthonormality relation (1), and that a given function f is to be approximated by $S_n(x) = c_1\phi_1(x) + \cdots + c_n\phi_n(x)$, where the coefficients c_i are not necessarily those of Eq. (9). Show that the mean square error R_n given by Eq. (6) may be written in the form

$$R_n = \int_0^1 r(x) f^2(x)\,dx - \sum_{i=1}^{n} a_i^2 + \sum_{i=1}^{n} (c_i - a_i)^2,$$

where the a_i are the Fourier coefficients given by Eq. (9). Show that R_n is minimized if $c_i = a_i$ for each i.

4. In this problem we show by examples that the (Riemann) integrability of f and f^2 are independent.

(a) Let $f(x) = \begin{cases} x^{-1/2}, & 0 < x \le 1, \\ 0, & x = 0. \end{cases}$

Show that $\int_0^1 f(x)\,dx$ exists as an improper integral, but $\int_0^1 f^2(x)\,dx$ does not.

(b) Let $f(x) = \begin{cases} 1, & x \text{ rational}, \\ -1, & x \text{ irrational}. \end{cases}$

Show that $\int_0^1 f^2(x)\,dx$ exists, but $\int_0^1 f(x)\,dx$ does not.

5. Suppose that it is desired to construct a set of polynomials $f_0(x), f_1(x), f_2(x), \dots, f_k(x), \dots$, where $f_k(x)$ is of degree k, that are orthonormal on the interval $0 \le x \le 1$. That is, the set of polynomials must satisfy

$$(f_j, f_k) = \int_0^1 f_j(x) f_k(x)\,dx = \delta_{jk}.$$

(a) Find $f_0(x)$ by choosing the polynomial of degree zero such that $(f_0, f_0) = 1$.
(b) Find $f_1(x)$ by determining the polynomial of degree one such that $(f_0, f_1) = 0$ and $(f_1, f_1) = 1$.
(c) Find $f_2(x)$.
(d) The normalization condition $(f_k, f_k) = 1$ is somewhat awkward to apply. Let $g_0(x), g_1(x), \dots, g_k(x), \dots$ be the sequence of polynomials that are orthogonal on $0 \le x \le 1$ and that are normalized by the condition $g_k(1) = 1$. Find $g_0(x)$, $g_1(x)$, and $g_2(x)$ and compare them with $f_0(x)$, $f_1(x)$, and $f_2(x)$.

6. Suppose that it is desired to construct a set of polynomials $P_0(x), P_1(x), \dots, P_k(x), \dots$, where $P_k(x)$ is of degree k, that are orthogonal on the interval $-1 \le x \le 1$; see Problem 5. Suppose further that $P_k(x)$ is normalized by the condition $P_k(1) = 1$. Find $P_0(x)$, $P_1(x)$, $P_2(x)$, and $P_3(x)$. Note that these are the first four Legendre polynomials (see Problem 11 of Section 4.2.1).

7. This problem develops some further results associated with mean convergence. Let $R_n(a_1, \ldots, a_n)$, $S_n(x)$, and a_i be defined by Eqs. (6), (2), and (9), respectively.
(a) Show that

$$R_n = \int_0^1 r(x) f^2(x)\, dx - \sum_{i=1}^{n} a_i^2.$$

Hint: Substitute for $S_n(x)$ in Eq. (6) and integrate, using the orthogonality relation (1).

(b) Show that $\sum_{i=1}^{n} a_i^2 \le \int_0^1 r(x) f^2(x)\, dx$. This result is known as *Bessel's inequality*.

(c) Show that $\sum_{i=1}^{\infty} a_i^2$ converges.

(d) Show that $\lim_{n\to\infty} R_n = \int_0^1 r(x) f^2(x)\, dx - \sum_{i=1}^{\infty} a_i^2$.

(e) Show that $\sum_{i=1}^{\infty} a_i \phi_i(x)$ converges to $f(x)$ in the mean if and only if

$$\int_0^1 r(x) f^2(x)\, dx = \sum_{i=1}^{\infty} a_i^2.$$

This result is known as *Parseval's equation*.

In Problems 8 through 10 let $\phi_1, \phi_2, \ldots, \phi_n, \ldots$ be the normalized eigenfunctions of the Sturm-Liouville problem (11), (12).

8. Show that if a_n is the nth Fourier coefficient of a square integrable function f, then $\lim_{n\to\infty} a_n = 0$.
Hint: Use Bessel's inequality, Problem 7(b).
9. Show that the series

$$\phi_1(x) + \phi_2(x) + \cdots + \phi_n(x) + \cdots$$

cannot be the eigenfunction series for any square integrable function.
Hint: See Problem 8.
10. Show that the series

$$\phi_1(x) + \frac{\phi_2(x)}{\sqrt{2}} + \cdots + \frac{\phi_n(x)}{\sqrt{n}} + \cdots$$

is not the eigenfunction series for any square integrable function.
Hint: Use Bessel's inequality, Problem 7(b).
11. Show that Parseval's equation in Problem 7(e) is obtained formally by squaring the series (10) corresponding to f, multiplying by the weight function r, and integrating term by term.

REFERENCES

The following books were mentioned in the text in connection with certain theorems about Sturm-Liouville problems:

Birkhoff, G., and Rota, G.-C., *Ordinary Differential Equations* (3rd ed.) (New York: Wiley, 1978).
Sagan, H., *Boundary and Eigenvalue Problems in Mathematical Physics* (New York: Wiley, 1961).

Weinberger, H., *A First Course in Partial Differential Equations* (New York: Wiley, 1965).
Yosida, K., *Lectures on Differential and Integral Equations* (New York: Wiley-Interscience, 1960).

The following books are convenient sources of numerical and graphical data about Bessel and Legendre functions:

Abramowitz, M., and Stegun, I. A. (Eds.), *Handbook of Mathematical Functions* (New York: Dover, 1965); originally published by the National Bureau of Standards, Washington, D.C., 1964.
Jahnke, E. and Emde, F., *Tables of Functions with Formulae and Curves* (Leipzig: Teubner, 1938); reprinted by Dover, New York, 1945.

The following books also contain much information about Sturm-Liouville problems:

Cole, R. H., *Theory of Ordinary Differential Equations* (New York: Irvington, 1968).
Hochstadt, H., *Differential Equations: A Modern Approach*, (New York: Holt, 1964); reprinted by Dover, New York, 1975.
Miller, R. K., and Michel, A. N., *Ordinary Differential Equations* (New York: Academic Press, 1982).
Tricomi, F. G., *Differential Equations* (New York: Hafner, 1961).

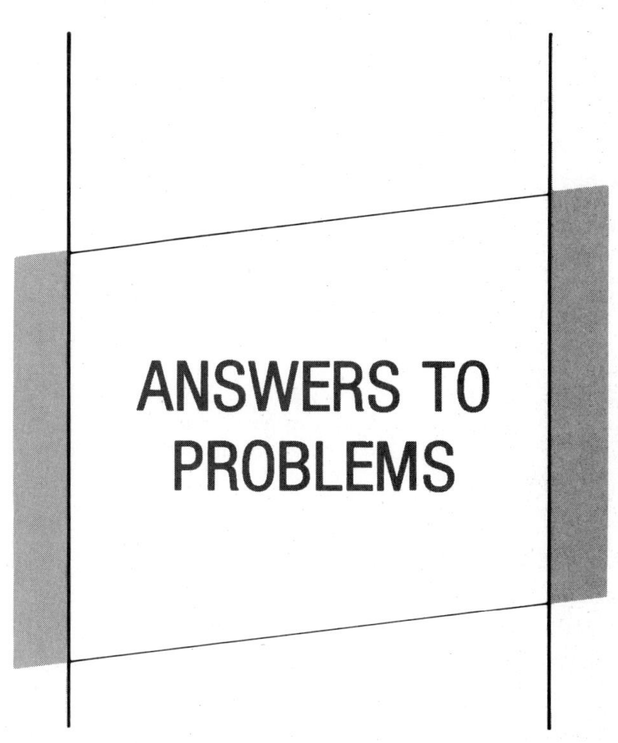

CHAPTER ONE

Section 1.1, Page 6

1. (a) 2nd order, linear (b) 2nd order, nonlinear
   (c) 4th order, linear (d) 1st order, nonlinear
   (e) 2nd order, nonlinear (f) 3rd order, linear
3. (a) $r = -2$ (b) $r = \pm 1$ (c) $r = 2, -3$ (d) $r = 0, 1, 2$
4. (a) $r = -1, -2$ (b) $r = 1, 4$
5. (a) 2nd order, linear (b) 2nd order, linear
   (c) 2nd order, linear (d) 2nd order, nonlinear
   (e) 4th order, linear (f) 2nd order, nonlinear

CHAPTER TWO

Section 2.1, Page 20

1. $y = ce^{-3x} + (x/3) - (1/9) + e^{-2x}$
2. $y = ce^{2x} + x^3e^{2x}/3$
3. $y = ce^{-x} + 1 + x^2e^{-x}/2$
4. $y = (c/x) + (3\cos 2x)/4x + (3\sin 2x)/2$
5. $y = ce^x + 2xe^x$
6. $y = (c - x\cos x + \sin x)/x^2$
7. $y = x^2e^{-x^2} + ce^{-x^2}$
8. $y = (\arctan x + c)/(1 + x^2)^2$
9. $y = 3e^x + 2(x - 1)e^{2x}$
10. $y = (x^2 - 1)e^{-2x}/2$
11. $y = (3x^4 - 4x^3 + 6x^2 + 1)/12x^2$
12. $y = (\sin x)/x^2$
13. $y = (x + 2)e^{2x}$
14. $y = x^{-2}[(\pi^2/4) - 1 - x\cos x + \sin x]$

15. $y = \text{arccosh}\, x$
20. (a) See Problem 2. (b) See Problem 4.

Section 2.2, Page 27

1. $y = (c + \sin x)/x - \cos x$
2. $y = (c - \cos x)/x^3$
3. $y = (c - 2x\cos x + 2\sin x)\cos x$
4. $y = [c + (x - 1)e^x]/x^2$
5. $y = (3x^2 - 4x + 6 + x^{-2})/12, \quad x > 0$
6. $y = x^{-1}(e^x + 1 - e), \quad x > 0$
7. $y = (2x + 1 - \pi)/\sin x, \quad 0 < x < \pi$
8. $y = x^{-2}(\sin x - x\cos x), \quad x > 0$
9. $y = (2x + \pi)\csc x - 2\cos x, \quad -\pi < x < 0$
10. $y = (x^3 + x + 1)/x(x + 2), \quad -2 < x < 0$
11. $y = e^{-x}\int_0^x \frac{e^t}{1 + t^2}\, dt, \quad -\infty < x < \infty$
12. $y = cx^{-2} + x^{-1}; \quad \lim_{x\to 0} |y| = \infty$ for all c
13. $y = cx + x^2; \quad \lim_{x\to 0} y = 0$ for all c
14. $y = cx + 2x^{3/2}; \quad \lim_{x\to 0} y = 0$ for all c, but $y'', y''', \ldots,$ all $\to \pm\infty$ as $x \to 0$
15. $y = \frac{c + \sin x}{x}; \quad \lim_{x\to 0} y = \begin{cases} \pm\infty, & c \neq 0 \\ 1, & c = 0 \end{cases}$
16. 0.671
17. (b) $-\sqrt{\pi}/2$
18. (a) $y_1(x) = \frac{1}{\mu(x)}; \quad y_2(x) = \frac{1}{\mu(x)}\int^x \mu(s)g(s)\, ds$
20. (a) $y = \frac{1}{2}(1 - e^{-2x})$ for $0 \leq x \leq 1; \quad y = \frac{1}{2}(e^2 - 1)e^{-2x}$ for $x > 1$
    (b) $y = e^{-2x}$ for $0 \leq x \leq 1$; $y = e^{-(x+1)}$ for $x > 1$
22. $y = \pm[5x/(2 + 5cx^5)]^{1/2}$
23. $y = r/(k + cre^{-rx})$
24. $y = \pm[\epsilon/(\sigma + c\epsilon e^{-2\epsilon x})]^{1/2}$
25. $y = \pm\{\mu(t)/[2\int^t \mu(s)\, ds + c]\}^{1/2}, \quad$ where $\mu(t) = \exp(2\Gamma\sin t + 2Tt)$

Section 2.3, Page 36

1. (a) $2x + 5y > 0$ or $2x + 5y < 0$
   (b) $x^2 + y^2 < 1$
   (c) Everywhere
   (d) $x + y > 0$ or $x + y < 0$
   (e) $1 - x^2 + y^2 > 0$, or $1 - x^2 + y^2 < 0, \quad x \neq 0, y \neq 0$
   (f) Everywhere
2. $|x| < 1$
3. $y = [2(x - \frac{7}{8})]^{-1/2}; \; x > \frac{7}{8}$
4. $x > -(1 - e^{-3/2})^{1/3}$
5. For given c, solutions are valid for $|x| < |c|/2$.
   $(0, 4)$: $y = (16 - 4x^2)^{1/2}; \quad (1, -1)$: $y = -(5 - 4x^2)^{1/2}$
6. (a) $y_1(x)$ is solution for $x \geq 2; \quad y_2(x)$ is solution for all x
   (b) f_y is not continuous at $(2, -1)$

Section 2.4, Page 41

1. $3y^2 - 2x^3 = c; \quad y \neq 0$
2. $3y^2 - 2\ln|1 + x^3| = c; \quad x \neq -1, y \neq 0$
3. $y^{-1} + \cos x = c$ if $y \neq 0$; also $y = 0$; everywhere
4. $\arctan y - x - x^2/2 = c$; everywhere
5. $2\tan 2y - 2x - \sin 2x = c$ if $\cos 2y \neq 0$; also $y = \pm(2n + 1)\pi/4$ for any integer n; everywhere
6. $y = \sin[\ln|x| + c]$ if $x \neq 0$ and $|y| < 1$; also $y = \pm 1$; $x \neq 0$ and $|y| < 1$

7. $y^2 - x^2 + 2(e^y - e^{-x}) = c; \quad y + e^y \neq 0$
8. $3y + y^3 - x^3 = c;$ everywhere
9. $y = \frac{1}{3}\arcsin(3\cos^2 x), \quad 0.95 < x < 2.19$ approximately
10. $y = [2(1 - x)e^x - 1]^{1/2}, \quad -1.68 < x < 0.77$ approximately
11. $r = 2/(1 - 2\ln\theta), \quad 0 < \theta < \sqrt{e}$
12. $y = -[2\ln(1 + x^2) + 4]^{1/2}, \quad -\infty < x < \infty$
13. $y = [3 - 2\sqrt{1 + x^2}]^{-1/2}, \quad |x| < \frac{1}{2}\sqrt{5}$
14. $y = -\frac{1}{2} + \frac{1}{2}\sqrt{4x^2 - 15}, \quad x > \frac{1}{2}\sqrt{15}$
15. $y = -\sqrt{(x^2 + 1)/2}, \quad -\infty < x < \infty$
16. $y^3 - 4y - x^3 = -1, \quad |x^3 - 1| < 16/3\sqrt{3}$ or $-1.28 < x < 1.60$
17. $2y^3 - 3(\arcsin x)^2 = c$
18. $y = \dfrac{a}{c}x + \dfrac{bc - ad}{c^2}\ln|cx + d| + k; \quad c \neq 0, \quad cx + d \neq 0$
19. $x = \dfrac{c}{a}y + \dfrac{ad - bc}{a^2}\ln|ay + b| + k; \quad a \neq 0, \quad ay + b \neq 0$
20. $|y + 2x|^3|y - 2x| = c$

Section 2.5, Page 50

1. (a) 13.20 years (b) 29.6 mg
2. $-11.7\ln 2/\ln\frac{2}{3} \cong 20.0$ days
3. $1620\ln\frac{4}{3}/\ln 2 \cong 672.4$ years
4. (a) $Q(t) = 35.36 + 64.64\exp(-0.02828t)$, Q in mg, t in days
   (b) 35.36 mg (c) 171.9 days (d) 2.828 mg/day
5. (a) 0.0001245 year^{-1} (b) $Q_0\exp(-0.0001245t)$, t in years
   (c) 12,927 years
6. (a) $(\ln 2)/r$ years (b) 9.90 years (c) 8.66%
7. (a) $k(e^{rt} - 1)/r$ (b) $k \cong \$3930$ (c) 9.77%
8. (a) \$549,119 (b) \$864,268
9. (a) \$74,119 (b) \$98,943 (c) \$123,767
   (d) \$148,591
10. (a) $(k/r) + [S_0 - (k/r)]e^{rt}$ (b) rS_0
    (c) $(1/r)\ln[k/(k - k_0)]$ years (d) $T \cong 8.66$ years
    (e) $rS_0e^{rT}/(e^{rT} - 1)$ (f) \$119,716
11. (a) $S(40) = \$624,601; \quad S(40)/I \cong 12.25$
    (b) $S(40) = \$456,233; \quad S(40)/I \cong 7.24$
    (c) $S(40) = \$662,935; \quad S(40)/I \cong 10.52$
12. $p = (6.0 \times 10^8)\exp(0.005135t)$; A.D. 2376
13. $t = \ln\frac{13}{8}/\ln\frac{13}{12}$ min $\cong 6.07$ min
14. 1 hr, 58 min
15. 11:12 P.M.
16. $r = 3 - 2t$ mm, $\quad 0 < t < \frac{3}{2}$
17. $t = 100\ln 100$ min $\cong 460.5$ min
18. $Q = 50e^{-0.2}(1 - e^{-0.2})$ lb $\cong 7.42$ lb
19. $Q = 200 + t - [100(200)^2/(200 + t)^2]$ lb, $\quad t < 300$;
    $c = 121/125$ lb/gal; $\lim_{t\to\infty} c = 1$ lb/gal
20. (a) $x = 0.04[1 - \exp(-t/12{,}000)]$ (b) $\tau \cong 36$ min
21. (a) $c = k + (P/r) + [c_0 - k - (P/r)]e^{-rt/V}; \quad \lim_{t\to\infty} c = k + (P/r)$
    (b) $T = (V\ln 2)/r; \quad T = (V\ln 10)/r$
    (c) Superior, $T = 430$ years; Michigan, $T = 71.4$ years; Erie, $T = 6.05$ years; Ontario, $T = 17.6$ years

Section 2.6, Page 65

1. $N = 0$ is unstable
2. $N = -a/b$ is stable, $N = 0$ is unstable
3. $N = 1$ is stable, $N = 0$ and $N = 2$ are unstable
4. $N = 0$ is unstable
5. $N = 0$ is stable
6. $N = 0$ is stable
7. (c) $N = [N_0 + (1 - N_0)kt]/[1 + (1 - N_0)kt]$
8. $N = 1$ is semistable
9. $N = -1$ is stable, $N = 0$ is semistable, $N = 1$ is unstable
10. $N = -1$ and $N = 1$ are stable, $N = 0$ is unstable
11. $N = 0$ is stable, $N = b^2/a^2$ is unstable
12. $N = 2$ is stable, $N = 0$ is semistable, $N = -2$ is unstable
13. $N = 0$ and $N = 1$ are semistable
15. (a) $\tau = (1/r)\ln 4$; 55.452 years
    (b) $T = (1/r)\ln[\beta(1 - \alpha)/(1 - \beta)\alpha]$; 175.78 years
16. (a) $N = 0$ is unstable, $N = K$ is stable
    (b) Concave up for $0 < N \le K/e$, concave down for $K/e \le N < K$
17. (a) $N(t) = K\exp\{[\ln(N_0/K)]e^{-rt}\}$
    (b) $N(2) \simeq 0.7153K \simeq 57.6 \times 10^6$ kg
    (c) $\tau \simeq 2.215$ years
18. (c) $Y = EN_2 = KE[1 - (E/r)]$ (d) $Y_m = Kr/4$ for $E = r/2$
19. (a) $N_{1,2} = K[1 \mp \sqrt{1 - (4h/rK)}\,]/2$
20. (a) $y = 0$ is unstable, $y = 1$ is stable (b) $y = y_0[y_0 + (1 - y_0)e^{-\alpha t}]$
21. (a) $y = y_0 e^{-\beta t}$ (b) $x = x_0\exp[-\alpha y_0(1 - e^{-\beta t})/\beta]$
    (c) $x_0\exp(-\alpha y_0/\beta)$
22. (b) $z = 1/[\nu + (1 - \nu)e^{\beta t}]$ (c) 0.0927
24. (a) $\lim_{t\to\infty} x(t) = \min(p, q);$ $x(t) = pq[e^{\alpha(q-p)t} - 1]/[qe^{\alpha(q-p)t} - p]$
    (b) $\lim_{t\to\infty} x(t) = p;$ $x = p^2\alpha t/(p\alpha t + 1)$

Section 2.7, Page 76

1. (a) 50.4 m (b) 5.25 sec
2. (a) 47.3 m (b) 5.14 sec
3. (a) $x_m = v_0^2/2g$ (b) $t_m = v_0/g$ (c) $t = 2v_0/g$
4. $v = (mg/k) + (v_0 - mg/k)e^{-kt/m};$ $v_l = mg/k$
5. $t = (m\ln 10)/k$
6. (a) $v = 5(1 - e^{-0.2t})$ ft/sec (b) $v_l = 5$ ft/sec
7. (a) 176.7 ft/sec (b) 1074.5 ft (c) 15 ft/sec (d) 256.6 sec
8. $2\frac{m}{k}(\sqrt{v_0} - \sqrt{v}) + 2\frac{m^2g}{k^2}\ln\left|\frac{mg - k\sqrt{v_0}}{mg - k\sqrt{v}}\right| = t;$ $v_l = \left(\frac{mg}{k}\right)^2$
9. $v = \sqrt{\frac{mg}{k}}\,\frac{e^{2\sqrt{kg/m}\,t} - 1}{e^{2\sqrt{kg/m}\,t} + 1};$ $v_l = \sqrt{\frac{mg}{k}}$
10. $v_l = \left(\frac{mg}{k}\right)^{1/r}$
11. (a) $x_m = -\frac{m^2g}{k^2}\ln\left(1 + \frac{kv_0}{mg}\right) + \frac{mv_0}{k}$ (b) $t_m = \frac{m}{k}\ln\left(1 + \frac{kv_0}{mg}\right)$
12. (a) $v = -(mg/k) + [v_0 + (mg/k)]\exp(-kt/m)$
    (b) $v = v_0 - gt;$ yes (c) $v = 0$ for $t > 0$
13. (a) $v_l = 2a^2g(\rho - \rho')/9\mu$ (b) $e = 4\pi a^3g(\rho - \rho')/3E$

14. $\dfrac{x_m}{R} = -1 + \left(1 - \dfrac{v_0^2}{v_e^2}\right)^{-1} = \dfrac{v_0^2}{v_e^2}\left(1 + \dfrac{v_0^2}{v_e^2} + \dfrac{v_0^4}{v_e^4} + \cdots\right)$

15. (a) $v = R\sqrt{2g/(R+x)}$ (b) 50.6 hr

16. $v_e = \left(\dfrac{2gR}{1+\xi}\right)^{1/2}$; altitude $\cong$ 1536 miles

Section 2.8, Page 83

1. $x^2 + 3x + y^2 - 2y = c$
2. Not exact
3. $x^3 - x^2y + 2x + 2y^3 + 3y = c$
4. $x^2y^2 + 2xy = c$
5. $ax^2 + 2bxy + cy^2 = k$
6. Not exact
7. $e^x \sin y + 2y\cos x = c$; also $y = 0$
8. Not exact
9. $e^{xy}\cos 2x + x^2 - 3y = c$
10. $y\ln x + 3x^2 - 2y = c$
11. Not exact
12. $x^2 + y^2 = c$
13. $y = [x + \sqrt{28 - 3x^2}]/2, \quad |x| < \sqrt{28/3}$
14. $y = [x - (24x^3 + x^2 - 8x - 16)^{1/2}]/4, \quad x > 0.986$
15. (a) $b = 3; \quad x^2y^2 + 2x^3y = c$ (b) $b = 1; \quad e^{2xy} + x^2 = c$
16. $\int^y N(x,t)\,dt + \int^x \left[M(s,y) - \int^y N_s(s,t)\,dt\right] ds = c$

Section 2.9, Page 87

1. $x^2 + 2\ln|y| - y^{-2} = c$; also $y = 0$
2. $e^x \sin y + 2y\cos x = c$
3. $xy^2 - (y^2 - 2y + 2)e^y = c$
4. $x^2e^x \sin y = c$
6. $\mu(t) = \exp\int^t R(s)\,ds$, where $t = xy$
7. $\mu(x) = e^{3x}; \quad (3x^2y + y^3)e^{3x} = c$
8. $\mu(x) = e^{-x}; \; y = ce^x + 1 + e^{2x}$
9. $\mu(y) = y; \; xy + y\cos y - \sin y = c$
10. $\mu(y) = e^{2y}/y; \quad xe^{2y} - \ln|y| = c$; also $y = 0$
11. $\mu(y) = \sin y; \quad e^x \sin y + y^2 = c$
12. $\mu(y) = y^2; \quad x^4 + 3xy + y^4 = c$
13. $\mu(x,y) = xy; \quad x^3y + 3x^2 + y^3 = c$

Section 2.10, Page 90

1. $y = cx + x\ln|x|$
2. $y = cx^2$
3. $\arctan(y/x) - \ln|x| = c$
4. $x^2 + y^2 - cx^3 = 0$
5. $|y - x| = c|y + 3x|^5$
6. $|y + x||y + 4x|^2 = c$
7. $-2x/(x + y) = \ln c|x + y|$
8. $x/(x + y) + \ln|x| = c$
9. (a) $|y - x| = c|y + x|^3$
   (b) $|y - x + 3| = c|y + x + 1|^3$
10. $|y + x + 4||y + 4x + 13|^2 = c$
11. $-2(x - 2)/(x + y - 3) = \ln c|x + y - 3|$
12. $x^2y^2 + 2x^3y = c$
15. (a) No (b) Yes (c) Yes (d) No

Section 2.11, Page 92

1. $y = (c/x^2) + (x^3/5)$
2. $\arctan(y/x) - \ln\sqrt{x^2 + y^2} = c$
3. $x^2 + xy - 3y - y^3 = 0$
4. $x = ce^y + ye^y$
5. $x^2y + xy^2 + x = c$
6. $y = x^{-1}(1 - e^{1-x})$
7. $(x^2 + y^2 + 1)e^{-y^2} = c$
8. $y = (4 + \cos 2 - \cos x)/x^2$
9. $x^2y + x + y^2 = c$
10. $(y^2/x^3) + (y/x^2) = c$
11. $x^3/3 + xy + e^y = c$
12. $y = ce^{-x} + e^{-x}\ln(1 + e^x)$
13. $2(y/x)^{1/2} - \ln|x| = c$
14. $x^2 + 2xy + 2y^2 = 34$
15. $y = c/\cosh^2(x/2)$
16. $(2/\sqrt{3})\arctan[(2y - x)/\sqrt{3}\,x] - \ln|x| = c$

17. $y = ce^{3x} - e^{2x}$
18. $y = cx^{-2} - x$
19. $3y - 2xy^3 - 10x = 0$
20. $e^x + e^{-y} = c$
21. $e^{-y/x} + \ln|x| = c$
22. $y^3 + 3y - x^3 + 3x = 2$
23. $\dfrac{1}{y} = -x\displaystyle\int^x \frac{e^{2t}}{t^2}\,dt + cx$; also $y = 0$
24. $\sin^2 x \sin y = c$
25. $x^2/y + \arctan(y/x) = c$
26. $x^2 + 2x^2y - y^2 = c$
27. $\sin x \cos 2y - \frac{1}{2}\sin^2 x = c$
28. $2xy + xy^3 - x^3 = c$
29. $2\arcsin(y/x) - \ln|x| = c$
30. $xy^2 - \ln|y| = 0$
31. $x + \ln|x| + x^{-1} + y - 2\ln|y| = c$; also $y = 0$
32. $x^3y^2 + xy^3 = -4$
33. (b) $x = \int G'(p)/p\,dp$
34. $x + e^{-y} = c$
36. (a) $y = x + (c - x)^{-1}$ (b) $y = x^{-1} + 2x(c - x^2)^{-1}$
(c) $y = \sin x + (c\cos x - \frac{1}{2}\sin x)^{-1}$
37. (a) $v' - [x(t) + b]v = b$
(b) $v = [b\int^t \mu(s)\,ds + c]/\mu(t)$, $\mu(t) = \exp[-(at^2/2) - bt]$
38. (a) $dy/dx = 2y/x$ (b) $dy/dx = -x/2y$ (c) $x^2 + 2y^2 = c$
39. (a) $y^2 - x^2 = c$ (b) $x^2 + (y - c)^2 = c^2$
(c) $x - y = c(x + y)^3$ (d) $x^2 - 2cy = c^2$; $c > 0$
40. (a) $y - 3x = c$ (b) $\ln\sqrt{x^2 + y^2} + \arctan(y/x) = c$, or $r = ke^{-\theta}$
41. $y - b = c(x - a)$
42. $(x - 2)^2 + y^2 = 9$
43. $x = cy^2$
44. In six months: \$4 million. After only one year, the fortune becomes unbounded.
45. (b) $(h/a)\sqrt{k/\alpha\pi}$; yes (c) $k/\alpha \le \pi a^2$
46. (b) $k^2/2g(\alpha a)^2$
47. (b) $W = [W_0^{1/3} + (\alpha t/3)]^3$

CHAPTER THREE

Section 3.1, Page 111

1. (a) $y = c_1x^{-1} + c_2 + \ln x$ (b) $y = c_1 \ln x + c_2 + x$
(c) If $c_1 = k^2 > 0$, $y = (1/k)\ln|(k - x)/(k + x)| + c_2$; if $c_1 = -k^2 < 0$, $y = (2/k)\arctan(x/k) + c_2$; if $c_1 = 0$, $y = -2x^{-1} + c_2$; also $y = c$
(d) $y = \pm\frac{2}{3}(x - 2c_1)\sqrt{x + c_1} + c_2$; also $y = c$
Hint: $\mu(v) = v^{-3}$ is an integrating factor
(e) $y = c_1e^{-x} + c_2 - xe^{-x}$;
(f) $c_1^2y = c_1x - \ln|1 + c_1x| + c_2$ if $c_1 \ne 0$; $y = (x^2/2) + c_2$ if $c_1 = 0$; also $y = c$
2. (a) $y^2 = c_1x + c_2$ (b) $y = c_1\sin(x + c_2) = k_1\sin x + k_2\cos x$
(c) $\frac{1}{3}y^3 - 2c_1y + c_2 = 2x$; also $y = c$
(d) $x + c_2 = \pm\frac{2}{3}(y - 2c_1)(y + c_1)^{1/2}$
(e) $y\ln|y| - y + c_1y + x = c_2$; also $y = c$
(f) $e^y = (x + c_2)^2 + c_1$
3. (a) $y = \frac{4}{3}(x + 1)^{3/2} - \frac{1}{3}$ (b) $y = 2(1 - x)^{-2}$
(c) $y = c_1\arctan x + c_2 + 3\ln x - \frac{3}{2}\ln(x^2 + 1)$
(d) $y = \frac{1}{2}x^2 + \frac{3}{2}$

4. (a) Any interval not containing $x = 0$ (b) Any interval
(c) Any interval not containing $x = 0$ or $x = 1$
(d) Any interval not containing $x = 0$ (e) Any interval
(f) Any interval not containing $x = (2n + 1)\pi/2$, $n = 0, \pm 1, \pm 2, \ldots$

5. $\phi''(0) = -p(0)a_1 - q(0)a_0$,
$\phi'''(0) = [p^2(0) - p'(0) - q(0)]a_1 + [p(0)q(0) - q'(0)]a_0$, yes

6. (a) $y'' - y = 0$ (b) $y'' + y = 0$
(c) $(1 - x\cot x)y'' - xy' + y = 0$ (d) $y'' - 2y' + y = 0$
(e) $x^2y'' - 2xy' + 2y = 0$ (f) $y'' - y = 0$

Section 3.2, Page 120

2. $y = \frac{2}{3}e^x + \frac{1}{3}e^{-2x}$, $y = 0$
8. (a) $b + cx$ (b) $(c - a)\sin x + b\cos x$
(c) $(ar^2 + br + c)e^{rx}$ (d) $ar(r - 1)x^{r-2} + brx^{r-1} + cx^r$
9. (a) $(2a + 2b + c)x^2$ (b) $(ar^2x^2 + brx + c)e^{rx}$
(c) $[ar^2 + (b - a)r + c]x^r$ (d) $-a + b + c\ln x$
10. (a) $(n - m)e^{(m+n)x}$ (b) -1 (c) x^2e^x (d) $-e^{2x}$ (e) 0
11. (a) $-\infty < x < \infty$ (b) $-\infty < x < \infty$
(c) $-\infty < x < \infty$ (d) $x > 0$ or $x < 0$
13. (a) $y = (e^x - e^{-x})/2$ (b) $y = \sinh x$
(c) $y = 4e^{-2x} - 3e^{-3x}$ (d) $y = 1 - e^{-(x-1)}$
14. 2/25
17. If x_1 is an inflection point and $y = \phi(x)$ is a solution, then from the differential equation $p(x_1)\phi'(x_1) + q(x_1)\phi(x_1) = 0$.
18. (a) Yes, $y = c_1e^{-x^2/2}\int^x e^{t^2/2}\,dt + c_2e^{-x^2/2}$
(b) No
(c) Yes, $y = \dfrac{1}{\mu(x)}\left[c_1\int^x \dfrac{\mu(t)}{t}\,dt + c_2\right]$, $\mu(x) = \exp\left[-\int^x\left(\dfrac{1}{t} + \dfrac{\cos t}{t}\right)dt\right]$
(d) Yes, $y = c_1x^{-1} + c_2x$
19. (a) $x^2\mu'' + 3x\mu' + (1 + x^2 - \nu^2)\mu = 0$
(b) $(1 - x^2)\mu'' - 2x\mu' + \alpha(\alpha + 1)\mu = 0$ (c) $\mu'' - x\mu = 0$
21. The Legendre equation and the Airy equation are self-adjoint.

Section 3.3, Page 126

5. Linearly independent
6. Linearly dependent
7. Linearly independent
8. Linearly dependent
9. $W(c_1y_1, c_2y_2) = c_1c_2W(y_1, y_2) \neq 0$
14. If y_1 and y_2 are linearly independent, then $W(y_1, y_2)$ cannot change sign. Show that $W(y_1, y_2)$ will change sign if y_1 has none or more than one zero between consecutive zeros of y_2.

Section 3.4, Page 130

1. $y_2(x) = e^{-2x}$
2. $y_2(x) = xe^{-x}$
3. $y_2(x) = x^{-1}$, $x > 0$ or $x < 0$
4. $y_2(x) = x^{-2}$, $x > 0$ or $x < 0$
5. $y_2(x) = x^{-1}\ln x$
6. $y_2(x) = xe^x$
7. $y_2(x) = \sin x$
8. $y_2(x) = \cos x^2$
9. $y_2(x) = x$
10. $y_2(x) = x^{1/4}e^{-2\sqrt{x}}$

11. $y_2(x) = \sinh x + x \operatorname{sech} x$
12. $y_2(x) = x^{-1/2} \cos x$
14. $f(\xi) = c_1 e^{-\delta\xi^2/2} \int_0^{\xi} e^{\delta s^2/2}\, ds + c_2 e^{-\delta\xi^2/2}$
16. $y_2(x) = x^{-2}$

Section 3.5, Page 136

1. $y = c_1 e^x + c_2 e^{-3x}$
2. $y = c_1 e^{-x/2} + c_2 x e^{-x/2}$
3. $y = c_1 e^{x/2} + c_2 e^{-x/3}$
4. $y = c_1 e^{x/2} + c_2 e^x$
5. $y = c_1 e^x + c_2 e^{-x}$
6. $y = c_1 e^x + c_2 x e^x$
7. $y = c_1 + c_2 e^{-5x}$
8. $y = c_1 e^{(9+3\sqrt{5})x/2} + c_2 e^{(9-3\sqrt{5})x/2}$
9. $y = c_1 e^{(1+\sqrt{3})x} + c_2 e^{(1-\sqrt{3})x}$
10. $y = c_1 e^{-x} + c_2 x e^{-x}$
11. $y = e^x$
12. $y = 2xe^{3x}$
13. $y = \frac{1}{10} e^{-9(x-1)} + \frac{9}{10} e^{x-1}$
14. $y = 7e^{-2(x+1)} + 5xe^{-2(x+1)}$

Section 3.5.1, Page 140

1. $y = c_1 e^x \cos x + c_2 e^x \sin x$
2. $y = c_1 e^x \cos \sqrt{5}\, x + c_2 e^x \sin \sqrt{5}\, x$
3. $y = c_1 e^{2x} + c_2 e^{-4x}$
4. $y = c_1 e^{-x} \cos x + c_2 e^{-x} \sin x$
5. $y = c_1 e^{x/3} + c_2 x e^{x/3}$
6. $y = e^{-3x}(c_1 \cos 2x + c_2 \sin 2x)$
7. $y = \frac{1}{2} \sin 2x$
8. $y = e^{-2x} \cos x + 2e^{-2x} \sin x$
9. $y = -e^{x-\pi/2} \sin 2x$
10. $y = (1 + 2\sqrt{3}) \cos x - (2 - \sqrt{3}) \sin x$
13. (b) $y_0 + (a/b) y_0'$
16. $\pm 2^{1/4} e^{i\pi/8}$; $\quad \pm 2^{1/4} e^{-i\pi/8}$; $\quad \pm e^{i\pi/4} = \pm(1 + i)/\sqrt{2}$
17. (a) $y = c_1 e^{ix} + c_2 e^{-2ix} = (c_1 \cos x + c_2 \cos 2x) + i(c_1 \sin x - c_2 \sin 2x)$, no
    (b) $y = c_1 e^{r_1 x} + c_2 e^{r_2 x}$ where $r_1 = -1 + 2^{1/4} e^{-i\pi/8}$ and $r_2 = -1 - 2^{1/4} e^{-i\pi/8}$, no
19. (a) Yes, $y = c_1 \cos z + c_2 \sin z$, $\quad z = \int^x e^{-t^2/2}\, dt$
    (b) No
    (c) Yes, $y = c_1 e^{-x^2/4} \cos(\sqrt{3}\, x^2/4) + c_2 e^{-x^2/4} \sin(\sqrt{3}\, x^2/4)$
21. (a) $y = c_1 \cos \ln x + c_2 \sin \ln x$
    (b) $y = c_1 x^{-1} + c_2 x^{-2}$
    (c) $y = c_1 x^2 + c_2 x^2 \ln x$
    (d) $y = c_1 x^6 + c_2 x^{-1}$
23. $v(x) = \exp\left[-\frac{1}{2} \int^x p(t)\, dt\right]$, $\quad f(x) = q(x) - \frac{1}{4} p^2(x) - \frac{1}{2} p'(x)$

Section 3.6.1, Page 154

1. $y = e^x - \frac{1}{2} e^{-2x} - x - \frac{1}{2}$
2. $y = c_1 e^{3x} + c_2 e^{-x} - \frac{1}{3} e^{2x}$
3. $y = \frac{7}{10} \sin 2x - \frac{19}{40} \cos 2x + \frac{1}{4} x^2 - \frac{1}{8} + \frac{3}{5} e^x$
4. $y = c_1 + c_2 e^{-2x} + \frac{3}{2} x - \frac{1}{2} \sin 2x - \frac{1}{2} \cos 2x$
5. $y = c_1 \cos 3x + c_2 \sin 3x + \frac{1}{18}(x^2 - \frac{2}{3} x + \frac{1}{9}) e^{3x} + \frac{2}{3}$
6. $y = 4xe^x - 3e^x + \frac{1}{6} x^3 e^x + 4$
7. $y = c_1 e^{-x} + c_2 e^{-x/2} + (x^2 - 6x + 14) - \frac{3}{10} \sin x - \frac{9}{10} \cos x$
8. $y = c_1 \cos x + c_2 \sin x - \frac{1}{3} x \cos 2x - \frac{5}{9} \sin 2x$
9. $y = c_1 e^{-x} + c_2 x e^{-x} + \frac{1}{25} e^x (3 \cos x + 4 \sin x)$
10. $u = c_1 \cos \omega_0 t + c_2 \sin \omega_0 t + (\omega_0^2 - \omega^2)^{-1} \cos \omega t$
11. $u = c_1 \cos \omega_0 t + c_2 \sin \omega_0 t + (1/2\omega_0) t \sin \omega_0 t$
12. $u = e^{-\mu t/2}(c_1 \cos \sqrt{\omega_0^2 - \mu^2/4}\ \ t + c_2 \sin \sqrt{\omega_0^2 - \mu^2/4t})$
    $$+ \frac{(\omega_0^2 - \omega^2) \cos \omega t + \mu\omega \sin \omega t}{(\omega_0^2 - \omega^2)^2 + \mu^2\omega^2}$$
13. $y = c_1 e^{-x/2} \cos(\sqrt{3}\, x/2) + c_2 e^{-x/2} \sin(\sqrt{3}\, x/2) + \frac{1}{2} - \frac{1}{13} \sin 2x + \frac{3}{26} \cos 2x$
14. $y = c_1 e^{-x/2} \cos(\sqrt{15}\, x/2) + c_2 e^{-x/2} \sin(\sqrt{15}\, x/2) + \frac{1}{6} e^x - \frac{1}{4} e^{-x}$

15. $y = c_1 e^{-x} + c_2 e^{2x} + \frac{1}{6}xe^{2x} + \frac{1}{8}e^{-2x}$
16. $y_p(x) = x(A_0x^4 + A_1x^3 + A_2x^2 + A_3x + A_4) + x(B_0x^2 + B_1x + B_2)e^{-3x}$
$+ D\sin 3x + E\cos 3x$
17. $y_p(x) = A_0x + A_1 + x(B_0x + B_1)\sin x + x(D_0x + D_1)\cos x$
18. $y_p(x) = e^x(A\cos 2x + B\sin 2x) + (D_0x + D_1)e^{2x}\sin x + (E_0x + E_1)e^{2x}\cos x$
19. $y_p(x) = Ae^{-x} + x(B_0x^2 + B_1x + B_2)e^{-x}\cos x + x(D_0x^2 + D_1x + D_2)e^{-x}\sin x$
20. $y_p(x) = A_0x^2 + A_1x + A_2 + x^2(B_0x + B_1)e^{2x} + (D_0x + D_1)\sin 2x$
$+ (E_0x + E_1)\cos 2x$
21. $y_p(x) = x(A_0x^2 + A_1x + A_2)\sin 2x + x(B_0x^2 + B_1x + B_2)\cos 2x$
22. $y_p(x) = (A_0x^2 + A_1x + A_2)e^x\sin 2x + (B_0x^2 + B_1x + B_2)e^x\cos 2x$
$+ e^{-x}(D\cos x + E\sin x) + Fe^x$
23. $y = c_1\cos\lambda x + c_2\sin\lambda x + \sum_{m=1}^{N}\frac{a_m}{\lambda^2 - m^2\pi^2}\sin m\pi x$
24. $y = \begin{cases} t, & 0 \le t \le \pi \\ -\left(1 + \frac{\pi}{2}\right)\sin t - \frac{\pi}{2}\cos t + \frac{\pi}{2}e^{(\pi - t)}, & t > \pi \end{cases}$
25. No

Section 3.6.2, Page 160

1. $y_p(x) = e^x$
2. $y_p(x) = -\frac{2}{3}xe^{-x}$
3. $y_p(x) = \frac{3}{2}x^2e^{-x}$
4. $y_p(x) = -(\cos x)\ln(\tan x + \sec x)$
5. $y_p(x) = (\sin 3x)\ln(\tan 3x + \sec 3x) - 1$
6. $y_p(x) = -e^{-2x}\ln x$
7. $y_p(x) = \frac{3}{4}(\sin 2x)\ln\sin 2x - \frac{3}{2}x\cos 2x$
8. $y = c_1x + c_2xe^x - 2x^2$
9. $y = c_1(1 + x) + c_2e^x + \frac{1}{2}e^{2x}(x - 1)$
10. $y = c_1x^{-1/2}\cos x + c_2x^{-1/2}\sin x - \frac{3}{2}x^{1/2}\cos x$
11. $y = c_1e^x + c_2x - \frac{1}{2}e^{-x}(2x - 1)$
12. $y_p(x) = \int^x [e^{3(x-t)} - e^{2(x-t)}]g(t)\,dt$
13. $y_p(x) = x^{-1/2}\int^x \frac{g(t)\sin(x - t)}{t^{3/2}}\,dt$
18. (a) $y = c_1x + c_2x^2 + 4x^2\ln x$ (b) $y = c_1x^{-1} + c_2x^{-5} + \frac{1}{12}x$

Section 3.7.1, Page 171

1. $u = \frac{1}{4}\cos 8t$ ft, $\frac{1}{4}$ ft, 8 rad/sec, $\pi/4$ sec; t in sec
2. $u = \frac{5}{7}\sin 14t$ cm; t in sec
3. $u = \frac{1}{4\sqrt{2}}\sin 8\sqrt{2}\,t - \frac{1}{12}\cos 8\sqrt{2}\,t$ ft, $\frac{\sqrt{22}}{24}$ ft, $8\sqrt{2}$ rad/sec, $\frac{\pi\sqrt{2}}{8}$ sec; t in sec
5. Approximately 1%
6. $c = \sqrt{10}/2$ lb sec/ft
7. $r = \sqrt{A^2 + B^2}$, $r\cos\theta = B$, $r\sin\theta = -A$; $R = r$;
$\delta = \theta + (4n + 1)\pi/2$, $n = 0, 1, 2, \ldots$
8. $\omega_0 = \sqrt{k/m}$; (a) $u = u_0\cos\omega_0 t$ (b) $u = (\dot{u}_0/\omega_0)\cos(\omega_0 t - \pi/2)$
(c) $u = \sqrt{u_0^2 + \left(\frac{\dot{u}_0}{\omega_0}\right)^2}\cos(\omega_0 t - \delta)$, $\delta = \arctan\frac{\dot{u}_0/\omega_0}{u_0}$
9. $2\pi\sqrt{l/g}$
10. $\rho l\ddot{x} + \hat{\rho}gx = 0$, $2\pi\sqrt{\rho l/\hat{\rho}g}$
11. $u = e^{-10t}(2\cos 4\sqrt{6}\,t + (5/\sqrt{6})\sin 4\sqrt{6}\,t)$ cm; t in sec

12. $u = (1/8\sqrt{31})e^{-2t}\sin 2\sqrt{31}\,t$ ft, t in sec; $t = n\pi/2\sqrt{31}$, $n = 1, 2, \ldots$
13. $c > 8$ lb sec/ft, overdamped; $c = 8$ lb sec/ft, critically damped;
$c < 8$ lb sec/ft, underdamped
14. $u \cong 0.0572e^{-0.15t}\cos(3.87t - 0.507)$ m; t in sec
15. $\mu = (4km - c^2)^{1/2}/2m$

(a) $u = u_0\sqrt{1 + (c/2m\mu)^2}\,e^{-ct/2m}\cos(\mu t - \delta)$, $\delta = \arctan(c/2m\mu)$

(b) $u = (\dot{u}_0/\mu)e^{-ct/2m}\cos(\mu t - \pi/2)$

(c) $u = \sqrt{u_0^2 + \left(\dfrac{\dot{u}_0}{\mu} + \dfrac{cu_0}{2m\mu}\right)^2}\,e^{-ct/2m}\cos(\mu t - \delta)$, $\delta = \arctan\dfrac{\dot{u}_0/\mu + cu_0/2\mu m}{u_0}$

17. $\dot{u}_0 < -cu_0/2m$
20. (a) $2\pi/\sqrt{31}$ (b) $c = 5$ lb sec/ft

Section 3.7.2, Page 178

1. $-2\sin 8t\sin t$
2. $2\sin\frac{1}{2}t\cos\frac{13}{2}t$
3. $u = \frac{151}{1482}\cos 16t + \frac{16}{247}\cos 3t$ ft; t in sec; $\omega = 16$ rad/sec
4. $u = \frac{64}{45}(\cos 7t - \cos 8t) = \frac{128}{45}\sin\frac{1}{2}t\sin\frac{15}{2}t$ ft; t in sec
5. $u = (\cos 8t + \sin 8t - 8t\cos 8t)/4$ ft; t in sec; $1/8$, $\pi/8$, $\pi/4$, $3\pi/8$ sec
6. $\omega_0 = \sqrt{k/m}$

(a) $u = \left[u_0 - \dfrac{F}{m(\omega_0^2 - \omega^2)}\right]\cos\omega_0 t + \dfrac{F}{m(\omega_0^2 - \omega^2)}\cos\omega t$

(b) $u = \dfrac{\dot{u}_0}{\omega_0}\sin\omega_0 t + \dfrac{F_0}{m(\omega_0^2 - \omega^2)}(\cos\omega t - \cos\omega_0 t)$

(c) $u = \left[u_0 - \dfrac{F}{m(\omega_0^2 - \omega^2)}\right]\cos\omega_0 t + \dfrac{\dot{u}_0}{\omega_0}\sin\omega_0 t + \dfrac{F}{m(\omega_0^2 - \omega^2)}\cos\omega t$

7. $u = \begin{cases} (F_0/m)(t - \sin t), & 0 \le t \le \pi \\ (F_0/m)[(2\pi - t) - 3\sin t], & \pi < t \le 2\pi \\ -(4F_0/m)\sin t, & 2\pi < t \end{cases}$

8. $u = \frac{8}{901}(30\cos 2t + \sin 2t)$ ft, t in sec
9. $m = 4$ slugs
10. $u = (\sqrt{2}/6)\cos(3t - 3\pi/4)$ m
12. $u = e^{-ct/2m}(c_1\cos\mu t + c_2\sin\mu t) + (F_0/\Delta)[m(\omega_0^2 - \omega^2)\sin\omega t - c\omega\cos\omega t]$
$\mu = (4km - c^2)^{1/2}/2m$, $\omega_0^2 = k/m$, $\Delta = m^2(\omega_0^2 - \omega^2)^2 + c^2\omega^2$

(a) $c_1 = u_0 + \dfrac{c\omega F_0}{\Delta}$, $c_2 = \dfrac{cu_0}{2m\mu} + \dfrac{F_0\omega}{\mu\Delta}\left[\dfrac{c^2}{2m} - m(\omega_0^2 - \omega^2)\right]$

(b) $c_1 = \dfrac{c\omega F_0}{\Delta}$, $c_2 = \dfrac{\dot{u}_0}{\mu} + \dfrac{F_0\omega}{\mu\Delta}\left[\dfrac{c^2}{2m} - m(\omega_0^2 - \omega^2)\right]$

(c) $c_1 = u_0 + \dfrac{c\omega F_0}{\Delta}$, $c_2 = \dfrac{\dot{u}_0}{\mu} + \dfrac{cu_0}{2m\mu} + \dfrac{F_0\omega}{\mu\Delta}\left[\dfrac{c^2}{2m} - m(\omega_0^2 - \omega^2)\right]$

Section 3.8, Page 182

4. (a) $Q = 10^{-6}(2e^{-500t} - e^{-1000t})$ coulombs; t in sec
(b) $Q = 10^{-6}(1 - 500t)e^{-500t}$ coulombs; t in sec
(c) $Q = 10^{-6}e^{-100t}(\cos 200t + \frac{1}{2}\sin 200t)$ coulombs; t in sec

5. $R = 10^3$ ohms
6. $Q = \psi(t) = 10^{-6}(e^{-4000t} - 4e^{-1000t} + 3)$ coulombs; t in sec
   $\psi(0.001) \cong 1.52 \times 10^{-6}$; $\psi(0.01) \cong 3 \times 10^{-6}$; $Q \to 3 \times 10^{-6}$ as $t \to \infty$
7. $I \cong 0.0273 \cos(120\pi t + \delta)$ amperes;
   $\delta = \arctan(-0.9)$, δ in the fourth quadrant; t in sec
8. $\omega = 1/\sqrt{LC}$
9. (d) $I = \dfrac{E_0}{\sqrt{R^2 + (\omega L - 1/\omega C)^2}} \sin(\omega t - \delta)$, $\delta = \arctan \dfrac{\omega L - 1/\omega C}{R}$

CHAPTER FOUR

Section 4.1, Page 192

1. (a) $\rho = 1$ (b) $\rho = 2$ (c) $\rho = \infty$
   (d) $\rho = \frac{1}{2}$ (e) $\rho = \frac{1}{2}$ (f) $\rho = 1$
   (g) $\rho = 3$ (h) $\rho = 2$ (i) $\rho = e$
2. (a) $\displaystyle\sum_{n=0}^{\infty} \frac{(-1)^n x^{2n+1}}{(2n+1)!}$, $\rho = \infty$ (b) $\displaystyle\sum_{n=0}^{\infty} \frac{x^n}{n!}$, $\rho = \infty$
   (c) $1 + (x - 1)$, $\rho = \infty$ (d) $1 - 2(x+1) + (x+1)^2$, $\rho = \infty$
   (e) $\displaystyle\sum_{n=1}^{\infty} (-1)^{n+1} \frac{(x-1)^n}{n}$, $\rho = 1$ (f) $\displaystyle\sum_{n=0}^{\infty} (-1)^n x^n$, $\rho = 1$
   (g) $\displaystyle\sum_{n=0}^{\infty} x^n$, $\rho = 1$ (h) $\displaystyle\sum_{n=0}^{\infty} (-1)^{n+1}(x-2)^n$, $\rho = 1$
3. $y' = 1 + 2^2 x + 3^2 x^2 + 4^2 x^3 + \cdots + (n+1)^2 x^n + \cdots$
   $y'' = 2^2 + 3^2 \cdot 2x + 4^2 \cdot 3x^2 + 5^2 \cdot 4x^3 + \cdots + (n+2)^2(n+1)x^n + \cdots$
4. $y' = a_1 + 2a_2 x + 3a_3 x^2 + 4a_4 x^3 + \cdots + (n+1)a_{n+1}x^n + \cdots$
   $\displaystyle = \sum_{n=1}^{\infty} n a_n x^{n-1} = \sum_{n=0}^{\infty} (n+1) a_{n+1} x^n$
   $y'' = 2a_2 + 6a_3 x + 12a_4 x^2 + 20a_5 x^3 + \cdots + (n+2)(n+1)a_{n+2}x^n + \cdots$
   $\displaystyle = \sum_{n=2}^{\infty} n(n-1) a_n x^{n-2} = \sum_{n=0}^{\infty} (n+2)(n+1) a_{n+2} x^n$
6. $a_n = (-2)^n a_0/n!$, $n = 1, 2, \ldots$; $a_0 e^{-2x}$

Section 4.2, Page 206

1. $a_{n+2} = \dfrac{a_n}{(n+2)(n+1)}$
   $\displaystyle y_1(x) = 1 + \frac{x^2}{2!} + \frac{x^4}{4!} + \frac{x^6}{6!} + \cdots = \sum_{n=0}^{\infty} \frac{x^{2n}}{(2n)!} = \cosh x$
   $\displaystyle y_2(x) = x + \frac{x^3}{3!} + \frac{x^5}{5!} + \frac{x^7}{7!} + \cdots = \sum_{n=0}^{\infty} \frac{x^{2n+1}}{(2n+1)!} = \sinh x$
2. $a_{n+2} = \dfrac{a_n}{n+2}$
   $\displaystyle y_1(x) = 1 + \frac{x^2}{2} + \frac{x^4}{2 \cdot 4} + \frac{x^6}{2 \cdot 4 \cdot 6} + \cdots = \sum_{n=0}^{\infty} \frac{x^{2n}}{2^n n!}$
   $\displaystyle y_2(x) = x + \frac{x^3}{3} + \frac{x^5}{3 \cdot 5} + \frac{x^7}{3 \cdot 5 \cdot 7} + \cdots = \sum_{n=0}^{\infty} \frac{2^n n! x^{2n+1}}{(2n+1)!}$

3. $(n+2)a_{n+2} - a_{n+1} - a_n = 0$
$y_1(x) = 1 + \frac{1}{2}(x-1)^2 + \frac{1}{6}(x-1)^3 + \frac{1}{6}(x-1)^4 + \cdots$
$y_2(x) = (x-1) + \frac{1}{2}(x-1)^2 + \frac{1}{2}(x-1)^3 + \frac{1}{4}(x-1)^4 + \cdots$

4. $a_{n+4} = -\dfrac{k^2 a_n}{(n+4)(n+3)};\qquad a_2 = a_3 = 0$

$$y_1(x) = 1 - \frac{k^2x^4}{3\cdot 4} + \frac{k^4x^8}{3\cdot 4\cdot 7\cdot 8} - \frac{k^6x^{12}}{3\cdot 4\cdot 7\cdot 8\cdot 11\cdot 12} + \cdots$$
$$= 1 + \sum_{m=0}^{\infty} \frac{(-1)^{m+1}(k^2x^4)^{m+1}}{3\cdot 4\cdot 7\cdot 8\cdots(4m+3)(4m+4)}$$
$$y_2(x) = x - \frac{k^2x^5}{4\cdot 5} + \frac{k^4x^9}{4\cdot 5\cdot 8\cdot 9} - \frac{k^6x^{13}}{4\cdot 5\cdot 8\cdot 9\cdot 12\cdot 13} + \cdots$$
$$= x\left[1 + \sum_{m=0}^{\infty} \frac{(-1)^{m+1}(k^2x^4)^{m+1}}{4\cdot 5\cdot 8\cdot 9\cdots(4m+4)(4m+5)}\right]$$

Hint: Let $n = 4m$ in the recurrence relation, $m = 1, 2, 3, \ldots$.

5. $(n+2)(n+1)a_{n+2} - n(n+1)a_{n+1} + a_n = 0, \qquad n \geq 1; \qquad a_2 = -\frac{1}{2}a_0$
$y_1(x) = 1 - \frac{1}{2}x^2 - \frac{1}{6}x^3 - \frac{1}{24}x^4 + \cdots, \qquad y_2(x) = x - \frac{1}{6}x^3 - \frac{1}{12}x^4 + \cdots$

6. $a_{n+2} = -\dfrac{(n^2-2n+4)a_n}{2(n+1)(n+2)}, \qquad n \geq 2; \qquad a_2 = -a_0, \qquad a_3 = -\frac{1}{4}a_1$
$y_1(x) = 1 - x^2 + \frac{1}{6}x^4 - \frac{1}{30}x^6 + \cdots$
$y_2(x) = x - \frac{1}{4}x^3 + \frac{7}{160}x^5 - \frac{19}{1920}x^7 + \cdots$

7. $a_{n+2} = -a_n/(n+1), \qquad n = 0, 1, 2, \ldots$

$$y_1(x) = 1 - \frac{x^2}{1} + \frac{x^4}{1\cdot 3} - \frac{x^6}{1\cdot 3\cdot 5} + \cdots = 1 + \sum_{n=1}^{\infty} \frac{(-1)^n x^{2n}}{1\cdot 3\cdot 5\cdots(2n-1)}$$
$$y_2(x) = x - \frac{x^3}{2} + \frac{x^5}{2\cdot 4} - \frac{x^7}{2\cdot 4\cdot 6} + \cdots = x + \sum_{n=1}^{\infty} \frac{(-1)^n x^{2n+1}}{2\cdot 4\cdot 6\cdots(2n)}$$

8. $a_{n+2} = -[(n+1)^2 a_{n+1} + a_n + a_{n-1}]/(n+1)(n+2), \qquad n = 1, 2, \ldots$
$a_2 = -(a_0 + a_1)/2$
$y_1(x) = 1 - \frac{1}{2}(x-1)^2 + \frac{1}{6}(x-1)^3 - \frac{1}{12}(x-1)^4 + \cdots$
$y_2(x) = (x-1) - \frac{1}{2}(x-1)^2 + \frac{1}{6}(x-1)^3 - \frac{1}{6}(x-1)^4 + \cdots$

9. $y = \displaystyle\sum_{n=0}^{\infty} \frac{2^n n! x^{2n+1}}{(2n+1)!}, \qquad y = 2\sum_{n=0}^{\infty} \frac{x^{2n}}{2^n n!} + \sum_{n=0}^{\infty} \frac{2^n n! x^{2n+1}}{(2n+1)!}$

10. $y_1(x) = 1 - \frac{1}{3}(x-1)^3 - \frac{1}{12}(x-1)^4 + \frac{1}{18}(x-1)^6 + \cdots$
$y_2(x) = (x-1) - \frac{1}{4}(x-1)^4 - \frac{1}{20}(x-1)^5 + \frac{1}{28}(x-1)^7 + \cdots$

Section 4.2.1, Page 212

1. (a) $\phi''(0) = -1, \qquad \phi'''(0) = 0, \qquad \phi^{iv}(0) = 3$
   (b) $\phi''(0) = 0, \qquad \phi'''(0) = -2, \qquad \phi^{iv}(0) = 0$
   (c) $\phi''(1) = 0, \qquad \phi'''(1) = -6, \qquad \phi^{iv}(1) = 42$
   (d) $\phi''(0) = 0, \qquad \phi'''(0) = -a_0, \qquad \phi^{iv}(0) = -4a_1$
2. (a) $\rho = \infty, \rho = \infty$ (b) $\rho = 1, \rho = 3, \rho = 1$
   (c) $\rho = 1, \rho = \sqrt{3}$ (d) $\rho = 1$
3. (a) $\rho = \infty$ (b) $\rho = \infty$ (c) $\rho = \infty$
   (d) $\rho = \infty$ (e) $\rho = 1$ (f) $\rho = \sqrt{2}$
   (g) $\rho = \infty$ (h) $\rho = 1$

4. (a) $y_1(x) = 1 - \frac{\alpha^2}{2!}x^2 - \frac{(2^2-\alpha^2)\alpha^2}{4!}x^4 - \frac{(4^2-\alpha^2)(2^2-\alpha^2)\alpha^2}{6!}x^6 - \cdots$
$- \frac{\left[(2m-2)^2-\alpha^2\right]\cdots(2^2-\alpha^2)\alpha^2}{(2m)!}x^{2m} - \cdots$

$y_2(x) = x + \frac{1-\alpha^2}{3!}x^3 + \frac{(3^2-\alpha^2)(1-\alpha^2)}{5!}x^5 + \cdots$
$+ \frac{\left[(2m-1)^2-\alpha^2\right]\cdots(1-\alpha^2)}{(2m+1)!}x^{2m+1} + \cdots$

(b) $y_1(x)$ or $y_2(x)$ terminates with x^n as $\alpha = n$ is even or odd.
(c) $n = 0, \quad y = 1; \qquad n = 1,\ y = x; \qquad n = 2, \quad y = 1 - 2x^2;$
$n = 3, \quad y = x - \frac{4}{3}x^3$

5. $y_1(x) = 1 - \frac{x^3}{3!} + \frac{x^5}{5!} + \frac{x^6}{2\cdot 3\cdot 5\cdot 6} + \cdots$

$y_2(x) = x - \frac{x^4}{3\cdot 4} + \frac{x^6}{2\cdot 3\cdot 5\cdot 6} + \cdots$

6. $y_1(x) = 1 - \frac{1}{6}x^3 + \frac{1}{12}x^4 - \frac{1}{40}x^5 + \cdots, \ y_2(x) = x - \frac{1}{12}x^4 + \frac{1}{20}x^5 + \cdots$
$\rho = \infty$

7. Cannot specify arbitrary initial conditions at $x = 0$; hence $x = 0$ is a singular point.

8. $f'(0) = \lim_{h\to 0}\frac{e^{-1/h^2} - 0}{h} = 0, \qquad f''(0) = \lim_{h\to 0}\frac{(2/h^3)e^{-1/h^2} - 0}{h} = 0,$ etc.

9. (a) $y = 1 + x + \frac{x^2}{2!} + \cdots + \frac{x^n}{n!} + \cdots = e^x$

(b) $y = 1 + \frac{x^2}{2} + \frac{x^4}{2\cdot 4} + \frac{x^6}{2\cdot 4\cdot 6} + \cdots + \frac{x^{2n}}{2^n\cdot n!} + \cdots$

(c) $y = 1 + x + \frac{1}{2}x^2 + \frac{1}{2}x^3 + \cdots$

(d) $y = 1 + x + x^2 + \cdots + x^n + \cdots = \frac{1}{1-x}$

(e) $y = a_0\left(1 + x + \frac{x^2}{2!} + \cdots + \frac{x^n}{n!} + \cdots\right) + 2!\left(\frac{x^3}{3!} + \frac{x^4}{4!} + \cdots + \frac{x^n}{n!} + \cdots\right)$

$= a_0e^x + 2\left(e^x - 1 - x - \frac{x^2}{2}\right) = ce^x - 2 - 2x - x^2$

(f) $y = a_0\left(1 - \frac{x^2}{2} + \frac{x^4}{2^2 2!} - \frac{x^6}{2^3 3!} + \cdots + \frac{(-1)^n x^{2n}}{2^n n!} + \cdots\right)$

$+\left(x + \frac{x^2}{2} - \frac{x^3}{3} - \frac{x^4}{2\cdot 4} + \frac{x^5}{3\cdot 5} + \cdots\right)$

$= a_0e^{-x^3/2} + \left(x + \frac{x^2}{2} - \frac{x^3}{3} - \frac{x^4}{2\cdot 4} + \frac{x^5}{3\cdot 5} + \cdots\right)$

Section 4.3, Page 219

	$x = -1$	$x = 0$	$x = 1$
1.	Ordinary	Regular singular	Ordinary
2.	Regular singular	Regular singular	Regular singular
3.	Regular singular	Irregular singular	Regular singular
4.	Regular singular	Irregular singular	Regular singular
5.	Irregular singular	Ordinary	Regular singular
6.	Ordinary	Ordinary	Irregular singular

	$x = -1$	$x = 0$	$x = 1$
7.	Ordinary	Ordinary	Ordinary
8.	Regular singular	Regular singular	Irregular singular

9. Regular singular point
10. Regular singular point
11. (a) Regular singular point
(b) Regular singular point
14. (a) Regular singular point
(b) Irregular singular point
(c) Irregular singular point
(d) Irregular singular point

Section 4.4, Page 225

1. $y = c_1 x^{-1} + c_2 x^{-2}$
2. $y = c_1|x + 1|^{-1/2} + c_2|x + 1|^{-3/2}$
3. $y = c_1 x^2 + c_2 x^2 \ln|x|$
4. $y = c_1 x^{-1} \cos(2\ln|x|) + c_2 x^{-1} \sin(2\ln|x|)$
5. $y = c_1 x + c_2 x \ln|x|$
6. $y = c_1(x - 1)^{-3} + c_2(x - 1)^{-4}$
7. $y = c_1|x|^{(-5+\sqrt{29})/2} + c_2|x|^{(-5-\sqrt{29})/2}$
8. $y = c_1|x|^{3/2} \cos(\frac{1}{2}\sqrt{3}\ln|x|) + c_2|x|^{3/2} \sin(\frac{1}{2}\sqrt{3}\ln|x|)$
9. $y = c_1 x^3 + c_2 x^3 \ln|x|$
10. $y = c_1(x - 2)^{-2} \cos(2\ln|x - 2|) + c_2(x - 2)^{-2} \sin(2\ln|x - 2|)$
11. $y = c_1|x|^{-1/2} \cos(\frac{1}{2}\sqrt{15}\ln|x|) + c_2|x|^{-1/2} \sin(\frac{1}{2}\sqrt{15}\ln|x|)$
12. $y = c_1 x + c_2 x^4$
14. $\alpha < 1$
15. $\beta > 0$
17. (a) $y = c_1 x^{-1} + c_2 x^2$
(b) $y = c_1 x^2 + c_2 x^2 \ln x + \frac{1}{4}\ln x + \frac{1}{4}$
(c) $y = c_1 x^{-1} + c_2 x^{-5} + \frac{1}{12}x$
(d) $y = c_1 x + c_2 x^2 + 3x^2 \ln x + \ln x + \frac{3}{2}$
(e) $y = c_1 \cos(2\ln x) + c_2 \sin(2\ln x) + \frac{1}{3}\sin(\ln x)$
(f) $y = c_1 x^{-3/2} \cos(\frac{3}{2}\ln x) + c_2 x^{-3/2} \sin(\frac{3}{2}\ln x)$
19. $x > 0$: $\quad c_1 = k_1, \quad c_2 = k_2; \qquad x < 0$: $\quad c_1(-1)^{r_1} = k_1, \quad c_2 = k_2$
20. (a) $y = c_1 x^{1/2 - i(1+\sqrt{3}/2)} + c_2 x^{1/2 - i(1-\sqrt{3}/2)}$

$$= c_1 x^{1/2}\left\{\cos\left[\left(1 + \tfrac{1}{2}\sqrt{3}\right)\ln x\right] - i\sin\left[\left(1 + \tfrac{1}{2}\sqrt{3}\right)\ln x\right]\right\}$$
$$+ c_2 x^{1/2}\left\{\cos\left[\left(1 - \tfrac{1}{2}\sqrt{3}\right)\ln x - i\sin\left[\left(1 - \tfrac{1}{2}\sqrt{3}\right)\ln x\right]\right\}$$

(b) $y = c_1 x^{2i} + c_2 x^{-i}$

$$= c_1[\cos(2\ln x) + i\sin(2\ln x)] + c_2[\cos(\ln x) - i\sin(\ln x)]$$

(c) $y = c_1 x^{1+i} + c_2 x^{-1-i}$

$$= c_1 x[\cos(\ln x) + i\sin(\ln x)] + c_2 x^{-1}[\cos(\ln x) - i\sin(\ln x)]$$

Section 4.5, Page 232

1. $r(2r - 1) = 0; \qquad a_n = -\dfrac{a_{n-2}}{(n + r)[2(n + r) - 1]}; \qquad r_1 = \frac{1}{2},\ r_2 = 0$

$$y_1(x) = x^{1/2}\left[1 - \frac{x^2}{2 \cdot 5} + \frac{x^4}{2 \cdot 4 \cdot 5 \cdot 9} - \frac{x^6}{2 \cdot 4 \cdot 6 \cdot 5 \cdot 9 \cdot 13} + \cdots \right.$$
$$\left. + \frac{(-1)^n x^{2n}}{2^n n! 5 \cdot 9 \cdot 13 \cdots (4n + 1)} + \cdots \right]$$
$$y_2(x) = 1 - \frac{x^2}{2 \cdot 3} + \frac{x^4}{2 \cdot 4 \cdot 3 \cdot 7} - \frac{x^6}{2 \cdot 4 \cdot 6 \cdot 3 \cdot 7 \cdot 11} + \cdots$$
$$+ \frac{(-1)^n x^{2n}}{2^n n! 3 \cdot 7 \cdot 11 \cdots (4n - 1)} + \cdots$$

2. $r^2 - \frac{1}{9} = 0; \quad a_n = -\dfrac{a_{n-2}}{(n+r)^2 - \frac{1}{9}}; \quad r_1 = \frac{1}{3},\ r_2 = -\frac{1}{3}$

$$y_1(x) = x^{1/3}\left[1 - \frac{1}{1!(1+\frac{1}{3})}\left(\frac{x}{2}\right)^2 + \frac{1}{2!(1+\frac{1}{3})(2+\frac{1}{3})}\left(\frac{x}{2}\right)^4 + \cdots + \frac{(-1)^m}{m!(1+\frac{1}{3})(2+\frac{1}{3})\cdots(m+\frac{1}{3})}\left(\frac{x}{2}\right)^{2m} + \cdots\right]$$

$$y_2(x) = x^{-1/3}\left[1 - \frac{1}{1!(1-\frac{1}{3})}\left(\frac{x}{2}\right)^2 + \frac{1}{2!(1-\frac{1}{3})(2-\frac{1}{3})}\left(\frac{x}{2}\right)^4 + \cdots + \frac{(-1)^m}{m!(1-\frac{1}{3})(2-\frac{1}{3})\cdots(m-\frac{1}{3})}\left(\frac{x}{2}\right)^{2m} + \cdots\right]$$

Hint: Let $n = 2m$ in the recurrence relation, $m = 1, 2, 3, \ldots$.

3. $r(r-1) = 0; \quad a_n = -\dfrac{a_{n-1}}{(n+r)(n+r-1)}; \quad r_1 = 1,\ r_2 = 0$

$$y_1(x) = x\left[1 - \frac{x}{1!2!} + \frac{x^2}{2!3!} + \cdots + \frac{(-1)^n}{n!(n+1)!}x^n + \cdots\right]$$

4. $r^2 = 0; \quad a_n = \dfrac{a_{n-1}}{(n+r)^2}; \quad r_1 = r_2 = 0$

$$y_1(x) = 1 + \frac{x}{(1!)^2} + \frac{x^2}{(2!)^2} + \cdots + \frac{x^n}{(n!)^2} + \cdots$$

5. $r(3r-1) = 0; \quad a_n = -\dfrac{a_{n-2}}{(n+r)[3(n+r)-1]}; \quad r_1 = \frac{1}{3},\ r_2 = 0$

$$y_1(x) = x^{1/3}\left[1 - \frac{1}{1!7}\left(\frac{x^2}{2}\right) + \frac{1}{2!7\cdot 13}\left(\frac{x^2}{2}\right)^2 + \cdots + \frac{(-1)^m}{m!7\cdot 13\cdots(6m+1)}\left(\frac{x^2}{2}\right)^m + \cdots\right]$$

$$y_2(x) = 1 - \frac{1}{1!5}\left(\frac{x^2}{2}\right) + \frac{1}{2!5\cdot 11}\left(\frac{x^2}{2}\right)^2 + \cdots + \frac{(-1)^m}{m!5\cdot 11\cdots(6m-1)}\left(\frac{x^2}{2}\right)^m + \cdots$$

Hint: Let $n = 2m$ in the recurrence relation, $m = 1, 2, 3, \ldots$.

6. $r^2 - 2 = 0; \quad a_n = -\dfrac{a_{n-1}}{(n+r)^2 - 2}; \quad r_1 = \sqrt{2}, \quad r_2 = -\sqrt{2}$

$$y_1(x) = x^{\sqrt{2}}\left[1 - \frac{x}{1(1+2\sqrt{2})} + \frac{x^2}{2!(1+2\sqrt{2})(2+2\sqrt{2})} + \cdots + \frac{(-1)^n}{n!(1+2\sqrt{2})(2+2\sqrt{2})\cdots(n+2\sqrt{2})}x^n + \cdots\right]$$

$$y_2(x) = x^{-\sqrt{2}}\left[1 - \frac{x}{1(1-2\sqrt{2})} + \frac{x^2}{2!(1-2\sqrt{2})(2-2\sqrt{2})} + \cdots + \frac{(-1)^n}{n!(1-2\sqrt{2})(2-2\sqrt{2})\cdots(n-2\sqrt{2})}x^n + \cdots\right]$$

7. $r^2 = 0; \quad r_1 = 0, \quad r_2 = 0$

$$y_1(x) = 1 + \frac{\alpha(\alpha+1)}{2 \cdot 1^2}(x-1) - \frac{\alpha(\alpha+1)[1 \cdot 2 - \alpha(\alpha+1)]}{(2 \cdot 1^2)(2 \cdot 2^2)}(x-1)^2 + \cdots$$
$$+(-1)^{n+1}\frac{\alpha(\alpha+1)[1 \cdot 2 - \alpha(\alpha+1)] \cdots [n(n-1) - \alpha(\alpha+1)]}{2^n(n!)^2}(x-1)^n + \cdots$$

8. (a) $r_1 = \frac{1}{2}, \quad r_2 = 0$ at both $x = \pm 1$.

(b) $y_1(x) = |x-1|^{1/2} \times$

$$\left[1 + \sum_{n=1}^{\infty} \frac{(-1)^n(1+2\alpha)\cdots(2n-1+2\alpha)(1-2\alpha)\cdots(2n-1-2\alpha)}{2^n(2n+1)!}(x-1)^n\right]$$

$y_2(x) = 1 +$

$$\sum_{n=1}^{\infty} \frac{(-1)^n\alpha(1+\alpha)\cdots(n-1+\alpha)(-\alpha)(1-\alpha)\cdots(n-1-\alpha)}{n!1 \cdot 3 \cdot 5 \cdots (2n-1)}(x-1)^n$$

9. $r^2 = 0; \quad r_1 = 0, \quad r_2 = 0; \quad a_n = \dfrac{(n-1-\lambda)a_{n-1}}{n^2}$

$$y_1(x) = 1 + \frac{-\lambda}{(1!)^2}x + \frac{(-\lambda)(1-\lambda)}{2!}x^2 + \cdots$$
$$+ \frac{(-\lambda)(1-\lambda)\cdots(n-1-\lambda)}{(n!)^2}x^n + \cdots$$

For $\lambda = n$, the coefficients of all terms past x^n are zero.

12. (b) $[(n-1)^2 - 1]b_n = -b_{n-2}$, and it is impossible to determine b_2.

Section 4.5.1, Page 238

1. $x = 0$; $\quad r(r-1) = 0$; $\quad r_1 = 1, r_2 = 0$
2. $x = 0$; $\quad r^2 - 3r + 2 = 0$; $\quad r_1 = 2, r_2 = 1$
3. $x = 0$; $\quad r(r-1) = 0$; $\quad r_1 = 1, r_2 = 0$
   $x = 1$; $\quad r(r+5) = 0$; $\quad r_1 = 0, r_2 = -5$
4. None
5. $x = 0$; $\quad r^2 + 2r - 2 = 0$; $\quad r_1 = -1 + \sqrt{3}, r_2 = -1 - \sqrt{3}$
6. $x = 0$; $\quad r(r - \frac{3}{4}) = 0$; $\quad r_1 = \frac{3}{4}, r_2 = 0$
   $x = -2$; $\quad r(r - \frac{5}{4}) = 0$; $\quad r_1 = \frac{5}{4}, r_2 = 0$
7. $x = 0$; $\quad r^2 + 1 = 0$; $\quad r_1 = i, r_2 = -i$
8. $x = -1$; $\quad r^2 - 7r + 3 = 0$; $\quad r_1 = (7 + \sqrt{37})/2, r_2 = (7 - \sqrt{37})/2$
9. $y_1(x) = x - \dfrac{x^3}{4!} + \dfrac{x^5}{6!} + \cdots$
10. $r_1 = \frac{1}{2}, r_2 = 0$
    $y_1(x) = (x-1)^{1/2}[1 - \frac{3}{4}(x-1) + \frac{53}{480}(x-1)^2 + \cdots], \qquad \rho = 1$
11. (c) *Hint:* $(n-1)(n-2) + (1 + \alpha + \beta)(n-1) + \alpha\beta$
    $= (n - 1 + \alpha)(n - 1 + \beta)$

    (d) *Hint:* $(n-\gamma)(n-1-\gamma) + (1 + \alpha + \beta)(n-\gamma) + \alpha\beta$
    $= (n - \gamma + \alpha)(n - \gamma + \beta)$

Section 4.7, Page 253

1. (a) $y_1(x) = \displaystyle\sum_{n=0}^{\infty} \frac{(-1)^n x^n}{n!(n+1)!}$

$$y_2(x) = -y_1(x)\ln x + \frac{1}{x}\left[1 - \sum_{n=1}^{\infty} \frac{H_n + H_{n-1}}{n!(n-1)!}(-1)^n x^n\right]$$

(b) $y_1(x) = \frac{1}{x}\sum_{n=0}^{\infty}\frac{(-1)^n x^n}{(n!)^2}$, $\quad y_2(x) = y_1(x)\ln x - \frac{2}{x}\sum_{n=1}^{\infty}\frac{(-1)^n H_n}{(n!)^2}x^n$

(c) $y_1(x) = \sum_{n=0}^{\infty}\frac{(-1)^n 2^n}{(n!)^2}x^n$, $\quad y_2(x) = y_1(x)\ln x - 2\sum_{n=1}^{\infty}\frac{(-1)^n 2^n H_n}{(n!)^2}x^n$

(d) $y_1(x) = \frac{1}{x}\sum_{n=0}^{\infty}\frac{(-1)^n}{n!(n+1)!}x^n$

$$y_2(x) = -y_1(x)\ln x + \frac{1}{x^2}\left[1 - \sum_{n=1}^{\infty}\frac{H_n + H_{n-1}}{n!(n-1)!}(-1)^n x^n\right]$$

2. $y_1(x) = x^{3/2}\left[1 + \sum_{m=1}^{\infty}\frac{(-1)^m}{m!(1+\frac{3}{2})(2+\frac{3}{2})\cdots(m+\frac{3}{2})}\left(\frac{x}{2}\right)^{2m}\right]$

$$y_2(x) = x^{-3/2}\left[1 + \sum_{m=1}^{\infty}\frac{(-1)^m}{m!(1-\frac{3}{2})(2-\frac{3}{2})\cdots(m-\frac{3}{2})}\left(\frac{x}{2}\right)^{2m}\right]$$

Hint: Let $n = 2m$ in the recurrence relation, $m = 1, 2, 3, \ldots$. For $r = -\frac{3}{2}$, $a_1 = 0$ and a_3 is arbitrary.

CHAPTER FIVE

Section 5.1, Page 258

1. $\phi^{(n)}(0) = -p_1(0)y_0^{(n-1)} - \cdots - p_n(0)y_0$
   $\phi^{(n+1)}(0) = p_1(0)[p_1(0)y_0^{(n-1)} + \cdots + p_n(0)y_0] - p_1'(0)y_0^{(n-1)} - \cdots - p_n(0)y_0' - p_n'(0)y_0$
2. (a) $y''' = -\cos x$ (b) $y''' + y' = 0$
   (c) $y''' - 2y'' - y' + 2y = 0$ (d) $x^3y''' - 3x^2y'' + 6xy' - 6y = 0$
   (e) $y''' + y' = 1$ (f) $y^{\text{iv}} - y'' = 0$
3. (a) $-\infty < x < \infty$ (b) $x > 0$ or $x < 0$
   (c) $x > 1$, or $0 < x < 1$, or $x < 0$ (d) $x > 0$

Section 5.2, Page 261

4. (a) 1 (b) 1 (c) $-6e^{-2x}$ (d) e^{-2x} (e) $6x$
5. $\sin^2 x = \frac{1}{10}(5) - \frac{1}{2}\cos 2x$
6. (a) $a_0[n(n-1)(n-2)\cdots 1] + a_1[n(n-1)\cdots 2]x + \cdots + a_n x^n$
   (b) $(a_0r^n + a_1r^{n-1} + \cdots + a_n)e^{rx}$
   (c) $e^x, e^{-x}, e^{2x}, e^{-2x}$; yes, $W(e^x, e^{-x}, e^{2x}, e^{-2x}) \neq 0$, $-\infty < x < \infty$
8. (a) $y = c_1x + c_2x^2 + c_3x^3$ (b) $y = c_1x^2 + c_2x^3 + c_3(x+1)$

Section 5.3, Page 268

1. (a) $\sqrt{2}\,e^{i[(\pi/4)+2m\pi]}$ (b) $2e^{i[(2\pi/3)+2m\pi]}$ (c) $3e^{i(\pi+2m\pi)}$
   (d) $e^{i[(3\pi/2)+2m\pi]}$ (e) $2e^{i[(11\pi/6)+2m\pi]}$ (f) $\sqrt{2}\,e^{i[(5\pi/4)+2m\pi]}$
2. (a) $1, \frac{1}{2}(-1+i\sqrt{3}), \frac{1}{2}(-1-i\sqrt{3})$ (b) $2^{1/4}e^{-\pi i/8}, 2^{1/4}e^{7\pi i/8}$
   (c) $1, i, -1, -i$ (d) $(\sqrt{3}+i)/\sqrt{2}, -(\sqrt{3}+i)/\sqrt{2}$
3. $y = c_1e^x + c_2xe^x + c_3e^{-x}$
4. $y = c_1e^x + c_2xe^x + c_3x^2e^x$
5. $y = c_1e^x + c_2e^{2x} + c_3e^{-x}$
6. $y = c_1 + c_2x + c_3e^{2x} + c_4xe^{2x}$
7. $y = c_1\cos x + c_2\sin x + e^{\sqrt{3}x/2}(c_3\cos\frac{1}{2}x + c_4\sin\frac{1}{2}x)$
   $+e^{-\sqrt{3}x/2}(c_5\cos\frac{1}{2}x + c_6\sin\frac{1}{2}x)$
8. $y = c_1e^x + c_2e^{-x} + c_3e^{2x} + c_4e^{-2x}$

9. $y = c_1e^x + c_2xe^x + c_3x^2e^x + c_4e^{-x} + c_5xe^{-x} + c_6x^2e^{-x}$
10. $y = c_1 + c_2x + c_3e^x + c_4e^{-x} + c_5\cos x + c_6\sin x$
11. $y = c_1 + c_2e^x + c_3e^{2x} + c_4\cos x + c_5\sin x$
12. $y = c_1 + c_2e^{2x} + e^{-x}(c_3\cos\sqrt{3}\,x + c_4\sin\sqrt{3}\,x)$
13. $y = e^x[(c_1 + c_2x)\cos x + (c_3 + c_4x)\sin x] + e^{-x}[(c_5 + c_6x)\cos x + (c_7 + c_8x)\sin x]$
14. $y = (c_1 + c_2x)\cos x + (c_3 + c_4x)\sin x$
15. $y = 2 - 2\cos x + \sin x$ 16. $y = \cos x$
17. $y = 2x - 3$ 18. $y = e^{(x-\pi/2)} + \sin x$
19. $y = \frac{1}{2}(\cosh x - \cos x) + \frac{1}{2}(\sinh x - \sin x)$
21. (a) $y = c_1x + c_2x^2 + c_3x^{-1}$ (b) $y = c_1x + c_2x\ln x + c_3x(\ln x)^2$
(c) $y = c_1x + c_2\cos(\ln x) + c_3\sin(\ln x)$
(d) $y = c_1 + c_2x^{-1} + x[c_3\cos(2\ln x) + c_4\sin(2\ln x)]$
(e) $y = c_1 + c_2x + c_3x\ln x$ (f) $y = c_1x^2 + c_2x^{-1} + c_3x^{-2}$
24. (a) Stable (b) Not stable (c) Not stable
(d) Not stable (e) Not stable (f) Stable

Section 5.4, Page 273

1. $y = c_1e^x + c_2xe^x + c_3e^{-x} + \frac{1}{2}xe^{-x} + 3$
2. $y = c_1e^x + c_2e^{-x} + c_3\cos x + c_4\sin x - 3x - \frac{1}{4}x\sin x$
3. $y = c_1e^{-x} + c_2\cos x + c_3\sin x + \frac{1}{2}xe^{-x} + 4(x - 1)$
4. $y = c_1 + c_2e^x + c_3e^{-x} + \cos x$
5. $y = c_1 + c_2x + c_3e^{-2x} + c_4e^{2x} - \frac{1}{3}e^x - \frac{1}{48}x^4 - \frac{1}{16}x^2$
6. $y = c_1\cos x + c_2\sin x + c_3x\cos x + c_4x\sin x + 3 + \frac{1}{9}\cos 2x$
7. $y = c_1 + c_2x + c_3x^2 + c_4e^{-x} + e^{x/2}\left(c_5\cos\frac{\sqrt{3}}{2}x + c_6\sin\frac{\sqrt{3}}{2}x\right) + \frac{x^4}{24}$
8. $y = c_1 + c_2x + c_3x^2 + c_4e^{-x} + \frac{1}{20}\sin 2x + \frac{1}{40}\cos 2x$
9. $y = \frac{3}{16}(1 - \cos 2x) + \frac{1}{8}x^2$
10. $y = (x - 4)\cos x - (\frac{3}{2}x + 4)\sin x + 3x + 4$
11. $y = 1 + \frac{1}{4}(x^2 + 3x) - xe^x$
12. $y_p(x) = x(A_0x^3 + A_1x^2 + A_2x + A_3) + Bx^2e^x$
13. $y_p(x) = x(A_0x + A_1)e^{-x} + B\cos x + C\sin x$
14. $y_p(x) = Ax^2e^x + B\cos x + C\sin x$
15. $y_p(x) = Ax^2 + (B_0x + B_1)e^x + x(C\cos 2x + D\sin 2x)$
16. $y_p(x) = x(A_0x^2 + A_1x + A_2) + (B_0x + B_1)\cos x + (C_0x + C_1)\sin x$
17. $y_p(x) = Ae^x + (B_0x + B_1)e^{-x} + xe^{-x}(C\cos x + D\sin x)$
18. $t_0 = a_0, \quad t_n = a_0\alpha^n + a_1\alpha^{n-1} + \cdots + a_{n-1}\alpha + a_n$

Section 5.5, Page 278

1. $y_p(x) = -\ln\cos x - (\sin x)\ln(\sec x + \tan x)$
2. $y_p(x) = -x^2/2$ 3. $y_p(x) = e^{4x}/30$ 4. $y_p(x) = x^4/15$
5. $y_p(x) = \frac{1}{2}\int^x [e^{x-t} - \sin(x - t) - \cos(x - t)]g(t)\,dt$
6. $y_p(x) = \frac{1}{2}\int^x [\sinh(x - t) - \sin(x - t)]g(t)\,dt$
7. $y_p(x) = \frac{1}{2}\int^x e^{(x-t)}(x - t)^2g(t)\,dt; \quad y_p(x) = -xe^x\ln|x|$
8. $y_p(x) = \frac{1}{2}\int^x \left(\frac{x}{t^2} - 2\frac{x^2}{t^3} + \frac{x^3}{t^4}\right)g(t)\,dt$

CHAPTER SIX

Section 6.1, Page 284

1. (a) Piecewise continuous (b) Neither
   (c) Continuous (d) Piecewise continuous
2. (a) $1/s^2$, $s > 0$ (b) $2/s^3$, $s > 0$ (c) $n!/s^{n+1}$, $s > 0$
3. $s/(s^2 + a^2)$, $s > 0$
4. (a) $\dfrac{s}{s^2 - b^2}$, $s > |b|$ (b) $\dfrac{b}{s^2 - b^2}$, $s > |b|$
   (c) $\dfrac{s - a}{(s - a)^2 - b^2}$, $s - a > |b|$ (d) $\dfrac{b}{(s - a)^2 - b^2}$, $s - a > |b|$
5. (a) $\dfrac{b}{s^2 + b^2}$, $s > 0$ (b) $\dfrac{s}{s^2 + b^2}$, $s > 0$
   (c) $\dfrac{b}{(s - a)^2 + b^2}$, $s > a$ (d) $\dfrac{s - a}{(s - a)^2 + b^2}$, $s > a$
6. (a) $\dfrac{1}{(s - a)^2}$, $s > a$ (b) $\dfrac{2as}{(s^2 + a^2)^2}$, $s > 0$
   (c) $\dfrac{s^2 + a^2}{(s - a)^2(s + a)^2}$, $s > |a|$ (d) $\dfrac{n!}{(s - a)^{n+1}}$, $s > a$
   (e) $\dfrac{2a(3s^2 - a^2)}{(s^2 + a^2)^3}$, $s > 0$ (f) $\dfrac{2a(3s^2 + a^2)}{(s^2 - a^2)^3}$, $s > |a|$
7. (a) Converges (b) Converges (c) Diverges
9. (c) $\Gamma(3/2) = \sqrt{\pi}/2$; $\Gamma(11/2) = 945\sqrt{\pi}/32$

Section 6.2, Page 294

1. $\frac{3}{2} \sin 2t$
2. $2t^2 e^t$
3. $\frac{2}{5} e^t - \frac{2}{5} e^{-4t}$
4. $\frac{9}{5} e^{3t} + \frac{6}{5} e^{-2t}$
5. $2e^{-t} \cos 2t$
6. $2 \cosh 2t - \frac{3}{2} \sinh 2t$
7. $2e^t \cos t + 3e^t \sin t$
8. $3 - 2 \sin 2t + 5 \cos 2t$
9. $-2e^{-2t} \cos t + 5e^{-2t} \sin t$
10. $2e^{-t} \cos 3t - \frac{5}{3} e^{-t} \sin 3t$
11. $y = \frac{1}{5}(e^{3t} + 4e^{-2t})$
12. $y = 2e^{-t} - e^{-2t}$
13. $y = e^t \sin t$
14. $y = e^{2t} - te^{2t}$
15. $y = 2e^t \cosh \sqrt{3}\, t - (2/\sqrt{3}) e^t \sinh \sqrt{3}\, t$
16. $y = 2e^{-t} \cos 2t + \frac{1}{2} e^{-t} \sin 2t$
17. $y = te^t - t^2 e^t + \frac{2}{3} t^3 e^t$
18. $y = \cosh t$
19. $y = \cos \sqrt{2}\, t$
20. $y = (\omega^2 - 4)^{-1}[(\omega^2 - 5) \cos \omega t + \cos 2t]$
21. $y = \frac{1}{5}(\cos t - 2 \sin t + 4e^t \cos t - 2e^t \sin t)$
22. $y = \frac{1}{5}(e^{-t} - e^t \cos t + 7e^t \sin t)$
23. $y = 2e^{-t} + te^{-t} + 2t^2 e^{-t}$
24. $Y(s) = \dfrac{s}{s^2 + 4} + \dfrac{1 - e^{-\pi s}}{s(s^2 + 4)}$
25. $Y(s) = \dfrac{1}{s^2(s^2 + 1)} - \dfrac{e^{-s}(s + 1)}{s^2(s^2 + 1)}$
26. $Y(s) = (1 - e^{-s})/s^2(s^2 + 4)$

29. (a) $1/(s-a)^2$ (b) $2b(3s^2-b^2)/(s^2+b^2)^3$
(c) $n!/s^{n+1}$ (d) $n!/(s-a)^{n+1}$
(e) $2b(s-a)/[(s-a)^2+b^2]^2$ (f) $[(s-a)^2-b^2]/[(s-a)^2+b^2]^2$

31. (a) $Y' + s^2Y = s$ (b) $s^2Y'' + 2sY' - [s^2+\alpha(\alpha+1)]Y = -1$

Section 6.3, Page 302

2. (a) $F(s) = 2e^{-2s}/s^3$ (b) $F(s) = e^{-s}(s^2+2)/s^3$
(c) $F(s) = \dfrac{e^{-\pi s}}{s^2} - \dfrac{e^{-2\pi s}}{s^2}(1+\pi s)$ (d) $F(s) = \dfrac{1}{s}(e^{-s} + 2e^{-3s} - 6e^{-4s})$
(e) $F(s) = s^{-2}[(1-s)e^{-2s} - (1+s)e^{-3s}]$
(f) $F(s) = (1-e^{-s})/s^2$

3. (a) $f(t) = t^3e^{2t}$ (b) $f(t) = \frac{1}{3}u_2(t)[e^{t-2} - e^{-2(t-2)}]$
(c) $f(t) = 2u_2(t)e^{t-2}\cos(t-2)$ (d) $f(t) = u_2(t)\sinh 2(t-2)$
(e) $f(t) = u_1(t)e^{2(t-1)}\cosh(t-1)$ (f) $f(t) = u_1(t) + u_2(t) - u_3(t) - u_4(t)$

5. (a) $f(t) = 2(2t)^n$ (b) $f(t) = \frac{1}{2}e^{-t/2}\cos t$
(c) $f(t) = \frac{1}{6}e^{t/3}(e^{2t/3} - 1)$ (d) $f(t) = \frac{1}{2}e^{t/2}u_2(t/2)$

6. (a) $F(s) = s^{-1}(1-e^{-s}), \quad s > 0$
(b) $F(s) = s^{-1}(1 - e^{-s} + e^{-2s} - e^{-3s}), \quad s > 0$
(c) $F(s) = \dfrac{1}{s}[1 - e^{-s} + \cdots + e^{-2ns} - e^{-(2n+1)s}] = \dfrac{1-e^{-(2n+2)s}}{s(1+e^{-s})}, \quad s > 0$
(d) $F(s) = \dfrac{1}{s}\sum_{n=0}^{\infty}(-1)^n e^{-ns} = \dfrac{1/s}{1+e^{-s}}, \quad s > 0$

8. (a) $\mathscr{L}\{f(t)\} = \dfrac{1/s}{1+e^{-s}}, \quad s > 0$ (b) $\mathscr{L}\{f(t)\} = \dfrac{1-e^{-s}}{s(1+e^{-s})}, \quad s > 0$
(c) $\mathscr{L}\{f(t)\} = \dfrac{1-(1+s)e^{-s}}{s^2(1-e^{-s})}, \quad s > 0$
(d) $\mathscr{L}\{f(t)\} = \dfrac{1+e^{-\pi s}}{(1+s^2)(1-e^{-\pi s})}, \quad s > 0$

9. (a) $\mathscr{L}\{f(t)\} = s^{-1}(1-e^{-s}), \quad s > 0$ (b) $\mathscr{L}\{g(t)\} = s^{-2}(1-e^{-s}), \quad s > 0$
(c) $\mathscr{L}\{h(t)\} = s^{-2}(1-e^{-s})^2, \quad s > 0$

10. (b) $\mathscr{L}\{p(t)\} = \dfrac{1-e^{-s}}{s^2(1+e^{-s})}, \quad s > 0$

Section 6.3.1, Page 309

1. $y = 1 - \cos t + \sin t - u_{\pi/2}(t)[1 - \sin t]$
2. $y = e^{-t}\sin t + \frac{1}{2}u_\pi(t)[1 + e^{-(t-\pi)}\cos t + e^{-(t-\pi)}\sin t]$
$\frac{1}{2}u_{2\pi}(t)[1 - e^{-(t-2\pi)}\cos t - e^{-(t-2\pi)}\sin t]$
3. $y = \frac{1}{6}[1 - u_{2\pi}(t)](2\sin t - \sin 2t)$
4. $y = \frac{1}{6}(2\sin t - \sin 2t) - \frac{1}{6}u_\pi(t)(2\sin t + \sin 2t)$
5. $y = 1 - u_1(t)[1 - e^{-(t-1)} - (t-1)e^{-(t-1)}]$
6. $y = e^{-t} - e^{-2t} + u_2(t)[\frac{1}{2} - e^{-(t-2)} + \frac{1}{2}e^{-2(t-2)}]$
7. $y = \cos t + u_\pi(t)(1 + \cos t)$
8. $y = h(t) - u_{\pi/2}(t)h(t - \pi/2)$,
$h(t) = \frac{4}{25}(-4 + 5t + 4e^{-t/2}\cos t - 3e^{-t/2}\sin t)$
9. $y = t - u_1(t)[t - 1 - \sin(t-1)]$

10. $y = h(t) + u_\pi(t)h(t - \pi)$,
$h(t) = \frac{4}{17}[-4\cos t + \sin t + 4e^{-t/2}\cos t + e^{-t/2}\sin t]$
11. $y = \cos 2t + u_\pi(t)h(t - \pi) - u_{2\pi}(t)h(t - 2\pi)$, $h(t) = (1 - \cos 2t)/4$
12. $y = u_1(t)h(t - 1) - u_2(t)h(t - 2)$, $h(t) = -1 + (\cos t + \cosh t)/2$
13. $y = h(t) - u_\pi(t)h(t - \pi)$, $h(t) = (3 - 4\cos t + \cos 2t)/12$
14. $y = 1 + \sum_{n=1}^{\infty} (-1)^n u_{n\pi}(t)[1 - \cos(t - n\pi)]$

Section 6.4, Page 314

1. $y = e^{-t}\cos t + e^{-t}\sin t - u_\pi(t)e^{-(t-\pi)}\sin t$
2. $y = \frac{1}{2}u_\pi(t)\sin 2t - \frac{1}{2}u_{2\pi}(t)\sin 2t$
3. $y = 2te^{-t} + u_{2\pi}(t)[1 - e^{-(t-2\pi)} - (t - 2\pi)e^{-(t-2\pi)}]$
4. $y = \cosh t + 2u_1(t)\sinh(t - 1)$
5. $y = \frac{1}{2}\sqrt{2}\,e^{-t}\sin\sqrt{2}\,t + \frac{1}{4}e^{-t}\cos\sqrt{2}\,t + \frac{1}{4}(\sin t - \cos t)$
$+ \frac{1}{2}\sqrt{2}\,u_\pi(t)e^{-(t-\pi)}\sin\sqrt{2}\,(t - \pi)$
6. $y = \cos\omega t - \omega^{-1}u_{\pi/\omega}(t)\sin\omega t$
7. $y = [1 + u_\pi(t)]\sin t$
8. $y = u_{\pi/4}(t)\sin 2(t - \pi/4)$
9. $y = u_{\pi/2}(t)[1 - \cos(t - \pi/2)] + u_\pi(t)\sin(t - \pi) - u_{3\pi/2}(t)[1 - \cos(t - 3\pi/2)]$
10. $y = u_{\pi/6}(t)\sin 2(t - \pi/6)$
11. $y = \frac{1}{5}\cos t + \frac{2}{5}\sin t - \frac{1}{5}e^{-t}\cos t - \frac{3}{5}e^{-t}\sin t + u_{\pi/2}(t)e^{-(t-\pi/2)}\sin(t - \pi/2)$
12. $y = u_1(t)[\sinh(t - 1) - \sin(t - 1)]/2$

Section 6.5, Page 319

3. $\sin t * \sin t = \frac{1}{2}(\sin t - t\cos t)$ is negative when $t = 2\pi$, for example.
4. (a) $F(s) = 2/s^2(s^2 + 4)$ (b) $F(s) = 1/(s + 1)(s^2 + 1)$
(c) $F(s) = 1/s^2(s - 1)$ (d) $F(s) = s/(s^2 + 1)^2$
5. (a) $f(t) = \dfrac{1}{6}\int_0^t (t - \tau)^3 \sin\tau\, d\tau$ (b) $f(t) = \int_0^t e^{-(t-\tau)}\cos 2\tau\, d\tau$
(c) $f(t) = \dfrac{1}{2}\int_0^t (t - \tau)e^{-(t-\tau)}\sin 2\tau\, d\tau$ (d) $f(t) = \int_0^t \sin(t - \tau)g(\tau)\, d\tau$
6. $y = \dfrac{1}{\omega}\sin\omega t + \dfrac{1}{\omega}\int_0^t \sin\omega(t - \tau)g(\tau)\, d\tau$
7. $y = \int_0^t e^{-(t-\tau)}\sin(t - \tau)\sin\alpha\tau\, d\tau$
8. $y = \dfrac{1}{8}\int_0^t e^{-(t-\tau)/2}\sin 2(t - \tau)g(\tau)\, d\tau$
9. $y = e^{-t/2}\cos t - \frac{1}{2}e^{-t/2}\sin t + \int_0^t e^{-(t-\tau)/2}\sin(t - \tau)[1 - u_\pi(\tau)]\, d\tau$
10. $y = 2e^{-2t} + te^{-2t} + \int_0^t (t - \tau)e^{-2(t-\tau)}g(\tau)\, d\tau$
11. $y = 2e^{-t} - e^{-2t} + \int_0^t [e^{-(t-\tau)} - e^{-2(t-\tau)}]\cos\alpha\tau\, d\tau$
12. $y = \dfrac{1}{2}\int_0^t [\sinh(t - \tau) - \sin(t - \tau)]g(\tau)\, d\tau$
13. $y = \frac{4}{3}\cos t - \frac{1}{3}\cos 2t + \dfrac{1}{6}\int_0^t [2\sin(t - \tau) - \sin 2(t - \tau)]g(\tau)\, d\tau$
14. $\Phi(s) = \dfrac{F(s)}{1 + K(s)}$
15. (c) $\phi(t) = \frac{1}{3}(4\sin 2t - 2\sin t)$ (d) $u(t) = \frac{1}{3}(2\sin t - \sin 2t)$

CHAPTER SEVEN

Section 7.1, Page 328

2. $x_1' - x_2 = 0, \qquad x_1(0) = u_0$
$x_2' + p(t)x_2 + q(t)x_1 = g(t), \qquad x_2(0) = u_0'$

Section 7.2, Page 338

1. (a) $\begin{pmatrix} 6 & -6 & 3 \\ 5 & 9 & -2 \\ 2 & 3 & 8 \end{pmatrix}$ (b) $\begin{pmatrix} -15 & 6 & -12 \\ 7 & -18 & -1 \\ -26 & -3 & -5 \end{pmatrix}$

(c) $\begin{pmatrix} 6 & -12 & 3 \\ 4 & 3 & 7 \\ 9 & 12 & 0 \end{pmatrix}$ (d) $\begin{pmatrix} -8 & -9 & 11 \\ 14 & 12 & -5 \\ 5 & -8 & 5 \end{pmatrix}$

2. (a) $\begin{pmatrix} 1 - i & -7 + 2i \\ -1 + 2i & 2 + 3i \end{pmatrix}$ (b) $\begin{pmatrix} 3 + 4i & 6i \\ 11 + 6i & 6 - 5i \end{pmatrix}$

(c) $\begin{pmatrix} -3 + 5i & 7 + 5i \\ 2 + i & 7 + 2i \end{pmatrix}$ (d) $\begin{pmatrix} 8 + 7i & 4 - 4i \\ 6 - 4i & -4 \end{pmatrix}$

3. (a) $\begin{pmatrix} -2 & 1 & 2 \\ 1 & 0 & -1 \\ 2 & -3 & 1 \end{pmatrix}$ (b) $\begin{pmatrix} 1 & 3 & -2 \\ 2 & -1 & 1 \\ 3 & -1 & 0 \end{pmatrix}$ (c), (d) $\begin{pmatrix} -1 & 4 & 0 \\ 3 & -1 & 0 \\ 5 & -4 & 1 \end{pmatrix}$

4. (a) $\begin{pmatrix} 3 - 2i & 2 - i \\ 1 + i & -2 + 3i \end{pmatrix}$ (b) $\begin{pmatrix} 3 + 2i & 1 - i \\ 2 + i & -2 - 3i \end{pmatrix}$

(c) $\begin{pmatrix} 3 + 2i & 2 + i \\ 1 - i & -2 - 3i \end{pmatrix}$

5. $\begin{pmatrix} 10 & 6 & -4 \\ 0 & 4 & 10 \\ 4 & 4 & 6 \end{pmatrix}$

6. (a) $\begin{pmatrix} 7 & -11 & -3 \\ 11 & 20 & 17 \\ -4 & 3 & -12 \end{pmatrix}$ (b) $\begin{pmatrix} 5 & 0 & -1 \\ 2 & 7 & 4 \\ -1 & 1 & 4 \end{pmatrix}$ (c) $\begin{pmatrix} 6 & -8 & -11 \\ 9 & 15 & 6 \\ -5 & -1 & 5 \end{pmatrix}$

8. (a) $4i$ (b) $12 - 8i$ (c) $2 + 2i$ (d) 16

10. $\begin{pmatrix} \frac{3}{11} & -\frac{4}{11} \\ \frac{2}{11} & \frac{1}{11} \end{pmatrix}$

11. $\begin{pmatrix} \frac{1}{6} & \frac{1}{12} \\ -\frac{1}{2} & \frac{1}{4} \end{pmatrix}$

12. $\begin{pmatrix} 1 & -3 & 2 \\ -3 & 3 & -1 \\ 2 & -1 & 0 \end{pmatrix}$

13. $\begin{pmatrix} \frac{1}{3} & \frac{1}{3} & 0 \\ \frac{1}{3} & -\frac{1}{3} & \frac{1}{3} \\ -\frac{1}{3} & 0 & \frac{1}{3} \end{pmatrix}$

14. Singular

15. $\begin{pmatrix} \frac{1}{2} & -\frac{1}{4} & \frac{1}{8} \\ 0 & \frac{1}{2} & -\frac{1}{4} \\ 0 & 0 & \frac{1}{2} \end{pmatrix}$

16. $\begin{pmatrix} \frac{1}{10} & \frac{3}{10} & \frac{1}{10} \\ -\frac{2}{10} & \frac{4}{10} & -\frac{2}{10} \\ -\frac{7}{10} & -\frac{1}{10} & \frac{3}{10} \end{pmatrix}$

17. Singular

18. $\begin{pmatrix} 1 & 1 & 0 & 1 \\ 1 & 0 & 1 & 1 \\ 1 & 1 & 1 & 1 \\ 0 & 1 & 0 & 1 \end{pmatrix}$

19. $\begin{pmatrix} 6 & \frac{13}{5} & -\frac{8}{5} & \frac{2}{5} \\ 5 & \frac{11}{5} & -\frac{6}{5} & \frac{4}{5} \\ 0 & -\frac{1}{5} & \frac{1}{5} & \frac{1}{5} \\ -2 & -\frac{4}{5} & \frac{4}{5} & -\frac{1}{5} \end{pmatrix}$

22. (a) $\begin{pmatrix} 7e^t & 5e^{-t} & 10e^{2t} \\ -e^t & 7e^{-t} & 2e^{2t} \\ 8e^t & 0 & -e^{2t} \end{pmatrix}$

(b) $\begin{pmatrix} 2e^{2t} - 2 + 3e^{3t} & 1 + 4e^{-2t} - e^t & 3e^{3t} + 2e^t - e^{4t} \\ 4e^{2t} - 1 - 3e^{3t} & 2 + 2e^{-2t} + e^t & 6e^{3t} + e^t + e^{4t} \\ -2e^{2t} - 3 + 6e^{3t} & -1 + 6e^{-2t} - 2e^t & -3e^{3t} + 3e^t - 2e^{4t} \end{pmatrix}$

(c) $\begin{pmatrix} e^t & -2e^{-t} & 2e^{2t} \\ 2e^t & -e^{-t} & -2e^{2t} \\ -e^t & -3e^{-t} & 4e^{2t} \end{pmatrix}$ (d) $(e-1)\begin{pmatrix} 1 & 2e^{-1} & \frac{1}{2}(e+1) \\ 2 & e^{-1} & -\frac{1}{2}(e+1) \\ -1 & 3e^{-1} & e+1 \end{pmatrix}$

Section 7.3, Page 350

1. $x_1 = -\frac{1}{3},\ x_2 = \frac{7}{3},\ x_3 = -\frac{1}{3}$
2. No solution
3. $x_1 = -c,\ x_2 = c + 1,\ x_3 = c,$ where c is arbitrary
4. $x_1 = c,\ x_2 = -c,\ x_3 = -c,$ where c is arbitrary
5. $x_1 = 0,\ x_2 = 0,\ x_3 = 0$
6. (a) Linearly independent (b) $\mathbf{x}^{(1)} - 5\mathbf{x}^{(2)} + 2\mathbf{x}^{(3)} = \mathbf{0}$
   (c) $2\mathbf{x}^{(1)} - 3\mathbf{x}^{(2)} + 4\mathbf{x}^{(3)} - \mathbf{x}^{(4)} = \mathbf{0}$
   (d) Linearly independent (e) $\mathbf{x}^{(1)} + \mathbf{x}^{(2)} - \mathbf{x}^{(4)} = \mathbf{0}$
8. (a) $3\mathbf{x}^{(1)}(t) - 6\mathbf{x}^{(2)}(t) + \mathbf{x}^{(3)}(t) = \mathbf{0}$ (b) Linearly independent

10. $\lambda_1 = 2,\ \mathbf{x}^{(1)} = \begin{pmatrix} 1 \\ 3 \end{pmatrix};\quad \lambda_2 = 4,\ \mathbf{x}^{(2)} = \begin{pmatrix} 1 \\ 1 \end{pmatrix}$

11. $\lambda_1 = 1 + 2i,\ \mathbf{x}^{(1)} = \begin{pmatrix} 1 \\ 1 - i \end{pmatrix};\quad \lambda_2 = 1 - 2i,\ \mathbf{x}^{(2)} = \begin{pmatrix} 1 \\ 1 + i \end{pmatrix}$

12. $\lambda_1 = -3,\ \mathbf{x}^{(1)} = \begin{pmatrix} 1 \\ -1 \end{pmatrix};\quad \lambda_2 = -1,\ \mathbf{x}^{(2)} = \begin{pmatrix} 1 \\ 1 \end{pmatrix}$

13. $\lambda_1 = 0,\ \mathbf{x}^{(1)} = \begin{pmatrix} 1 \\ i \end{pmatrix};\quad \lambda_2 = 2,\ \mathbf{x}^{(2)} = \begin{pmatrix} 1 \\ -i \end{pmatrix}$

14. $\lambda_1 = 2,\ \mathbf{x}^{(1)} = \begin{pmatrix} \sqrt{3} \\ 1 \end{pmatrix};\quad \lambda_2 = -2,\ \mathbf{x}^{(2)} = \begin{pmatrix} 1 \\ -\sqrt{3} \end{pmatrix}$

15. $\lambda_1 = \lambda_2 = -3,\ \mathbf{x}^{(1)} = \begin{pmatrix} 1 \\ 1 \end{pmatrix}$

16. $\lambda_1 = 1,\ \mathbf{x}^{(1)} = \begin{pmatrix} 2 \\ -3 \\ 2 \end{pmatrix};\quad \lambda_2 = 1 + 2i,\ \mathbf{x}^{(2)} = \begin{pmatrix} 0 \\ 1 \\ -i \end{pmatrix};\quad \lambda_3 = 1 - 2i,\ \mathbf{x}^{(3)} = \begin{pmatrix} 0 \\ 1 \\ i \end{pmatrix}$

17. $\lambda_1 = 1,\ \mathbf{x}^{(1)} = \begin{pmatrix} 1 \\ 0 \\ -1 \end{pmatrix};\quad \lambda_2 = 2,\ \mathbf{x}^{(2)} = \begin{pmatrix} -2 \\ 1 \\ 0 \end{pmatrix};\quad \lambda_3 = 3,\ \mathbf{x}^{(3)} = \begin{pmatrix} 0 \\ 1 \\ -1 \end{pmatrix}$

18. $\lambda_1 = \lambda_2 = \lambda_3 = 2,\ \mathbf{x}^{(1)} = \begin{pmatrix} 0 \\ 1 \\ -1 \end{pmatrix}$

19. $\lambda_1 = -1,\ \mathbf{x}^{(1)} = \begin{pmatrix} 1 \\ -4 \\ 1 \end{pmatrix};\quad \lambda_2 = -1,\ \mathbf{x}^{(2)} = \begin{pmatrix} 1 \\ 0 \\ -1 \end{pmatrix};\quad \lambda_3 = 8,\ \mathbf{x}^{(3)} = \begin{pmatrix} 2 \\ 1 \\ 2 \end{pmatrix}$

20. (a) $\mathbf{T} = \begin{pmatrix} 1 & 1 \\ 3 & 1 \end{pmatrix}$ (b) $\mathbf{T} = \begin{pmatrix} 1 & 1 \\ 1-i & 1+i \end{pmatrix}$ (c) $\mathbf{T} = \begin{pmatrix} 1 & 1 \\ i & -i \end{pmatrix}$

Section 7.4, Page 356

2. (c) $W(t) = c \exp \int [p_{11}(t) + p_{22}(t)]\, dt$
6. (a) $W(t) = t^2$
   (b) $\mathbf{x}^{(1)}$ and $\mathbf{x}^{(2)}$ are linearly independent at each point except $t = 0$; they are linearly independent on every interval.
   (c) At least one coefficient must be discontinuous at $t = 0$.
   (d) $\mathbf{x}' = \begin{pmatrix} 0 & 1 \\ -2t^{-2} & 2t^{-1} \end{pmatrix}\mathbf{x}$
7. (a) $W(t) = t(t-2)e^t$
   (b) $\mathbf{x}^{(1)}$ and $\mathbf{x}^{(2)}$ are linearly independent at each point except $t = 0$ and $t = 2$; they are linearly independent on every interval.
   (c) There must be at least one discontinuous coefficient at $t = 0$ and $t = 2$.
   (d) $\mathbf{x}' = \begin{pmatrix} 0 & 1 \\ \dfrac{2-2t}{t^2-2t} & \dfrac{t^2-2}{t^2-2t} \end{pmatrix}\mathbf{x}$

Section 7.5, Page 365

1. $\mathbf{x} = c_1\begin{pmatrix} 1 \\ 2 \end{pmatrix}e^{-t} + c_2\begin{pmatrix} 2 \\ 1 \end{pmatrix}e^{2t}$
2. $\mathbf{x} = c_1\begin{pmatrix} 1 \\ 1 \end{pmatrix}e^{-t} + c_2\begin{pmatrix} 2 \\ 3 \end{pmatrix}e^{-2t}$
3. $\mathbf{x} = c_1\begin{pmatrix} 1 \\ 1 \end{pmatrix}e^{t} + c_2\begin{pmatrix} 1 \\ 3 \end{pmatrix}e^{-t}$
4. $\mathbf{x} = c_1\begin{pmatrix} 1 \\ -4 \end{pmatrix}e^{-3t} + c_2\begin{pmatrix} 1 \\ 1 \end{pmatrix}e^{2t}$
5. $\mathbf{x} = c_1\begin{pmatrix} 1 \\ -1 \end{pmatrix}e^{-3t} + c_2\begin{pmatrix} 1 \\ 1 \end{pmatrix}e^{-t}$
6. $\mathbf{x} = c_1\begin{pmatrix} 1 \\ -1 \end{pmatrix}e^{t/2} + c_2\begin{pmatrix} 1 \\ 1 \end{pmatrix}e^{2t}$
7. $\mathbf{x} = c_1\begin{pmatrix} 3 \\ 4 \end{pmatrix} + c_2\begin{pmatrix} 1 \\ 2 \end{pmatrix}e^{-2t}$
8. $\mathbf{x} = c_1\begin{pmatrix} 5 \\ -1 \end{pmatrix} + c_2\begin{pmatrix} 2 \\ -1 \end{pmatrix}e^{-3t}$
9. $\mathbf{x} = c_1\begin{pmatrix} -2 \\ 1 \end{pmatrix} + c_2\begin{pmatrix} -3 \\ 1 \end{pmatrix}e^{t}$
10. $\mathbf{x} = c_1\begin{pmatrix} 1 \\ i \end{pmatrix} + c_2\begin{pmatrix} 1 \\ -i \end{pmatrix}e^{2t}$
11. $\mathbf{x} = c_1\begin{pmatrix} 2+i \\ -1 \end{pmatrix}e^{t} + c_2\begin{pmatrix} 1 \\ -1 \end{pmatrix}e^{-it}$
12. $\mathbf{x} = c_1\begin{pmatrix} 1 \\ 1 \\ 1 \end{pmatrix}e^{4t} + c_2\begin{pmatrix} 1 \\ -2 \\ 1 \end{pmatrix}e^{t} + c_3\begin{pmatrix} 1 \\ 0 \\ -1 \end{pmatrix}e^{-t}$
13. $\mathbf{x} = c_1\begin{pmatrix} 1 \\ -4 \\ 1 \end{pmatrix}e^{-t} + c_2\begin{pmatrix} 1 \\ 0 \\ -1 \end{pmatrix}e^{-t} + c_2\begin{pmatrix} 2 \\ 1 \\ 2 \end{pmatrix}e^{8t}$
14. $\mathbf{x} = c_1\begin{pmatrix} 4 \\ -5 \\ -7 \end{pmatrix}e^{-2t} + c_2\begin{pmatrix} 3 \\ -4 \\ -2 \end{pmatrix}e^{-t} + c_3\begin{pmatrix} 0 \\ 1 \\ -1 \end{pmatrix}e^{2t}$
15. $\mathbf{x} = c_1\begin{pmatrix} 1 \\ -4 \\ -1 \end{pmatrix}e^{t} + c_2\begin{pmatrix} 1 \\ -1 \\ -1 \end{pmatrix}e^{-2t} + c_3\begin{pmatrix} 1 \\ 2 \\ 1 \end{pmatrix}e^{3t}$
16. $\mathbf{x} = -\dfrac{3}{2}\begin{pmatrix} 1 \\ 3 \end{pmatrix}e^{2t} + \dfrac{7}{2}\begin{pmatrix} 1 \\ 1 \end{pmatrix}e^{4t}$
17. $\mathbf{x} = \dfrac{1}{2}\begin{pmatrix} 1 \\ 1 \end{pmatrix}e^{-t} + \dfrac{1}{2}\begin{pmatrix} 1 \\ 5 \end{pmatrix}e^{3t}$

18. $\mathbf{x} = \begin{pmatrix} 0 \\ -2 \\ 1 \end{pmatrix} e^{t} + 2\begin{pmatrix} 1 \\ 1 \\ 0 \end{pmatrix} e^{2t}$

19. $\mathbf{x} = 6\begin{pmatrix} 1 \\ 2 \\ -1 \end{pmatrix} e^{t} + 3\begin{pmatrix} 1 \\ -2 \\ 1 \end{pmatrix} e^{-t} - \begin{pmatrix} 2 \\ 1 \\ -8 \end{pmatrix} e^{4t}$

21. $\mathbf{x} = c_1\begin{pmatrix} 1 \\ 1 \end{pmatrix} t + c_2\begin{pmatrix} 1 \\ 3 \end{pmatrix} t^{-1}$

22. $\mathbf{x} = c_1\begin{pmatrix} 1 \\ 3 \end{pmatrix} t^{2} + c_2\begin{pmatrix} 1 \\ 1 \end{pmatrix} t^{4}$

23. $\mathbf{x} = c_1\begin{pmatrix} 3 \\ 4 \end{pmatrix} + c_2\begin{pmatrix} 1 \\ 2 \end{pmatrix} t^{-2}$

24. $\mathbf{x} = c_1\begin{pmatrix} 1 \\ 2 \end{pmatrix} t^{-1} + c_2\begin{pmatrix} 2 \\ 1 \end{pmatrix} t^{2}$

28. $\mathbf{x} = c_1\begin{pmatrix} 1 \\ 1 \end{pmatrix} e^{-3t} + c_2\left[\begin{pmatrix} 1 \\ 1 \end{pmatrix} te^{-3t} - \begin{pmatrix} 0 \\ \frac{1}{4} \end{pmatrix} e^{-3t}\right]$

29. $\mathbf{x} = c_1\begin{pmatrix} 2 \\ 1 \end{pmatrix} e^{t} + c_2\left[\begin{pmatrix} 2 \\ 1 \end{pmatrix} te^{t} + \begin{pmatrix} 1 \\ 0 \end{pmatrix} e^{t}\right]$

30. (a) $\begin{pmatrix} I \\ V \end{pmatrix} = c_1\begin{pmatrix} 1 \\ 3 \end{pmatrix} e^{-2t} + c_2\begin{pmatrix} 1 \\ 1 \end{pmatrix} e^{-t}$

31. (a) $\left(\dfrac{1}{CR_2} - \dfrac{R_1}{L}\right)^2 - \dfrac{4}{CL} > 0$

Section 7.6, Page 372

1. $\mathbf{x} = c_1 e^{t}\begin{pmatrix} \cos 2t \\ \cos 2t + \sin 2t \end{pmatrix} + c_2 e^{t}\begin{pmatrix} \sin 2t \\ -\cos 2t + \sin 2t \end{pmatrix}$

2. $\mathbf{x} = c_1 e^{-t}\begin{pmatrix} 2\cos 2t \\ \sin 2t \end{pmatrix} + c_2 e^{-t}\begin{pmatrix} -2\sin 2t \\ \cos 2t \end{pmatrix}$

3. $\mathbf{x} = c_1\begin{pmatrix} 5\cos t \\ 2\cos t + \sin t \end{pmatrix} + c_2\begin{pmatrix} 5\sin t \\ -\cos t + 2\sin t \end{pmatrix}$

4. $\mathbf{x} = c_1 e^{t/2}\begin{pmatrix} 5\cos \frac{3}{2}t \\ 3(\cos \frac{3}{2}t + \sin \frac{3}{2}t) \end{pmatrix} + c_2 e^{t/2}\begin{pmatrix} 5\sin \frac{3}{2}t \\ 3(-\cos \frac{3}{2}t + \sin \frac{3}{2}t) \end{pmatrix}$

5. $\mathbf{x} = c_1 e^{-t}\begin{pmatrix} \cos t \\ 2\cos t + \sin t \end{pmatrix} + c_2 e^{-t}\begin{pmatrix} \sin t \\ -\cos t + 2\sin t \end{pmatrix}$

6. $\mathbf{x} = c_1\begin{pmatrix} -2\cos 3t \\ \cos 3t + 3\sin 3t \end{pmatrix} + c_2\begin{pmatrix} -2\sin 3t \\ \sin 3t - 3\cos 3t \end{pmatrix}$

7. $\mathbf{x} = c_1\begin{pmatrix} 2 \\ -3 \\ 2 \end{pmatrix} e^{t} + c_2 e^{t}\begin{pmatrix} 0 \\ \cos 2t \\ \sin 2t \end{pmatrix} + c_3 e^{t}\begin{pmatrix} 0 \\ \sin 2t \\ -\cos 2t \end{pmatrix}$

8. $\mathbf{x} = c_1\begin{pmatrix} 2 \\ -2 \\ 1 \end{pmatrix} e^{-2t} + c_2 e^{-t}\begin{pmatrix} -\sqrt{2}\sin\sqrt{2}\,t \\ \cos\sqrt{2}\,t \\ -\cos\sqrt{2}\,t - \sqrt{2}\sin\sqrt{2}\,t \end{pmatrix} + c_3 e^{-t}\begin{pmatrix} \sqrt{2}\cos\sqrt{2}\,t \\ \sin\sqrt{2}\,t \\ \sqrt{2}\cos\sqrt{2}\,t - \sin\sqrt{2}\,t \end{pmatrix}$

9. $\mathbf{x} = e^{-t}\begin{pmatrix} \cos t - 3\sin t \\ \cos t - \sin t \end{pmatrix}$

10. $\mathbf{x} = e^{-2t}\begin{pmatrix} \cos t - 5\sin t \\ -2\cos t - 3\sin t \end{pmatrix}$

11. $\mathbf{x} = c_1 t^{-1}\begin{pmatrix} \cos(\sqrt{2}\ln t) \\ \sqrt{2}\sin(\sqrt{2}\ln t) \end{pmatrix} + c_2 t^{-1}\begin{pmatrix} \sin(\sqrt{2}\ln t) \\ -\sqrt{2}\cos(\sqrt{2}\ln t) \end{pmatrix}$

12. $\mathbf{x} = c_1\begin{pmatrix} 5\cos(\ln t) \\ 2\cos(\ln t) + \sin(\ln t) \end{pmatrix} + c_2\begin{pmatrix} 5\sin(\ln t) \\ -\cos(\ln t) + 2\sin(\ln t) \end{pmatrix}$

13. (b) $\begin{pmatrix} I \\ V \end{pmatrix} = c_1 e^{-t/2}\begin{pmatrix} \cos t/2 \\ 4\sin t/2 \end{pmatrix} + c_2 e^{-t/2}\begin{pmatrix} \sin t/2 \\ -4\cos t/2 \end{pmatrix}$

(c) $c_1 = 2, \quad c_2 = -\frac{3}{4}$ in answer to part (b).

(d) $\lim_{t\to\infty} I(t) = \lim_{t\to\infty} V(t) = 0$; No

14. (b) $\begin{pmatrix} I \\ V \end{pmatrix} = c_1 e^{-t}\begin{pmatrix} \cos t \\ -\cos t - \sin t \end{pmatrix} + c_2 e^{-t}\begin{pmatrix} \sin t \\ -\sin t + \cos t \end{pmatrix}$

(c) Use $c_1 = 2$ and $c_2 = 3$ in answer to part (b).

(d) $\lim_{t\to\infty} I(t) = \lim_{t\to\infty} V(t) = 0$; No

Section 7.7, Page 380

1. $\mathbf{x} = c_1\begin{pmatrix} 2 \\ 1 \end{pmatrix}e^t + c_2\left[\begin{pmatrix} 2 \\ 1 \end{pmatrix}te^t + \begin{pmatrix} 1 \\ 0 \end{pmatrix}e^t\right]$

2. $\mathbf{x} = c_1\begin{pmatrix} 1 \\ 2 \end{pmatrix} + c_2\left[\begin{pmatrix} 1 \\ 2 \end{pmatrix}t - \begin{pmatrix} 0 \\ \frac{1}{2} \end{pmatrix}\right]$

3. $\mathbf{x} = c_1\begin{pmatrix} 2 \\ 1 \end{pmatrix}e^{-t} + c_2\left[\begin{pmatrix} 2 \\ 1 \end{pmatrix}te^{-t} + \begin{pmatrix} 0 \\ 2 \end{pmatrix}e^{-t}\right]$

4. $\mathbf{x} = c_1\begin{pmatrix} 1 \\ 1 \end{pmatrix}e^{-t/2} + c_2\left[\begin{pmatrix} 1 \\ 1 \end{pmatrix}te^{-t/2} + \begin{pmatrix} 0 \\ 2/5 \end{pmatrix}e^{-t/2}\right]$

5. $\mathbf{x} = c_1\begin{pmatrix} -3 \\ 4 \\ 2 \end{pmatrix}e^{-t} + c_2\begin{pmatrix} 0 \\ 1 \\ -1 \end{pmatrix}e^{2t} + c_3\left[\begin{pmatrix} 0 \\ 1 \\ -1 \end{pmatrix}te^{2t} + \begin{pmatrix} 1 \\ 0 \\ 1 \end{pmatrix}e^{2t}\right]$

6. $\mathbf{x} = c_1\begin{pmatrix} 1 \\ 1 \\ 1 \end{pmatrix}e^{2t} + c_2\begin{pmatrix} 1 \\ 0 \\ -1 \end{pmatrix}e^{-t} + c_3\begin{pmatrix} 0 \\ 1 \\ -1 \end{pmatrix}e^{-t}$

7. $\mathbf{x} = \begin{pmatrix} 3 + 4t \\ 2 + 4t \end{pmatrix}e^{-3t}$

8. $\mathbf{x} = \begin{pmatrix} -1 \\ 2 \\ -33 \end{pmatrix}e^t + 4\begin{pmatrix} 0 \\ 1 \\ -6 \end{pmatrix}te^t + 3\begin{pmatrix} 0 \\ 0 \\ 1 \end{pmatrix}e^{2t}$

9. $\mathbf{x} = c_1\begin{pmatrix} 0 \\ 1 \\ -1 \end{pmatrix}e^{2t} + c_2\left[\begin{pmatrix} 0 \\ 1 \\ -1 \end{pmatrix}te^{2t} + \begin{pmatrix} 1 \\ 0 \\ 1 \end{pmatrix}e^{2t}\right]$

$+ c_3\left[\begin{pmatrix} 0 \\ 1 \\ -1 \end{pmatrix}t^2e^{2t} + 2\begin{pmatrix} 1 \\ 0 \\ 1 \end{pmatrix}te^{2t} + 2\begin{pmatrix} 1 \\ 0 \\ 2 \end{pmatrix}e^{2t}\right]$

10. $\mathbf{x} = c_1\begin{pmatrix} 2 \\ 1 \end{pmatrix}t + c_2\left[\begin{pmatrix} 2 \\ 1 \end{pmatrix}t\ln t + \begin{pmatrix} 1 \\ 0 \end{pmatrix}t\right]$

11. $\mathbf{x} = c_1\begin{pmatrix} 1 \\ 1 \end{pmatrix}t^{-3} + c_2\left[\begin{pmatrix} 1 \\ 1 \end{pmatrix}t^{-3}\ln t - \begin{pmatrix} 0 \\ \frac{1}{4} \end{pmatrix}t^{-3}\right]$

12. (a) $\mathbf{x}^{(1)}(t) = \begin{pmatrix} 1 \\ 0 \\ 2 \end{pmatrix}e^t, \quad \mathbf{x}^{(2)}(t) = \begin{pmatrix} 0 \\ 2 \\ -3 \end{pmatrix}e^t$ (d) $\mathbf{x}^{(3)}(t) = \begin{pmatrix} 2 \\ 4 \\ -2 \end{pmatrix}te^t + \begin{pmatrix} 0 \\ 0 \\ -1 \end{pmatrix}e^t$

14. (b) $\begin{pmatrix} I \\ V \end{pmatrix} = -\begin{pmatrix} 1 \\ -2 \end{pmatrix}e^{-t/2} + \left[\begin{pmatrix} 1 \\ -2 \end{pmatrix}te^{-t/2} + \begin{pmatrix} 2 \\ 0 \end{pmatrix}e^{-t/2}\right]$

Section 7.8, Page 386

1. $\mathbf{\Phi}(t) = \begin{pmatrix} -\frac{1}{3}e^{-t} + \frac{4}{3}e^{2t} & \frac{2}{3}e^{-t} - \frac{2}{3}e^{2t} \\ -\frac{2}{3}e^{-t} + \frac{2}{3}e^{2t} & \frac{4}{3}e^{-t} - \frac{1}{3}e^{2t} \end{pmatrix}$

2. $\mathbf{\Phi}(t) = \begin{pmatrix} 3 - 2e^{-2t} & -\frac{3}{2} + \frac{3}{2}e^{-2t} \\ 4 - 4e^{-2t} & -2 + 3e^{-2t} \end{pmatrix}$

3. $\mathbf{\Phi}(t) = \begin{pmatrix} \frac{3}{2}e^t - \frac{1}{2}e^{-t} & -\frac{1}{2}e^t + \frac{1}{2}e^{-t} \\ \frac{3}{2}e^t - \frac{3}{2}e^{-t} & -\frac{1}{2}e^t + \frac{3}{2}e^{-t} \end{pmatrix}$

4. $\mathbf{\Phi}(t) = \begin{pmatrix} \frac{1}{5}e^{-3t} + \frac{4}{5}e^{2t} & -\frac{1}{5}e^{-3t} + \frac{1}{5}e^{2t} \\ -\frac{4}{5}e^{-3t} + \frac{4}{5}e^{2t} & \frac{4}{5}e^{-3t} + \frac{1}{5}e^{2t} \end{pmatrix}$

5. $\mathbf{\Phi}(t) = \begin{pmatrix} \cos t + 2\sin t & -5\sin t \\ \sin t & \cos t - 2\sin t \end{pmatrix}$

6. $\mathbf{\Phi}(t) = \begin{pmatrix} e^{-t}\cos 2t & -2e^{-t}\sin 2t \\ \frac{1}{2}e^{-t}\sin 2t & e^{-t}\cos 2t \end{pmatrix}$

7. $\mathbf{\Phi}(t) = \begin{pmatrix} -\frac{1}{2}e^{2t} + \frac{3}{2}e^{4t} & \frac{1}{2}e^{2t} - \frac{1}{2}e^{4t} \\ -\frac{3}{2}e^{2t} + \frac{3}{2}e^{4t} & \frac{3}{2}e^{2t} - \frac{1}{2}e^{4t} \end{pmatrix}$

8. $\mathbf{\Phi}(t) = \begin{pmatrix} e^t + 2te^t & -4te^t \\ te^t & e^t - 2te^t \end{pmatrix}$

9. $\mathbf{\Phi}(t) = \begin{pmatrix} -2e^{-2t} + 3e^{-t} & -e^{-2t} + e^{-t} & -e^{-2t} + e^{-t} \\ \frac{5}{2}e^{-2t} - 4e^{-t} + \frac{3}{2}e^{2t} & \frac{5}{4}e^{-2t} - \frac{4}{3}e^{-t} + \frac{13}{12}e^{2t} & \frac{5}{4}e^{-2t} - \frac{4}{3}e^{-t} + \frac{1}{12}e^{2t} \\ \frac{7}{2}e^{-2t} - 2e^{-t} - \frac{3}{2}e^{2t} & \frac{7}{4}e^{-2t} - \frac{2}{3}e^{-t} - \frac{13}{12}e^{2t} & \frac{7}{4}e^{-2t} - \frac{2}{3}e^{-t} - \frac{1}{12}e^{2t} \end{pmatrix}$

10. $\mathbf{\Phi}(t) = \begin{pmatrix} \frac{1}{6}e^t + \frac{1}{3}e^{-2t} + \frac{1}{2}e^{3t} & -\frac{1}{3}e^t + \frac{1}{3}e^{-2t} & \frac{1}{2}e^t - e^{-2t} + \frac{1}{2}e^{3t} \\ -\frac{2}{3}e^t - \frac{1}{3}e^{-2t} + e^{3t} & \frac{4}{3}e^t - \frac{1}{3}e^{-2t} & -2e^t + e^{-2t} + e^{3t} \\ -\frac{1}{6}e^t - \frac{1}{3}e^{-2t} + \frac{1}{2}e^{3t} & \frac{1}{3}e^t - \frac{1}{3}e^{-2t} & -\frac{1}{2}e^t + e^{-2t} + \frac{1}{2}e^{3t} \end{pmatrix}$

Section 7.9, Page 394

1. $\mathbf{x} = c_1\begin{pmatrix} 1 \\ 1 \end{pmatrix}e^t + c_2\begin{pmatrix} 1 \\ 3 \end{pmatrix}e^{-t} + \frac{3}{2}\begin{pmatrix} 1 \\ 1 \end{pmatrix}te^t - \frac{1}{4}\begin{pmatrix} 1 \\ 3 \end{pmatrix}e^t + \begin{pmatrix} 1 \\ 2 \end{pmatrix}t - \begin{pmatrix} 0 \\ 1 \end{pmatrix}$

2. $\mathbf{x} = c_1\begin{pmatrix} \sqrt{3} \\ 1 \end{pmatrix}e^{2t} + c_2\begin{pmatrix} 1 \\ -\sqrt{3} \end{pmatrix}e^{-2t} - \begin{pmatrix} 2/3 \\ 1/\sqrt{3} \end{pmatrix}e^t + \begin{pmatrix} -1 \\ 2/\sqrt{3} \end{pmatrix}e^{-t}$

3. $\mathbf{x} = \frac{1}{5}(2t - \frac{3}{2}\sin 2t - \frac{1}{2}\cos 2t + c_1)\begin{pmatrix} 5\cos t \\ 2\cos t + \sin t \end{pmatrix}$
$+ \frac{1}{5}(-t - \frac{1}{2}\sin 2t + \frac{3}{2}\cos 2t + c_2)\begin{pmatrix} 5\sin t \\ -\cos t + 2\sin t \end{pmatrix}$

4. $\mathbf{x} = c_1\begin{pmatrix} 1 \\ -4 \end{pmatrix}e^{-3t} + c_2\begin{pmatrix} 1 \\ 1 \end{pmatrix}e^{2t} - \begin{pmatrix} 0 \\ 1 \end{pmatrix}e^{-2t} + \frac{1}{2}\begin{pmatrix} 1 \\ 0 \end{pmatrix}e^t$

5. $\mathbf{x} = c_1\begin{pmatrix} 1 \\ 2 \end{pmatrix} + c_2\left[\begin{pmatrix} 1 \\ 2 \end{pmatrix}t - \frac{1}{2}\begin{pmatrix} 0 \\ 1 \end{pmatrix}\right] - 2\begin{pmatrix} 1 \\ 2 \end{pmatrix}\ln t + \begin{pmatrix} 2 \\ 5 \end{pmatrix}t^{-1} - \begin{pmatrix} \frac{1}{2} \\ 0 \end{pmatrix}t^{-2}$

6. $\mathbf{x} = c_1\begin{pmatrix} 1 \\ 2 \end{pmatrix} + c_2\begin{pmatrix} -2 \\ 1 \end{pmatrix}e^{-5t} + \begin{pmatrix} 1 \\ 2 \end{pmatrix}\ln t + \frac{8}{5}\begin{pmatrix} 1 \\ 2 \end{pmatrix}t + \frac{4}{25}\begin{pmatrix} -2 \\ 1 \end{pmatrix}$

7. $\mathbf{x} = c_1\begin{pmatrix} 1 \\ 2 \end{pmatrix}e^{3t} + c_2\begin{pmatrix} 1 \\ -2 \end{pmatrix}e^{-t} + \frac{1}{4}\begin{pmatrix} 1 \\ -8 \end{pmatrix}e^t$

8. $\mathbf{x} = c_1\begin{pmatrix} 1 \\ 1 \end{pmatrix}e^t + c_2\begin{pmatrix} 1 \\ 3 \end{pmatrix}e^{-t} + \begin{pmatrix} 1 \\ 0 \end{pmatrix}e^t + 2\begin{pmatrix} 1 \\ 1 \end{pmatrix}te^t$

9. $\mathbf{x} = c_1\begin{pmatrix} 1 \\ 1 \end{pmatrix}e^{-t/2} + c_2\begin{pmatrix} 1 \\ -1 \end{pmatrix}e^{-2t} + \begin{pmatrix} \frac{5}{2} \\ \frac{3}{2} \end{pmatrix}t - \begin{pmatrix} \frac{17}{4} \\ \frac{15}{4} \end{pmatrix} + \begin{pmatrix} \frac{1}{6} \\ \frac{1}{2} \end{pmatrix}e^t$

10. $\mathbf{x} = c_1\begin{pmatrix} 1 \\ \sqrt{2} \end{pmatrix}e^{-t} + c_2\begin{pmatrix} -\sqrt{2} \\ 1 \end{pmatrix}e^{-4t} - \frac{1}{3}\begin{pmatrix} \sqrt{2} - 1 \\ 2 - \sqrt{2} \end{pmatrix}te^{-t} + \frac{1}{9}\begin{pmatrix} 2 + \sqrt{2} \\ -1 - \sqrt{2} \end{pmatrix}e^{-t}$

11. $\mathbf{x} = (\frac{1}{2}\sin^2 t + c_1)\begin{pmatrix} 5\cos t \\ 2\cos t + \sin t \end{pmatrix} + (-\frac{1}{2}t - \frac{1}{2}\sin t\cos t + c_2)\begin{pmatrix} 5\sin t \\ -\cos t + 2\sin t \end{pmatrix}$

12. $\mathbf{x} = [\frac{1}{5}\ln(\sin t) - \ln(-\cos t) - \frac{2}{5}t + c_1]\begin{pmatrix} 5\cos t \\ 2\cos t + \sin t \end{pmatrix}$

$+[\frac{2}{5}\ln(\sin t) - \frac{4}{5}t + c_2]\begin{pmatrix} 5\sin t \\ -\cos t + 2\sin t \end{pmatrix}$

13. (a) $\mathbf{\Psi}(t) = \begin{pmatrix} e^{-t/2}\cos\frac{1}{2}t & e^{-t/2}\sin\frac{1}{2}t \\ 4e^{-t/2}\sin\frac{1}{2}t & -4e^{-t/2}\cos\frac{1}{2}t \end{pmatrix}$ (b) $\mathbf{x} = e^{-t/2}\begin{pmatrix} \sin\frac{1}{2}t \\ 4 - 4\cos\frac{1}{2}t \end{pmatrix}$

14. $\mathbf{x} = c_1\begin{pmatrix} 1 \\ 1 \end{pmatrix}t + c_2\begin{pmatrix} 1 \\ 3 \end{pmatrix}t^{-1} - \begin{pmatrix} 2 \\ 3 \end{pmatrix} + \frac{1}{2}\begin{pmatrix} 1 \\ 3 \end{pmatrix}t - \begin{pmatrix} 1 \\ 1 \end{pmatrix}t\ln t - \frac{1}{3}\begin{pmatrix} 4 \\ 3 \end{pmatrix}t^2$

15. $\mathbf{x} = c_1\begin{pmatrix} 2 \\ 1 \end{pmatrix}t^2 + c_2\begin{pmatrix} 1 \\ 2 \end{pmatrix}t^{-1} + \begin{pmatrix} 3 \\ 2 \end{pmatrix}t + \frac{1}{10}\begin{pmatrix} -2 \\ 1 \end{pmatrix}t^4 - \frac{1}{2}\begin{pmatrix} 2 \\ 3 \end{pmatrix}$

CHAPTER EIGHT

Section 8.2, Page 404

1. (a) 1.1, 1.22, 1.364, 1.537 (b) 1.105, 1.232, 1.386, 1.572
   (c) $y = \phi(x) = (1 + e^{2x})/2$; 1.111, 1.246, 1.411, 1.613
2. (a) 1.25, 1.54, 1.878, 2.274 (b) 1.26, 1.564, 1.922, 2.344
   (c) $y = \phi(x) = e^{2x} + 0.5x$; 1.271, 1.592, 1.972, 2.426
3. (a) 1.1, 1.222, 1.375, 1.573 (b) 1.105, 1.236, 1.404, 1.627
4. (a) 1.576, 1.249, 1.014, 0.8618 (b) 1.600, 1.293, 1.072, 0.9302
5. (a) 3.2, 3.407, 3.622, 3.844 (b) 3.202, 3.411, 3.628, 3.851
6. (a) 1.1, 1.210, 1.328, 1.455 (b) 1.102, 1.214, 1.335, 1.464
7. (a) − 0.1661, −0.4109, −0.8047, 4.159
   (b) − 0.1747, −0.4342, −0.8891, −3.098
9. 1.595, 2.464
12. (a) $\phi_3(x) = 1 + x + x^2 + \frac{2}{3}x^3$; $\phi_2(0.4) = 1.56$, $\phi_3(0.4) = 1.603$
   (b) $\phi_3(x) = 1 + \frac{5}{2}x + 2x^2 + \frac{4}{3}x^3 - \frac{1}{6}x^4$; $\phi_2(0.4) = 2.299$, $\phi_3(0.4) = 2.401$
   (c) $\phi_2(x) = 1 + x + x^2 + \frac{2}{3}x^3 + \frac{1}{6}x^4 + \frac{2}{15}x^5 + \frac{1}{63}x^7$; $\phi_2(0.4) = 1.608$

Section 8.3, Page 410

1. $e_{n+1} = [2\phi(\bar{x}_n) - 1]h^2$, $|e_{n+1}| \le \left[1 + 2\max_{0\le x\le 1}|\phi(x)|\right]h^2$,
   $e_{n+1} = e^{2\bar{x}_n}h^2$, $|e_1| \le 0.012$, $|e_4| \le 0.022$
2. $e_{n+1} = [2\phi(\bar{x}_n) - \bar{x}_n]h^2$, $|e_{n+1}| \le \left[1 + 2\max_{0\le x\le 1}|\phi(x)|\right]h^2$,
   $e_{n+1} = 2e^{2\bar{x}_n}h^2$, $|e_1| \le 0.024$, $|e_4| \le 0.045$
3. $e_{n+1} = [\bar{x}_n + \bar{x}_n^2\phi(\bar{x}_n) + \phi^3(\bar{x}_n)]h^2$
4. $e_{n+1} = [19 - 15\bar{x}_n\phi^{-1/2}(\bar{x}_n)]h^2/4$
5. $e_{n+1} = \{1 + [\bar{x}_n + \phi(\bar{x}_n)]^{-1/2}\}h^2/4$
6. $e_{n+1} = \{2 - [\phi(\bar{x}_n) + 2\bar{x}_n^2]\exp[-\bar{x}_n\phi(\bar{x}_n)] - \bar{x}_n\exp[-2\bar{x}_n\phi(\bar{x}_n)]\}h^2/2$
7. (a) $\phi(x) = 1 + (1/5\pi)\sin 5\pi x$ (b) 1.2, 1.0, 1.2
   (c) 1.1, 1.1, 1.0, 1.0 (d) $h < 1/\sqrt{50\pi} \cong 0.08$

10. (a) $y_1(h = 0.2) = 1.4$, $y_2(h = 0.1) = 1.44$; $\phi(0.2) \cong 1.48$, exact 1.4918
(b) $y_1(h = 0.2) = 1.2$, $y_2(h = 0.1) = 1.22$; $\phi(0.2) \cong 1.24$, exact 1.246
(c) $y_1(h = 0.2) = 1.5$, $y_2(h = 0.1) = 1.54$; $\phi(0.2) \cong 1.58$, exact 1.592
(d) $y_1(h = 0.2) = 1.2$, $y_2(h = 0.1) = 1.222$; $\phi(0.2) \cong 1.244$
12. (a) 56.75310 (b) 62.027322
13. (a) 1.05, 1.11, 1.17, 1.24 (b) 1.13, 1.27, 1.42, 1.58
(c) 1.05, 1.11, 1.17, 1.24
14. (a) 0 (b) 60 (c) −92.16
15. $0.224 \neq 0.225$

Section 8.4, Page 416

1. 1.11, 1.244, 1.408, 1.608
2. 1.27, 1.588, 1.966, 2.415
3. 1.111, 1.252, 1.436, 1.688
4. 1.625, 1.338, 1.133, 1.001
5. 3.204, 3.415, 3.633, 3.859
6. 1.105, 1.219, 1.341, 1.473
8. $e_{n+1} = 38e^{4\bar{x}_n}(h^3/3)$, $|e_{n+1}| \leq 691.6h^3$ on $0 \leq x \leq 1$, $|e_1| \leq 0.0189$
9. (a) $e_{n+1} = 2e^{2\bar{x}_n}(h^3/3)$, $|e_{n+1}| \leq 4.926h^3$ on $0 \leq x \leq 1$, $|e_1| \leq 0.00081$
(b) $e_{n+1} = 4e^{2\bar{x}_n}(h^3/3)$, $|e_{n+1}| \leq 9.852h^3$ on $0 \leq x \leq 1$, $|e_1| \leq 0.0016$
11. (a) 1.11, 1.244, 1.408, 1.608 (b) 1.27, 1.588, 1.966, 2.415
(c) 1.111, 1.250, 1.434, 1.683 (d) 1.624, 1.337, 1.131, 0.9993

Section 8.5, Page 419

1. 1.11, 1.244, 1.408, 1.608
2. 1.27, 1.588, 1.966, 2.415
3. 1.11, 1.249, 1.431, 1.678
4. 1.623, 1.336, 1.130, 0.9979
5. 3.204, 3.415, 3.633, 3.859
6. 1.105, 1.219, 1.342, 1.473
8. 2.464, 3.737, 5.610
10. $y_{n+1} = y_n + hy_n' + (h^2/2)y_n'' + (h^3/6)y_n'''$
$y_n''' = f_{xx} + 2f_{xy}f + f_{yy}f^2 + f_xf_y + f_y^2f$
$e_{n+1} = \phi^{\text{iv}}(\bar{x}_n)h^4/24$, $x_n < \bar{x}_n < x_n + h$
11. $\phi(1) \cong 64.586856$, exact 64.897803

Section 8.6, Page 423

1. 1.246, 1.613
2. 1.592, 2.425
3. 1.253, 1.696
4. 1.334, 0.9941
5. 3.415, 3.859
6. 1.218, 1.473
7. 5.778
10. $\phi(1) \simeq 64.885876$, exact 64.897803

Section 8.7, Page 430

1. (b) $\phi_2(x) - \phi_1(x) = 0.001e^x \to \infty$ as $x \to \infty$
2. (b) $\phi_1(x) = \ln[e^x/(2 - e^x)]$; $\phi_2(x) = \ln[1/(1 - x)]$
3. (a) $y = x$
4. (a) $y = x^2$

Section 8.8, Page 437

1. $y_{4p} = 1.61267$, $y_{4c} = 1.61276$
2. $y_{4p} = 2.42536$, $y_{4c} = 2.42553$
3. $y_{4p} = 1.69396$, $y_{4c} = 1.69613$
4. $y_{4p} = 0.991185$, $y_{4c} = 0.991055$
5. *Hint:* $\phi'(x) \cong y_n' + \dfrac{y_n' - y_{n-1}'}{h}(x - x_n)$.
7. (a) $y_{4p} = 1.61269$, $y_{4c} = 1.61276$
(b) $y_{4p} = 2.42539$, $y_{4c} = 2.42553$
(c) $y_{4p} = 1.69412$, $y_{4c} = 1.69604$
(d) $y_{4p} = 0.994789$, $y_{4c} = 0.993360$

Section 8.9, Page 440

1. (a)$x_1 = 1.1, \quad x_2 = 1.26;$ $\quad y_1 = 0.4, \quad y_2 = 0.76$
   (b) $x_1 = 1.2, \quad x_2 = 1.451;$ $\quad y_1 = 1.1, \quad y_2 = 1.232$
   (c) $x_1 = 0.8, \quad x_2 = 0.582;$ $\quad y_1 = 1.1, \quad y_2 = 1.18$
2. 1.30, $\quad -0.24$
3. (a) $x_1 = 1.32, \quad y_1 = 0.72$ $\quad$ (b) $x_1 = 1.50, \quad y_1 = 1.26$
   (c) $x_1 = 0.56, \quad y_1 = 1.16$
4. $x_1 = 1.32007, \quad y_1 = -0.250667$
5. $x_{4p} = 1.99423, \quad y_{4p} = -0.662181; \quad x_{4c} = 1.99521, \quad y_{4c} = -0.662442$
6. $x' = y, \quad y' = -t^2 y - 3x + t; \quad x(0) = 1, \quad y(0) = 2$
   $x_1 = 1.2, \quad x_2 = 1.37; \quad y_1 = 1.7, \quad y_2 = 1.348$

CHAPTER NINE

Section 9.1, Page 454

1. (a) $x = 4e^{-t}, \quad y = 2e^{-2t}; \quad y = x^2/8$
   (b) $x = 4e^{-t}, \quad y = 2e^{2t}; \quad y = 32x^{-2}; \quad x = 4e^{-t}, \quad y = 0$
   (c) $x = 4\cos t, \quad y = 4\sin t; \quad x^2 + y^2 = 16;$
   $x = -4\sin t, \quad y = 4\cos t; \quad x^2 + y^2 = 16$
   (d) $x = \sqrt{a}\cos t, \quad y = -\sqrt{b}\sin t; \quad (x^2/a) + (y^2/b) = 1$
2. (a) $(0,0)$ $\quad$ (b) $(0,0), \quad (-\frac{1}{2}, 1)$
   (c) $(0,0), \quad (1,0), \quad (0,2), \quad (\frac{1}{2}, \quad \frac{1}{2})$ $\quad$ (d) $(0,0)$
   (e) $(\pm n\pi, 0), \quad n = 0, 1, 2, \ldots$ $\quad$ (f) $(0,0)$
3. (a) $x = Ae^{-t}, \quad y = Be^{-2t}; \quad y = Bx^2/A^2, \quad A \neq 0$
   (b) $x = Ae^{-t}, \quad y = Be^{2t}; \quad y = BA^2/x^2, \quad A \neq 0$
   (c) $x = Ae^{t}, \quad y = Be^{-2t}; \quad y = BA^2/x^2, \quad A \neq 0$
   (d) $x = Ae^{-t}, \quad y = Be^{2t} - Ae^{-t}; \quad y = BA^2/x^2 - x, \quad A \neq 0$
8. $2x^2 - \alpha x^4 + 2y^2 = c$

Section 9.2, Page 467

1. $r_1 = -1, \quad r_2 = 2;$ saddle point, unstable
2. $r_1 = 2, \quad r_2 = 4;$ improper node, unstable
3. $r_1 = -1, \quad r_2 = 1;$ saddle point, unstable
4. $r_1 = -3, \quad r_2 = -3;$ improper node, asymptotically stable
5. $r_1, r_2 = -1 \pm i;$ spiral point, asymptotically stable
6. $r_1, r_2 = \pm i;$ center, stable
7. $r_1, r_2 = 1 \pm 2i;$ spiral point, unstable
8. $r_1 = -1, \quad r_2 = -\frac{1}{4};$ improper node, asymptotically stable
9. $r_1 = 1, \quad r_2 = 1;$ improper node, unstable
10. $r_{1,2} = \pm 3i;$ center, stable
11. $r_1 = -1, \quad r_2 = -1;$ proper node, asymptotically stable
12. $r_{1,2} = (1 \pm 3i)/2;$ spiral point, unstable
13. $x_0 = 1, \quad y_0 = 1; \quad r_1 = \sqrt{2}, \quad r_2 = -\sqrt{2};$ saddle point, unstable
14. $x_0 = -1, \quad y_0 = 0; \quad r_1 = -1, \quad r_2 = -3;$
    improper node, asymptotically stable
15. $x_0 = -2, \quad y_0 = 1; \quad r_{1,2} = -1 \pm \sqrt{2}\, i;$ spiral point, asymptotically stable
16. $x_0 = \gamma/\delta, \quad y_0 = \alpha/\beta; \quad r_{1,2} = \pm\sqrt{\beta\delta}\, i;$ center, stable

17. $c^2 > 4km$, improper node, asymptotically stable; $c^2 = 4km$, improper node, asymptotically stable; $c^2 < 4km$, spiral point, asymptotically stable

Section 9.3, Page 482

1. Spiral point, asymptotically stable
2. PN, IN, or SpP, unstable
3. Spiral point, asymptotically stable
4. Spiral point, unstable
5. Saddle point, unstable
6. Saddle point, unstable
7. $\mu < 2$, spiral point, unstable; $\mu = 2$, PN, IN, or SpP, unstable; $\mu > 2$, improper node, unstable
8. Spiral point, unstable
9. C or SpP, indeterminate
10. Improper node, asymptotically stable
11. (a) $(0,0)$, PN, IN, or SpP, unstable; $(-1,1)$, saddle point, unstable
    (b) $(1,1)$, PN, IN, or SpP, asymptotically stable; $(-1,-1)$, saddle point, unstable
    (c) $(0,0)$, improper node, unstable; $(0,\frac{3}{2})$, improper node, asymptotically stable; $(1,0)$, saddle point, unstable; $(-1,2)$, saddle point, unstable
    (d) $(1,1)$, spiral point, asymptotically stable; $(-1,1)$, saddle point, unstable
15. (b) Refer to Table 9.2.
18. (a) $dx/dt = y, \quad dy/dt = -g(x) - c(x)y$
    (b) The linear system is $dx/dt = y, \quad dy/dt = -g'(0)x - c(0)y$
    (c) The eigenvalues satisfy $r^2 + c(0)r + g'(0) = 0$
19. (a) $X = (\epsilon_1\sigma_2 - \epsilon_2\alpha_1)/(\sigma_1\sigma_2 - \alpha_1\alpha_2), \quad Y = (\epsilon_2\sigma_1 - \epsilon_1\alpha_2)/(\sigma_1\sigma_2 - \alpha_1\alpha_2)$
    (c) Asymptotically stable improper node
20. (a) $X = (\epsilon_1\sigma_2 - \epsilon_2\alpha_1)/(\sigma_1\sigma_2 - \alpha_1\alpha_2), \quad Y = (\epsilon_1\sigma_1 - \epsilon_1\alpha_2)/(\sigma_1\sigma_2 - \alpha_1\alpha_2)$
    (c) Unstable saddle point.

Section 9.4, Page 498

1. (a, b) $(0,0)$: $r_1 = 1$, $r_2 = \frac{5}{2}$, improper node, unstable
   $(0,\frac{5}{3})$: $r_1 = \frac{11}{6}$, $r_2 = -\frac{5}{2}$, saddle point, unstable
   $(1,0)$: $r_1 = \frac{11}{4}$, $r_2 = -1$, saddle point, unstable
   $(2,2)$: $r_{1,2} = (-5 \pm \sqrt{3})/2$, improper node, asymptotically stable
2. (a, b) $(0,0)$: $r_1 = 1$, $r_2 = \frac{1}{2}$, improper node, unstable
   $(0,2)$: $r_1 = \frac{1}{2}$, $r_2 = -2$, saddle point, unstable
   $(\frac{3}{2},0)$: $r_1 = -\frac{3}{2}$, $r_2 = \frac{7}{8}$, saddle point, unstable
   $(\frac{4}{5},\frac{7}{5})$: $r_{1,2} = (-2.2 \pm \sqrt{2.04})/2$, improper node, asymptotically stable
3. (a, b) $(0,0)$: $r_1 = 1$, $r_2 = -\frac{1}{4}$, saddle point, unstable
   $(2,0)$: $r_1 = \frac{3}{4}$, $r_2 = -1$, saddle point, unstable
   $(\frac{1}{2},\frac{3}{2})$: $r_{1,2} = (-1 \pm \sqrt{11}\,i)/8$, spiral point, asymptotically stable
4. (a, b) $(0,0)$: $r_1 = \frac{3}{2}$, $r_2 = 2$, improper node, unstable
   $(0,4)$: $r_1 = -\frac{1}{2}$, $r_2 = -2$, improper node, asymptotically stable
   $(\frac{3}{2},0)$: $r_1 = -\frac{1}{4}$, $r_2 = -\frac{3}{2}$, improper node, asymptotically stable
   $(1,1)$: $r_{1,2} = (-3 \pm \sqrt{13})/4$, saddle point, unstable
5. (a, b) $(0,0)$: $r_1 = \frac{9}{8}$, $r_2 = -1$, saddle point, unstable
   $(\frac{9}{8},0)$: $r_1 = -\frac{9}{8}$, $r_2 = \frac{1}{8}$, saddle point, unstable
   $(1,\frac{1}{4})$: $r_{1,2} = (-1 \pm \sqrt{0.5})/2$, improper node, asymptotically stable
7. (a) Critical points are $x = 0$, $y = 0$; $x = \epsilon_1/\sigma_1$, $y = 0$; $x = 0$, $y = \epsilon_2/\sigma_2$. Mixed state (X, Y) is not possible since one or the other of X and Y is negative. $x \to 0$, $y \to \epsilon_2/\sigma_2$ as $t \to \infty$, the redear survive.
   (a) Same as part (a) except $x \to \epsilon_1/\sigma_1$, $y \to 0$ as $t \to \infty$, the bluegill survive.

8. (a) $X = (B - \gamma_1 R)/(1 - \gamma_1\gamma_2)$, $Y = (R - \gamma_2 B)/(1 - \gamma_1\gamma_2)$.
   (b) X is reduced, Y is increased; yes, if B becomes less than $\gamma_1 R$, then $x \to 0$ and $y \to R$ as $t \to \infty$.
9. $t = 0, T, 2T, \ldots$: H is a max., dP/dt is a max.
   $t = T/4, 5T/4, \ldots$: dH/dt is a min., P is a max.
   $t = T/2, 3T/2, \ldots$: H is a min., dP/dt is a min.
   $t = 3T/4, 7T/4, \ldots$: dH/dt is a max., P is a min.
11. Trap foxes in half cycle when $dP/dt > 0$, trap rabbits in half cycle when $dH/dt > 0$, trap rabbits and foxes in quarter cycle when $dP/dt > 0$ and $dH/dt > 0$, trap neither in quarter cycle when $dP/dt < 0$ and $dH/dt < 0$.
12. $dH/dt = aH - \alpha HP - \beta H$, $dP/dt = -cP + \gamma HP - \delta P$, where β and δ are constants of proportionality. New center is $H = (c + \delta)/\gamma > c/\gamma$ and $P = (a - \beta)/\alpha < a/\alpha$ so equilibrium value of prey is increased and equilibrium value of predator is decreased!
13. Let $A = a/\sigma - c/\gamma > 0$. The critical points are $(0,0)$, $(a/\sigma, 0)$, and $(c/\gamma, \sigma A/\alpha)$, where $(0,0)$ is a saddle point, $(a/\sigma, 0)$ is a saddle point, and $(c/\gamma, \sigma A/\alpha)$ is an asymptotically stable node if $(c\sigma/\gamma)^2 - 4c\sigma A \geq 0$ or an asymptotically stable spiral point if $(c\sigma/\gamma)^2 - 4c\sigma A < 0$. $(H, P) \to (c/\gamma, \sigma A/\alpha)$ as $t \to \infty$.

CHAPTER TEN

Section 10.2, Page 521

1. $\alpha^2 u_{xx} = u_t$, $\quad 0 < x < 1,\ t > 0$ $\qquad \alpha^2 = 0.000114\ \text{m}^2/\text{sec}$
   $u(x,0) = 20$, $\quad 0 \leq x \leq 1$ $\qquad x$ in meters
   $u(0,t) = 60$, $\quad u(1,t) = 20$, $\quad t > 0$ $\qquad t$ in seconds
2. $\alpha^2 u_{xx} = u_t$, $\quad 0 < x < 2$, $\quad t > 0$ $\qquad \alpha^2 = 0.000171\ \text{m}^2/\text{sec}$
   $u(x,0) = 30 + 50x - 20x^2$, $\quad 0 \leq x \leq 2$ $\qquad x$ in meters
   $u(0,t) = 30$, $\quad u(2,t) = 50$, $\quad t > 0$ $\qquad t$ in seconds
4. $u(x,t) = e^{-400\pi^2 t}\sin 2\pi x - 2e^{-2500\pi^2 t}\sin 5\pi x$
5. $u(x,t) = 2e^{-\pi^2 t/16}\sin(\pi x/2) - e^{-\pi^2 t/4}\sin \pi x + 4e^{-\pi^2 t}\sin 2\pi x$
6. $(\text{length})^{-2}$
7. (a) $xX'' - \lambda X = 0$, $\quad T' + \lambda T = 0$
   (b) $X'' - \lambda x X = 0$, $\quad T' + \lambda t T = 0$
   (c) $X'' - \lambda(X' + X) = 0$, $\quad T' + \lambda T = 0$
   (d) $[p(x)X']' + \lambda r(x)X = 0$, $\quad T'' + \lambda T = 0$
   (e) Not separable
9. (a) $aw_{xx} - bw_t + (c - b\delta)w = 0$ $\qquad$ (b) $\delta = c/b$ if $b \neq 0$
10. $X'' + \mu^2 X = 0$, $\quad Y'' + (\lambda^2 - \mu^2)Y = 0$, $\quad T' + \alpha^2\lambda^2 T = 0$
11. $r^2R'' + rR' + (\lambda^2 r^2 - \mu^2)R = 0$, $\quad \Theta'' + \mu^2\Theta = 0$, $\quad T' + \alpha^2\lambda^2 T = 0$

Section 10.3, Page 530

1. (a) $T = 2l$ $\qquad$ (b) $T = 1$ $\qquad$ (c) Not periodic
   (d) $T = 1$ $\qquad$ (e) Not periodic $\qquad$ (f) $T = 2\pi/5$
   (g) $T = 2\pi/|m|$, $m \neq 0$ $\qquad$ (h) Not periodic $\qquad$ (i) $T = 2$
   (j) $T = 4$
3. $\int_0^x f(t)\,dt$ may not be periodic; for example, let $f(t) = 1 + \cos t$.

5. $f(x) = \frac{2l}{\pi} \sum_{n=1}^{\infty} \frac{(-1)^n}{n} \sin \frac{n\pi x}{l}$ 6. $f(x) = \frac{1}{2} - \frac{2}{\pi} \sum_{n=1}^{\infty} \frac{\sin[(2n-1)\pi x/l]}{2n-1}$

7. $f(x) = \frac{2l}{\pi} \sum_{n=1}^{\infty} \frac{1}{n} \sin \frac{n\pi x}{l}$ 8. $f(x) = \frac{2}{\pi} \sum_{n=1}^{\infty} \frac{(-1)^{n+1}}{n} \sin n\pi x$

9. $f(x) = \frac{1}{2} - \frac{1}{\pi} \sum_{n=1}^{\infty} \frac{\sin 2n\pi x}{n}$ 10. $f(x) = \frac{4}{\pi} \sum_{n=1}^{\infty} \frac{\sin[(2n-1)\pi x/2]}{2n-1}$

11. $f(x) = -\frac{\pi}{4} + \sum_{n=1}^{\infty} \left[\frac{2\cos(2n-1)x}{\pi(2n-1)^2} + \frac{(-1)^{n+1}\sin nx}{n} \right]$

12. $f(x) = \frac{1}{2} + \frac{4}{\pi^2} \sum_{n=1}^{\infty} \frac{\cos(2n-1)\pi x}{(2n-1)^2}$

13. $f(x) = \frac{3l}{4} + \sum_{n=1}^{\infty} \left[\frac{2l\cos[(2n-1)\pi x/l]}{(2n-1)^2\pi^2} + \frac{(-1)^{n+1} l \sin(n\pi x/l)}{n\pi} \right]$

14. $f(x) = \sum_{n=1}^{\infty} \left[-\frac{2}{n\pi} \cos \frac{n\pi}{2} + \left(\frac{2}{n\pi}\right)^2 \sin \frac{n\pi}{2} \right] \sin \frac{n\pi x}{2}$

15. $f(x) = 2l - x$ in $l < x < 2l$; $f(x) = -2l - x$ in $-3l < x < -2l$

16. $f(x) = x - 1$ in $1 < x < 2$; $f(x) = x - 8$ in $8 < x < 9$

17. $f(x) = -l - x$ in $-l < x < 0$

Section 10.4, Page 538

1. $f(x) = \frac{4}{\pi} \sum_{n=1}^{\infty} \frac{\sin(2n-1)\pi x}{2n-1}$

2. $f(x) = \frac{\pi}{4} - \sum_{n=1}^{\infty} \left[\frac{2}{(2n-1)^2\pi} \cos(2n-1)x + \frac{(-1)^n}{n} \sin nx \right]$

3. $f(x) = \frac{1}{2} - \frac{1}{2}\cos 2x$ 4. $f(x) = s + \frac{2}{\pi} \sum_{n=1}^{\infty} \frac{\sin n\pi s}{n} \cos n\pi x$

5. $f(x) = \frac{l}{2} + \frac{4l}{\pi^2} \sum_{n=1}^{\infty} \frac{\cos[(2n-1)\pi x/l]}{(2n-1)^2}$

6. $f(x) = \frac{2}{3} + \frac{4}{\pi^2} \sum_{n=1}^{\infty} \frac{(-1)^{n+1}}{n^2} \cos n\pi x$

7. $f(x) = \frac{1}{2} + \frac{2}{\pi} \sum_{n=1}^{\infty} \frac{(-1)^{n-1}}{2n-1} \cos(2n-1)x$

8. $f(x) = \frac{a_0}{2} + \sum_{n=1}^{\infty} (a_n \cos n\pi x + b_n \sin n\pi x)$;

$$a_0 = \tfrac{1}{3}, \qquad a_n = \frac{2(-1)^n}{n^2\pi^2}, \qquad b_n = \begin{cases} -1/n\pi, & n \text{ even} \\ 1/n\pi - 4/n^3\pi^3, & n \text{ odd} \end{cases}$$

9. $y = (\omega \sin nt - n \sin \omega t)/\omega(\omega^2 - n^2)$, $\omega^2 \neq n^2$

$y = (\sin nt - nt\cos nt)/2n^2$, $\omega^2 = n^2$

10. $y = \sum_{n=1}^{\infty} b_n(\omega \sin nt - n \sin \omega t)/\omega(\omega^2 - n^2)$, $\omega \neq 1, 2, 3, \ldots$

$$y = \sum_{\substack{n=1 \\ n \neq m}}^{\infty} b_n (m \sin nt - n \sin mt)/m(m^2 - n^2) + b_m (\sin mt - mt \cos mt)/2m^2,$$

$\omega = m$

11. $y = \dfrac{4}{\pi} \displaystyle\sum_{n=1}^{\infty} \dfrac{1}{\omega^2 - (2n-1)^2} \left[\dfrac{1}{2n-1} \sin(2n-1)t - \dfrac{1}{\omega} \sin \omega t \right]$

12. $y = \cos \omega t + \dfrac{1}{2\omega^2}(1 - \cos \omega t) + \dfrac{4}{\pi^2} \displaystyle\sum_{n=1}^{\infty} \dfrac{\cos(2n-1)\pi t - \cos \omega t}{(2n-1)^2 [\omega^2 - (2n-1)^2]}$

Section 10.5, Page 546

1. (a) Odd (b) Odd (c) Neither (d) Odd
   (e) Even (f) Even (g) Neither (h) Even
   (i) Even (j) Even (k) Neither (1) Neither
2. (a) 0 (b) $\frac{2}{5}$ (c) $2\pi(-1)^{n+1}/n$
   (d) 0 (e) 0 (f) 0
4. $f(x) = \dfrac{1}{4} + \dfrac{4}{\pi^2} \displaystyle\sum_{n=1}^{\infty} \dfrac{1 - \cos(n\pi/2)}{n^2} \cos(n\pi x/2)$

   $f(x) = \dfrac{4}{\pi^2} \displaystyle\sum_{n=1}^{\infty} \dfrac{(n\pi/2) - \sin(n\pi/2)}{n^2} \sin(n\pi x/2)$
5. $f(x) = \dfrac{1}{2} + \dfrac{2}{\pi} \displaystyle\sum_{n=1}^{\infty} \dfrac{(-1)^{n-1}}{2n-1} \cos \dfrac{(2n-1)\pi x}{2}$
6. $f(x) = \displaystyle\sum_{n=1}^{\infty} \dfrac{2}{n\pi} \left(-\cos n\pi + \dfrac{2}{n\pi} \sin \dfrac{n\pi}{2} \right) \sin \dfrac{n\pi x}{2}$
7. $f(x) = 1$
8. $f(x) = \dfrac{4}{\pi} \displaystyle\sum_{n=1}^{\infty} \dfrac{\sin(2n-1)x}{2n-1}$
9. $f(x) = \displaystyle\sum_{n=1}^{\infty} \dfrac{2}{n\pi} \left(\cos \dfrac{n\pi}{3} + \cos \dfrac{2n\pi}{3} - 2\cos n\pi \right) \sin \dfrac{nx}{3}$
10. $f(x) = \dfrac{1}{2} - \dfrac{1}{\pi} \displaystyle\sum_{n=1}^{\infty} \dfrac{\sin 2n\pi x}{n}$
11. $f(x) = \dfrac{l}{2} + \dfrac{4l}{\pi^2} \displaystyle\sum_{n=1}^{\infty} \dfrac{\cos[(2n-1)\pi x/l]}{(2n-1)^2}$
12. $f(x) = \dfrac{2l}{\pi} \displaystyle\sum_{n=1}^{\infty} \dfrac{\sin(n\pi x/l)}{n}$
13. $f(x) = \dfrac{\pi}{4} + \dfrac{1}{\pi} \displaystyle\sum_{n=1}^{\infty} \left[\dfrac{2\pi}{n} \sin \dfrac{n\pi}{2} + \dfrac{4}{n^2} \left(\cos \dfrac{n\pi}{2} - 1 \right) \right] \cos \dfrac{nx}{2}$
14. $f(x) = 2 \displaystyle\sum_{n=1}^{\infty} \dfrac{(-1)^n \sin nx}{n}$
19. 2.895 20. (b) 3.061
24. Extend $f(x)$ antisymmetrically into $(l, 2l]$; that is, so that $f(2l - x) = -f(x)$ for $0 \le x < l$. Then extend this function as an even function into $(-2l, 0)$.

Section 10.6, Page 556

1. $u(x,t) = \sum_{n=1}^{\infty} c_n e^{-n^2\pi^2\alpha^2 t/l^2} \sin(n\pi x/l)$; $\alpha^2 = 1.14 \text{ cm}^2/\text{sec}$, $l = 100$ cm
   (a) $c_n = 100(1 - \cos n\pi)/n\pi$ (b) $c_n = 400 \sin(n\pi/2)/n^2\pi^2$
   (c) $c_n = 100[\cos(n\pi/4) - \cos(3n\pi/4)]/n\pi$
2. $u(x,t) = \dfrac{400}{\pi} \sum_{n=1}^{\infty} \dfrac{1}{2n-1} e^{-(2n-1)^2\pi^2\alpha^2 t/400} \sin \dfrac{(2n-1)\pi x}{20}$
   (a) $u(10,30) \cong 35.9°\text{C}$ (b) $u(10,30) \cong 67.2°\text{C}$ (c) $u(10,30) \cong 97.4°\text{C}$
3. $t \cong 400(\alpha\pi)^{-2} \ln(400/25\pi)$
   (a) $t \cong 38.6$ sec (b) $t \cong 76.7$ sec (c) $t \cong 550$ sec
4. (a) $u(x,t) = \dfrac{60x}{l} + \sum_{n=1}^{\infty} \left(\dfrac{50}{n\pi} + \dfrac{70}{n\pi} \cos n\pi\right) e^{-0.86n^2\pi^2 t/l^2} \sin \dfrac{n\pi x}{l}$
   (b) $u(5,30) \cong 12.6°\text{C}$; $u(5,60) \cong 13.7°\text{C}$
   (c) $u(5,30) \cong 14.1°\text{C}$; 12%; third term $\cong -0.005°\text{C}$
   (d) $t \cong 160$ sec
5. (a) $f(x) = 2x,\ 0 \le x \le 50$; $f(x) = 200 - 2x,\ 50 < x \le 100$
   (b) $u(x,t) = 20 - \dfrac{x}{5} + \sum_{n=1}^{\infty} c_n e^{-1.14n^2\pi^2 t/(100)^2} \sin \dfrac{n\pi x}{100}$,
   $$c_n = \frac{1}{50}\int_0^{100}\left[f(x) - \left(20 - \frac{x}{5}\right)\right] \sin \frac{n\pi x}{100}\,dx = \frac{800}{n^2\pi^2} \sin \frac{n\pi}{2} - \frac{40}{n\pi}$$
6. $u(x,t) = \dfrac{2}{\pi} + \sum_{n=1}^{\infty} c_n e^{-n^2\pi^2\alpha^2 t/l^2} \cos \dfrac{n\pi x}{l}$,
   $$c_n = \frac{2}{l}\int_0^l \cos\frac{n\pi x}{l} \sin \frac{\pi x}{l}\,dx = \begin{cases} 0, & n \text{ odd}, \\ -4/(n^2 - 1)\pi, & n \text{ even}; \end{cases}$$
   $\lim_{t\to\infty} u(x,t) = 2/\pi$
8. $v(x) = 0$
9. $v(x) = T$
10. (b) $u(x,t) = \sum_{n=1}^{\infty} c_n e^{-(2n-1)^2\pi^2\alpha^2 t/4l^2} \sin \dfrac{(2n-1)\pi x}{2l}$,
   $$c_n = \frac{2}{l}\int_0^l f(x) \sin \frac{(2n-1)\pi x}{2l}\,dx$$
11. $u(x,t) = T + \sum_{n=1}^{\infty} c_n e^{-(2n-1)^2\pi^2\alpha^2 t/4l^2} \sin \dfrac{(2n-1)\pi x}{2l}$,
   $$c_n = \frac{2}{l}\int_0^l [f(x) - T] \sin \frac{(2n-1)\pi x}{2l}\,dx$$
12. $v(x) = T(1 + x)/(1 + l)$
13. $$u(x) = \begin{cases} T\dfrac{x}{a}\left[\dfrac{\xi}{\xi + (l/a) - 1}\right], & 0 \le x \le a \\ T\left[1 - \dfrac{l-x}{a}\,\dfrac{1}{\xi + (l/a) - 1}\right], & a \le x \le l \end{cases}$$ where $\xi = \kappa_2 A_2/\kappa_1 A_1$
14. (e) $u_n(x,t) = e^{-\lambda_n^2\alpha^2 t} \sin \lambda_n x$

Section 10.7, Page 567

1. $u(x,t) = \dfrac{4Al}{\pi^2} \sum_{n=1}^{\infty} \dfrac{1}{n^2} \sin \dfrac{n\pi}{2} \sin \dfrac{n\pi x}{l} \cos \dfrac{n\pi at}{l}$

2. $u(x,t) = \dfrac{2Al}{\pi^2} \displaystyle\sum_{n=1}^{\infty} \frac{1}{n^2}\left(\sin\frac{n\pi}{4} + \sin\frac{3n\pi}{4}\right) \sin\frac{n\pi x}{l} \cos\frac{n\pi at}{l}$

3. $u(x,t) = \displaystyle\sum_{n=1}^{\infty} c_n \sin\frac{n\pi x}{l} \sin\frac{n\pi at}{l}, \qquad c_n = \frac{2}{n\pi a}\int_0^l g(x)\sin\frac{n\pi x}{l}\,dx$

4. $u(x,t) = \dfrac{4Al^2}{a\pi^3} \displaystyle\sum_{n=1}^{\infty} \frac{1}{n^3} \sin\frac{n\pi}{2} \sin\frac{n\pi x}{l} \sin\frac{n\pi at}{l}$

6. $u(x,t) = \displaystyle\sum_{n=1}^{\infty} c_n \sin\frac{(2n-1)\pi x}{2l} \cos\frac{(2n-1)\pi at}{2l},$

$c_n = \dfrac{2}{l}\displaystyle\int_0^l f(x)\sin\frac{(2n-1)\pi x}{2l}\,dx$

9. $\phi(x + at)$ represents a wave moving in the negative x direction with speed $a > 0$.

10. (a) 248 ft/sec (b) 49.6 πn rad/sec (c) Frequencies increase; modes are unchanged

11. $u(x,t) = \displaystyle\sum_{n=1}^{\infty} c_n \cos\beta_n t \sin(n\pi x/l), \qquad \beta_n^2 = (n\pi a/l)^2 + \alpha^2,$

$c_n = \dfrac{2}{l}\displaystyle\int_0^l f(x)\sin(n\pi x/l)\,dx$

17. $r^2R'' + rR' + (\lambda^2 r^2 - \mu^2)R = 0, \quad \Theta'' + \mu^2\Theta = 0, \quad T'' + \lambda^2 a^2 T = 0$

Section 10.8, Page 577

1. (a) $u(x,y) = \displaystyle\sum_{n=1}^{\infty} c_n \sin\frac{n\pi x}{a} \sinh\frac{n\pi y}{a}, \qquad c_n = \frac{2/a}{\sinh(n\pi b/a)}\int_0^a g(x)\sin\frac{n\pi x}{a}\,dx$

(b) $u(x,y) = \dfrac{4a}{\pi^2} \displaystyle\sum_{n=1}^{\infty} \frac{1}{n^2} \frac{\sin(n\pi/2)}{\sinh(n\pi b/a)} \sin\frac{n\pi x}{a} \sinh\frac{n\pi y}{a}$

2. $u(x,y) = \displaystyle\sum_{n=1}^{\infty} c_n \sin\frac{n\pi x}{a} \sinh\frac{n\pi(b-y)}{a},$

$c_n = \dfrac{2/a}{\sinh(n\pi b/a)}\displaystyle\int_0^a h(x)\sin\frac{n\pi x}{a}\,dx$

3. $u(x,y) = v(x,y) + w(x,y)$, where $v(x,y)$ is found in Problem 2 and $w(x,y)$ is given by Eq. (12) in the text.

5. $u(r,\theta) = \dfrac{c_0}{2} + \displaystyle\sum_{n=1}^{\infty} r^{-n}(c_n\cos n\theta + k_n\sin n\theta);$

$c_n = \dfrac{a^n}{\pi}\displaystyle\int_0^{2\pi} f(\theta)\cos n\theta\,d\theta, \qquad k_n = \frac{a^n}{\pi}\int_0^{2\pi} f(\theta)\sin n\theta\,d\theta$

6. $u(r,\theta) = \displaystyle\sum_{n=1}^{\infty} c_n r^n \sin n\theta, \qquad c_n = \frac{2}{\pi a^n}\int_0^{\pi} f(\theta)\sin n\theta\,d\theta$

7. $u(r,\theta) = \displaystyle\sum_{n=1}^{\infty} c_n r^{n\pi/\alpha}\sin(n\pi\theta/\alpha), \qquad c_n = (2/\alpha)a^{-n\pi/\alpha}\int_0^{\alpha} f(\theta)\sin(n\pi\theta/\alpha)\,d\theta$

8. $u(x,y) = \displaystyle\sum_{n=1}^{\infty} c_n e^{-n\pi y/a}\sin(n\pi x/a), \qquad c_n = \frac{2}{a}\int_0^a f(x)\sin(n\pi x/a)\,dx$

10. (b) $u(x,y) = c_0 + \displaystyle\sum_{n=1}^{\infty} c_n\cosh(n\pi x/b)\cos(n\pi y/b),$

$c_n = \dfrac{2/n\pi}{\sinh(n\pi a/b)}\displaystyle\int_0^b f(y)\cos(n\pi y/b)\,dy$

11. $u(r,\theta) = c_0 + \sum_{n=1}^{\infty} r^n(c_n \cos n\theta + k_n \sin n\theta)$,

$c_n = \dfrac{1}{n\pi a^{n-1}} \int_0^{2\pi} g(\theta) \cos n\theta \, d\theta, \qquad k_n = \dfrac{1}{n\pi a^{n-1}} \int_0^{2\pi} g(\theta) \sin n\theta \, d\theta$

Necessary condition is $\int_0^{2\pi} g(\theta)\, d\theta = 0$.

12. $u(x, y) = \sum_{n=1}^{\infty} c_n \cosh(n\pi x/b) \sin(n\pi y/b)$,

$c_n = \dfrac{2/b}{\cosh(n\pi a/b)} \int_0^b f(y) \sin(n\pi y/b)\, dy$

13. $u(x, y) = \sum_{n=1}^{\infty} c_n \sinh[(2n-1)\pi x/2b] \sin[(2n-1)\pi y/2b]$,

$c_n = \dfrac{2/b}{\sinh[(2n-1)\pi a/2b]} \int_0^b f(y) \sin[(2n-1)\pi y/2b]\, dy$

14. $u(x, y) = c_0 y + \sum_{n=1}^{\infty} c_n \cos(n\pi x/a) \sinh(n\pi y/a)$,

$c_0 = \dfrac{1}{ab} \int_0^a f(x)\, dx, \qquad c_n = \dfrac{2/a}{\sinh(n\pi b/a)} \int_0^a f(x) \cos(n\pi x/a)\, dx$

CHAPTER ELEVEN

Section 11.1, Page 592

1. (a) Homogeneous (b) Nonhomogeneous
   (c) Nonhomogeneous (d) Homogeneous
   (e) Nonhomogeneous (f) Homogeneous
3. (a) $\mu(x) = e^{-x^2}$ (b) $\mu(x) = 1/x$
   (c) $\mu(x) = e^{-x}$ (d) $\mu(x) = (1 - x^2)^{-1/2}$
4. $a^2X'' + (\lambda - k)X = 0, \qquad T'' + cT' + \lambda T = 0$

Section 11.2, Page 599

1. $\lambda_n = [(2n-1)\pi/2]^2; \quad \phi_n(x) = \sin[(2n-1)\pi x/2]; \quad n = 1, 2, 3, \ldots$
2. $\lambda_n = [(2n-1)\pi/2]^2; \quad \phi_n(x) = \cos[(2n-1)\pi x/2]; \quad n = 1, 2, 3, \ldots$
3. $\lambda_0 = 0; \quad \phi_0(x) = 1; \quad \lambda_n = n^2\pi^2; \quad \phi_n(x) = \cos n\pi x; \quad n = 1, 2, 3, \ldots$
4. $\lambda_n = -[(2n-1)\pi/2]^2; \quad \phi_n(x) = \sin[(2n-1)\pi x/2]; \quad n = 1, 2, 3, \ldots$
5. $\lambda_0 = 0; \quad \phi_0(x) = 1 - x$
   For $n = 1, 2, 3, \ldots$, $\phi_n(x) = \sin \mu_n x - \mu_n \cos \mu_n x$ and $\lambda_n = -\mu_n^2$, where μ_n are roots of $\mu = \tan \mu$.
   $\lambda_1 \cong -20.2; \quad \lambda_n \cong -(2n+1)^2\pi^2/4$ for large n
6. $\phi_n(x) = \sin\sqrt{\lambda_n}\, x$ where $\sqrt{\lambda_n}$ are roots of $\sqrt{\lambda} = -\tan\sqrt{\lambda}\,\pi; \quad \lambda_1 \cong 0.621;$
   $\lambda_n \cong (2n-1)^2/4$ for large n
7. $\phi_n(x) = \cos\sqrt{\lambda_n}\, x$ where $\sqrt{\lambda_n}$ are roots of $\sqrt{\lambda} = \cot\sqrt{\lambda}; \quad \lambda_1 \cong 0.740;$
   $\lambda_n \cong (n-1)^2\pi^2$ for large n
8. $\phi_n(x) = \sin\sqrt{\lambda_n}\, x + \sqrt{\lambda_n} \cos\sqrt{\lambda_n}\, x$ where $\sqrt{\lambda_n}$ are roots of
   $(\lambda - 1)\sin\sqrt{\lambda} - 2\sqrt{\lambda}\cos\sqrt{\lambda} = 0; \quad \lambda_1 \cong 1.69; \quad \lambda_n \cong (n-1)^2\pi^2$ for large n
9. (a) $s(x) = e^x$ (b) $\lambda_n = n^2\pi^2$, $\phi_n(x) = e^x \sin n\pi x; \quad n = 1, 2, 3, \ldots$

10. Positive eigenvalues $\lambda = \lambda_n$ where $\sqrt{\lambda_n}$ are roots of $\sqrt{\lambda} = \frac{2}{3}\tan 3\sqrt{\lambda}\, l$; corresponding eigenfunctions are $\phi_n(x) = e^{-2x}\sin 3\sqrt{\lambda_n}\, x$. If $l = \frac{1}{2}$, $\lambda_0 = 0$ is eigenvalue, $\phi_0(x) = xe^{-2x}$ is eigenfunction; if $l \neq \frac{1}{2}$, $\lambda = 0$ is not eigenvalue. If $l \leq \frac{1}{2}$ there are no negative eigenvalues; if $l > \frac{1}{2}$ there is one negative eigenvalue $\lambda = -\mu^2$ where μ is a root of $\mu = \frac{2}{3}\tanh 3\mu l$; corresponding eigenfunction is $\phi_{-1}(x) = e^{-2x}\sinh 3\mu x$.
11. No real eigenvalues
12. Only eigenvalue is $\lambda = 0$; eigenfunction is $\phi(x) = x - 1$.
13. (a) $2\sin\sqrt{\lambda} - \sqrt{\lambda}\cos\sqrt{\lambda} = 0$
    (b) $2\sinh\sqrt{\mu} - \sqrt{\mu}\cosh\sqrt{\mu} = 0, \qquad \mu = -\lambda$
16. (a) $\lambda_n = \mu_n^4$, where μ_n is a root of $\sin\mu l \sinh\mu l = 0$, hence $\lambda_n = (n\pi/l)^4$;
    $\phi_n(x) = \sin(n\pi x/l)$
    (b) $\lambda_n = \mu_n^4$, where μ_n is a root of $\sin\mu l\cosh\mu l - \cos\mu l\sinh\mu l = 0$;
    $$\phi_n(x) = \frac{\sin\mu_n x\sinh\mu_n l - \sin\mu_n l\sinh\mu_n x}{\sinh\mu_n l}$$
    (c) $\lambda_n = \mu_n^4$, where μ_n is a root of $1 + \cosh\mu l\cos\mu l = 0$;
    $$\phi_n(x) = \frac{\begin{array}{c}[(\sin\mu_n x - \sinh\mu_n x)(\cos\mu_n l + \cosh\mu_n l)\\ +(\sin\mu_n l + \sinh\mu_n l)(\cosh\mu_n x - \cos\mu_n x)]\end{array}}{\cos\mu_n l + \cosh\mu_n l}$$
17. (c) $\phi_n(x) = \sin\sqrt{\lambda_n}\, x$, where λ_n satisfies $\cos\sqrt{\lambda_n}\, l - \gamma\sqrt{\lambda_n}\, l\sin\sqrt{\lambda_n}\, l = 0$

Section 11.3, Page 612

1. (a) $\phi_n(x) = \sqrt{2}\sin(n - \frac{1}{2})\pi x; \qquad n = 1, 2, \ldots$
   (b) $\phi_n(x) = \sqrt{2}\cos(n - \frac{1}{2})\pi x; \qquad n = 1, 2, \ldots$
   (c) $\phi_0(x) = 1, \qquad \phi_n(x) = \sqrt{2}\cos n\pi x; \qquad n = 1, 2, \ldots$
   (d) $\phi_n(x) = \dfrac{\sqrt{2}\cos\sqrt{\lambda_n}\, x}{(1 + \sin^2\sqrt{\lambda_n})^{1/2}}$, where λ_n satisfies $\cos\sqrt{\lambda_n} - \sqrt{\lambda_n}\sin\sqrt{\lambda_n} = 0$
   (e) $\phi_n(x) = (2\sqrt{1 + n^2\pi^2}/n\pi\sqrt{e^2 - 1})e^x \sin n\pi x; \qquad n = 1, 2, \ldots$
2. (a) $a_n = \dfrac{2\sqrt{2}}{(2n-1)\pi}; \qquad n = 1, 2, \ldots$
   (b) $a_n = \dfrac{4\sqrt{2}(-1)^{n-1}}{(2n-1)^2\pi^2}; \qquad n = 1, 2, \ldots$
   (c) $a_n = \dfrac{2\sqrt{2}}{(2n-1)\pi}(1 - \cos[(2n-1)\pi/4]) \qquad n = 1, 2, \ldots$
   (d) $a_n = \dfrac{2\sqrt{2}\sin(n - \frac{1}{2})(\pi/2)}{(n - \frac{1}{2})^2\pi^2}; \qquad n = 1, 2, \ldots$
3. In all parts $\alpha_n = (1 + \sin^2\sqrt{\lambda_n})^{1/2}$ and $\cos\sqrt{\lambda_n} - \sqrt{\lambda_n}\sin\sqrt{\lambda_n} = 0$.
   (a) $a_n = \dfrac{\sqrt{2}\sin\sqrt{\lambda_n}}{\sqrt{\lambda_n}\,\alpha_n}; \qquad n = 1, 2, \ldots$
   (b) $a_n = \dfrac{\sqrt{2}(2\cos\sqrt{\lambda_n} - 1)}{\lambda_n\alpha_n}; \qquad n = 1, 2, \ldots$

(c) $a_n = \dfrac{\sqrt{2}\left(1 - \cos\sqrt{\lambda_n}\right)}{\lambda_n \alpha_n}$; $\quad n = 1, 2, \ldots$

(d) $a_n = \dfrac{\sqrt{2}\,\sin\left(\sqrt{\lambda_n}/2\right)}{\alpha_n\sqrt{\lambda_n}}$; $\quad n = 1, 2, \ldots$

4. (a) Not self-adjoint (b) Self-adjoint
(c) Not self-adjoint (d) Self-adjoint
(e) Self-adjoint

7. If $a_2 = 0$ or $b_2 = 0$, then the corresponding boundary term is missing.

12. (a) $\lambda_1 = \pi^2/l^2$; $\quad \phi_1(x) = \sin(\pi x/l)$
(b) $\lambda_1 \cong (4.49)^2/l^2$; $\quad \phi_1(x) = \sin\sqrt{\lambda_1}\,x - \sqrt{\lambda_1}\,x\cos\sqrt{\lambda_1}\,l$
(c) $\lambda_1 = (2\pi)^2/l^2$; $\quad \phi_1(x) = 1 - \cos(2\pi x/l)$

13. $\lambda_1 = \pi^2/4l^2$; $\quad \phi_1(x) = 1 - \cos(\pi x/2l)$

Section 11.4, Page 625

1. (a) $y = 2\displaystyle\sum_{n=1}^{\infty} \frac{(-1)^{n+1}\sin n\pi x}{(n^2\pi^2 - 2)\,n\pi}$

(b) $y = 2\displaystyle\sum_{n=1}^{\infty} \frac{(-1)^{n+1}\sin\left(n - \frac{1}{2}\right)\pi x}{\left[\left(n - \frac{1}{2}\right)^2\pi^2 - 2\right]\left(n - \frac{1}{2}\right)^2\pi^2}$

(c) $y = -\frac{1}{4} - 4\displaystyle\sum_{n=1}^{\infty} \frac{\cos(2n-1)\pi x}{\left[(2n-1)^2\pi^2 - 2\right](2n-1)^2\pi^2}$

(d) $y = 2\displaystyle\sum_{n=1}^{\infty} \frac{\left(2\cos\sqrt{\lambda_n} - 1\right)\cos\sqrt{\lambda_n}\,x}{\lambda_n(\lambda_n - 2)\left(1 + \sin^2\sqrt{\lambda_n}\right)}$

(e) $y = 8\displaystyle\sum_{n=1}^{\infty} \frac{\sin(n\pi/2)\sin n\pi x}{(n^2\pi^2 - 2)\,n^2\pi^2}$

2. For each part the solution is

$$y = \sum_{n=1}^{\infty} \frac{c_n}{\lambda_n - \mu}\phi_n(x), \qquad c_n = \int_0^1 f(x)\phi_n(x)\,dx, \qquad \mu \neq \lambda_n$$

where $\phi_n(x)$ is given in parts (a), (b), (c), and (d), respectively, of Problem 1 in Section 11.3, and λ_n is the corresponding eigenvalue. In part (c) summation starts at $n = 0$.

3. (a) $a = -\dfrac{1}{2}$, $\quad y = \dfrac{1}{2\pi^2}\cos\pi x + \dfrac{1}{\pi^2}\left(x - \dfrac{1}{2}\right) + c\sin\pi x$
(b) No solution (c) a is arbitrary, $y = c\cos\pi x + a/\pi^2$
(d) $a = 0$, $\quad y = c\sin\pi x - (x/2\pi)\sin\pi x$

7. $v(x) = a + (b - a)x$ 8. $v(x) = 1 - \frac{3}{2}x$

9. (a) $u(x,t) = \sqrt{2}\left[-\dfrac{4c_1}{\pi^2} + \left(\dfrac{4c_1}{\pi^2} + \dfrac{1}{\sqrt{2}}\right)e^{-\pi^2 t/4}\right]\sin\dfrac{\pi x}{2}$

$$- \sqrt{2}\sum_{n=2}^{\infty} \frac{4c_n}{(2n-1)^2\pi^2}\left(1 - e^{-(n-1/2)^2\pi^2 t}\right)\sin\left(n - \tfrac{1}{2}\right)\pi x,$$

$$c_n = \frac{4\sqrt{2}(-1)^{n+1}}{(2n-1)^2\pi^2}, \qquad n = 1, 2, \ldots$$

(b) $u(x,t) = \sqrt{2}\sum_{n=1}^{\infty}\left[\frac{c_n}{\lambda_n - 1}(e^{-t} - e^{-\lambda_n t}) + \alpha_n e^{-\lambda_n t}\right]\frac{\cos\sqrt{\lambda_n}\,x}{\left(1 + \sin^2\sqrt{\lambda_n}\right)^{1/2}}$,

$$c_n = \frac{\sqrt{2}\sin\sqrt{\lambda_n}}{\sqrt{\lambda_n}\left(1 + \sin^2\sqrt{\lambda_n}\right)^{1/2}}, \qquad \alpha_n = \frac{\sqrt{2}\left(1 - \cos\sqrt{\lambda_n}\right)}{\lambda_n\left(1 + \sin^2\sqrt{\lambda_n}\right)^{1/2}},$$

and λ_n satisfies $\cos\sqrt{\lambda_n} - \sqrt{\lambda_n}\sin\sqrt{\lambda_n} = 0$

(c) $u(x,t) = 8\sum_{n=1}^{\infty}\frac{\sin(n\pi/2)}{n^4\pi^4}(1 - e^{-n^2\pi^2 t})\sin n\pi x$

(d) $u(x,t) = \sqrt{2}\sum_{n=1}^{\infty}\frac{c_n\left(e^{-t} - e^{-(n-1/2)^2\pi^2 t}\right)\sin\left(n - \frac{1}{2}\right)\pi x}{\left(n - \frac{1}{2}\right)^2\pi^2 - 1}$,

$$c_n = \frac{2\sqrt{2}(2n-1)\pi + 4\sqrt{2}(-1)^n}{(2n-1)^2\pi^2}$$

10. (a) $r(x)w_t = [p(x)w_x]_x - q(x)w$
$w(0,t) = 0,\ w(1,t) = 0, \qquad w(x,0) = f(x) - v(x)$

11. (a) $u(x,t) = x^2 - 2x + 1 + \frac{4}{\pi}\sum_{n=1}^{\infty}\frac{e^{-(2n-1)^2\pi^2 t}\sin(2n-1)\pi x}{2n-1}$

(b) $u(x,t) = -\cos\pi x + e^{-9\pi^2 t/4}\cos(3\pi x/2)$

17. In all cases solution is $y = \int_0^1 G(x,s)f(s)\,ds$, where $G(x,s)$ is given below.

(a) $G(x,s) = \begin{cases} 1 - x, & 0 \le s \le x \\ 1 - s, & x \le s \le 1 \end{cases}$

(b) $G(s,x) = \begin{cases} s(2-x)/2, & 0 \le s \le x \\ x(2-s)/2, & x \le s \le 1 \end{cases}$

(c) $G(x,s) = \begin{cases} \cos s\sin(1-x)/\cos 1, & 0 \le s \le x \\ \sin(1-s)\cos x/\cos 1, & x \le s \le 1 \end{cases}$

(d) $G(x,s) = \begin{cases} s, & 0 \le s \le x \\ x, & x \le s \le 1 \end{cases}$

Section 11.5, Page 637

1. $y = \sum_{n=1}^{\infty}\frac{c_n}{\lambda_n - \mu}J_0(\sqrt{\lambda_n}\,x), \qquad c_n = \int_0^1 f(x)J_0\left(\sqrt{\lambda_n}\,x\right)dx\Big/\int_0^1 xJ_0^2\left(\sqrt{\lambda_n}\,x\right)dx$,
$\sqrt{\lambda_n}$ are roots of $J_0(\sqrt{\lambda}) = 0$

2. (c) $y = -\frac{c_0}{\mu} + \sum_{n=1}^{\infty}\frac{c_n}{\lambda_n - \mu}J_0\left(\sqrt{\lambda_n}\,x\right)$
$c_0 = 2\int_0^1 f(x)\,dx;\ c_n = \int_0^1 f(x)J_0\left(\sqrt{\lambda_n}\,x\right)dx\Big/\int_0^1 xJ_0^2\left(\sqrt{\lambda_n}\,x\right)dx,\ n = 1, 2, \ldots$
$\sqrt{\lambda_n}$ are roots of $J_0'(\sqrt{\lambda}) = 0$

3. (d) $a_n = \int_0^1 xJ_k\left(\sqrt{\lambda_n}\,x\right)f(x)\,dx\Big/\int_0^1 xJ_k^2\left(\sqrt{\lambda_n}\,x\right)dx$

(e) $y = \sum_{n=1}^{\infty}\frac{c_n}{\lambda_n - \mu}J_k\left(\sqrt{\lambda_n}\,x\right), \quad c_n = \int_0^1 f(x)J_k\left(\sqrt{\lambda_n}\,x\right)dx\Big/\int_0^1 xJ_k^2\left(\sqrt{\lambda_n}\,x\right)dx$

4. (b) $y = \sum_{n=1}^{\infty}\frac{c_n}{\lambda_n - \mu}P_{2n-1}(x), \qquad c_n = \int_0^1 f(x)P_{2n-1}(x)\,dx\Big/\int_0^1 P_{2n-1}^2(x)\,dx$

Section 11.6, Page 643

1. (b) $u(\xi, 2) = f(\xi + 1), \quad u(\xi, 0) = 0, \quad 0 \le \xi \le 2$
   $u(0, \eta) = u(2, \eta) = 0, \quad 0 \le \eta \le 2$
2. $u(r, t) = \sum_{n=1}^{\infty} k_n J_0(\lambda_n r) \sin \lambda_n at, \quad k_n = \dfrac{1}{\lambda_n a} \int_0^1 r J_0(\lambda_n r) g(r)\, dr \Big/ \int_0^1 r J_0^2(\lambda_n r)\, dr$
3. Superpose the solution of Problem 2 and the solution [Eq. (21)] of the example in the text.
6. $u(r, z) = \sum_{n=1}^{\infty} c_n e^{-\lambda_n z} J_0(\lambda_n r), \qquad c_n = \int_0^1 r J_0(\lambda_n r) f(r)\, dr \Big/ \int_0^1 r J_0^2(\lambda_n r)\, dr,$
   and λ_n are roots of $J_0(\lambda) = 0$.
7. (b) $v(r, \theta) = \frac{1}{2} c_0 J_0(kr) + \sum_{m=1}^{\infty} J_m(kr)(b_m \sin m\theta + c_m \cos m\theta),$
   $b_m = \dfrac{1}{\pi J_m(kc)} \int_0^{2\pi} f(\theta) \sin m\theta\, d\theta; \qquad m = 1, 2, \ldots$
   $c_m = \dfrac{1}{\pi J_m(kc)} \int_0^{2\pi} f(\theta) \cos m\theta\, d\theta; \qquad m = 0, 1, 2, \ldots$
8. $c_n = \int_0^1 r f(r) J_0(\lambda_n r)\, dr \Big/ \int_0^1 r J_0^2(\lambda_n r)\, dr$
10. $u(\rho, s) = \sum_{n=0}^{\infty} c_n \rho^n P_n(s)$, where $c_n = \int_{-1}^1 f(\cos^{-1} s) P_n(s)\, ds \Big/ \int_{-1}^1 P_n^2(s)\, ds;$
    P_n is the nth Legendre polynomial and $s = \cos \phi$.

Section 11.7, Page 651

2. (a) lub = 1; no max. (b) lub = 1; no max.
   (c) lub = 1; no max. (d) lub = max. = 1
   (e) no lub or max.
5. (a) $f_0(x) = 1$ (b) $f_1(x) = \sqrt{3}(1 - 2x)$
   (c) $f_2(x) = \sqrt{5}(-1 + 6x - 6x^2)$
   (d) $g_0(x) = 1, \quad g_1(x) = 2x - 1, \quad g_2(x) = 6x^2 - 6x + 1$
6. $P_0(x) = 1, \quad P_1(x) = x, \quad P_2(x) = (3x^2 - 1)/2, \quad P_3(x) = (5x^3 - 3x)/2$

INDEX